MICROWAVE SPECTROSCOPY

MICROWAVE SPECTROSCOPY

C. H. TOWNES

University Professor of Physics
University of California

A. L. SCHAWLOW

Professor of Physics
Stanford University

DOVER PUBLICATIONS, INC.

NEW YORK

Published in Canada by General Publishing Com-
pany, Ltd., 30 Lesmill Road, Don Mills, Toronto,
Ontario.
Published in the United Kingdom by Constable
and Company, Ltd., 10 Orange Street, London
WC 2.

This Dover edition, first published in 1975, is an
unabridged and corrected republication of the work
first published in 1955 by the McGraw-Hill Book
Company, Inc.

International Standard Book Number: 0-486-61798-X
Library of Congress Catalog Card Number: 74-83620

Manufactured in the United States of America
Dover Publications, Inc.
180 Varick Street
New York, N.Y. 10014

PREFACE

This volume is concerned primarily with a relatively new field, the microwave spectroscopy of gases. The origins of microwave spectroscopy might be said to lie in the early high-frequency measurements of dielectric constants, or much more directly in the 1933 experiment of Cleeton and Williams on absorption of centimeter radiation by ammonia gas. However, a generally useful and accurate spectroscopy in the microwave region has come only after development of microwave oscillators and techniques and after demonstration in 1946 of the high resolution obtainable with low-pressure gases. Since 1946 microwave spectroscopy of gases has developed very rapidly and yielded a wide range of useful information in fields as far separated as nuclear physics, molecular structure, chemical kinetics, quantum electrodynamics, and astronomy. The purpose of this book is to present a systematic and fairly complete account of the large amount of theory, experimental data, and experimental know-how which has been developed, and to make this information more available to students, prospective research workers, and those interested in the many practical problems to which microwave spectroscopy may be applied.

The furious activity in microwave spectroscopy has made a book in this field both very desirable, because of the plethora of research work and results which need digestion and coordination, and at the same time very difficult because of the rapid development of ideas and outmoding of techniques. It is only now that microwave spectroscopy appears to have matured to the point where one can attempt a book of some perspective and lasting value.

Most microwave spectra are associated with molecules, although some important atomic microwave spectra occur and are treated here. Molecular spectra have previously been well described from the point of view of infrared spectroscopy; but the different frequency range, higher resolution, and greater accuracy of microwave spectroscopy makes available for study rather different types of phenomena such as hyperfine structure, pressure broadening, and Stark and Zeeman effects. The discussion is particularly full for those phases of the theory of molecular spectra which older types of spectroscopy have not been able to test adequately or which have been developed by persons interested in microwave spectra.   In addition, some attention is directed toward obtaining information about nuclear and molecular properties from the interpretation of molecular spectra.

Although a fairly complete account of microwave techniques is included, only those parts which are especially useful or somewhat peculiar to microwave spectroscopy are discussed in detail.   The reader can obtain further elaboration on standard microwave techniques from the large number of excellent books on microwaves which have been published in the last few years.   An attempt has been made to present not only the basic theory and properties of various types of microwave spectroscopes, but also some details of construction and specific operating characteristics which would be useful to the person faced with the problem of making and operating this type of instrument.

This volume is not primarily intended as a text, but rather as a readable reference which can be used both by students interested in one of the many aspects of microwave spectroscopy, and by those interested in research in this field.   The authors have endeavored to discuss the material critically, systematically, and in the simplest way consistent with some completeness in a single volume.   It is hoped that the simplicity of wording and mathematics (no group theory, as such, is used) will allow most of the discussion to be profitably read by those with a very elementary knowledge of quantum mechanics and atomic physics.

Considerable effort has been directed at making this volume valuable as a reference to a variety of users.   Although the treatment is a continuously developed one, attention has been directed at making each chapter and section as independent of other sections as is practical.   For example, certain expressions and definitions of symbols are often repeated in order to reduce the need for reference to other parts of the book. Appendixes give most of the tables and information needed in doing research and in interpreting microwave spectra.   They also contain extensive data on nuclear and molecular constants, including essentially all those determined by microwave techniques.   A complete list of publications concerning microwave spectroscopy and a subject index are included.   The tables and appendixes include material published up to January 1, 1955.

Microwave spectra of solids and other closely related types of spectroscopy are not discussed because it appeared that justice to such a wide variety of fields could hardly be done in one volume.   However, much of the material will be of use to those interested in types of microwave and radio-frequency spectroscopy not specifically discussed here.

The authors are grateful to their many colleagues in microwave physics who have provided information, criticism, and encouragement for this volume.   These include the following, who have given helpful comments on certain sections of the manuscript:

P. W. Anderson, T. S. Benedict, R. Beringer, D. G. Burkhard, D. K. Coles, B. P. Dailey, H. M. Foley, R. A. Frosch, R. H. Hughes, C. K. Jen, C. M. Johnson, W. C. King, D. R. Lide, H. Lyons, A. H. Nethercot, R. Novick, W. V. Smith, M. W. P. Strandberg, and E. B. Wilson, Jr. In addition, they wish to thank the many students and members of the Columbia Radiation Laboratory who have uncovered errors and provided valuable discussion.

The appendixes and tables have profited from the cooperation and assistance of a number of others whose help it is a pleasure to acknowledge. Inclusion of the extensive table in Appendix IV was made both possible and convenient through the efforts of T. E. Turner and G. Reitwiesner. J. A. Klein and G. C. Dousmanis did a major part of the work involved in compiling the bibliography and Appendix VI, respectively. L. C. Aamodt, J. F. Lotspeich, M. McDermott, and B. Herzog performed many of the computations necessary for the appendixes. J. Kraitchman assisted with Appendix III, in addition to checking many of the formulas and derivations.

One of the authors (C.H.T.) would like to thank Columbia University for grant of the E. K. Adams Fellowship which assisted in preparation of this work. The other (A.L.S.) would also like to thank this University for its hospitality during the time when much of the book was written.

CHARLES H. TOWNES
ARTHUR L. SCHAWLOW

CONTENTS

ix

LIST OF SYMBOLS

English letters

a magnetic hyperfine-structure interval factor

a_0 $a_0 = h^2/4\pi^2\mu e^2$ = radius of first Bohr orbit for hydrogen

A largest rotational constant of an asymmetric rotor

$\mathcal{A}$ dyadic (tensor) connecting angular momentum and magnetic hyperfine structure

b impact parameter, distance of closest approach of molecules

B rotational constant; intermediate rotational constant (usually in cycles/sec.)

B_0 rotational constant in ground vibrational state

B_e rotational constant in absence of zero-point vibrations

$\mathbf{B}$ magnetic induction

c velocity of light

C smallest rotational constant of an asymmetric rotor

C $C = F(F + 1) - I(I + 1) - J(J + 1)$

D dissociation energy

D_0, D_J, D_{JK} centrifugal distortion constants

D spectroscopic symbol for state with $l = 2$

$\mathbf{D}$ electric displacement

e electron charge or, in some cases, the proton charge

e subscript indicating equilibrium distance or conditions

E electric field intensity

$\mathbf{E}$ electric field

f fraction of molecules in state of interest

f_v fraction of molecules in vibrational state of interest

F quantum number for total angular momentum including nuclear spin

g g factor = magnetic moment (in Bohr magnetons or sometimes in nuclear magnetons) divided by angular momentum (in units $h/2\pi$). This is negative for electron spin or orbital motion

g_I nuclear g factor = nuclear magnetic moment (in nuclear magnetons) divided by angular momentum (in units $h/2\pi$)

h Planck's constant

$\hbar$ $\hbar = h/2\pi$, where h is Planck's constant

H   magnetic field intensity

H Hamiltonian

i   electric current

I moment of inertia of molecule

I spin of nucleus in units of $h/2\pi$

J rotational quantum number

J   total angular momentum excluding nuclear spin

k Boltzmann constant

K quantum number of component of angular momentum along molecular axis

l angular-momentum quantum number for polyatomic molecule with excited degenerate vibration

l electronic orbital angular-momentum quantum number for a single electron

L   electronic orbital angular momentum of an entire atom or molecule

m mass of electron

m atomic magnetic quantum number

M mass of nucleus

M projection of **J** on fixed direction, magnetic quantum number

M molecular weight

M_F projection of total angular momentum, including nuclear spin, on space-fixed axis

M_I projection of nuclear spin angular momentum on space-fixed axis

M_J projection of total angular momentum, excluding nuclear spin, on space-fixed axis

M_L projection of electronic orbital angular momentum on space-fixed axis

M_S projection of electron spin on space-fixed axis

$\mathfrak{M}$ dyadic (tensor) connecting angular momentum and magnetic moment

n atomic principal quantum number

N number of molecules per cubic centimeter

N   total orbital angular momentum including rotation of molecule

N_0 Avogadro's number; number of molecules in one molecular weight

O   orbital angular momentum due to rotation of molecule

P associated Legendre function

P   total angular momentum

P spectroscopic symbol for state with $l = 1$

q negative electric field gradient at nucleus $= \partial^2 V/\partial z^2$

q_J negative electric field gradient in the direction of J, or $(\partial^2 V/\partial z_J^2)_{av}$

q_l l-type doubling constant

q_Λ Λ-type doubling constant

Q nuclear quadrupole moment

Q resonator figure of merit

r internuclear distance

r intermolecular distance

r_e equilibrium distance between nuclei

R dipole moment matrix element

R Rydberg constant

s electron spin quantum number

s quantum number for internal torsional states

S degree of degeneracy for a symmetric top

S transition strength

S spectroscopic symbol for state with $l = 0$

S electron spin angular momentum

T absolute temperature

U potential energy

U_p number of unbalanced P electrons oriented along the bond

v vibrational quantum number

v linear velocity of molecule

V potential energy

W energy

x fractional importance of covalent bond

x_e anharmonicity vibrational constant

Y Dunham's molecular energy constant

Z nuclear charge, in units of the proton charge

Z characteristic impedance of a transmission line

Z_i effective nuclear charge

Greek letters

α attenuation constant

α rotation-vibration interaction constant

α fine-structure constant $2\pi e^2/hc$

α torsion angle of one part of a molecule with respect to another part

γ absorption coefficient

γ propagation constant

δ skin depth

$\Delta\nu$ line-breadth parameter

$\Delta\nu$ frequency bandwidth from which noise is passed by amplifier

ϵ dielectric constant

λ_g wavelength in waveguide

λ_0 wavelength in free space

η field gradient asymmetry

θ coordinate polar angle

κ Ray's asymmetry parameter $= (2B - A - C)/(A - C)$

Λ projection of electronic angular momentum on molecular axis

μ reduced mass

μ molecular electric dipole moment, or in some cases molecular magnetic dipole moment

μ magnetic permeability

$|\mu_{ij}|^2$ square of dipole moment matrix element, summed over directions

μ_n nuclear magneton $= he/4\pi Mc$, where M is proton mass

μ_0 Bohr magneton $= he/4\pi mc$, where m is electron mass

ν frequency

ν_0 resonant frequency; frequency absorbed by undisturbed molecule $= \omega_0/2\pi$

ν_1, ν_2, ν_3 frequencies of vibration of polyatomic molecule

π Zeeman components of line for transitions with $\Delta M = 0$

Π molecular state having one unit of electronic orbital angular momentum

ρ electric charge density

σ effective cross section area of molecule

σ Zeeman components of line for transitions with $\Delta M = \pm 1$

Σ molecular state having zero electronic orbital angular momentum

Σ projection of total electron spin on molecular axis

τ volume of space

τ integer used in specifying levels of asymmetric rotor

τ mean lifetime between collisions

τ resistivity

ψ wave function

ω angular frequency $= 2\pi\nu$

ω_e vibrational frequency (at equilibrium)

ω_0 2π times the natural molecular frequency $= 2\pi\nu_0$

Ω projection of total angular momentum, excluding nuclear spin, on molecular axis (absolute value)

Ω_I projection of nuclear spin angular momentum on molecular axis

Ω_F projection of total angular momentum, including nuclear spin, on molecular axis

INTRODUCTION

Some low-pressure gases selectively absorb electromagnetic radiation of particular wavelengths in the millimeter and centimeter range. This type of absorption can be observed in an experiment broadly represented by Fig. 1.

The source of microwaves (electromagnetic radiation of wavelength between 1 and 1000 mm) is usually an electronic tube, which emits radiation through a hollow metal pipe called a waveguide. The microwaves are detected after passage through a region of low-pressure gas (10 mm to 10^{-4} mm Hg pressure) by a silicon "crystal" or other detecting device. This detector produces an electrical signal proportional to the

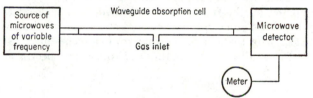

FIG. 1. Experiment for measuring microwave absorption.

microwave power which, after possible amplification, is observed on a meter or oscilloscope. As the frequency of the microwaves is varied, absorption appears as a sudden decrease in the voltage output of the detector.

Electronic techniques are characteristic of microwave spectroscopy, being involved in the production, detection, and amplification of microwaves. In some cases very sensitive electronic circuits are needed for proper detection and amplification, since the fractional power decrease may be quite small—as small as one part in 10^8 in an absorption path of 1 meter. In a few cases the absorption may be as much as 90 per cent in 1 meter path, and very easily detectable.

At gas pressures near 1 atm, a small microwave absorption may occur over a wide range of frequency. As the pressure is lowered, the range of frequency absorbed decreases proportionally down to pressures near 10^{-3} mm Hg, where the range is so small that the term absorption "line" is well merited. Very significantly, and contrary to experience in most other types of spectroscopy, the intensity of absorption in the center of the line does not appreciably decrease with this enormous decrease in pressure.

Because of the narrowness of absorption lines at low pressures, and the flexibility and sensitivity of electronic techniques, this type of experiment and its many refinements and ramifications form a basis for the precise, widely applicable microwave spectroscopy of gases which is the subject of this volume.

Consider now the frequencies absorbed. These must be interpretable in terms of the structure and behavior of the absorbing molecules. The motions (or transitions) of electrons in atoms and molecules are known to produce characteristically spectra in the optical and ultraviolet region. The slower vibrational motions of atoms in molecules are primarily responsible for the rich infrared spectra. It is the still slower end-over-end rotation of molecules which have characteristic frequencies so low that they lie in the microwave range and dominate microwave spectra.

Discussion of the interpretation of microwave spectra will begin with the rather simple diatomic molecules and progress in following chapters to successively more complex cases of linear polyatomic molecules, symmetric-top molecules, and asymmetric-top molecules.

Superimposed on the frequencies associated with molecular rotation are many interesting fine and hyperfine effects, some of which have been observed clearly for the first time by microwave techniques. These will be discussed after the broader outlines of rotational spectra have been treated.

CHAPTER 1

ROTATIONAL SPECTRA OF DIATOMIC MOLECULES

1-1. The Rigid Rotor. If the distance between nuclei in a diatomic molecule is considered fixed, the possible frequencies of the end-over-end rotation of this "rigid rotor" can be rather simply obtained. Using assumptions of the "old" quantum mechanics, the angular momentum must be some integral multiple of $h/2\pi$, so that

$$2\pi\nu I = \frac{Jh}{2\pi}$$

where h is Planck's constant, I is the molecular moment of inertia about axes perpendicular to the internuclear axis, ν is the frequency of rotation, and J is a positive integer giving the angular momentum in units of $h/2\pi$. Hence the frequencies expected from such a system are

$$\nu = \frac{Jh}{4\pi^2 I} \tag{1-1}$$

The moment of inertia I comes largely from the nuclei, where most of the molecular mass is concentrated, and for diatomic molecules of ordinary masses is of such size that for small integral values of J, the frequency ν is of the order 10,000 to 100,000 Mc, or the wavelength in the region 3 cm to 3 mm.

On this simple basis one might expect a rotation about the molecular axis to occur also and to have characteristic frequencies a few thousand times greater because the moment of inertia about this axis is produced by electrons, which are very much lighter than the nuclei. These frequencies lie then near the optical region, and in a very rough way the electronic frequencies may be regarded as due to this type of rotation about the molecular axis. Since these frequencies are very high, they lie far beyond the microwave range and are not ordinarily excited at room temperature. They will therefore be neglected in most of the following treatment. A somewhat more sophisticated and rigorous determination of the frequencies produced by a rigid diatomic molecule can be obtained by finding the permitted energy levels from wave mechanics (see [62], p. 271, or [305], p. 60). As the molecule rotates about its center of gravity, its orientation in space may be specified by the spherical

3

polar coordinates θ and ϕ. The wave equation may then be written

$$\frac{h^2}{8\pi^2 I} \left[\frac{1}{\sin\theta} \frac{\partial}{\partial\theta} \left(\sin\theta \frac{\partial\psi}{\partial\theta} \right) + \frac{1}{\sin^2\theta} \frac{\partial^2\psi}{\partial\phi^2} \right] + W\psi = 0 \qquad (1\text{-}2)$$

where ψ is the wave function and W the rotational energy of the molecule. The variables θ and ϕ may be separated by substituting

$$\psi = \Theta(\theta)\Phi(\phi)$$

which gives

$$\frac{d^2\Phi}{d\phi^2} = -M^2\Phi \qquad (1\text{-}3)$$

and

$$\frac{h^2}{8\pi^2 I} \left[\frac{1}{\sin\theta} \frac{d}{d\theta} \left(\sin\theta \frac{d\Theta}{d\theta} \right) - \frac{M^2\Theta}{\sin^2\theta} \right] + W\Theta = 0 \qquad (1\text{-}4)$$

where M^2 is an arbitrary constant.

Solutions of these equations which are single-valued and normalized can be obtained only when

$$W = \frac{h^2}{8\pi^2 I} J(J+1)$$

where J is a positive integer and M is an integer such that $|M| \leq J$. Such solutions are

$$\Phi_M = \frac{1}{\sqrt{2\pi}} e^{iM\phi} \qquad (1\text{-}5)$$

$$\Theta_{MJ} = \left[\frac{(2J+1)(J-|M|)!}{2(J+|M|)!} \right]^{\frac{1}{2}} P_J^{|M|}(\cos\theta) \qquad (1\text{-}6)$$

where $P_J^{|M|}(\cos\theta)$ is an associated Legendre function. $[J(J+1)](h^2/4\pi^2)$ is the square of the total angular momentum, so that the angular momentum may for convenience be designated by J. Similarly the projection of the angular momentum on the polar axis is given by $M(h/2\pi)$, or simply by the integer M.

The frequency observed when the molecule makes a transition between a lower state of energy W_1 and an upper state of energy W_2 is given by

$$\nu = \frac{W_2 - W_1}{h} = \frac{h}{8\pi^2 I} [J_2(J_2+1) - J_1(J_1+1)] \qquad (1\text{-}7)$$

From the correspondence principle, these frequencies may be expected to approximately equal the frequencies given by expression (1-1); hence J_2 should equal $J_1 + 1$, and

$$\nu = 2B(J+1) \qquad (1\text{-}8)$$

where J is the angular-momentum quantum number for the lower state (J_1), and $B = (h/8\pi^2 I)$ is called the rotational constant. The quantity

B is often expressed in units of cm^{-1} for infrared spectroscopy. In that case $B = (h/8\pi^2 Ic)$. For microwave spectroscopy, B will generally be given in cycles per second, or $B = h/8\pi^2 I$. However, numerical values will usually be quoted in megacycles, or 10^6 cycles/sec. The selection rule that $J_2 = J_1 + 1$ or $\Delta J = \pm 1$ for dipole radiation of a diatomic molecule will be more rigorously demonstrated in the discussion of intensities later in this chapter.

1-2. Energy Levels of the Diatomic Molecule. From Eq. (1-8) it is seen that the spectrum of a rigid rotor consists of absorption lines equally spaced in frequency with an interval $2B$. Although the rigid rotor is an idealization to which actual molecules conform to a good approximation, accurate spectroscopic measurement reveals many deviations from this approximation. As J increases and the molecule rotates faster, it stretches so that the moment of inertia increases. Moreover, the nuclei vibrate back and forth along the line joining them even in the lowest vibrational state. A much greater difficulty from the point of view of obtaining a complete theoretical treatment is that the entire molecular system, composed of interacting electrons as well as nuclei, is so complicated that an exact quantum-mechanical solution is impossible.

However, since the electrons are very much lighter than the nuclei and move in electric fields of approximately the same intensity, the electron motion is very much faster than that of the nuclei; *i.e.*, many cycles of the electronic motion occur during a small portion of a cycle of the nuclear motion. It is therefore reasonable to treat first the electronic motion, considering the nuclei as fixed. Then the internuclear distance r appears as a parameter. In this way the electrons are found to be capable of occupying several states, each giving the molecule a particular value of the energy U, for each internuclear distance. Generally in microwave spectroscopy only the lowest of these electronic states is important.

As the internuclear distance is slowly varied, the electronic energy varies. Because the electronic motion is so fast in comparison with the nuclear motion, at each instant the electronic energy may be considered to have reached its equilibrium value corresponding to that distance. Thus we are justified in treating the vibration and rotation of the nuclei separately from the electronic motion. In this treatment $U(r)$, which is the sum of the electron energy plus energy of electrostatic interaction between the two nuclei, appears as the potential energy. The validity of this approximation was discussed by Born and Oppenheimer ([8]; see also [62], pp. 259–274, and [21], Chap. I). They showed that the entire molecular energy, including that due to electronic motion, can be expanded in powers of $(m/M)^{\frac{1}{4}}$, where m is the electronic mass and M an average nuclear mass. Separation of nuclear and electronic motions hence corresponds to selecting the larger terms of the series expansion and neglecting those which are smaller by $(m/M)^{\frac{1}{2}}$ or more. In some cases the neglected terms lead to observable effects, but they can only with difficulty be taken into account.

Using the approximation that the variation in electron energy with nuclear motion may be included in the potential $U(r)$, the wave equation for vibration and rotation of a diatomic molecule becomes

$$\frac{1}{M_1} \nabla_1^2 \psi + \frac{1}{M_2} \nabla_2^2 \psi + \frac{8\pi^2}{h^2} [W - U(r)]\psi = 0 \qquad (1\text{-}9)$$

in which ψ is the wave function for the nuclear motion, M_1 and M_2 are the nuclear masses, and

$$\nabla_i^2 = \frac{\partial^2}{\partial x_i^2} + \frac{\partial^2}{\partial y_i^2} + \frac{\partial^2}{\partial z_i^2} \qquad \text{where } i = 1 \text{ or } 2 \qquad (1\text{-}10)$$

x_i, y_i, and z_i being Cartesian coordinates of the ith nucleus relative to axes fixed in space.

Transforming to spherical polar coordinates r, θ, ϕ of the second nucleus relative to the first as origin (cf. [62], p. 264),

$$\frac{1}{r^2} \frac{\partial}{\partial r} \left(r^2 \frac{\partial \psi}{\partial r} \right) + \frac{1}{r^2 \sin \theta} \frac{\partial}{\partial \theta} \left(\sin \theta \frac{\partial \psi}{\partial \theta} \right) + \frac{1}{r^2 \sin^2 \theta} \frac{\partial^2 \psi}{\partial \phi^2}$$
$$+ \frac{8\pi^2 \mu}{h^2} [W - U(r)]\psi = 0 \quad (1\text{-}11)$$

where μ is the reduced mass, $M_1 M_2/(M_1 + M_2)$. The variables may be separated by the substitution

$$\Psi = R(r)\Theta(\theta)\Phi(\phi) \qquad (1\text{-}12)$$

$\Theta(\theta)$ and $\Phi(\phi)$ turn out to be the same as the wave functions found above for the rigid rotor.

The radial wave function $R(r)$ obtained by the separation process is given by

$$\frac{1}{r^2} \frac{d}{dr} \left(r^2 \frac{dR}{dr} \right) + \left\{ \frac{8\pi^2 \mu}{h^2} [W - U(r)] - \frac{J(J+1)}{r^2} \right\} R = 0 \quad (1\text{-}13)$$

The term $J(J + 1)/r^2$ may be regarded as a potential energy associated with the centrifugal force due to the rotational angular momentum J. Substituting the expression

$$R(r) = \frac{1}{r} S(r) \qquad (1\text{-}14)$$

we get

$$\frac{d^2 S}{dr^2} + \left\{ -\frac{J(J+1)}{r^2} + \frac{8\pi^2 \mu}{h^2} [W - U(r)] \right\} S = 0 \qquad (1\text{-}15)$$

The solutions of Eq. (1-15) will obviously depend on the form of $U(r)$. Since it is seldom possible actually to solve the electronic wave equation, it is customary to use an empirical expression for $U(r)$.

From experimental studies of molecular spectra and from calculations on simple molecules, the general form of $U(r)$ is known to be that of Fig. 1-1 (see [471]). At large distances the atoms are independent, and the force between them is negligible. Their energy is then just the sum of the energies of the individual atoms. At very small distances, when the atoms are "in contact," they must repel each other. At some intermediate distance there must be a potential minimum, corresponding to the equilibrium distance of the atoms.

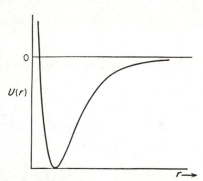

FIG. 1-1. Variation of molecular potential energy $U(r)$ with internuclear distance r.

Solution for Morse Potential. A potential which fulfills these requirements is the Morse function [16]

$$U(r) = D(1 - e^{-a(r-r_e)})^2 \tag{1-16}$$

where D = dissociation energy of the molecule
r_e = equilibrium distance between nuclei
a = a constant

The Morse function differs from the true potential at $r = 0$, where the actual potential would be extremely large. However, the Morse potential is also quite large at $r = 0$ and this is a region where the wave function of the vibrating rotor is expected to be small so that the discrepancy is not serious.

Using the Morse potential function, the radial equation (1-15) becomes

$$\frac{d^2S}{dr^2} + \left[-\frac{J(J+1)}{r^2} + \frac{8\pi^2\mu}{h^2}(W - D - De^{-2a(r-r_e)} + 2De^{-a(r-r_e)}) \right]S = 0 \tag{1-17}$$

The solution of this equation for $J = 0$ has been given by Morse [16] and for any J by Pekeris [52]. Substituting

$$y = e^{-a(r-r_e)} \quad \text{and} \quad A = J(J+1)\frac{h^2}{8\pi^2\mu r_e^2} \tag{1-18}$$

in Eq. (1-17), we obtain

$$\frac{d^2S}{dy^2} + \frac{1}{y}\frac{dS}{dy} + \frac{8\pi^2\mu}{a^2h^2}\left(\frac{W-D}{y^2} + \frac{2D}{y} - D - \frac{Ar_e^2}{y^2r^2}\right)S = 0 \tag{1-19}$$

For $A \neq 0$, it is necessary to expand r_e^2/r^2 in terms of y:

$$\frac{r_e^2}{r^2} = \frac{1}{[1 - (\ln y)/ar_e]^2} = 1 + \frac{2}{ar_e}(y - 1)$$

$$+ \left(-\frac{1}{ar_e} + \frac{3}{a^2 r_e^2}\right)(y - 1)^2 + \cdots \quad (1\text{-}20)$$

If the first three terms of this Taylor expansion are retained, Eq. (1-19) becomes

$$\frac{d^2 S}{dy^2} + \frac{1}{y}\frac{dS}{dy} + \frac{8\pi^2 \mu}{a^2 h^2}\left(\frac{W - D - c_0}{y^2} + \frac{2D - c_1}{y} - D - c_2\right)S = 0 \quad (1\text{-}21)$$

in which

$$c_0 = A\left(1 - \frac{3}{ar_e} + \frac{3}{a^2 r_e^2}\right)$$

$$c_1 = A\left(\frac{4}{ar_e} - \frac{6}{a^2 r_e^2}\right) \quad (1\text{-}22)$$

$$c_2 = A\left(-\frac{1}{ar_e} + \frac{3}{a^2 r_e^2}\right)$$

Eq. (1-21) can be further simplified by the substitutions

$$S(y) = e^{-z/2}z^{b/2}F(z) \qquad z = 2dy$$

$$d^2 = \frac{8\pi^2 \mu}{a^2 h^2}(D + c_2) \qquad b^2 = -\frac{32\pi^2 \mu}{a^2 h^2}(W - D - c_0) \quad (1\text{-}23)$$

so that it becomes

$$\frac{d^2 F}{dz^2} + \left(\frac{b + 1}{z} - 1\right)\frac{dF}{dz} + \frac{v}{z}F = 0 \quad (1\text{-}24)$$

where

$$v = \frac{4\pi^2 \mu}{a^2 h^2 d}(2D - c_1) - \tfrac{1}{2}(b + 1) \quad (1\text{-}25)$$

As in the usual quantum-mechanical treatment of the simple harmonic oscillator or of the hydrogen atom (*cf.* [62]), for the solution of Eq. (1-24) to be finite and vanish at the ends of its range, it must be given by a terminating series, *i.e.*, a polynomial. In fact, Eq. (1-24) is identical in form with the equation for Laguerre polynomials found in the solution of the hydrogen atom. This requirement can be shown to restrict v to the values 0, 1, 2, ... Strictly speaking, the solutions thus obtained satisfy the boundary condition $S \to 0$ as $r \to -\infty$ rather than the proper condition $S \to 0$ as $r \to 0$. Ter Haar [156] has examined this approximation and shown that it is usually a good one.

It is possible to solve for W using Eqs. (1-25), (1-23), (1-22), and

(1-18), which give

$$W_{Jv} = D + c_0 - \frac{(D - \frac{1}{2}c_1)^2}{D + c_2} + \frac{ah(D - \frac{1}{2}c_1)}{\pi \sqrt{2\mu} \sqrt{D + c_2}} \left(v + \frac{1}{2}\right)$$
$$- \frac{a^2 h^2}{8\pi^2 \mu} \left(v + \frac{1}{2}\right)^2 \quad (1\text{-}26)$$

Expanding Eq. (1-26) in powers of c_1/D and c_2/D, it takes the form:

$$\frac{W_{Jv}}{h} = \omega_e(v + \tfrac{1}{2}) - x_e\omega_e(v + \tfrac{1}{2})^2 + J(J + 1)B_e - D_e J^2(J + 1)^2$$
$$- \alpha_e(v + \tfrac{1}{2})J(J + 1) \quad (1\text{-}27)$$

where

$$\omega_e = \frac{a}{2\pi} \sqrt{\frac{2D}{\mu}} \qquad x_e = \frac{h\omega_e}{4D} \qquad B_e = \frac{h}{8\pi^2 I_e}$$
$$D_e = \frac{h^3}{128\pi^6 \mu^3 \omega_e^2 r_e^6} = \frac{4B_e^3}{\omega_e^2} \qquad (1\text{-}28)$$
$$\alpha_e = \frac{3h^2\omega_e}{16\pi^2 \mu r_e^2 D} \left(\frac{1}{ar_e} - \frac{1}{a^2 r_e^2}\right) = 6\sqrt{\frac{x_e B_e^3}{\omega_e}} - \frac{6B_e^2}{\omega_e}$$

ω_e, α_e, B_e in (1-27) and (1-28) are expressed in cycles per second. The terms in Eq. (1-27) can be identified with the solutions of more specialized problems, so that each can be given a physical significance. Thus the first term involving $(v + \frac{1}{2})$ has the form of the solution of the wave equation of a pure vibrator with a harmonic potential. The second term is obtained when the vibrator potential is made anharmonic by the addition of a cubic term in the potential energy. A term of the form $BJ(J + 1)$ is just that obtained in Eq. (1-4), the solution of the rigid rotor problem, while the next to last term comes from centrifugal stretching of the rotating molecule. The last term allows for the change in average moment of inertia due to vibration and the consequent change in rotational energy.

Dunham's Solution for Energy Levels. Some other more refined potentials have been used for problems in optical spectra involving excited rotational or vibrational states ([471], pp. 102, 108). Dunham [34] has calculated the energy levels of a vibrating rotor, by a Wentzel-Kramers-Brillouin method, for any potential which can be expanded as a series of powers of $(r - r_e)$ in the neighborhood of the potential minimum. This treatment shows that the energy levels can be written in the form

$$F_{vJ} = \sum_{l,j} Y_{lj}(v + \tfrac{1}{2})^l J^j(J + 1)^j \quad (1\text{-}29)$$

where l and j are summation indices, v and J are, respectively, vibrational and rotational quantum numbers, and Y_{lj} are coefficients which depend on molecular constants. The effective potential function of the vibrating rotor may be written in the form

$$U = a_0\xi^2(1 + a_1\xi + a_2\xi^2 + \cdots) + B_eJ(J + 1)(1 - 2\xi + 3\xi^2 - 4\xi^3 + \cdots) \quad (1\text{-}30)$$

where $\xi = (r - r_e)/r_e$, $B_e = h/8\pi^2\mu r_e^2$. The term involving $B_eJ(J + 1)$ allows for the influence of the rotation on the effective potential.

Dunham [34] shows that the first 15 Y_{lj}'s are

$$
\begin{aligned}
Y_{00} &= B_e/8(3a_2 - 7a_1^2/4) \\
Y_{10} &= \omega_e[1 + (B_e^2/4\omega_e^2)(25a_4 - 95a_1a_3/2 - 67a_2^2/4 \\
&\quad + 459a_1^2a_2/8 - 1155a_1^4/64)] \\
Y_{20} &= (B_e/2)[3(a_2 - 5a_1^2/4) + (B_e^2/2\omega_e^2)(245a_6 - 1365a_1a_5/2 \\
&\quad - 885a_2a_4/2 - 1085a_3^2/4 + 8535a_1^2a_4/8 + 1707a_2^3/8 \\
&\quad + 7335a_1a_2a_3/4 - 23{,}865a_1^3a_3/16 - 62{,}013a_1^2a_2^2/32 \\
&\quad + 239{,}985a_1^4a_2/128 - 209{,}055a_1^6/512)] \\
Y_{30} &= (B_e^2/2\omega_e)(10a_4 - 35a_1a_3 - 17a_2^2/2 + 225a_1^2a_2/4 \\
&\quad - 705a_1^4/32) \\
Y_{40} &= (5B_e^3/\omega_e^2)(7a_6/2 - 63a_1a_5/4 - 33a_2a_4/4 - 63a_3^2/8 \\
&\quad + 543a_1^2a_4/16 + 75a_2^3/16 + 483a_1a_2a_3/8 - 1953a_1^3a_3/32 \\
&\quad - 4989a_1^2a_2^2/64 + 23{,}265a_1^4a_2/256 - 23{,}151a_1^6/1024)
\end{aligned} \quad (1\text{-}31)
$$

$$
\begin{aligned}
Y_{01} &= B_e\{1 + (B_e^2/2\omega_e^2)[15 + 14a_1 - 9a_2 + 15a_3 - 23a_1a_2 \\
&\quad + 21(a_1^2 + a_1^3)/2]\} \\
Y_{11} &= (B_e^2/\omega_e)\{6(1 + a_1) + (B_e^2/\omega_e^2)[175 + 285a_1 - 335a_2/2 \\
&\quad + 190a_3 - 225a_4/2 + 175a_5 + 2295a_1^2/8 - 459a_1a_2 \\
&\quad + 1425a_1a_3/4 - 795a_1a_4/2 + 1005a_2^2/8 - 715a_2a_3/2 \\
&\quad + 1155a_1^3/4 - 9639a_1^2a_2/16 + 5145a_1^2a_3/8 \\
&\quad + 4677a_1a_2^2/8 - 14{,}259a_1^3a_2/16 \\
&\quad + 31{,}185(a_1^4 + a_1^5)/128]\} \\
Y_{21} &= (6B_e^3/\omega_e^2)[5 + 10a_1 - 3a_2 + 5a_3 - 13a_1a_2 \\
&\quad + 15(a_1^2 + a_1^3)/2] \\
Y_{31} &= (20B_e^4/\omega_e^3)[7 + 21a_1 - 17a_2/2 + 14a_3 - 9a_4/2 + 7a_5 \\
&\quad + 225a_1^2/8 - 45a_1a_2 + 105a_1a_3/4 - 51a_1a_4/2 + 51a_2^2/8 \\
&\quad - 45a_2a_3/2 + 141a_1^3/4 - 945a_1^2a_2/16 + 435a_1^2a_3/8 \\
&\quad + 411a_1a_2^2/8 - 1509a_1^3a_2/16 + 3807(a_1^4 + a_1^5)/128]
\end{aligned} \quad (1\text{-}32)
$$

$$
\begin{aligned}
Y_{02} &= -(4B_e^3/\omega_e^2)\{1 + (B_e^2/2\omega_e^2)[163 + 199a_1 - 119a_2 + 90a_3 \\
&\quad - 45a_4 - 207a_1a_2 + 205a_1a_3/2 - 333a_1^2a_2/2 + 693a_1^2/4 \\
&\quad + 46a_2^2 + 126(a_1^3 + a_1^4/2)]\} \\
Y_{12} &= -(12B_e^4/\omega_e^3)(\tfrac{19}{2} + 9a_1 + 9a_1^2/2 - 4a_2) \\
Y_{22} &= -(24B_e^5/\omega_e^4)[65 + 125a_1 - 61a_2 + 30a_3 - 15a_4 \\
&\quad + 495a_1^2/4 - 117a_1a_2 + 26a_2^2 + 95a_1a_3/2 - 207a_1^2a_2/2 \\
&\quad + 90(a_1^3 + a_1^4/2)]
\end{aligned} \quad (1\text{-}33)
$$

$$Y_{03} = 16B_e^5(3 + a_1)/\omega_e^4$$
$$Y_{13} = (12B_e^6/\omega_e^5)(233 + 279a_1 + 189a_1^2 + 63a_1^3 - 88a_1a_2$$
$$\qquad - 120a_2 + 80a_3/3)$$
$$Y_{04} = -(64B_e^7/\omega_e^6)(13 + 9a_1 - a_2 + 9a_1^2/4) \tag{1-34}$$

It should be noted that B_e is generally much smaller than ω_e. For most molecules the ratio B_e^2/ω_e^2 is of the order of 10^{-6}, although for light molecules such as H_2 it approaches more nearly to 10^{-3}. In such cases more terms are required in the expressions for the various coefficients.

If B_e/ω_e is small, the Y's can be related to the ordinary band spectrum constants as follows:

$$
\begin{array}{llll}
Y_{10} \approx \omega_e & Y_{20} \approx -\omega_e x_e & Y_{30} \approx \omega_e y_e & \\
Y_{01} \approx B_e & Y_{11} \approx -\alpha_e & Y_{21} \approx \gamma_e & (1\text{-}35) \\
Y_{02} \approx -D_e & Y_{12} \approx -\beta_e & Y_{40} \approx \omega_e z_e & \\
Y_{03} \approx H_e & & &
\end{array}
$$

where these symbols refer to the coefficients in the Bohr theory expansion for the molecular energy levels:

$$F_{vJ} = \omega_e(v + \tfrac{1}{2}) - \omega_e x_e(v + \tfrac{1}{2})^2 + \omega_e y_e(v + \tfrac{1}{2})^3 + \omega_e z_e(v + \tfrac{1}{2})^4$$
$$\qquad + B_v J(J + 1) - D_e J^2(J + 1)^2 + H_e J^3(J + 1)^3 + \cdots \tag{1-36}$$

where $B_v = B_e - \alpha_e(v + \tfrac{1}{2}) + \gamma_e(v + \tfrac{1}{2})^2 \ldots$ (cf. [471], p. 92, pp. 107–108).

Sandeman [103] has extended Dunham's treatment to include other terms of the same order of magnitude which involve higher powers of the vibrational quantum number.

For the special case of the Morse potential function, Dunham shows that all the Y_{l0}'s except Y_{10} and Y_{20} vanish and all but the first terms in the expressions for Y_{10} and Y_{20} are zero. Because of the simplicity of the expressions obtained with the Morse function, and because it does give a quite good fit to the actual potential in the region of $r = r_e$, the Morse function has been widely used.

Dependence of Energy on Isotopic Masses. Since the frequencies of lines in microwave spectra can be measured with great precision, and since they can be used to evaluate the molecular moment of inertia, they permit an accurate evaluation of atomic or nuclear masses, or rather the mass ratios of isotopic nuclei.

To a good approximation we can use the Morse potential solution. The usual expansion for energy levels, appropriate to the Morse potential or other similar potentials, is given by (1-27), from which the frequency of a microwave rotational transition, where J changes by one unit, is easily shown to be

$$\nu = \frac{W_{J+1} - W_J}{h} = 2B_e(J + 1) - 2\alpha_e(v + \tfrac{1}{2})(J + 1) - 4D_e(J + 1)^3$$
$$\qquad\qquad = 2B_v(J + 1) - 4D_e(J + 1)^3 \tag{1-37}$$

The constants B_e, α_e, and D_e are usually expressed in cm^{-1} in optical work. In the above formula they, and therefore the frequency, are in cycles per second, which may be divided by 10^6 to convert to megacycles, the most usual unit for microwave work.

B_e and α_e can be evaluated directly from microwave spectra if two lines can be measured with different values of v; for instance, the same rotational transition in the ground vibrational state and the first excited vibrational state. The term in $(J + 1)^3$ is often negligible because $D_e = (4B_e^3/\omega_e^2)$ is smaller in magnitude than B_e by $4(B_e/\omega_e)^2$, or approximately 10^{-5} for most molecules. However, for very light molecules or large J this term may be rather prominent. When required it can be calculated with sufficient accuracy from $B_0 \approx B_e$ and ω_e, which is usually obtainable from optical spectra.

If the nuclear masses are known from mass spectrographic or other measurements, a determination of B_e allows an evaluation of the internuclear distance r_e, since B_e is related to the moment of inertia I_e.

$$r_e = \sqrt{\frac{I_e}{\mu}} = \sqrt{\frac{h}{8\pi^2 B_e \mu}} \qquad (1\text{-}38)$$

where $\mu = M_1 M_2/(M_1 + M_2)$ is the reduced mass. The accuracy with which r_e can be determined for a diatomic molecule is limited mainly by the error in Planck's constant h, which is required to calculate I_e from B_e. The best available value of this constant is

$$h = (6.6252 \pm 0.0005) \times 10^{-27} \text{ erg-sec}$$

[795] so that r_e can be determined to an accuracy of about 1 part in 6000. It is often convenient to have B_e in megacycles, r_e in angstroms, and μ in atomic mass units. In these units

$$I_e = \frac{5.055 \times 10^5}{B_e} = \frac{I_e \text{ in cgs units}}{1.6598 \times 10^{-40}} \qquad (1\text{-}39)$$

and

$$r_e = \sqrt{\frac{5.055 \times 10^5}{\mu B_e}} \qquad \text{angstrom units} \qquad (1\text{-}40)$$

Table 1a gives the constants of a number of representative diatomic molecules. Table 1b lists certain constants of one isotopic species of all diatomic molecules whose microwave rotational spectra have been studied.

If the spectroscopic constants have been measured for one isotopic species of a molecule, their values for other species may be found from the following relations which are deducible from Eq. (1-28):

$$\omega_e \propto \frac{1}{\sqrt{\mu}} \qquad B_e \propto \frac{1}{\mu} \qquad \alpha_e \propto \frac{1}{\mu^{\frac{3}{2}}} \qquad D_e \propto \frac{1}{\mu^2} \qquad (1\text{-}41)$$

The values in Table 1a have been calculated with the aid of these relations in some cases.

TABLE 1-1a. MOLECULAR CONSTANTS OF SOME REPRESENTATIVE DIATOMIC MOLECULES

Molecule	Y_{01} (approx. B_e), Mc	Y_{01} (approx. B_e), cm^{-1}	α_e, Mc	I_e, A$^2 \times$ atomic mass units	r_e, A	ω_e, cm^{-1}	$D_e = \dfrac{4B_e^3}{\omega_e^2}$, Mc	μ, debyes	Reference
H^{1}Cl^{35}	317,510	10.591	9050	1.592	1.275	2989.74	15.94	1.18	[336a] [471]
DI^{127}	($B_0 = 97,537.2$)	($B_0 = 3.25348$)	1840	($I_0 = 5.183$)	1.604	1630	1.56	0.38	[755a] [827a] [782b]
C^{12}O^{16}	57,897.5	1.93124	524.0	8.731	1.128	2170.21	0.1834	0.10	[336a] [457]
C^{13}O^{16}	55,344.9	1.84610	488.3	9.134	1.128	2074.81	0.1753	0.10	[457]
Cl^{35}F^{19}	15,483.69	0.516479	130.67	32.65	1.628	793.2	0.02626	0.88	[366]
Cl^{37}F^{19}	15,189.22	0.506657	126.96	33.28	1.628	778.6	0.02527	0.88	[366]
Br^{79}F^{19}	10,706.9	0.357143	156.3	47.21	1.759	671	0.0121	1.29	[534]
Br^{81}F^{19}	10,655.7	0.355435	155.8	47.44	1.759	670	0.0121	1.29	[534]
K^{44}Cl^{35}	3,767.394	0.125667	22.865	134.2	2.667	300	0.003	10.48	[835] [938]
K^{39}Cl^{35}	3,856.370	0.128634	23.680	131.1	2.667	300	0.003	10.48	[835] [938]
K^{39}Cl^{37}	3,746.583	0.124972	22.676	134.9	2.667	300	0.003	10.48	[835] [938]
I^{127}Cl^{35}	3,422.300	0.114155	16.06	147.7	2.321	384.2	0.00121	0.65	[330]
I^{127}Cl^{37}	3,277.365	0.109320	15.05	154.2	2.321	376	0.00111	0.65	[330]

TABLE 1-1b. MOLECULAR CONSTANTS OF DIATOMIC MOLECULES WHOSE
ROTATIONAL SPECTRA HAVE BEEN MEASURED IN THE
MICROWAVE REGION

Molecule	Y_{01} (approx B_e), Mc	α_e, Mc	r_e, 10^{-8} cm or angstroms	μ, 10^{-18} esu or debyes	Reference
DBr^{79}	($B_0 = 127{,}358.2$)	2500	1.414	0.79	[336a] [676c] [927]
DI	($B_0 = 97{,}537.2$)	1840	1.604	0.38	[755a] [782b] [827a]
FCl^{35}	15,483.69	130.67	1.628	0.881	[366]
$Br^{79}Cl^{35}$	4,570.92	23.22	2.138	0.57	[535]
ICl^{35}	3,422.300	16.06	2.321	0.65	[330]
FBr^{79}	10,706.9	156.3	1.759	1.29	[534]
$C^{12}O^{16}$	57,897.5	524.0	1.128	0.10	[336a] [457]
$C^{12}S^{32}$	24,584.352	177.544	1.535	2.0	[777]
$N^{14}O^{16}$	51,084.5	534	1.151	0.16	[336a] [782b] [924]
Li^6Br^{79}	19,161.51	208.8	2.170	6.19	[938]
Li^6I	15,381.45	152.6	2.392	6.25	[938]
$NaCl^{35}$	6,536.86	48.28	2.361	8.5	[751] [987]
$NaBr^{79}$	4,534.51	28.25	2.502	......	[938]
NaI	3,531.76	19.44	2.712	......	[938]
$K^{39}Cl^{35}$	3,856.370	23.680	2.667	10.48	[835] [987]
$K^{39}Br^{79}$	2,434.947	12.136	2.821	10.41	[799]
$K^{39}I$	1,825.01	8.034	3.048	11.05	[938]
$Rb^{85}Cl^{35}$	2,627.400	13.601	2.787	......	[938]
$Rb^{85}Br^{79}$	1,424.83	5.578	2.945	......	[938]
$Rb^{85}I$	984.31	3.281	3.177	......	[938]
CsF	5,527.27	33.13	2.345	7.874	[423a] [938]
$CsCl^{35}$	2,161.20	10.085	2.906	10.40	[751] [606a] [938]
$CsBr^{79}$	1,081.34	3.718	3.072	......	[938]
CsI	708.36	2.044	3.315	12.1	[938]

1-3. Mass Measurements. Once B_e has been measured for two isotopic species of a molecule, the reduced mass ratio is given directly by the
ratio of the B_e's. That is,

$$\frac{\mu^{(1)}}{\mu^{(2)}} = \frac{B_e^{(2)}}{B_e^{(1)}} = \frac{M_1(M + M_2)}{M_2(M + M_1)} \qquad (1\text{-}42)$$

or

$$\frac{M_1}{M_2} = \frac{(M/M_2)(B_e^{(2)}/B_e^{(1)})}{1 + M/M_2 - B_e^{(2)}/B_e^{(1)}} \qquad (1\text{-}43)$$

where M_1 and M_2 are the masses of the two isotopes; M is the mass of the
other nucleus in the molecule. From (1-43) and microwave measurements of $B_e^{(2)}/B_e^{(1)}$, the mass ratio M_1/M_2 can be obtained with great
precision. The mass ratio M/M_2 need be known only moderately accurately since it enters into both the numerator and denominator. Planck's
constant and the other constants required to compute r_e from B_e do not
enter at all. By this procedure the mass ratio of the chlorine 35 to the

chlorine 37 isotope has been found from the spectra of ICl and FCl. The values obtained are compared in Table 1-2 with another microwave measurement in the triatomic molecule ClCN and with values from mass-spectroscopic and nuclear-disintegration work. It may be seen from this table that they agree well with other determinations and represent very accurate determinations of the Cl^{35}/Cl^{37} mass ratio.

TABLE 1-2. VALUE OF Cl^{35}/Cl^{37} MASS RATIO

Method	Value	Reference
Mass spectroscopy	0.9459777 ± 20	[562a]
Nuclear reaction	0.9459893 ± 110	[110a]
Microwave (ICl)	0.9459801 ± 50	[330]
Microwave (FCl)	0.9459775 ± 40	[366]
Microwave-molecular beam (KCl)	0.9459803 ± 15	[835]
Microwave (CsCl)	0.9459781 ± 30	[938]
Microwave (ClCN)	0.9459906 ± 120	[490]

Within the limits of error given in Table 1-2 for the microwave measurements, there seem to be no theoretical uncertainties which should affect mass ratios. However, microwave measurements can be made considerably more precise, so that it is important to examine even small effects which might possibly cause errors. Such effects have been observed in optical spectra of hydrides where the large mass ratio between H^1 and H^2, the rapid rotation, and vibration with large amplitudes make them unusually large (see, for example, [57], [58], and [71]).

The more important errors in measuring mass ratios from rotational spectra of diatomic molecules (other than simply inaccurate measurements of B_e) may be grouped under three causes:

1. Anharmonicity of the potential function
2. Uncertainties in electronic behavior, including L uncoupling
3. Inaccurate knowledge of the mass of the other atom in the molecule, or in M/M_2

Anharmonicity has been partly taken into account by the use of the Morse function, but if this is not a sufficiently good approximation to the potential curve, Dunham's method must be used.

The Dunham coefficient corresponding most nearly to B_e is Y_{01}, but actually

$$Y_{01} = B_e\{1 + B_e^2/2\omega_e^2[15 + 14a_1 - 9a_2 + 15a_3 - 23a_1a_2 \\ + 21(a_1^2 + a_1^3)/2]\} \quad (1\text{-}31)$$

where $a_1, a_2, \ldots$ are the coefficients in the expansion for the potential curve in terms of powers of $(r - r_e)/r_e$ [cf. (1-30)]. Then

$$Y_{01} = B_e[1 + (B_e^2/\omega_e^2)\beta_{01}] \quad (1\text{-}44)$$

where β_{01} does not depend on M_1, M_2

$$Y_{01}(\mu_1)/Y_{01}(\mu_2) = \mu_2/\mu_1 \left[1 + \beta_{01} \left(\frac{B_e^2}{\omega_e^2} \right) \left(\frac{\mu_2 - \mu_1}{\mu_2} \right) \right] \quad (1\text{-}45)$$

β_{01} can be calculated from spectroscopically observable quantities, as Dunham shows.

$$\beta_{01} = Y_{10}^2 Y_{21}/4Y_{01}^3 + 16a_1 Y_{20}/3Y_{01} - 8a_1 - 6a_1^2 + 4a_1^3 \quad (1\text{-}46)$$
$$a_1 = Y_{11}Y_{10}/6Y_{01}^2 - 1$$

Since β_{01} is multiplied by B_e^2/ω_e^2 in the expression for Y_{01}, it will enter as a small correction, and it is sufficiently accurate to replace the coefficients entering into it by their approximate values from (1-35), *i.e.*,

$$Y_{10} \approx \omega_e \qquad Y_{01} \approx B_e \qquad Y_{21} \approx \gamma_e \qquad Y_{20} \approx -\omega_e x_e \qquad Y_{11} \approx -\alpha_e$$
$$a_1 \approx (-\alpha_e \omega_e/6B_e^2) - 1$$

For example, in the case of ICl, β_{01} is approximately 50, so that the fractional correction to the mass ratio deduced from the microwave measurements is $50(B_e^2/\omega_e^2)[(\mu^{37} - \mu^{35})/\mu^{35}] = 2 \times 10^{-7}$. Since the accuracy of the present measurements is about 2×10^{-6}, an improvement by a factor of 10 would make this correction appreciable.

It is customary in calculating the moment of inertia of a molecule to assume each atom has the proper number of electrons to make it neutral, and that the entire mass of the atom is concentrated at a point. Thus for diatomic molecules the moment of inertia is generally written $I_e = [M_1 M_2/(M_1 + M_2)]r_e^2$, where M_1 and M_2 are the masses of neutral atoms. Such an approximation is good only because the electrons are very light compared with the atomic nuclei which are, indeed, concentrated within a very small radius. However, uncertainties in the location and behavior of electrons in a rotating molecule do appreciably affect the moment of inertia. In NaCl, for example, it would be more correct to associate with Cl the mass of one electron more than the neutral atom and with Na one electron less since the molecule is largely ionic or $\overset{+}{\text{Na}}\overset{-}{\text{Cl}}$. In the case of LiBr and LiI, it has been shown [938] that the rotational spectrum gives a mass ratio for the two Li isotopes which is in agreement with other measurements of this ratio only if the Li is assumed to have lost one electron and be $\overset{+}{\text{Li}}$.

Electrons are certainly not concentrated at the nuclei, but are arranged more or less spherically about their respective nuclei. Hence the moments of inertia might be expected to be greater than those calculated from the point mass assumption by an amount approximately equal to the moments of inertia of the electrons about their respective nuclei. This last contribution to the moment of inertia would be rather large, but fortunately it is not really present because a completely spherical shell of electrons around an atom would not, in fact, rotate as the molecule rotates. This is known as the "slip effect," for the spherical shells of electrons appear to remain fixed in orientation or slip as the molecule rotates (*cf.* [339] and discussion of this effect in Chap. 8). The valence shell of electrons, however, is not completely spherical, and part of it must be considered to rotate with the molecule, giving a contribution of approximately nmr^2, where n is the number of rotating valence electrons, m the electron mass, and r some average of their distance from the nuclei with which they are supposed to be associated. This is of approximately the same magnitude as the uncertainty in moment of inertia due to an uncertainty of the position in the molecule of one electron. If n is taken as 1 and $r = r_e$, the error in moment of inertia is of the order mr_e^2, or a fractional error of $m(M_1 + M)/M_1 M$. This is less than 1 part in 10,000 for almost all atoms, and hence would not affect a calculation of r_e from B_e. On the other hand, it does affect a determination of mass ratios, giving a fractional error in the mass ratio M_1/M_2 of $m(M_1 - M_2)/M_1 M_2$. For ICl, this fractional error is 8×10^{-7}, which is of the same general magnitude as the other errors discussed above. However, for the light nucleus of Li in LiI, such an effect would be as large as 10^{-5}, and easily detectable.

L uncoupling also involves the behavior of electrons during rotation and is very closely related to the above effects, although it may be described in somewhat different language. The rotational momentum of the molecule can to a very small extent be transferred to the molecular electrons. Rotation tends to excite the valence electrons from their normal $^1\Sigma$ state of zero angular momentum to excited $^1\Pi$ states, which have unit angular momentum, and hence slightly change the rotational energy. This process, known as L uncoupling, is very difficult to evaluate quantitatively from theory, since little exact knowledge of the electronic wave functions and excited states is available. However, it can be approximately evaluated from experimental results. Since a Π state has an electronic angular momentum and magnetic moment, even a small excitation of this state contributes a considerable part of the magnetic moment of the rotating molecule, which is of the order of a nuclear magneton, or one two-thousandth that of an electron. Hence a measurement of the molecular magnetic moment allows a rough estimate of the extent of L uncoupling or of its effect on the rotational energy.

Electrons in a Π state also produce a large magnetic field at the positions of the nuclei and hence a magnetic hyperfine interaction with the nuclei. Although this is not the only source of magnetic hyperfine interactions in a rotating molecule, it is probably a major contributor, so that measurement of the magnetic hyperfine interactions allow an estimate of the L uncoupling.

It is estimated that L uncoupling in ICl or FCl produces uncertainties in the mass ratio Cl^{35}/Cl^{37} of about 1 part in 10^6. This is again large enough to be of importance in accurate microwave measurements. Lighter molecules, which rotate faster, would in general involve larger errors from L uncoupling.

These electronic effects and their interrelations will be discussed in some detail in Chap. 8. That chapter shows that the entire contribution of electrons to the kinetic energy of rotation of a $^1\Sigma$ molecule can be evaluated by a measurement of the rotational magnetic moment of the molecule. The magnetic moment of a $^1\Sigma$ molecule is due to the rotation of both nuclear and electronic charges. The nuclear effect can be calculated by assuming that the nuclei form a rigid rotating frame. If their effect is subtracted from the measured magnetic moment, the electronic contribution to the moment can be determined. The change in rotational energy due to electron motion is, from (8-29) and (11-15),

$$W_R = g_e J (J + 1) h B_e \qquad (1\text{-}47)$$

where $g_e J$ is the magnetic moment in Bohr magnetons due to the motion of all the electrons. This expression allows the possibility of precise corrections for all effects due to electron motion [type (2) above].

Finally, it is of interest to examine how accurately the mass M must be known for the atom whose mass is not being measured. In determining the ratio M_1/M_2, it may be seen from Eq. (1-42) that M/M_2 is assumed known. A fractional error ϵ in this ratio will give a fractional error δ in the determination of M_1/M_2 which is

$$\delta = \frac{(M_2 - M_1)\epsilon}{M + M_2} \qquad (1\text{-}48)$$

It is evident that, when the fractional change in weight of the molecule, $(M_2 - M_1)/(M + M_2)$, is small, the ratio M/M_2 need not be known with high accuracy. The error produced by uncertainties in M/M_2 is not, of course, due to theoretical uncertainties in the behavior of the molecule as are errors (1) and (2) on page 15. However, inaccurate knowledge of M/M_2 may often give errors in M_1/M_2 of the same order as those of type (1) and (2).

So far microwave mass measurements are just at the threshold of difficulties of the types discussed here. Since the measurements of B_e with microwaves can be rather easily improved by another factor of 10, these difficulties will provide an ultimate limit to the accuracy of most mass ratio measurements of about 1 part in 10^6. This limit, of course, represents a very good accuracy and one which cannot always be obtained by other methods, i.e., an error of 10^{-4} mass unit for nuclei of atomic mass 100.

1-4. Absorption Intensities and Selection Rules. A molecule interacts appreciably with a microwave electromagnetic field to emit or absorb radiation only if it has an electric or magnetic dipole moment μ. Usually the dipole is an electric moment due to the positive and negative charges in the molecule. In the ICl molecule, for example, the Cl has an excess negative charge and the I an excess positive charge, so that the molecule is a small rotating dipole which acts in many ways like a small antenna in radiating or receiving electromagnetic waves whose frequency coincides with its frequency of rotation. The rate of radiation is small because the molecule is so small (approximately 10^{-8} cm) compared with the wavelengths radiated (approximately 1 cm).

As will be discussed in some detail in Chap. 13, the intensity of a narrow microwave absorption line in a gas may usually be written

$$\gamma = \frac{8\pi^2 N f |\mu_{ij}|^2 \nu^2 \, \Delta\nu}{3ckT[(\nu - \nu_0)^2 + (\Delta\nu)^2]} \qquad (1\text{-}49)$$

where N = the number of molecules per cc in absorption cell

f = fraction of these molecules in the lower of the two states involved in the transition

$|\mu_{ij}|^2$ = square of the dipole moment matrix element for the transition, summed over the three perpendicular directions in space

ν = frequency

ν_0 = resonant frequency or, to a good approximation, the center frequency of the absorption line

$\Delta\nu$ = half width of the line at half maximum, or line-breadth parameter

c = velocity of light

k = Boltzmann constant

T = absolute temperature

The peak absorption of the line occurs very near to $\nu = \nu_0$, and is

$$\gamma_{\text{max}} = \frac{8\pi^2 N f |\mu_{ij}|^2 \nu_0^2}{3ckT \, \Delta\nu} \qquad (1\text{-}50)$$

The fraction of molecules in a particular vibrational state of energy $h\omega_e(v + \frac{1}{2})$ is given by

$$f_v = \frac{e^{-h\omega_e(v+\frac{1}{2})/kT}}{\displaystyle\sum_{n=0}^{\infty} e^{(h\omega_e/kT)(n+\frac{1}{2})}} = e^{-vh\omega_e/kT}(1 - e^{-h\omega_e/kT}) \qquad (1\text{-}51)$$

since

$$1 + e^{-h\omega_e/kT} + e^{-2h\omega_e/kT} + e^{-3h\omega_e/kT} + \cdots = \frac{1}{1 - e^{-h\omega_e/kT}} \qquad (1\text{-}52)$$

Of the molecules in a particular vibrational state, we must find the fraction f_J in a particular rotational state J in order to obtain the fraction $f = f_v f_J$ in the particular state of interest for expression (1-50). The angular momentum J may be oriented in space $2J + 1$ different ways, corresponding to the different values of the magnetic quantum number $M = J, J - 1, J - 2, \ldots, -J$. The fraction having angular momentum J is then

$$f_J = \frac{(2J + 1)e^{-hBJ(J+1)/kT}}{\sum\limits_{n=0}^{\infty} (2n + 1)e^{-hBn(n+1)/kT}} \tag{1-53}$$

If B/kT is sufficiently small, the sum, often called the partition function, may be replaced by an integral,

$$\sum_{n=0}^{\infty} (2n + 1)e^{-hBn(n+1)/kT} = \int_0^{\infty} (2x + 1)e^{-hBx(x+1)/kT}\, dx = \frac{kT}{hB} \tag{1-54}$$

In case B/kT is large enough that a more accurate value of the rotational partition function is needed, it may be written as an expansion

$$\sum_{n=0}^{\infty} (2n + 1)e^{-hBn(n+1)/kT} = \frac{kT}{hB} + \frac{1}{3} + \frac{1}{15}\frac{hB}{kT} + \frac{4}{315}\left(\frac{hB}{kT}\right)^2$$
$$+ \frac{1}{315}\left(\frac{hB}{kT}\right)^3 + \cdots \tag{1-55}$$

Using only the first term of this expansion,

$$f_J = \frac{(2J + 1)hBe^{-hBJ(J+1)/kT}}{kT} \tag{1-56}$$

In most cases of microwave rotational spectra, $hB/kT \approx \frac{1}{200}$ at room temperature, so that not only is hB/kT small, but $e^{-hBJ(J+1)/kT}$ can usually be approximated as unity for low values of J. Hence

$$f = (2J + 1)\frac{hB}{kT}\, e^{-vh\omega_e/kT}(1 - e^{-h\omega_e/kT}) \tag{1-57}$$

Dipole Moment Matrix Elements. The dipole moment of a macroscopic linear array of charges oriented along the z axis would be defined as

$$\mu = \sum_i e_i z_i \text{ or } \int \rho(z)z\, dz \tag{1-58}$$

where e_i is the size of the ith charge and z_i is its coordinate. In the integral form, $\rho(z)$ is the charge density per unit length. A linear molecule may be thought of as having an inherent or permanent dipole moment of the same type oriented along its axis. However, the molecular orientation is not fixed in space, so that its average dipole moment in any one direction is zero unless it is subjected to external electric fields or other constraining forces.

A measure of the effectiveness of an electric field along the z axis in exerting a torque on a rotating molecule, or in inducing a transition between states JM and $J'M'$ is given by the dipole moment matrix elements

$$\mu_z(JMJ'M') = \int \psi_{JM}^* \mu_z \psi_{J'M'} \, d\tau \tag{1-59}$$

where μ_z is the projection of the molecular dipole moment on the z axis. ψ_{JM} represents the molecular wave function for a total angular momentum J and magnetic quantum number M. Similarly, for electric fields in the x or y directions the dipole matrix elements

$$\mu_x(JMJ'M') = \int \psi_{JM}^* \mu_x \psi_{J'M'} \, d\tau \tag{1-60}$$
$$\mu_y(JMJ'M') = \int \psi_{JM}^* \mu_y \psi_{J'M'} \, d\tau \tag{1-61}$$

are important. The intensity of absorption or emission of radiation polarized with the electric vector in x, y, or z directions due to a transition between the states JM and $J'M'$, is just proportional to $|\mu_x(JMJ'M')|^2$, $|\mu_y(JMJ'M')|^2$, or $|\mu_z(JMJ'M')|^2$, respectively. In expression (1-50) giving the intensity of an absorption line, in which either the radiation is unpolarized or the molecules are randomly distributed in various M states, the quantity $|\mu_{ij}|^2$ is given by

$$|\mu_{ij}|^2 = \sum_{M'} |\mu_x(JMJ'M')|^2 + |\mu_y(JMJ'M')|^2 + |\mu_z(JMJ'M')|^2 \tag{1-62}$$

In terms of spherical polar coordinates fixed in space, the components of the dipole moment are

$$\mu_x = \mu \sin \theta \cos \phi \qquad \mu_y = \mu \sin \theta \sin \phi \qquad \mu_z = \mu \cos \theta \tag{1-63}$$

In these coordinates the matrix elements become:

$$\mu_x(JMJ'M') = \mu \int \psi_{JM}^* \sin \theta \cos \phi \, \psi_{J'M'} \, d\tau$$
$$\mu_y(JMJ'M') = \mu \int \psi_{JM}^* \sin \theta \sin \phi \, \psi_{J'M'} \, d\tau \tag{1-64}$$
$$\mu_z(JMJ'M') = \mu \int \psi_{JM}^* \cos \theta \, \psi_{J'M'} \, d\tau$$

where ψ_{JM} is an eigenfunction for the rotating molecule.

For the rigid rotator, on substituting the eigenfunctions from (1-5) and (1-6) and using $d\tau = \sin \theta \, d\phi \, d\theta$, we get

$$\mu_z(JMJ'M') = \mu N_{JM} N_{J'M'} \int_0^\pi P_J^{|M|}(\cos \theta) \cos \theta \, P_{J'}^{|M'|}(\cos \theta) \sin \theta \, d\theta$$
$$\int_0^{2\pi} e^{-iM\phi} e^{iM'\phi} \, d\phi \tag{1-65}$$

where N_{JM} and $N_{J'M'}$ are the normalization factors for ψ_{JM} and $\psi_{J'M'}$, or

$$N_{JM} = \frac{1}{\sqrt{2\pi}} \left[\frac{(2J+1)(J-|M|)!}{2(J+|M|)!} \right]^{\frac{1}{2}} \tag{1-66}$$

The second integral in (1-65) has the value of 2π if $M = M'$; otherwise it is zero. The first integral may be obtained from the properties of the

Legendre polynomials ([471], p. 73; [62], p. 307; [537], p. 136)

$$\cos \theta \, P_J^{|M|}(\cos \theta) = \frac{J + |M|}{2J + 1} P_{J-1}^{|M|}(\cos \theta)$$
$$+ \frac{J - |M| + 1}{2J + 1} P_{J+1}^{|M|}(\cos \theta) \quad (1\text{-}67)$$

so that, remembering that M must equal M',

$$\mu_z(JMJ'M) = 2\pi\mu N_{JM} N_{J'M} \left[\frac{J + M}{2J + 1} \int P_{J-1}^{|M|}(\cos \theta) P_{J'}^{|M|}(\cos \theta) \sin \theta \, d\theta \right.$$
$$\left. + \frac{J - |M| + 1}{2J + 1} \int P_{J+1}^{|M|}(\cos \theta) P_{J'}^{|M|}(\cos \theta) \sin \theta \, d\theta \right] \quad (1\text{-}68)$$

These integrals vanish unless $J' = J \pm 1$, giving the selection rule that for a transition $\Delta J = \pm 1$. Taking J as the lower state so that

$$J' = J + 1$$

the first integral vanishes and

$$\mu_z(JMJ'M) = 2\pi\mu \frac{N_{JM}}{N_{J+1,M}} \frac{J - |M| + 1}{2J + 1}$$
$$\int N_{J+1,m}^2 [P_{J+1}^{|M|}(\cos \theta)]^2 \sin \theta \, d\theta \quad (1\text{-}69)$$

The integral remaining is just the normalization integral and equals $1/2\pi$ with our choice of normalization factor. Putting in values of N_{JM} and $N_{J+1,M}$,

$$\mu_z(J,M,J + 1,M) = \mu \sqrt{\frac{(J + 1)^2 - M^2}{(2J + 1)(2J + 3)}} \quad (1\text{-}70)$$

Similarly, μ_x and μ_y are zero unless $J = J' \pm 1$ and $M = M' \pm 1$. Then

$$\mu_x(J,M,J + 1,M + 1) = -i\mu_y(J,M,J + 1,M + 1)$$
$$= \frac{-\mu}{2} \sqrt{\frac{(J + M + 2)(J + M + 1)}{(2J + 1)(2J + 3)}} \quad (1\text{-}71)$$
$$\mu_x(J,M,J + 1,M - 1) = i\mu_y(J,M,J + 1,M - 1)$$
$$= \frac{\mu}{2} \sqrt{\frac{(J - M + 1)(J - M + 2)}{(2J + 1)(2J + 3)}} \quad (1\text{-}72)$$

The signs or "phases" of the matrix elements (1-71) and (1-72) are usually not important and may be positive or negative according to one's definition of $P_J^M(\cos \theta)$. The signs given are those adopted by Condon and Shortley [56] but are not used by all authors.

The probability of inducing an absorption in a particular molecule in a state JM by radiation with the electric field in the z direction is then proportional to

$$|\mu_z|^2 = \mu^2 \frac{(J + 1)^2 - M^2}{(2J + 1)(2J + 3)} \tag{1-73}$$

and only the transition $J + 1,M \leftarrow JM$ can occur. Here the arrow indicates that the absorption process produces a transition from the state JM to the state $J + 1,M$.* Probability of emission of radiation of the same polarization due to the transition $J + 1,M \rightarrow JM$ is proportional to the same quantity. For radiation with the electric vector in the x or y direction, the absorption probability is proportional to

$$|\mu_x|^2 = \frac{\mu^2}{4} \frac{(J + M + 2)(J + M + 1)}{(2J + 1)(2J + 3)} \tag{1-74}$$

for a transition $J + 1,M + 1 \leftarrow JM$ and

$$|\mu_x|^2 = \frac{\mu^2}{4} \frac{(J - M + 1)(J - M + 2)}{(2J + 1)(2J + 3)} \tag{1-75}$$

for $J + 1,M - 1 \leftarrow JM$.

For an electric vector in the z direction, the result that M, the angular momentum about the z axis, cannot change is easily understood because there can then be no torque on the molecule about the z axis. For electric fields in the x or y directions, however, there is a torque about the z axis and M changes by one unit.

For any particular initial state JM, it can be shown from (1-73), (1-74), and (1-75) that

$$|\mu_{ij}|^2 = \sum_{M'} |\mu_x(JMJ'M')|^2 + |\mu_y(JMJ'M')|^2 + |\mu_z(JMJ'M')|^2$$

$$= \mu^2 \frac{(J + 1)}{2J + 1} \qquad \text{for the transition } J + 1 \leftarrow J$$

$$= \mu^2 \frac{J + 1}{2J + 3} \qquad \text{for the transition } J + 1 \rightarrow J \tag{1-76}$$

The expression is independent of M, as it should be since it represents the probability of absorption of unpolarized radiation, which should be independent of molecular orientation.

It should be noted that, although the individual matrix elements are identical for reverse transitions $J'M' \leftarrow JM$ and $JM \leftarrow J'M'$, the average matrix element given by (1-76) for a transition $J + 1 \leftarrow J$ is greater than that for the reverse transition $J + 1 \rightarrow J$. This must be true in

---

* In spectroscopy it is conventional to denote transitions by writing first the state having higher energy. For an absorption process, which is the common type of transition in microwave work, this involves writing the final state before the initial one. We follow the common convention for the sake of uniformity with other branches of spectroscopy, although the lower state is written first in much of the previous microwave literature.

order to maintain thermal equilibrium when transitions take place, since there are $2J + 3$ states of angular momentum $J + 1$, but only $2J + 1$ states of the lower angular momentum J.

Peak Intensities of Absorption Lines.   Combining (1-57) and (1-76) for an absorption in which $J + 1 \leftarrow J$

$$|\mu_{ij}|^2 f = \mu^2 \frac{hB(J + 1)}{kT} f_v = \frac{\mu^2 f_v}{2kT} h\nu_0$$

where f_v is the fraction of molecules in the vibrational state being considered.   The peak intensity of an absorption line of a diatomic molecule given by (1-50) becomes

$$\gamma_{max} = \frac{4\pi^2 h N f_v \mu^2 \nu_0^3}{3c(kT)^2 \Delta\nu} \tag{1-77}$$

Since the line-breadth parameter $\Delta\nu$ is proportional to the molecular density N at low pressures, $N/\Delta\nu$ and therefore γ_{max} is quite constant for pressures from about 1 to 10^5 microns or 10^{-3} to 100 mm Hg.   As a standard procedure, the value $\Delta\nu$ is often given for 1 mm Hg pressure. Then the universal constants of (1-77) may be evaluated and rewritten

$$\gamma_{max} = 5.48 \times 10^{-17} f_v \frac{\mu^2 \nu_0^3}{\Delta\nu} \tag{1-78}$$

where μ is measured in debye units, or 10^{-18} esu, ν_0 and $\Delta\nu$ are measured in megacycles and T is taken as 300°K.   Typical values would be $f_v \approx 1$, $\mu = 1$, $\nu_0 = 30,000$ ($\lambda = 1$ cm), $\Delta\nu = 15$, giving $\gamma_{max} = 10^{-4}$/cm. This value of γ_{max} represents a conveniently strong absorption for microwave spectroscopic work, *i.e.*, 1 per cent absorption in one meter path length.   Because of the rapid variation of intensity with ν_0, absorption-line intensities for wavelengths longer than 10 cm are usually too weak to be readily observed, while those for wavelengths as short as 1 mm are quite intense.

Measurements of intensity, combined with expression (1-78), can be used in some cases to evaluate $\Delta\nu$ if μ is known.   Although dipole moments are usually measured most accurately by Stark effects (Chap. 10), they may be determined to an accuracy of a few per cent from (1-78) by a measurement of both intensity γ_{max} and the half width of the line $\Delta\nu$.

Expression (1-77) indicates that the absorption intensity γ_{max} increases rapidly with decreasing temperature T.   For this reason it is often advantageous to strengthen absorption lines by decreasing the temperature of the gas to -78°C (dry ice) or lower if there is sufficient vapor pressure.   The exact dependence of γ_{max} on T depends on how $\Delta\nu$ varies with T. $\Delta\nu$ will be shown in Chap. 13 to vary as T^n, where $-1 \leq n \leq -\frac{1}{2}$. Even when $n = -1$, γ_{max} can be increased by decreasing the gas temperature T.

CHAPTER 2

LINEAR POLYATOMIC MOLECULES

2-1. Pure Rotational Spectra—General Considerations. Except for the complication of more possible modes of vibration, the spectrum of a linear polyatomic molecule is much like that of a diatomic molecule. There is a very small moment of inertia about the molecular axis, and hence angular momentum about this axis is not easily excited, and the linear molecule rotates end over end with energies of the same form as for a diatomic molecule

$$W = \frac{h^2}{8\pi^2 I} J(J + 1)$$

where I is the moment of inertia and vibrations are neglected. Wave functions and intensities for linear molecules are also similar to those for a diatomic molecule [Eqs. (1-5), (1-6), and (1-77)] if vibrations are neglected. Vibrations do, however, add considerably to the actual complexity of the spectra as well as introduce some new phenomena.

As an obvious extension of the effects of vibrations seen in diatomic molecules, the rotational constant $B = h/8\pi^2 I$ for a polyatomic linear molecule should be written

$$B = B_e - \sum_i \alpha_i(v_i + \tfrac{1}{2}) - J(J + 1)D \tag{2-1}$$

where $-\alpha_i$ represents the change in the equilibrium value B_e due to excitation of the ith vibration, D the change due to centrifugal stretching, and v_i is the quantum number for the vibrational excitation. Even in the ground state, where all v_i are zero, the zero-point vibrations change the rotational constant by $-\sum_i \alpha_i/2$. A linear molecule has one or more degenerate modes of vibration, $i.e.$, modes which have the same frequency ω_i and the same value of α_i. If these are counted as a single vibration, then (2-1) is written:

$$B = B_e - \sum_i \alpha_i \left(v_i + \frac{d_i}{2}\right) - J(J + 1)D$$

where d_i is the degree of the degeneracy, or the number of degenerate modes with the same value α_i. For the diatomic molecule where there

25

is only one vibrational mode, α can be fairly simply treated. We shall see that the addition of just one more atom to the diatomic molecule very greatly complicates the vibrational effects.

The commonest type of linear polyatomic molecule is triatomic, such as carbonyl sulfide or OCS, where in equilibrium the atoms are arranged in a straight line in the order of this chemical formula. Any arbitrary relative vibration of the atoms in OCS may be described as a sum of four types or normal modes of vibrations. For each normal mode, the displacement of each atom is periodic with a definite frequency and proportional to a variable called the normal coordinate q_i. Then the general motion of one atom, which we shall label s, is given by

$$x_s = \sum_i l_{is}q_i \qquad y_s = \sum_i m_{is}q_i \qquad z_s = \sum_i n_{is}q_i \qquad (2\text{-}2)$$

Two of these normal modes are degenerate, or are similar and have the same frequency. The three different vibrations are shown in Fig. 2-1, where the arrows indicate the relative directions and magnitudes of the

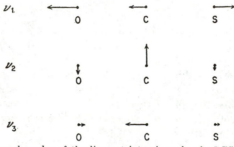

FIG. 2-1. The normal modes of the linear triatomic molecule OCS. Directions and lengths of arrows indicate relative motions of the three nuclei for the three different vibrations. The mode ν_2 is degenerate, a vibration of the same frequency with displacements perpendicular to the page being also possible.

displacements of each atom. The lowest frequency of vibration, which is indicated by ν_2, corresponds to a bending of the molecule. This bending can take place in either of two perpendicular planes, corresponding to the two degenerate modes of oscillation of this type. Of the two stretching modes, by convention the one of lower frequency in which the two outer atoms move in opposite directions is called ν_1 and the higher-frequency vibration in which these atoms move in the same direction is ν_3.

Allowing for the four modes of vibration, two of which are degenerate,

$$B_v = B_e - \alpha_1(v_1 + \tfrac{1}{2}) - \alpha_2(v_2 + 1) - \alpha_3(v_3 + \tfrac{1}{2}) \qquad (2\text{-}3)$$

The α's as before must be evaluated from the molecular potential function. The potential function for a linear triatomic molecule is considerably more complex than that for a diatomic molecule. Taking a coordinate system with origin at the center of mass and for which the z axis lies along the molecular axis, the potential for small displacements of the atoms may be written

$$V = \frac{K_1}{2}(z_2 - z_1)^2 + \frac{K_2}{2}(z_3 - z_2)^2 + \frac{K_3}{2}(z_3 - z_1)^2$$
$$+ K_{12}(z_2 - z_1)(z_3 - z_2) + K_{13}(z_2 - z_1)(z_3 - z_1)$$
$$+ K_{23}(z_3 - z_2)(z_3 - z_1) + \frac{K_4}{2}(x^2 + y^2) \quad (2\text{-}4)$$

where the subscripts 1, 2, 3 refer to changes of the coordinates of the first, second, and third atoms from the equilibrium values. The K's are potential constants, and

$$x = (2m_2 x_2 - m_1 x_1 - m_3 x_3)\frac{m_1 + m_2 + m_3}{3m_2(m_1 + m_3)}$$
$$y = (2m_2 y_2 - m_1 y_1 - m_3 y_3)\frac{m_1 + m_2 + m_3}{3m_2(m_1 + m_3)}$$

which represent relative displacements of the central and two end atoms perpendicular to the molecular axis. Here m_1, m_2, and m_3 are the masses of the three atoms. This potential is so complex that frequently it is simplified by assuming that forces occur only through the conventionally recognized molecular bonds corresponding to their stretching or bending. This "valence bond" approximation, which makes all constants in V zero except K_1, K_2, and K_4, is usually a moderately good approximation. It tends to be rather poor if a stretching of one bond tends to affect the nature of a second bond as in bonds said to be "conjugated."

Evaluation of the α's involves not only the terms in the potential given above, but also the anharmonic potential constants. These anharmonic constants are most conveniently expressed if the potential is written in terms of the normal coordinates (see [130], p. 70)

$$V_{\text{anharmonic}} = k_{111}q_1^3 + k_{113}q_1^2 q_3 + k_{133}q_1 q_3^2 + k_{122}q_1 q_2^2 + k_{322}q_3 q_2^2 + k_{333}q_3^3$$
$$(2\text{-}5)$$

For these normal coordinates q_i, the constants l_{is}, m_{is}, and n_{is} of (2-2) have been chosen so that the kinetic energy due to vibration comes out simply

$$\text{KE} = \tfrac{1}{2}\sum_i \left(\frac{dq_i}{dt}\right)^2$$

Because of symmetry about the molecular axis, odd powers of q_2 do not appear in (2-5). A. H. Nielsen [119] has obtained for the α's

$$\alpha_1 = \frac{2B_e^2}{c\omega_1}\left(1 - 4\xi_{21}^2\frac{\omega_1^2}{\omega_1^2 - \omega_2^2} - 4\xi_{23}^2 + 6I_e^{\frac{1}{2}}\frac{\xi_{23}k_{111}}{4\pi^2c^2\omega_1^2} + 2I_e^{\frac{1}{2}}\frac{\xi_{21}k_{113}}{4\pi^2c^2\omega_3^2}\right)$$

$$\alpha_2 = \frac{B_e^2}{c\omega_2}\left[-4I_e^{\frac{1}{2}}\frac{\xi_{21}k_{223}}{4\pi^2c^2\omega_3^2} + 4I_e^{\frac{1}{2}}\frac{\xi_{23}k_{122}}{4\pi^2c^2\omega_1^2} - (3\omega_2^2 + \omega_1^2)\frac{\xi_{21}^2}{\omega_2^2 - \omega_1^2}\right.$$

$$\left. - (3\omega_2^2 + \omega_3^2)\frac{\xi_{23}^2}{\omega_2^2 - \omega_3^2}\right]$$

(2-6)

α_3 is similar to α_1 but with the indices 1 and 3 interchanged throughout. Even though these expressions appear somewhat complex, they have been abbreviated by using the quantities

$$\xi_{21} = \left[\frac{m_1 m_3}{(m_1 + m_3)I_e}\right]^{\frac{1}{2}}(z_{e_1} - z_{e_3})\cos\gamma$$

$$+ \left[\frac{m_2(m_1 + m_2 + m_3)}{(m_1 + m_3)I_e}\right]^{\frac{1}{2}} z_{e_2}\sin\gamma$$

$$\xi_{23} = -\left[\frac{m_1 m_3}{(m_1 + m_3)I_e}\right]^{\frac{1}{2}}(z_{e_1} - z_{e_3})\sin\gamma$$

(2-7)

$$+ \left[\frac{m_2(m_1 + m_2 + m_3)}{(m_1 + m_3)I_e}\right]^{\frac{1}{2}} z_{e_2}\cos\gamma$$

$$\sin\gamma = \frac{1}{\sqrt{2}}\left[1 + \frac{(k_1' - k_3')^2}{4(k_4')^2 + (k_1' - k_3')^2}\right]^{\frac{1}{2}}$$

The subscript e in z_{e_1}, z_{e_2}, and z_{e_3} indicates the equilibrium value of z_1, z_2, and z_3.

$$k_1' = \frac{m_1 + m_3}{m_1 m_3}\left[\frac{K_1 m_3^2}{(m_1 + m_3)^2} + \frac{K_2 m_1^2}{(m_1 + m_3)^2} + K_3 + \frac{2K_{12}m_1 m_3}{(m_1 + m_3)^2}\right.$$

$$\left. + \frac{2K_{13}m_3}{m_1 + m_3} + \frac{2K_{23}m_1}{m_1 + m_3}\right]$$

$$k_3' = \frac{m_1 + m_2 + m_3}{m_2(m_1 + m_3)}(K_1 + K_2 - 2K_{12})$$

$$k_4' = \left(\frac{m_1 + m_2 + m_3}{m_1 m_2 m_3}\right)^{\frac{1}{2}}\left[-\frac{K_1 m_3}{m_1 + m_3} + \frac{K_2 m_1}{m_1 + m_3} - \frac{K_{12}(m_1 - m_3)}{m_1 + m_3}\right.$$

$$\left. - K_{13} + K_{23}\right]$$

The frequencies are written as ω_1, ω_2, and ω_3 to indicate as usual the ideal frequencies for infinitesimal vibrations. These do not in fact differ significantly from the observed frequencies of the lowest vibrational states, ν_1, ν_2, and ν_3. The frequencies ω are all expressed in cm^{-1}, and B_e and the α's are in cycles per second.

There are so many potential constants involved in (2-6) that for no molecule of the form XYZ (such as OCS) have they yet been all evaluated so that the α's can be theoretically determined. However, if the triatomic molecule is symmetric such as OCO (CO_2), the potential constants and formulas simplify considerably [119], and one or two cases have been completely worked out from infrared measurements. There is, of course, no pure rotational spectrum for these symmetric linear molecules because the dipole moment is zero.

As in the case of diatomic molecules, the contributions to α from the anharmonic force constants are usually larger than those from the harmonic potential terms. Thus the experimental values of α_1 and α_3 in all known cases are positive as a result of anharmonicities instead of negative as would be expected from a purely harmonic potential. From Table 2-2 it may be seen that α_2 is negative in cases which have been measured.

TABLE 2-1. MOLECULAR DIMENSIONS AND VIBRATION PARAMETERS OF SOME LINEAR TRIATOMIC MOLECULES*

l_1 is the distance between the first two atoms and l_2 that between the last two in the chemical formula. K_1 and K_2, the force constants corresponding to stretching l_1 or l_2, respectively, are evaluated under the assumption of valence forces only.

Molecule	l_1, 10^{-8} cm	l_2, 10^{-8} cm	K_1, 10^5 dynes/cm	K_2, 10^5 dynes/cm	ν_1, cm^{-1}	ν_2, cm^{-1}	ν_3, cm^{-1}
$H^1C^{12}N^{14}$	1.068	1.156	5.8	17.9	2089	712	3312
$Cl^{35}C^{12}N^{14}$	1.629	1.163	5.2	16.7	729	397	2201
$Br^{79}C^{12}N^{14}$	1.789	1.160	4.2	16.9	580	368	2187
$I^{127}C^{12}N^{14}$	1.995	1.159	3.0	16.7	470	321	2158
$O^{16}C^{12}S^{32}$	1.161	1.561	14.2	8.0	859	527	2079
$N^{14}N^{14}O^{16}$	1.126	1.191	14.6	13.7	1285	589	2224

* All data on force constants and frequencies are from Herzberg [130]. Internuclear distances come from microwave work (see Appendix VI).

The centrifugal stretching constant D has also been evaluated by Nielsen [119] as

$$D = 4B_e^3\left(\frac{\xi_{21}^2}{c^2\omega_3^2} + \frac{\xi_{23}^2}{c^2\omega_1^2}\right) \tag{2-8}$$

where the ξ's are defined in (2-7), ω_1 and ω_3 are in cm^{-1}, B_e and D are in cycles per second. The centrifugal stretching of linear molecules is very small for most rotational lines in the microwave region, but not too small to be detected and measured. The calculated values of D assuming valence bond forces for several linear molecules are listed in Table 2-2 and compared with measured values.

Coriolis Forces. It may be at first surprising to note from (2-6) some terms which are "resonant." For example, the term in α_2 which is

$$-\frac{B_e^2(3\omega_2^2 + \omega_3^2)}{c\omega_2(\omega_2^2 - \omega_3^2)}\xi_{23}^2$$

would become very large if ω_2 were close to ω_3. These terms are generally thought of as due to Coriolis forces, and represent a coupling of the modes of vibration ω_2 and ω_3 by Coriolis forces in the rotating molecule. The Coriolis force is a fictitious force which must be introduced if mechanical motion is studied in a rotating coordinate system and the rotation is otherwise overlooked. It has a value $\mathbf{F} = 2\mathbf{v} \times \boldsymbol{\omega}$, where $\boldsymbol{\omega}$ is the vector angular velocity of rotation of the coordinate system and $\mathbf{v}$ is the vector velocity of motion through the coordinate system.

A Coriolis force occurs in the case of a rotating and vibrating diatomic molecule as well as in the more complex case discussed here. As the rotating diatomic molecule stretches, its rotation is slowed down and as the molecule contracts its rotation is speeded up by Coriolis forces. Such changes in rotational velocity are often attributed simply to the law of conservation of angular momentum—as the molecule expands its moment of inertia increases and hence it must slow down in order to conserve angular momentum. This is not all, but it is a part of the origin of Coriolis forces, and the diatomic molecule may be regarded as subject to Coriolis forces which slightly change its rotation and give the "harmonic" contribution to the rotation-vibration interaction α_e which is $-6B_e^2/\omega_e$ [Eq. (1-28)].

TABLE 2-2. ROTATIONAL CONSTANTS AND DIPOLE MOMENTS OF SOME LINEAR POLYATOMIC MOLECULES

All rotational constants are given in megacycles.

Molecule	B_0	α_1	α_2	α_3	D (Calculated)	D (Observed)	q_l	μ, debyes or 10^{-18} esu	Reference
$H^1C^{12}N^{14}$	44,315.80	279	−21	324	0.065	0.10	224.47	3.00	[130] [532] [733] [803] [911] [997]
$Cl^{35}C^{12}N^{14}$	5,970.82	......	−16.39	....	0.00159	....	7.50	2.80	[329] [528]
$Br^{79}C^{12}N^{14}$	4,120.19	15.54	−11.49	....	0.000842	0.00091	3.91	2.94	[329] [531]
$I^{127}C^{12}N^{14}$	3,225.53	9.33	− 9.50	....	0.000626	0.00088	2.69	3.72	[329] [531]
$O^{16}C^{12}S^{32}$	6,081.48	20.5	−10.59	36.4	0.00128	0.00131	6.39	0.709	[329] [530] [946] [968]
$O^{16}C^{12}Se^{80}$	4,017.68	13.27	− 6.92	....	....	0.00076	3.15	0.754	[418]
$S^{32}C^{12}Se^{80}$	2,017(?)	......	......	...	....	....	....	.....	[432]
$S^{32}C^{12}Te^{130}$	1,559.93	......	− 3.245	....	....	....	0.660	0.17	[932]
$N^{14}N^{14}O^{16}$	12,561.64	51.6	−12.6	104	0.00524	0.0057	....	0.166	[184] [528] [755] [915]
$H^1C^{12}C^{12}Cl^{35}$	5,684.24	......	....	...	....	....	....	0.44	[425]
$H^1C^{12}C^{12}C^{12}N^{14}$	4,549.07	......	....	...	....	....	....	3.6	[548]

For diatomic molecules, introduction of Coriolis forces seems an unnecessary complication. It is only for polyatomic molecules that the Coriolis-force approach is generally used, for there it is of real value in simplifying one's view of rotation-vibration interactions. Figure 2-2 shows the effect of Coriolis forces on a rotating linear triatomic molecule. It is evident that the vibration ω_3 excites some motion of the type ω_2, and vice versa. This is the reason the resonance-type terms involving the difference between two frequencies are often called Coriolis terms. The other terms not involving anharmonic force constants or resonance denominators may also properly be called Coriolis terms, although they may be just as simply and correctly regarded as due to effects of harmonic vibration on the moment of inertia.

2-2. l-Type Doubling. Bending or perpendicular modes of oscillation in a linear polyatomic molecule are rather different from anything that occurs in a diatomic molecule, and they introduce a new phenomenon known as l-type doubling. If the molecule is not rotating, then it may bend in two perpendicular planes, say the xz plane and the yz plane, with exactly the same frequencies of oscillation. These are the two degenerate modes which have been considered to have the same rotation-vibration constant α. However, if the molecule is rotating about the x axis, then bending in the xz plane is not quite equivalent to bending in the yz plane, the effective moments of inertia about the axis of rotation being different for the two cases. In addition it may be seen from Fig. 2-2 that, when the bending vibration is perpendicular to the angular momentum $\mathbf{J}$ of the molecule, the vibrational frequencies ν_1 and ν_3 are excited by Coriolis forces. However, when the bending motion is parallel to $\mathbf{J}$, the Coriolis force $2\mathbf{v} \times \boldsymbol{\omega}$ is zero and other vibration modes are not excited. Hence the two directions of vibration coupled with rotation give different energies also because of the different Coriolis forces. As a result of this vibration-rotation interaction, the two degenerate energy levels are slightly split, the splitting being called l-type doubling.

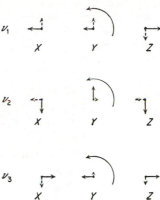

FIG. 2-2. Coriolis forces in a linear XYZ molecule. The curved arrows indicate direction of rotation. Solid straight arrows give the normal motion of the different modes of vibration, and dashed arrows indicate the Coriolis forces.

A more accurate description of l-type doubling and a calculation of the magnitude of the splitting must come from a quantum-mechanical treatment [120]. We start by considering a two-dimensional simple harmonic oscillator in the xy plane representing the two bending modes of equal frequency ω. The oscillator may be studied in Cartesian coordinates and found to have various allowed energy levels $(n_x + \frac{1}{2})h\nu$ for vibration along the x axis and similarly allowed energies $(n_y + \frac{1}{2})h\nu$

for vibrations along the y axis. Here n_x and n_y are, of course, positive integers. The total energy $W = (n_x + n_y + 1)h\nu$ does not uniquely determine the state of the oscillator since various combinations of n_x and n_y can give the same energy.

Considering the oscillator from a classical point of view, a proper phasing of the x and y motions will make the oscillator travel in a circle or ellipse and hence have an angular momentum. To discuss this angular momentum quantum-mechanically, it is appropriate to use cylindrical coordinates, specifying the state of the oscillator by its distance from the origin r and the angle χ between r and the x axis. The wave equation may be solved in these coordinates, giving wave functions

$$\psi_{vl} = N_{vl}\rho^{|l|}e^{-\rho^2/2+il\chi}F^{|l|}_{\frac{1}{2}(v+|l|)}(\rho^2) \qquad (2\text{-}9)$$

where $N_{vl} = 2 \dfrac{\left[\left(\dfrac{v-|l|}{2}\right)!\right]^{\frac{1}{2}}}{\left[\left(\dfrac{v+|l|}{2}\right)!\right]^{\frac{3}{2}}} \left(\dfrac{\pi m \nu}{h}\right)^{\frac{1}{2}}$

$F^l_{\frac{1}{2}(v+l)}$ = an associated Laguerre polynomial

$\rho = 2\pi \sqrt{m\nu/h}\, r$

m = mass of oscillator

The energy is given by $h\nu(v + 1)$, and the angular momentum l (in units of $h/2\pi$) can take on only the values $v, v - 2, v - 4, \ldots, -v$.

Similarly the linear molecule may have angular momentum about its axis as a result of vibration in one or more degenerate modes. This angular momentum affects the energies of rotation, making the molecule very similar to the symmetric-top molecules described in the next chapter with angular momentum about the symmetry axis. For a linear triatomic molecule where there is only one degenerate mode of vibration, the wave function for the molecule becomes

$$\psi_{vlJ} = \psi_{vl}(\rho,\chi) R_{Jl}(\theta,\phi) \qquad (2\text{-}10)$$

where R_{Jl} = symmetric-top wave function discussed in Chap. 3

$\rho = 2\pi \sqrt{\nu/h}\, (q_x^2 + q_y^2)^{\frac{1}{2}}$, where q_x and q_y are normal coordinates for the two degenerate vibrations

$l = v, v - 2, v - 4, \ldots, -v$

The total angular momentum J cannot be less than the angular momentum l about the axis, or $J \geq |l|$. Except for the l-type doubling energy, the rotational energy is very similar to that for a symmetric top. However, the energy associated with rotation around the symmetry axis is normally attributed to vibration in this case,

$$W = h\nu_2(v + 1) + B_v[J(J + 1) - l^2] - D_v[J(J + 1) - l^2]^2 \qquad (2\text{-}11)$$

Rotational frequencies are, according to (2-11),

$$\nu = 2B_v(J + 1) - 4D_v(J + 1)[(J + 1)^2 - l^2] \qquad (2\text{-}12)$$

The energies given by (2-11), which are similar to those of a symmetric top, still do not include l-type doubling, but indicate a degeneracy between $+l$ and $-l$. The l-type doubling is in many ways similar to the splitting of the symmetric-top energy levels by a slight asymmetry. Behavior of the vibrating linear molecule's wave functions and energy levels is exactly parallel to that of the slightly asymmetric rotor to be discussed in Chap. 4. The magnitude of splitting may even be approximately calculated from the crude model that the molecule is permanently bent by an amount equal to its average vibrational displacement, and by then treating it as an asymmetric rotor. The rather complete treatment of H. H. Nielsen ([402] and [624]) gives for the energy-level splitting when $|l| = 1$

$$\frac{B_e^2}{\omega_2} \left(1 + 4 \sum_i \xi_{2i}^2 \frac{\omega_2^2}{\omega_i^2 - \omega_2^2} \right) (v_2 + 1) J(J + 1) \qquad (2\text{-}13)$$

where $v_2 =$ the quantum number for the degenerate vibration ω_2

$\quad \omega_i' =$ a molecular vibrational frequency other than ω_2

$\quad \xi_{2i} =$ certain Coriolis parameters of the molecule which are dependent on the masses, dimensions, and harmonic force constants of the molecule. For a linear triatomic molecule, they are given by (2-7).

The quantity

$$2 \frac{B_e^2}{\omega_2} \left(1 + 4 \sum_i \xi_{2i}^2 \frac{\omega_2^2}{\omega_i^2 - \omega_2^2} \right)$$

may be abbreviated by q_l, the l-type doubling constant. In most cases $4 \sum_i \xi_{2i}^2 [\omega_2^2/(\omega_i^2 - \omega_2^2)]$ is near 0.3; so q_l is very approximately $2.6 B_e^2/\omega_2$. In the few cases where q has been fairly accurately calculated from the above expression, the values agree within a few per cent with measured values [505]. For $l = 2$ or greater, the splitting is usually too small to observe, being of the order $B(B/\omega_2)^l$.

Allowing for the above splitting, expression (2-12) for the rotational frequencies is modified to

$$\nu = \left[2B_v \pm \frac{q_l}{2} (v_2 + 1) \right] (J + 1) - 4D(J + 1)[(J + 1)^2 - l^2] \qquad (2\text{-}14)$$

where, of course,

$$B_v = B_e - \alpha_1(v_1 + \tfrac{1}{2}) - \alpha_2(v_2 + 1) - \alpha_3(v_3 + \tfrac{1}{2})$$

The above discussion may be readily extended to linear molecules involving more than three atoms, and hence more than one pair of degenerate vibrations.

Introduction of an angular momentum l about the molecular axis due to vibration affects not only the energy levels but also the intensities.

The dipole matrix elements are similar to those for symmetric and slightly asymmetric rotors to be discussed in following chapters. For any value of J, the energy level of these degenerate states may be specified by l and a subscript 1 or 2 indicating, respectively, the lower or upper of the two split states. Allowed transitions of the following types occur

$$\left. \begin{array}{l} J + 1, l_1 \leftarrow J, l_1 \\ J + 1, l_2 \leftarrow J, l_2 \end{array} \right\} \quad |\mu_{ij}|^2 = \mu^2 \frac{(J + 1)^2 - l^2}{(J + 1)(2J + 1)} \tag{2-15}$$

$$J, l_2 \leftarrow J, l_1 \quad |\mu_{ij}|^2 = \mu^2 \frac{l^2}{(J + 1)(2J + 1)} \tag{2-16}$$

where μ_{ij} is the dipole matrix element between the two states from which intensities may be computed according to (1-49) and μ is the molecular dipole moment. If the molecular states are designated by asymmetric-top rotation (Chap. 4), the state J, l_1 becomes $J_{|l|, J+1-|l|}$ and J, l_2 is $J_{|l|, J-|l|}$. The dipole matrix elements for transitions between these levels are identical with those for a slightly asymmetric top.

A type of transition with $\Delta J = 0$ is indicated by (2-16) which cannot occur when $l = 0$. The frequency of such transitions, from (2-13), is

$$\nu = \frac{q_l}{2} (v_2 + 1) J(J + 1) \tag{2-17}$$

which is a rather low frequency for many molecules unless J is very large. However, Shulman and Townes [505] showed that for HCN q_l is so large that these transitions occur in the microwave region for moderate values of J. The observed frequencies for a series of these lines are given in Table 2-3. It may be noted that q_l is not accurately

TABLE 2-3. OBSERVED LINES IN HCN AND VALUES OF
l-TYPE DOUBLING CONSTANT q_l
(After Westerkamp [997])

J	Frequency, Mc	$q_l = \text{frequency}/J(J + 1)$, Mc
6	9,423.3	224.365
8	16,147.8	224.274
9	20,181.4	224.238
10	24,660.4	224.185
11	29,585.1	224.129
12	34,953.5	224.061

constant but decreases with increasing J by an amount which is of the order $q_l(B/\omega_2)^2 J(J + 1)$ ([529], [547], [997]). A small variation in q_l of this type may be expected in analogy with the expansion of all rotational constants in powers of B/ω, and has been justified by H. H. Nielsen [505].

Major features of the rotation-vibration spectrum of a linear molecule have now been described. They are illustrated in Fig. 2-3, which is the

$J = 2 \leftarrow 1$ transition of OCS. Each vibrational mode produces a series of lines of exponentially decreasing intensity. The degenerate mode shows l-type doubling when $|l| = 1$. Larger values of $|l|$ cannot occur in this transition, since $|l|$ must be less than J.

2-3. Perturbations between Vibrational States—Fermi Resonance. Ordinarily, the values of α may be determined from the separation between lines of two adjacent vibrational states. Thus the frequency difference between the ground state (000) and the excited state (100) in the $J = 2 \leftarrow 1$ transition shown in Fig. 2-3 should equal $4\alpha_1$. Similarly

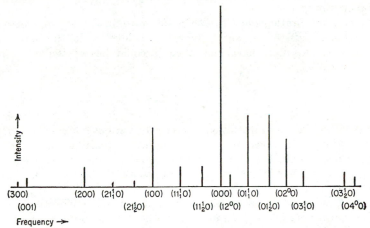

FIG. 2-3. Rotational transition $J = 2 \leftarrow 1$ in OCS showing excited vibrational states and l-type doubling. The vibrational state is given by vibrational quantum numbers in brackets $(v_1 v_2 v_3)$, v_2 having a superscript $|l|$. In case $|l| = 1$, a subscript 1 is applied to the lower-frequency component of the l-type doublet, and 2 to the higher-frequency component. Intensities of excited states in this figure are much larger than normal, being appropriate for a temperature of 800°C.

the frequency difference between (000) and the center of the two l-type doublets $(01^1_1 0)$ and $(01^1_2 0)$ should equal $4\alpha_2$. In most cases this is a satisfactory method of determining the α's. If higher excited vibrational states are observed, they also allow determination of the α's. The frequency difference between (000) and $(02^0 0)$ for a $J = 2 \leftarrow 1$ transition should according to (2-14) give $8\alpha_2$, except for very small correction terms similar to Y_{21} in (1-29) or (1-32). However, in all polyatomic molecules there are perturbations between vibrational states which shift the energy levels, destroy the regularity predicted by (2-14), and make somewhat inaccurate the values of the α's obtained from simple application of this formula. Perturbations between vibrational states were first noticed in CO_2, and are generally called "Fermi resonance" effects because they were explained by Fermi as due to interaction between two states of nearly the same energy.

"Fermi resonance" effects on the rotational spectrum of OCS are rather prominent, and may be seen from Fig. 2-4, which illustrates the $J = 3 \leftarrow 2$ transition. In this figure, the (02^00) line is displaced from the ground state (000) less than twice as much as the center of the two lines (01^110) and (01^120). In addition, the (02^00) line does not coincide with the center of the unsplit doublet (02^20). According to (2-14), these should differ only by the very small quantity $16D(J + 1)$. Separation between (02^00) and (02^20) illustrates the fact that these perturbation interactions depend not just on the energies of vibration, but also on symmetry properties.

The lower vibrational levels of OCS are shown in Fig. 2-5 with possible interactions between adjacent vibrational levels. If the molecular potential were purely harmonic, then it could be written in normal coordinates

$$V = \frac{k_1}{2} q_1^2 + \frac{k_2}{2} q_2^2 + \frac{k_3}{2} q_3^2$$

and there would be no interaction between the various normal coordinates or vibrations. However, there are anharmonic terms in the potential (2-5) which couple the various normal modes. Thus a term such as $k_{122}q_1q_2^2$ from (2-5) allows a variation of q_1 to affect the behavior of q_2 or vice versa.

Let ψ_n^0 represent the molecular wave function for a vibrational level with quantum numbers v_1, v_2, v_3, and $|l|$ when the anharmonic potential terms causing interactions between modes are omitted. Then these interactions may be taken into account by the usual quantum-mechanical perturbation theory. If the initial energies are W_n^0, the perturbed energies W are found by solving a determinant of the form:

$$\begin{vmatrix} W_1^0 - W & W_{21'} & W_{31} & \cdots \\ W_{12} & W_2^0 - W & W_{32} & \cdots \\ W_{13} & W_{23} & W_3^0 - W & \cdots \\ \cdots & \cdots & \cdots & \cdots \end{vmatrix} = 0 \qquad (2\text{-}18)$$

where $W_{ni} = \int \psi_n^* V_{anh} \psi_i \, d\tau$

$V_{anh} = k_{113}q_1^2q_3 + k_{133}q_1q_3^2 + k_{122}q_1q_2^2 + k_{322}q_3q_2^2 + \cdots$

is the anharmonic perturbing potential. Since V_{anh} does not contain any of the angular coordinates θ, ϕ, or χ indicating the molecular orientation, W_{ni} will be zero unless states n and i have the same dependence on θ, ϕ, and χ, or hence have the same angular-momenta quantum numbers. This is connected with the fact that internal motions of the molecule cannot change its angular momentum. For a given vibrational state, any arbitrary total angular momentum J and magnetic quantum number M may be found (unless $J < |l|$). However, any value of $|l|$ cannot be

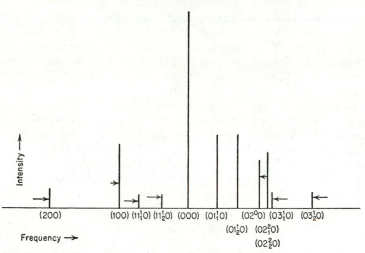

FIG. 2-4. Rotational transition $J = 3 \leftarrow 2$ of OCS showing shifts in rotational frequencies due to perturbations. The arrows indicate the effects of "Fermi resonance" perturbations. Notation is the same as for Fig. 2-3.

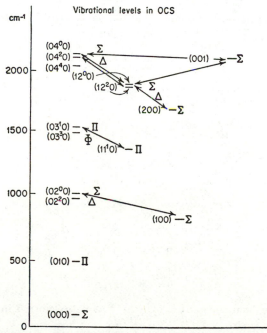

FIG. 2-5. Vibrational levels in OCS. Lines indicate possible interaction between nearby vibrational levels.

found, since a vibrational state is characterized by a particular value $|l|$ of the angular momentum about the symmetry axis. The value of $|l|$ is indicated in Fig. 2-5 by superscripts. It is also given in standard molecular notation by Greek letters, where Σ, Π, Δ, Φ, and Γ represent values of $|l| = 0, 1, 2, 3, 4$, respectively. Only states of the same l can perturb each other, since otherwise W_{ni} is zero.

The perturbations which are most effective in changing the energies and wave functions are those between states which have nearly the same energy. The arrows in Fig. 2-5 indicate such nearby vibrational states which may perturb each other. These are mostly pairs of levels of the type $(v_1, v_2^{|l|}, v_3)$ and $(v_1 - 1, v_2 + 2^{|l|}, v_3)$. Indicating such a pair of states by subscripts 1 and 2, respectively, the determinant (2-18) may be factored into a number of determinants of the type

$$\begin{vmatrix} W_1^0 - W & W_{21} \\ W_{12} & W_2^0 - W \end{vmatrix} \tag{2-19}$$

where $W_{12} = \int \psi_{v_1 v_2 v_3}^* V_{\text{anh}} \psi_{v_1-1, v_2+2, v_3} \, d\tau$

The only nonvanishing term of V_{anh} after integration is

$$W_{12} = k_{122} \int \psi_{v_1 v_2 v_3}^* q_1 q_2^2 \psi_{v_1-1, v_2+2, v_3} \, d\tau \tag{2-20}$$

Since vibrational wave functions for simple harmonic motion are well known ([62], p. 74), (2-20) may be evaluated as

$$W_{12} = - \frac{h^{\frac{3}{2}} v_1^{\frac{1}{2}} [(v_2 + 2)^2 - l^2]^{\frac{1}{2}}}{16 \sqrt{2} \, \pi^3 c^3 \omega_1^{\frac{1}{2}} \omega_2} k_{122} \tag{2-21}$$

The perturbed energies are, from (2-19),

$$W = \frac{W_1^0 + W_2^0 \pm \sqrt{\delta^2 + 4|W_{12}|^2}}{2} \tag{2-22}$$

where $\delta = W_1^0 - W_2^0$. The perturbed wave functions ψ_1 and ψ_2 are combinations of the unperturbed wave functions ψ_1^0 and ψ_2^0

$$\psi_1 = a\psi_1^0 - b\psi_2^0 \qquad \psi_2 = b\psi_1^0 + a\psi_2^0 \tag{2-23}$$

where

$$a = \left(\frac{\sqrt{\delta^2 + 4|W_{12}|^2} + \delta}{2\sqrt{\delta^2 + 4|W_{12}|^2}} \right)^{\frac{1}{2}}$$

$$b = \left(\frac{\sqrt{\delta^2 + 4|W_{12}|^2} - \delta}{2\sqrt{\delta^2 + 4|W_{12}|^2}} \right)^{\frac{1}{2}} \tag{2-24}$$

If the rotational constant is expressed as an expansion in the normal coordinates,

$$B = B_e + \sum_i q_i B_i' + \sum_{i,j} q_i q_j B_{ij}'' + \cdots$$

The effective value of B for a particular vibrational state $(v_1 v_2 v_3)$ may be evaluated as

$$B_v = \int \psi^*_{v_1 v_2 v_3} B \psi_{v_1 v_2 v_3} \, d\tau = B_e + B''_{11} \int \psi^*_{v_1 v_2 v_3} q_1^2 \psi_{v_1 v_2 v_3} \, d\tau$$
$$+ B''_{22} \int \psi^*_{v_1 v_2 v_3} q_2^2 \psi_{v_1 v_2 v_3} \, d\tau + B''_{33} \int \psi^*_{v_1 v_2 v_3} q_3^2 \psi_{v_1 v_2 v_3} \, d\tau$$

since all other terms of second or lower order in the q's are identically zero. Now $\int \psi^*_{v_1 v_2 v_3} q_i^2 \psi_{v_1 v_2 v_3} \, d\tau$ is proportional to the energy of oscillation of the ith mode; hence one may also write

$$B_v = B_e - \alpha_1(v_1 + \tfrac{1}{2}) - \alpha_2(v_2 + 1) - \alpha_3(v_3 + \tfrac{1}{2})$$

where, for example,

$$-\alpha_1(v_1 + \tfrac{1}{2}) = B''_{11} \int \psi^*_{v_1 v_2 v_3} q_1^2 \psi_{v_1 v_2 v_3} \, d\tau$$

The effective value of B may be similarly evaluated for the perturbed wave function $\psi_1 = a\psi_1^0 + b\psi_2^0$ and shown to be

$$B_1 = a^2 B_1^0 + b^2 B_2^0 = B_e + a^2(B_1^0 - B_e) + b^2(B_2^0 - B_e) \qquad (2\text{-}25)$$

where B_1^0 and B_2^0 are the appropriate values of B for the unperturbed states ψ_1^0 and ψ_2^0, respectively. The deviation of B from the equilibrium value B_e due to vibrations is thus intermediate between the deviation for the two unperturbed states. The sum of the B values for the two states does not change, that is,

$$B_1 + B_2 = B_1^0 + B_2^0 \qquad (2\text{-}26)$$

since $a^2 + b^2 = 1$.

For various excited vibrational levels of OCS, Table 2-3 indicates the importance of perturbations in the rotational transitions $J = 2 \leftarrow 1$ and $J = 3 \leftarrow 2$. Since α_1 and α_2 are rather different, even a small perturbation between pairs of vibrational states $(v_1, v_2^{|l|}, v_3)$ and $(v_1 - 1, v_2 + 2^{|l|}, v_3)$ can appreciably affect the rotational frequencies. From infrared measurements of vibrational frequencies [54] the separation between the ideal unperturbed states (100) and (020) may be determined as 165 cm^{-1}. The unperturbed value of α_2 may be obtained from the separation between rotational levels for the (000) and (01^{1}0) states in Table 2-3. The frequency change due to perturbation of the (02^{0}0) state may be then obtained from the known value of α_2. The frequency change of the (10^{0}0) state must be just equal to this but of opposite sign to satisfy (2-26) so that the unperturbed value of α_1 may be determined. From (2-25) the value of a^2 and b^2 for the (10^{0}0) state can then be obtained as 0.944 and 0.056, respectively. The interaction energy $W_{12}(v_1 = 1, v_2 = 0)$ is then evaluated from (2-24) as 43 cm^{-1}. This quantity, combined with δ,

α_1, and α_2, allows prediction of the shift for all other excited states listed in Table 2-3.

Perturbation effects have also been found in CO_2, OCSe, and BrCN, for which $W_{12}(v_1 = 1, v_2 = 0)$ is 50.4, 46, and 60.5 cm^{-1}, respectively.

TABLE 2-4. "FERMI RESONANCE" PERTURBATIONS IN THE ROTATIONAL
SPECTRUM OF OCS

Perturbation corrections are calculated from the values $\delta = 165$ cm^{-1},

$$W_{12}(v_1 = 1, v_2 = 0) = 43 \text{ cm}^{-1},$$

$\alpha_1 = 20.5$ Mc, $\alpha_2 = -10.59$ Mc.   Pairs of interacting vibrational states are bracketed.

Rotational transition	Vibrational state $v_1 v_2{}^{|l|} v_3$	Observed frequency (center of l-type doublets where doubling is present), Mc	Correction for perturbations, Mc
$J = 2 \leftarrow 1$	0 0⁰ 0	24,325.92	0
	0 1¹ 0	24,368.17	0
	⎰1 0⁰ 0	24,253.44	−9.42
	⎱0 2⁰ 0	24,401.0	+9.42
	⎰2 0⁰ 0	24,179.62	−17.54
	⎱1 2⁰ 0	....	+17.54
	⎰1 1¹ 0	24,303.4	−17.15
	⎱0 3¹ 0	24,435	+17.15
$J = 3 \leftarrow 2$	0 0⁰ 0	36,488.82	0
	0 1¹ 0	36,551.7	0
	0 2² 0	36,615.3	0
	⎰1 0⁰ 0	....	−14.83
	⎱0 2⁰ 0	36,600.8	+14.83

2-4. Moments of Inertia and Internuclear Distances. The most obvious quantity determined by measurement of the rotational spectrum of a linear molecule is its rotational constant B, and consequently its effective moment of inertia $J = h/8\pi^2 B$. If vibrational motions of the molecule are negligible, as they are to a certain approximation in the ground vibrational state, and if all atomic masses are considered as concentrated at the atomic nuclei, then the molecular moment of inertia is dependent only on internuclear distances and atomic masses. A linear molecule of three atoms with masses m_1, m_2, and m_3 has a moment of inertia about its center of mass

$$I = \frac{m_1 m_2 l_{12}^2 + m_1 m_3 l_{13}^2 + m_2 m_3 l_{23}^2}{m_1 + m_2 + m_3} \tag{2-27}$$

where l_{ij} represents the distance between masses m_i and m_j. Moments of inertia of the more general linear molecule of an arbitrary number of point masses may be found from a similar formula

$$I = \frac{\frac{1}{2} \sum_j \sum_i m_i m_j l_{ij}^2}{\sum_i m_i} \tag{2-28}$$

Expression (2-28) is correct, in fact, for the moment of inertia of any planar molecule about an axis through its center of mass and perpendicular to the plane of the molecule.

If the atomic masses are assumed known, the moment of inertia of a linear molecule of n atoms contains $n - 1$ unknown distances (there are obvious relations between some of the l_{ij}'s for a linear molecule). For a diatomic molecule, a measurement of I would immediately allow determination of the one unknown distance. In the case of a linear molecule of n atoms, rotational spectra from $n - 1$ different isotopic species, giving values of I for which the known masses are different but the unknown distances are not varied, are needed to determine all the internuclear distances.

Fortunately, it is usually easy to obtain moments of inertia of several isotopic species of polyatomic molecules. Measurements of effective moments of inertia can be made to very high precision, so that the measurements themselves usually offer no bar to very accurate determination of internuclear distances. The one serious limitation to easy attainment of extremely accurate internuclear distances for polyatomic molecules from microwave measurements is the occurrence of rotation-vibration interactions which were provisionally neglected above. Expressions (2-27) and (2-28) apply simply to the molecule if all atoms are at rest in their equilibrium positions, in which case they give the equilibrium moments of inertia I_e. What is actually measured is a reciprocal of the moment of inertia averaged over the ground vibrational state, or B_0. B_0 differs from the equilibrium value B_e by $-\frac{1}{2} \sum_i \alpha_i$. The α_i may depend in a rather complex way on the potential constants and masses, as is evident for the triatomic molecule from (2-6).

Although for the diatomic molecule evaluation of α and allowance for it are easily made, in polyatomic molecules it is rarely practical to measure all the α's and determine their dependence on the various isotopic masses. Usually various isotopic values of B_0 are measured and assumed to be equivalent to values of B_e, so that internuclear distances may be evaluated from (2-28). The errors resulting from neglect of the α's, or zero-point vibration, must be accepted as the primary limit on the accuracy of distances obtained. In the case of OCS, a large number of isotopic species have been measured, so that the two internuclear distances may be determined from various pairs of isotopic combinations.

The results of this procedure are shown in Table 2-5. Discrepancies between the various determinations in Table 2-5 are due primarily to the neglected zero-point vibrations.

Although Table 2-5 illustrates the usual size of errors due to neglect of zero-point vibrations, they can in some cases be considerably larger or considerably smaller. The linear molecule NNO affords an example where serious errors in internuclear distances can be made if the wrong isotopic species are used for their determination. The central nitrogen is very near the center of gravity of the molecule, and hence changing its mass from N^{14} to N^{15} might be expected to have a very small effect on the moment of inertia about the center of mass, or on B_e. However, a change in mass of the central nitrogen can rather markedly affect the vibrational frequencies and hence change the rotation-vibration interaction $-\frac{1}{2}\sum_i \alpha_i$. In fact, experimentally the rotational transitions of $N^{14}N^{15}O^{16}$ are found to occur at higher frequencies than those for $N^{14}N^{14}O^{16}$ [357]. If rotation-vibration interactions are neglected, this would indicate that an increase in mass of the central nitrogen had decreased the moment of inertia, which is clearly impossible from (2-27). A solution for the internuclear distances from the measured frequencies for $N^{14}N^{15}O^{16}$ and $N^{14}N^{14}O^{16}$ yields the interesting result that one of the internuclear distances is imaginary. This is obviously an extreme case, and one which can be remedied by using isotopic species involving a change of mass of the end nitrogen or of the oxygen. Then, even though the rotation-vibration interactions are changed by these isotopic substitutions, the equilibrium moment of inertia is changed appreciably and the effects of vibrations represent only small fractional errors in the observed isotopic effects. The internuclear distances given for N_2O in Table 2-1 were obtained from the pair $N^{14}N^{14}O^{16}$ and $N^{15}N^{14}O^{16}$.

TABLE 2-5. INTERNUCLEAR DISTANCES IN OCS CALCULATED FROM VARIOUS ISOTOPIC PAIRS [329]

Zero-point vibrations are neglected, and are the main source of discrepancies between the different values.

Pair of isotopic molecules used	O—C distance, angstroms	C—S distance, angstroms
$O^{16}C^{12}S^{32}$, $O^{16}C^{12}S^{34}$	1.1647	1.5576
$O^{16}C^{12}S^{32}$, $O^{16}C^{13}S^{32}$	1.1629	1.5591
$O^{16}C^{12}S^{34}$, $O^{16}C^{13}S^{34}$	1.1625	1.5594
$O^{16}C^{12}S^{32}$, $O^{18}C^{12}S^{32}$	1.1552	1.5653

2-5. Determination of Nuclear Masses. For a polyatomic linear molecule of n atoms, $n - 1$ isotopic species must be measured to determine the internuclear distances. If additional isotopic species are measured, the measurements should allow accurate determination of mass ratios, assuming zero-point vibrations can be neglected. In OCS, for example, at least 11 different isotopic species have been measured.

It might at first be thought that two different sulfur isotopes can be used to determine the two unknown distances and that then measurement of moments of inertia for more than one C or O isotope should allow their mass ratios to be determined. Since frequency changes due to isotopic substitution are characteristically of the order of a few hundred megacycles, and frequencies can be measured to an accuracy of a few kilocycles, an accuracy of approximately 10^{-5} mass unit might be expected. A look at Table 2-5, however, provides a quick disillusionment. Because of zero-point vibrations, the internuclear distances are not known to better than 1 part in 10^3, so that the accuracy in mass ratios obtainable by this general method would not be very worthwhile.

In spite of the inaccurate determination of internuclear distances, there is a fairly general way of using the very accurate measurements of microwave spectroscopy to obtain accurate mass information. If rotational frequencies for three isotopes of the same element in one molecule are measured, and if masses of two of these isotopes are known, then the mass of the third isotope can usually be determined with an accuracy comparable with or better than that achieved by more standard methods. Use of two isotopes of known masses might be regarded as a "calibration" of the isotopic shift in the molecule and the effect of zero-point vibrations so that the mass of a third isotope may be determined.

This process of "calibration" with two known isotopic masses can be made exact if zero-point vibrations are negligible, without requiring detailed knowledge of the structure and internuclear distances of the molecule.

The moment of inertia through the center of gravity of a molecule and along a particular direction defined as the z axis of the molecule may be written

$$I_0 = \sum_i m_i(x_i^2 + y_i^2) \tag{2-29}$$

where m_i is the mass of the ith atom, and $x_i^2 + y_i^2$ is the square of its distance from the axis. If the mass of the nth atom is changed an amount Δm, by making an isotopic substitution, the new moment of inertia through the new center of mass about an axis parallel to z is

$$I_1 = I_0 + \frac{\Delta m_1 M_0(x_n^2 + y_n^2)}{\Delta m_1 + M_0} \tag{2-30}$$

where M_0 is $\sum_i m_i$, the total mass of the original molecule before isotopic substitution and $x_n^2 + y_n^2$ is the square of the distance from the nth atom to the center of mass of the original molecule. Equation (2-30) results from the fact that the moment of inertia of an extended body about any axis is the sum of its moment of inertia about a parallel axis through the center of gravity and the moment of inertia of a point mass of equal magnitude placed at its center of gravity. If the mass of the nth atom had been changed by an amount Δm_2, the new moment of inertia would have been

$$I_2 = I_0 + \frac{\Delta m_2 M_0}{\Delta m_2 + M_0} (x_n^2 + y_n^2) \qquad (2\text{-}31)$$

Combining (2-29), (2-30), and (2-31),

$$\frac{I_1 - I_0}{I_2 - I_0} = \frac{M_2}{M_1} \left(\frac{\Delta m_1}{\Delta m_2} \right) \qquad (2\text{-}32)$$

where M_1 and M_2 are the total masses $\Delta m_1 + M_0$ and $\Delta m_2 + M_0$, respectively, for the two isotopic substitutions. Since the moments of inertia are inversely proportional to the rotational constants, which are the quantities actually measured, (2-32) may more conveniently be written

$$\frac{\Delta m_1}{\Delta m_2} = \frac{m_1 - m_0}{m_2 - m_0} = \frac{M_1}{M_2} \frac{B^{(2)}}{B^{(1)}} \frac{(B^{(0)} - B^{(1)})}{(B^{(0)} - B^{(2)})} \qquad (2\text{-}33)$$

Here m_0 represents the original mass of the elements being isotopically replaced, m_1 and m_2 the first and second replacements, $B^{(0)}$, $B^{(1)}$ and $B^{(2)}$ the corresponding values of the rotational constants. Evidently if m_1 and m_2 are known, m_0 may be determined from a measurement of the rotational frequencies. About all that need be known of the molecule is the total mass of other atoms which it contains. It should be clear from the derivation that (2-32) or (2-33) holds not only for linear molecules, but for any type as long as the moments of inertia or rotational constants are appropriate for an axis fixed in direction with respect to the molecule. If two isotopic masses are not accurately known, but their mass difference is known, then mass differences between them and other isotopes may be obtained from (2-33).

Expression (2-33) is strictly correct only if the equilibrium values of the B's are used. However, if the rotational constants B_0 for the ground state are used in (2-33), it still gives the mass difference ratios to very good accuracy. Since the equilibrium values for B are seldom obtainable, masses are generally obtained from (2-33) by using the ground-state rotational constants. Various mass difference ratios obtained from (2-33) are compared with those obtained by other methods in Table 2-6. The errors listed for the microwave data are primarily due to zero-point vibrations or to the use of B_0 rather than B_e, and their magnitudes will be discussed below [456]. Figure 2-6 shows a curve of Se masses found by microwave measurements, and calculated by assuming values for the two masses, Se^{76} and Se^{80}. This curve shows quite clearly the odd-even mass variation, or the difference in nuclear mass behavior between odd and even isotopes. Although some errors may occur in the shape of the curve due to zero-point vibrations, these errors are unimportant in determining the odd-even mass variation. The even masses establish a smooth curve, and the odd-even mass variation is obtained simply by observing how far off of this curve the odd isotopes Se^{75}, Se^{77}, and Se^{79} occur.

We turn now to an examination of the errors in mass determinations caused by zero-point vibrations. The rotational constants usually

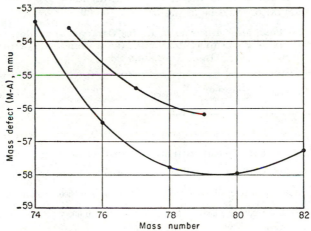

FIG. 2-6. Variation of masses of the stable Se isotopes as a function of mass number. The experimental masses are determined after assuming values for the two masses Se^{76} and Se^{80}. Note that Se^{75}, Se^{77}, and Se^{79}, having odd mass numbers, fall considerably above the curve established by the even isotopes. (*From Geschwind, Gunther-Mohr, and Townes* [924a].)

inserted in (2-33) are those for the ground vibrational state. For a particular isotopic species, this corresponds to $B_0 = B_e - \frac{1}{2} \sum_i \alpha_i$, where α_i is the rotation-vibration interaction for the ith vibrational mode. This sum is abbreviated in the subsequent discussion to $\alpha = \sum_i \alpha_i$.

From (2-33) it is evident that, if α has the same dependence on mass variation as does B_e, then no error is introduced in the mass difference ratio $(m_1 - m_0)/(m_2 - m_0)$, for then the ratios of B's are the same regardless of whether the B_e's or the B_0's are used. Since α and B_e both vary in an approximately linear way for small fractional changes in mass, and since α is much smaller than B_e in any case, the error introduced by α is usually not large. If B_e is changed an amount $\Delta\nu$ due to an isotopic change, then the zero-point vibration effects will be changed by an amount roughly proportional, or $\alpha\Delta\nu/2B$. As long as this change is linearly proportional to $\Delta\nu$ or to the change in mass Δm, no great errors will result. However, an error due to nonlinear dependence of α on Δm will occur of magnitude approximately $\delta = (\alpha_e/2B)\,\Delta\nu\,(\Delta m/m)$. The fractional error which this will produce in the mass difference ratio is simply $\delta/\Delta\nu$, or

$$\frac{\delta}{\Delta\nu} = \frac{\alpha_e}{2B}\frac{\Delta m}{m} \tag{2-34}$$

This expression represents, of course, only a very rough estimate. For the sulfur masses in OCS, $\Delta m/m \approx \frac{1}{16}$ and $\alpha_e/2B \approx \frac{1}{1000}$, so that the error in mass ratio would be of the order 1/16,000.

TABLE 2-6. SOME MASS DIFFERENCE RATIOS OBTAINED FROM ROTATIONAL SPECTRA OF POLYATOMIC MOLECULES AND A COMPARISON WITH OTHER DETERMINATIONS

Ratio of mass differences	Molecule	Microwave measurement	Reference	Other measurements	Method
$\dfrac{S^{33} - S^{32}}{S^{34} - S^{32}}$	OCS	0.500714 ± 0.00003	[574]	0.500727 ± 0.00002	Mass spectra [855a]
$\dfrac{Se^{77} - Se^{76}}{Se^{80} - Se^{77}}$	OCSe	0.33395 ± 0.00002	[456]	0.33394 ± 0.00003	Mass spectra [911a]
$\dfrac{Si^{30} - Si^{29}}{Si^{30} - Si^{28}}$	SiD$_3$F	0.49938 ± 0.00003	[574] [888]	0.49943 ± 0.00001	Mass spectra [855a]
$\dfrac{O^{17} - O^{16}}{O^{18} - O^{16}}$	OCS	0.501042 ± 0.00008	[924a[	0.501044 ± 0.00007	Nuclear reaction [924a]
$\dfrac{Cl^{36} - Cl^{35}}{Cl^{37} - Cl^{36}}$	CH$_3$Cl	1.0018 ± 0.0004	[924a]	1.00179 ± 0.00007	Nuclear reaction [924a]

A more exact expression may be obtained for the error due to neglect of zero-point vibration by expanding α and B_e about their values when $m = m_0$ in powers of the change in isotopic mass Δm

$$\begin{aligned}
\alpha &= \alpha^{(0)} + \alpha'\,\Delta m + \alpha''\frac{(\Delta m)^2}{2} + \cdots \\
B_e &= B_e^{(0)} + B'\,\Delta m + B''\frac{(\Delta m)^2}{2} + \cdots
\end{aligned} \tag{2-35}$$

The "experimental" value of the ratio of mass differences obtained by using (2-33) and neglecting vibrational effects can then be related to the "true" value which would be obtained by using equilibrium rotational constants as follows:

$$\begin{aligned}
\left(\frac{m_1 - m_0}{m_2 - m_0}\right)_{\exp} = \left(\frac{m_1 - m_0}{m_2 - m_0}\right)_{\text{true}} \Bigg[&1 - \frac{1}{4}\frac{\alpha'}{B'}\left(\frac{B''}{B'} - \frac{\alpha''}{\alpha'}\right)(m_2 - m_1) \\
&+ \frac{1}{2}\frac{\alpha}{B}\left(\frac{B'}{B} - \frac{\alpha'}{\alpha}\right)(m_2 - m_1) + \cdots \Bigg]
\end{aligned} \tag{2-36}$$

This expression is not a final answer because usually an evaluation of α' and α'' is very difficult. In fact, if they could be properly evaluated, then corrections could be applied to eliminate these particular errors. However, it does allow an estimation of errors.

B and α are positive in all known cases, B' and α' are negative, and B'' and α'' positive. In this respect α and B are similar, if not identical, functions of m. Hence the two terms in each bracket, $(B''/B' - \alpha''/\alpha')$ and $(B'/B - \alpha'/\alpha)$, tend to cancel. In addition, the two error terms

$$\frac{1}{4}\frac{\alpha'}{B'}\left(\frac{B''}{B'} - \frac{\alpha''}{\alpha'}\right)(m_2 - m_1) \qquad \text{and} \qquad -\frac{1}{2}\frac{\alpha}{B}\left(\frac{B'}{B} - \frac{\alpha'}{\alpha}\right)(m_2 - m_1)$$

may be expected to be of opposite sign and to cancel partially. Hence an upper limit for the fractional error in $(m_1 - m_0)/(m_2 - m_0)$ may be taken as the biggest of the four terms which multiply $m_2 - m_1$. However, the actual error to be expected should be considerably less than this upper limit. Detailed estimates [573][924a] indicate that errors of the type given by (2-36) in the ratio $\dfrac{S^{33} - S^{32}}{S^{34} - S^{32}}$ obtained from the spectrum of OCS are less than 1 part in 15,000. This corresponds to an uncertainty of about 0.03 millimass unit in determining the mass of S^{33} if masses of S^{32} and S^{34} are assumed to be known.

It is, of course, possible to find cases for which the errors given by (2-36) would be very serious. These would be primarily cases where the masses being measured are located near the center of gravity of the molecule—such as the central nitrogen in NNO. Location near the center of gravity would make B' very small, but α' and α'' would not necessarily be small, so that the error terms of (2-36) may be large. However, if these unfavorable cases are avoided, nuclear masses of the medium and heavy atoms may be measured in polyatomic molecules to an accuracy of one or two ten-thousandths of a mass unit. Perhaps the best assurance that the zero-point vibration errors are usually not serious and that no unforeseen errors are present is given by comparing the ratios in Table 2-6 which have been measured both by microwave spectroscopy and by other well-accepted techniques. Where accurate ratios are available from other techniques, they usually agree with microwave results very well and indicate that the errors due to zero-point vibrations are no larger than those which have been estimated.

CHAPTER 3

SYMMETRIC-TOP MOLECULES

3-1. Introduction and General Features of Rotational Spectra. For the normal rotation of a linear molecule, no angular momentum occurs in the direction of the molecular axis because the moment of inertia about this axis is so small that one quantum of angular momentum represents a large excitation energy. For a more general type of molecule, however, there is no axis about which the moment of inertia is extremely small, and the normal rotational states of nonlinear molecules may involve rotation about any molecular axis.

The moments of inertia of a molecule (or of any system of masses) may be represented by an ellipsoid whose orientation is fixed in the molecule and whose center coincides with the center of mass. The shape of the ellipsoid is such that the molecular moment of inertia about any axis through the center of mass is just equal to half the distance between intersections of this axis and the ellipsoid. Every ellipsoid has three perpendicular principal axes, and if the coordinate system is oriented so that x, y, and z are along the principal axes of the ellipsoid of inertia, then the equation of the ellipsoid of inertia may be written simply

$$\frac{x^2}{I_x^2} + \frac{y^2}{I_y^2} + \frac{z^2}{I_z^2} = 1 \tag{3-1}$$

where I_x, I_y, and I_z are the moments of inertia along the directions of the principal axes, and are hence called the principal moments of inertia.

A molecular rotation can usually most simply be described in terms of motions about the principal axes. The special case of a linear molecule has an ellipsoid of inertia which is a flat disk, since the moment of inertia along the molecular axis, which we shall take as the z direction, is very small, and the other two principal moments of inertia are equal. The general rotating body where all three principal moments of inertia are different is called an asymmetric rotor or asymmetric top, and usually the principal moments of inertia are indicated by I_A, I_B, and I_C in increasing order of size. For some molecules, two moments of inertia, such as I_A and I_B or I_B and I_C, may be equal, in which case the molecule is called a symmetric rotor or symmetric top. The linear molecule is a special case of a symmetric rotor, since for it the two largest moments of inertia I_B and I_C are equal.

In many cases it is very easy to pick out the principal axes of a molecule and to see whether or not two principal moments of inertia are equal. If the molecule has an axis of symmetry, then this is always a principal axis of the molecule. An axis of symmetry is recognized if the distribution of atoms in space is unchanged when the molecule is rotated about some axis by an angle of $2\pi/n$, in which case the molecule is said to have an n-fold axis. For example, the water molecule, H_2O, has a configuration as follows

$$H \diagdown \overset{\displaystyle\uparrow}{\underset{\displaystyle\vert}{O}} \diagup H$$

Symmetry
axis

An axis in the plane of all three nuclei, passing through the oxygen nucleus lying halfway between the two hydrogens, is a twofold axis of symmetry, since the molecule will have just the same arrangement of atoms if it is rotated by π radians, or 180°, about this axis. Since the orientation of the ellipsoid of inertia must also remain unchanged as a result of this rotation, it is easy to see that this symmetry axis is a principal axis of inertia. The water molecule is not, however, a symmetric top, because its three principal moments of inertia are all different. If a molecule has an axis of three- or more fold symmetry, then it is always a symmetric top. An example would be NH_3, which is a pyramid-shaped molecule with the N at the apex and the three hydrogens equidistant from the nitrogen. An axis through the N and midway between the three hydrogens is a threefold axis of symmetry. It is also from the above discussion a principal axis of the ellipsoid of inertia and is usually taken as the z axis. If the molecule is rotated through an angle of $2\pi/3$ radians, or 120°, the ellipsoid of inertia must be unchanged. This is possible only if I_x and I_y are equal, so that the cross section of the ellipsoid of inertia in a plane perpendicular to the axis of symmetry z degenerates into a circle. The same argument would hold for any axis of symmetry greater than threefold. A linear molecule, for example, has an infinityfold axis of symmetry.

By far the most common varieties of symmetric tops are linear molecules, which have already been treated, and those with a threefold axis of symmetry. It can conceivably happen that a molecule with less symmetry than a threefold axis is a symmetric top, which would be called an "accidental" symmetric top. Since the high resolution and accuracy of microwave spectroscopy can detect even a very slight deviation from equality of two moments of inertia, it is extremely unlikely that two moments of inertia would be nearly enough equal by "accident" to make the molecule appear to be a symmetric top. Hence this variety of symmetric top with less than a threefold axis of symmetry will be ignored. Cases of this type which are almost symmetric will be treated in Chap. 4 as "slightly asymmetric" rotors.

Symmetric-rotor Spectra—Semiclassical Discussion. Much of the behavior—energy levels and selection rules—of a symmetric top can be deduced from classical mechanics and the correspondence principle. Figure 3-1 illustrates this classical motion. The axis of the molecule precesses around the total angular momentum P with a frequency that can be shown to be $P/2\pi I_B$. At the same time it may spin about its axis (see [130], p. 22). The energy of rotation would be given by

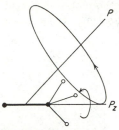

FIG. 3-1. Classical motion of a symmetric top. This is a combined rotation around the molecular axis associated with P_z and a precession of this axis around the total angular momentum P. The molecule represented is methyl chloride,

$$Cl— C \overset{\nearrow H}{\underset{\searrow H}{—H}}$$

$$W = \tfrac{1}{2}I_x\omega_x^2 + \tfrac{1}{2}I_y\omega_y^2 + \tfrac{1}{2}I_z\omega_z^2 = \frac{P_x^2}{2I_x} + \frac{P_y^2}{2I_y} + \frac{P_z^2}{2I_z} \tag{3-2}$$

where x, y, and z are directions along the principal axes of inertia, z being the symmetry axis of the molecule. Now since the molecule is a symmetric top, I_x and I_y are equal and will be both called I_B, which is the normal symbol for the moment of inertia of intermediate size. I_z will be properly designated either I_A or I_C according to whether it is smaller or larger than I_B. If the relative sizes of the moments of inertia are unknown or unimportant, then I_z will always be designated as I_C. Using the fact that $I_x = I_y = I_B$ and that $P^2 = P_x^2 + P_y^2 + P_z^2$, (3-2) becomes

$$W = \frac{P^2}{2I_B} + P_z^2\left(\frac{1}{2I_C} - \frac{1}{2I_B}\right) \tag{3-3}$$

The square of the total angular momentum P^2 is quantized and must equal $J(J+1)h^2/4\pi^2$, where J is an integer. Similarly, the component of the angular momentum along some direction, say the z axis, is quantized, so that $P_z^2 = K^2h^2/4\pi^2$, where K is an integer. Hence (3-3) becomes

$$W = \frac{J(J+1)h^2}{8\pi^2 I_B} + \left(\frac{h^2}{8\pi^2 I_C} - \frac{h^2}{8\pi^2 I_B}\right)K^2$$

or, defining the rotational constants

$$A = \frac{h}{8\pi^2 I_A} \qquad B = \frac{h}{8\pi^2 I_B} \qquad C = \frac{h}{8\pi^2 I_C} \tag{3-4}$$

$$\frac{W}{h} = BJ(J+1) + (C-B)K^2 \tag{3-5}$$

The expression (3-5) gives the correct allowed energy levels for a symmetric top, which are illustrated in Fig. 3-2. If K is zero, the energy levels are just those previously found for a linear molecule. For a given value of J, however, K may have a number of values. K cannot, of course, be larger than J since K represents a component of J. It can therefore be one of the integers

$$K = J, J - 1, \ldots, -J \tag{3-6}$$

or have $2J + 1$ different values. Since the energy is independent of the sign of K, levels with the same absolute magnitude of K coincide, so that all levels for which K is greater than zero are doubly degenerate, and there

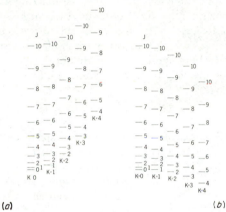

(a)   (b)

FIG. 3-2. Energy levels of typical symmetric-top molecules. (a) prolate; (b) oblate symmetric top.

are only $J + 1$ different energy values for each possible value of J. For each particular K, there is an infinite series of levels with different values of J. These are identical in relative spacing with the linear molecule levels except that the series must start with $J = K$ rather than $J = 0$.

A prolate symmetric top is one in which the molecule is more or less elongated like a cigar, and for which the moment of inertia about the symmetry axis I_z is smaller than the moment of inertia about the other principal axes. In that case $h/8\pi^2 I_z - h/8\pi^2 I_B$, which is the coefficient of K^2 in (3-5), is positive, so that the energy levels for the same J increase with increasing K as in Fig. 3-2a. An oblate symmetric top, which might be represented by a pancake shape, would have I_z larger than I_B, and hence the coefficient of K^2 in (3-5) would be negative. In this case energy levels for the same J decrease in energy with increasing K as in Fig. 3-2b. For a spherical top, with all moments of inertia equal, the coefficient of K^2 in (3-5) is zero and the energy depends only on the total angular momentum J.

In order to determine the spectrum to be expected, the selection rules in addition to the energy levels are needed. Because of the symmetry, there can be no dipole moment perpendicular to the axis of a symmetric top, and hence no torque along the axis due to electric fields associated with radiation. This indicates from the correspondence principle that the angular momentum along the molecular axis cannot change due to radiation, or $\Delta K = 0$. The dipole moment lies along the molecular axis and this axis precesses around the total angular momentum, which is fixed in direction, with a frequency $P/2\pi I_B$ as mentioned above. Hence the frequency to be expected classically is just $P/2\pi I_B$, which is identical with what would be expected from a linear molecule. This frequency can be obtained approximately with the selection rule $\Delta J = \pm 1$, which is identical with the rigorous result of a quantum-mechanical calculation.

It is important to note that, because of the above selection rules, as long as centrifugal stretching and other small effects are neglected, the frequencies observed for a symmetric top do not depend in any way on K or the moment of inertia about the symmetry axis. They are given simply by

$$\nu = \frac{2h(J + 1)}{8\pi^2 I_B} = 2B(J + 1) \tag{3-7}$$

The fact that the frequencies observed do not depend on I_C is an advantage in simplifying the spectrum; it is a disadvantage, however, in preventing any direct determination from the spectrum of the moment of inertia about the axis of a symmetric top.

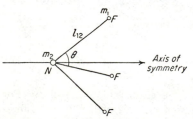

The simplest type of symmetric top (other than a linear molecule or an "accidental" symmetric top, which never is exactly symmetric) is composed of three identical atoms arranged in an equilateral triangle,

FIG. 3-3. A simple pyramidal symmetric-top molecule, NF_3.

and another atom equidistant from these three. The fourth atom may be in the plane of the three identical atoms, or out of this plane so that the molecule is pyramidal. Planar molecules of this type include the trihalides of elements of the third column of the periodic table such as BF_3, BCl_3, and $AlCl_3$. Symmetric planar molecules give no pure rotational spectrum because they have no permanent dipole moment. Simple pyramidal symmetric molecules include the trihydrides and trihalides of the fifth-column elements such as NH_3, NF_3, PH_3, PCl_3, or AsF_3 (see Fig. 3-3). In order for these molecules to be symmetric rotors, the three like atoms must of course be of the same isotopic mass—NH_2D, where one hydrogen has been replaced with deuterium, is by no means a symmetric rotor.

Moments of Inertia.   The moment of inertia about the symmetry axis is given by

$$I_C = 2m_1 l_{12}^2 (1 - \cos \theta) \tag{3-8}$$

where m is the mass of one of the three identical atoms, l_{12} the distance from one of them to the fourth atom, and θ is the angle between two lines joining the fourth atom with two of the other atoms, *i.e.*, the angle of one face of the pyramid at its apex.   The angle θ is usually called the bond

TABLE 3-1. ROTATIONAL CONSTANT AND STRUCTURE OF SYMMETRIC
PYRAMIDAL MOLECULES

l_{12} and θ are defined as in Fig. 3-2.   Where errors in l_{12} and θ are not listed, the errors have not been estimated.

Molecule	B_0, Mc	l_{12}, angstroms	θ	Reference
NH_3	298,000	1.014	106°47′	[130] [662]
NF_3	10,680.96	1.371	102°9′	[527]
PH_3	133,478.3	1.421	93°27′	[606] [736] [866] [909]
PF_3	7,819.90	1.55	102°	[367] [947]
$PCl_3{}^{35}$	2,617.1	2.043 ± 0.003	100°6′ ± 20′	[481]
$PBr_3{}^{79}$	996.8	....	....	[551]
AsH_3	111,620	1.523	92°0′	[606] [736]
AsF_3	5,878.971	1.712 ± 0.006	102° ± 2°	[266] [824]
$AsCl_3{}^{35}$	2,147.2	2.161 ± 0.004	98°25′ ± 30′	[481]
$Sb^{121}H_3$	88,000	1.712	91°30′	[606] [736]
$Sb^{121}Cl_3{}^{35}$	1,754	2.325 ± 0.005	99°30′ ± 1°30′	[597] [947]

angle because the chemical bonds are represented as straight lines between each of the three identical atoms and the fourth atom.   The two equal moments of inertia perpendicular to the axis of symmetry are given by

$$I_B = m_1 l_{12}^2 (1 - \cos \theta) + \frac{m_1 m_2 l_{12}^2}{3m_1 + m_2} (1 + 2 \cos \theta) \tag{3-9}$$

Since the frequencies of allowed rotational transitions depend only on I_B, observation of the rotational spectrum of this type of molecule does not allow determination of both parameters l_{12} or θ, which give the complete molecular configuration.   If, however, spectra of two different isotopic species of the same molecule are observed, such as $N^{14}F_3$ and $N^{15}F_3$, then two different moments I_B are measured giving two equations of the type (3-9), and both molecular parameters l_{12} and θ may be determined.   Structural information about these molecules may also be obtained from their nonsymmetric isotopic forms, *i.e.*, NH_2D, $AsCl_2^{35}Cl^{37}$, etc.   These asymmetric molecules will be discussed in Chap. 4.   Table 3-1 includes the best data available on the structure of symmetric pyramidal molecules. The internuclear distances and angles are subject to the same type of uncertainties due to zero-point vibrations as are the internuclear distances in linear molecules. Where estimates of these or other errors have been made, they are included in Table 3-1.

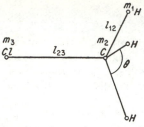

FIG. 3-4. A common type of symmetric-top molecule illustrated by methyl chloride.

Another common type of symmetric-top molecule involves elements of the fourth column of the periodic table bonded to three like atoms and a fourth different atom or group of atoms. Methyl chloride, an example of this type, is shown in Fig. 3-4. The moment of inertia I_C for this type of molecule is of course also given by (3-8), and the moment of inertia perpendicular to the axis of symmetry is

$$I_B = m_1 l_{12}^2 (1 - \cos \theta) + \frac{m_1(m_2 + m_3)l_{12}^2}{3m_1 + m_2 + m_3} (1 + 2 \cos \theta)$$
$$+ \frac{m_3 l_{23}}{3m_1 + m_2 + m_3} \times \left[(3m_1 + m_2)l_{23} + 6m_1 l_{12} \left(\frac{1 + 2 \cos \theta}{3} \right)^{\frac{1}{2}} \right] \quad (3\text{-}10)$$

For these molecules three isotopic species must be measured in order to determine the three structural parameters l_{12}, l_{23}, and θ. In the rather

TABLE 3-2. AN EXAMPLE OF SERIOUS VARIATIONS IN STRUCTURAL PARAMETERS FROM VARIOUS ISOTOPIC COMBINATIONS AS A RESULT OF ZERO-POINT VIBRATIONS

The best values are those in the last row obtained by using the asymmetric species CHD$_2$Cl. (Data from Miller *et al.* [729].)

Isotopic species used	Molecular parameters obtained		
	l_{12} (CH)	l_{23} (CCl)	θ (HCH)
$C^{12}H_3Cl^{35}$, $C^{12}H_3Cl^{37}$, $C^{13}H_3Cl^{37}$	1.123	1.7813	110°57′
$C^{12}H_3Cl^{35}$, $C^{12}H_3Cl^{37}$, $C^{12}D_3Cl^{37}$	1.128	1.7872	112°31′
$C^{12}H_3Cl^{37}$, $C^{13}H_3Cl^{37}$, $C^{12}D_3Cl^{37}$	0.949	1.7850	104°09′
$C^{12}H_3Cl^{35}$, $C^{13}H_3Cl^{35}$, $C^{12}HD_2Cl^{35}$	1.101	1.7815	110°13′

common case where the three identical atoms of mass m are hydrogen, effects of zero-point vibrations can give rather large uncertainties in the positions of these hydrogens (l_{12} and θ). Variations in structural parameters of CH$_3$Cl as a result of using various combinations of effective moments of inertia of three isotopic species are indicated in Table 3-2. In this case variations in the hydrogen positions obtained are especially large. It has been shown that for methyl chloride the average C—H distance appears to be 0.009 A greater than the average C—D distance of the deuterated compound, and that the HCH angle is smaller than the DCD angle by about $\frac{1}{5}°$ [729]. Symmetric-top molecules of the methyl chloride type (containing five atoms) which have so far been measured are listed in Table 3-3. In some cases all structural parameters have not been determined from microwave measurements because a sufficient

TABLE 3-3. SYMMETRIC TOPS OF FIVE ATOMS FOR WHICH MICROWAVE SPECTRA ARE KNOWN

For definition of l_{12}, l_{23}, and θ see Fig. 3-4.

Molecule	B_0, Mc	l_{12}, angstroms	l_{23}, angstroms	θ	Reference
CH_3F	25,536.12	1.11	1.39	110°	[367] [946]
CH_3Cl^{35}	13,292.95	1.113	1.781	110°31'	[280] [412] [496] [729]
CH_3Br^{79}	9,568.19	1.113	1.939	111°14'	[280] [531] [533] [729]
CH_3I	7,501.31	1.113	2.1392	111°25'	[531] [729]
$Si^{28}H_3F$	14,327.9	1.46	1.5946	109°20'	[522] [772]
$Si^{28}H_3Cl^{35}$	6,673.8	1.44	2.050	110°	[315] [362] [727] [772]
$Si^{28}H_3Br^{79}$	4,321.72	1.57 ± 0.03	2.209 ± 0.001	111°20' ± 1°	[409] [521]
$Ge^{74}H_3Cl^{35}$	4,333.91	1.52	2.148	111°	[362] [727]
$Ge^{74}H_3Br^{79}$	2,375.88	1.55 ± 0.05	2.297 ± 0.001	112° ± 1°	[521]
CF_3H	10,348.74	1.332	1.098	108°48'	[367] [690]
CF_3Cl^{35}	3,335.56	1.32	1.77	109°	[358]
CF_3Br^{79}	2,098.06	1.33	1.91	108°	[520] [743]
CF_3I	1,523.23	1.33	2.13	108°	[743]
$CCl_3^{35}H$	3,301.94	1.767	1.073	110°24'	[545] [690]
$CBr_3^{79}H$	1,247.61	1.930 ± 0.003	1.07	110°48' ± 16'	[764]
SiF_3H	7,207.98	1.46	1.565	108°17'	[525]
$Si^{28}F_3Cl^{35}$	2,477.7	1.560	1.989	108°30'	[525] [641]
$Si^{28}F_3Br^{79}$	1,549.9	1.56	2.15	109°	[525] [641]
$Ge^{74}F_3Cl^{35}$	2,166.60	1.69 ± 0.02	2.067 ± 0.005	107°40' ± 1°30'	[555]
PF_3O	4,594.25	1.52	1.45 ± 0.03	102°30' ± 2°	[518] [765]
PF_3S	2,657.63	1.53	1.87	100°20'	[686] [765]
$PCl_3^{35}O$	2,015.20	1.99	1.45 ± 0.03	103°30' ± 2°	[765]
$PCl_3^{35}S$	1,402.64	2.02	1.85 ± 0.02	100°30' ± 2°	[765]
MnO_3F	4,129.11	1.586 ± 0.005	1.724 ± 0.005	108°27' ± 7'	[943a]
ReO_3F	3,566.75	....	....	....	[943a]
ReO_3Cl^{35}	2,094.20	1.761	2.230	108°20' ± 1°	[665] [943a]

number of isotopic species were not measured. In these cases one or two structural parameters have been estimated by other means.

More complex symmetric tops which have been studied in the microwaves are listed in Table 3-4. In many of these cases some structural parameters are estimated.

Isotopic mass ratios may be determined from microwave measurements on the moments of inertia of symmetric-top molecules. As in

TABLE 3-4. SYMMETRIC-TOP MOLECULES OF MORE THAN FIVE ATOMS

Molecule	B_0	Structure	Reference
CH_3CN	9,198.83		[446] [480] [543]
CH_3NC	10,052.90		[480] [543]
CH_3CCH	8,545.84		[544]
CH_3CCBr^{79}	1,561.11		[744]
CH_3CCI	1,259.02		[744]
$CH_3Hg^{202}Cl^{35}$	2,076.20		[462] [928]
$CH_3Hg^{202}Br^{79}$	1,139.88		[462] [928]
$CH_3Hg^{202}I$	788.0		[462]

TABLE 3-4. SYMMETRIC-TOP MOLECULES OF MORE THAN FIVE ATOMS
(*Continued*)

Molecule	B_0	Structure	Reference
$CH_3Hg^{202}CN$	1,747	H—C—Hg—C≡N (with H's on C)	[712b]
CH_3CF_3	5,185	H—C—C—F (with H's on C, F's on C)	[268]
CH_3SiH_3	10,968.96	H—C—Si—H (with H's on C, H's on Si)	[604]
CH_3SiF_3	3,715.63	H—C $\xrightarrow{1.88}$ Si—F; $109°$, 1.10, 1.55, $109°$	[498] [641]
$CH_3SiCl_3^{35}$	1,769.84	H—C $\xrightarrow{1.88}$ Si—Cl; $109°$, 1.09, 2.02, $109°$	[847]
$CH_3Sn^{120}H_3$	6,890.2	H—C $\xrightarrow{2.143}$ Sn—H; $109°$, 1.09, 1.70, $109°$	[603]
CF_3CN	2,945.54	F—C $\xrightarrow{1.46}$ C $\xrightarrow{1.16}$ N; $108°$, 1.335	[743]
CF_3CCH	2,877.95	F—C $\xrightarrow{1.49}$ C $\overset{1.21}{\equiv}$ C $\xrightarrow{1.06}$ H; $109°$, 1.33	[557]

TABLE 3-4. SYMMETRIC-TOP MOLECULES OF MORE THAN FIVE ATOMS
(*Continued*)

Molecule	B_0	Structure	Reference
CF_3SF_5	1,097.6		[713]
$B^{11}H_3CO$	8,657.22		[461]
$(CN)_3P$	2,326		[712b]
$(CH_3)_3CCl^{35}$	3,016		[550]
$(CH_3)_3CBr^{79}$	2,044		[550]
$(CH_3)_3CI$	1,562		[550]

TABLE 3-4. SYMMETRIC-TOP MOLECULES OF MORE THAN FIVE ATOMS
(Continued)

Molecule	B_0	Structure	Reference
$(CH_3)_3SiCl^{35}$	2,197.44		[847]
$C_3H_6O_3$	5,273.6		[553]
$C_8H_{13}Cl^{35}$	1,090.90		[851]
$C_8H_{13}Br^{79}$	725.9		[851]
$B_5{}^{11}H_9*$	7,002.85		[939]

* Angle between plane containing two equivalent borons and the apical boron, and that containing the same equivalent borons and the hydrogen bonded to them is 196°.

polyatomic linear molecules, accurate determination of all the rotation-vibration effects and their dependence on mass is almost impossible, but mass difference ratios as given by (2-32) may be obtained

$$\frac{m_1 - m_0}{m_2 - m_0} = \frac{M_1}{M_2} \frac{I_1 - I_0}{I_2 - I_0} \tag{2-32}$$

It was shown in Chap. 2 that this expression is valid for any type of molecule as long as the moments of inertia I represent the various values for an axis of fixed orientation with respect to the molecule and for the several isotopic masses m_0, m_1, and m_2. If a molecule is to be a symmetric top with changes of the mass of one atom, this atom must necessarily be on the molecular axis, and hence the desired moments of inertia are simply given by the measured values of B, so that (2-32) becomes

$$\frac{m_1 - m_0}{m_2 - m_0} = \frac{M_1}{M_2} \frac{B^{(2)}}{B^{(1)}} \frac{(B^{(0)} - B^{(1)})}{(B^{(0)} - B^{(2)})} \tag{3-11}$$

Zero-point vibrations may be expected to give approximately the same types and magnitudes of errors in evaluation of masses from (3-11) as in the case of linear molecules. Mass difference ratios which have been measured from observations on symmetric tops are listed in Table 3-5.

TABLE 3-5. MASS-RATIO DETERMINATIONS FROM MEASUREMENTS OF
ROTATIONAL CONSTANTS OF SYMMETRIC TOPS
(From Geschwind, Gunther-Mohr, and Townes [924a])

Molecule	Ratio of mass differences	Other determinations
GeH_3Cl^{35}	$\dfrac{Ge^{72} - Ge^{70}}{Ge^{74} - Ge^{70}} = 0.49985 \pm 0.00003$	0.49978 ± 0.00002
	$\dfrac{Ge^{76} - Ge^{74}}{Ge^{74} - Ge^{70}} = 0.50013 \pm 0.00003$	0.50011 ± 0.00002
SiH_3Cl^{35}	$\dfrac{Si^{30} - Si^{29}}{Si^{30} - Si^{28}} = 0.49941 \pm 0.00005$	0.49934 ± 0.00020 0.49943 ± 0.00003
SiD_3F	$\dfrac{Si^{30} - Si^{29}}{Si^{30} - Si^{28}} = 0.49934 \pm 0.00003$	

3-2. Symmetric-top Wave Functions. Discussion of energy levels and selection rules has been so far on a semiclassical basis. A quantum-mechanical treatment of course starts with the Hamiltonian and hence the wave equation for a symmetric top. The motion of a top is usually described in terms of Euler's angles, which are illustrated in Fig. 3-5. θ and ϕ are equivalent to the usual polar angles between an axis fixed in space and some axis fixed in the molecule, and χ (often designated ψ where there is no chance of confusion with the wave function) is the angle of rotation around the axis fixed in the molecule. For a symmetric top, this chosen axis is naturally the molecular or symmetry axis. Eulerian angles are specified in several ways by various authors. Here they are defined as follows in agreement with Casimir [24]. Axes x, y, and z are fixed in the body; X, Y, and Z in space. The position of the body is specified from a starting position in which the two sets of axes coincide. The body is first rotated an angle ϕ about the Z axis, then through an angle θ about the x axis, and finally through an angle χ about the z axis. It can be shown ([81], p. 230) that the wave equation can be written in the above coordinates

$$\frac{1}{\sin\theta}\frac{\partial}{\partial\theta}\left(\sin\theta\frac{\partial\psi}{\partial\theta}\right) + \frac{1}{\sin^2\theta}\frac{\partial^2\psi}{\partial\phi^2}$$

$$+ \left(\frac{\cos^2\theta}{\sin^2\theta} + \frac{C}{B}\right)\frac{\partial^2}{\partial\chi^2}\psi - \frac{2\cos\theta}{\sin^2\theta}\frac{\partial^2\psi}{\partial\chi\,\partial\phi}$$

$$+ \frac{W}{hB}\psi = 0 \quad (3\text{-}12)$$

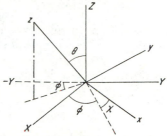

FIG. 3-5. Diagram showing Euler's angles for specifying the position of a rotating body. One dashed line is the line of nodes, or intersection between the xy and XY planes. Another is the projection of the z axis on the XY plane.

where C is the rotational constant for the symmetry axis and B is the rotational constant $h/8\pi^2 I_B$ for an axis perpendicular to the symmetry axis. The variables in (3-12) may be separated, and the solutions written in the form

$$\psi = \Theta(\theta)e^{iM\phi}e^{iK\chi} \quad (3\text{-}12a)$$

where M and K must be integers 0, ± 1, ± 2, . . . in order to make the wave function single-valued. Θ satisfies the equation

$$\frac{1}{\sin\theta}\frac{d}{d\theta}\left(\sin\theta\frac{\partial\Theta}{\partial\theta}\right) - \left[\frac{M^2}{\sin^2\theta} + \left(\frac{\cos^2\theta}{\sin^2\theta} + \frac{C}{B}\right)K^2\right.$$

$$\left. - 2\frac{\cos\theta}{\sin^2\theta}KM - \frac{W}{hB}\right]\Theta = 0 \quad (3\text{-}13)$$

by introducing the variables

$$x = \tfrac{1}{2}(1 - \cos\theta) \quad (3\text{-}14)$$

and letting

$$\Theta(\theta) = x^{\frac{1}{2}|K-M|}(1-x)^{\frac{1}{2}|K+M|}F(x) \quad (3\text{-}15)$$

the equation for F may be found to be

$$x(1-x)\frac{d^2F}{dx^2} + (\alpha - \beta x)\frac{dF}{dx} + \gamma F = 0 \quad (3\text{-}16)$$

where

$\alpha = |K - M| + 1$
$\beta = |K + M| + |K - M| + 2$
$\gamma = \dfrac{W}{hB} - \dfrac{CK^2}{B} + K^2 - (\tfrac{1}{2}|K+M| + \tfrac{1}{2}|K-M|)$ $\quad (3\text{-}17)$
$\qquad \times (\tfrac{1}{2}|K+M| + \tfrac{1}{2}|K-M| + 1)$

This is a well-known form of equation called the hypergeometric equation. Its solution, known as a hypergeometric function, can be obtained as a power series

$$F(x) = \sum_{n=0}^{\infty} a_n x^n \qquad (3\text{-}18)$$

where

$$a_{n+1} = \frac{n(n-1) + \beta n - \gamma}{(n+1)(n+\alpha)} a_n \qquad (3\text{-}19)$$

For ψ to be a satisfactory normalizable wave function, the series must terminate and become just a polynomial, which requires that the energy W is

$$\frac{W}{h} = BJ(J+1) + (C-B)K^2 \qquad (3\text{-}20)$$

with

$$J = n_{\max} + \tfrac{1}{2}|K+M| + \tfrac{1}{2}|K-M| \qquad (3\text{-}21)$$

$n_{\max} =$ the largest value of n in Eq. (3-18) for which a_n does not vanish.

In order for ψ to be normalized and to give matrix elements with signs or "phases" consistent with those of Condon and Shortley [56], the first term a_0 of the series (3-18) for $F(x)$ must be taken as (cf. [931])

$$e^{\frac{i\pi}{2}|K-M|} \left[\frac{(2J+1)(J+\tfrac{1}{2}|K+M| + \tfrac{1}{2}|K-M|)!(J-\tfrac{1}{2}|K+M| + \tfrac{1}{2}|K-M|)!}{8\pi^2(J-\tfrac{1}{2}|K+M| + \tfrac{1}{2}|K-M|)!\,|K-M|!\,(J+\tfrac{1}{2}|K+M| - \tfrac{1}{2}|K-M|)!} \right]^{1/2}$$
$$(3\text{-}22)$$

This expression may be regarded as a normalization and phase factor for ψ.

From (3-21) J must be a positive integer which is equal to or larger than $|K|$ or $|M|$, so that

$$\begin{aligned} J &= 0, 1, 2, \ldots \\ K &= 0, \pm 1, \pm 2, \ldots, \pm J \\ M &= 0, \pm 1, \pm 2, \ldots, \pm J \end{aligned} \qquad (3\text{-}23)$$

As the reader may suspect, $J(J+1)h^2/4\pi^2$ can be shown to be the square of the total angular momentum; $Kh/2\pi$ is its projection on the molecular axis, and $Mh/2\pi$ its projection on the polar axis fixed in space. The energy (3-20) can be seen to be identical with Eq. (3-5) which was obtained from a semiclassical approach.

3-3. Symmetry and Inversion. The energy or behavior of a rotating molecule remains unchanged after certain types of changes of coordinates or symmetry operations, and hence the quantum-mechanical wave functions describing the molecule might be expected to remain unchanged. The wave equation may be written

$$H_{op}\psi = W\psi \qquad (3\text{-}24)$$

where H_{op} represents the Hamiltonian operator for the energy which, for a symmetric top, takes the form (3-12) if Eulerian angles are used as coordinates. If Cartesian coordinates are used, H_{op} is easily seen to be unchanged when the coordinate system is inverted about the origin, *i.e.*, when x is replaced by $-x$, y by $-y$, and z by $-z$. This is shown by the fact that terms in H_{op} do not involve odd powers of the coordinates, but only terms of the type $\partial^2/\partial x^2$, $x(\partial/\partial y)$, etc.

If the coordinate change $x \to -x'$, $y \to -y'$, and $z \to -z'$ is made on (3-24), then H_{op} is unchanged, and the new ψ' must be a solution of (3-24), for the same energy W. If this energy does not represent a degenerate level, for which there are several different solutions of (3-24), then the new ψ' must be the same as the old ψ, or differ only by a multiplicative constant. Let this constant be c. If now another such transformation is made, $x' \to -x''$, $y \to -y''$, $z' \to -z''$, the new ψ'' is $\psi'' = c\psi' = c^2\psi$, but it also must be identical with ψ since this is just the reverse of the original transformation. Hence $c^2 = 1$, or $c = \pm 1$. If $c = +1$, ψ is unchanged by inversion about the origin, is said to be symmetric with respect to this operation, and is designated as an even ($+$) level. If $c = -1$, ψ simply changes sign on inversion, is said to be antisymmetric, and is designated as an odd ($-$) level.

One reason for the importance of the symmetry of the wave function with respect to inversion is its connection with the dipole matrix elements which determine the intensities of transitions. These have the form

$$\int \psi_1 x \psi_2 \, d\tau, \quad \int \psi_1 y \psi_2 \, d\tau, \text{ and } \int \psi_1 z \psi_2 \, d\tau \qquad (3\text{-}25)$$

where ψ_1 and ψ_2 are wave functions for the two states between which transitions occur and $d\tau$ is a volume element.

The integrals are to be taken over all values of the coordinates. Each integral may be equated to the sum of two integrals, the first over all positive values of x, y, z and the second over all negative values. The second integral is obtained from the first by the transformation $x \to -x'$, $y \to -y'$, and $z \to -z'$. In this transformation, the integrals will not of course change value, but may be seen to change sign unless ψ_1 and ψ_2 have opposite symmetries. Thus if ψ_1 and ψ_2 have the same symmetry, the integral over all space is the sum of two integrals which are equal in magnitude and opposite in sign, and so is zero. This establishes the selection rule

$$+ \longleftrightarrow - \qquad + \leftarrow\!\!\!|\!\!\!\to + \qquad - \leftarrow\!\!\!|\!\!\!\to -$$

The transitions of a diatomic molecule, for which the matrix elements vanish unless $J \pm 1 \leftarrow J$ (page 22), must of course obey the above selection rule and will be found to do so. We shall examine the symmetry of symmetric-top wave functions, but before doing so, the coordinate system for a symmetric top must be more completely described.

Consider a planar symmetric top such as BF_3 with three like atoms and a fourth atom equidistant from them. The orientation of this molecule can be specified in terms of an axis through the center of mass and perpendicular to the plane of the molecule. In order to define the positive direction of this axis, the three like (fluorine) atoms must be labeled with the numbers 1, 2, 3. The positive direction is then taken as the direction of advance of a right-handed screw rotated in the direction of successive nuclei 1, 2, 3. This is illustrated in Fig. 3-6a, or in Fig. 3-6b where the positive direction is into the page. θ and ϕ are the usual polar angles between the molecular axis and a fixed polar axis (z), and χ, illustrated in Fig. 3-6b, is the angle of rotation around the molecular axis. If the coordinates of each atom in the molecule are inverted about the origin, then Fig. 3-6b shows that the positive direction of the molecular axis does not change, so that θ and ϕ are unaffected. However, χ is changed into $\chi + \pi$.

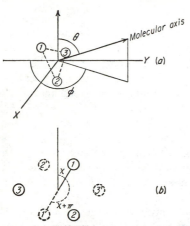

FIG. 3-6. Coordinates for a symmetric-top molecule. (a) The positive direction of the molecular axis of a symmetric-top molecule (e.g., BF_3) and the polar angles θ and ϕ. The three identical nuclei are labeled 1, 2, and 3. (b) A symmetric-top molecule viewed along the positive direction of the molecular axis showing the angle χ which indicates rotation around the molecular axis. The vertical line is the "line of nodes." The solid circles represent the nuclear positions before an inversion through the origin, and the dotted circles are their positions after such an inversion.

The rotational wave functions for a symmetric top are from (3-12a), of the form

$$\psi_{JKM} = e^{iM\phi}e^{iK\chi}\Theta_{JKM}(\theta) \quad (3\text{-}26)$$

so that when the coordinates undergo an inversion about the origin, the new wave function becomes

$$\psi = \psi e^{iK\pi} = (-1)^K\psi \quad (3\text{-}27)$$

Hence the rotational wave function is even $(+)$ or odd $(-)$ with respect to inversion according to whether K is even or odd.

In addition to the rotational part of the wave function, the electronic, vibrational, and spin coordinates must be considered. The complete wave function may be indicated by a product

$$\psi_{\text{total}} = \psi_e\psi_v\psi_R\psi_I \quad (3\text{-}28)$$

where ψ_e, ψ_v, ψ_R, and ψ_I represent the parts of the wave function dependent, respectively, on electronic, vibrational, rotational, and spin coordinates. The behavior of ψ_{total} with respect to any symmetry operation depends on the behavior of each of these four parts. The electronic wave functions for almost all polyatomic molecules in the ground state are symmetric, so they may be neglected in considering the symmetry of ψ_{total}. The spin wave functions ψ_I may usually be either symmetric or antisymmetric, but consideration of them will be postponed until later.

The vibrational wave function for a molecule in the ground vibrational state is always symmetric, so in this case the symmetry of ψ_{total} depends only on ψ_R (if ψ_I is neglected). Consider excited states of the vibration for which the central boron atom in the symmetric top illustrated by Fig. 3-6 moves perpendicularly to the plane of the three fluorines. The coordinate for this motion will be called h. It indicates the distance moved by the boron from the center of mass and will be called positive when the boron has moved along the positive direction of the molecular axis, negative if the boron moves in the opposite direction. The wave functions for a harmonic vibration of this type are ([62], p. 74)

$$\psi_v(h) = c_1 e^{-c_2{}^2 h^2/2} H_v(c_2 h) \tag{3-29}$$

where c_1 and c_2 are constants, and H_v is a Hermite polynomial of order n. The lowest energy level corresponds to $v = 0$, with higher integral values $v = 1, 2, \ldots$ being successively higher in energy. H_v involves only odd or even powers of h according to whether v is odd or even. Hence when an inversion about the origin occurs, which changes h into $-h'$, the new vibrational wave function is

$$\psi_v' = (-1)^v \psi_v \tag{3-30}$$

The rotation-vibration wave functions hence have symmetries with respect to inversion as shown in Fig. 3-7, where the ground and first excited vibration states are indicated.

There are many symmetric tops such as CH_3Cl or NH_3 which are not normally considered planar molecules, but which have the same symmetry with respect to inversion as BF_3. NH_3 may actually be considered planar, but with a potential function which differs so much from a simple harmonic potential that as a result of vibration the nitrogen spends most of its time some distance from the plane of the three hydrogens. The energy levels of a particle moving in a parabolic potential are equally spaced as on the left side of Fig. 3-8. If the potential is distorted by a hill being gradually raised in the center, the energy levels approach each other in pairs as shown in Fig. 3-8. For a very high potential hill, the particle again has equally spaced energy levels, but two sets of them corresponding to oscillations on either one side of the hill or the other. However, even when the hill is so high that the particle does not have enough energy to go to its top, a quantum-mechanical "tunnel effect" occurs which allows the particle to oscillate slowly from one side of the potential barrier to the other. (This will be discussed more fully in

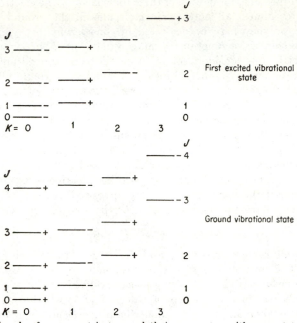

FIG. 3-7. Levels of a symmetric top and their symmetry with respect to inversion. The vibration is a nondegenerate mode in which atoms on the molecular axis move parallel to this axis.

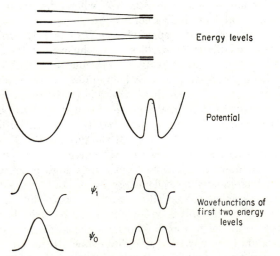

FIG. 3-8. Behavior of a vibration with introduction of a potential barrier.

Chap. 12 on the ammonia spectrum.) The rise of the central potential hill modifies the wave functions as is shown in Fig. 3-8, but does not destroy their symmetry. For BF_3 there is no potential hill and the potential minimum occurs when the boron is in the plane of the fluorines, so the energy levels are those on the left of Fig. 3-8. NH_3, on the other hand, is a pyramidal molecule. Potential minima occur when the nitrogen is on either side of the plane of the hydrogens, and the energy levels correspond to those on the right-hand side of Fig. 3-8. In NH_3, the central potential hill is only moderately high, and the two lowest vibrational levels are separated by an energy gap such that a transition between them falls in the microwave region. In the case of NF_3, CH_3Cl, or almost any other nonplanar top, the potential hill is so high that the two lowest vibrational levels have almost coalesced, their separation corresponding to frequencies which are so low that usually many years are required for an oscillation period.

For nonplanar symmetric tops, the transition between these two lowest levels is often referred to as inversion, for the classical motion corresponds to the molecule being turned inside out. This is not identical, however, with inversion of the molecule about the center of mass. Let the position of nitrogen with respect to the plane of the hydrogens in NH_3 be given by wave functions ψ_0 and ψ_1 as shown on the right side of Fig. 3-8. The energy for ψ_0 is W_0, so that it varies with time as $\psi_0 e^{2\pi i W_0 t/h}$, and similarly ψ_1 varies as

$$\psi_1 e^{2\pi i(W_0+\Delta)t/h}$$

where Δ is the energy separating the two lowest levels. If at time $t = 0$ the nitrogen is on the negative side of the hydrogens, then the wave function for the system may be written

$$\psi = \frac{1}{\sqrt{2}}(\psi_0 + \psi_1 e^{2\pi i\Delta t/h})e^{2\pi i W_0 t/h} \tag{3-31}$$

which at time $t = 0$ is $(\psi_0 + \psi_1)/\sqrt{2}$, and at time $h/2\Delta$ is $(\psi_0 - \psi_1)/\sqrt{2}$, corresponding to the nitrogen being on the positive side. Hence the NH_3 molecule inverts itself with a frequency (for a complete cycle) $\nu = \Delta/h$, which happens to be about 2.4×10^{10} times per second. CH_3Cl, on the other hand, exists for some time with Cl on one particular side of the molecular axis, and inverts only very slowly.

Symmetry and spacing of the rotation and inversion levels of NH_3 are indicated in Fig. 3-9. Possible transitions, allowed by the rules $+\longleftrightarrow-$, $\Delta J = 0, \pm 1$, $\Delta K = 0$, are also shown. Figure 3-10 shows the same things for CH_3Cl, where the inversion levels are so close together as to be indistinguishable, and the rule $+\longleftrightarrow-$ is no longer of importance since $+$ and $-$ levels coincide in pairs. Some levels in Fig. 3-9 for which $K = 0$ are drawn as dashed lines. These levels cannot occur

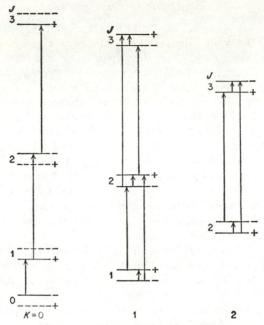

FIG. 3-9. Levels and possible transitions in the rotation-inversion spectrum of NH_3. The dotted levels for $K = 0$ are forbidden by the exclusion principle.

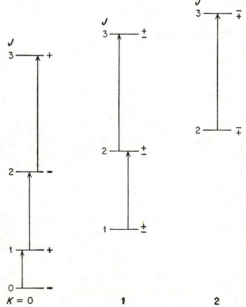

FIG. 3-10. Levels and possible transitions in the rotation-inversion spectrum of CH_3Cl, where the inversion frequency is negligibly small.

because of certain properties of the spin wave functions to be discussed below.

For NH_3, the doubling of each level due to inversion produces a doubling of the rotational lines with a separation between doublet components of twice the inversion frequency. This rotational spectrum lies in the infrared region. In addition, transitions between inversion levels with $\Delta J = 0$, $\Delta K = 0$ occur and produce lines which occur in the normal microwave range near 1 cm. In the case of CH_3Cl, the pure rotational spectrum is observed in the microwave range, and although each rotational line is split by twice the inversion frequency, this splitting is so small that it is completely unobservable even with the high resolution of microwave spectroscopy.

3-4. Effects of Nuclear Spins and Statistics. Other types of symmetry operations in addition to inversion about the center of mass may also be considered. For a symmetric top with a threefold axis of symmetry such as NH_3 or BF_3, a rotation of 120° about the symmetry axis should leave the molecule essentially unchanged, and reasoning similar to that applied above to inversion about the origin shows that this rotation must either leave the wave function unchanged or change only its sign if the state is not degenerate. Similarly an interchange of two H nuclei in NH_3 or two F nuclei in BF_3 is another permissible symmetry operation which must affect the wave function in the same way.

Symmetry considerations may also be applied to the case of interchanging two identical particles in any type of system. It is found experimentally that H^1, F^{19}, and any other nucleus of odd nuclear mass always occur in antisymmetric wave functions. These nuclei are said to obey Fermi-Dirac statistics. Nuclei of even mass always occur in symmetric wave functions and are said to obey Einstein-Bose statistics. Hence it is essential that any true and permissible wave function for NH_3 changes sign when two H nuclei are exchanged.

Consider just rotation of 120° around the symmetry axis of NH_3. This is equivalent to exchanging two pairs of H nuclei, say first numbers 1 and 2, then 2 and 3. Since there are two exchanges, the wave function must be unchanged by a rotation of 120° if H obeys either Fermi-Dirac or Bose-Einstein statistics. The only one of Euler's angles which changes with such a rotation is χ, which enters the wave function as $e^{iK\chi}$ or $e^{-iK\chi}$. Hence after a 120°, or $2\pi/3$, rotation,

$$\psi' = \psi e^{\pm(2\pi/3)Ki} \qquad (3\text{-}32)$$

If K is a multiple of 3, then the exponential in (3-32) equals 1, and $\psi' = \psi$, so that the wave function is symmetric. If K is not a multiple of 3, then ψ is neither symmetric nor antisymmetric. This indicates that the state is degenerate, which is true since the same energy is obtained for $+K$ as for $-K$. In order to make wave functions of the correct symmetry when K is not a multiple of 3, the spin wave function ψ_I must be considered, for when nuclei are exchanged not only their spatial coordinates are changed, but also their spins.

Before discussing spin wave functions, consider the symmetry operation of exchanging only one pair of nuclei, e.g., 2 and 3. An interchange of these two nuclei changes the positive direction of the molecular axis as defined above, since it changes the relative order of the nuclei. The molecular variables are transformed as follows:

$$\theta' \to \pi - \theta \qquad \phi' \to \phi + \pi$$
$$\chi' \to \pi - \chi \qquad h' \to -h \qquad (3\text{-}33)$$

Detailed examination of the wave functions given by (3-12a) shows that for $\theta \to \theta' - \pi$, $\chi \to \pi - \chi'$, and $\phi \to \phi' - \pi$, the wave function changes as

$$\psi_{JKM} \to (-1)^J \psi_{J,-K,M} \qquad (3\text{-}34)$$

For this change of variables, ψ is neither symmetric nor antisymmetric but the wave functions are degenerate. They can be put in symmetric or antisymmetric form by choosing a wave function $\psi_{JKM} + \psi_{J,-K,M}$ or $\psi_{JKM} - \psi_{J,-K,M}$. In exchanging nuclei 2 and 3, χ changed to $\pi - \chi$. If, instead of interchanging 2 and 3, 1 and 3 had been interchanged, then the variables would have transformed as

$$\theta' \to \pi - \theta \qquad \phi' \to \phi + \pi$$
$$\chi' \to \pi - \chi + \frac{2\pi}{3} \qquad h' \to -h \qquad (3\text{-}35)$$

The new symmetrized forms $e^{iK\chi} \pm e^{-iK\chi}$ would then for this exchange no longer be symmetric or antisymmetric unless K is a multiple of 3. Spin wave functions are needed to produce ψ_{total} which has the correct type of symmetry for all possible interchanges of nuclei.

Spin Wave Functions. The spin wave function for a nucleus needs to tell the projection of the spin on some fixed direction, which in the case of hydrogen can take on only two values, $+\frac{1}{2}$ and $-\frac{1}{2}$. The $+\frac{1}{2}$ value will be indicated graphically by a vector pointed up, and $-\frac{1}{2}$ by a vector pointed down. Various possible spin functions which exactly specify the spin orientation of three H nuclei are then represented by Fig. 3-11.

The spin wave functions I and VIII of Fig. 3-11 are clearly symmetric with respect to interchange of any two nuclei, since they do not change. Spin function II, however, is symmetric with respect to interchange of 2 and 3, but it changes into III if 1 and 2 are interchanged, or into IV if 1 and 3 are interchanged. All spin functions shown in Fig. 3-11 are of this type except I and VIII, and hence are degenerate. Functions which are symmetric or antisymmetric with respect to interchange of any nuclei can be formed by combining the variation with χ and spin. Such functions automatically are symmetric for a 120° rotation, which corresponds to successive interchange of two pairs of nuclei. If K is not a multiple of 3, they are

$$\psi_{JKM}(\text{II} + e^{2\pi Ki/3}\text{III} + e^{4\pi Ki/3}\text{IV}) \pm \psi_{J,-K,M}(\text{II} + e^{-2\pi Ki/3}\text{III} + e^{-4\pi Ki/3}\text{IV})$$
$$\psi_{JKM}(\text{V} + e^{2\pi Ki/3}\text{VI} + e^{4\pi Ki/3}\text{VII}) \quad (3\text{-}36)$$
$$\pm \psi_{J,-K,M}(\text{V} + e^{-2\pi Ki/3}\text{VI} + e^{-4\pi Ki/3}\text{VII})$$

If K is a multiple of 3, these reduce to

$$(\psi_{JKM} \pm \psi_{J,-K,M})(\text{II} + \text{III} + \text{IV})$$
$$(\psi_{JKM} \pm \psi_{J,-K,M})(\text{V} + \text{VI} + \text{VII}) \quad (3\text{-}37)$$

and two additional functions have the correct symmetry,

$$(\psi_{JKM} \pm \psi_{J,-K,M})\text{I}$$
$$(\psi_{JKM} \pm \psi_{J,-K,M})\text{VIII} \quad (3\text{-}38)$$

In all these expressions, when J is even the $+$ sign produces a function which is symmetric with respect to interchange of two identical nuclei, and the $-$ sign gives an antisymmetric function [$cf.$ Eq. (3-34)]. When J is odd, the symmetries are reversed.

Not all the functions (3-36) or (3-37) and (3-38) are permitted for a particular rotation-inversion state. In the lowest inversion state, the vibrational (inversion) part of the wave function does not change sign for a transformation of the type (3-33) or (3-35), which reverses the sign of h. The parts of the wave function involving h, θ, and ϕ are symmetric with respect to interchange of two nuclei if $J + K$ is even [$cf.$ (3-34)]. For these cases antisymmetric forms, with the $-$ sign, must be chosen from (3-36), or (3-37) and (3-38) to make the total wave function change sign with exchange of two nuclei, $i.e.$, if J is odd, the $-$ sign is chosen in these equations, or if J is even, the $+$ sign is needed. In the upper inversion state, the vibrational part of the wave function is antisymmetric for interchange of two nuclei, and the choice of $+$ or $-$ signs in (3-36), or (3-37) and (3-38) must be just reversed.

When $K = 0$, the functions of the form (3-37) or (3-38) become zero when the $-$ sign is used, and hence no such wave function exists. This is the reason why half the levels are nonexistent when $K = 0$ as indicated in Fig. 3-9. In the lowest inversion state, when $K = 0$ and $J = 0$, a $-$ sign in (3-37) or (3-38) would be called for, but this makes the wave function zero. In the upper inversion state, however, when $K = 0$ and $J = 0$, a $+$ sign is called for and such a wave function is not zero. When $K = 0$ and J is odd, however, the ground inversion level requires the $+$ sign and hence is the state in which molecules can exist.

Nucleus	1	2	3
I	↑	↑	↑
II	↓	↑	↑
III	↑	↓	↑
IV	↑	↑	↓
V	↑	↓	↓
VI	↓	↑	↓
VII	↓	↓	↑
VIII	↓	↓	↓

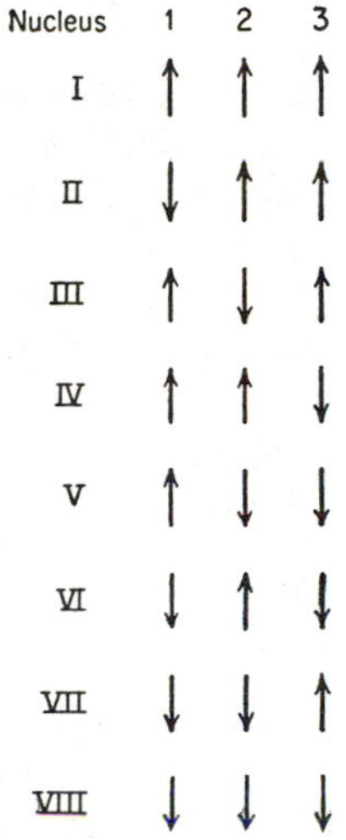

FIG. 3-11. The eight possible spin states for three nuclei with spin $\frac{1}{2}$.

Statistical Weights. It is evident that, with the exception of the case $K = 0$, there are twice as many acceptable wave functions from (3-37) and (3-38) for the case where K is a multiple of 3 as can be obtained from (3-36) for the cases where K is not a multiple of 3. This gives the states for which K is a multiple of 3 twice the statistical weight and hence approximately twice the population of otherwise similar states for which K is not a multiple of 3.

The above discussion has been specialized to the case of three identical atoms with spins $\frac{1}{2}$, which is by far the most common case. If there are three identical atoms with spin I, the statistical weights are given by Table 3-6 [25], [43]. Regardless of the type of statistics, the levels with K a multiple of 3 always have a greater weight, the ratio between these and other values of K being $2:1$ in the case of spin $\frac{1}{2}$ as was found above.

If $K = 0$, we have seen that alternate inversion levels on the level diagram of Fig. 3-9 are missing, beginning with the lowest inversion level for $J = 0$. If the spin I of the identical nuclei had been zero, Bose-Einstein statistics would apply instead of Fermi-Dirac, and for $K = 0$ the role of permitted and forbidden levels on Fig. 3-9 would be just

TABLE 3-6. STATISTICAL WEIGHTS

Statistical weights due to nuclear spin for rotational levels of a symmetric-top molecule with three identical nuclei of spin I. These apply to molecules in a non-degenerate vibrational state. For a degenerate vibrational state, K should be replaced by $K - l$.

	Statistical weights	Nuclear spin, I
K a multiple 3, but not 0 K not a multiple of 3	$\frac{1}{3}(2I + 1)(4I^2 + 4I + 3)$ $\frac{1}{3}(2I + 1)(4I^2 + 4I)$	$0 \quad \frac{1}{2} \quad 1 \quad \frac{3}{2}$
Ratio	$\dfrac{4I^2 + 4I + 3}{4I^2 + 4I}$	$\infty \quad \frac{2}{1} \quad \frac{11}{8} \quad \frac{6}{5}$
$K = 0$, J even lower inversion level or J odd upper inversion level, Fermi-Dirac statistics	$\frac{1}{3}(2I + 1)(2I - 1)I$	
$K = 0$, J odd, lower inversion level or J even, upper inversion level, Fermi-Dirac statistics	$\frac{1}{3}(2I + 1)(2I + 3)(I + 1)$	
Ratio Fermi-Dirac statistics	$\dfrac{(2I - 1)I}{(2I + 3)(I + 1)}$	$0 \qquad \frac{1}{5}$
Ratio Bose-Einstein statistics	$\dfrac{(2I + 3)(I + 1)}{(2I - 1)I}$	$\infty \qquad \frac{10}{1}$

reversed. If the spin is greater than $\frac{1}{2}$, none of these $K = 0$ levels is forbidden: their statistical weights for the case of Fermi-Dirac statistics are given in Table 3-6. For Bose-Einstein statistics, these formulas for statistical weights of alternate levels are just reversed.

For symmetric-top molecules with four identical nuclei equidistant around the axis, the statistical weights are as follows [43]:

For $K \neq 0$:

$(I + 1)(2I + 1)(2I^2 + I + 1)$
    for K a multiple of 4, Bose-Einstein statistics or for K even, not a multiple of 4 and Fermi-Dirac statistics.
$I(2I + 1)(2I^2 + 3I + 2)$
    for K a multiple of 4, Fermi-Dirac statistics or for K even, not a multiple of 4 and Bose-Einstein statistics.
$I(I + 1)(2I + 1)^2$
    for K odd.

For $K = 0$:

$\frac{1}{2}(I + 1)(2I + 1)(2I^2 + 3I + 2)$
    for J even, Bose-Einstein statistics.
$\frac{1}{2}I(I + 1)(2I - 1)(2I + 1)$
    for J odd, Bose-Einstein statistics.
$\frac{1}{2}I(2I + 1)(2I^2 + I + 1)$
    for J even, Fermi-Dirac statistics.
$\frac{1}{2}I(I + 1)(2I + 1)(2I + 3)$
    for J odd, Fermi-Dirac statistics.

In this case, the inversion levels are considered to coincide, so statistical weights of the $K = 0$ levels refer to the sum of both inversion levels.

The above considerations about statistical weights apply only to molecules in nondegenerate vibrational states. When a degenerate vibrational mode is excited, a new angular momentum l is introduced (*cf.* l-type doubling in Chap. 2 and below). In this case similar statistical weights apply, but with K replaced by $K - l$ [943a]. Thus levels with $K - l$ a multiple of 3 have greater statistical weights than do other levels for molecules with a threefold axis.

Nuclear spins are important in a wide variety of other types of molecules. Some of these are discussed by Placzek and Teller [43], Wilson ([64] and [90]), and Minden [730].

3-5. Intensities of Symmetric-top Transitions. Intensities of symmetric-top absorption lines may be calculated from the basic formula (1-49). Some of the quantities in this expression, however, such as the matrix element μ and the fraction of molecules f in a given state, must be evaluated for the symmetric-top case.

Selection rules for dipole radiation of a nonplanar symmetric top (dipole moment taken only along the molecular axis since it is a truly symmetric top) are

$$\Delta J = 0, \pm 1 \qquad \Delta K = 0 \qquad + \rightarrow - \qquad (3\text{-}39)$$

The last selection rule, taken from symmetry considerations above, is needed to specify the inversion levels involved in a transition, and may be applied by referring to Fig. 3-9.   Matrix elements may be calculated as indicated in (1-59), (1-60), and (1-61).   However, since wave functions for a symmetric top are considerably more complex than those for a linear molecule, actual evaluation of these integrals is more difficult.   The matrix elements are of course nonzero only for transitions given by the selection rules (3-39).   The nonzero matrix elements are as follows:

$J + 1 \leftarrow J, K \leftarrow K$:

$$|\mu_{ij}|^2 = \mu^2 \frac{(J + 1)^2 - K^2}{(J + 1)(2J + 1)} \tag{3-40}$$

$J \leftarrow J, K \leftarrow K$:

$$|\mu_{ij}|^2 = \frac{\mu^2 K^2}{J(J + 1)} \tag{3-41}$$

$J - 1 \leftarrow J, K \leftarrow K$:

$$|\mu_{ij}|^2 = \mu^2 \frac{J^2 - K^2}{J(2J + 1)} \tag{3-42}$$

These are the matrix elements appropriate for substitution into (1-49), and represent the sum of the components $|R_x|^2$, $|R_y|^2$, and $|R_z|^2$ for a particular molecule orientation specified by M, the projection of J on the z axis, to all possible final M states.   The components $|R_x|^2$, $|R_y|^2$, and $|R_z|^2$ and their dependence on M are given in Table 4-4.

The quantity μ in the above equations is the usual dipole moment of the molecule.   One might question whether a symmetric top has a dipole moment, since we have been considering inversion as a type of vibration and even a pyramidal molecule as an oscillating planar molecule.   This dipole moment μ, however, is to be evaluated without considering inversion, but taking the symmetric top in its normal pyramidal configuration.   Thus although NH_3 is inverting approximately 3×10^{10} times per second, μ is called a "permanent" dipole moment and is to be calculated when N is on one side only of the three hydrogens.   This is the NH_3 dipole moment which commonly enters into other physical and chemical measurements.   Quantum-mechanically, μ should be evaluated in the state $(\psi_0 + \psi_1)/\sqrt{2}$, where ψ_0 and ψ_1 represent wave functions for the two inversion levels.

The fraction of molecules f in a particular initial state is also needed to calculate absorption intensities from (1-49).   This fraction f is the product of the fraction f_v in the vibrational state of interest and the fraction of these f_{JK} in a particular rotational state.   If statistical weight due to nuclear spin is neglected, the probability of a molecule's being in a state J, K is proportional to

$$(2J + 1)e^{-[BJ(J+1)+(C-B)K^2]h/kT} \tag{3-43}$$

$2J + 1$ is the statistical weight due to the different orientations of J.

The fraction of molecules in this rotational state would be

$$f_{JK} = \frac{(2J+1)e^{-[BJ(J+1)+(C-B)K^2]h/kT}}{\displaystyle\sum_{J=0}^{\infty}\sum_{K=-J}^{J}(2J+1)e^{-[BJ(J+1)+(C-B)K^2]h/kT}} \tag{3-44}$$

Here B and C are in cycles per second as in (3-4). When Bh and Ch are small compared with kT the sums may be replaced by integrals, giving

$$f_{JK} = (2J+1)e^{-[BJ(J+1)+(C-B)K^2]h/kT}\sqrt{\frac{B^2Ch^3}{\pi(kT)^3}} \tag{3-45}$$

It should be noted that (3-45) applies to one particular value of K, and does not allow for K degeneracy. A more accurate evaluation of the sum has been made ([130], p. 506) but is not of much interest to us because the error in approximating the sum as an integral is usually very small and in addition (3-44) does not allow for the statistical weights introduced by the spins of the like particles which must always be considered in a symmetric top.

The usual type of symmetric top has threefold symmetry about the axis and the separation between inversion levels is negligible. In that case, from Table 3-6 the degeneracy due to spin and inversion levels (or spin and K degeneracy) for each value of J and K is proportional to (omitting a constant factor $(2I+1)/3$):

For K a multiple of 3, but not 0,

$$S(I,K) = 2(4I^2+4I+3) \tag{3-46a}$$

For $K = 0$,

$$S(I,K) = (4I^2+4I+3) \tag{3-46b}$$

For K not a multiple of 3,

$$S(I,K) = 2(4I^2+4I) \tag{3-46c}$$

Allowing for this degeneracy, (3-44) becomes

$$f_{JK} = \frac{S(I,K)(2J+1)e^{-[BJ(J+1)+(C-B)K^2]h/kT}}{\displaystyle\sum_{J=0}^{\infty}\sum_{K=0}^{J}S(I,K)(2J+1)e^{-[BJ(J+1)+(C-B)K^2]h/kT}} \tag{3-47}$$

It is evident that (3-47) is unaffected by the omission of a constant factor $(2I+1)/3$ in the degeneracy $S(I,K)$ associated with spin and inversion as given in (3-46). Again assuming B and C to be much smaller than kT, the sums in (3-47) may be changed to integrals, so that one obtains

$$f_{JK} = \frac{S(I,K)(2J+1)}{4I^2+4I+1}\sqrt{\frac{B^2Ch^3}{\pi(kT)^3}}\,e^{-[BJ(J+1)+(C-B)K^2]h/kT} \tag{3-48}$$

For low values of J and K, the exponential in (3-48) is very close to 1, so that

$$f_{JK} = \frac{S(I,K)(2J + 1)}{4I^2 + 4I + 1} \sqrt{\frac{B^2Ch^3}{\pi(kT)^3}} \qquad (3\text{-}49)$$

The fraction of molecules in a given vibrational state may be obtained analogously to (1-51) as

$$f_v = e^{-W_v/kT} \prod_n (1 - e^{-h\omega_n/kT})^{d_n} \qquad (3\text{-}50)$$

where d_n is the degeneracy of a vibrational mode of frequency ω_n and $\prod_n$ represents the product

$$(1 - e^{-h\omega_1/kT})^{d_1}(1 - e^{-h\omega_2/kT})^{d_2}(1 - e^{-h\omega_3/kT})^{d_3} \cdots$$

for all vibrational modes. Since a symmetric top has a number of vibrational modes, the product $\prod_n$ is sometimes appreciably less than 1, but in simple symmetric tops not usually less than about 0.5. Substituting (3-49) and (3-40) into (1-49), and setting $2B(J + 1) = h\nu_0$, the intensity for a transition $J + 1 \leftarrow J, K \leftarrow K$ is

$$\gamma = \frac{4\pi hNf_vS(I,K)}{(4I^2 + 4I + 1)3c(kT)^2} \sqrt{\frac{\pi Ch}{kT}} \mu^2 \left[1 - \frac{K^2}{(J + 1)^2}\right] \frac{\nu_0\nu^2\,\Delta\nu}{(\nu - \nu_0)^2 + (\Delta\nu)^2} \qquad (3\text{-}51)$$

It may be seen that γ increases, as in a linear molecule, approximately as ν^3. However, for a symmetric top γ is more strongly dependent on the temperature T than it is for the linear molecule. The expression (3-51) might be summed over all possible K for a particular J transition, since transitions for the various values of K coincide in frequency (approximately—see discussion of centrifugal distortions in Sec. 3-6). Summing over the K values, assuming J is large, and letting $\dfrac{S(I,K)}{4I^2 + 4I + 1}$ be approximately 2,

$$\gamma_{\text{total}} = \frac{2\pi h^2Nf_v}{9c(kT)^2B} \sqrt{\frac{\pi Ch}{kT}} \mu^2 \frac{(4J + 3)(J + 2)}{(J + 1)^2} \frac{\nu_0^2\nu^2\,\Delta\nu}{(\nu - \nu_0)^2 + (\Delta\nu)^2} \qquad (3\text{-}52)$$

This shows that the entire intensity for a transition $J + 1 \leftarrow J$ summed over all K increases approximately as ν^4, even more rapidly than intensities of transitions for linear molecules.

The maximum absorption coefficient of a symmetric-top transition $J + 1 \leftarrow J$ may also be written from (3-51), after some numerical evaluation,

$$\gamma_{\max} = \frac{1.23 \times 10^{-20} f_v S(I,K) \sqrt{C}}{4I^2 + 4I + 1} \mu^2 \left[1 - \frac{K^2}{(J+1)^2} \right] \frac{\nu_0^3}{\Delta\nu} \quad \text{cm}^{-1} \quad (3\text{-}53)$$

where C = rotational constant about the symmetry axis, Mc

μ = dipole moment, debye units (10^{-18} esu)

$\Delta\nu$ = half width at half maximum at a pressure of 1 mm Hg, Mc

ν_0 = resonant frequency, Mc

the temperature is assumed to be 300°K

For symmetric tops with more than three identical nuclei, the statistical weight factor $\dfrac{S(I,K)}{4I^2 + 4I + 1}$ in (3-53) must be adjusted, but this factor is of the order of 1 and may be set equal to 2 if an approximate evaluation of $\gamma_{\max}$ is all that is required. The only other type of transition, $J \leftarrow J$, which occurs for a symmetric top is rather rarely of interest since it is a transition between inversion levels which usually corresponds to a negligibly small frequency. However, these inversion transitions are important in the case of NH_3 and will be discussed in more detail in Chap. 13-

It may be noted that (3-51) or (3-53) does not reduce to the corresponding expression for a linear molecule when $K = 0$ and C is infinite. This is because C has been assumed much smaller than kT in deriving these expressions, and hence they are no longer valid when C is allowed to be large. Because of the additional possible rotation around the symmetry axis, symmetric-top molecules are generally distributed throughout more states than are linear molecules and individual transitions are usually less intense by a factor of approximately 10. Typical intensity of a symmetric-top absorption at 1 cm wavelength is $\gamma_{\max} = 10^{-6}$ cm^{-1}.

Since variations in intensity with K depend on the nuclear spins (cf. [3-53] and Table 3-6), they may be used as a means of determining nuclear spins when several identical nuclei occur in the same molecule. This method of spin measurement has very limited utility, however, since molecules of this type are normally found only for abundant isotopes of common elements such as H and the halogens whose spins are already well known [233].

3-6. Centrifugal Stretching in Symmetric Tops. The discussion so far has assumed a rigid symmetric top, with no effects of vibration or centrifugal distortion. Centrifugal stretching of symmetric tops is somewhat more complicated than that for linear molecules because it involves both angular momentum quantum numbers J and K. It is fairly obvious that the amount of centrifugal distortion of the molecule cannot depend on the sign of the angular rotation (e.g., whether it is clockwise or counterclockwise). Hence the change in rotational energy must involve only even powers of the momentum, such as the square of the total angular momentum $J(J + 1)$ or of the component of momentum along the symmetry axis K^2. The rotational energy W including centrifugal effects may therefore be written

$$W(J,K) = BJ(J + 1) + (C - B)K^2 - D_J J^2(J + 1)^2$$
$$- D_{JK}J(J + 1)K^2 - D_K K^4 \quad (3\text{-}54)$$

plus terms of higher order in $J(J + 1)$ and K^2 where the centrifugal distortion constants D_J, D_{JK}, and D_K are of the order of B^2/ω, ω being a molecular vibrational frequency. The distortion constants are hence very small compared with the rotational constants B and C. The frequency due to a rotational transition $J + 1 \leftarrow J$, $\Delta K = 0$ is, from (3-54),

$$\nu = 2(J + 1)(B - D_{JK}K^2) - 4D_J(J + 1)^3 \quad (3\text{-}55)$$

Without centrifugal distortion, the rotational transitions for a symmetric top are all equally spaced, and for a given transition of J, the various possible values of K all give identical frequencies. From expression (3-55), it may be seen that centrifugal distortion destroys both these simple relations, although the previous approximation of a rigid rotor is still an extremely good one. Because of the term D_{JK}, molecules with different values of K have effectively slightly different values of the rotational constant B, so that their rotational transitions are not precisely superimposed. A high-resolution microwave spectroscope usually resolves separate rotational lines due to molecules with different values of K, so that D_{JK} is easily evaluated.

Centrifugal distortion for molecules of this type was first observed in the infrared rotational spectrum of NH_3 and PH_3 [44a]. In this case the lines due to different values of K were not resolved, but the center of gravity of successive J transitions could be fitted by an expression of the form

$$\nu = 2(J + 1)B - 4D(J + 1)^3$$

The constants D_J, D_{JK}, and D_K depend of course on the various molecular force constants and the moments of inertia. The calculation from observed vibrational frequencies and rotational constants is both tedious and uncertain, since some of the force constants needed cannot always be accurately determined. However, Slawsky and Dennison [94a] obtained theoretical values of the centrifugal distortion constants of NH_3, ND_3, and PH_3 which fit experimental measurements of infrared spectra. These values are listed in Table 3-7. Chang and Dennison [783a] have succeeded in calculating D_J and D_{JK} for CH_3Cl to an accuracy of a few per cent. Nielsen [624] has given general expressions for the centrifugal stretching constants of symmetric molecules of the type XY_3. Appropriate force constants must be evaluated, however, before numerical results can be obtained from these expressions.

Values of D_J and D_{JK} for a number of molecules are given in Table 3-7.

It can be expected that D_J will always be positive as is illustrated in the table, since centrifugal forces due to rotation about any given axis will always tend to increase the moment of inertia about that axis or decrease the effective rotational constant. The sign of D_{JK} may in principle be either positive or negative. It is striking that all molecules of the type XY_3 which have so far been examined have D_{JK} negative, while molecules involving a methyl group or its derivatives have a positive D_{JK}.

TABLE 3-7. VALUES OF CENTRIFUGAL STRETCHING CONSTANTS FOR REPRESENTATIVE SYMMETRIC-TOP MOLECULES

Molecule	D_J, Mc	D_{JK}, Mc	Reference
NH_3	19	-28	[94a]
ND_3	5.2	-7.8	[94a]
PH_3	3.7	-4.6	[94a]
AsF_3	....	0.009 ± 0.002	[824]
CH_3Cl^{35}	0.0181	0.189	[531] [913]
CH_3I	0.0080	0.0994	[531]
H_3CCCl	....	0.0072	[744]
F_3CCCH	0.00024	0.0063	[557]
F_3GeCl	$\sim$0.0006	$|D_{JK}| < 0.001$	[555]
$H_3B^{11}CO$	....	0.00036	[461]

3-7. Rotation-Vibration Interactions and l-Type Doubling in Symmetric Tops. The rotation-vibration constants α for symmetric tops have been discussed theoretically by Shaffer [109a] and an extensive and systematic treatment has been given by Nielsen [624]. However, theoretical evaluation of these rotation-vibration constants involves knowing so many different force constants that as yet essentially no comparison between theory and experimental values of α has been possible. There are a number of experimentally determined values of the α's in various symmetric tops. In many cases, however, the multiplicity of vibrational modes makes it difficult to assign uniquely a given observed rotational line due to an excited state to a particular vibrational mode, and hence to assign uniquely measured values of α.

There are always a number of degenerate vibrational modes in symmetric tops. Even the simplest types (of the form XY_3) have two sets of doubly degenerate vibrational modes, and each of these modes can produce an angular momentum which reacts in an interesting way with the angular momentum of rotation. If a degenerate vibrational mode involves only motion perpendicular to the molecular axis, then as in the case of the degenerate modes of a linear molecule, the vibration produces an angular momentum $l\hbar$ about the molecular axis, where l is an integer and $\hbar = h/2\pi$. In general, the vibrational motion may not be entirely perpendicular to the axis, and it produces angular momentum $\zeta l\hbar$ about the axis of symmetry, where $|\zeta| \leq 1$ (cf. [130]). The molecule may at the same time be rotating about the axis of symmetry, and it is the sum of the vibrational angular momentum plus that due to rotation of the molecular frame which is quantized and equal to $K\hbar$, where K is an integer. These two types of motions are illustrated in Fig. 3-12.

A given value of the angular momentum may be produced in a variety of ways. For example, if the molecule is excited to the first excited state of a degenerate mode for which $\zeta = 1$, then a momentum $K = 1$ may be formed entirely of vibrational momentum ($l = 1$) or it may be a combination of vibrational momentum $l = -1$ and an angular momentum of the molecular frame of two units in the positive direction.

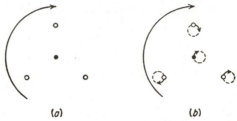

(a) (b)

Fig. 3-12. Angular momentum along symmetry axis of XY_3 molecule. (a) Rotation of molecular frame only; (b) rotation of molecular frame plus angular momentum produced by degenerate vibration.

Similarly $K = -1$ can be formed in two ways. The angular momentum due to rotation of the molecular frame is $K - \zeta l$, and the energy associated with this momentum is hence not CK^2, but $C(K - \zeta l)^2$ or

$$C(K^2 - 2\zeta lK + \zeta^2 l^2)$$

Since $\zeta^2 l^2$ does not vary with rotational state, it may be omitted so that the rotational energy is written

$$W_R = BJ(J + 1) + (C - B)K^2 - 2K\zeta lC \qquad (3\text{-}56)$$

The energy levels with $l = \pm 1$ are compared in Fig. 3-13 with those for $l = 0$, assuming $\zeta = 1$.

When $K = l = \pm 1$, the molecule may be regarded as having no overall rotational motion about the symmetry axis. This is strictly true only if $\zeta = 1$, when the entire angular momentum K is provided by vibrational motion very much as in the case of a linear molecule with an angular momentum $l = \pm 1$ due to a degenerate vibration. In this case, it is not hard to understand why l-type doubling occurs just as in a linear molecule [503], [624].

If the symmetric top is the common variety with a threefold symmetry axis, then levels for which $K = -l = \pm 1$ are not split, nor is there appreciable splitting when $|K| > 1$. Since these cases correspond to the

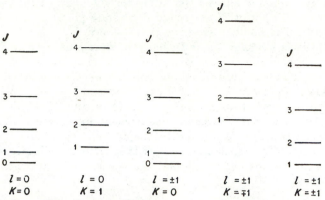

Fig. 3-13. Comparison of rotational energy levels for a symmetric top in the ground state with $l = 0$ and those for an excited degenerate vibrational mode with $l = \pm 1$. The vibrational energy has been disregarded. Each level indicated by $l = \pm 1$ is actually two which may be slightly different in energy due to l-type doubling. ζ is assumed to be positive.

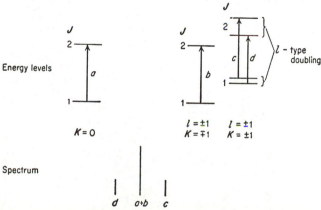

Fig. 3-14. l-type doubling in a symmetric rotator with three-fold symmetry axis. Energy levels and allowed transitions are shown above, and the resulting spectrum below. ζ is assumed to be negative. Note that the transitions a and b are between doubly degenerate levels and are superimposed to a good approximation, although some splitting due to centrifugal distortion may occasionally be observed.

framework of the molecule rotating as a whole, any change in the effective values of B due to vibration may be thought of as equivalent for all orientations of the angular momentum J because of the averaging effect of the rotation. However, this qualitative explanation cannot be taken too seriously, since de Heer has shown [568] that, for a molecule with a fourfold axis of symmetry, the levels $K = -l = \mp 1$ are split by an amount similar to those for which $K = l = \pm 1$.

Energy levels and allowed transitions for a $J = 2 \leftarrow 1$ rotational line of a symmetric top with an excited degenerate vibration are shown in Fig. 3-14. The resulting spectrum is also shown.

3-8. Dipole Moment Due to Degenerate Vibrations. There are many symmetric molecules which have zero dipole moment because of their symmetry and hence one would not normally expect microwave absorption due to pure rotational transitions of these molecules. Mizushima and Venkateswarlu [846] have shown, however, that if certain types of symmetric molecules with zero dipole moment are excited to degenerate vibrational states, it is possible to observe pure rotational transitions due to an effective dipole moment resulting from the vibrations. Such an effect should occur in molecules like allene, C_3H_4, or spherical tops such as CF_4.

CHAPTER 4

ASYMMETRIC-TOP MOLECULES

An asymmetric top is a rotor with no two principal moments of inertia equal. General principles involved in the motion of such rotors are of course the same as for symmetric tops, but the details turn out to be much more complex. This complexity shows up not only in the quantum-mechanical behavior of an asymmetric top, but also in its classical motion. The classical motion is well known and closely parallels the quantum-mechanical behavior; but it is not simple enough to afford any generally useful model for the quantum-mechanical case. For this reason the discussion below will begin immediately with a quantum-mechanical approach. Much of the discussion follows the recent extensive work of King, Hainer, and Cross on asymmetric rotors [118], [122], [215], [372].

4-1. Energy Levels of Asymmetric and Slightly Asymmetric Rotors. A general picture of the behavior of the energy levels of an asymmetric top may be had from examining their behavior as the rotor begins to deviate from the two simple extremes, the prolate and the oblate symmetric top. As in (3-2), the energy is

$$W = \frac{P_x^2}{2I_x} + \frac{P_y^2}{2I_y} + \frac{P_z^2}{2I_z} = \frac{4\pi^2 A}{h} P_x^2 + \frac{4\pi^2 B}{h} P_y^2 + \frac{4\pi^2 C}{h} P_z^2$$

where for an asymmetric rotor the constants A, B, and C are all different. If the three rotational constants are, in decreasing order of size, A, B, and C, a prolate symmetric top corresponds to $B = C$, and an oblate symmetric top to $B = A$. The range of values of B between A and C correspond to various conditions of asymmetry. If B differs from A or from C by only a small amount, the rotor may be called a slightly asymmetric top. Figure 4-1 shows in a qualitative way how the energy levels vary as B varies from C to A so that on the left-hand side the levels are just those of a prolate symmetric top $(B = C)$ and on the right they correspond to an oblate symmetric top $(B = A)$. A slight asymmetry splits the levels $\pm K$ which are degenerate for symmetric tops. Note that, as the value of B changes, no two levels of the same J cross. Levels of different J values may cross, however.

Various parameters may be used to indicate the degree of asymmetry.

Ray's asymmetry parameter [37] is

$$\kappa = \frac{2B - A - C}{A - C} \qquad (4\text{-}1)$$

which becomes -1 for a prolate symmetric top $(B = C)$ and 1 for an oblate symmetric top $(B = A)$, varying between these two values for asymmetric cases. Another parameter, especially appropriate for a slightly asymmetric prolate top, is

$$b_P = \frac{C - B}{2A - B - C} = \frac{\kappa + 1}{\kappa - 3} \qquad (4\text{-}2)$$

b_P is zero for a prolate symmetric top, and increases in size as the top becomes more asymmetric. The analogous asymmetry parameter for a slightly asymmetric oblate top is

$$b_O = \frac{A - B}{2C - B - A} = \frac{\kappa - 1}{\kappa + 3} \qquad (4\text{-}3)$$

For an asymmetric top, the total angular momentum J and its projection M on an axis fixed in space are constants of the motion and "good" quantum numbers which can be used to specify the state of the rotor. However, neither in the classical motion (see [130], p. 42) nor for the quantum-mechanical solution is the component of the angular momentum constant along any direction in the rotating asymmetric molecule. This means that the quantum number K, which in a symmetric top is the projection of J on the symmetry axis, is no longer a "good" quantum number and cannot very well be used to specify the rotational state. In fact there is no set of convenient quantum numbers which can specify the state and also have simple physical meaning. Regardless of the fact that K is not a good quantum number for an asymmetric top, the energy levels may be specified by giving the value of J, and the value of K_{-1} for the limiting prolate and K_1 for the limiting oblate symmetric top. The subscripts -1 and 1 used here are the asymmetry parameter κ. Thus a level may be designated as $J_{K_{-1}K_1}$ or 5_{32}, indicating a J of 5 and a level which in the limiting cases connects on the left-hand side of Fig. 4-1 with a K of 3, and on the right-hand side with a K of 2. Another method of designating the levels is by J_τ, where J is the total angular momentum and τ is an integer between $-J$ and J which indicates simply the order of the energy levels of a given J. Thus J_{-J} represents the lowest energy state of total angular momentum, J_{-J+1} the next, and J_J the highest. It may be seen from Fig. 4-1 that $\tau = K_{-1} - K_1$, so that these two common ways of designating the states are easily related.

If a molecule is a slightly asymmetric prolate top, the energy may be conveniently written in the form

$$\frac{W}{h} = \frac{B + C}{2} J(J + 1) + \left(A - \frac{B + C}{2}\right) w \qquad (4\text{-}4)$$

It may be seen by comparison with the energy of a prolate symmetric top that, since $B \approx C$, w must approximately equal K^2. Exact expressions for the various possible values of w, regardless of the size of the asymmetry parameter b, are as follows:

$J = 0$: $w = 0$

$J = 1$: $w = 0$
         $w - 1 - b = 0$
         $w - 1 + b = 0$

$J = 2$: $w - 4 = 0$
         $w - 1 + 3b = 0$
         $w - 1 - 3b = 0$
         $w^2 - 4w - 12b^2 = 0$

$J = 3$: $w - 4 = 0$
         $w^2 - 4w - 60b^2 = 0$
         $w^2 - (10 - 6b)w + (9 - 54b - 15b^2) = 0$
         $w^2 - (10 + 6b)w + (9 + 54b - 15b^2) = 0$

$J = 4$: $w^2 - 10(1 - b)w + (9 - 90b - 63b^2) = 0$
         $w^2 - 10(1 + b)w + (9 + 90b - 63b^2) = 0$
         $w^2 - 20w + (64 - 28b^2) = 0$
         $w^3 - 20w^2 + (64 - 208b^2)w + 2880b^2 = 0$

$J = 5$: $w^2 - 20w + 64 - 108b^2 = 0$
         $w^3 - 20w^2 + (64 - 528b^2)w + 6720b^2 = 0$
         $w^3 - w^2(35 - 15b) + w(259 - 510b - 213b^2)$
                 $- (225 - 3375b - 4245b^2 + 675b^3) = 0$
         $w^3 - w^2(35 + 15b) + w(259 + 510b - 213b^2)$
                 $- (225 + 3375b - 4245b^2 - 675b^3) = 0$

$J = 6$: $w^3 - w^2(35 - 21b) + w(259 - 714b - 525b^2)$
                $- 225 + 4725b + 9165b^2 - 3465b^3 = 0$
         $w^3 - w^2(35 + 21b) + w(259 + 714b - 525b^2)$
                $- 225 - 4725b + 9165b^2 + 3465b^3 = 0$
         $w^3 - 56w^2 + w(784 - 336b^2) - 2304 + 9984b^2 = 0$
         $w^4 - 56w^3 + w^2(784 - 1176b^2)$
            $- w(2304 - 53{,}664b^2) - 483{,}840b^2 + 55{,}440b^4 = 0$

$$(4\text{-}5)$$

Similar expressions for J up to 11 have been developed by Randall, Dennison, Ginsburg, and Weber [84] and equations for still higher values of J may be derived from Wang's general expressions [18]. If the molecule is only slightly asymmetric, w may conveniently be expanded

$$w = K^2 + c_1 b_P + c_2 b_P^2 + c_3 b_P^3 + \cdots \tag{4-6}$$

where b_P is the asymmetry parameter for a prolate top, which is given

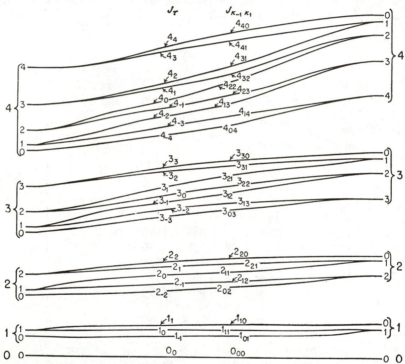

Fig. 4-1. Qualitative behavior of the asymmetric-top energy levels. The rotational constant B varies from left to right, equaling C and giving a prolate symmetric top on the left, and equaling A to give an oblate symmetric top on the right.

by (4-2). The coefficients c_1, c_2, and c_3 are given in Appendix III for a slightly asymmetric prolate top.

For the oblate case, similar formulas are good. The energy is

$$\frac{W}{h} = \frac{A+B}{2} J(J+1) + \left(C - \frac{A+B}{2}\right) w \tag{4-7}$$

with

$$w = K^2 + c_1 b_o + c_2 b_o^2 + c_3 b_o^3 + \cdots \tag{4-8}$$

and b_o is defined by (4-3). K is now the appropriate value for an oblate rotor, and c_1, c_2, c_3 may be obtained from the table in Appendix III used for the prolate case if the values of K_{-1} and K_1 are interchanged or the sign of τ reversed.

It may be noted from Eqs. (4-5) or Appendix III that levels with $K = 1$ which are degenerate for the symmetric case are split an amount proportional to the asymmetry b and to $J(J + 1)$. For levels with

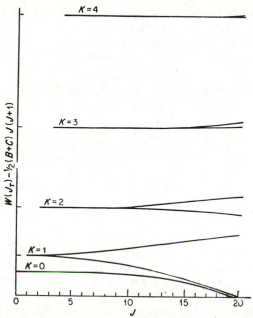

Fig. 4-2. Rotational energy of a slightly asymmetric top (b about 0.01) as a function of J. [The term $\frac{1}{2}(B + C)J(J + 1)$ is subtracted from the energy, *i.e.*, the deviations of the curves from horizontal lines represent the deviations from the levels of the symmetric top.] (*From Dieke and Kistiakowsky* [47a].)

higher K, the splitting due to asymmetry is very much less and proportional to b^K, which is just analogous to the case of l-type doubling described in Chap. 2 for linear molecules. Wang [18] has shown that for small asymmetries the splitting of these levels which are degenerate in the symmetric case is given approximately by

$$\Delta w = \frac{b^K(J + K)!}{8^{K-1}(J - K)![(K - 1)!]^2} \tag{4-9}$$

where b and K are the appropriate prolate- or oblate-top values.

The variation of energy levels with J and K for a molecule of slight asymmetry is shown in Fig. 4-2. In addition to a splitting of the degenerate levels which increases with J, there is in most cases a deviation of the center of the two degenerate levels from their symmetric-top values. This deviation is usually proportional to b^2. An approximation for the splitting which is good to the $(K + 2)$nd power of the asymmetry b has been given by Kivelson [825]. He showed that (4-9) should be multiplied by a factor of the form $\{1 + [C_1 + C_2 J(J + 1) + C_3 J^2(J + 1)^2]b^2\}$, where the constants C_1, C_2, and C_3 are tabulated [825] for various values of K.

The molecule PCl_3 might ordinarily be classed as a symmetric top, as it is when all three chlorines are of the same isotopic species, Cl^{35} or Cl^{37}. However, common species of this molecule contain two atoms of Cl^{35} and one of Cl^{37}, or one of Cl^{35} and two of Cl^{37}, making them slightly

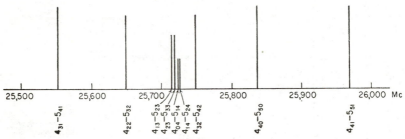

FIG. 4-3. The $5 \leftarrow 4$ transition of a slightly asymmetric rotor $PCl_2^{35}Cl^{37}$. ($b_0 = -0.037$).

asymmetric. This asymmetry affects the energy levels in a very noticeable way, since the observed frequencies are measured accurately. However, the effect of a slight asymmetry on the selection rules and intensities of transitions is generally negligible since intensities are not usually measured to high accuracy. The matrix elements and intensity relations given in Chap. 3 for a symmetric top may therefore be applied. The spectrum of the $J = 5 \leftarrow 4$ transition of $PCl_2^{35}Cl^{37}$ is shown in Fig. 4-3. Even though the asymmetry parameter

$$b_o = \frac{A - B}{2C - B - A}$$

is only -0.037, most of the lines which would coincide for the symmetric molecule PCl_3^{35} are split by the asymmetry.

Slightly asymmetric tops are commonly and naturally formed by a mixture of isotopes as in $PCl_2^{35}Cl^{37}$ or asymmetric methyl chloride CH_2DCl, and by adding a light off-axis atom to an otherwise symmetric structure as in methyl alcohol

$$H_3C\!-\!O\diagup^{\displaystyle H} \qquad \text{or in} \qquad {}^{\displaystyle H}\diagdown N\!\!=\!\!C\!\!=\!\!S$$

Even the molecule $O\!\!=\!\!N$ is approximately a prolate symmetric top
 Cl

with an asymmetry parameter b as small as -0.0002 [632]. Slight asymmetries may also occur in what may seem more accidental ways. For any bent triatomic molecule (as NOCl), there is some value of the bond angle which makes the molecule an oblate symmetric top, and therefore a range of bond angles (in addition to those near 0 and 180°) for which the molecule is only slightly asymmetric.

In case a molecule is very asymmetric and the asymmetry parameter b is large, an expansion for the energy of the type (4-6) or (4-8) is no longer appropriate. Equations (4-5) may of course be solved completely for any arbitrary value of b and hence the energies obtained from (4-4). However, it is generally better to express the energy in the form

$$\frac{W}{h} = \tfrac{1}{2}(A + C)J(J + 1) + \tfrac{1}{2}(A - C)E_\tau \tag{4-10}$$

where E_τ has replaced w in (4-4) as a numeric to be evaluated for the particular case and amount of asymmetry. τ is the integer used above to specify the order of the energy among the different levels of the same J. Certain values of E_τ or $E_{K_{-1}K_1}$ may be evaluated explicitly by solution of linear or quadratic equations. They are given in Table 4-1.

The quantity E_τ in (4-10) is a function only of the asymmetry parameter, and if Ray's asymmetry parameter $\kappa = (2B - A - C)/(A - C)$ is used, then

$$E_\tau(\kappa) = -E_{-\tau}(-\kappa) \tag{4-11a}$$

or in the $J_{K_{-1}K_1}$ notation,

$$E_{mn}(\kappa) = -E_{nm}(-\kappa) \tag{4-11b}$$

Since E_τ is a complicated function which must be computed and tabulated, relations (4-11) are of real value. Only positive or only negative values of κ need be tabulated, and from these the other values may be obtained from (4-11). The table in Appendix IV, computed by Turner, Hicks, and Reitwiesner [877a], gives values of E_τ for all levels with J less than 13, and for values of κ from 0 to $+1$ in steps of 0.01. These values of κ are not so closely spaced as might be desired for microwave work but, with care in interpolation between tabulated values of κ, energies may be obtained for any κ with sufficient accuracy for most microwave work.

It may also be noted that when $\kappa = 0$, which is sometimes called the most asymmetric case, (4-11) leads to the equation $E_\tau(0) = E_{-\tau}(0)$, so that the energy levels are symmetrically spaced [about $E_0(0)$] when κ is zero. This relation might be guessed from examination of Fig. 4-1, since $\kappa = 0$ is just halfway between the two limiting types of symmetric tops.

A number of sum rules for energy levels have been found by Mecke ([130], p. 50). They are of help in checking the correctness of energy-level computations. The most easily interpreted of these sum rules is

$$\frac{\sum_\tau \dfrac{W_{J\tau}}{h}}{2J + 1} = \tfrac{1}{3}(A + B + C)J(J + 1) \qquad (4\text{-}12)$$

which states that the average rotational energy of all levels of a particular J is given by $J(J + 1)$ times a rotational constant which is the average of A, B, and C.

TABLE 4-1. SOLUTIONS OF $E(\kappa)$, THE ENERGY PARAMETER FOR ASYMMETRIC ROTORS [cf. EQ. (4-10)] FROM LINEAR OR QUADRATIC EQUATIONS

$J_{K_{-1}K_1}$	$E_{K_{-1}K_1}$
0_{00}	0
1_{10}	$\kappa + 1$
1_{11}	0
1_{01}	$\kappa - 1$
2_{20}	$2[\kappa + (\kappa^2 + 3)^{\frac{1}{2}}]$
2_{21}	$\kappa + 3$
2_{11}	4κ
2_{12}	$\kappa - 3$
2_{02}	$2[\kappa - (\kappa^2 + 3)^{\frac{1}{2}}]$
3_{30}	$5\kappa + 3 + 2(4\kappa^2 - 6\kappa + 6)^{\frac{1}{2}}$
3_{31}	$2[\kappa + (\kappa^2 + 15)^{\frac{1}{2}}]$
3_{21}	$5\kappa - 3 + 2(4\kappa^2 + 6\kappa + 6)^{\frac{1}{2}}$
3_{22}	4κ
3_{12}	$5\kappa + 3 - 2(4\kappa^2 - 6\kappa + 6)^{\frac{1}{2}}$
3_{13}	$2[\kappa - (\kappa^2 + 15)^{\frac{1}{2}}]$
3_{03}	$5\kappa - 3 - 2(4\kappa^2 + 6\kappa + 6)^{\frac{1}{2}}$
4_{40}	$\cdots$
4_{41}	$5\kappa + 5 + 2(4\kappa^2 - 10\kappa + 22)^{\frac{1}{2}}$
4_{31}	$10\kappa + 2(9\kappa^2 + 7)^{\frac{1}{2}}$
4_{32}	$5\kappa - 5 + 2(4\kappa^2 + 10\kappa + 22)^{\frac{1}{2}}$
4_{22}	$\cdots$
4_{23}	$5\kappa + 5 - 2(4\kappa^2 - 10\kappa + 22)^{\frac{1}{2}}$
4_{13}	$10\kappa - 2(9\kappa^2 + 7)^{\frac{1}{2}}$
4_{14}	$5\kappa - 5 - 2(4\kappa^2 + 10\kappa + 22)^{\frac{1}{2}}$
4_{04}	$\cdots$
5_{42}	$10\kappa + 6(\kappa^2 + 3)^{\frac{1}{2}}$
5_{24}	$10\kappa - 6(\kappa^2 + 3)^{\frac{1}{2}}$

Applicability of Various Approximate Methods. Many useful methods for approximate evaluation of the rotational energy of asymmetric rotors have been developed. The discussion below is largely concerned with the applicability or range of usefulness of approximate methods. The reader is advised to consult original papers for detailed descriptions of the actual methods of computation (*cf.* also [871]).

Approximations are of most interest for large values of J, since many of the equations (4-5) for small J's are easily solved, and tabulations (Appendix IV) are available for J's up to 12.   For any large J, the various types of solutions for energy levels needed may be indicated on the two-dimensional diagram of Fig. 4-4.   Various approximations are applicable in various parts of this diagram, but unfortunately in the

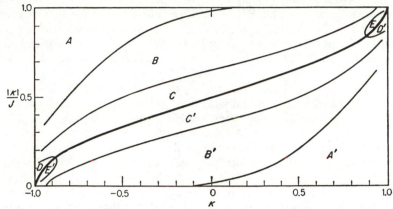

FIG. 4-4. Sketch of the regions of error expected in approximating E_τ.   (*From Hainer, Cross, and King* [372].)

regions C and C', no known approximate method is very good because the energy levels change very rapidly across the heavy line

$$\frac{E_\tau(\kappa)}{J(J+1)} = \kappa$$

shown in this figure.

One type of approximation already discussed is a power-series expansion in the asymmetry parameter b [*cf.* (4-6) and (4-8)].   This expansion yields good results in the regions A and A', that is, when $|K|/J \approx 1$. In the regions D and D', that is, when $|K|/J$ is small, the expansion is good only for very small asymmetries.   Inaccuracy of such an expansion in these regions may be guessed from the very large size of the coefficients given in Appendix III when $|K|/J$ is small, indicating the rapid change in energy near the line $[E_\tau(\kappa)]/[J(J+1)] = \kappa$.

Another power-series expansion about $\kappa = 0$ of the form

$$E_\tau = a_0 + a_1\kappa + a_2\kappa^2 + \cdots \tag{4-13}$$

has been used by King, Hainer, and Cross [118].   The coefficients a_0, a_1, and a_2 are tabulated for all levels with J less than 13.   This expansion tends to be poor where the $\kappa = 0$ line crosses the $[E_\tau(\kappa)]/[J(J+1)] = \kappa$ line in Fig. 4-4, or hence for small E_τ.

The "correspondence principle approximation" [215] uses the techniques of early quantum mechanics by setting the integrals of angular momentum components for each of the Eulerian angles integrated over a complete cycle of the angle equal to some simple numerical multiple of $h/2\pi$. This leads to elliptic integrals which must be solved to obtain E_r. This approximation allows no difference in the energy between the two levels which are degenerate in the symmetric-top cases, and in general a good estimate of its error is the amount of actual splitting of these levels as given in (4-9). Hence for large values of J it yields very good results in regions A and A' of Fig. 4-4, is considerably worse in B and B', bad in C and C', and very bad in D, D', E, and E' (for numerical illustrations of the errors due to this and the following approximation, see [372]).

Another method of approximating energy levels for large J takes advantage of the similarity between the matrix for the energy of an asymmetric rotor and a matrix arising from Mathieu's differential equation [274]. Characteristic values from Mathieu's equation with some corrections by perturbation methods may hence be used to approximate the asymmetric rotor energies. This "Mathieu's equation approximation" supplements the "correspondence principle approximation" by yielding good values for the energy in regions D and D' in Fig. 4-4 where the latter method is very poor. Elsewhere it is not appreciably better than the "correspondence principle approximation" and is less convenient because the amount of computation is somewhat greater. This approximation is therefore particularly useful when $|K|/J \ll 1$.

A similar type of approach may be used when $|K|/J \approx 1$. In this case, the energy matrix for large J becomes similar to one obtained from a harmonic oscillator. This "harmonic oscillator approximation" [369] is similar to the "correspondence principle approximation" in giving no splitting between the levels which are degenerate in the symmetric-rotor case. These two approximate methods are good in the same regions, that is, A and A'.

4-2. Symmetry Considerations and Intensities. The spectrum of an asymmetric rotor is complicated not only by the irregular distribution of energy levels, but also because the selection rules and transition probabilities between these levels are more complex than in the symmetric case. The selection rules are complicated partly by the increased number of separate levels, and partly because of the arbitrary direction of the dipole moment. It may be remembered that the dipole moment for a symmetric rotor must lie along the symmetry axis (if the possible, but nonexistent, accidentally symmetric rotor is excluded). In an asymmetric rotor the dipole moment may lie in any arbitrary direction with respect to the principal axes of inertia. However, the dipole moment is not uncommonly parallel to one of the three principal axes, and selection rules in such cases will be considered first.

General selection rules are usually the result of some symmetry property, so we turn to an examination of symmetry relations. The rotational behavior of a molecule can be deduced from its ellipsoid of inertia, which is symmetric with respect to a rotation of 180° about any principal axis even though the molecule itself may not be symmetric with respect to such a rotation. Hence the wave function ψ must be either symmetric, antisymmetric, or degenerate with respect to such a rotation. Since degeneracy of the pure rotational energy levels is usually removed by asymmetry of the rotor, the only cases which need be considered are when ψ is symmetric $(+)$ or antisymmetric $(-)$.

For the limiting case of a symmetric prolate rotor, symmetry of the wave functions with respect to a rotation of 180° around the axis of least moment of inertia, which is the molecular axis, is easily determined. Since the wave function depends on this angle χ as $e^{\pm iK_{-1}\chi}$, it is symmetric when K_{-1} is even, antisymmetric when K_{-1} is odd. Now this symmetry property of the wave function is not changed as a result of a perturbation of the same symmetry as the initial Hamiltonian; i.e., although the wave function changes somewhat, it maintains the same symmetry or antisymmetry when the moments of inertia are slightly changed and the molecule becomes asymmetric. Hence any asymmetric wave function $\psi_{J_{K_{-1}K_1}}$ is symmetric with respect to rotation of 180° around the axis of smallest moment of inertia when K_{-1} is even, antisymmetric when K_{-1} is odd. Similar consideration of the limiting oblate symmetric rotor shows that $\psi_{J_{K_{-1}K_1}}$ is symmetric with respect to rotation around the axis of greatest moment of inertia when K_1 is even, antisymmetric when K_1 is odd.

Suppose the axes are labeled a, b, and c in increasing order of size of the moments of inertia. The symmetry for axes a and c has already been determined. The symmetry for the intermediate axis b is derivable from them. Since successive rotations of 180° about each of the three axes brings the molecule back to its original orientation and the coordinates back to their original values, the symmetry for the rotation about b must be just such that it will counteract the effect of rotations about a and c. Hence the wave function is symmetric for a 180° rotation about b if K_{-1} and K_1 are both odd or both even; otherwise it is antisymmetric.

If the dipole moment lies along the a axis, the dipole moment will reverse direction because of a rotation of 180° about either b or c. The matrix element, on which intensities depend, is of the form

$$\mu_g = \int \psi_{J_{K_{-1}K_1}M} \, \mu \cos (ag) \, \psi_{J'_{K'_{-1}K'_1}M'} \, d\tau \qquad (4\text{-}14)$$

where $J_{K_{-1}K_1}M$ and $J'_{K'_{-1}K'_1}M'$ represent the quantum numbers of initial and final states, and $\cos (ag)$ is the cosine of the angle between a and some axis fixed in space. If $\cos (ag)$ changes sign for a rotation of 180° around c, the $\psi_{J_{K_{-1}K_1}}\psi_{J'_{K'_{-1}K'_1}}$ must also change sign if μ_g is not zero. Otherwise

μ_g would appear to change sign, and since the matrix element μ_g cannot change with this symmetry operation on the coordinates, it would have to equal zero. Hence transitions can occur only when K_1 and K'_1 are of different parity (one even and the other odd). A similar argument for rotation around b shows that K_{-1} and K'_{-1} must be of the same parity. This type of procedure may be applied to a molecule with dipole moment along the b or c direction to establish selection rules for these cases.

Wave-function symmetries are summarized in Table 4-2 and selection rules in Table 4-3.

TABLE 4-2. SYMMETRY PROPERTIES OF ASYMMETRIC-TOP WAVE FUNCTIONS

Designation			Behavior with 180° rotation about principal axes		
Cross, Hainer, and King K_{-1} K_1		Dennison c a	a (least moment)	b (intermediate moment)	c greatest moment
e	e	$+$ $+$	$+$	$+$	$+$
e	o	$-$ $+$	$+$	$-$	$-$
o	o	$-$ $-$	$-$	$+$	$-$
o	e	$+$ $-$	$-$	$-$	$+$

TABLE 4-3. SELECTION RULES FOR ASYMMETRIC TOPS
In all cases $\Delta J = 0, \pm 1$.

Axes parallel to dipole moment	Allowed transitions	
	Cross, Hainer, and King	Dennison
a (least)	$ee \longleftrightarrow eo$ $oo \longleftrightarrow oe$	$++ \longleftrightarrow -+$ $-- \longleftrightarrow +-$
b (intermediate)	$ee \longleftrightarrow oo$ $eo \longleftrightarrow oe$	$++ \longleftrightarrow --$ $-+ \longleftrightarrow +-$
c (greatest)	$ee \longleftrightarrow oe$ $oo \longleftrightarrow eo$	$++ \longleftrightarrow +-$ $-- \longleftrightarrow -+$

Symmetry properties have been discussed in terms of the evenness or oddness of K_{-1} and K_1, which are indicated by e or o in the tables. Since the symmetries for all three axes are not independent, it is sufficient to give the symmetry for the two axes only, and in the K notation they are given in the order a, c, the letters e and o being used for even and odd, respectively. The symmetry or antisymmetry of the wave function may also be indicated by $+$ or $-$. When this notation is used, the two axes are designated in the order c, a. This is the older notation, and unfortunately the notation e, o above uses just the reverse order of axes. The usual selection rules $\Delta J = 0$, ± 1 for dipole radiation of a rotating body also apply to asymmetric rotors.

If the molecular dipole moment does not lie along any principal axis, it may be resolved into components along the three axes and the allowed transitions are the sum of all those allowed by Table 4-3 for each component. Thus if the dipole moment has nonzero components along all three axes, all possible transitions consistent with the general selection rule $\Delta J = 0$, ± 1 are allowed.

Dipole Matrix Elements. In order to evaluate intensities, the dipole matrix elements must be obtained. The z component of the matrix element for a transition $j \leftarrow i$ is

$$\mu_z = \mu_a \int \cos (az) \, \psi_i \psi_j^* \, d\tau + \mu_b \int \cos (bz) \, \psi_i \psi_j^* \, d\tau$$
$$+ \, \mu_c \int \cos (cz) \, \psi_i \psi_j^* \, d\tau \quad (4\text{-}15)$$

where μ_a, μ_b, μ_c are the components of the dipole moment along the three principal axes of the molecule and cos (az), cos (bz), cos (cz) represent the cosines of angles between the principal axes and the z axis fixed in space. To obtain the integrals in (4-15), the wave functions for an asymmetric rotor are needed. The general form of these wave functions is discussed below, although the functions are not explicitly given.

The wave functions ψ_{JKM} for either a prolate or an oblate symmetric top form a complete set of functions in terms of which the wave functions for an asymmetric rotor may be expanded as follows:

$$\psi_{J_{K_{-1}K_1}M} = \sum_{J'} \sum_K \sum_{M'} a_{J'KM'} \psi_{J'KM'} \quad (4\text{-}16)$$

where $a_{J'KM'}$ is an appropriate numerical coefficient. Since the total angular momentum J and its projection M on a fixed axis are good quantum numbers for any asymmetry and more than one value cannot be involved in a given state, $J' = J$ and $M' = M$, so that (4-16) reduces to

$$\psi_{J_{K_{-1}K_1}M} = \sum_K a_{JKM} \psi_{JKM} \quad (4\text{-}17)$$

Prolate symmetric-top wave functions are, of course, most appropriate when the rotor approximates a prolate symmetric top, and oblate top functions when it is nearly an oblate symmetric top.

Since the function $\psi_{J_{K_{-1}K_1}M}$ must be either symmetric or antisymmetric with respect to a 180° rotation as pointed out above, only odd or only even K will appear in the sum; hence for an expansion in prolate wave functions,

$$\psi_{J_{K_{-1}K_1}M} = \sum_{K = K_{-1}\pm 2n} a_{JKM}\psi_{JKM} \tag{4-18}$$

where n is an integer. The energy depends on the a_{JKM}, since they give the probability of rotation with the particular angular momenta J and K. However, since the energy cannot depend on M, the projection of J on some arbitrary direction in space, the a_{JKM} must be independent of the quantum number M. The a_{JKM} in (4-17) or (4-18) may hence be indicated simply by a_{JK}.

The coefficients a_{JK} can be evaluated (see, for example, [379]), but no simple closed expression may be found for them, except in special cases.

From (4-18) it may be seen that the matrix elements for an asymmetric rotor can be derived from those for a symmetric rotor. In cases of slightly asymmetric rotors, a sum of type (4-18) reduces to essentially one term, and matrix elements are, to good accuracy, the same as those for symmetric tops.

The dipole matrix elements for a symmetric top may be broken up into several factors [122]

$$\mu_g = \mu\phi_{JJ'}\phi_{JKJ'K'}\phi_{JMJ'M'} \tag{4-19}$$

where μ is the molecular dipole moment, or its component along some principal axis. The ϕ's, which might be called factors of the direction-cosine matrix from (4-15), are each dependent on the rotational quantum numbers indicated by subscripts. The ϕ's are also dependent on the particular component μ_g which is being evaluated and on the molecular axis along which the dipole moment μ lies. Table 4-4 gives expressions for the various ϕ's.

From Table 4-4 the dipole matrix elements of symmetric tops for

TABLE 4-4. VALUES OF FACTORS OF THE DIRECTION-COSINE MATRIX ELEMENTS

The dipole moment matrix element is $\mu\phi_{JJ'}\phi_{JKJ'K'}\phi_{JMJ'M'}$. Subscript a applies to cases where μ is along the molecular axis, b or c to cases where μ is perpendicular to this axis. Subscripts x, y, or z apply for μ_x, μ_y, or μ_z, which are the appropriate elements for polarization along the x, y, or z directions, respectively. The phases chosen are consistent with reference [56]. Matrix elements listed are appropriate for a prolate symmetric top (with a the symmetry axis). For an oblate symmetric top, ϕ_a should be replaced by ϕ_c, ϕ_c by ϕ_b, and ϕ_b by ϕ_a.

Matrix element factor	Value of J'		
	$J+1$	J	$J-1$
$\phi_{JJ'}$	$[4(J+1)\sqrt{(2J+1)(2J+3)}]^{-1}$	$[4J(J+1)]^{-1}$	$[4J\sqrt{4J^2-1}]^{-1}$
$(\phi_a)_{JKJ'K}$	$2\sqrt{(J+1)^2-K^2}$	$2K$	$+2\sqrt{J^2-K^2}$
$(\phi_b \text{ or } \pm i\phi_c)_{J,K,J',K\pm1}$	$\mp\sqrt{(J\pm K+1)(J\pm K+2)}$	$\sqrt{(J\mp K)(J\pm K+1)}$	$\pm\sqrt{(J\mp K)(J\mp K-1)}$
$(\phi_z)_{JMJ'M}$	$2\sqrt{(J+1)^2-M^2}$	$2M$	$+2\sqrt{J^2-M^2}$
$(\phi_x \text{ or } \mp i\phi_y)_{J,M,J',M\pm1}$	$\mp\sqrt{(J\pm M+1)(J\pm M+2)}$	$\sqrt{(J\mp M)(J\pm M+1)}$	$\pm\sqrt{(J\mp M)(J\mp M-1)}$

transitions between individual M values may be obtained, and from these the average square of the matrix element, or $|\mu_{ij}|^2$ as given in (3-40),

(3-41), and (3-42). A rather unreliable, but sometimes useful, estimate of the transition intensity for an asymmetric rotor may be made by interpolating between the intensities of the corresponding transitions for a prolate and an oblate symmetric top. For slightly asymmetric rotors, Lide [720] has given the dipole matrix elements as those of the symmetric rotors plus a correction term proportional to the asymmetry. Numerical values for the intensities of the more important transitions involving levels with J less than 13 may be obtained as a function of asymmetry in Appendix V [122].

A general view of the many transitions which can occur in an asymmetric rotor is given by Table 4-5. The changes in pseudo quantum numbers K_{-1} and K_1 are indicated by numbers, a minus sign being put before the numbers when a change in K_{-1} or K_1 is negative. Superscripts a, b, and c indicate components of the molecular dipole moment along directions of the least, intermediate, and greatest principal moments of inertia, respectively. For an arbitrary direction of the dipole moment, all possible changes ΔK_{-1} and ΔK_1 are allowed except that both these changes cannot be even (cf. Table 4-3). Any particular transition is due to only one of the components of the dipole moment, that is, μ_a, μ_b, or μ_c.

Intensities of absorption lines of asymmetric rotors are given by the basic formula (1-59) or (13-19). The matrix element $|\mu_{ij}|^2$ used in these expressions is the sum $\sum_{M'} (\mu_x^2 + \mu_y^2 + \mu_z^2)$ for any arbitrary orientation of the molecule indicated by M, the projection of J on a fixed axis. As was found in Chap. I, this quantity is independent of M, and furthermore the sum of $|\mu_{ij}|^2$ for all states M of the transition $J' \leftarrow J$, must just equal the sum for all states M' of the transition $J' \rightarrow J$ in order to maintain thermal equilibrium. Since there are $2J + 1$ different M states,

$$(2J + 1)|\mu_{J' \leftarrow J}|^2 = (2J' + 1)|\mu_{J' \rightarrow J}|^2 \qquad (4\text{-}20)$$

Here states designated properly by $J_{K_{-1}K_1}$ and $J'_{K'_{-1}K'_1}$ have been indicated more briefly by J and J'. The quantity tabulated in Appendix V is what may be called the transition strength

$$^xS_{JJ'} = (2J + 1) \frac{|\mu_{J' \leftarrow J}|^2}{\mu^2} \qquad (4\text{-}21)$$

The superscript x indicates the principal axis parallel to the dipole moment μ which produces the transition, and hence takes on the values a, b, or c. The quantity $|\mu_{ij}|^2$ which is needed to obtain the absorption coefficient of a microwave line from expression (13-19) can hence be easily obtained from the entries in Appendix V, since

$$|\mu_{ij}|^2 = \frac{\mu^2 S}{2J + 1} \qquad (4\text{-}22)$$

TABLE 4-5. PERMITTED TRANSITIONS BETWEEN ASYMMETRIC ROTOR LEVELS OF LOW J VALUES

Numbers indicate changes in K_{-1} and K_1; the letter superscript indicates the axis along which the molecular dipole moment must have a nonzero component for the transition to occur. Thus $^a2,-1$ indicates a dipole moment component along the principal axis of smallest moment of inertia and $\Delta K_{-1} = 2$, $\Delta K_1 = -1$.

	$0_{0,0}$	$1_{0,1}$	$1_{1,1}$	$1_{1,0}$	$2_{0,2}$	$2_{1,2}$	$2_{1,1}$	$2_{2,1}$	$2_{2,0}$	$3_{0,3}$	$3_{1,3}$	$3_{1,2}$	$3_{2,2}$	$3_{2,1}$	$3_{3,1}$	$3_{3,0}$
$0_{0,0}$	—	$^a0,1$	$^b1,1$	$^c1,0$												
$1_{0,1}$	$^a0,-1$				$^a0,1$	$^b1,1$	$^c1,0$	—	$^a2,-1$							
$1_{1,1}$	$^b-1,-1$				$^b-1,1$	$^a0,1$	—	$^c1,0$	$^b1,-1$							
$1_{1,0}$	$^c-1,0$				$^c-1,2$	—	$^a0,1$	$^b1,1$	$^c1,0$							
$2_{0,2}$		$^a0,-1$	$^b1,-1$	$^c1,-2$						$^a0,1$	$^b1,1$	$^c1,0$	—	$^a2,-1$	$^b3,-1$	$^c3,-2$
$2_{1,2}$		$^b-1,-1$	$^a0,-1$	—						$^b-1,1$	$^a0,1$	—	$^c1,0$	$^b1,-1$	$^a2,-1$	—
$2_{1,1}$		$^c-1,0$	—	$^a0,-1$						$^c-1,2$	—	$^a0,1$	$^b1,1$	$^c1,0$	—	$^a2,-1$
$2_{2,1}$		—	$^c-1,0$	$^b-1,-1$						—	$^c-1,2$	$^b-1,1$	$^a0,1$	—	$^c1,0$	$^b1,-1$
$2_{2,0}$		$^a-2,1$	$^b-1,1$	$^c-1,0$						$^a-2,3$	$^b-1,3$	$^c-1,2$	—	$^a0,1$	$^b1,1$	$^c1,0$
$3_{0,3}$					$^a0,-1$	$^b1,-1$	$^c1,-2$	—	$^a2,-3$							
$3_{1,3}$					$^b-1,-1$	$^a0,-1$	—	$^c1,-2$	$^b1,-3$							
$3_{1,2}$					$^c-1,0$	—	$^a0,-1$	$^b1,-1$	$^c1,-2$							
$3_{2,2}$					—	$^c-1,0$	$^b-1,-1$	$^a0,-1$	—							
$3_{2,1}$					$^a-2,1$	$^b-1,1$	$^c-1,0$	—	$^a0,-1$							
$3_{3,1}$					$^b-3,1$	$^a-2,1$	—	$^c-1,0$	$^b-1,-1$							
$3_{3,0}$					$^c-3,2$	—	$^a-2,1$	$^b-1,1$	$^c-1,0$							

where μ is the component of the dipole moment responsible for the transition and J refers to the lower state.

Because of relation (4-20) if $^x S_{JJ}$ for the transition $J_{K_{-1}K_1} \leftarrow J'_{K'_{-1}K'_1}$ is tabulated, the similar quantity for the reverse transition is immediately available, since it has the same value.

$$^x S_{J_{kl}J'_{mn}}(\kappa) = {}^x S_{J'_{mn}J_{kl}}(\kappa) \tag{4-23}$$

where k, l, m, and n represent the values of K_{-1} and K_1, and the dependence of S on the asymmetry parameter κ is indicated. In addition, it can be shown that

$$^x S_{J_{kl}J'_{mn}}(\kappa) = {}^{x'} S_{J_{lk}J'_{nm}}(-\kappa) \tag{4-24}$$

or that the strength of a transition $J'_{mn} \leftarrow J_{kl}$ for a rotor with asymmetry parameter κ is just equal to that of what might be called the "inverse" transition $J'_{nm} \leftarrow J_{lk}$ of a rotor with an equal and opposite asymmetry parameter. The relation (4-23) is rather similar and closely related to Eqs. (4-11) connecting energies for positive and negative values of κ. Note that in (4-24) the component of the dipole moment involved may be different for the inverse transitions. If the component is required to be along a in the first transition, it will be along c in the inverse transition, and vice versa. However, if the dipole moment required for the first transition lies along the b axis, it will also lie along the b axis for the inverse transition.

For rotation-vibration spectra, which occur in the infrared region, the groups of transitions of types $J - 1 \leftarrow J$, $J \leftarrow J$, and $J + 1 \leftarrow J$ often give somewhat separate parts or branches of the spectra, and are called the P, Q, and R branches, respectively. These three types of transitions are intermingled in the pure rotational spectra which occur in the microwave region. However, the designation of branches P, Q, and R for transitions with $\Delta J = -1$, 0, and $+1$, respectively, is still useful as an aid in classifying transitions. Types of transitions, approximately in order of intensity, are listed in Table 4-6. Transitions called "forbidden" are forbidden only for symmetric rotors but tend to be weak even in the asymmetric cases. Still weaker or more highly "forbidden" transitions occur, involving larger changes in K_{-1} or K_1, but they are not included in Table 4-6.

Appendix V lists transition strengths for the various types of transitions shown in Table 4-6, and in the same order. Because of relations (4-23) and (4-24), strengths of four types of transitions are given by the same entry in the table, i.e.,

$$^x S_{J_{kl}J'_{mn}}(\kappa), \quad {}^x S_{J'_{mn}J_{kl}}(\kappa), \quad {}^{x'} S_{J_{lk}J'_{nm}}(-\kappa), \quad {}^{x'} S_{J'_{nm}J_{lk}}(-\kappa)$$

TABLE 4-6. SUMMARY OF THE STRONGER TRANSITIONS IN AN ASYMMETRIC
ROTOR SPECTRUM

"Branches" or groups of transitions indicated by P, Q, and R involve absorption transitions $J - 1 \leftarrow J$, $J \leftarrow J$, or $J + 1 \leftarrow J$, respectively. Notation such as $^cP_{1,-2}$ indicates a transition for which $\Delta J = -1$, $\Delta K_{-1} = 1$, $\Delta K_1 = -2$. The superscript, which in this case is c, indicates the principal axis along which a component of the dipole moment must lie in order for such a transition to occur. The forbidden transitions cannot occur in symmetric rotors and tend to be weak in asymmetric rotors. (After Cross, Hainer, and King [122].)

	"Symmetric-rotor" subbranches		
	a and c subbranches		
	Prolate and oblate		
$^cQ_{1,0}$	$^cQ_{-1,0}$	$^aQ_{0,1}$	$^aQ_{0,-1}$
$^cR_{1,0}$	$^cP_{-1,0}$	$^aR_{0,1}$	$^aP_{0,-1}$
Prolate only (c)		Oblate only (a)	
$^cQ_{-1,2}$	$^cQ_{1,-2}$	$^aQ_{2,-1}$	$^aQ_{-2,1}$
$^cR_{-1,2}$	$^cP_{1,-2}$	$^aR_{2,-1}$	$^aP_{-2,1}$
	b subbranches		
$^bQ_{-1,1}$	$^bQ_{1,-1}$	$^bQ_{1,-1}$	$^bQ_{-1,1}$
$^bR_{1,1}$	$^bP_{-1,-1}$	$^bR_{1,1}$	$^bP_{-1,-1}$
Prolate only		Oblate only	
$^bR_{-1,3}$	$^bP_{1,-3}$	$^bR_{3,-1}$	$^bP_{-3,1}$
	First-order forbidden subbranches		
	a and c subbranches		
$^cQ_{-3,2}$	$^cQ_{3,-2}$	$^aQ_{2,-3}$	$^aQ_{-2,3}$
$^cQ_{-3,4}$	$^cQ_{3,-4}$	$^aQ_{4,-3}$	$^aQ_{-4,3}$
$^cR_{3,-2}$	$^cP_{-3,2}$	$^aR_{-2,3}$	$^aP_{2,-3}$
$^cR_{-3,4}$	$^cP_{3,-4}$	$^aR_{4,-3}$	$^aP_{-4,3}$
$^cR_{-3,4}$	$^cP_{3,-4}$	$^aR_{4,-3}$	$^aP_{-4,3}$
	b subbranches		
$^bQ_{-3,3}$	$^bQ_{3,-3}$	$^bQ_{3,-3}$	$^bQ_{-3,3}$
$^bR_{-3,3}$	$^bP_{3,-3}$	$^bR_{3,-3}$	$^bP_{-3,3}$
$^bR_{-3,5}$	$^bP_{3,-5}$	$^bR_{5,-3}$	$^bP_{-5,3}$

Transition strengths are listed for κ values of 1, 0.5, 0, -0.5, and -1. Strengths for intermediate values of κ must be obtained by interpolation.

Intensities of Absorption Lines. Absorption intensities involve not only $|\mu_{ij}|^2$ but also the fraction f of molecules in the ground state of the transition. Neglecting the effects of nuclear spin, f may be written as previously (pages 19 and 20)

$$f = f_{J_{K_{-1}K_1}} f_v$$

where

$$f_{J_{K_{-1}K_1}} = \frac{(2J+1)e^{-\frac{W_{J_{K_{-1}K_1}}}{kT}}}{\sum_J (2J+1)e^{-\frac{W_{J_{K_{-1}K_1}}}{kT}}} \tag{4-25}$$

$$f_v = e^{-W_v/kT}\prod_n (1 - e^{-h\omega_n/kT})^{d_n} \tag{4-26}$$

where $W_{J_{K_{-1}K_1}}$ is the rotational energy and W_v the vibrational energy. If the temperature is high enough so that $kT/h \gg A$,

$$\sum_J (2J+1)e^{-\frac{W_{J_{K_{-1}K_1}}}{kT}} = \sqrt{\frac{\pi}{ABC}}\left(\frac{kT}{h}\right)^3 \tag{4-27}$$

where A, B, and C are the rotational constants in cycles per second. Better approximations for the partition function may be found [48]. However, when T is greater than $100°K$, the approximation (4-27) for the partition function has an error less than 2 per cent for all known cases.

In the discussion of asymmetric rotors thus far, centrifugal stretching effects have been neglected. A precise evaluation of the partition function must take such effects into account. Their contribution to the partition function may in some cases be as large as 1 per cent. Hence if more accuracy than that given by (4-27) is needed, a rather detailed evaluation of the partition function must be made, taking into account centrifugal distortion.

Although the high-temperature approximation may be used for the denominator of (4-25), it cannot always be assumed that the exponential in the numerator of (4-25) is approximately equal to 1. For a symmetric molecule, only the lower rotation states give transitions in the microwave region, $W_{JK} \ll kT$, and the Boltzmann factor $e^{-W_{JK}/kT}$ may usually be safely set equal to unity. For asymmetric rotors, however, microwave transitions may occur between states each of which has a very large rotational energy, so that the corresponding factor is sometimes considerably smaller than 1 and must be retained.

Assuming no effect of nuclear spins, the maximum absorption coefficient for an asymmetric transition in the microwave region from (4-25), (4-27), and (13-19), is

$$\gamma_{\max} = \frac{8\pi hNf_v}{3c(kT)^2}\sqrt{\frac{\pi hABC}{kT}}\,e^{-\frac{W_{J_{K_{-1}K_1}}}{kT}}(2J+1)|\mu_{ij}|^2\frac{\nu^2}{\Delta\nu} \tag{4-28}$$

where all quantities are in cgs or electrostatic units. $(2J+1)|\mu_{ij}|^2$ is just $\mu_x^2\,{}^xS_{J_{kl}J'_{mn}}$, the square of the appropriate component of the dipole moment times the number tabulated in Appendix V. Letting $T = 300°K$

and substituting values for the universal constants,

$$\gamma_{\text{max}} = 2.46 \times 10^{-20} f_v \sqrt{ABC} \; e^{-\frac{W_{J_{K_{-1}K_1}}}{kT}} (2J+1)|\mu_{ij}|^2 \frac{\nu^2}{\Delta \nu} \quad (4\text{-}29)$$

The Effects of Nuclear Spins and Statistics. If there are two equivalent nuclei in a molecule, *i.e.*, identical isotopes of the same element which have exactly the same molecular environment, the nuclear spins and statistics will affect the population of molecular states and hence the transition intensities. If two equivalent nuclei occur, the molecule has a twofold axis of symmetry, and coordinates of the two equivalent nuclei may be interchanged by a 180° rotation about this symmetry axis, or by combinations of inversion and 180° rotations about various axes. In an asymmetric rotor, there cannot be more than two equivalent nuclei, because if there were, the molecule would have at least a threefold axis and would be a symmetric top (*cf.* Sec. 3-1). However, there may be more than one pair of equivalent nuclei.

In order to avoid the complication of possible inversion of the molecule, we consider first a planar molecule. The molecules H_2O, NO_2, and H_2CO are examples of this type. If a planar molecule has two equivalent nuclei, it has a symmetry axis which is the intersection between the plane in which the molecule lies and the perpendicular bisector of the line joining the two equivalent nuclei. This symmetry axis must be a principal axis of the ellipsoid of inertia (Sec. 3-1) and a rotation of 180° about this axis interchanges the positions of the two equivalent nuclei. If the nuclei obey Bose-Einstein statistics, an interchange of both the position and spin coordinates must leave the wave function unchanged; if they obey Fermi-Dirac statistics, the wave function must change sign (*cf.* page 69).

The wave function can be written as a product of rotational and nuclear spin functions

$$\psi = \psi_{J_{K_{-1}K_1}} \psi_N \quad (4\text{-}30)$$

The symmetry of the rotational function has already been discussed.

Let the spin function of the first nucleus of spin I be written $\sigma_m(1)$, where m is the projection of I on an axis fixed in space, and may have the $2I + 1$ values $I, I - 1, I - 2, \ldots, -I$. Similarly, the spin functions for the second nucleus may be written $\sigma_{m'}(2)$, and spin functions for the two nuclei $\sigma_m(1)\sigma_{m'}(2)$. There are in all $(2I + 1)^2$ such combinations. For the $2I + 1$ cases when $m = m'$, the function $\sigma_m(1)\sigma_{m'}(2)$ is clearly symmetric with respect to an interchange of the spin coordinates of nucleus 1 and 2. If $m \neq m'$, then this function is neither symmetric nor antisymmetric, but equal numbers of symmetric and antisymmetric combinations can be formed from these of the type

$$\sigma_m(1)\sigma_{m'}(2) + \sigma_m(2)\sigma_{m'}(1) \quad \text{(symmetric)}$$
$$\sigma_m(1)\sigma_{m'}(2) - \sigma_m(2)\sigma_{m'}(1) \quad \text{(antisymmetric)}$$

There are $[(2I + 1)^2 - (2I + 1)]/2$ of each of these types, so the total number of symmetric spin functions becomes

$$n_{sym} = (2I + 1)(I + 1) \qquad (4\text{-}31a)$$

and of antisymmetric spin functions

$$n_{antisym} = (2I + 1)I \qquad (4\text{-}31b)$$

A molecular rotation of 180° about the symmetry axis and an exchange of spin coordinates of the two equivalent nuclei amounts to an exchange of all coordinates of these two nuclei. Hence if Bose-Einstein statistics apply to the two nuclei, the symmetric spin functions must be used with rotational functions which are symmetric with respect to a rotation around the symmetry axis, and antisymmetric spin functions with antisymmetric rotational functions. The ratio of the number of spin states or the statistical weights of the levels which are symmetric to those which are antisymmetric with respect to a 180° rotation about the symmetry axis is then, from (4-31),

$$\text{For Bose-Einstein statistics: } \frac{I + 1}{I} \qquad (4\text{-}32a)$$

$$\text{For Fermi-Dirac statistics: } \frac{I}{I + 1} \qquad (4\text{-}32b)$$

Since the molecular symmetry axis in question is a principal axis of inertia, Table 4-2 gives the behavior of $\psi_{J_{K_{-1}K_1}}$ with respect to 180° rotations about this axis. Statistical weights due to nuclear spins for a number of cases are given in Table 4-7. Expression (4-25) giving the fraction of molecules in a given rotational state must accordingly be modified for molecules with two equivalent atoms by multiplying the probability for each state by the nuclear spin statistical weight factor from (4-31) or Table 4-7. Actually only the ratio of statistical weights for the two types of states is of importance. The statistical weight due to nuclear spins may be considered unity for the more populated states and $I/(I + 1)$ for the others. If this is done, the partition function [the denominator of (4-25)] is multiplied by $(2I + 1)/[2(I + 1)]$.

If there are more than one pair of equivalent nuclei in a molecule, the symmetry properties of each pair may be taken into account. An example is $CH_2Cl_2^{35}$, which is not planar, but the inversion, which will be discussed below, may be neglected. This molecule has a twofold axis of symmetry which exchanges the positions of the two hydrogens and the two chlorines at the same time. Let the hydrogens have S_1 symmetric spin functions and A_1 antisymmetric spin functions. Similarly the chlorines will have S_2 symmetric and A_2 antisymmetric spin functions. The product of hydrogen and chlorine functions will give $S_1S_2 + A_1A_2$ symmetric and $S_1A_2 + S_2A_1$ antisymmetric total spin functions. Hence from (4-31), letting the spins of the two different types of nuclei be I_1

and I_2,

$$n_{\mathrm{sym}} = (2I_1 + 1)(2I_2 + 1)(2I_1I_2 + I_1 + I_2 + 1) \qquad (4\text{-}33a)$$
$$n_{\mathrm{antisym}} = (2I_1 + 1)(2I_2 + 1)(2I_1I_2 + I_1 + I_2) \qquad (4\text{-}33b)$$

Both H and Cl^{35} obey Fermi-Dirac statistics, so that an interchange of both pairs of nuclei must leave the wave function unchanged, and symmetric total spin functions must be paired with symmetric rotational functions. The resulting statistical weights are given in Table 4-7.

TABLE 4-7. EXAMPLES OF STATISTICAL WEIGHTS DUE TO SPINS OF
EQUIVALENT NUCLEI

Molecule	Symmetric levels*	Statistical weight	Antisymmetric levels*	Statistical weight
H_2O	ee,oo	1	eo,oe	3
D_2O	ee,oo	6	eo,oe	3
H_2CO, H_2C_2O	ee,eo	1	oe,oo	3
D_2CO, D_2C_2O	ee,eo	6	oe,oo	3
NO_2^{16}, SO_2^{16}	ee,oo	1	eo,oe	0
H_2C^{12}——$C^{12}H_2$ O	ee,oo	10	eo,oe	6
CH_2F_2, $H_2C = CF_2$	ee,eo	10	oe,oo	6
D_2C^{12}——$C^{12}D_2$ O	ee,oo	45	eo,oe	36
CD_2F_2, $D_2C = CF_2$	ee,eo	15	oe,oo	21
H_2C^{13}——$C^{13}H_2$ O	ee,oo	28	eo,oe	36
$CH_2Cl_2^{35}$	ee,oo	36	eo,oe	28
Cl^{35} Cl^{35} $C^{12}=C^{12}$ H H	ee,oo	36	eo,oe	28
$CD_2Cl_2^{35}$	ee,oo	66	eo,oe	78
NDH_2	ee,eo (lower inversion level) or oo,oe (upper inversion level)	1	oo,oe (lower inversion level) or ee,eo (upper inversion level)	3

* Rotational levels specified by evenness or oddness of $K_{-1}K_1$.

Other cases of more than one pair of equivalent nuclei may be similarly treated. In general for a molecule with n pairs of identical nuclei, the number of symmetric spin functions which can be formed is

$$n_{\mathrm{sym}} = \tfrac{1}{2}\left[\prod_{i=1}^{n}(2I_i + 1)\right]\left[\prod_{i=1}^{n}(2I_i + 1) + 1\right] \qquad (4\text{-}34a)$$

and the number of antisymmetric functions

$$n_{\text{antisym}} = \tfrac{1}{2}\left[\prod_{i=1}^{n}(2I_i + 1)\right]\left[\prod_{i=1}^{n}(2I_i + 1) - 1\right] \qquad (4\text{-}34b)$$

If the molecule has a twofold axis of symmetry and the n pairs of nuclei are interchanged by a 180° rotation around this axis, the ratio of intensities of rotation levels of odd and even K is given by Eqs. (4-34).

4-3. Centrifugal Distortion. Centrifugal distortion is enormously more important in the microwave spectra of asymmetric rotors than in the spectra of symmetric tops. In the latter it produces very small shifts of the order of 1 Mc or less, whereas in the microwave spectra of some asymmetric rotors, centrifugal distortions change the rotational frequencies many hundreds of megacycles. This is because microwave transitions may occur in asymmetric rotors between states of rather large angular momentum and of very large rotational energies. In light symmetric tops, transitions between states of rather small J are generally observed. Furthermore, in the heavier symmetric molecules which give microwave spectra for transitions between states of larger J, the moment of inertia is so large that rotational energies in these states are still rather small.

Consider as an example the asymmetric rotor SO_2, whose rotational constants lie between 8000 and 80,000 Mc. If this molecule were linear, a transition involving J as small as 2 would fall in the "K-band" region near 24,000 Mc. However, in the actual spectrum between 20,000 and 30,000 Mc transitions involving J values from 3 to 35 have been identified [647]. Many other transitions occur in this region which are thought to involve still higher values of J. The rotational energy for $J = 35$ is of the order of 1000 cm^{-1}, or 3×10^7 Mc. Although the centrifugal distortion is a small fraction of the rotational energy, for $J = 35$ it is as large as about 0.3 cm^{-1}, or 10^4 Mc, and hence corresponds to an enormous shift of the observed microwave lines. Hence for an accurate understanding of the microwave spectrum of SO_2, rather accurate knowledge of centrifugal effects is necessary.

Still more extreme cases are the light molecules H_2O and HDO. The $6_{-5} \leftarrow 5_{-1}$ transition of H_2O which lies at 22,235 Mc involves levels with rotational energies near 500 cm^{-1}, or 1.5×10^7 Mc, even though J is only 5 or 6. For $J = 11$ states of H_2O, centrifugal distortion corrections occur which are as large as 9 per cent of the entire rotational energy, or 280 cm^{-1} [84]. Transitions involving these particular states of H_2O lie in the infrared region, however.

Lest the reader infer that centrifugal distortions produce even more difficulties in the interpretation of microwave spectra than they really do, it should be pointed out that, for all but the lightest molecules, it is usually possible to find rotational transitions in the microwave region of low angular momentum J which are not shifted by centrifugal effects more than a few megacycles. Hence moments of inertia and parameters for asymmetric rotors may often be obtained with sufficient accuracy without allowing for centrifugal distortion. Furthermore, in some cases where centrifugal distortion is large, microwave transitions occur between states of rather similar angular momentum, so that the net shift in frequency due to centrifugal distortion is not so large.

In order to understand qualitatively what variables are important to centrifugal distortion, consider first a molecule which rotates about only one axis, so that classically the energy of rotation may be written

$$W = \frac{1}{2I} P^2 \qquad (4\text{-}35)$$

where I is the moment of inertia and P the angular momentum. Assume now that I depends on one coordinate R. Then the centrifugal force tending to increase R is

$$F = -\frac{\partial W}{\partial R} = -\tfrac{1}{2}P^2 \frac{\partial(1/I)}{\partial R} \qquad (4\text{-}36)$$

Because of this force, R is modified by a small amount ΔR such that the restoring force $k\,\Delta R$ equals F. Hence

$$\Delta R = -\tfrac{1}{2}\frac{P^2}{k}\frac{\partial(1/I)}{\partial R} \qquad (4\text{-}37)$$

There is a consequent change in the energy of rotation

$$\Delta W_P = \tfrac{1}{2}P^2 \frac{\partial(1/I)}{\partial R}\,\Delta R = -\tfrac{1}{4}\frac{P^4}{k}\left[\frac{\partial(1/I)}{\partial R}\right]^2 \qquad (4\text{-}38)$$

In addition, the potential energy stored as a result of the displacement ΔR is

$$\Delta W_k = \tfrac{1}{2}k(\Delta R)^2 = \tfrac{1}{8}\frac{P^4}{k}\left[\frac{\partial(1/I)}{\partial R}\right]^2 \qquad (4\text{-}39)$$

Combining (4-38) and (4-39), the total energy change due to centrifugal distortion is

$$\Delta W = -\tfrac{1}{8}\frac{P^4}{k}\left[\frac{\partial(1/I)}{\partial R}\right]^2 \qquad (4\text{-}40)$$

From (4-40) it may be seen that centrifugal distortion always decreases the energy by an amount dependent on the fourth power of the angular momentum, and the inverse of a molecular force-constant.

In the general case, angular momenta about all three principal axes of the molecule will be involved, and the energy due to centrifugal distortion must be written in the more general form (4-41) given by Wilson and Howard [76], [715].

$$\Delta W = \tfrac{1}{4}\sum_{\alpha\beta\gamma\delta} \tau_{\alpha\beta\gamma\delta}P_\alpha P_\beta P_\gamma P_\delta \qquad (4\text{-}41)$$

where P_α, P_β, etc., is each an angular momentum about some principal axis of the molecule (they are not all different, and all may be the same). The molecular constant $\tau_{\alpha\beta\gamma\delta}$ is

$$\tau_{\alpha\beta\gamma\delta} = -\tfrac{1}{2} \sum_{ij} \frac{\partial \mu_{\alpha\beta}}{\partial R_i} \frac{\partial \mu_{\gamma\delta}}{\partial R_j} (k^{-1})_{ij} \qquad (4\text{-}42)$$

Here $\mu_{\alpha\beta}$ and $\mu_{\gamma\delta}$ correspond to $1/I$ in (4-40). In this generalized form, they are elements of the matrix which is inverse to the inertia matrix or dyadic. Before displacements δR_i and δR_j are made, the moments of inertia are with respect to principal axes, so that this matrix is simply

$$(\mu) = \begin{pmatrix} 1/I_{xx} & 0 & 0 \\ 0 & 1/I_{yy} & 0 \\ 0 & 0 & 1/I_{zz} \end{pmatrix} \qquad (4\text{-}43)$$

The derivative $\partial\mu_{\alpha\beta}/\partial R_i$ is more complicated. The matrix element $(k^{-1})_{ij}$ replaces the factor $1/k$ in (4-40). It is an element of the matrix which is inverse to the matrix of force constants k_{ij} obtained from the potential energy

$$V = \tfrac{1}{2} \sum_{ij} k_{ij} R_i R_j \qquad (4\text{-}44)$$

The above expressions assume that the potential may be taken as harmonic, *i.e.*, of the form (4-44). This is probably a sufficiently good approximation since the effects of centrifugal distortion are usually not extremely large, and if they are very large the potential constants are not usually known so well that anharmonic terms in the potential would greatly improve the accuracy. However, it is possible that in the future the effects of anharmonic potential constants on centrifugal distortion may be determined and used as a method of evaluating these constants. An additional approximation is involved in omitting from (4-41) terms proportional to the sixth power of momenta and higher. These may be of importance in some extreme cases [346].

If R_i, R_j are taken as normal coordinates Q_i, Q_j, then by definition of normal coordinates the potential has the particularly simple form

$$V = \tfrac{1}{2} \sum_i k_{ii} Q_i^2 \qquad (4\text{-}45)$$

The coordinates Q_i are usually taken such that the vibrating mass associated with each coordinate may be assumed to be unity, and the molecular vibrational frequency $\nu_i = \left(\dfrac{1}{2\pi}\right) \sqrt{k_{ii}}$. Hence (4-42) becomes

$$\tau_{\alpha\beta\gamma\delta} = -\tfrac{1}{2} \sum_i \frac{\partial \mu_{\alpha\beta}}{\partial Q_i} \frac{\partial \mu_{\gamma\delta}}{\partial Q_i} \left(\frac{1}{4\pi^2 \nu_i^2}\right) \qquad (4\text{-}46)$$

Fortunately, many of the total of 81 constants $\tau_{\alpha\beta\gamma\delta}$ are usually zero, and others are not independent. In a molecule such as H_2O, there are only four independent coefficients of this type. They can often be evaluated empirically by fitting the observed spectra. However, in some cases it is important to calculate the constants from the known geometry and force constants of the molecule.

The coefficients $\tau_{\alpha\beta\gamma\delta}$ are probably most easily calculated by using expression (4-42), although this depends of course on what information about the molecule is available. If α, β, γ, and δ are directions of principal axes of inertia as assumed above,

$$\frac{\partial \mu_{\alpha\beta}}{\partial R_i} = -\frac{1}{I_{\alpha\alpha}I_{\beta\beta}}\frac{\partial I_{\alpha\beta}}{\partial R_i} \tag{4-47}$$

where $I_{\alpha\beta}$ represents an element of the moment of inertia matrix or dyadic. In obtaining the derivatives $\partial \mu_{\alpha\beta}/\partial R_i$, the variation in the coordinates R_i must be taken in such a way that the center of gravity of the molecule is not changed, nor the orientation of its principal axes (cf. "Eckart conditions" [715]). If small variations $\Delta\alpha_j$, $\Delta\beta_j$, etc., are found for the Cartesian coordinates α_j, β_j of each atom j of the molecule such that R_i is changed by ΔR_i, and other internal molecular coordinates remain unchanged, then the above requirements give [826]

$$\frac{\partial I_{\alpha\beta}}{\partial R_i} = \frac{\Delta I_{\alpha\beta}}{\Delta R_i} = \frac{-2}{I_{\gamma\gamma}\Delta R_i}\left(\sum_l m_l\alpha_l^2 \sum_j m_j\beta_j\Delta\alpha_j + \sum_l m_l\beta_l^2 \sum_j m_j\alpha_j\Delta\beta_j\right) \tag{4-48}$$

Kivelson and Wilson [826] have given sample calculations of this type and formulas for several common cases.

A calculation of the effects of centrifugal distortion on the frequencies of rotational spectra depends on evaluation not only of the constants $\tau_{\alpha\beta\gamma\delta}$, but also of the effect of the operators P_α, P_β, P_γ, and P_δ. A closed general expression for the effect of such operators is not possible. However, since the energy due to centrifugal distortion is almost always a small part of the total rotational energy, the energy contributions due to it may usually be treated with sufficient accuracy by first-order perturbation theory. The Hamiltonian for the rotational energy without centrifugal distortion is as in (3-2)

$$H_0 = \frac{P_x^2}{2I_{xx}} + \frac{P_y^2}{2I_{yy}} + \frac{P_z^2}{2I_{zz}} \tag{4-49}$$

The perturbing centrifugal terms may be taken as

$$H^1 = \tfrac{1}{4}\sum_{\alpha\beta}\tau_{\alpha\alpha\beta\beta}P_\alpha^2 P_\beta^2 \tag{4-50}$$

Matrix elements for a Hamiltonian of the form (4-50) are given by Nielsen [108]. The technique of first-order perturbation theory may be described as that of obtaining wave functions appropriate to the large part H_0 of the Hamiltonian, and then averaging the perturbing energy H^1 over these states. All terms involving odd powers of any component of P have been omitted from (4-50), since they are zero for a large class of molecules and in any case their average is zero so that they contribute nothing in this first-order approximation.

It can be shown [715] that the energy resulting from H_0 and the first-order effects of H^1 is

$$W = W_0 + A_1 W_0^2 + A_2 W_0 J(J+1) + A_3 J^2 (J+1)^2 \\ + A_4 J(J+1)(P_z^2)_{av} + A_5(P_z^4)_{av} + A_6 W_0 (P_z^2)_{av} \quad (4\text{-}51)$$

where the A's are constants of the molecule which are given explicitly in terms of the moments of inertia and the $\tau_{\alpha\alpha\beta\beta}$ by Kivelson and Wilson [715], W_0 is the rotational energy assuming no centrifugal distortion, and $(P_z^2)_{av}$ represents the average or expectation value of P_z^2.

For a symmetric or nearly symmetric rotor, $(P_z^2)_{av} = K^2$ and $(P_z^4)_{av} = K^4$ so that (4-51) reduces to a form similar to that given for a symmetric rotor in (3-54). In other cases, $(P_z^2)_{av}$ may be obtained from the relation [352]

$$(P_z^2)_{av} = \frac{\partial W_0}{\partial (1/I_{zz})} \quad (4\text{-}52)$$

The quantity $\partial W_0/\partial(1/I_{zz})$ may be evaluated by methods given in the earlier part of this chapter for obtaining the rotational energy of a rigid rotor. Evaluation of $(P_z^4)_{av}$ is a bit complex when the approximation $(P_z^4)_{av} \approx (P_z^2)_{av}^2$ is not sufficiently accurate. A method for its evaluation is discussed by Kivelson and Wilson [715] (cf. also [59] and [601]).

Lawrance and Strandberg ([601]; cf. also [59]) have developed an expression similar to (4-51) and applied it to H_2CO by evaluating the constants empirically from the observed spectral frequencies. Certain transitions were observed involving values of J as high as 31 and centrifugal distortion corrections as large as 600 Mc. These fitted the expected form (4-51) within a few megacycles.

A rather complete calculation of centrifugal distortion effects from molecular geometry and force constants has been carried out for SO_2 [647], for PH_2D and PHD_2 [866], and for HDS [587]. Matrix elements and helpful details of this type of calculation are included in the discussion of HDS by Hillger and Strandberg [587]. In each of the above cases the fit to experimental data obtained by the calculation of centrifugal distortion from known force constants appears to be satisfactory, although there is some minor disagreement in the case of HDS.

4-4. Structures of Asymmetric Rotors. Because of the complexities of the spectra of asymmetric rotors, the rotational structure of the spectra of only two (H_2O and HDO) had been fairly completely solved before the

TABLE 4-8. STRUCTURES OF ASYMMETRIC ROTORS WHICH HAVE BEEN
OBTAINED FROM MICROWAVE SPECTRA

Molecule	Structure	Reference
B_2BrH_5		[447]
CHNO (HNCO)		[479]
CHNS (HNCS)		[434] [794]
CH_2Cl_2		[732]
CH_2F_2		[722]
CH_2O		[353] [601]

TABLE 4-8. STRUCTURES OF ASYMMETRIC ROTORS WHICH HAVE BEEN
OBTAINED FROM MICROWAVE SPECTRA (*Continued*)

Molecule	Structure	Reference
CH₄O (CH₃OH)	 (Oxygen lies 0.079 A above symmetry axis of methyl group)	[817]
CH₄S (CH₃SH)		[640] [870]
C₂H₂F₂ (CH₂CF₂)		[406]
C₂H₂O (H₂C₂O)		[708]
C₂H₃NS (CH₃SCN)		[345]
C₂H₃NS (CH₃NCS)		[345]

TABLE 4-8. STRUCTURES OF ASYMMETRIC ROTORS WHICH HAVE BEEN
OBTAINED FROM MICROWAVE SPECTRA (*Continued*)

Molecule	Structure	Reference
C_2H_4O	(Angle between C—C bond and plane containing carbon and its two hydrogens is 159°25′)	[316] [566]
C_2H_4S	(Angle between C—C bond and plane containing carbon and its two hydrogens is 151°43′)	[566]
C_2H_5N (ethylenimine)	(Angle between N—H bond and CCN plane is 112°. Angle between C—C bond and CH_2 plane is 159°25′)	[877]
C_3H_3N (vinyl cyanide)		[763a]
C_4H_5N (pyrrole)	(Molecule is entirely planar)	[763]
C_5H_5N (pyridine)		[893] [914]

TABLE 4-8. STRUCTURES OF ASYMMETRIC ROTORS WHICH HAVE BEEN
OBTAINED FROM MICROWAVE SPECTRA (*Continued*)

Molecule	Structure	Reference
C_6H_5F (fluorobenzene)		[796]
ClF_3		[867]
FNO (NOF)		[609]
FNO_2 (NO_2F)		[748]
F_2OS		[802] [922]
F_2SO_2		[684]

TABLE 4-8. STRUCTURES OF ASYMMETRIC ROTORS WHICH HAVE BEEN
OBTAINED FROM MICROWAVE SPECTRA (*Concluded*)

Molecule	Structure	Reference
HN_3	H—112°39′—N≡N≡N; 1.021, 1.240, 1.134	[427]
H_2S	S, 1.323, 92°6′, H, H	[783]
O_2S (SO_2)	S, 1.432, 119°32′, O, O	[647] [565]
O_3	O, 1.278, 116°49′, O, O	[875]

advent of high-resolution microwave spectroscopy. A large number have
now been worked out by microwave techniques, and it seems feasible to
solve the rotational spectrum of any asymmetric rotor which does not
have serious complications due to internal motions (*cf.* Chap. 12) or
an exceptionally complex hyperfine structure (*cf.* Chap. 6). The Stark
effect has been extremely valuable in identifying and working out this type
of spectrum (*cf.* Chap. 10). The structures of asymmetric rotors which
have been obtained from microwave results are given in Table 4-8.

CHAPTER 5

ATOMIC SPECTRA

While most microwave spectra have their origin in absorption by molecules, certain types of atomic spectra may fall in the microwave region. Atomic theory is important even for molecular spectroscopy because it is often convenient to consider a molecule as being composed of atoms whose properties are not too greatly different from their properties in the free state. Moreover, many molecular phenomena are sufficiently analogous to phenomena in atoms so that it is worthwhile to study first the simpler atomic case.

This chapter will present a summary of those parts of the theory of atomic spectra which are needed for microwave spectroscopy. More extensive treatments are given in the several books devoted to the subject (*e.g.*, A. C. Candler [79], G. Herzberg [124], L. Pauling and S. Goudsmit [23], H. E. White [53], and for a more advanced, quantum-mechanical treatment, E. U. Condon and G. H. Shortley [56]).

5-1. The Hydrogen Atom. The simplest atom is that of hydrogen, consisting of a single proton and an electron. It is described by the wave equation

$$\nabla^2\psi + \frac{8\pi^2\mu}{h^2}(W - V)\psi = 0$$

or, in spherical coordinates,

$$\frac{1}{r^2}\frac{\partial}{\partial r}\left(r^2\frac{\partial\psi}{\partial r}\right) + \frac{1}{r^2\sin^2\theta}\frac{\partial^2\psi}{\partial\phi^2} + \frac{1}{r^2\sin\theta}\frac{\partial}{\partial\theta}\left(\sin\theta\frac{\partial\psi}{\partial\theta}\right)$$
$$+ \frac{8\pi^2\mu}{h^2}(W - V)\psi = 0 \quad (5\text{-}1)$$

where the nucleus, or more exactly the center of mass of the electron and nucleus, is taken as the origin of coordinates, and $\mu = mM/(M + m)$ is the reduced mass of the atom. W is the total energy of the atom, $V = -Ze^2/r$ is the potential energy, Z is the nuclear charge in units of the proton charge, and e is the proton charge.

By a process of separating the variables similar to that used for the diatomic molecule (Chap. 1), the wave equation may be solved [62], giving

$$\psi = R(r)\Theta(\theta)\Phi_m(\phi) \quad (5\text{-}2)$$

where

$$\Phi_M = \frac{1}{\sqrt{2\pi}} e^{im\phi} \tag{5-3}$$

$$\Theta_{Ml} = \left[\frac{(2l+1)(l-|m|)!}{2(l+|m|)!} \right]^{\frac{1}{2}} P_l^{|m|} (\cos \theta) \tag{5-4}$$

$$R_{nl} = \sqrt{\frac{4(n-l-1)!Z^3}{[(n+l)!]^3 n^4 a_1^3}} \left(\frac{2Zr}{na_0} \right)^l e^{-Zr/na_0} L_{n+l}^{2l+1} \left(\frac{2Zr}{na_0} \right) \tag{5-5}$$

and where $n = 1, 2, \ldots$ is the principal quantum number

$l = 0, 1, 2, \ldots, (n-1)$ is the orbital quantum number

$m = -l, -l+1, \ldots, l-1, l$ is the magnetic quantum number and should not be confused with the same symbol used for the mass of the electron

$P_l^{|m|}$ = associated Legendre polynomial

L_{n+l}^{2l+1} = associated Laguerre polynomial

$a_0 = h^2/4\pi^2\mu e^2$ is the radius of the first orbit of the hydrogen atom in the Bohr theory

It may be observed that Eq. (5-1) for the hydrogen atom is exactly the same as the wave equation (1-11) for a diatomic molecule with the potential $V(r)$ replacing the molecular potential $U(r)$. The hydrogen atom may, in fact, be regarded as a diatomic molecule with the proton and electron as the two atoms. The parts of the wave function (5-3) and (5-4) which depend on angle are identical with those for a diatomic or linear molecule (1-5) and (1-6). They are the same for any spherically symmetric potential, since these functions represent conservation of total angular momentum (l or J) and of the projection of the angular momentum (m or M) on a chosen axis. Unlike the diatomic molecule, the dependence of the potential on r in this atomic case is very simple and the radial wave function (5-5) can be determined. In more complex atoms, the potential for a single electron may often be considered spherically symmetric, so that the angular parts of the wave function are unchanged. However, the dependence of the potential on r is usually very difficult to determine, so that the radial wave function and energy cannot be exactly obtained.

Figures 5-1, 5-2, and 5-3 show the radial and angular distribution of the electrons. The s electron wave function is seen to be the only one which does not vanish at the center of the nucleus. The s electron is also the only one which has a spherical charge distribution.

The allowed energy levels for the hydrogen atom are given by

$$W = -\frac{2\pi^2\mu e^4 Z^2}{n^2 h^2} = -Rhc \frac{Z^2}{n^2} \tag{5-6}$$

where $R = 2\pi^2\mu e^4/ch^3$ is the Rydberg constant in cm^{-1}. W is in ergs; to convert to cm^{-1}, divide by hc. On this model of the hydrogen atom,

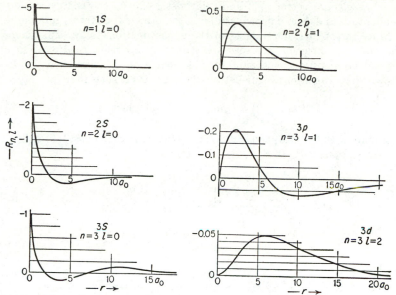

Fig. 5-1. The radial part of the hydrogen electronic wave function $R_{n,l}$ plotted as a function of the distance between the electron and the nucleus.

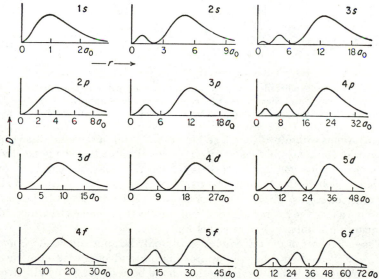

Fig. 5-2. The electronic density distribution $r^2(R_{n,l})^2$ is plotted as a function of the electron-nuclear distance for several states of the hydrogen atom. The ordinate shows the probability of finding the electron between spherical shells of radii r and $r + dr$.

which neglects electron spin and relativistic effects, the energy is independent of l and m, and depends only on the principal quantum number n.

Since the angular dependence of the wave functions is the same as for the diatomic molecule, the selection rules and intensity relations which depend on angular momentum are identical. As in Chap. 1, it can be shown that transitions occur between energy levels such that

$$l' = l \pm 1 \qquad \text{and} \qquad m' = m, m \pm 1$$

A state with $l = 0$ is called an S state; states for which $l = 1, 2, 3, \ldots$ are, respectively, $P, D, F, G, H, \ldots$ states.

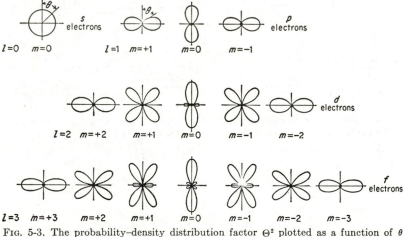

Fig. 5-3. The probability–density distribution factor Θ^2 plotted as a function of θ for s, p, d, f electrons. For states with $m = 0$ the scale is approximately $l(l + 1)$ times that of the other states having the same value of l. (*After White* [53].)

5-2. Atoms with More Than One Electron.

While a wave equation can be written for many-electron atoms, it is not possible to solve it exactly. To a good approximation, however, each electron may be considered to move in a spherically symmetrical field produced by the nucleus and all the other electrons. Then the angular part of the solution will be just the same as for the hydrogen atom, and the electron can be characterized by quantum numbers l and m as before.

The radial part of the wave function can only be approximated. Some useful methods of making the approximation and obtaining numerical or sometimes analytical wave functions have been discussed by Hartree, Fock, Fermi, Thomas, Slater, and others [305].

Particularly simple are the alkali-type atoms which have one electron outside a spherically symmetrical core of "closed electron shells." As long as the valence electron stays out beyond all other electrons in the atom, it moves in a Coulomb field with $Z_{eff} = 1$. To this extent the wave functions and energy levels resemble those of the hydrogen atom, and each state is characterized by quantum numbers n, l, m. The energy is still independent of m and for large values of l does not depend greatly on l. But for small values of l, especially for S states where $l = 0$, the wave functions (Figs. 5-1 and 5-2) show a relatively large probability of the electron's being near the nucleus. In these states, the electron

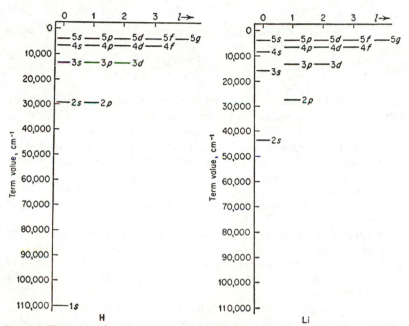

FIG. 5-4. Energy levels of hydrogen and an alkali atom. The term value, or energy, of the atom is referred to a zero which represents the energy after the valence electron has been completely removed or the atom ionized.

spends considerable time inside the closed electron shells, in a region of larger Z_{eff}, so that its binding energy is increased. Figure 5-4 shows the energy levels of a hydrogen atom and those of an alkali atom.

The energies for levels in a complex atom are often given by a formula analogous to Eq. (5-6) for the hydrogenic case. However, since Eq. (5-6) no longer applies exactly, it can be made to agree with the observed energies only by modifying the value of n or the value of Z, or both. Thus for the first case,

$$W = -\frac{RhcZ_o^2}{n^{*2}} = -\frac{RhcZ_o^2}{(n-\sigma)^2} \tag{5-7}$$

where R = Rydberg constant

Z_o = effective nuclear charge in the outer region of the atom. $Z_o = 1$ for alkali atoms, but may be 2, 3, 4, . . . for ionized alkalilike atoms such as Be^+, B^{++}, C^{+++}, etc.

n^* = effective principal quantum number; not an integer

n = principal quantum number, and is an integer

σ = quantum defect

From Eq. (5-7) it is seen that

$$n^* = n - \sigma \tag{5-8}$$

If n is retained and Z is modified, the term value or energy depends on an effective Z, Z_{eff}, such that

$$W = - \frac{RhcZ_{\text{eff}}^2}{n^2} \tag{5-9}$$

Equations (5-7) and (5-9) are written down by purely formal analogy with the expression which applies to a hydrogenic atom. Since n^* in Eq. (5-7) or Z_{eff} in Eq. (5-9) are not integers but are empirical constants, these equations are useful only when the same constant can be used for several purposes. Equation (5-7) is quite useful for describing the term values of alkalilike atoms because the quantum defect, $\sigma = n - n^*$, varies only slowly with n and l. Thus if n^* is known for one level, its value, and hence the term value for another level, can be found at least approximately.

An effective value of Z is often used in connection with atomic fine structure, or for the energy levels of different atoms in an isoelectronic sequence. It is then sometimes convenient to express Z_{eff} in terms of Z_{inner} and Z_{outer}, which are effective nuclear charges in the inner and outer region of the atom.

5-3. Fine Structure, Electron Spin, and the Vector Model. When spectral lines are examined with instruments of moderate resolving power, they are found to have a structure, *i.e.*, to consist of several components. This "fine structure" is explained by attributing to the electron a spin angular momentum and a magnetic moment. The angular momentum is related to a spin quantum number s, and its magnitude is given by $\sqrt{s(s + 1)}$ in units $h/2\pi$. For a single electron s always has the value $\frac{1}{2}$. This s should not be confused with the same letter used to denote a state with $l = 0$. The corresponding spin magnetic moment is $1 + (\alpha/2\pi)$, in units of the Bohr magneton* $he/4\pi mc$, where m is the electron mass. The fine structure constant α is $2\pi e^2/hc$ and approximately equals $\frac{1}{137}$. A further quantum number $m_s = \pm\frac{1}{2}$ is required to describe the projec-

---

* The Bohr magneton is usually taken as a positive quantity even though the electron charge is negative. Regardless of this convention, it should be remembered that the electron magnetic moment is opposite to the direction of its spin.

tion of s on some fixed direction, just as m_l gives the projection of the orbital angular momentum on a fixed direction.

The electron's orbital angular momentum and spin angular momentum are vectors and will therefore add in some way similar to the ordinary rules of vector addition to form a resultant denoted by $\mathbf{j}$. The magnitude of $\mathbf{j}$ is also quantized. It is specified by the total-angular-momentum quantum number j which has the values $j = l + s$ and $j = |l - s|$ (when $s = \frac{1}{2}$). The corresponding vector equation is

$$\mathbf{j} = \mathbf{l} + \mathbf{s} \tag{5-10}$$

where the two possible values of $\mathbf{j}$ correspond to $\mathbf{l}$ and $\mathbf{s}$ being nearly parallel or nearly antiparallel.

The magnitudes of $\mathbf{j}$, $\mathbf{l}$, and $\mathbf{s}$ are, respectively, $\sqrt{j(j+1)}$, $\sqrt{l(l+1)}$, and $\sqrt{s(s+1)}$. However, sometimes equations are written as if the magnitudes were j, l, and s. This convention is frequently employed because of its conciseness (cf. [23]). Using it, one may speak of $\mathbf{l}$ and $\mathbf{s}$ being parallel or antiparallel for then the magnitude of their resultant would just be $l + s$ or $|l - s|$. The method leads to equations which must finally be corrected by replacing j^2 by $j(j + 1)$, l^2 by $l(l + 1)$, and s^2 by $s(s + 1)$.

The two orientations of $\mathbf{s}$ relative to $\mathbf{l}$ produce levels with slightly different energies. They are distinguished by appending the value of j as a subscript to the term symbol, for example, $S_{\frac{1}{2}}$, $P_{\frac{1}{2}}$, $P_{\frac{3}{2}}$, $D_{\frac{3}{2}}$, $D_{\frac{5}{2}}$. Since there are in general two possible orientations of $\mathbf{s}$ relative to $\mathbf{l}$, for any given l there are two possible j values. The *multiplicity* of the term is therefore said to be 2, and this is denoted by a superscript 2 to the left of the term symbol. The doublet symbol is used even for S states where only one value of j is possible. Some typical states in a one-electron (hydrogenic or alkalilike) spectrum are $^2S_{\frac{1}{2}}$, $^2P_{\frac{1}{2}}$, $^2P_{\frac{3}{2}}$, $^2D_{\frac{3}{2}}$, $^2D_{\frac{5}{2}}$, ... The two levels $^2P_{\frac{1}{2}}$ and $^2P_{\frac{3}{2}}$ form a fine-structure doublet term.

The splitting between the two levels with different values of j which occurs when l is not zero is caused primarily by the magnetic interaction between the electron spin and electron orbital magnetic moments. This interaction can be derived from the magnitude of the magnetic field at the electron caused by the relative motion of the nucleus with respect to the electron. The splitting, however, is reduced by another contribution to the energy exactly half as large and in the opposite direction which is connected with a precession of the electron axis due to relativistic effects [7]. When both are taken into account, the energy level for a hydrogenic atom with nuclear charge Z is displaced by an amount

$$W_s = \frac{1}{2} \frac{e^2 h^2 Z}{4\pi^2 m^2 c^2} \left(\frac{1}{r^3}\right)_{\text{av}} sl \cos(\mathbf{s},\mathbf{l}) \tag{5-11}$$

where e = electron charge

m = mass of the electron

c = velocity of light

h = Planck's constant

$(1/r^3)_{av}$ = average of the inverse cube of the distance between nucleus of charge Z and the electron.

Two values of cos (s,l) are possible corresponding to l and s being "parallel" or "antiparallel"; *i.e.*, the quantum number j can equal $l + s$ or $l - s$. The quantity sl cos (s,l) appearing in (5-11) can be evaluated from the vector model. Figure 5-5 shows how the vectors l and s com-

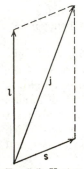

bine to form a resultant j. From the trigonometric relation between the sides and angles of a triangle in the diagram

$$sl \cos (s,l) = \frac{|j|^2 - |s|^2 - |l|^2}{2}$$

where $|j|^2$, $|s|^2$, and $|l|^2$ represent, respectively, squares of the magnitudes of the vectors j, s, and l. From quantum mechanics, these are to be replaced by $j(j + 1)$, $s(s + 1)$, and $l(l + 1)$, so that

$$sl \cos (s,l) = \frac{j(j + 1) - s(s + 1) - l(l + 1)}{2} \quad (5\text{-}12)$$

Fig. 5-5. Vector addition of l and s to form resultant j.

This method of adding vectors trigonometrically and then correcting the square of their magnitudes can be applied to any quantum vectors. It will be found useful for the addition of nuclear angular momenta to atomic or molecular angular momenta.

The vector model, which treats angular momenta as classical vectors except for quantum conditions imposed on their magnitudes, is very helpful because it permits easy calculation and visualization of many quantum-mechanical results. It always gives correctly the results of a true quantum-mechanical calculation if only the cosines between vectors or their projections on arbitrary directions are of importance. It is not directly applicable, however, when more complicated functions are required, such as the squares of cosines of angles between vectors.

Substituting the expression (5-12) for sl cos (s,l) into (5-11), the displacements of the two levels $j = l \pm s = l \pm \frac{1}{2}$ are given by

$$W_s = \frac{1}{2} \frac{e^2 h^2}{4\pi^2 m^2 c^2} \left(\frac{Z}{r^3}\right)_{av} \frac{\pm (l + \frac{1}{2}) - \frac{1}{2}}{2}$$

The $+$ sign corresponds to $j = l + \frac{1}{2}$. The splitting or energy difference between the two levels is then

$$\Delta \nu = R \alpha^2 a_0^3 \left(\frac{Z}{r^3}\right)_{av} (l + \frac{1}{2}) \quad \text{cm}^{-1} \quad (5\text{-}13a)$$

For a hydrogenic orbit, Z is constant and

$$\left(\frac{1}{r^3}\right)_{av} = \frac{Z^3}{n^3 a_0^3 l(l + \frac{1}{2})(l + 1)} \tag{5-13b}$$

so that

$$\Delta\nu = \frac{R\alpha^2 Z^4}{n^3 l(l + 1)} \tag{5-13c}$$

where $R = 2\pi^2 m e^4/ch^3$ cm^{-1} [Eq. (5-6)]

$\alpha = 2\pi e^2/hc$ is the fine-structure constant

$a_0 = h^2/4\pi^2 m e^2$ is the Bohr radius of hydrogen

For an alkali-type atom a similar formula might be expected to hold approximately. The quantum number n in (5-13) must again be modified to an effective value $n^* = n - \sigma$. In addition, the Z^4 is not a simply determinable quantity since the electron is affected by Z_o, or Z_{outer}, when it is very far from the nucleus, and Z_i, or Z_{inner}, when it is inside the shells of other electrons and close to the nucleus. A good empirical expression, which can also be justified theoretically, is

$$\Delta\nu = \frac{R\alpha^2 Z_i^2 Z_o^2}{(n^*)^3 l(l + 1)} = \frac{5.83 Z_i^2 Z_o^2}{(n^*)^3 l(l + 1)} \qquad \text{cm}^{-1} \tag{5-14}$$

n^* is the effective principal quantum number which may be evaluated from the term energy and from (5-7). Z_o is the same Z_o as in (5-7), which is the net charge on the atom after removing the valence electron. Z_i, the effective value of Z near the nucleus, may be taken as approximately 4 less than the nuclear charge Z for p electrons. Methods of approximating $\Delta\nu$ and appropriate values of Z_i are discussed by White [53].

Although (5-14) represents as accurate an expression as can usually be obtained for many-electron atoms, the hydrogen case can be treated much more completely. A more exact theory for the hydrogenic case will be discussed below.

5-4. Atoms with More Than One Valence Electron. Many atoms have more than one valence electron, each one of which is characterized by quantum numbers n, l, and s. The angular momenta $\mathbf{l}$ and $\mathbf{s}$ may be coupled in several different ways by the interactions between the electrons. The most common coupling scheme, and one that applies for all light atoms, is known as LS or Russell-Saunders coupling. For this scheme the individual $\mathbf{l}$'s are coupled so that they add vectorially to form a resultant $\mathbf{L}$, and the spins couple to form a resultant $\mathbf{S}$. Finally $\mathbf{L}$ and $\mathbf{S}$ add vectorially to form a total angular momentum $\mathbf{J}$. To the vectors $\mathbf{L}$, $\mathbf{S}$, and $\mathbf{J}$ correspond quantum numbers L, S, and J, while their magnitudes are, respectively, $\sqrt{L(L + 1)}$, $\sqrt{S(S + 1)}$, and $\sqrt{J(J + 1)}$. The state of the atom as a whole is represented by a term symbol quite analogous to that used for the hydrogen atom. In fact the hydrogenlike atoms may be regarded as especially simple cases of LS coupling. A capital letter $S, P, D, F, \ldots$ gives the value of L, corresponding, respectively, to $L = 0, 1, 2, 3, \ldots$. A subscript to the right gives the value of J, while $2S + 1$ is given by a superscript to the left. $2S + 1$ is called the "multiplicity" of the state. This is because, as long as $L \geq S$, J may assume any of the values $L + S, L + S - 1, \ldots, L - S$, or $2S + 1$ different J states are possible. For example, the states ${}^3P_2$, ${}^3P_1$, ${}^3P_0$ have the same L and S, but different J's; they constitute a triplet.

In compounding the angular momenta of the individual electrons, it is usually necessary to consider only those electrons outside closed shells, because for all closed shells L, S, and J are zero. The number of electrons in a closed shell is governed by the Pauli exclusion principle.

In a strong electric or magnetic field, the interactions of the individual electrons with the field are stronger than their interactions with each other. Then in addition to n, l, and s, each electron has quantum numbers m_l and m_s which are, respectively, the projections of l and s on the field direction. m_l can have the values $l, l - 1, \ldots, -l$, while $m_s = +\frac{1}{2}, -\frac{1}{2}$. Pauli's exclusion principle states that only one electron in any one atom can have a given set of the five quantum numbers n, l, s, m_l, m_s. From this it follows that $2(2l + 1)$ electrons may have a given n and l. They form a closed subshell, with zero angular momentum and a spherical distribution of charge. For a given value of n, l may assume the values $0, 1, 2, \ldots, n - 1$. Including all these possible values of l, $2n^2$ electrons have the same value of n, and they constitute a closed shell.

5-5. Selection Rules and Intensities. The intensity of a transition between two states of an atom is proportional to the square of the matrix element of the electric dipole moments between the states, that is,

$$I \propto |\int \psi_2^* ez\psi_1 \, d\tau|^2 \tag{5-15}$$

and to the population of the initial state.

The value of the matrix element involves both the radial and the angular parts of the wave function. The radial part of the wave function depends on the quantum numbers n and l and on the particular atom in question. It is usually difficult to evaluate exactly. However, the angular part of the wave function depends only on the angular momentum quantum numbers and not on the details of the particular atom as long as the electron can be considered to move in a spherical potential. General selection rules may hence be found for the changes in angular momentum quantum numbers allowed during a radiative transition. They are very similar to the selection rules of Sec. 1-4 and may be written

$$L' = L \pm 1, L$$
$$J' = J \pm 1, J \text{ (but } J' \text{ and } J \text{ cannot both be zero)} \tag{5-16}$$
$$\Sigma l_i \text{ changes from odd to even or vice versa}$$
$$\Delta m_J = 0, \pm 1$$

Although these selection rules are very useful in predicting which transitions may occur, they give little information about the absolute or relative intensities of transitions, which often depend on the radial parts of the wave function.

In cases involving a group of transitions with the various initial states and the various final states differing among themselves only in the relative orientation of the angular momentum vectors **L**, **S**, and **J**, the radial part of the matrix element (5-15) is constant. The relative intensities of such transitions then depend only on the angular momentum quantum numbers. A typical example of this type of case is a group of fine-structure components of "one" transition. Their relative intensities are given by [23], [53]:

For transitions $L \leftarrow L - 1$:

$$J \leftarrow J - 1, I = \frac{B(L + J + S + 1)(L + J + S)(L + J - S)(L + J - S - 1)}{J}$$

$$J \leftarrow J, \quad I = -\frac{B(L+J+S+1)(L+J-S)(L-J+S)(L-J-S-1)(2J+1)}{J(J+1)}$$

$$J \leftarrow J + 1, I = \frac{B(L - J + S)(L - J + S - 1)(L - J - S - 1)(L - J - S - 2)}{(J + 1)}$$

$$(5\text{-}17)$$

For transitions $L \leftarrow L$:

$$J \leftarrow J - 1, I = -\frac{A(L + J + S + 1)(L + J - S)(L - J + S + 1)(L - J - S)}{J}$$

$$J \leftarrow J, \quad I = \frac{A[L(L + 1) + J(J + 1) - S(S + 1)]^2(2J + 1)}{J(J + 1)}$$

$$J \leftarrow J + 1, I = -\frac{A(L + J + S + 2)(L + J - S + 1)(L - J + S)(L-J-S-1)}{(J + 1)}$$

$$(5\text{-}18)$$

Expressions (5-17) and (5-18) for the relative intensities take into account both the squares of the matrix element (5-15) and the relative populations of the different levels, assuming that the populations are proportional to the statistical weights or M degeneracy.

In Appendix I, these relative intensities have been tabulated. It should be noted that the lines whose intensities are to be compared must be close together in frequency for Eqs. (5-17) and (5-18) to apply. Otherwise the intensity should be multiplied by a frequency factor which is ν^4 for optical emission lines, or ν^2 for microwave absorption.

These formulas and tables depend only on the magnitude and directions of the vectors which have been added, and not on the specific form of interaction between the vectors. They may therefore be used for transitions in which angular momentum vectors other than L, S, and J are involved if appropriate changes are made in the symbols.

For example, if the nucleus has an angular momentum I, it can interact with the electronic angular momentum J to give a new total angular momentum quantum number F. The energy separations between levels with different values of F give rise to hyperfine structure, which will be considered shortly. Then the relative intensities of different transitions in a hyperfine-structure pattern are given by the same equations (5-17) and (5-18), provided we replace J by F, L by J, and S by I. Here the symbol I should not be confused with the same letter used to denote intensity in Eqs. (5-17) and (5-18).

The general behavior of the relative intensities of fine-structure components may be readily understood. The radiation fields act primarily on the orbital angular momentum and change L by one unit without changing S. This is because the force of the electric component of the field on the electron charge is much greater than magnetic forces acting on the electron spin. If L is increased during the transition to $L + 1$ and S is unaffected in magnitude or direction, then J would be expected to change in approximately the same way as L, or to increase to $J + 1$. It will be observed that the stronger transitions in Appendix I tend to be those for which the change in J has the same value as the change in L. This is most notably true for large values of L, where quantum-mechanical results most closely approximate classical expectations.

5-6. Fine Structure—More Exact Treatment.   Although most atoms are so complex that a theoretical calculation of fine structure which is much more exact than the expressions (5-13) and (5-14) given above is very difficult and even uncertain, the hydrogen atom and hydrogenlike one-electron ions are sufficiently simple to allow a considerably more exact treatment. Because of this simplicity, the hydrogen spectrum, and the hydrogen fine structure in particular, has been much used as a testing ground for atomic theory. Several different treatments of the hydrogen fine structure have met with temporary success, but each in turn has required modification as experimental techniques and theoretical knowledge advanced. The basis of the modern treatment of the hydrogen atom and its fine structure was given in 1928 by Dirac. He proposed a relativistic form of quantum mechanics which inherently provides the electron with the properties previously postulated separately as a spin and magnetic moment.

According to the Dirac theory, the energy levels of a hydrogenlike atom are given by

$$W = mc^2(\{1 + (\alpha Z)^2[n - |K| + (K^2 - \alpha^2 Z^2)^{\frac{1}{2}}]^{-2}\}^{-\frac{1}{2}} - 1) \quad (5\text{-}19)$$

where $|K| = j + \frac{1}{2}$. It may be seen from Eq. (5-19) that the levels with the same value of n and j are degenerate. If this equation is expanded in powers of αZ, it gives

$$\frac{W_{nj}}{hc} = -\frac{Z^2 R}{n^2} - \frac{\alpha^2 Z^4 R}{n^3}\left[(j + \tfrac{1}{2})^{-1} - \left(\frac{3}{4n}\right)\right] + \cdots \quad (5\text{-}20)$$

so that the fine-structure doublet separation between the $^2P_{\frac{3}{2}}$ and $^2P_{\frac{1}{2}}$ levels is

$$\Delta\nu = \frac{R\alpha^2 Z^4}{n^3 l(l + 1)} \quad (5\text{-}21)$$

which agrees with the previous result [Eq. (5-13)], although here it is evident that terms in higher orders of αZ have been neglected. Figure

FIG. 5-6. Fine structure of $n = 2$ levels of hydrogen (Dirac theory). The dotted line indicates the position the levels would have according to the Bohr theory, which does not consider fine structure.

5-6 shows the $n = 2$ levels of the hydrogen atom according to the Dirac theory.

Optical spectroscopic measurements of the fine structure of the hydrogen H_α line ($n = 3$ to $n = 2$), indicated that the structure of the $n = 2$ level of hydrogen was in reasonable, although not precise, agreement with the predictions of the Dirac theory. However, because hydrogen is such a light gas, the atoms have large thermal velocities which, through the Doppler effect, cause considerable broadening of the lines. The resultant blurring of the pattern was sufficient even at liquid-air temperatures to make the disagreement uncertain. There was some indication that the $^2S_{\frac{1}{2}}$ level was displaced upward with respect to the P levels by about 0.03 cm^{-1}, or 1000 Mc [80], [88], [89]. On the other hand, the apparent shifts were attributed by some [97] to impurities in the light source.

As early as 1928 Grotrian [9] had pointed out that the selection rules permitted transitions between states with the same value of n, and that radio waves might be used to induce such transitions. Several attempts were made [31], [60] to observe transitions between the $2s\ ^2S_{\frac{1}{2}}$ and $2p\ ^2P_{\frac{1}{2}}$ levels which on the Dirac theory should be separated by $\Delta\nu = 0.365$ cm^{-1}, or 10,950 Mc. Spark-gap oscillators were used, and the radiation was passed through a hydrogen gas discharge tube. Betz reported absorptions of about 25 per cent at wavelengths of 3, 9, and 27 cm, but these absorptions seem to be impossibly large. Haase [60] found no resonant absorption at all.

It does appear possible to detect this fine-structure splitting by direct microwave absorption. Hydrogen atoms may be excited to the $2s$ $^2S_{\frac{1}{2}}$ or $2p$ $^2P_{\frac{3}{2}}$ states in an electrical discharge in wet hydrogen. From the $^2P_{\frac{3}{2}}$ state they decay rapidly, since this state has a natural lifetime τ_P of only 1.6×10^{-9} sec. The $2s$ $^2S_{\frac{1}{2}}$ state is metastable, however, so that atoms tend to accumulate in it and they may absorb microwave energy by making a transition to the $2p$ $^2P_{\frac{3}{2}}$ state. On the other hand, a relatively small electric field may be expected to produce a Stark effect in the $^2S_{\frac{1}{2}}$ state which would make transitions to the ground state possible and drastically reduce its lifetime. Reesor [631] has looked for direct microwave absorption due to the $2p \leftarrow 2s$ transition in a discharge tube and failed to find it.

The expression (1-49) giving the absorption coefficient in a gas for which both the upper and lower states are populated according to the Boltzmann distribution law needs only a slight modification to apply to this case where all the atoms considered are in the lower state of the transition. It becomes

$$\gamma = \frac{8\pi^2 N}{3hc} \frac{|e\mathbf{r}|^2 \nu \, \Delta\nu}{(\nu - \nu_0)^2 + (\Delta\nu)^2} \tag{5-22}$$

where γ is the absorption coefficient for microwave radiation of frequency ν, ν_0 is the resonant frequency, and $|e\mathbf{r}|$ is the dipole moment matrix element, or e times the matrix element of the coordinate vector $\mathbf{r}$ of the atomic electron for the transition $2s$ $^2S_{\frac{1}{2}} \rightarrow 2p$ $^2P_{\frac{3}{2}}$. The square of the matrix element $|e\mathbf{r}|^2$ for the transition can be calculated as $6a_0^2 e^2$, where $a_0 = h^2/4\pi^2 me^2$ is the radius of the first Bohr orbit for hydrogen. If broadening of the resonance absorption due to collisions can be ignored then there is no appreciable broadening of the $^2S_{\frac{1}{2}}$ state and the half width is just $1/(4\pi\tau_P)$, where τ_P is the natural lifetime of the $^2P_{\frac{3}{2}}$ state, 1.6×10^{-9} sec. This short lifetime gives a half width of about 50 Mc. The number of atoms N in the $2s$ $^2S_{\frac{1}{2}}$ state is very difficult to calculate. However, a rough estimate gives $N = 5 \times 10^{10}$ atoms/cm³ under optimum conditions [484], so that the absorption coefficient becomes, from (5-22),

$$\gamma = 1.6 \times 10^{-4} \text{ cm}^{-1} \tag{5-23}$$

This is a large absorption coefficient by microwave spectroscopic standards, but its detection would be made more difficult by the large half width of the line, which is $1/(4\pi\tau_P) \approx 50$ Mc. The effective absorption might be reduced by transitions in the reverse direction if the $^2P_{\frac{3}{2}}$ state is appreciably populated. A further complication is the background of continuous absorption by free electrons in the discharge tube. This might be of the order of 10^{-4} cm^{-1} in a typical case.

While it seems that the hydrogen fine structure could probably be detected by direct absorption, no experiment of this type has yet been done. An alternative experiment, which is very suitable for an accurate measurement of this resonance, is the atomic-beam method of Lamb and Retherford [484]. Figure 5-7 is a block diagram of their apparatus.

Hydrogen was dissociated in an oven, and collimated by slits to form an atomic beam. Some of the atoms were excited to the metastable $2s\ ^2S_{\frac{1}{2}}$ state by an electron bombarder. Then the beam passed through a region in which a radio-frequency field was applied and finally hit a

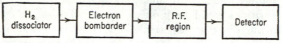

Fig. 5-7. Block diagram of Lamb and Retherford's apparatus for measurement of hydrogen fine structure.

detector. This detector was a tungsten ribbon, from which metastable atoms can eject electrons by giving up their excitation energy. Atoms in the ground state are not detected at all.

When the radio-frequency field of the proper frequency is applied, transitions are induced from the $2s\ ^2S_{\frac{1}{2}}$ to the $2p\ ^2P_{\frac{3}{2}}$ state. From there the atoms quickly decay to the ground state. Since the density in the beam is low there is little imprisonment of resonance radiation to increase the $2p\ ^2P_{\frac{3}{2}}$ population.

To avoid the necessity for varying the radio frequency over a wide range which would make difficult the maintenance of constant radio-frequency power, the transition region was placed in a variable magnetic field. Then the Zeeman effect of the transition was observed and the frequency for zero field was found by extrapolation.

The fine structure of singly ionized helium was also studied by Lamb and Skinner [381], [485]. In this case the decay of the metastable atoms on application of the microwave field was detected by observation of the ultraviolet photons emitted in the transition to the ground state.

In both hydrogen and helium the $2s\ ^2S_{\frac{1}{2}}$ level was found to be higher than the $2p\ ^2P_{\frac{1}{2}}$. However the separation between the levels $2p\ ^2P_{\frac{1}{2}}$ and $2p\ ^2P_{\frac{3}{2}}$ in hydrogen agreed fairly well with expression (5-21) [790]. For hydrogen the measured separation between $2s\ ^2S_{\frac{1}{2}}$ and $2p\ ^2P_{\frac{1}{2}}$ was 1057.777 $\pm$ 0.10 Mc [876]; while in ionized helium the shift was 14,020 $\pm$ 100 Mc for the corresponding level [485]. To achieve this precision in the measurement of the center frequency of a broad line requires a careful consideration of the factors affecting the shape and position of the line [790], [876].

The apparent upward shifts of the $2s\ ^2S_{\frac{1}{2}}$ levels are now explained fairly well by the interaction between the atomic electron and its radiation field [177]. The calculation is very difficult to carry out accurately, but the calculated shift for hydrogen agrees within about 0.5 Mc with the observed shift [876].

Probably the simplest atom from the theoretical point of view is positronium. It consists of a positron and electron only, so that there is no complication from the short-range forces associated with heavy nucleons. This atom is not stable, decaying in about 10^{-8} sec with the annihilation of the particles and emission of either two or three γ rays. Consequently, it is not easy to study experimentally, and was discovered only recently by Deutsch [568a]. Two γ rays are emitted in opposite directions from states with $J = 0$. States with $J = 1$ must emit three photons to conserve angular momentum, and hence have a longer lifetime. If transitions could be induced by a radio-frequency field from the $J = 1$ to the $J = 0$ states, triplet positronium would be converted to singlet, and the transition detected by the increase in double quantum annihilation. For this direct experiment, a frequency near 2×10^5 Mc would be needed.

However, lower frequencies can be used to measure the fine structure of positronium with the aid of the Zeeman effect. In a magnetic field the $J = 1$ state is split into $M = 0$ and $M = \pm 1$ components. The $M = 0$ state acquires some singlet character, so that double quantum annihilation can occur from it. Weinstein, Deutsch, and Brown [997a] have used the annihilation radiation to detect microwave induced transitions from the $J = 1$, $M = \pm 1$ to the $M = 0$ levels. Since this Zeeman splitting depends in part on the ratio of magnetic field to the singlet-triplet separation, this measurement determined the fine structure splitting between the $J = 0$ and 1 states as $(2.0338 \pm 0.0004) \times 10^5$ Mc. The calculated value of 2.0337×10^5 Mc [712a] is in excellent agreement.

A few other fine-structure splittings in atoms can probably be studied by microwave techniques. However, in many cases the fine structure is so large that transition frequencies do not lie in the microwave region, or in other cases the lifetimes of both states are so small that application of microwave techniques is difficult.

5-7. Hyperfine Structure. Atomic nuclei have radii near 10^{-12} cm, and hence are very small compared with the size of electron orbits, which are approximately 10^{-8} cm. Nuclei are also some 10^4 times heavier than electrons. To a good approximation electronic energies can therefore be obtained by considering nuclei to be positive point charges of infinite mass. However, effects on electronic energy levels due to the finite size and mass of nuclei, although small, often appear on careful observation of atomic spectra. They are called hyperfine structure because they produce a very small splitting of atomic lines, usually much smaller than the fine structure.

If the nucleus is to be considered other than a point charge, it must be recognized that the nucleus involves a charge distribution and that this charge distribution may be in motion, producing magnetic fields and giving the nucleus an angular momentum. As for an atom or any other quantum-mechanical system, the angular momentum of the nucleus must be $Ih/2\pi$, where I is an integer or half integer and is usually called the nuclear spin.

A number of types of hyperfine interactions between nuclei and electrons are independent of the relative orientation of nuclear spin I and electronic angular momentum J. These include the small shifts due to the finite nuclear mass, variation of the electron potential from a coulomb potential when electrons are within the nuclear radius, and isotropic (*i.e.*, independent of nuclear orientation) polarization of the nucleus by electron fields. These effects slightly change each electronic energy level but can usually be detected only by examining their variation between two or more isotopes, and hence are called "isotope effects." Thus a given chemical element may produce a number of slightly different superimposed spectra, each associated with a particular isotope of the element. Since the "isotope effects" do not represent small splittings of the energy levels of any one atomic system, but rather small differences between the spectra of different systems, they are not generally observed by microwave spectroscopy.

On the other hand, hyperfine interactions which vary with nuclear orientation give small splittings of electronic energy levels and are often observed with microwave or radio-frequency techniques. These effects may be either electric or magnetic in origin. Although the magnetic effects are usually most prominent in atoms, electric effects predominate in molecules. The electric interactions will be discussed first.

Hyperfine Structure Due to Electric Charge Distribution in the Nucleus. Motion of the center of mass of the nucleus is unchanged for the various possible nuclear orientations; hence it is the natural origin in considering a nucleus of finite size. Let V_0 be the electrostatic potential produced at the nuclear center of mass by all electronic charges in the atom, and $\partial V_0/\partial x$ represent its derivative evaluated at the same point. The electrostatic energy of a charge $\rho(x,y,z)\,\Delta x\,\Delta y\,\Delta z$, where ρ represents the nuclear charge density, is $\Delta W = \rho\,\Delta x\,\Delta y\,\Delta z\,V(x,y,z)$.

Expanding V as a series and writing the volume element $\Delta x\,\Delta y\,\Delta z = \Delta\tau$

$$\Delta W = \rho\,\Delta\tau\left[\, V_0 + x\frac{\partial V_0}{\partial x} + y\frac{\partial V_0}{\partial y} + z\frac{\partial V_0}{\partial z} + \tfrac{1}{2}x^2\frac{\partial^2 V_0}{\partial x^2} + \tfrac{1}{2}y^2\frac{\partial^2 V_0}{\partial y^2} \right.$$
$$+ \tfrac{1}{2}z^2\frac{\partial^2 V_0}{\partial z^2} + xy\frac{\partial^2 V_0}{\partial x\,\partial y} + yz\frac{\partial^2 V_0}{\partial y\,\partial z} + zx\frac{\partial^2 V_0}{\partial x\,\partial z}$$
$$\left. + \cdots \frac{x^n y^m z^p}{n!\,m!\,p!}\frac{\partial^{n+m+p}V_0}{\partial x^n\,\partial y^m\,\partial z^p} + \cdots \right] \qquad (5\text{-}24)$$

Integrating over the entire nuclear volume,

$$W = \int \rho(x,y,z) \left[V_0 + x \frac{\partial V_0}{\partial x} + y \frac{\partial V_0}{\partial y} + z \frac{\partial V_0}{\partial z} + \tfrac{1}{2}x^2 \frac{\partial^2 V_0}{\partial x^2} + \tfrac{1}{2}y^2 \frac{\partial^2 V_0}{\partial y^2} \right.$$
$$\left. + \tfrac{1}{2}z^2 \frac{\partial^2 V_0}{\partial z^2} + xy \frac{\partial^2 V_0}{\partial x \, \partial y} + yz \frac{\partial^2 V_0}{\partial y \, \partial z} + zx \frac{\partial^2 V_0}{\partial z \, \partial x} + \cdots \right] dv \quad (5\text{-}25)$$

The first term may be easily integrated to give ZeV_0, where Z is the atomic number of the nucleus and Ze its total charge. This is the term which is independent of nuclear size or shape. The second term may be written

$$\frac{\partial V_0}{\partial x} \int \rho(x,y,z)x \, dv$$

where the integral is the nuclear dipole moment in the x direction. If this nuclear dipole is not produced by an external field, such as that of the electrons, but is a property only of the nucleus, it can be shown to be zero except in very rare cases. Suppose the wave function and hence the charge distribution for a nucleus is known and the dipole moment in the x direction, $\int \rho x \, dx \, dy \, dz$, has the value μ_x. If the positive directions of the nuclear coordinates x, y, and z are now reversed, a new wave function can be found and a new charge density which is just the same function of the new coordinates x', y', z' as it was of the old coordinates x, y, z. This is possible because, for all known forces within the nucleus, the Hamiltonian or wave equation turns out to depend only on even powers of the coordinates and hence remains unchanged when the signs of all coordinates are reversed. The charge density at x will be replaced by a similar charge density at x' or $-x$. However, the direction of the angular momentum does not change on reversing all coordinates. In the new coordinate system, the dipole moment $\mu_{x'}$ has the same value as before but is oppositely directed, that is, $\mu_{x'} = -\mu_x$. Other nuclear properties, however, including the nuclear angular momentum, will have remained unchanged. We must conclude, therefore, that if the nucleus has a dipole moment in one direction with respect to its angular momentum, there must be a degenerate nuclear state, or one of the same energy, with an oppositely directed dipole. Normally, such identical or degenerate states of the nucleus are not encountered, and hence the nucleus has no inherent dipole moment.* If a nucleus has angular momentum I, there are

---

* A similar proof may be applied to any system, showing that no dipole moments may exist in nature without degeneracy. What is ordinarily referred to as the permanent dipole moment of a molecule in fact does not give a molecule an average dipole moment in one direction unless there is degeneracy or an external field. The dipole moment of a large macroscopic collection of charges may be regarded as existing only because of the very close spacing, and hence effective degeneracy, of the rotational energy levels of such a large system.

$2I + 1$ different possible states having the same energy, corresponding to the different values of M_I, the projection of I on a fixed direction. It might be thought that this is a degeneracy which allows a dipole moment. However, since the angular momentum operator is similar to the Hamiltonian in that it does not change sign when all coordinates are reversed, an argument similar to that above shows that no dipole moment can exist unless the system has two states of the same energy and the same value of M_I. It can thus be shown that all terms of (5-25) involving odd powers of the coordinates will normally be zero. However, terms such as $\int \frac{1}{2}\rho x^2 \, dv$ and $\int \rho xy \, dv$ are not necessarily zero because they do not change sign with reversal of direction of all coordinates. These terms are associated with the quadrupole moment of the nucleus.

Before reexpressing these terms in a more convenient form, it is interesting to note their approximate magnitude. The potential V produced by the electron is e/r_e, where r_e is the distance between electron and nucleus. Hence $\partial^2 V/\partial x^2$ is roughly e/r_e^3. The integral $\int \frac{1}{2}\rho x^2 \, dv$ is of the order $\int \rho r_n^2 \, dv = Zer_n^2$, where r_n is the nuclear radius. Hence the term $\partial^2 V/\partial x^2 \int \frac{1}{2}\rho x^2 \, dv \approx Ze^2 r_n^2/r_e^3$. This might be compared with the first term in our expansion, $ZeV = Ze^2/r_e$, giving the electrostatic energy for a point nucleus. The ratio of the two is r_n^2/r_e^2, or 10^{-8} if r_n is 10^{-12} cm and an average value of 10^{-8} cm is taken for r_e. The usual electrostatic energy is of the order 10^5 cm^{-1}, so the energy associated with the small correction terms of this type is expected to be 0.001 cm^{-1}, or 30 Mc. Still higher-order terms in the expansion (5-24) which are nonzero involve fourth powers of the coordinates [that is, $x^4(\partial^4 V/\partial x^4)$, etc.] They are associated with the nuclear hexadecapole (16-pole) and are expected to be still smaller than the quadrupole terms by a factor of roughly 10^8. In most cases this makes them only a few cycles per second, and too small for present experimental accuracy to detect.

Part of the energy due to terms of (5-25) containing second derivatives of the potential does not vary with nuclear orientation. To eliminate this part we subtract

$$\int \frac{1}{6}\rho(x^2 + y^2 + z^2)\left(\frac{\partial^2 V}{\partial x^2} + \frac{\partial^2 V}{\partial y^2} + \frac{\partial^2 V}{\partial z^2}\right) dv \text{ or } \nabla^2 V \int \frac{1}{6}\rho r^2 \, dv$$

If electrons do not penetrate the nucleus, then $\nabla^2 V$ is zero everywhere ρ is not zero, and this energy integral vanishes. If the electrons do penetrate the nucleus, then this energy represents a deviation from a Coulomb field within the nuclear radius, and is an important part of the atomic isotope shift.

The Electric Quadrupole Moment. Eliminating then the parts of (5-25) which are independent of nuclear orientation and the dipole terms, which have been shown to vanish, the remaining terms are attributable to a nuclear electric quadrupole and may be written

$$W_Q = \tfrac{1}{6} \int \rho \left[(3x^2 - r^2) \frac{\partial^2 V}{\partial x^2} + (3y^2 - r^2) \frac{\partial^2 V}{\partial y^2} + (3z^2 - r^2) \frac{\partial^2 V}{\partial z^2} \right.$$
$$\left. + 6xy \frac{\partial^2 V}{\partial x \, \partial y} + 6yz \frac{\partial^2 V}{\partial y \, \partial z} + 6zx \frac{\partial^2 V}{\partial z \, \partial x} \right] dv \quad (5\text{-}26)$$

or

$$W_Q = -\tfrac{1}{6} \mathbb{Q} : \nabla \mathbf{E} \quad (5\text{-}27)$$

which is the inner product between the quadrupole moment dyadic

$$\mathbb{Q} = \int (3\mathbf{rr} - r^2 \mathbf{1}) \rho \, dx \, dy \, dz \quad (5\text{-}28)$$

and the gradient of the electric field due to the electrons.

The properties of dyadics may be found in [105] or [63]. A dyadic $\mathbf{AB}$ is formed from the two vectors $\mathbf{A} = A_x \mathbf{i} + A_y \mathbf{j} + A_z \mathbf{k} = \sum_n A_n \mathbf{e}_n$. $\mathbf{B} = B_x \mathbf{i} + B_y \mathbf{j} + B_z \mathbf{k} = \sum_n B_n \mathbf{e}_n$, where $\mathbf{e}_n$ represents one of the three unit vectors $\mathbf{i}$, $\mathbf{j}$, or $\mathbf{k}$. The dyadic has nine components and may be written $\sum_{nm} A_n B_m \mathbf{e}_n \, \mathbf{e}_m$. The unit dyadic $\mathbf{1}$ is $\mathbf{ii} + \mathbf{jj} + \mathbf{kk}$, and is said to be diagonal because no "cross terms" of the type $\mathbf{ij}$ or $\mathbf{jk}$ occur. The inner product of two dyadics $\mathbf{AB} : \mathbf{CD}$ is the scalar quantity $\sum_{nm} A_m B_n C_n D_m$, which is analogous to the scalar product of two vectors.

By a proper choice of axes any symmetric dyadic such as the quadrupole moment dyadic may be diagonalized. This eliminates all terms except those multiplying $\mathbf{ii}$, $\mathbf{jj}$, or $\mathbf{kk}$.

The charges in the nucleus are rotating very rapidly about the direction of the nuclear spin. If an average is made over a time long enough for the nuclear particles to rotate many times, but so short that the electrons or charges outside the nucleus have not appreciably changed position, the electric field gradient may be considered constant and the nuclear charge distribution cylindrical. Using a new coordinate system with z_n in the direction of the nuclear spin, all nondiagonal terms of $\mathbb{Q}$ become zero, and the diagonal terms are simply related;

$$\int \rho (3x_n^2 - r^2) \, dv = \int \rho (3y_n^2 - r^2) \, dv = -\tfrac{1}{2} \int \rho (3z_n^2 - r^2) \, dv \quad (5\text{-}29)$$

The entire quadrupole moment dyadic may hence be expressed in terms of one constant, called "the" nuclear quadrupole moment

$$Q = \frac{1}{e} \int \rho (3z_n^2 - r^2) \, dx \, dy \, dz \quad (5\text{-}30)$$

where e is the charge of one proton. From (5-30) it can be seen that a nucleus whose charge distribution is spherical has zero quadrupole moment, for then the average value of $3z_n^2$ is just equal to the average value of $r^2 = x_n^2 + y_n^2 + z_n^2$. The quadrupole moment may be considered then a measure of the deviation of the nuclear charge from spherical shape. If the charge distribution is somewhat elongated along the nuclear axis z_n, then Q is positive; if it is flattened along the nuclear axis, Q is negative. From (5-26) the quadrupole energy becomes

$$W_Q = \frac{e}{6} Q \left[\frac{\partial^2 V}{\partial z_n^2} - \frac{1}{2} \left(\frac{\partial^2 V}{\partial x_n^2} + \frac{\partial^2 V}{\partial y_n^2} \right) \right] \tag{5-31}$$

If the potential V is due entirely to charges outside the nucleus, then

$$\frac{\partial^2 V}{\partial x_n^2} + \frac{\partial^2 V}{\partial y_n^2} = - \frac{\partial^2 V}{\partial z_n^2}$$

from Laplace's equation, and

$$W_Q = \frac{e}{4} Q \frac{\partial^2 V}{\partial z_n^2} \tag{5-32}$$

The potential V is produced by electrons which are in rapid motion, so rapid that the nuclear axis z_n may be considered stationary during the time that the electrons traverse their entire orbits, or take up all possible positions. Hence, (5-31) may be averaged over all possible electron positions

$$W_Q = \frac{e}{6} Q \left[\frac{\partial^2 V}{\partial z_n^2} - \frac{1}{2} \left(\frac{\partial^2 V}{\partial x_n^2} + \frac{\partial^2 V}{\partial y_n^2} \right) \right]_{\text{av}} \tag{5-33}$$

or, using Laplace's equation again

$$W_Q = \frac{e}{4} Q \left(\frac{\partial^2 V}{\partial z_n^2} \right)_{\text{av}} \tag{5-34}$$

If the average electron charge density is spherical, then

$$\frac{\partial^2 V}{\partial x_n^2} = \frac{\partial^2 V}{\partial y_n^2} = \frac{\partial^2 V}{\partial z_n^2} \qquad \text{and} \qquad W_Q = 0$$

Since only s electrons, which have spherically symmetric distributions, have large probabilities of being found within the nucleus, it is customary to set

$$W_Q = \frac{e}{4} Q \left(\frac{\partial^2 V'}{\partial z_n^2} \right)_{\text{av}} \tag{5-35}$$

where V' is the potential due only to the electronic distribution outside a small sphere surrounding the nucleus. This gives a small error, because p or d electrons, which are not spherically distributed, have a finite, though small, probability of being inside the nucleus. The density of a nonspherically distributed p electron must, however, be zero at the center of the nucleus, and its average density within the nucleus is given in order of magnitude by $e(r^2/r_e^5)$, where r is the distance from the center of the nucleus and r_e is the radius of the electron orbit. Hence from Poisson's equation $(\partial^2 V/\partial x_n^2)$ or $(\partial^2 V/\partial y_n^2)$ due to this p electron cannot be much greater than $e(r_n^2/r_e^5)$. The neglected contribution to the energy is therefore of the order

$$eQ\,\frac{er_n^2}{r_e^5} \approx \frac{e^2 r_n^4}{r_e^5}$$

which is of the size of the hexadecapole energy discussed above and usually unobservably small. We shall henceforth define the quadrupole energy without this small contribution as

$$\frac{eQ}{4}\left(\frac{\partial^2 V'}{\partial z_n^2}\right)_{\mathrm{av}}$$

or, omitting the prime,

$$\frac{eQ}{4}\left(\frac{\partial^2 V}{\partial z_n^2}\right)_{\mathrm{av}}$$

We turn now to an examination of the gradient of the electric field, $(\partial^2 V/\partial z_n^2)_{\mathrm{av}}$, along the nuclear axis produced by the electrons. About an axis parallel to the electronic angular momentum, $\mathbf{J}(h/2\pi)$ or J, the average electric field is cylindrically symmetric, and the dyadic or tensor $(-\nabla\mathbf{E})_{\mathrm{av}}$ is hence diagonal with components

$$\left(\frac{\partial^2 V}{\partial x_1^2}\right)_{\mathrm{av}} = \left(\frac{\partial^2 V}{\partial y_1^2}\right)_{\mathrm{av}} = \left(-\frac{1}{2}\frac{\partial^2 V}{\partial z_1^2}\right)_{\mathrm{av}}$$

Here z_1^{-} is along the axis of electronic angular momentum. These quantities, which are elements of a tensor, can easily be transformed to the nuclear coordinate system. If θ is the angle between I and J or z_n and z_1, and if x_1 and x_n are chosen parallel to each other,

$$\begin{aligned}
\left(\frac{\partial^2 V}{\partial z_n^2}\right)_{\mathrm{av}} &= \sin^2\theta\left(\frac{\partial^2 V}{\partial y_1^2}\right)_{\mathrm{av}} + \cos^2\theta\left(\frac{\partial^2 V}{\partial z_1^2}\right)_{\mathrm{av}} \\
&= \frac{3\cos^2\theta - 1}{2}\left(\frac{\partial^2 V}{\partial z_1^2}\right)_{\mathrm{av}}
\end{aligned} \tag{5-36}$$

defining $(\partial^2 V/\partial z_1^2)_{\mathrm{av}} \equiv q_J$, the quadrupole energy becomes

$$W_Q = \frac{eq_J Q}{4}\frac{3\cos^2\theta - 1}{2} \tag{5-37}$$

Expression (5-37) is the correct classical expression for quadrupole energy, but since the quantum numbers I and J are usually small, only a quantum-mechanical derivation, first given by Casimir [70], can be relied on for an accurate expression. To treat the energy quantum-mechanically, the Hamiltonian may be taken from Eq. (5-27) as

$$H = -\frac{1}{6}\,\mathbf{Q}\!:\!\nabla\mathbf{E}$$

Q and ∇E can be replaced by operators having known eigenvalues for the usual nuclear and electronic wave functions by the following method. The operator $\frac{3}{2}(\mathbf{II} + \widetilde{\mathbf{II}}) - I^2\mathbf{1}$ has the same angular dependence with respect to nuclear orientation as Q, which is proportional to $3\mathbf{rr} - r^2\mathbf{1}$. Hence its matrix elements between states of various orientations must be identical with Q except for a proportionality constant.*

$$(Im_I|Q|Im_I) = \text{const } [Im_I|\tfrac{3}{2}(\mathbf{II} + \widetilde{\mathbf{II}}) - I^2\mathbf{1}|Im_I] \qquad (5\text{-}38)$$

where I, m_I are quantum numbers for the nuclear spin and the component of the nuclear spin in a direction z fixed in space. The symmetrized expression $\frac{3}{2}(\mathbf{II} + \widetilde{\mathbf{II}})$, where $\widetilde{\mathbf{II}}$ indicates the transpose of $\mathbf{II}$, is used because Q is a symmetric operator. To evaluate the proportionality constant, consider the zz component for the state $m_I = I$

$$\begin{aligned}(II|Q_{zz}|II) &= \text{const } (II|3I_z^2 - I^2|II) \\ &= \text{const } [3I^2 - I(I+1)]\end{aligned} \qquad (5\text{-}39)$$

from the known expectation values for I_z^2 and I^2. The quantity

$$(II|Q_{zz}|II)$$

corresponds at least approximately to the classical quadrupole moment eQ, and will be so defined. Hence $eQ = \text{const } I(2I - 1)$, or

$$Q_{op} = \frac{eQ}{I(2I-1)}\left[\tfrac{3}{2}(\mathbf{II} + \widetilde{\mathbf{II}}) - I^2\mathbf{1}\right] \qquad (5\text{-}40)$$

Similarly, it can be shown that

$$(\nabla E)_{op} = \frac{q_J}{J(2J-1)}\left[\tfrac{3}{2}(\mathbf{JJ} + \widetilde{\mathbf{JJ}}) - J^2\mathbf{1}\right] \qquad (5\text{-}41)$$

where $q_J = \left(JJ\left|\dfrac{\partial^2 V}{\partial z^2}\right|JJ\right)$. Expression (5-41) applies only to the common case where J is a "good" quantum number.

Since the electronic potential at the nucleus due to charge e is

$$\frac{e}{r} = \frac{e}{\sqrt{x^2 + y^2 + z^2}}$$

where r is the distance from nucleus to charge,

$$\frac{\partial^2 V}{\partial z^2} = e\,\frac{3z^2 - r^2}{r^5} = e\,\frac{3\cos^2\theta - 1}{r^3}$$

where θ is the angle between the z axis fixed in space and the radius vector $\mathbf{r}$. This gives

$$q_J = \int \rho_{JJ}\,\frac{3\cos^2\theta - 1}{r^3}\,d\tau = \left(\frac{\partial^2 V}{\partial z_1^2}\right)_{av} \qquad (5\text{-}42)$$

* For a more complete discussion, see [969a], p. 16.

where ρ_{JJ} indicates the electron charge density for a state $m_J = J$. If only one electron of wave function ψ is included,

$$q_J = e \int \psi_{JJ}^* \frac{3\cos^2\theta - 1}{r^3} \psi_{JJ}\, d\tau \qquad (5\text{-}43)$$

The operator for the quadrupole energy is $W_{op} = -\frac{1}{6}\mathsf{Q}_{op} : (\nabla\mathbf{E})_{op}$. In case J is a constant or "good" quantum number, this becomes from (5-40) and (5-41) and use of the commutation rules for the components of angular momenta

$$\frac{1}{2} \frac{eq_J Q}{I(2I - 1)J(2J - 1)} [3(\mathbf{I}\cdot\mathbf{J})^2 + \tfrac{3}{2}(\mathbf{I}\cdot\mathbf{J}) - I^2 J^2] \qquad (5\text{-}44)$$

If terms off diagonal in J must be considered, the basic form of W_{op} above must be used. It is easily shown that, if I and J are allowed to be very large, (5-44) becomes equivalent to the classical expression (5-37), since $\mathbf{I}\cdot\mathbf{J}$ becomes $IJ\cos\theta$.

The total angular momentum of the system in units $h/2\pi$ may be written $\mathbf{F} = \mathbf{I} + \mathbf{J}$ and is of course constant, as is its projection

$$m_F = m_I + m_J$$

on the fixed axis. A representation F, m_F is hence appropriate, and in this representation I^2, J^2, and $\mathbf{I}\cdot\mathbf{J}$ are diagonal so that the expectation values of (5-44) can be easily determined. The diagonal elements for these operators are $I(I + 1)$, $J(J + 1)$, and $C/2$, respectively, where

$$C = F(F + 1) - I(I + 1) - J(J + 1)$$

Hence

$$W_Q = \frac{1}{2} \frac{eq_J Q}{I(2I - 1)J(2J - 1)} [\tfrac{3}{4}C(C + 1) - I(I + 1)J(J + 1)] \qquad (5\text{-}45)$$

From expression (5-45) hyperfine splitting of energy levels due to nuclear quadrupole effects may be calculated. The magnitude of the quadrupole effects determined by the quadrupole coupling constant eqQ, which involves both the nuclear quadrupole moment Q, and q_J, the second derivative of the potential along the axis of electronic angular momentum. This coupling constant may vary from zero to a few tenths of a wave number, or thousands of megacycles, and of course can be either positive or negative. Evaluation of the quantity q_J will be discussed at some length in Chap. 9. If q_J is known, then determination of the quadrupole coupling constant allows a determination of the nuclear constant Q.

Nuclear Polarizability. Another type of electrostatic interaction is an electric polarization of the nucleus. The strong electrostatic fields due to atomic electrons induce a small electric dipole moment on the nucleus which slightly increases the force of attraction between electron and nucleus, thus lowering the electronic energy levels by an amount which depends on the nuclear polarizability and the square of the electric field strength.

The nuclear polarizability α_z along the spin axis is not necessarily the same as its polarizability α_x perpendicular to this axis. Hence the energy depends on the orientation of the spin with respect to the electric field. Since the polarizability is a symmetric tensor just as is the quadrupole moment, it is not difficult to show that the energy of polarization has the same dependence on nuclear orientation as that due to a quadrupole moment. Gunther-Mohr, Geschwind, and Townes [583] have given the variation of energy of polarization with angle as

$$W_A = \frac{e}{3} \frac{(\alpha_z - \alpha_x)p_J}{I(2I-1)J(2J-1)} \left[\tfrac{3}{4}C(C+1) - I(I+1)J(J+1)\right] \quad (5\text{-}46)$$

This expression is very similar to the energy (5-45) due to a quadrupole moment. Here p_J corresponds to q_J, and is defined in analogy with (5-42) as

$$p_J = \int \rho_{JJ} \frac{3\cos^2\theta - 1}{r^4} d\tau = \frac{2}{e}(E_z^2 - E_x^2)_{\text{av}} \quad (5\text{-}47)$$

The coupling constant $ep_J(\alpha_z - \alpha_x)$ is equivalent to $\tfrac{3}{2}eq_JQ$, and in many cases these two are experimentally indistinguishable. The difference $\alpha_z - \alpha_x$ in classical polarizabilities can be expressed quantum-mechanically as

$$\alpha_z - \alpha_x = \frac{2I(I+1)}{2I-1} \sum_n \frac{|\mu_{0n}|^2_{M=I} - |\mu_{0n}|^2_{M=I-1}}{W_n - W_0} \quad (5\text{-}48)$$

where $|\mu_{0n}|$ is the z component of the electric dipole matrix element between the ground state of energy W_0 and an excited state of energy W_n. Subscripts $M = I$ and $M = I - 1$ indicate that the matrix elements are states with the projection M of $\mathbf{I}$ on the z direction equal to I and $I - 1$, respectively.

If the nuclear matrix elements and energy levels were known, the polarizabilities could be easily calculated. However, generally only rough magnitudes for these quantities are known. $|\mu_{0n}|$ may be taken approximately equal to one protonic charge times one nuclear radius, and $W_n - W_0$ as 1 Mev to roughly evaluate $\alpha_z - \alpha_x$. The quantity $\alpha_z - \alpha_x$ may also be taken as approximately equal to the nuclear volume. From such estimates and evaluation of p from Hartree wave functions, it may be shown that the anisotropic polarization effects are usually about 1 per cent of the nuclear quadrupole effects, hence generally not larger than a few megacycles. They can be experimentally distinguished from nuclear quadrupole effects only because they depend in a different way on the electron-nuclear distance r. Thus the relative size of quadrupole coupling and polarization energy will be different in different electronic states. If precision measurements of hyperfine-structure splittings could be made in several electronic states of an atom with two isotopes, the ratio of the splittings for the two isotopes would vary from state to state. Because it is not usually possible to measure precisely the hyperfine structure of excited states, this method is not very practical.

However, in different types of molecules the electronic configuration about a given atom may be quite different. Then if nuclear polarization is appreciable the apparent quadrupole coupling ratio between two isotopes would depend on the molecule in which it is measured.   Certain variations in the ratios of quadrupole moments for Cl^{35} and Cl^{37} have been observed but cannot clearly be attributed to nuclear polarizability ([575], [760]).

Magnetic Hyperfine Structure. Another type of hyperfine structure which depends on the nuclear orientation is connected with magnetic interactions between the nucleus and atomic electrons. If the structure of the nucleus and the possibility of having circulating charges within the nucleus is considered, it is not surprising that this magnetic hyperfine structure shows that a magnetic dipole moment μ must be attributed to the nucleus. Each possible orientation of the nuclear spin has a slightly different energy due to interaction between the nuclear magnetic moment and the magnetic field at the nucleus produced by the spins and orbital motions of surrounding electrons.

The electrons precess about their total angular momentum $\mathbf{J}$ so that the currents and magnetic fields must on the average be cylindrically symmetric about the direction of $\mathbf{J}$, and hence the magnetic field they produce at the nucleus is parallel to $\mathbf{J}$.   For similar reasons the magnetic moment μ_I of the nucleus is parallel to its spin $\mathbf{I}$. The energy of interaction is $\mu_I H \cos \theta$ or $\mathbf{\mu}_I \cdot \mathbf{H}$, which may be written

$$W = a\mathbf{I} \cdot \mathbf{J} \qquad (5\text{-}49)$$

since $\mathbf{\mu}_I$ is parallel to $\mathbf{I}$ and $\mathbf{H}$ to $\mathbf{J}$. The quantity a is a constant for a given electronic state and nucleus, and is known as the interval factor.

The quantity $\mathbf{I} \cdot \mathbf{J}$ involves the cosine of the angle between $\mathbf{I}$ and $\mathbf{J}$, and can easily be obtained from the vector model [*cf.* Eq. (5-12)], so that

$$W = \frac{a}{2} [F(F + 1) - J(J + 1) - I(I + 1)] \qquad (5\text{-}50)$$

where F is the magnitude of the vector $\mathbf{F} = \mathbf{I} + \mathbf{J}$. F may take on the values $I + J, I + J - 1, \ldots, |I - J|$. The total number of different values of F is $2I + 1$ if $I < J$, or otherwise $2J + 1$.

If the angular momenta of a number of electrons add up vectorially to a resultant angular momentum of zero, as in the case of a closed shell of electrons, then the average magnetic field at the nucleus is zero. Hence in evaluating the constant a, it is necessary to take into account only the unfilled shell of electrons, which in many common cases may consist of only one electron, or a closed shell minus one electron. Although a nonrelativistic treatment of the magnetic field H at the nucleus gives a good approximation for hyperfine structure produced by electrons with orbital angular momentum, for s electrons relativistic theory (summarized in [67]) is necessary. Hyperfine structure is particularly important and large for s electrons, since they penetrate most closely to the nucleus. For non-s electrons (*i.e.*, with $l > 0$) in a hydrogenic atom,

$$a = \frac{2\mu_I\mu_0}{I}\left(\frac{1}{r^3}\right)_{av}\frac{l(l+1)}{j(j+1)} = \frac{g(I)}{1836}\frac{e^2h^2}{8\pi^2m^2c^2}\left(\frac{1}{r^3}\right)_{av}\frac{l(l+1)}{j(j+1)} \quad (5\text{-}51)$$

where $g(I)$ is the "nuclear g factor," *i.e.*, the ratio of the nuclear magnetic moment in nuclear magnetons to its angular momentum in units of $h/2\pi$. One nuclear magneton $= he_P/4\pi M_Pc$, where M_P is the mass of the proton and e_P its charge. Since M_P is 1836 times the mass of the electron, the nuclear magneton is 1836 times smaller than the Bohr magneton.

Substituting the quantum-mechanical value of $\left(\dfrac{1}{r^3}\right)_{av}$ [Eq. (5-13b)]

$$
\begin{aligned}
a &= \frac{g(I)}{1836}\frac{R\alpha^2Z^3}{n^3(l+\frac{1}{2})j(j+1)} \\
&= \frac{g(I)}{1836}\frac{\Delta\nu\,l(l+1)}{Z(l+\frac{1}{2})j(j+1)}
\end{aligned} \quad (5\text{-}52)
$$

where $\Delta\nu$ is the fine-structure doublet splitting given by Eq. (5-13c).

Expression (5-52) follows either from a nonrelativistic calculation which treats the electrons as point particles with electric charge and a magnetic dipole moment, or from a semirelativistic calculation which neglects the electron's binding energy in comparison with its rest mass. A more exact relativistic treatment [19], [28] gives

$$a = \frac{g(I)}{1836}\frac{\Delta\nu\,l(l+1)}{Z(l+\frac{1}{2})j(j+1)}\frac{\kappa}{\lambda} \quad (5\text{-}53)$$

where $\kappa = \dfrac{4j(j+\frac{1}{2})(j+1)}{(4\rho^2-1)\rho}$

$\rho = \sqrt{(j+\frac{1}{2})^2 - (\alpha Z_i)^2}$

$\lambda = \left[\dfrac{2l(l+1)}{(\alpha Z_i)^2}\right]\{[(l+1)^2 - (\alpha Z_i)^2]^{\frac{1}{2}} - 1 - [l^2 - (\alpha Z_i)^2]^{\frac{1}{2}}\}$

The values of λ and κ are tabulated by Goudsmit [41].

Since ρ depends on j, κ is quite different for $p_{\frac{1}{2}}$ and $p_{\frac{3}{2}}$ states. Thus the relativistic correction makes the constant larger for a $p_{\frac{1}{2}}$ electron than for a $p_{\frac{3}{2}}$ electron. This is because, in a relativistic treatment, spin and orbital angular momentum are not sharply separated. Thus a $p_{\frac{1}{2}}$ electron has some of the character of an $s_{\frac{1}{2}}$ electron (the only other type for which

$j = \frac{1}{2}$). We shall see that an $s_{\frac{1}{2}}$ electron has a large interaction constant. The relativistic theory shows that for an s electron

$$a = \frac{g(I)}{1836} \frac{e^2 h^2}{3\pi m^2 c^2} \psi^2(0)\kappa \qquad (5\text{-}54)$$

$\psi^2(0)$ is the electron density at the nucleus, or the square of the non-relativistic Schrödinger wave function at $r = 0$, the center of the nucleus. For a hydrogenic atom $\psi^2(0) = Z^3/\pi a_o^3 n^3$, so that

$$a = \frac{g(I)}{1836} \frac{e^2 h^2 Z^3 \kappa}{3\pi^2 m^2 c^2 a_o^3 n^3} \qquad (5\text{-}55)$$

$$a = \frac{g(I)}{1836} \frac{8 R \alpha^2 Z^3 \kappa}{3n^3} \qquad (5\text{-}56)$$

It is interesting that this expression for an s electron is just what would be obtained by more or less arbitrarily using expression (5-52) for a and setting $l = 0$, $j = \frac{1}{2}$, assuming the relativistic correction κ is unity.

Interaction between atomic magnetic fields and a nuclear magnetic octupole moment has recently been detected by Jaccarino, King, Satten, and Stroke [942]. A rough estimate of the magnitude of magnetic octupole interaction would indicate that it is smaller than the dipole interaction by the square of the ratio of nuclear radius to the distance between nucleus and electron [cf. discussion of hexadecapole moment above]. However, the magnetic octupole actually gives effects somewhat larger than such an estimate. In the case of atomic I and In the effects are many kilocycles in size and have been detected by molecular beam techniques [942] [953].

General Considerations about the Existence of Nuclear Moments. If a nucleus has spin I, the pole of highest order which can occur is given by 2^{2I}. Thus, for $I = 0$, no dipole or quadrupole moment may exist, but only a monopole (charge). If $I = \frac{1}{2}$, a dipole moment may exist, but not a quadrupole moment, which occurs only if $I \geq 1$. This limit to the order of poles which may occur can be proved quite generally, but we shall only attempt an indication of why it occurs. In an external field, a nucleus of spin I can have $2I + 1$ different orientations or states and hence $2I + 1$ different energies. In order to specify these energies completely, only $2I + 1$ different constants of the nucleus need be given. Thus, if $I = 0$, there is no need for more than one constant, the monopole strength or electric charge (no magnetic "charge" exists). If $I = \frac{1}{2}$, two states occur, and one need specify only the monopole and dipole strength. When $I \geq 1$, a quadrupole moment is needed, etc. Additional discussion of this limit can be found in [969a].

Poles of various orders alternate between electric and magnetic type for reasons of symmetry. Thus, as shown above, electric monopoles and quadrupoles exist, but ordinarily electric dipoles do not occur. However, magnetic dipoles and octupoles are permitted.

5-8. Penetrating Orbits. If the state of an electron is not well approximated by a hydrogenlike wave function, as in the cases of valence electrons which penetrate a closed shell of electrons around the nucleus, an exact expression for the interval factor a is very difficult to calculate. However, on the basis of some approximate models, the following expressions for a may be obtained for these cases.

For a non-s electron [41]:

$$a = \frac{g(I)}{1836} \frac{\Delta\nu}{Z_i} \frac{l(l+1)\kappa}{(l+\frac{1}{2})j(j+1)\lambda} \tag{5-57}$$

and for an s electron [40]

$$a = \frac{g(I)}{1836} \frac{8R\alpha^2 Z_i Z_o^2}{3n^{*3}} \kappa \left(1 - \frac{d\sigma}{dn}\right) \tag{5-58}$$

where n^* is the effective principal quantum number and $\sigma = n - n^*$ is the quantum defect. If the energy levels of the atom satisfy a Rydberg-Ritz equation,

$$T = \frac{RZ_o^2}{(n - \alpha - \beta T)^2} \tag{5-59}$$

where α and β are constants, T is the term value, and R is the Rydberg constant, then [359]

$$\frac{d\sigma}{dn} = \frac{\beta}{\beta - n^*/2T} \tag{5-60}$$

In the above equations, for s electrons $Z_i = Z$, the nuclear charge. The equation then works well for elements of medium weight but gives values of a which may be 10 to 20 per cent high for very light or very heavy elements. For heavy elements an additional correction for the finite radius of the nucleus should be made, and with this addition the formula is quite accurate [359].

For p electrons, it is usual to put $Z_i = Z - 4$ in place of Z in (5-52). This works well in the equation for $\Delta\nu$, the fine-structure splitting, but is not so good for hyperfine structure where the average of a different power of Z is involved (*cf.* [364]). If the nuclear moment and so $g(I)$ is known, it is also possible to use the observed hyperfine-structure interval factor, a, to evaluate $(1/r^3)_{\mathrm{av}}$ [67]. For a non-s electron, from Eq. (5-53), with relativistic corrections,

$$\left(\frac{1}{r^3}\right)_{\mathrm{av}} = \frac{1836}{g(I)} \frac{8\pi^2 m^2 c^2}{e^2 h^2} \frac{j(j+1)}{l(l+1)} \frac{\lambda}{\kappa} a \tag{5-61}$$

This value of $(1/r^3)_{\mathrm{av}}$ may then be used in evaluation of nuclear quadrupole moments from observed quadrupole coupling energies.

5-9. Zeeman Effects for Atoms. When an atom is placed in a magnetic field, the energy levels undergo a splitting known as the Zeeman effect (*cf.* Chap. 11 for Zeeman effects on molecules).   It is convenient to distinguish three cases: (a) a weak magnetic field, where this splitting is considerably less than the hyperfine structure, (b) a strong field, where the splitting is much larger than the hyperfine structure, and (c) intermediate fields.

In the weak-field case, the nuclear spin I remains coupled to the electronic angular momentum J, and their resultant F has $2F + 1$ possible values of the component, M_F, along the field direction.   Then the energy due to magnetic hyperfine structure and interactions with the magnetic field is [*cf.* (11-13)]

$$W(F,M_F) = \frac{a}{2}[F(F + 1) - I(I + 1) - J(J + 1)]$$
$$- \left\{\frac{\mu_I}{I}[I(I + 1) + F(F + 1) - J(J + 1)]\right.$$
$$\left. + \frac{\mu_J}{J}[F(F + 1) + J(J + 1) - I(I + 1)]\right\} \frac{M_F H}{2F(F + 1)} \quad (5\text{-}62)$$

where a is a constant which gives the strength of the magnetic hyperfine structure, μ_I is the nuclear magnetic moment, μ_J the atomic (electron spin and orbital combined) magnetic moment, and H the applied magnetic field.

In a very strong field, I and J interact more strongly with the field than with each other.   Then

$$W(I,J,M_I,M_J) = aM_I M_J - \frac{\mu_J}{J} H M_J - \frac{\mu_I}{I} H M_I \quad (5\text{-}63)$$

where M_I and M_J are the quantum numbers for the projection of I and J, respectively, on H.   The intermediate field case is generally more complicated.   For the important special case of $J = \frac{1}{2}$ (*e.g.*, hydrogen, the alkalies, silver, gold, indium, thallium) the energies are given by Breit and Rabi [19a]

$$W(F,M_F) = -\frac{\Delta W}{2(2I + 1)} - \frac{\mu_I}{I} H M_F \pm \frac{\Delta W}{2}\sqrt{1 + \frac{4M_F}{2I + 1}x + x^2}$$
$$(5\text{-}64)$$

where $\Delta W \equiv (a/2)(2I + 1) \equiv h\,\Delta\nu$, and $\Delta\nu$ is the zero-field hyperfine structure splitting

$$x = \frac{(-\mu_J/J + \mu_I/I)}{\Delta W} H$$

Zeeman effects in atoms are often relatively large (several megacycles per gauss).   It is then possible to measure transitions between Zeeman components of several hyperfine levels by varying the applied magnetic field until the transitions coincide with a given microwave frequency.

When a resonant cavity spectrograph is employed, this possibility is especially useful.

Beringer and Heald [898] obtained an accurate measurement of the Zeeman splitting in atomic hydrogen, using a frequency near 9500 Mc and a variable magnetic field. The hydrogen was dissociated by a discharge just before passing through the cavity resonator. From their measurements, and the molecular-beam measurement of the zero field hyperfine structure splitting [737a], the electron-spin magnetic moment g factor is found to be $g_s = -2(1.001148 \pm 0.000006)$. The precision of this value was not limited by the microwave spectrum, but by the absolute calibration of magnetic fields in terms of proton resonance frequencies. A similar technique has been applied to the atoms O [739a], N [932a], and P [913b].

5-10. Microwave Studies of Atomic Hyperfine Structure. Since the separations between hyperfine-structure levels often lie in the microwave range, it is possible to use microwaves to induce transitions between these levels. The electric dipole matrix element between these states vanishes because they belong to the same electronic configuration. However, the magnetic dipole matrix element is not zero and it permits the transition to take place. The peak intensity for a transition in which $\Delta F = \pm 1$ is then given by Eq. (1-50), where in this case μ_{ij} is the appropriate matrix element for the magnetic dipole moment of the atom.

The transitions are most likely to be found in the microwave region for an atom in a $^2S_\frac{1}{2}$ ground state, because the largest number of atoms would occur in the ground state, and a $^2S_\frac{1}{2}$ state has a relatively large hyperfine structure. In that case, the matrix elements are ([56], pp. 64–72), for $F = I + \frac{1}{2} \leftarrow F = I - \frac{1}{2}$,

$$|\mu_{ij}|^2 = \frac{4[(I + \frac{1}{2})^2 - m_F^2]}{(2I + 1)^2} \mu_0^2 \qquad \text{when } \Delta m_F = 0 \qquad (5\text{-}65)$$

$$|\mu_{ij}|^2 = \frac{2(I + \frac{1}{2} \pm m_F)(I + \frac{3}{2} \pm m_F)}{(2I + 1)^2} \mu_0^2 \qquad \text{when } \Delta m_F = \pm 1 \qquad (5\text{-}66)$$

where m_F = projection of the total angular momentum, F, on a fixed direction

μ_0 = the Bohr magneton, $he/4\pi mc$.

The transitions with $\Delta m_F = \pm 1$ are polarized so that the magnetic vector is perpendicular to the fixed direction and those with $\Delta m_F = 0$ are polarized with the magnetic vector parallel to the fixed direction.

If the above matrix elements are substituted into (1-50) to obtain the absorption coefficient for each component of the transition ($\Delta m_F = 0$, $\Delta m_F = +1$, or $\Delta m_F = -1$), then $Nf/3$ in this formula must be interpreted as the number of atoms in the ground state of each component. In case the atom is in an external magnetic field, each value of m_F has a slightly different energy, and the intensity of individual components is needed.   If, however, there is no external magnetic field for one polarization of the incident radiation, then what is wanted is the average of all transitions for which $\Delta m_F = 0$ or $\Delta m_F = \pm 1$.   This average, after multiplication by 3 to obtain the sum of the squares of the dipole matrix elements for all three directions of polarization, is

$$|\mu_{ij}|^2_{\text{av}} = \frac{4(I + 1)}{2I + 1}\, \mu_0^2 \tag{5-67}$$

Expression (5-67) is the appropriate quantity to insert in the customary way into (1-50) if Nf is taken to be the fraction of atoms in the state $F = I - \frac{1}{2}$.

The hyperfine structure of the $^2S_{\frac{1}{2}}$ ground state of cesium was investigated by Roberts, Beers, and Hill [405].   In that case, using the notation of Eq. (1-50) and calculating the intensity of the entire transition,

$Nf = 2.5 \times 10^{14}$ atoms/cm^3 corresponding to a pressure of 3×10^{-2} mm

$T = 500°$K

$\Delta\nu = 1.5 \times 10^5$ cycles/sec (estimated roughly from kinetic theory)

$\nu_0 = 9.2 \times 10^9$ cycles/sec

$I = \frac{7}{2}$

$|\mu_{ij}|^2_{\text{av}} = \frac{9}{4}\mu_0^2$

so that $\gamma_{\text{max}} = 6 \times 10^{-9}$ cm^{-1} for an average component.   The cesium was placed in a microwave resonant cavity which was used to control the frequency of a klystron oscillator.   The cavity was in a variable magnetic field so that each component could be brought in turn to the resonant frequency of the cavity.   As the magnetic field was varied to make a component of the line approach the resonant frequency of the cavity, this resonant frequency was slightly changed by the anomalous dispersion associated with the cesium resonance, and the consequent variation in frequency of the controlled oscillator was detected.

Few other atoms have been investigated by microwave absorption spectroscopy because of the relatively weak absorptions and the difficulty of obtaining many materials in atomic form.   However, Shimoda and Nishikawa [642] obtained a measurement of transitions between hyperfine components of Na^{23} at 1772 Mc.   Hyperfine interactions in H [737a], N [932a], and P [913b] have been obtained from microwave transitions between Zeeman components (see Sec. 5.9).   A large number of atoms have been investigated by molecular-beam techniques (see [969a]), which are particularly suited for this purpose.

5-11. Microwave Spectra from Astronomical Sources.   Microwave radiation due to transitions between the hyperfine components of atomic hydrogen in interstellar space was first detected by Ewen and Purcell [571] and independently discovered by Muller and Oort [621].   This radiation has a wavelength near 21 cm and penetrates the earth's ionosphere and gas and dust particles in interstellar space rather readily. The frequency corresponding to the transition between hyperfine components of hydrogen has been measured in the laboratory as 1420.405 Mc [737a]. Near this frequency, hydrogen gas has a large enough absorption coefficient to be opaque for certain directions through our own galaxy, the Milky Way. Hence it also radiates an intensity corresponding to a black body at about 100°K, which is the effective temperature for the hyperfine levels of H in interstellar space.

In interstellar space, hydrogen atoms occur with a density near 1 atom/cm³, or a pressure of less than 10^{-19} atm. Since a collision between atoms is very rare, occurring only once in a number of years, the dominating source of broadening is the Doppler effect. The various parts of the Milky Way have random velocities as large as about ± 10 km/sec with respect to each other so that the Doppler effect gives a line-width parameter $\Delta\nu$ of about $\dfrac{\nu}{3 \times 10^4}$, or 50 kc, for $\nu = 1420$ Mc.

The total amount of power received by an antenna from the radiation by interstellar hydrogen is approximately $kT\,\Delta\nu$, where T is the temperature 100°K, and $\Delta\nu$ the line width of 50 kc. This amounts to slightly less than 10^{-16} watt but is enough to give a signal which is a few hundred times background noise in a carefully constructed radiometer of the type described in Chap. 15.

If the temperature of an object is measured by the intensity of radiation at a given frequency, the apparent temperature is given by

$$T = T_o(1 - e^{-\gamma L}) \tag{5-68}$$

where T_o = temperature of the object
L = thickness of the object
γ = absorption coefficient at the frequency of measurement
For an opaque object, $T = T_o$. For less opaque material the apparent temperature T of emission is reduced. It must be at least as large as about 1° for an emission line to be detectable. Temperature changes which are this small correspond to observation of a gas which is almost transparent, that is, $\gamma L \ll 1$. In this case the observed temperature is, from (5-68),

$$T \approx T_o\gamma L \tag{5-69}$$

Now the absorption coefficient γ for a gas is given by Eq. (1-50). For the ground atomic state of hydrogen, $f = 1$ and μ is approximately one Bohr magneton, since the transition involves a magnetic rather than an electric dipole moment. Inserting values of the constants, expression (5-69) becomes

$$T \approx 5 \times 10^{-19} \frac{NL}{\lambda} \tag{5-70}$$

Here $\nu/\Delta\nu = 3 \times 10^4$ due to Doppler effect has been used and λ is the wavelength. For a temperature change as large as 1°C, the number of molecules NL which must be in the path of observation is

$$NL = 2 \times 10^{18}\lambda \qquad (5\text{-}71)$$

Since the longest dimension of the Milky Way is approximately 10^{23} cm, an interstellar gas must have a density as high as about $N = 2 \times 10^{-5}\lambda$ to be detected by microwave emission if the transition involved is due to a magnetic dipole moment. Since hydrogen has a density of approximately 1 atom/cm³ in interstellar space, its radiation is clearly observable. In fact, in this case γL is larger than unity in most directions through our galaxy, and the galaxy is opaque at the center of the hydrogen line. On the other hand, deuterium would be very difficult to observe, since it probably has a density of only 10^{-3} or 10^{-4} atom/cm³, and the transition between hyperfine levels falls at longer wavelengths.

If the transition is due to an electric dipole moment, then the dipole matrix element is approximately 1 debye, which is 100 times larger than the matrix element due to a magnetic dipole transition, and the minimum observable density of interstellar gas would be approximately

$$N = 2 \times 10^{-9}\lambda$$

A few molecules with dipole moments, such as the radical OH, are thought to have densities as high as $N = 10^{-6}$, and hence their spectra may possibly be observed.

Transitions between several other hyperfine levels in atoms may eventually be observed from astronomical objects. The hyperfine structure of N^{14} in several states of ionization may be sufficiently intense for observation since the density of nitrogen in interstellar gas is approximately 10^{-3} atom/cm³. However, frequencies for the hyperfine structure of N^{14} are known experimentally only for the ground state of the neutral atom. The hyperfine structure of N^{14} in high states of ionization may possibly also be observed in the sun's atmosphere.

The microwave line emitted by interstellar hydrogen has been particularly valuable for astronomy. For example, its observation has shown that, in certain directions through our galaxy, there are several strips of gas each moving systematically at velocities appropriate to the successive arms of a rotating spiral nebula. This seems to give the clearest evidence so far available that our galaxy is a spiral nebula.

CHAPTER 6

QUADRUPOLE HYPERFINE STRUCTURE IN MOLECULES

6-1. Introduction. In most atoms, the predominant hyperfine structure is due to interaction between a nuclear magnetic moment and magnetic fields of the atomic electrons. Effects of a nuclear quadrupole moment are smaller and give small deviations from the expected magnetic hyperfine intervals. However, for most molecules in the ground state, the magnetic fields due to various electrons almost completely cancel, giving zero or only very small magnetic fields at the nucleus. Electric quadrupole effects in molecules may still be sizable, however, and they become the dominating source of hyperfine structure.

The cancellation of magnetic fields in molecules due to electronic motions is simply because the electrons are paired; *i.e.*, for each electron with an angular momentum and hence magnetic field, there is another electron in a similar state but with oppositely directed angular momentum. The net electronic angular momentum for the electrons in the ground state of most molecules is indicated by the spectroscopic term $^1\Sigma_0$, which signifies that the net electronic spin and orbital angular momentum are both zero. It is not surprising that electronic momenta are paired off in a molecule if the nature of molecular bonds is considered. Generally an atom forms chemical bonds with each of its unpaired electrons, each one pairing with an electron from another atom in the molecule to give a net zero angular momentum. For the rare molecules such as NO, ClO_2 and NO_2 having an odd number of electrons, a complete pairing of electron spins is impossible. These molecules are hence paramagnetic and have large magnetic hyperfine structures. There are in addition a few cases of molecules having an even number of electrons in which the chemical bonds are unusual and the electron spins are not paired. The most notable example of this case is O_2, which is in a $^3\Sigma_1$ state, having two parallel electron spins.

For the overwhelming majority of molecules, however, magnetic hyperfine effects are extremely small, and it is electric quadrupole hyperfine structure that is evident when molecular spectra are examined with high resolution.

The discussion given in the preceding chapter of interaction between a nuclear electric quadrupole moment and a surrounding charge distribution is of course as valid for a molecular system as for an atomic system since the charge distribution assumed is a general one. Expression (5-45), which is

$$W_Q = \frac{1}{2} \frac{eq_J Q}{I(2I - 1)J(2J - 1)} \left[\tfrac{3}{4}C(C + 1) - I(I + 1)J(J + 1)\right] \quad (6\text{-}1)$$

may be applied to the molecular case if J is taken as the angular momentum of the molecule, and q_J, given by (5-42), is

$$q_J = \int \rho_{JJ} \frac{(3 \cos^2 \theta - 1)}{r^3} d\tau = \left(\frac{\partial^2 V}{\partial z_J^2}\right)_{\mathrm{av}}$$

where the charge density ρ applies to all charges in the molecule outside of a small sphere around the nucleus considered. It may be seen that the integral is just the average value of the second derivative of the potential at the nucleus due to all extranuclear charges ρ_{JJ}, taken along the direction of J, which is fixed in space and labeled z_J. The only problem peculiar to the molecular case is the particular evaluation of q_J, which will depend not only on the charge distribution in the molecule, but also on an average of the orientation of the molecule with respect to J. For a molecule of little symmetry, such as an asymmetric top, evaluation of q_J in terms of the various molecular axes and rotational quantum numbers can be rather tedious. We therefore begin by considering the much simpler and fortunately common case of a linear molecule.

6-2. Quadrupole Hyperfine Structure in Linear Molecules. In a linear molecule, the charge distribution is symmetric around the molecular axis, and hence if z_m indicates the direction of the molecular axis,

$$\frac{\partial^2 V}{\partial x_m^2} = \frac{\partial^2 V}{\partial y_m^2} = -\frac{1}{2} \frac{\partial^2 V}{\partial z_m^2}$$

using Laplace's equation and the equivalence of the x and y directions. A transformation of coordinates allows a reexpression of q_J.

$$\begin{aligned}
q_J &= \left(\frac{\partial^2 V}{\partial z_J^2}\right)_{\mathrm{av}} = \left(\frac{\partial^2 V}{\partial z_m^2} \cos^2 \theta_{mJ} + \frac{\partial^2 V}{\partial x_m^2} \sin^2 \theta_{mJ}\right)_{\mathrm{av}} \\
&= \frac{\partial^2 V}{\partial z_m^2} \left(\frac{3 \cos^2 \theta_{mJ} - 1}{2}\right)_{\mathrm{av}}
\end{aligned} \quad (6\text{-}2)$$

where θ_{mJ} is the angle between the molecular axis and J. The quantity $\partial^2 V/\partial z_m^2$ is the second derivative of the potential at the nucleus under discussion along the direction of the molecular axis due to all charges outside a small sphere surrounding the nucleus. It is a property of the molecule independent of the rotational state of the molecule, and will be designated as q_m or simply q in analogy to q_J which is a similar quantity referred to the direction of J, and hence dependent on the rotational state.

In order to evaluate $[(3 \cos^2 \theta_{mJ} - 1)/2]_{av}$ we must use the molecular wave functions which have already been given in Chap. 1 as

$$\psi_{J,M=J} = \sqrt{\frac{2J+1}{4\pi(2J)!}} \, P_J^J(\cos \theta) e^{iJ\phi}$$

then

$$\left(\frac{3 \cos^2 \theta_{mJ} - 1}{2}\right)_{av}$$

$$= \frac{2J+1}{4\pi(2J)!} \int_0^\pi \int_0^{2\pi} [P_J^J(\cos \theta)]^2 \frac{(3 \cos^2 \theta - 1)}{2} \sin \theta \, d\theta \, d\phi$$

$$= \frac{-J}{2J+3} \qquad (6\text{-}3)$$

If J becomes very large, this result approaches the classical expectation that $[(3 \cos^2 \theta_{mJ} - 1)/2]_{av} = -\frac{1}{2}$, since classically the molecular axis should be perpendicular to the angular momentum and $\cos \theta_{mJ} = 0$. Hence from (6-1), (6-2), and (6-3), the quadrupole energy is

$$W_Q = -\frac{eq_m Q}{2I(2I-1)(2J-1)(2J+3)} [\tfrac{3}{4}C(C+1) - I(I+1)J(J+1)] \qquad (6\text{-}4)$$

where $C = F(F+1) - I(I+1) - J(J+1)$. F is the quantum number for the total angular momentum, which takes on the values $I+J$, $I+J-1, \ldots, |I-J|$.

The expression (6-4) gives the quadrupole energy for a single nucleus in a linear molecule in terms of the molecular constant q_m, the nuclear constant Q, and angular momentum quantum numbers I, J, F. It might be written $W = -eq_m Q f(I,J,F)$, where $eq_m Q$ or simply eqQ is called the quadrupole coupling constant and $f(I,J,F)$ might be called Casimir's function since it comes rather directly from theory developed by Casimir. This function is given in Appendix I for all values of I up to $\frac{11}{2}$ (excluding 0 and $\frac{1}{2}$, for which Q is necessarily zero, and 5 which is a rare case) and for values of J up to 10. It may be noted that, when F has its maximum or minimum values, corresponding to I and J parallel or antiparallel, respectively, Casimir's function is positive, whereas for intermediate values of F, the function is negative. This behavior corresponds roughly to the classically expected variation of $3 \cos^2 \theta_{IJ} - 1$.

From Appendix I and a knowledge of the quadrupole coupling constant eqQ the hyperfine energy levels can be easily determined. The constant eqQ may have an extremely wide range of values, but a representative value would be 100 Mc. To predict the hyperfine structure of a molecular transition, we need in addition some information about the selection rules and intensities. Selection rules for hyperfine structure are exactly the same as for fine structure assuming either type of interaction is very small compared with the separation between major energy levels. Thus if a microwave tends to change the angular momentum J of rotation of a molecule without acting on the nuclear spin I, this is an exact parallel to a light wave changing the orbital momentum L of electrons in an atom without acting on the electron spins S. The selection rules of Chap. 5 for fine structure then become for hyperfine structure

$$\Delta J = 0, \pm 1 \qquad \Delta F = 0, \pm 1 \qquad \Delta I = 0 \qquad (6\text{-}5)$$

Relative intensities for the different possibilities are given by appropriate substitution of quantum numbers in expression (5-17) and (5-18).

For transitions $J \leftarrow J - 1$:

$$F \leftarrow F - 1: \frac{B(J + F + I + 1)(J + F + I)(J + F - I)(J + F - I - 1)}{F}$$

$$F \leftarrow F: \quad -\frac{B(J + F + I + 1)(J + F - I)(J - F + 1)(J - F - I - 1)(2F + 1)}{F(F + 1)}$$

$$F \leftarrow F + 1: \frac{B(J - F + I)(J - F + I - 1)(J - F - I - 1)(J - F - I - 2)}{F + 1}$$

$$(6\text{-}6a)$$

For transitions $J \leftarrow J$:

$$F \leftarrow F - 1: -\frac{A(J + F + I + 1)(J + F - I)(J - F + I + 1)(J - F - I)}{F}$$

$$F \leftarrow F: \quad \frac{A[J(J + 1) + F(F + 1) - I(I + 1)]^2(2F + 1)}{F(F + 1)}$$

$$F \leftarrow F + 1: -\frac{A(J + F + I + 2)(J + F - I + 1)(J - F + I)(J - F - I - 1)}{F + 1}$$

$$(6\text{-}6b)$$

Since the probability of exciting a transition between two states is independent of the direction of the transition, relative intensities for the components of the transition $J - 1 \leftarrow J$ may be obtained simply by reversing all arrows in the first group of three equations.

Appendix I gives relative intensities of the various possible hyperfine component transitions for low values of J and I up to $\frac{11}{2}$. Values given in this appendix are simply calculated from the formulas above, with their absolute values adjusted so that the sum of all hyperfine components of a particular J transition is one hundred. This is convenient because the sum of the intensities of hyperfine components of a transition should just equal the intensity of the transition if it had not been split.

Only very rarely is the hyperfine structure of rotational lines involving J greater than 10 of interest, since it is usually not prominent enough to be observed. For such large values of J, the more intense components of the hyperfine structure are always those for which F changes in the same way as J ($\Delta F = \Delta J$), and for these the hyperfine splitting is very small. Relative intensities of these components of a transition involving large J are approximately proportional to the statistical weights $2F + 1$ or, therefore, to F. For each of the weaker components, when J is larger than 10 the intensity is a small fraction of that of the entire transition. This fraction is given within a factor of about 2 by the following expressions:

When $J + 1 \leftarrow J$, $F \leftarrow F$, fraction of intensity $\approx 1/2J^2$

$J + 1 \leftarrow J$, $F - 1 \leftarrow F$, fraction of intensity $\approx 1/10J^4$

$J \leftarrow J, F \pm 1 \leftarrow F$, fraction of intensity $\approx 1/2J^2$

Changes in quadrupolar energies may be similarly approximated for large J. For the stronger components of the transition where $\Delta F = \Delta J$, the change in quadrupole energy is a small fraction of the quadrupole coupling constant, being in almost all cases smaller in size than $eqQ/4J^2$. For the other, very much weaker, components, the change in energy is larger and can be approximated by

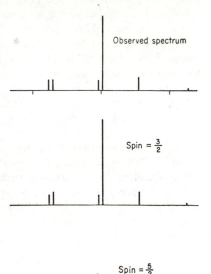

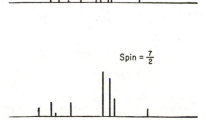

FIG. 6-1. Quadrupole hyperfine structure in the $J = 2 \leftarrow 1$ transition of $O^{16}C^{12}S^{33}$ due to S^{33}, showing the different patterns expected theoretically for various assumed values of the S^{33} spin. It is clear that the pattern for spin $\frac{3}{2}$ agrees well with the observed spectrum, and that other values of spin do not agree. For this case $eqQ = -29.2$ Mc.

$$\Delta W_Q (\Delta F = \Delta J \pm 1) = \mp \frac{3[2(F - J) \pm 1]eqQ}{8I(2I - 1)} + \text{terms of order } \frac{eqQ}{2J}$$

The quantities given in Appendix I allow a ready calculation of quadrupole hyperfine structure such as is shown in Fig. 6-1. In this figure the observed hyperfine structure in $O^{16}C^{12}S^{33}$ is compared with theoretically expected patterns assuming various values of the S^{33} spin. OCS is a linear molecule, and both O^{16} and C^{12} are known to have zero spins, and so produce no hyperfine structure. The quadrupole coupling constant assumed for S^{33} in OCS is chosen so that the computed hyperfine patterns will agree as closely as possible with the observed spectrum. It is evident that the observed structure agrees very well with the theory if the S^{33} spin is assumed to be $\frac{3}{2}$, whereas there is clear disagreement with the other calculated spectra. Thus a comparison of this type allows a determination of the S^{33} nuclear spin, its quadrupole coupling in OCS, and in addition the values of J involved in the transition, since both J and I determine the exact structure observed.

6-3. Quadrupole Hyperfine Structure in Symmetric Tops. For molecules which are not linear, the general theory of quadrupole coupling is unchanged, but the quantity q_J must be reevaluated. For the case of a nucleus on the axis of a symmetric top, this quantity is still rather simple in form. Because of the symmetry q_J may again be written, as in (6-2),

$$q_J = \frac{\partial^2 V}{\partial z_m^2} \left(\frac{3 \cos^2 \theta_{mJ} - 1}{2} \right)_{\text{av}}$$

where the direction z_m is as before along the molecular axis. For a symmetric top $[(3 \cos^2 \theta_{mJ} - 1)/2]_{\text{av}}$ has a somewhat different form, however.

$$\left(\frac{3 \cos^2 \theta_{mJ} - 1}{2} \right)_{\text{av}} = \int \psi^*_{J,K,M=J} \left(\frac{3 \cos^2 \theta_{mJ} - 1}{2} \right) \psi_{J,K,M=J} \, d\tau$$
$$= \left[\frac{3K^2}{J(J+1)} - 1 \right] \frac{J}{2J+3} \quad (6\text{-}7)$$

where $\psi_{J,K,M=J}$ is a symmetric-top wave function such as is given in Chap. 3. For J and K large, it can be seen that (6-7) gives the classically expected behavior, for then the cosine of the angle between J and the molecular axis ($\cos \theta_{mJ}$) is easily shown to be K/J or $K/\sqrt{J(J+1)}$ by use of the vector model.

The nuclear quadrupole energy for a nucleus on the axis of a symmetric top is, using (6-1), (6-2), and (6-7),

$$W_Q = \frac{eqQ \left[3 \dfrac{K^2}{J(J+1)} - 1 \right]}{2I(2I-1)(2J-1)(2J+3)} \left[\tfrac{3}{4}C(C+1) - I(I+1)J(J+1) \right]$$
$$(6\text{-}8)$$

where q, or q_m, is the second derivative of the potential (excluding charges in a small sphere around the nucleus) along the direction of the molecular axis [143], [145]. This expression is identical with that for a linear molecule except for a factor $1 - 3K^2/J(J + 1)$. The linear molecule is of course a special case of the symmetric top ($K = 0$). Appendix I may be used for the quadrupole energy levels of a symmetric top if energy values are multiplied by the factor $1 - 3K^2/J(J + 1)$. Relative intensities of hyperfine components, which are also given by Appendix I, apply to this case.

It should be noted that, in deriving (6-8), the electric field was assumed to be symmetric about the molecular axis ($\partial^2 V/\partial x_m^2 = \partial^2 V/\partial y_m^2$ on the axis). This will be true in all cases for a nucleus on the axis of a symmetric top, since to make the moments of inertia about x_m and y_m exactly equal, a symmetric arrangement of atoms is required. However, there may occur very rare cases of "accidentally" nearly symmetric tops for which this condition is not fulfilled, and the quadrupole levels must be described by the somewhat more complex theory developed in the latter part of this chapter for asymmetric molecules rather than by (6-8). If the nucleus is in a symmetric top, but not on the axis, then there is always present the complication of other like nuclei with quadrupole coupling, which will be treated later in this chapter.

The hyperfine structure of a symmetric rotor transition $J + 1 \leftarrow J$ is usually more complex than that for a linear molecule because such a transition involves a number of different K values, each with its own hyperfine structure. An example is the structure of the $J = 2 \leftarrow 1$ transition of CH_3I, shown in Fig. 6-2. Iodine has a spin of $\frac{5}{2}$ and a rather large quadrupole coupling constant, whereas C and H have spins of 0 and $\frac{1}{2}$, respectively, and hence give no quadrupole effects.

6-4. Second-order Quadrupole Effects. The quadrupole hyperfine structure has so far been considered small by comparison with the frequency of the rotational transition. The quantity $[(3 \cos^2 \theta_{mJ} - 1)/2]_{av}$ has been calculated, for example, in (6-3) and (6-7) with so-called first-order perturbation theory, or by using rotational wave functions which are assumed to be unchanged by the existence of quadrupole effects. If the quadrupole coupling is not small compared with the rotational frequencies, however, the molecular rotational wave functions will be modified and the energies for quadrupole effects given above will not be exactly correct.

A strong quadrupole interaction produces some exchange of angular momentum between the nucleus and the molecule; so the state of the rotating molecule can no longer be accurately specified by a fixed angular momentum J. Allowing for quadrupole interactions which are not small, the state of the molecule can only approximately be described by J, and the wave functions and quadrupole energy must be calculated from second-order perturbation theory. However, the total angular momentum F and its projection on a fixed axis M_F are quantized and cannot be changed by interactions within the molecule.

For a general symmetric top the energy given by second-order perturbation theory is of the form

$$W_Q = (IJKFM_F|H_Q|IJKFM_F) + \sum_{J'K'} \frac{|(IJKFM_F|H_Q|IJ'K'FM_F)|^2}{W_{JK} - W_{J'K'}} \quad (6\text{-}9)$$

where H_Q is the part of the Hamiltonian which represents the quadrupole energy. It is given by $-\frac{1}{6}\mathbf{Q}:\boldsymbol{\nabla}\mathbf{E}$, as discussed in Chap. 5. The quantities

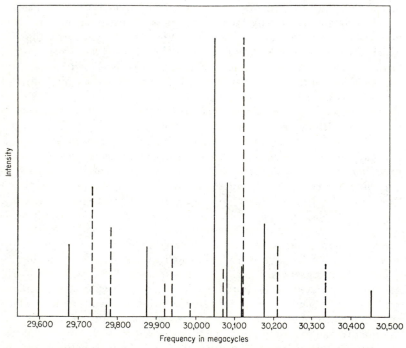

FIG. 6-2. Hyperfine structure in the $J = 2 \leftarrow 1$ transition of the symmetric rotor CH_3I due to the nuclear quadrupole moment of I^{127}. The spin of I^{127} is $\frac{5}{2}$ and its quadrupole coupling is -1934 Mc. Solid lines represent transitions due to molecules with $k = 0$, dashed lines those of molecules with $k = \pm1$.

$(IJKFM_F|H_Q|IJ'K'FM_F)$ are matrix elements of H_Q for the symmetric-rotor wave functions specified by the quantum numbers I, J, K, F, M_F and I, J', K', F, M_F. The quantum numbers I, F, and M_F are not summed over since they are unaffected by the perturbation, or in other words the matrix of H_Q is diagonal in I, F, and M_F. The first term of (6-9) is just the previously calculated first-order quadrupole energy (6-8) which gives quite accurately the entire quadrupole energy when $W_{JK} - W_{J'K'}$ is much larger than eqQ. The only matrix elements in the sum of (6-9) which are not identically zero are those for which $K' = K$ and either $J' = J \pm 1$ or $J' = J \pm 2$. They are [253]

$$(I,J,K,F,M_F|H_Q|I,J+1,K,F,M_F) = \frac{3eqQK[F(F+1) - I(I+1) - J(J+2)]}{8I(2I-1)J(J+2)}$$

$$\times \left\{ \frac{\left[\left(1 - \frac{K^2}{(J+1)^2}\right)(I+J+F+2)(J+F-I+1)\times(I+F-J)(I+J-F+1)\right]}{(2J+1)(2J+3)} \right\}^{\frac{1}{2}}$$

$$(I,J,K,F,M_F|H_Q|I,J+2,K,F,M_F) = \frac{3eqQ}{16I(2I-1)(2J+3)}$$

$$\times \left\{ \left[1 - \frac{K^2}{(J+1)^2}\right]\left[1 - \frac{K^2}{(J+2)^2}\right](I+J+F+3)(I+J+F+2) \right.$$

$$\times (I+J-F+2)(I+J-F+1)(J+F-I+2)$$

$$\times \left. \frac{(J+F-I+1)(I+F-J)(I+F-J-1)}{(2J+1)(2J+5)} \right\}^{\frac{1}{2}} \qquad (6\text{-}10)$$

From the matrix elements (6-10) and expression (6-9) the modifications in the quadrupole energy due to second-order perturbations may be taken into account. They are usually rather small, for they are less than first-order energies (6-8) by a factor usually somewhat smaller than $eqQ/(W_{JK} - W_{J'K'}) \approx eqQ/\nu$, where ν is the frequency of the observed rotational transition. Usually eqQ is a few hundred megacycles, while ν is a few ten thousands of megacycles. However, in some cases, such as a large molecule with the atoms I or Hg in it, eqQ/ν may not be small; in fact it may possibly be so large that a still better approximation than that given by (6-9) will be needed in order to fit the experimentally measured hyperfine structure.

A linear molecule in the ground vibrational state is of course a special case of a symmetric top for which K equals zero. When $K = 0$, the matrix element given by (6-10) which connects J and $J + 1$ is zero, and the matrix element connecting J and $J + 2$ simplifies somewhat. If the linear molecule is excited in a degenerate vibrational mode so that there is an angular momentum $|l|$ around the axis of the molecule, then as seen in Chap. 2, its rotational wave functions are similar to those for a symmetric top with $|K|$ equal to $|l|$. Hence matrix elements (6-10) apply to a linear molecule excited to any vibrational mode if the value of l is substituted for K.

The molecule ICN is a case where eqQ/ν is as large as is normally encountered. For I^{127}, $eqQ = -2420$ Mc, and the transition $J = 4 \leftarrow 3$ occurs near 25,800 Mc. The N^{14} nucleus in ICN has a quadrupole coupling of -3.7 Mc which is so small that, unless the spectrum is observed under very high resolution, the quadrupole effects due to N may be neglected. The effect of second-order corrections on the observed spectrum for the ground vibrational state is indicated in Table 6-1, which gives the first-order and second-order energies due to the I^{127} quadrupole coupling. It should be noted that the transitions $F = \frac{9}{2} \leftarrow \frac{7}{2}$ and $F = \frac{9}{2} \leftarrow \frac{9}{2}$ should, according to the usually first-order theory of quadrupole effects, exactly coincide. Second-order effects split these two lines by 6.32 Mc. Actual measurements of the hyperfine structure of the $J = 4 \leftarrow 3$ transition in ICN agree very well with the predicted combined first- and second-order effects.

TABLE 6-1. QUADRUPOLE SPLITTING OF THE $J = 4 \leftarrow 3$ TRANSITION OF ICN
IN THE GROUND VIBRATIONAL STATE

Without quadrupole effects, the transition would occur at 25,804 Mc. The frequency contributions to the various possible hyperfine transitions due to an I quadrupole coupling of -2420 Mc are listed. F' represents the initial value of F associated with a J of 3, and F is the final value.

Transition $F \leftarrow F'$	Quadrupole coupling contribution, Mc	
	First order	Second order
$\frac{13}{2} \leftarrow \frac{11}{2}$	18.33	0.55
$\frac{11}{2} \leftarrow \frac{11}{2}$	-410.65	0.10
$\frac{11}{2} \leftarrow \frac{9}{2}$	33.02	0.49
$\frac{9}{2} \leftarrow \frac{9}{2}$	-18.84	4.52
$\frac{9}{2} \leftarrow \frac{7}{2}$	-18.84	-1.80
$\frac{7}{2} \leftarrow \frac{7}{2}$	150.87	-0.83
$\frac{7}{2} \leftarrow \frac{5}{2}$	-74.98	-0.49
$\frac{5}{2} \leftarrow \frac{5}{2}$	189.01	-0.83
$\frac{5}{2} \leftarrow \frac{3}{2}$	-93.32	0.65
$\frac{3}{2} \leftarrow \frac{3}{2}$	165.95	-0.60
$\frac{3}{2} \leftarrow \frac{1}{2}$	-51.84	0.37

Transitions of ICN molecules in the first excited bending mode are split by l-type doubling, and each of the two l-type doublets is further split by the I quadrupole effects. First- and second-order quadrupole effects for these doublets are listed in Table 6-2. In this case $|l| = 1$, and the value of the corresponding quantum number K for the angular momentum around the symmetry axis is set equal to 1 in calculating second-order effects from the expressions (6-10).

The second-order effects of quadrupole coupling on the hyperfine structure as given in Tables 6-1 and 6-2 are not large, even though the quadrupole coupling constant for I^{127} in ICN is a rather large one. If the quadrupole coupling constant had been ten times smaller, or about 240 Mc, the second-order effects on frequency would have been 100 times smaller, since they depend on $(eqQ)^2$, and would then be detectable only by very accurate microwave frequency measurements.

In addition to changing the frequencies of hyperfine components, second-order quadrupole effects may affect their intensities.   The quadrupole coupling modifies the molecular rotational wave functions, adding to the wave function of rotational quantum number J a small component of the wave functions for $J \pm 1$ (if $K \neq 0$) and $J \pm 2$.   Because of this modification of the wave function, the matrix elements for the transitions and hence the intensities are slightly changed.   This change is usually too small to be of significance for the normal transitions unless extremely accurate intensity measurements are made.   More noticeable and important, however, is that the selection rules are modified to allow the occurrence of new transitions.   Thus since a small component of the wave function for $J + 2$ is mixed with the wave function for J, a weak transi-

TABLE 6-2. QUADRUPOLE SPLITTING OF THE $J = 4 \leftarrow 3$ TRANSITION OF ICN
IN THE FIRST EXCITED BENDING MODE, FOR WHICH $[l] = 1$
Notation is as in Table 6-1. l-type doubling is not included in this table but produces two sets of hyperfine components separated by the l-type doubling.

Transition $F \leftarrow F'$	Quadrupole coupling contribution, Mc	
	First order	Second order
$\frac{13}{2} \leftarrow \frac{11}{2}$	54.08	1.61
$\frac{11}{2} \leftarrow \frac{9}{2}$	22.18	-0.72
$\frac{9}{2} \leftarrow \frac{7}{2}$	-21.90	0.28
$\frac{7}{2} \leftarrow \frac{5}{2}$	-47.03	-0.20
$\frac{5}{2} \leftarrow \frac{3}{2}$	-34.40	-0.45

tion from this modified wave function to a $J + 3$ state may occur.   If the modified wave function is still identified (approximately) by the quantum number J, this effect would allow weak transitions corresponding to $\Delta J = \pm 3$.   Similarly, when K is not equal to zero and the quadrupole coupling is large, weak transitions corresponding to $\Delta J = \pm 2$ may be expected.   These types of transitions have not yet been observed but could probably be found in the case of molecules with quadrupole couplings as large as is found in ICN.   Their intensities may be calculated by evaluating the amounts of perturbation of the wave functions using the matrix elements given by (6-10).

6-5. Asymmetric Tops.   In principle, evaluation of nuclear quadrupole effects in asymmetric rotors is straightforward.   Expression (6-1) gives the energy, which is of the same form as for symmetric molecules, and q_J, the average second derivative of the potential along the direction of the angular momentum, is

$$q_J = \int \psi^*_{J_{K_{-1}K_1}, M=J} \left(\frac{\partial^2 V}{\partial z_J^2} \right) \psi_{J_{K_{-1}K_1}, M=J} \, d\tau \qquad (6\text{-}11)$$

where $\psi_{J_{K_{-1}K_1}, M=J}$ is an asymmetric-rotor wave function. Evaluation of the integral in (6-11), however, is considerably more complex than evaluating the similar expression for symmetric rotors. We shall follow the method of Bragg [258].

For an evaluation of (6-11), we shall express $\partial^2 V / \partial z_J^2$ in terms of the second derivatives of the potential along the principal axes of inertia x_m, y_m, and z_m of the molecule and the direction cosines α between this set of axes and z_J which is fixed in space.

$$
\begin{aligned}
\frac{\partial^2 V}{\partial z_J^2} = {} & \alpha_{z_J x_m}^2 \left(\frac{\partial^2 V}{\partial x_m^2} \right) + \alpha_{z_J y_m}^2 \left(\frac{\partial^2 V}{\partial y_m^2} \right) + \alpha_{z_J z_m}^2 \left(\frac{\partial^2 V}{\partial z_m^2} \right) \\
& + 2\alpha_{z_J x_m} \alpha_{z_J y_m} \left(\frac{\partial^2 V}{\partial x_m \, \partial y_m} \right) + 2\alpha_{z_J x_m} \alpha_{z_J z_m} \left(\frac{\partial^2 V}{\partial x_m \, \partial z_m} \right) \\
& + 2\alpha_{z_J y_m} \alpha_{z_J z_m} \left(\frac{\partial^2 V}{\partial y_m \, \partial z_m} \right) \quad (6\text{-}12)
\end{aligned}
$$

By an argument similar to that used to prove certain matrix elements of the dipole moment are zero (page 63), the integrals of the form

$$
\int \psi^* \alpha_{z_J x_m} \alpha_{z_J y_m} \frac{\partial^2 V}{\partial x_m \, \partial y_m} \psi \, d\tau \quad (6\text{-}13)
$$

may be shown to be zero. Because wave functions are either symmetric or antisymmetric with respect to a rotation around the principal axes, $\psi^* \psi$ is unchanged as a result of a 180° rotation about the axis x_m. However, $\alpha_{z y_m}$ is reversed in sign since the y_m direction has been reversed. Hence the integrand of (6-13) undergoes a reversal of sign due to this rotation and must be zero. Expression (6-11) becomes, therefore,

$$
q_J = (\alpha_{z x_m}^2)_{\mathrm{av}} \frac{\partial^2 V}{\partial x_m^2} + (\alpha_{z y_m}^2)_{\mathrm{av}} \frac{\partial^2 V}{\partial y_m^2} + (\alpha_{z z_m}^2)_{\mathrm{av}} \frac{\partial^2 V}{\partial z_m^2} \quad (6\text{-}14)
$$

where

$$
(\alpha_{z x_m}^2)_{\mathrm{av}} = \int \psi_{J_{K_{-1}K_1}, M=J}^* \alpha_{z x_m}^2 \psi_{J_{K_{-1}K_1}, M=J} \, d\tau \quad (6\text{-}15)
$$

Matrix elements of the direction cosines have already been discussed and the line strengths $S_{J_{K_{-1}K_1} J'_{K'_{-1}K'_1}}$ derived from their squares are tabulated in Appendix V. Matrix elements of the squares of direction cosines such as (6-15) may be obtained by squaring the direction cosine matrices. Some manipulation (cf. [258]) shows that $(\alpha_{z x_m}^2)_{\mathrm{av}}$ may hence be expressed in terms of the quantities tabulated in Appendix V as follows:

$$
(\alpha_{z z_m}^2)_{\mathrm{av}} = \frac{2J}{(2J+1)(2J+3)} \sum_{K'_{-1}K'_1} {}^x S_{J_{K_{-1}K_1} J_{K'_{-1}K'_1}} + \frac{1}{2J+3} \quad (6\text{-}16)
$$

where x or x_m may refer to the direction of any one of the three principal axes a, b, or c. This gives

$$q_J = \frac{2J}{(2J+1)(2J+3)} \sum_{K'_{-1}K'_1} \frac{\partial^2 V}{\partial a^2} \, {}^a S_{J_{K_{-1}K_1}J_{K'_{-1}K'_1}}$$
$$+ \frac{\partial^2 V}{\partial b^2} \, {}^b S_{J_{K_{-1}K_1}J_{K'_{-1}K'_1}} + \frac{\partial^2 V}{\partial c^2} \, {}^c S_{J_{K_{-1}K_1}J_{K'_{-1}K'_1}} \quad (6\text{-}17)$$

where second derivatives of the potential along the three principal axes of inertia are indicated by $\partial^2 V/\partial a^2$, $\partial^2 V/\partial b^2$, and $\partial^2 V/\partial c^2$. This potential as before (page 135) is due to all charges outside a small sphere around the nucleus. The quantities S in (6-17) are tabulated in Appendix V for values of the asymmetry parameter $\kappa = -1.0, -0.5, 0, 0.5$, and 1.0 only. Interpolation must be used for other values of κ.

Another form in which q_J may be expressed is [352]

$$q_J = \frac{1}{(J+1)(2J+3)} \frac{\partial^2 V}{\partial a^2} \left[J(J+1) + E(\kappa) - (\kappa+1)\frac{\partial E(\kappa)}{\partial \kappa} \right]$$
$$+ \frac{2}{(J+1)(2J+3)} \frac{\partial^2 V}{\partial b^2} \frac{\partial E(\kappa)}{\partial \kappa}$$
$$+ \frac{1}{(J+1)(2J+3)} \frac{\partial^2 V}{\partial c^2} \left[J(J+1) - E(\kappa) + (\kappa-1)\frac{\partial E(\kappa)}{\partial \kappa} \right] \quad (6\text{-}18)$$

where $E(\kappa)$ is the energy parameter for an asymmetric rotor of asymmetry κ as defined in (4-10). In (6-18) $E_{J_{K_{-1}K_1}}(\kappa)$ appropriate to the particular state for which q_J is being evaluated is of course used. $E(\kappa)$ and $\partial E(\kappa)/\partial \kappa$ may be obtained from Appendix IV. Although interpolation must still be used to evaluate $E(\kappa)$ and $\partial E(\kappa)/\partial \kappa$, the tabulation in Appendix IV uses smaller steps of κ than does Appendix V, so that more accuracy may often be obtained from (6-18) than from (6-17).

Expression (6-18) is also very useful if q_J must be evaluated for states not tabulated in Appendices IV or V ($J > 12$), since all the approximate methods discussed in Chap. 4 for evaluating $E(\kappa)$ for large J are available. [See [352] for expression of (6-18) in terms of approximations for $E(\kappa)$.]

If better accuracy is needed than can be obtained by interpolation, the integral in (6-11) may be evaluated by expanding the wave function as a sum of symmetric-top functions as in (4-18). This leads to the expression [258]

$$q_J = \frac{q_m}{(J+1)(2J+3)} \sum_K a_{JK}^2 [3K^2 - J(J+1)]$$
$$- 2a_{JK}a_{JK+2}[f'(J,K+1)]^{\frac{1}{2}}\eta \quad (6\text{-}19)$$

where $f'(J,n) = \frac{1}{4}(J^2 - n^2)[(J+1)^2 - n^2]$ \hfill (6-20)

$\qquad q_m = \partial^2 V/\partial z_m^2$ is the second derivative of the potential along the principal axis which most nearly represents a symmetry axis of the moment of inertia ellipsoid

$$\eta = \frac{\partial^2 V/\partial x_m^2 - \partial^2 V/\partial y_m^2}{q_m} \quad (6\text{-}21)$$

If the molecule is considered an oblate top, z_m is the c axis, and x_m and y_m must be taken as the a and b axes, respectively. If it is prolate, z_m is the a axis, x_m the b axis, and y_m the c axis.

Expression (6-19) can give q_J exactly, but only after evaluation of the a_{JK}'s, which is troublesome. For small values of the asymmetry parameter b [for definition of b, see (4-2) and (4-3)] q_J can be satisfactorily obtained for various values of K from the expressions below. Terms of order b^3 or higher are omitted.*

For $K = 0$:

$$q_J = \frac{q_m}{(J+1)(2J+3)} \left\{ -J(J+1) + (\tfrac{3}{2}b^2 - b\eta)f'(J,1) \right\} \qquad (6\text{-}22a)$$

For $K = 1$:

$$q_J = \frac{q_m}{(J+1)(2J+3)} \left\{ 3 - J(J+1) \mp \frac{\eta}{2}J(J+1) + (\tfrac{3}{2}b^2 - b\eta)\frac{f'(J,2)}{4} \right.$$
$$\left. \pm \tfrac{3}{128}b^2\eta J(J+1)f'(J,2) \right\} \qquad (6\text{-}22b)$$

For $K = 2$:

$$q_J = \frac{q_m}{(J+1)(2J+3)} \left\{ 12 - J(J+1) + (\tfrac{3}{2}b^2 - b\eta) \right.$$
$$\left. \left[\frac{f'(J,3)}{6} - \frac{f'(J,1)}{2} \mp \frac{f'(J,1)}{2} \right] \right\} \qquad (6\text{-}22c)$$

For $K = 3$:

$$q_J = \frac{q_m}{(J+1)(2J+3)} \left\{ 27 - J(J+1) + (\tfrac{3}{2}b^2 - b\eta) \right.$$
$$\left. \left[\frac{f'(J,4)}{8} - \frac{f'(J,2)}{4} \right] \mp \tfrac{3}{128}f'(J,2)J(J+1)b^2\eta \right\} \qquad (6\text{-}22d)$$

For $K > 3$:

$$q_J = \frac{q_m}{(J+1)(2J+3)} \left\{ 3K^2 - J(J+1) + (\tfrac{3}{2}b^2 - b\eta) \right.$$
$$\left. \left[\frac{f'(J,K+1)}{2(K+1)} - \frac{f'(J,K-1)}{2(K-1)} \right] \right\} \qquad (6\text{-}22e)$$

where $f'(J,n) = \tfrac{1}{4}(J^2 - n^2)[(J+1)^2 - n^2]$

The upper signs apply to the upper-energy level of a K-type doublet for a prolate rotor or the lower doublet of an oblate rotor. The lower signs apply to the lower-energy level of a doublet for a prolate rotor, or the upper doublet of an oblate rotor. Since quadrupole effects are not

---

* Expressions (6-22) were first given by G. Knight and B. T. Feld [379] but have been corrected by J. Kraitchman and A. Javan.

usually measured to high fractional accuracy, the approximate expressions (6-22) (neglecting terms in b^3 and higher) are satisfactory in many cases, and when b is small they are more accurate than expressions (6-17) or (6-18) with interpolation.

There can be no asymmetry in $\nabla \mathbf{E}$ about the axis for a nucleus on the axis of a real symmetric top. Nuclei off the axis of a symmetric top always occur as three or more equivalent nuclei, for the sum of which the asymmetry of the field, or of $\nabla \mathbf{E}$, cancels out. However, for a molecule which is accidentally very close to symmetric, an asymmetry of the electric field about the axis can give a sizable contribution to the quadrupole energy when $K = 1$. This shows up as the terms $\pm q_m J \eta / 2J + 3$ in Eq. (6-22). In the case of $H_2C=CHCl$, for example, where the asymmetry $b = -0.006$, hyperfine structure of transitions between levels with $K = 1$ can be fitted rather well by omitting all terms involving b, but these terms dependent on the asymmetry of the field η must be kept. When $K \neq 1$, an asymmetry of the field affects the quadrupole energy only if at the same time the molecule is asymmetric ($b \neq 0$).

Hyperfine structure in an asymmetric rotor must be fitted by the two parameters $\partial^2 V / \partial z_m^2$ and $\partial^2 V / \partial x_m^2 - \partial^2 V / \partial y_m^2$, or q_m and η instead of the one parameter q_m which is needed for symmetric tops. However, in many cases these two parameters can to a very good approximation be expressed in terms of a single property of the electric field, $\partial^2 V / \partial z_b^2$, the second derivative of the electrostatic potential along the direction of the chemical bond which binds the nucleus in question to the molecule. This is because the electrostatic fields are in many cases almost symmetric about the bond axis.

A clear example of this is the field at the Cl nucleus in the asymmetric varieties of methyl chloride. For CH_3Cl, ∇E at the chlorine nucleus is symmetric about the C—Cl axis because of the threefold-symmetry axis of the molecule. In CH_2DCl, ∇E must still be symmetrical about the C—Cl axis, but the molecule has become an asymmetric top, and no principal axis coincides with the C—Cl bond. In such case the various second derivatives of the potential along principal axes may be readily obtained.

$$\frac{\partial^2 V}{\partial z_m^2} = \frac{\partial^2 V}{\partial z_b^2} \frac{3\alpha_{z_b z_m}^2 - 1}{2}$$

$$\frac{\partial^2 V}{\partial x_m^2} = \frac{\partial^2 V}{\partial z_b^2} \frac{3\alpha_{z_b x_m}^2 - 1}{2}$$

$$\frac{\partial^2 V}{\partial y_m^2} = \frac{\partial^2 V}{\partial z_b^2} \frac{3\alpha_{z_b y_m}^2 - 1}{2} \qquad (6\text{-}23a)$$

where $\alpha_{z_b z_m}$, etc., represent the cosines of angles between the various axes.

More generally, if the second derivatives of the potential with respect to one set of Cartesian coordinates x_1, x_2, x_3 are known, those along any other set of axes x_1', x_2', x_3' may be obtained from the relations

$$\frac{\partial^2 V}{\partial x_1' \, \partial x_j'} = \sum_{kl} \alpha_{x'_1 x_k} \alpha_{x'_j x} \frac{\partial^2 V}{\partial x_k \, \partial x_l} \qquad (6\text{-}23b)$$

Here $\alpha_{x'_1 x_k}$ represents the cosine of the angle between the two axes x_i' and x_k.

Very frequently, the electric field at a nucleus is to good accuracy symmetric about a bond, as in the case of CH_2DCl above. In CH_2Cl_2 it has been shown by Myers and Gwinn [732] that expressions (6-23) give the correct values for q_m and η to within the experimental accuracy of 1 per cent of $\partial^2 V / \partial z_1^2$. On the other hand, in cases where double bonds are involved, the fields may not be at all symmetric about the bond (see Chap. 9). $H_2C{=}CHCl$ is a molecule in which apparently the double-bond character of the C—Cl bond is sufficient to make the field appreciably asymmetric.

The quadrupole energy so far discussed for asymmetric rotors has been of the first-order type. As in symmetric rotors, the quadrupole energy is sometimes large enough compared with the separation between rotational energy levels that second-order perturbations are important. Corresponding to Eq. (6-9), the quadrupole energy is then

$$W_Q = (I, J_{K_{-1}K_1}, F | H_Q | I, J_{K_{-1}K_1}, F)$$
$$+ \sum_{J'K'_{-1}K'_1} \frac{|(I, J_{K_{-1}K_1}, F | H_Q | I, J'_{K'_{-1}K'_1}, F)|^2}{W_{J_{K_{-1}K_1}} - W_{J'_{K'_{-1}K'_1}}} \qquad (6\text{-}24)$$

where $W_{J_{K_{-1}K_1}}$ is the rotational energy. The quantities in brackets are the matrix elements of the Hamilton for the quadrupole interaction H_Q, the first term being the first-order energy which has been discussed above. The sum is, of course, over all rotational states except the unperturbed state $J_{K_{-1}K_1}$.

It can be shown that all matrix elements involved in the sum of (6-24) are zero except those for which $J' = J \pm 1$ or $J' = J \pm 2$. These matrix elements have been discusssed by Bragg [258] but have not yet been evaluated in detail. In cases of near degeneracy where two asymmetric-top levels of appropriate symmetry lie close together, second-order quadrupole effects will, however, be of some importance.

6-6. Hyperfine Structure from Two or More Nuclei in the Same Molecule. A molecule may contain more than one nucleus which produces an observable hyperfine structure in its spectrum. This almost always occurs when more than one nucleus in the molecule has a spin greater than $\frac{1}{2}$ and hence is coupled by its quadrupole moment to the rotational motion of the molecule. In such cases the quadrupole energies are no longer given by expressions like (6-8), since the interaction between one nucleus and the molecule will affect the interaction between the second nucleus and the molecule, and vice versa. We shall consider first the case of two such nuclei, which is the one of most importance. The treatment follows that of Bardeen and Townes [252].

If nucleus 1 is coupled to the molecule much more strongly than nucleus 2, the system can be fairly well described by the vector model. In that case, the spin I_1 of nucleus 1 adds vectorially to the molecular angular momentum J to form a resultant F_1, which is quantized with the possible values $J + I_1$, $J + I_1 - 1$, . . . , $|J - I_1|$. The spin I_2 of the second, more weakly coupled nucleus then adds vectorially to F_1 to form the total angular momentum F which may have the values $F_1 + I_2$, $F_1 + I_2 - 1$, . . . , $|F_1 - I_2|$. The two angular momenta I_1 and J may be regarded as precessing around the vector F_1 with a precessional frequency approximately equal to the energy difference between F_1 and $F_1 + 1$ divided by h. Similarly the vectors I_2 and F_1 precess about the vector F, which is fixed in space. If the first nucleus is coupled to the molecule much more strongly than the second nucleus, then I_1 precesses so much more rapidly than I_2 that F_1 and I_2 may be thought of as stationary during a complete cycle of the motion of I_1 and J; hence the interaction between I_2 and J is averaged over this motion. If the interaction between I_2 and J is proportional to the cosine of the angle between them, then the vector model allows a rather simple and accurate calculation of this interaction energy, as will be seen from the discussion of magnetic hyperfine structure in Chap. 8. However, if the interaction is proportional to the square of the cosine between I_2 and J as is the quadrupole interaction, a calculation from more rigorous quantum mechanics must be used. In addition, if the coupling of nucleus 1 and nucleus 2 is not widely different, then averaging over the precession of nucleus 1 is no longer a good approximation, and the vector model must be abandoned for a more sophisticated treatment such as that below.

First consider wave functions which are formed by combining the vectors J and I_1 to produce F_1, and then combining F_1 and I_2 to produce F, and let wave functions of this type be represented by $\psi_1(F_1,F)$. The Hamiltonian for the two nuclear interactions may be written

$$H = H_1(I_1,J) + H_2(I_2,J) \qquad (6\text{-}25)$$

The energy due to H_1 can be readily evaluated for wave functions of the type ψ_1. In the case of quadrupole interactions this energy is obtained simply by letting F_1 take the place of F in expression (6-9). (Here second-order quadrupole interaction will be neglected, so that J is a good quantum number.) There will be a number of wave functions $\psi_1(F_1,F)$ having the same F but different F_1 and hence different energies associated with the interaction H_1. There may also be several wave functions of the same F_1 and different F, which will have the same energy if H_2 is negligibly small. This degeneracy is removed if H_2 is appreciable, but the energy associated with H_2 is not immediately calculable, since the wave function ψ_1 is not an eigenfunction of H_2.

Consider next wave functions formed by combining first the vectors I_2 and J to make the vector which will be designated F_2, then combining F_2 and I_1 to make the total angular momentum F. These are eigenfunctions of H_2 and may be designated $\psi_2(F_2,F)$. The number of different wave functions with the same F will be the same as before, and the two sets of wave functions are linearly related. Their relation may be written

$$\psi_1(F_1,F) = \sum_{F_2} c(F_1,F_2)\psi_2(F_2,F) \tag{6-26}$$

The matrix $c(F_1,F_2)$ is unitary, and the phases of the wave functions may be chosen so that these coefficients are all real. Hence the reverse transformation is

$$\psi_2(F_2,F) = \sum_{F_1} c(F_1,F_2)\psi_1(F_1,F) \tag{6-27}$$

In case both interactions, H_1 and H_2, are appreciable, the eigenfunctions are not given by either $\psi_1(F_1,F)$ or $\psi_2(F_2,F)$ but by appropriate linear combinations of either set. Let the correct wave function be given by the general expansion

$$\psi(F) = \sum_{F_1} a(F_1)\psi_1(F_1,F) \tag{6-28}$$

The Hamiltonian equation $H\psi = W\psi$ becomes

$$\sum_{F_1} H_1(I_1,J)a(F_1)\psi_1(F_1,F) + \sum_{F_1} H_2(I_2,J)a(F_1) \sum_{F_2} c(F_1,F_2)\psi_2(F_2,F)$$
$$= W \sum_{F_1} a(F_1)\psi_1(F_1,F) \tag{6-29}$$

in which use has been made of (6-25). Now ψ_1 is an eigenfunction of H_1 and ψ_2 of H_2, or

$$\begin{aligned} H_1(I_1,J)\psi_1(F_1,F) &= W_1(F_1)\psi_1(F_1,F) \\ H_2(I_2,J)\psi_2(F_2,F) &= W_2(F_2)\psi_2(F_2,F) \end{aligned} \tag{6-30}$$

Using these relations and replacing ψ_2 by (6-27), Eq. (6-29) becomes

$$\sum_{F_1} \Big\{ [A(F_1,F_1) + W(F_1) - W]a(F_1) $$
$$+ \sum_{F_1' \neq F_1} A(F_1,F_1')a(F_1') \Big\} \psi_1(F_1,F) = 0 \tag{6-31}$$

where

$$A(F_1,F_1') = \sum_{F_2} c(F_1,F_2)c(F_1',F_2)W_2(F_2) \tag{6-32}$$

Since all the ψ_1's are orthogonal, (6-31) may be written as a group of homogeneous equations of the form

$$[A(F_1,F) + W(F_1) - W]a(F_1) + \sum_{F_1' \neq F_1} A(F_1,F_1')a(F_1') = 0 \quad (6\text{-}33)$$

for each value of F_1 which, when added to I_2, gives the same value F for the total angular momentum. In order for these equations to have a solution, the determinant of their coefficients must be zero. A solution of this secular determinant gives the possible values of the energy W. If the interaction H_2 is much smaller than H_1 and there is no degeneracy in W_1, then the energy values are given to the first order in H_2 by

$$W = W(F_1) + A(F_1,F_1)$$

or

$$W = W(F_1) + \sum_{F_2} [c(F_1,F_2)]^2 W_2(F_2) \quad (6\text{-}34)$$

This is the case when the eigenfunctions $\psi_1(F_1,F)$ are essentially correct. Then the energy is given by the sum of the energy $W(F_1)$ and the various possible energies $W_2(F_2)$ weighted by the probability $[c(F_1,F_2)]^2$ of finding $\psi_2(F_2,F)$ in the transformation (6-26) of the wave function $\psi_1(F_1,F)$.

The quantities $c(F_1,F_2)$ are given in terms of I_1, I_2, J_1, F_1, F_2, and F by Tables 6-3, 6-4, and 6-5. Table 6-3 gives these coefficients for an arbitrary I_1, J, F_1, F_2, and F when $I_2 = \frac{1}{2}$, and Tables 6-4 and 6-5 give them when $I_2 = 1$ and $\frac{3}{2}$, respectively.

TABLE 6-3. TRANSFORMATION COEFFICIENTS
$c(F_1,F_2)$ for $I_1 = I$, $I_2 = \frac{1}{2}$, $\Sigma = I + J + F + \frac{1}{2}$

F_2	$F_1 = F - \frac{1}{2}$	$F_1 = F + \frac{1}{2}$
$J - \frac{1}{2}$	$\left[\dfrac{(\Sigma - 2F)(\Sigma - 2J)}{(2F + 1)(2J + 1)} \right]^{\frac{1}{2}}$	$\left[\dfrac{(\Sigma + 1)(\Sigma - 2I)}{(2F + 1)(2J + 1)} \right]^{\frac{1}{2}}$
$J + \frac{1}{2}$	$-\left[\dfrac{(\Sigma + 1)(\Sigma - 2I)}{(2F + 1)(2J + 1)} \right]^{\frac{1}{2}}$	$\left[\dfrac{(\Sigma - 2F)(\Sigma - 2J)}{(2F + 1)(2J + 1)} \right]^{\frac{1}{2}}$

These coefficients are related as follows to certain W functions defined by Racah [113]:

$$c(F_1,F_2) = (-1)^{F+J-I_1-I_2}[(2F_1 + 1)(2F_2 + 1)]^{\frac{1}{2}}W(F_1FJF_2;I_2I_1)$$

The W functions (W is not to be confused with the energy) have now been tabulated for most values of the variables which are of interest [674] [854a]. Hence the coefficients $c(F_1,F_2)$ for $I_2 > \frac{3}{2}$ may be obtained from them.

An Example of Two Nuclei with Hyperfine Structure. Calculation of a specific case may be helpful. Let $I_1 = \frac{3}{2}$, $I_2 = 1$, and $J = 2$. F_1 may

Table 6-4. Transformation Coefficients $c(F_1, F_2)$ for $I_1 = I$, $I_2 = 1$. $\Sigma = I + J + F + 1$
(From Bardeen and Townes [252])

F_2	$F_1 = F - 1$	$F_1 = F$	$F_1 = F + 1$
$J - 1$	$\left(\dfrac{(\Sigma - 2F - 1)(\Sigma - 2F)(\Sigma - 2J - 1)(\Sigma - 2J)}{2J(2J + 1)2F(2F + 1)}\right)^{\frac{1}{2}}$	$\left(\dfrac{2(\Sigma - 2F - 1)(\Sigma - 2I - 1)\Sigma(\Sigma - 2J)}{2J(2J + 1)2F(2F + 2)}\right)^{\frac{1}{2}}$	$\left(\dfrac{(\Sigma - 2I - 1)(\Sigma - 2I)\Sigma(\Sigma + 1)}{2J(2J + 1)(2F + 1)(2F + 2)}\right)^{\frac{1}{2}}$
J	$\left(\dfrac{2(\Sigma - 2F)(\Sigma - 2I - 1)\Sigma(\Sigma - 2J)}{2J(2J + 1)2F(2F + 2)}\right)^{\frac{1}{2}}$	$\dfrac{2F(F + 1) + 2J(J + 1) - 2I(I + 1)}{[2J(2J + 1)2F(2F + 2)]^{\frac{1}{2}}}$	$-\left(\dfrac{2(\Sigma - 2I)\Sigma(\Sigma + 1)(\Sigma - 2J - 1)}{2J(2J + 1)(2F + 1)(2F + 2)}\right)^{\frac{1}{2}}$
$J + 1$	$\left(\dfrac{(\Sigma - 2I - 1)(\Sigma - 2I)\Sigma(\Sigma + 1)}{(2J + 1)(2J + 2)2F(2F + 1)}\right)^{\frac{1}{2}}$	$-\left(\dfrac{2(\Sigma - 2I)\Sigma(\Sigma + 1)(\Sigma - 2J - 1)}{(2J + 1)(2J + 2)2F(2F + 2)}\right)^{\frac{1}{2}}$	$\left(\dfrac{(\Sigma - 2J - 1)(\Sigma - 2J)(\Sigma + 1)(\Sigma - 2I)}{(2J + 1)(2J + 2)(2F + 1)(2F + 2)}\right)^{\frac{1}{2}}$

Table 6-5. Transformation Coefficients $c(F_1, F_2)$ for $I_1 = I$, $I_2 = \tfrac{3}{2}$. $\Sigma = I + J + F + \tfrac{3}{2}$.
(From Bardeen and Townes [252])

F_2	$F_1 = F - \frac{3}{2}$	$F_1 = F - \frac{1}{2}$
$J - \frac{3}{2}$	$\left(\dfrac{(\Sigma - 2F - 2)(\Sigma - 2F - 1)(\Sigma - 2F)(\Sigma - 2J - 2)(\Sigma - 2J - 1)(\Sigma - 2J)}{(2F - 1)2F(2F + 1)(2J - 1)2J(2J + 1)}\right)^{\frac{1}{2}}$	$\left(\dfrac{3(\Sigma - 1)(\Sigma - 2F - 2)(\Sigma - 2F - 1)(\Sigma - 2J - 1)(\Sigma - 2J)(\Sigma - 2I - 2)}{(2F - 1)2F(2F + 1)(2J - 1)2J(2J + 1)}\right)^{\frac{1}{2}}$
$J - \frac{1}{2}$	$\left(\dfrac{3(\Sigma - 1)(\Sigma - 2F - 2)(\Sigma - 2F - 1)(\Sigma - 2J - 2)(\Sigma - 2J - 1)(\Sigma - 2I - 2)}{(2F - 1)2F(2F + 1)(2J - 1)2J(2J + 2)}\right)^{\frac{1}{2}}$	$\dfrac{(\Sigma - 2J - 1)(\Sigma - 2J)(-3I + 3J + F - \tfrac{3}{2}) + (\Sigma - 2F)(-3I + 3J + 3J - F - \tfrac{3}{2})[(\Sigma - 2F - 1)(\Sigma - 2J + 1)]^{\frac{1}{2}}}{[(2F - 1)2F(2F + 1)(2J - 1)2J(2J + 2)]^{\frac{1}{2}}}$
$J + \frac{1}{2}$	$\left(\dfrac{3(\Sigma - 1)(\Sigma - 2F - 2)(\Sigma - 2J - 1)(\Sigma - 2J)(\Sigma - 2I - 1)(\Sigma - 2I - 2)}{(2F - 1)2F(2F + 1)2J(2J + 1)(2J + 3)}\right)^{\frac{1}{2}}$	$\dfrac{[(2F - 1)2F + (\Sigma - 2I) + (\Sigma - 2J - 2I) + (\Sigma - 2I - 3J - F + \tfrac{3}{2})][(\Sigma - 2F - 1)(\Sigma - 2J - 1)]^{\frac{1}{2}}}{[(2F - 1)2F(2F + 1)2J(2J + 1)(2J + 3)]^{\frac{1}{2}}}$
$J + \frac{3}{2}$	$\left(\dfrac{(\Sigma - 2I - 3)(\Sigma - 2I - 2)(\Sigma - 2I - 1)(\Sigma - 2I)\Sigma(\Sigma + 1)}{(2F - 1)2F(2F + 1)(2J + 1)(2J + 2)(2J + 3)}\right)^{\frac{1}{2}}$	$-\left(\dfrac{3(\Sigma - 1)(\Sigma - 2F - 2)(\Sigma - 2J - 1)(\Sigma - 2I - 1)(\Sigma - 2I)\Sigma}{(2F - 1)2F(2F + 1)(2J + 1)(2J + 2)(2J + 3)}\right)^{\frac{1}{2}}$

F_2	$F_1 = F + \frac{1}{2}$	$F_1 = F + \frac{3}{2}$
$J - \frac{3}{2}$	$\left(\dfrac{3(\Sigma - 1)(\Sigma - 2F)(\Sigma - 2J - 2)(\Sigma - 2J - 1)(\Sigma - 2J)(\Sigma - 2I - 2)}{(2F + 1)(2F + 2)(2F + 3)(2J - 1)2J(2J + 1)}\right)^{\frac{1}{2}}$	$\left(\dfrac{(\Sigma - 2I - 1)(\Sigma - 2I)\Sigma(\Sigma + 1)(\Sigma - 2J - 1)(\Sigma - 2J)}{(2F + 1)(2F + 2)(2F + 3)(2J - 1)2J(2J + 1)}\right)^{\frac{1}{2}}$
$J - \frac{1}{2}$	$\dfrac{(\Sigma - 1)(\Sigma - 2F)(\Sigma - 2I - 2)(\Sigma - 2J - 1)(\Sigma - 2J) - 2I(\Sigma - 2F - 2)(\Sigma - 2J - 2I - 1)}{(2F + 1)(2F + 2)(2F + 3)(2J - 1)2J(2J + 1)}$	$\left(\dfrac{3(\Sigma + 1)(\Sigma - 2F)(\Sigma - 2I - 1)(\Sigma - 2J - 1)(\Sigma - 2J)(\Sigma - 2I)}{(2F + 1)(2F + 2)(2F + 3)(2J - 1)2J(2J + 1)}\right)^{\frac{1}{2}}$
$J + \frac{1}{2}$	$\dfrac{(\Sigma - 1)(\Sigma - 2F) + 3(\Sigma - 2F)(\Sigma - 2I) - 1][(\Sigma - 2F - 1)(\Sigma - 2J + 3)]^{\frac{1}{2}}}{(2F + 1)(2F + 2)(2F + 3)2J(2J + 1)(2J + 2)}$	$\left(\dfrac{3(\Sigma + 1)(\Sigma - 2F)(\Sigma - 2I - 1)(\Sigma - 2J - 1)(\Sigma - 2I)(\Sigma - 2J)}{(2F + 1)(2F + 2)(2F + 3)2J(2J + 1)(2J + 2)}\right)^{\frac{1}{2}}$
$J + \frac{3}{2}$	$\left(\dfrac{3(\Sigma + 1)(\Sigma - 2F)(\Sigma - 2J - 2)(\Sigma - 2J - 1)(\Sigma - 2J)(\Sigma - 2I)}{(2F + 1)(2F + 2)(2F + 3)(2J + 1)(2J + 2)(2J + 3)}\right)^{\frac{1}{2}}$	$-\left(\dfrac{(\Sigma - 2J - 2)(\Sigma - 2F)(\Sigma + 1)(\Sigma - 2I)(\Sigma - 2J - 1)(\Sigma - 2J)}{(2F + 1)(2F + 2)(2F + 3)(2J + 1)(2J + 2)(2J + 3)}\right)^{\frac{1}{2}}$

then have the values $\frac{1}{2}$, $\frac{3}{2}$, $\frac{5}{2}$, or $\frac{7}{2}$, F_2 the values 1, 2, or 3, and F may have values ranging from $\frac{1}{2}$ to $\frac{9}{2}$. There is only one possible wave function $\psi_1(F_1,F)$ or $\psi_2(F_2,F)$ which gives $F = \frac{9}{2}$, namely, $\psi_1(\frac{7}{2},\frac{9}{2})$ or $\psi_2(3,\frac{9}{2})$, which two functions must hence be identical. Therefore for $F = \frac{9}{2}$, all c's are zero except $c(\frac{7}{2},3)$ which is unity, as may be found from Table 6-4 or Table 6-5. The secular determinant for $F = \frac{9}{2}$ reduces to the single equation

$$W = W_1(\tfrac{7}{2}) + W_2(3)$$

There are two wave functions corresponding to $F = \frac{7}{2}$, linear combinations of $\psi_1(\frac{7}{2},\frac{7}{2})$ and $\psi_1(\frac{5}{2},\frac{7}{2})$, or of $\psi_2(3,\frac{7}{2})$ and $\psi_2(2,\frac{7}{2})$. The coefficients are, from Table 6-4 (or Table 6-5),

$$c(\tfrac{7}{2},3) = -\sqrt{\tfrac{1}{7}} \qquad c(\tfrac{7}{2},2) = \sqrt{\tfrac{6}{7}} \qquad c(\tfrac{5}{2},3) = \sqrt{\tfrac{6}{7}} \qquad c(\tfrac{5}{2},2) = \sqrt{\tfrac{1}{7}}$$

The secular determinant becomes, therefore

$$\begin{vmatrix} [W_1(\tfrac{7}{2}) + \tfrac{1}{7}W_2(3) + \tfrac{6}{7}W_2(2) - W] & \dfrac{\sqrt{6}}{7}[W_2(2) - W_2(3)] \\[2ex] \dfrac{\sqrt{6}}{7}[W_2(2) - W_2(3)] & [W_1(\tfrac{5}{2}) + \tfrac{6}{7}W_2(3) + \tfrac{1}{7}W_2(2) - W] \end{vmatrix}$$

$$= 0 \qquad (6\text{-}35)$$

which gives two possible values of the energy E corresponding to the two different states with $F = \frac{7}{2}$. For $H_2 \ll H_1$, the first-order solutions derived from the diagonal terms are sufficiently accurate. They are

$$\begin{aligned} W &= W_1(\tfrac{7}{2}) + \tfrac{1}{7}W_2(3) + \tfrac{6}{7}W_2(2) \\ \text{and} \qquad W &= W_1(\tfrac{5}{2}) + \tfrac{6}{7}W_2(3) + \tfrac{1}{7}W_2(2) \end{aligned} \qquad (6\text{-}36)$$

If H_2 is not very small, then the quadratic equation (6-35) must be solved.

The energies given by (6-36), which are correct when the coupling of nucleus 2 is much less than that of nucleus 1, are the energies for nucleus 1 coupled to the molecule and slightly perturbed by nucleus 2. If the coupling of nucleus 2 were much greater than that of nucleus 1, then the energies would be given primarily by coupling nucleus 2 to the molecule and these energies would be slightly perturbed by nucleus 1. For intermediate coupling cases, where neither of these approximations holds, description of the energies is much more complex. The energies of the various states have been computed for the case $I_1 = \frac{3}{2}$, $I_2 = 1$, $J = 2$, and their behavior is shown in Fig. 6-3. Assuming the couplings between the nuclei and the molecule are due to quadrupole moments, the energy levels are plotted as a function of the ratio of the quadrupole couplings of the two nuclei. This ratio, $\alpha = (eqQ)_{I=1}/(eqQ)_{I=\frac{3}{2}}$, is measured along the axis of abscissas, the function $(1 + \alpha)/(1 + \alpha^2)^{\frac{1}{2}}$ being plotted linearly for positive α and $(1 - \alpha)/(1 + \alpha^2)^{\frac{1}{2}}$ linearly for negative α. Energy is measured along the ordinate axis, $W/[(-eqQ)_{I=\frac{3}{2}}(1+\alpha^2)^{\frac{1}{2}}]$ being plotted linearly. Such a plot produces smooth curves and allows α a range from $-\infty$ to $+\infty$. Values of $1/\alpha$ rather than α are marked off in the region where $|\alpha| > 1$.

Near the line where $\alpha = 0$, the first-order approximation holds with the levels being close to those expected for a nucleus of spin $\frac{3}{2}$. Near the lines where $\alpha = \infty$, or $1/\alpha = 0$, the first-order approximation again holds, the levels being close to those expected for a nucleus of spin 1. It may

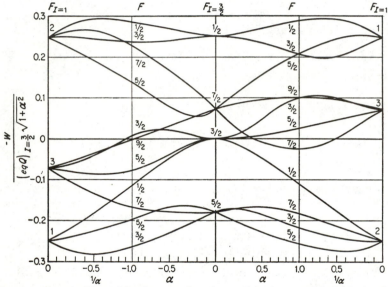

FIG. 6-3. Energies W resulting from quadrupole coupling of two nuclei of spin 1 and $\frac{3}{2}$, when $J = 2$. Parameter $\alpha = (eqQ_{I=1}/(eqQ)_{I=\frac{3}{2}}$. *(From Bardeen and Townes* [252].)

be seen that the first-order approximation is fairly accurate for $|\alpha| < 0.1$ or $|1/\alpha| < 0.1$, but that in intermediate ranges it can be very inaccurate, so that the secular equations must be solved in full. In these intermediate cases, the quantum numbers F_1 and F_2 do not have definite values for each state and the energy levels are designated in the figure only by the values of the total angular momentum F.

An example of a rotational spectrum with two quadrupole moments is given in Fig. 6-4. This is the $J = 2 \leftarrow 1$ transition of the ground state of the linear molecule $Cl^{35}C^{12}N^{14}$. In this case Cl has a spin of $\frac{3}{2}$ and a quadrupole coupling constant of -83.5 Mc, whereas N^{14}, with a spin of 1, has a coupling constant -3.83 Mc. Thus $\alpha \approx 0.05$ and the first-order theory is rather accurate, although some small deviations from it are noticeable.

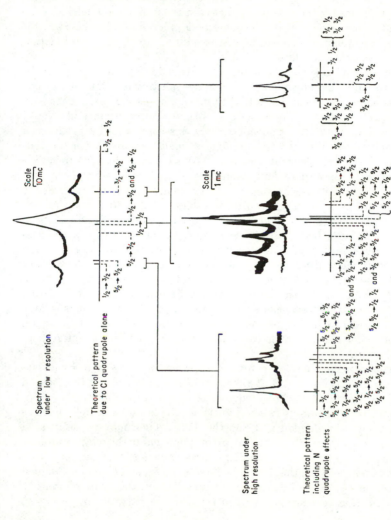

FIG. 6-4. Hyperfine structure of the $J = 2 \leftarrow 1$ transition of $Cl^{35}CN^{14}$ (ground vibrational state) with low and high dispersion and comparison with theoretical pattern. Spins for Cl^{35} and N^{14} are $\frac{3}{2}$ and 1, respectively. *(From Townes, Holden, and Merritt [329].)*

It should be noted that the technique for calculating the energies for the case of two nuclei given above is not restricted to quadrupole coupling but may be used for any case in which the energies W_1 and W_2, that is, the eigenvalues for the interaction between the individual nuclei and the molecule, can be obtained. The methods may thus also be used for magnetic coupling, which will be discussed in Chap. 8, or a combination of magnetic and quadrupole coupling.

Intensities of Hyperfine Components. In addition to energy levels, intensities are generally required. For the splitting of a line due to one nucleus alone, the fractional intensity in each hyperfine component may be formed from the fine- (or hyperfine-) structure intensity formulas (6-6), which are tabulated in Appendix I. If a second nucleus produces a hyperfine interaction, and its coupling is small compared with the coupling of the first nucleus, expressions (6-6) and the tables of Appendix I may again be applied to find intensities of all components of the still more finely split hyperfine structure. It is only necessary to replace J by F_1 and I by I_2 in order to use the tables. Appendix I does not list half-integral values of J, which will be needed when I_1 is half integral and J is replaced by F_1. However, values of intensities for such cases can usually be obtained to sufficient accuracy by interpolation between the nearest integral values. If the couplings of the two nuclei are not widely different, then true intensities cannot be found so directly, but a fair approximation to the intensities may be obtained by interpolating between the two extreme cases when coupling of the first nucleus is large compared with coupling of the second nucleus, and when it is small. Intensities for these extreme cases may be readily obtained from Appendix I.

Exact intensities may, of course, be obtained in cases of intermediate coupling by making use of the energy values obtained by the method described above and solving equations of the type (6-33) for $a(F_1)$. The matrix elements between hyperfine states for a single nucleus may be found in Condon and Shortley [56]. They may be designated $(I_1JF_1|\mu_J|I_1J'F_1')$ and their squares are proportional to the relative intensities given in Appendix I. A similar matrix element $(I_2F_1F|\mu_{F_1}|I_2F_1'F')$ for the transition between the states specified by I_2, F_1, and F can be obtained in the same way. For intermediate coupling cases, the relative intensity for a transition between states i and j would then be given by

$$\left| \sum_{F_1'} \sum_{F_1} a_i(F_1)a_j(F_1')(I_1JF_1|\mu_J|I_1J'F_1')(I_2F_1F|\mu_{F_1}|I_2F_1'F') \right|^2$$

Careful attention must be paid to the phases (see [56], p. 277). This procedure would at best be tedious and in most cases could be replaced by the more rapid approximate method of the preceding paragraph.

Hyperfine Structure Due to More Than Two Nuclei. The case of three nuclei coupled by quadrupole effects to a molecule is uncommon except in symmetric tops involving three halogens such as $AsCl_3$, $HCBr_3$, etc. The resulting hyperfine structure is so complex that so far only one example has been solved, the $J = 1 \leftarrow 0$ transition of $HCBr_3$ [620], [717]. Mizushima and Ito [620] give calculated patterns for the $J = 1 \leftarrow 0$ transition with a nuclear spin of 1, $\frac{3}{2}$, 2, or $\frac{5}{2}$.

For more than two nuclei coupled to the same molecule the method discussed above does not apply. Bersohn [437] has given a satisfactory technique for dealing with these more complicated cases by the use of a formulation due to Racah [113]. The state of a molecular system of angular momentum J with three nuclei of spins I_1, I_2, and I_3 may be designated by the vectors $\mathbf{I} = \mathbf{I}_1 + \mathbf{I}_2$, $\mathbf{I}^0 = \mathbf{I} + \mathbf{I}_3$, and $\mathbf{F} = \mathbf{I}^0 + \mathbf{J}$, or by the corresponding quantum numbers I, I^0, J, and F. Bersohn [437] obtains for the matrix elements of the quadrupole interactions H_Q for the three nuclei between states designated by the above quantum numbers, and when perturbations between different values of J are not important,

$$
\begin{aligned}
&(II^0JF|H_{Q_1} + H_{Q_2} + H_{Q_3}|I'I^{0'}J'F) \\
&= \frac{(J|eQ_1\,\partial^2 V_1/\partial z^2|J)(-1)^{I_2+I_3-I_1-I-I'+J'-F}}{8[J(2J-1)I_1(2I_1-1)]^{\frac{1}{2}}} \times [(2I+1)(2I'+1)(2I^0+1) \\
&\qquad (2I^{0'}+1)(2I_1+1)(2I_1+2)(2I_1+3)(2J+1)(2J+2)(2J+3)]^{\frac{1}{2}} \\
&\times W(I_1I_1II';2I_2)W(II'I^0I^{0'};2I_3)W(II'JJ';2F) \\
&+ \frac{(J|eQ_2\,\partial^2 V_2/\partial z^2|J)(-1)^{I_1+I_3-I_2-I-I'+J'-F}}{8[J(2J-1)I_2(2I_2-1)]^{\frac{1}{2}}} \times [(2I+1)(2I'+1)(2I^0+1) \\
&\qquad (2I^{0'}+1)(2I_2+1)(2I_2+2)(2I+3)(2J+1)(2J+2)(2J+3)]^{\frac{1}{2}} \\
&\times W(I_2I_2II';2I_1)W(II'I^0I^{0'};2I_3)W(II'JJ';2F) \\
&+ \frac{(J|eQ_3\,\partial^2 V_3/\partial z^2|J)(-1)^{I-I_3+I^0-I^{0'}+J'-F}}{[J(2J-1)I_3(2I_3-1)]^{\frac{1}{2}}}\,\delta_{II'} \times [(2I^0+1)(2I^{0'}+1) \\
&\qquad (2I_3+1)(2I_3+2)(2I_3+3)(2J+1)(2J+2)(2J+3)]^{\frac{1}{2}} \\
&\qquad\qquad\qquad \times W(I_3I_3I^0I^{0'};2I)W(I^0I^{0'}JJ';2F) \quad (6\text{-}37)
\end{aligned}
$$

$\delta_{II'}$ is unity for $I = I'$, and is otherwise zero. The matrix elements in (6-37) of the form $(J|eQ_1\,\partial^2 V_1/\partial z^2|J')$ depend on matrix elements of the direction cosines as may be seen from (6-12). If the rotational state is not much perturbed so that J is a good quantum number, these take the form $e(q_JQ)_1$, where q_J is given by (6-17) or (6-18). The W functions of the particular form used in (6-37) may be found in tables by Biedenharn, Blatt, and Rose [674].

In order to solve a specific problem, secular equations must be set up by the use of matrix elements (6-37) and solved. These matrix elements may of course also be used for the case of two nuclei. However, techniques already discussed for handling this case are usually simpler to understand and to apply.

CHAPTER 7

MOLECULES WITH ELECTRONIC ANGULAR MOMENTUM

7-1. Introduction. Molecules so far discussed have been in $^1\Sigma$ electronic states. That is, the sum of the orbital angular momenta of their electrons is zero, as is the sum of the electron spins. The $^1\Sigma$ case is usually the only one which needs consideration since it is the lowest electronic energy level of the vast majority of molecules. However, there are a few gaseous molecules, approximately 0.1 per cent of the total, which do have electronic angular momentum in the ground state, and hence are normally in states other than $^1\Sigma$. These include O_2 and the rare molecules with an odd number of electrons, such as NO, NO_2, and ClO_2. Molecules with an odd number of electrons cannot have zero electronic spin, and hence are never in $^1\Sigma$ states. Furthermore, if microwave spectroscopy is done on gases at high temperature or excited in an electrical discharge, molecules will be found in excited or dissociated states which are not $^1\Sigma$. Many molecules dissociate under high excitation into smaller parts which have an odd number of electrons and hence possess electronic angular momentum. These dissociated parts are of considerable importance in chemical reactions and gaseous discharges, and are usually called free radicals since they are free, chemically active groups of atoms or radicals.

Since the electrons move much more rapidly than the nuclei of a molecule, to a good approximation the nuclei may be assumed stationary while electron motions in the molecule are being discussed (*cf.* Born-Oppenheimer approximation mentioned in Chap. 1). The electrons in a molecule may have both orbital and spin angular momentum, and the resulting electronic states can be described in a way quite analogous to the electronic states of atoms discussed in Chap. 5. There is a basic difference between the molecular and atomic cases, however, because the electric fields due to an atomic nucleus are spherically symmetric, while those due to the two or more nuclei in a molecule are by no means spherically symmetric. Since a molecular electron does not move in a spherically symmetric field, torques are exerted on it by the field, and its angular momentum cannot be constant as it is in an atom. The simplest molecular case is a diatomic or linear molecule, where the fields are symmetric about the molecular axis. Because of this symmetry, no torque about the axis is exerted on a molecular electron, and the component of its angular momentum in the direction of the molecular axis is constant.

The diatomic molecule is somewhat similar to an atom subjected to a very large electric field along the direction of the molecular axis. This field produces a large Stark effect (see Chap. 10) which interferes with the orbital motion of the electrons. Although the orbital angular momentum is not constant and the quantum number L loses its significance, in many cases the projection M_L of L on the axis is constant. For molecules, the symbol Λ is written for M_L, since L itself is not of significance, and Λ is the Greek substitute for L. The energy is dependent on the value of Λ, which is, of course, integral and may equal L, $L - 1, \ldots, -L$. However, positive and negative values of Λ have the same energy (see Chap. 10), so that, unless $\Lambda = 0$, the levels are doubly degenerate. This degeneracy may be removed by the rotation-electron interaction discussed below called Λ-type doubling.

There is a close parallelism between the various angular momenta involved in molecules and those in atoms and the notation used in each case is outlined in Table 7-1. Molecules involve more types of angular momenta because of the possibility of end-over-end rotation of a molecule. In addition, for linear and symmetric-top molecules the projection of various angular momenta on the symmetry axis may be of interest as well as their projections on a direction fixed in space. Only the latter occur in discussions of atoms. It must be noted that not all the quantum numbers listed in Table 7-1 are normally used for any one case. For example, if the projection Λ of L on the molecular axis is constant, then its projection M_L on a space-fixed axis is not constant, and usually not of interest.

In diatomic or linear molecules where spin-orbit effects are not enormously large, the projection Λ of L on the molecular axis is constant and replaces L in importance. If the linear molecule is bent as in a bending vibrational mode, or if the molecule is not linear, then the electric fields are not symmetrical about the axis and Λ is no longer fixed or well defined. In nonlinear molecules, the orbital motion of electrons is almost completely "quenched" or suppressed, and a spin momentum is the only angular momentum in the molecule of distinctly electronic origin.

If the electronic orbital angular momentum Λ along the internuclear axis of a linear molecule is $0, \pm 1, \pm 2, \pm 3, \ldots$, the molecule is said to be in a $\Sigma, \Pi, \Delta, \Phi, \ldots$ state, respectively, in analogy with the atomic S, P, D, F states having $L = 0, 1, 2,$ or 3. In this and several other respects, notation for molecular spectra can be regarded as the Greek translation of notation for atomic spectra. If the electronic spin is $0, \frac{1}{2}, 1,$ $\ldots$, the state is said to be singlet, doublet, or triplet—again in analogy with the atomic case—and a corresponding superscript is applied to the left of the state designation. Thus $^2\Pi$ indicates $S = \frac{1}{2}$ and $\Lambda = 1$.

The component of the total angular momentum along the molecular axis may have the values $\Lambda + S$, $\Lambda + S - 1$, . . . , $\Lambda - S$, and is written as a right-hand subscript, for example, $^2\Pi_{\frac{3}{2}}$, $^2\Pi_{\frac{1}{2}}$. Its absolute value is called Ω, that is, $\Omega = |\Lambda + \Sigma|$.

Van Vleck [657] has discussed the general problem of coupling angular-momentum vectors in molecules. He shows that, if the internal angular

TABLE 7-1. EXPLANATION OF NOTATION FOR ANGULAR MOMENTA INVOLVED IN MOLECULAR SPECTRA AND COMPARISON WITH NOTATION FOR ATOMIC SPECTRA

When the nuclear spin is zero or unimportant, J is used instead of **F**. When there is no electronic spin, J is identical with **N** and is used in its place. Similarly, when the electronic spin is zero, the projection of J on the molecular axis is equal to K, which is used instead of Ω.

	Molecule	Atom
Nuclear spin angular momentum	**I**	**I**
Projection on space-fixed axis	M_I	M_I
Projection on molecular axis	Ω_I	
Electron spin angular momentum	**S**	**S**
Projection on space-fixed axis	M_S	M_S
Projection on molecular axis	Σ	
Electronic orbital angular momentum	**L**	**L**
Projection on space-fixed axis	M_L	M_L
Projection on molecular axis	Λ	
Sum of spin and electronic orbital momentum (**L + S**)	$\mathbf{J}_a$	**J**
Orbital angular momentum due to nuclear motion (rotation of molecule)	**O**	
Total orbital angular momentum including rotation of molecule	**N**	**L**
Projection on molecular axis	K	
Total angular momentum excluding nuclear spin	**J**	**J**
Projection on space-fixed axis	M_J	M_J
Projection on molecular axis (absolute value)	Ω	
Total angular momentum including nuclear spin	**F**	**F**
Projection on space-fixed axis	M_F	M_F
Projection on molecular axis	Ω_F	

momenta (all momenta except J, the total) are reversed in sign, they and the total angular-momentum vectors follow the same commutation rules as angular-momentum vectors in an atom. Hence for every problem of coupling angular-momentum vectors in a molecule there is a corresponding situation in atoms, and the matrix elements given by Condon and Shortley [56] for atoms can be applied in the molecular problem. Van Vleck's discussion is highly recommended to the advanced student, but it will not be needed below since no actual solution of intermediate coupling or other complex cases will be carried out.

7-2. Hund's Coupling Cases.* The notation of the last section is not always appropriate because the molecular angular momenta may interact or couple together in a variety of ways. The coupling schemes or cases were first systematically treated by Hund, who described five ideal cases. Molecules do not fit such ideal descriptions exactly, but Hund's cases are very good approximations to the actual states of many linear molecules. (For nonlinear molecules Hund's considerations are not useful.) Which case applies most closely depends on the relative strength of the various couplings, or the relative energy of interaction between the vectors. In all known cases, coupling between the nuclear spin and other vectors by hyperfine interactions is much smaller than other couplings. Interactions whose relative magnitude needs to be considered occur between any two of the vectors **L, S, N, O**, and **A**, where **A** is a vector along the molecular axis. Interactions between two of these vectors will be represented by a notation like **SA**, which indicates an interaction between the electronic spin and the molecular axis.

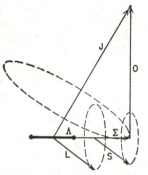

Fig. 7-1. Vector diagram of Hund's case (a). L and S precess rapidly around the molecular axis, which precesses more slowly around the total angular momentum J. Circles indicate these precessional motions.

Hund's Case (a). In Hund's case (a) the strongest interactions are between the molecular axis and **L** and between the axis and **S**. That is,

$$LA \gg LS \text{ or } LO$$
$$SA \gg SN$$

The vector model of this case is shown in Fig. 7-1. **L** interacts strongly with the axial field of the molecule and hence precesses about the molecular axis, so that its projection Λ is constant. Similarly **S** precesses with constant projection Σ, so that the total angular momentum along the molecular axis is $\Omega = |\Lambda + \Sigma|$. $\Lambda + \Sigma$ adds vectorially to the angular momentum **O** of the end-over-end molecular rotation to form the total angular momentum (excluding nuclear spin) **J**. The angular momenta Ω and **O** hence precess around the vector **J** which is fixed in space.

The relations between **J** and Ω in case (a) are just the same as those between **J** and the angular momentum K around the axis of a symmetric-top molecule. The quantity Ω is integral if the molecule contains an even number of electrons, but it is half integral if the number of electrons is odd. This is because Σ takes on only the values $S, S - 1, \ldots, -S$, and the sum of electron spins S is half integral if the number of electrons

---

* For an extensive treatment of Hund's coupling cases see also Herzberg [471].

is odd.   As in the case of a symmetric top, the total angular momentum
J cannot be less than its projection Ω on the axis so that J has values
$\Omega, \Omega + 1, \Omega + 2, \ldots$

Hund's Case (b).   In Hund's case (b) the electron spin is coupled more
strongly to $\mathbf{N} = \mathbf{\Lambda} + \mathbf{O}$ than to the molecular axis.   L, however, is still
strongly coupled to the molecular axis.

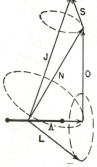

$$\mathbf{LA} \gg \mathbf{LS} \text{ or } \mathbf{LO}$$
$$\mathbf{SN} \gg \mathbf{SA}$$

The appropriate vector diagram for this case is shown
in Fig. 7-2.   **L** precesses rapidly around the molecular
axis.   **Λ** adds to **O** to form the total orbital angular
momentum **N**.   **N** and **S** add to form **J**, about which
they precess.

The spin is usually coupled to the axis by spin-orbit
coupling, *i.e.*, the spin is coupled to **Λ** rather than to the
axis itself.   Hence for molecules with $\Lambda = 0$, coupling
between the spin and molecular axis is small, and it is
these molecules with $\Lambda = 0$ which typically fall in
Hund's case (b).   When $\Lambda = 0$, the orbital angular
momentum is entirely due to molecular rotation
($N \equiv O$), and is perpendicular to the molecular axis.

There are molecules, however, with Λ not equal to
zero which fall approximately in case (b).   These are
usually very light molecules such as hydrides which
rotate rapidly and for large values of O produce a coupling between **O** and **S** which is greater than the spin-
axis interaction.   The **SA** interaction tends to be weak
in these cases anyhow, because the nuclear charge Z is
small (page 122), and spin-orbit coupling is small.
The free radical OH is in a $^2\Pi$ state and is a molecule
of this type.

Fig. 7-2. Vector
diagram for Hund's
case (b).   Preces-
sion of the molec-
ular axis, repre-
sented by the
largest ellipse, is
slower than the pre-
cession of **L** about
the axis, but faster
than the precessions
of **S** and **N** about **J**.
When $\Lambda = 0$, **N** is
identical with **O**
and is perpendic-
ular to the molec-
ular axis.

Hund's cases (a) and (b) are by far the most important.   However,
other cases are possible.   More detailed discussion of the following three
rare or nonexistent cases is given by Mulliken [22], [26], [100] and Weizel
[29].

Hund's Case (c).   For heavy nuclei, the atomic spin-orbit interaction
becomes very large.   Similarly in molecules involving heavy nuclei, the
interaction **LS** may be larger than the interaction between **L** and the
molecular axis.   This gives Hund's case (c), when

$$\mathbf{LS} \gg \mathbf{LA}$$

In this case Λ and S are not good quantum numbers, but **L** and **S** add
vectorially to form a resultant $\mathbf{J}_a$ which is then coupled to the inter-

nuclear axis with a projection Ω on this axis. Ω adds vectorially to the angular momentum **O** of end-over-end rotation to form the total angular moment **J**. Figure 7-3 is a vector diagram for this case.

Hund's Case (*d*). If the coupling between **L** and the angular momentum **O** is much larger than that between **L** and the molecular axis, Hund's case (*d*) occurs.

$$\mathbf{LO} \gg \mathbf{LA}$$

A vector diagram of this case is shown in Fig. 7-4.

Hund's Case (*e*). **L** and **S** may conceivably be strongly coupled and their resultant coupled to **O** rather than to the internuclear axis. This would give Hund's case (*e*) which has not yet been observed.

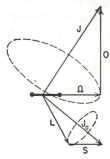

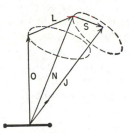

FIG. 7-3. Vector diagram for Hund's case (*c*). Precession of **L** and **S** about $\mathbf{J}_a$ is much faster than the precession (not shown) of $\mathbf{J}_a$ about the internuclear axis.

FIG. 7-4. Vector diagram for Hund's case (*d*). **L** is coupled to **O** rather than to the internuclear axis. **S** adds to the resultant **N** to form **J**.

Hund's coupling cases are idealizations to which many molecules approximately conform. However, noticeable deviations from these idealizations often occur and may be expected to be particularly evident if the spectra of these types of molecules are accurately measured by microwave spectroscopy. These deviations represent a partial uncoupling of two vectors by the effect of a third vector. In some cases, too, the deviations are very large, for a molecule may fall approximately into one coupling case for low rotational states, but into another case for high rotational states. For intermediate rotational states, such a molecule does not fall into any of Hund's coupling cases but is said to have intermediate coupling.

Some of the most interesting examples of uncoupling phenomena are those in which end-over-end rotation of the molecule uncouples the electronic momenta from the molecular axis, as it tends to do especially for high rotational states where the rotation is rapid. In extreme cases, rapid rotation may almost completely uncouple **S** from the molecular axis, producing a transition from Hund's case (*a*) for low rotational states to case (b) for higher states. Molecular rotation also interacts with the electronic orbital angular momentum Λ and removes the degeneracy corresponding to the two different possible orientations of Λ or Ω along the molecular axis which are degenerate when the molecule is not rotating. The effect is called Λ-type doubling, and may be regarded as an incipient uncoupling of L from the molecular axis by the rotation which, in extreme cases, would lead to Hund's case (d).

Another phenomenon associated with uncoupling is the magnetic field produced by end-over-end molecular rotation. This rotation slightly uncouples L from the molecular axis even in Σ states and orients L with respect to the angular momentum of rotation. Thus L uncoupling results in fields which interact with the electron spin moments or, when hyperfine structure can occur, with nuclear moments.

7-3. Rotational Energies. The rotational energies of molecules with electronic angular momentum will be treated first with the assumption of pure Hund's coupling cases, and then the additional energy terms due to uncoupling phenomena will be discussed.

Case (a). Because of the similarity of J and Ω in Hund's case (a) to J and K for an ordinary symmetric top, the energy levels in this case must be of the same form as (3-5), letting B_v be in energy units rather than in cycles per second.

$$W = B_v[J(J + 1) - \Omega^2] + A\Omega^2 \qquad (7\text{-}1)$$

In this case, however, the "rotational constant" A is extremely large, and $A\Omega^2$ represents electronic energy. Since any transition involving electronic energy would not lie in the microwave range, and such energy is not usually called rotational energy, terms of this type are omitted, so that

$$W(J) = B_v[J(J + 1) - \Omega^2] \qquad (7\text{-}2)$$

In addition, since changes in Ω would almost always produce frequencies higher than the microwave range, $B_v\Omega^2$ must be constant and may be neglected. Hence for the purposes of microwave spectroscopy the still simpler form

$$W(J) = B_vJ(J + 1) \qquad (7\text{-}3)$$

may be used.

Rotational energy levels for case (a) $^2\Pi$ and $^3\Delta$ levels are shown in Fig. 7-5. Except for the nonexistence of values of J less than Ω and for the added energy $B_v\Omega^2$, which is independent of J, the energy levels are much the same as those for a normal molecule in a $^1\Sigma$ state. However, half-integral values of J may occur instead of only integral values, since $J = \Omega,\ \Omega + 1,\ \Omega + 2,\ \ldots$. The levels also show an additional splitting, the Λ-doubling, which will be discussed more fully below.

Case (b).   If the electron spin is neglected, then the rotational energy of a case (b) molecule is of the same simple form as (7-3), that is,

$$W_R = B_v N(N + 1)$$

or $B_v J(J + 1)$, since $J \equiv N$ when spin is neglected.

The simplest type of case (b) molecules are in $^2\Sigma$ states, having electron spin of $\frac{1}{2}$.   The magnetic moment associated with this spin interacts

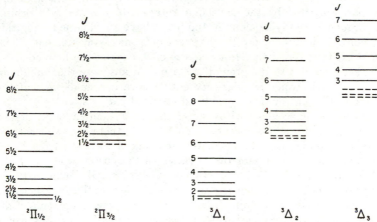

Fig. 7-5. The lower rotational energy levels of a case (a) molecule in $^2\Pi$ or $^3\Delta$ state. Λ-type doubling is too small to be shown on this scale.   The dotted levels cannot occur, since J must be $\geq \Omega$.

with the magnetic fields produced by molecular rotation, thus giving an interaction energy proportional to the cosine of the angle between $\mathbf{S}$ and $\mathbf{N}$, or of the form

$$W_M = \gamma \mathbf{S} \cdot \mathbf{N} \tag{7-4}$$

From the vector model,

$$\mathbf{S} \cdot \mathbf{N} = \frac{J(J + 1) - S(S + 1) - N(N + 1)}{2}$$

Hence for $J = N + \frac{1}{2}$, $W_M = (\gamma/2)N$, and for $J = N - \frac{1}{2}$,

$$W_M = -\frac{\gamma}{2}(N + 1)$$

Thus the total rotational energy including the electron spin interaction is [22]

$$\begin{aligned}
W &= B_v N(N + 1) + \frac{\gamma}{2} N & \text{when } J = N + \tfrac{1}{2} \\
W &= B_v N(N + 1) - \frac{\gamma}{2}(N + 1) & \text{when } J = N - \tfrac{1}{2}
\end{aligned} \tag{7-5}$$

These expressions neglect small centrifugal distortion terms of the form $D_vN^2(N+1)^2$, which occasionally may be needed.

The constant γ is a measure of the strength of the magnetic field produced by molecular rotation. This field can be regarded as partly due to the simple rotation of charges distributed in the molecule, but its largest part is due to L uncoupling [17]. L uncoupling (also discussed in Chaps. 1 and 8) is a transfer of angular momentum from the end-over-end motion of the molecule to electronic orbital motion. Electrons are thus partly excited by the molecular rotation to a state with orbital angular momentum lying along N and hence have a magnetic field in this direction which interacts with the electron's magnetic moment. The ease of excitation, and therefore the size of γ, depends on how near the excited Π state lies to the ground Σ state. In any case the amount of excitation of electronic angular momentum is rather small, so that the electronic state may still be considered essentially a $^2\Sigma$ state.

For case (b) molecules with S larger than $\frac{1}{2}$, $i.e.$, in $^3\Sigma$ or $^4\Sigma$ states, other types of interactions occur. In a $^3\Sigma$ state, there are two electrons with spins parallel, and the magnetic moments of these electrons interact. This "spin-spin interaction" varies as the $\cos^2 \theta$, where θ is the angle between the direction of the two parallel spins and the line between the two electrons, which when averaged is equivalent to the angle between S and the molecular axis [15]. In addition, the magnetic moments of the electrons magnetically polarize the molecule. Their fields partly excite the electrons into Π orbits, and these excited states produce fields which react back on the electron moments. Hebb [73] showed that this energy of magnetic polarization varies also with the $\cos^2 \theta$ and is not easily distinguishable from the spin-spin interaction. This type of interaction is also an L uncoupling due to $S \cdot L$ type interaction, and is large if there is a low-lying Π state which can easily be excited. It does not occur when $S = \frac{1}{2}$, for the same reasons that no nuclear quadrupole interactions occur when the nuclear spin is $\frac{1}{2}$, since, like the quadrupole interaction, it depends on $\cos^2 \theta$. In fact, this type of effect is sometimes called a pseudo quadrupole interaction because of the formal similarity.

To a first approximation, energies for a $^3\Sigma$ molecule are given by

$$W(J = N + 1) = B_vN(N + 1) - \frac{2\lambda(N + 1)}{2N + 3} + \gamma(N + 1)$$
$$W(J = N) = B_vN(N + 1) \qquad\qquad (7\text{-}6)$$
$$W(J = N - 1) = B_vN(N + 1) - \frac{2\lambda N}{2N - 1} - \gamma N$$

where γ is a constant which is a measure of the magnetic interaction of the type described above for the $^2\Sigma$ state and λ is a constant which measures the magnetic interaction of the spin-spin and polarization types.

The term involving λ represents "pseudo quadrupole" interactions, and except for the constants involved it may be obtained from the expression (5-45) for quadrupole energy in first-order perturbation theory. N takes the place of J, S that of I, and J that of F in using (5-45). As might be expected, this energy is not strongly dependent on N, and for large N it has the same value for $J = N + 1$ and $J = N - 1$, since these two states both have the same value of $\cos^2 \theta$ in the classical limit of large N. The interaction involving λ is usually so big that the approximate formulas (7-6) are too inaccurate to be of much value in interpreting microwave spectra. This interaction perturbs the molecular rotational motion in much the same way as a large nuclear quadrupole coupling (cf. Chap. 6). A given energy state cannot have a precisely defined rotational angular momentum N but involves small admixtures of the states $N + 2$ and $N - 2$. In the case of O_2, for which the ground state is $^3\Sigma$, λ is approximately 60,000 Mc, while B is near 40,000 Mc, so that the pseudo quadrupole interaction represents a large perturbation of the rotational energy levels. The exact energies may be obtained by solving a secular equation, as was first done by Schlapp [85]. A more precise form of Schlapp's equations given by Miller and Townes [841] is

$$W(J = N - 1) = B_v N(N + 1) - B_v(2N - 1) - \lambda - \frac{\gamma}{2}$$
$$+ \left[\lambda^2 - 2\lambda \left(B_v - \frac{\gamma}{2} \right) + (2N - 1)^2 \left(B_v - \frac{\gamma}{2} \right)^2 \right]^{\frac{1}{2}}$$
$$W(J = N) = B_v N(N + 1) \tag{7-7}$$
$$W(J = N + 1) = B_v N(N + 1) + B_v(2N + 3) - \lambda - \frac{\gamma}{2}$$
$$- \left[\lambda^2 - 2\lambda \left(B_v - \frac{\gamma}{2} \right) + (2N + 3)^2 \left(B_v - \frac{\gamma}{2} \right)^2 \right]^{\frac{1}{2}}$$

Constant terms $2\lambda - \gamma$ have been subtracted from each of the expressions (7-7), since they do not contribute to the frequency of transitions. The fine-structure separations $W(J = N) - W(J = N + 1)$ and

$$W(J = N) - W(J = N - 1)$$

are plotted in Fig. 7-6 for the O_2 molecule.

The O_2 molecule has no electric dipole moment, but since it is in a $^3\Sigma$ state, there is a magnetic dipole moment which allows transitions between rotational levels and their fine-structure components. Since B_v for O_2 is 43,100 Mc, and the rotational transitions obey the selection rule $\Delta N = \pm 2$ [990], they occur at wavelengths too short for present microwave techniques. However, transitions between fine-structure components without a change in N occur at wavelengths near 5 mm and have been fairly extensively studied.   Matrix elements for these transitions are [56]

$$|\mu_{J=N+1\leftarrow N}|^2 = 4\mu_0^2 \frac{N(2N+3)}{(2N+1)(N+1)}$$
$$|\mu_{J=N-1\leftarrow N}|^2 = 4\mu_0^2 \frac{(N+1)(2N-1)}{(2N+1)N} \tag{7-8}$$

Even though the Bohr magneton μ_0 which is involved is rather small (0.9×10^{-20} emu as compared with 10^{-18} esu for typical electric dipole

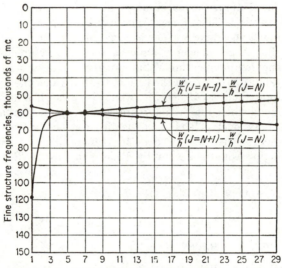

FIG. 7-6. Separation of spin triplets in $^3\Sigma_g^-$ ground state of O_2. (*From Artman* [892b].)

moments), some of these transitions have intensities as large as 10^{-5} cm^{-1} because of the large fraction of O_2 molecules in each state of low N.

Approximately 25 lines have been measured for $O^{16}O^{16}$ ([441], [554], [576], and [963]) and fitted to expression (7-7). From this fit, B_0 is found to be 43,101 Mc, $\lambda = 59,501$ Mc, and $\gamma = 252.7$ Mc. However, it is necessary to replace B_v by

$$B_{ef} = B_0 - N(N+1)D_v \tag{7-9}$$

for an exact fit.   Since N varies from 1 to about 25, and $D_v = 4B^3/\omega_e^2$, $N(N+1)D_v$ becomes as large as $0.0025B_v$ in (7-9). In addition, centrifugal stretching of the molecule appears to vary λ a small amount for large N.   The experimental measurements can be fitted with

$$\lambda = 59,501.6 + 0.0575N(N+1) \text{ Mc}$$

The spectrum of other isotopic species of O_2 may be predicted from the determination of B_v, λ, and γ for $O^{16}O^{16}$, since B_v and γ are inversely proportional to the reduced mass, while λ is approximately independent of the isotopic mass. Frequencies for $O^{16}O^{18}$ and $O^{18}O^{18}$ have been found to agree rather well with such predictions [841].

Pseudo quadrupole interactions also occur in $^4\Sigma$ states, another example of Hund's case (b). Since a quartet state has $S = \frac{3}{2}$, J may have the values $N + \frac{3}{2}$, $N + \frac{1}{2}$, $N - \frac{1}{2}$, and $N - \frac{3}{2}$. A good approximation to the rotational energy levels of a $^4\Sigma$ molecule has been given by Budó [78].

$$
\begin{aligned}
W(J = N - \tfrac{3}{2}) &= B_vN(N + 1) - B_v(2N - 1) \\
&\quad + B_v[4(N - \tfrac{1}{2})^2 - 6\lambda(N + 1) + 9\lambda^2]^{\frac{1}{2}} - 3(\gamma/2)(N + 1) \\
W(J = N - \tfrac{1}{2}) &= B_vN(N + 1) - B_v(2N - 1) \\
&\quad + B_v[4(N - \tfrac{1}{2})^2 + 6\lambda(N - 2) + 9\lambda^2]^{\frac{1}{2}} - (\gamma/2)(N + 4) \\
W(J = N + \tfrac{1}{2}) &= B_vN(N + 1) + B_v(2N + 3) \\
&\quad - B_v[4(N + \tfrac{3}{2})^2 - 6\lambda(N + 3) + 9\lambda^2]^{\frac{1}{2}} + (\gamma/2)(N - 3) \\
W(J = N + \tfrac{3}{2}) &= B_vN(N + 1) + B_v(2N + 3) \\
&\quad - B_v[4(N + \tfrac{3}{2})^2 + 6\lambda N + 9\lambda^2]^{\frac{1}{2}} + 3(\gamma/2)N
\end{aligned}
\tag{7-10}
$$

where λ and γ are coupling constants similar to those of expressions (7-6).

The general trend of fine-structure energy for $^4\Sigma$ levels is shown in Fig. 7-7. For large N, the levels with $J = N \pm \frac{3}{2}$ approach the same

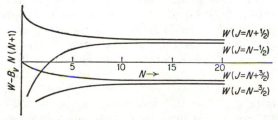

Fig. 7-7. Behavior of fine-structure energy for a $^4\Sigma$ state. *(From Budó [78].)*

energy value except for the small terms dependent on γ. Similarly, the levels with $J = N \pm \frac{1}{2}$ tend toward the same value.

Case (c). Rotational energies of molecules following case (c) are identical in form with those for case (a) and hence are given by (7-2) or (7-3).

Case (d). Rotational energies for case (d) molecules are given to a first approximation by

$$
W(O) = B_vO(O + 1)
\tag{7-11}
$$

Each of the rotational levels is split into several components by the interaction **LO** and the smaller interaction **SN**.

7-4. Spin Uncoupling. A common case of intermediate coupling is a transition from case (a) to case (b). For low rotational states, **S** is coupled to Λ, or to the molecular axis as in case (a). However, when the rotational frequency becomes larger than the frequency of precession of S about Λ, S is uncoupled from the molecular axis and couples instead to the total orbital angular momentum N as in case (b). Energy levels in the intermediate conditions where S is not simply coupled to Λ or N must be obtained from solution of a secular equation. For a $^2\Pi$ state with intermediate coupling, Hill and Van Vleck [10] obtain

$$
\begin{aligned}
W_1 &= B_v[(J + \tfrac{1}{2})^2 - \Lambda^2] - \left[(J + \tfrac{1}{2})^2 \left(B_v - \frac{\gamma}{2} \right)^2 \right. \\
&\qquad \left. + \Lambda^2(A - \gamma) \left(\frac{A + \gamma}{4} - B \right) \right]^{\frac{1}{2}} - \frac{\gamma}{2} - D_v J^4 \\
W_2 &= B_v[(J + \tfrac{1}{2})^2 - \Lambda^2] + \left[(J + \tfrac{1}{2})^2 \left(B_v - \frac{\gamma}{2} \right)^2 \right. \\
&\qquad \left. + \Lambda^2(A - \gamma) \left(\frac{A + \gamma}{4} - B \right) \right]^{\frac{1}{2}} - \frac{\gamma}{2} - D_v(J + 1)^4
\end{aligned}
\tag{7-12}
$$

where B_v is the usual rotational constant, A is the interaction constant between S and Λ (energy $= A\mathbf{S} \cdot \mathbf{\Lambda}$), and γ is the same interaction constant between S and the molecular rotation given in Eqs. (7-4) or (7-5). The energy W_1 applies to the state for which $J = N + \tfrac{1}{2}$ when J is very large, and W_2 to that for which $J = N - \tfrac{1}{2}$. The centrifugal distortion terms $D_v J^4$ and $D_v(J + 1)^4$ which are given are only approximately correct [77] but in most cases are so small that they are sufficiently accurate. Equations (7-12) are somewhat more complex than those usually given, because terms in γ have been included. γ is always much smaller than A and may for approximate results be omitted. For large J, where the transition to case (b) is complete, these equations reduce to the same form as (7-5). It is interesting to note that the molecular energies follow case (b) approximately not only when the rotational frequency $2B_v J$ is much larger than the precessional frequency $|A\Lambda|$ but also, rather unexpectedly, when $(A + \gamma)/4$ is nearly equal to B_v.

For small spin uncoupling ($2BJ \ll |A\Lambda|$), expressions (7-12) take the following form if terms independent of J are omitted.

$$
\begin{aligned}
W_1 &= B_v \left(1 - \frac{B_v}{|A\Lambda|} \right) J(J + 1) - D_v J^4 \\
W_2 &= B_v \left(1 + \frac{B_v}{|A\Lambda|} \right) J(J + 1) - D_v(J + 1)^4
\end{aligned}
\tag{7-13}
$$

Connections between energy levels in case (a), case (b), and intermediate coupling conditions are shown in Fig. 7-8. Since light molecules tend to have large rotational constants B and small fine-structure interaction constants A, they generally approach case (b) even for moderately low values of J. The free radical OH and other hydrides represent extremes

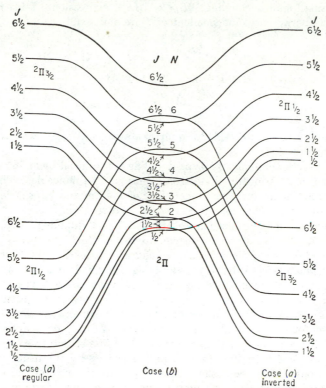

FIG. 7-8. Transition of $^2\Pi$ states from case (a) to case (b). On the left is a regular case (a) with the constant A of the interaction $A\,\mathbf{L}\cdot\mathbf{S}$ positive. On the right A is negative.

of this type. Heavier molecules tend to be close to case (a) for any rotational states of interest to microwave spectroscopy.

Similar formulas have been given for triplet states $(S = 1)$ [55], [69], [72]

$$W_1 = B_v[J(J + 1) - \sqrt{Z_1} - 2Z_2] - D_v(J - \tfrac{1}{2})^4$$
$$W_2 = B_v[J(J + 1) + 4Z_2] - D_v(J + \tfrac{1}{2})^4 \qquad (7\text{-}14)$$
$$W_3 = B_v[J(J + 1) + \sqrt{Z_1} - 2Z_2] - D_v(J + \tfrac{3}{2})^4$$

where $Z_1 = \Lambda^2 A/B_v(A/B_v - 4) + \tfrac{4}{3} + 4J(J + 1)$

$$Z_2 = \frac{1}{3Z_1}[\Lambda^2 A/B_v(A/B_v - 1) - \tfrac{4}{9} - 2J(J + 1)]$$

The term dependent on γ has been omitted from these expressions. Here W_1 is the energy of the state which for large J has $J = N + 1$, W_2 that for $J = N$, and W_3 that for $J = N - 1$.

For small spin uncoupling $(2B_vJ \ll |A\Lambda|)$, expressions (7-14) become (7-15) if energy terms independent of J are omitted.

$$W_1 = B_v\left(1 - \frac{2B_v}{A\Lambda}\right)J(J+1) - D_v(J - \tfrac{1}{2})^4$$

$$W_2 = B_vJ(J+1) - D_v(J + \tfrac{1}{2})^4 \tag{7-15}$$

$$W_3 = B_v\left(1 + \frac{2B_v}{A\Lambda}\right)J(J+1) - D_v(J + \tfrac{3}{2})^4$$

Formulas similar to (7-14) for quartet states $(S = \tfrac{3}{2})$ are also available [68], [78], [83], [87].

7-5. Λ-type Doubling. Λ-type doubling is produced by an interaction between the rotational and electronic motions in a molecule. This may be regarded as an incipient uncoupling of L from the internuclear axis which, under extreme conditions, would lead to Hund's case (d) where L is coupled to the molecular rotation. It is a doubling because this effect splits the two otherwise degenerate levels which are always present when $\Lambda \neq 0$. In general form and characteristics Λ-type doubling is entirely analogous to the l-type doubling described in Chap. 2, and in fact is the prototype of l-type doubling. As in the case of l-type doubling, the two perturbed states do not correspond simply to projections $+\Lambda$ and $-\Lambda$ of L on the internuclear axis, but are linear combinations of wave functions for positive and negative values of Λ. Λ-type doubling energies are in general smaller than rotational or fine-structure energies and hence have not been studied with high accuracy by the usual types of spectroscopy. However, they should be accurately measurable by microwave spectroscopy and may prove to be a particularly interesting feature of the microwave spectra of paramagnetic molecules.

The simplest cases of Λ-type doubling to discuss are singlet states, where the electron spin gives no complications since it is zero. In l-type doubling it may be remembered that the splitting was proportional to $B(B/\omega_e)^l$, where B/ω_e is the ratio of rotational energy to vibrational energy. Similarly for Λ-type doubling, when the electronic motion rather than a vibration provides the angular momentum about the axis, the Λ-type splitting is proportional to $B(B/\nu_e)^\Lambda$, where ν_e is the energy required to raise the electron from the ground state to some nearby excited state. Since the ratio B/ν_e is generally small $(\tfrac{1}{1000})$, Λ-type doubling for singlet states with Λ greater than 1 is almost always negligible, even for high-resolution microwave spectroscopy. When $\Lambda = 1$, the splitting of the two otherwise degenerate levels is given by

$$W = q_\Lambda J(J+1) \tag{7-16}$$

Theoretical evaluation of q_Λ is in general quite complicated. However, if an electron of orbital angular momentum is assumed to precess so that the projection of l on the molecular axis is 1, and the only low-lying excited Σ and Δ states correspond to the projection of this same l being 0 and 2, respectively, then q_Λ can be simply expressed [17], [27]. (This is the postulate of "pure precession.")

$$q_\Lambda = 2B_v^2 \frac{l(l + 1)}{h\nu_e[\Pi\,\Sigma]} \tag{7-17}$$

where $h\nu_e[\Pi\,\Sigma]$ is the energy difference between the Π and Σ electronic levels. Expression (7-17) allows at least a rough estimate of q_Λ in more complex cases.

When the electron spin is not zero, Λ-type doubling may be modified, especially if Hund's case (a) applies. However, when case (b) occurs, S does not affect the interaction between N and Λ, and (7-16) applies with the simple replacement of J by N

$$W = q_\Lambda N(N + 1) \tag{7-18}$$

A general summary of Λ-type doubling effects in pure cases (a) or (b) is given in Table 7-2. Rough magnitudes of the constants in Table 7-2

TABLE 7-2. SUMMARY OF Λ-TYPE DOUBLING EFFECTS
(After Van Vleck [17])

State	Coupling case	W (splitting of levels due to Λ-type doubling effects)
$^1\Pi$		$q_\Lambda J(J + 1)$
$^2\Pi$	Case (b)	$q_\Lambda N(N + 1)$
$^2\Pi$	Case (a) $\Omega = \frac{1}{2}$	$a(J + \frac{1}{2})$
	$\Omega = \frac{3}{2}$	$b(J^2 - \frac{1}{4})(J + \frac{3}{2})$
$^3\Pi$	Case (b)	$q_\Lambda N(N + 1)$
$^3\Pi$	Case (a) $\Omega = 0$	f
	$\Omega = 1$	$q_\Lambda J(J + 1)$
	$\Omega = 2$	~ 0
$^1\Delta$		$dJ(J^2 - 1)(J + 2)$

can be obtained from the pure-precession assumption as

$$q_\Lambda = \frac{4B_v^2}{h\nu_e} \qquad b = \frac{8B_v^3}{Ah\nu_e} \qquad d = \frac{48B^4}{(h\nu_e)^3}$$

$$a = \frac{4AB_v}{h\nu_e} \qquad f = \frac{2A^2}{h\nu_e}$$

where $h\nu_e$ is the separation between the ground and first excited electronic energy level.

For intermediate cases of $^2\Pi$ states the amount of Λ-type doubling has been given by Van Vleck [17].

$$W = \frac{q_\Lambda}{2}(J + \tfrac{1}{2})\left[\left(2 + \frac{A'}{B'}\right)\left(1 + \frac{2 - A/B}{X}\right)\right.$$
$$\left. + \frac{4(J + \tfrac{3}{2})(J - \tfrac{1}{2})}{X}\right] \quad (7\text{-}19)$$

where $X = \pm\left[\dfrac{A}{B}\left(\dfrac{A}{B} - 4\right) + 4(J + \tfrac{1}{2})^2\right]^{\frac{1}{2}}$ and A'/B' is a quantity which is approximately equal to A/B for the $^2\Pi$ state of the molecule in question, but which actually depends on matrix elements of A and B between this and other electronic states. When the interaction constant A is positive (regular fine structure), then a positive sign for X gives the state corresponding to $^2\Pi_{\frac{3}{2}}$, and a negative sign for X the state corresponding to $^2\Pi_{\frac{1}{2}}$. When A is negative (inverted fine structure), a negative X gives the $^2\Pi_{\frac{3}{2}}$ state, and a positive X the $^2\Pi_{\frac{1}{2}}$ state.

Matrix elements for transitions between Λ-type doublets are similar to those between l-type doublets, and hence for case (b) from (2-16) are

$$|\mu_{ij}|^2 = \frac{\mu^2\Lambda^2}{(J + 1)(2J + 1)} \quad (7\text{-}20)$$

where μ is the dipole moment along the molecular axis. For case (a), Λ in (7-20) is replaced by Ω. Although many of the diatomic hydrides have rotational frequencies too high for ordinary microwave spectroscopy, Λ-type doubling in these hydrides should afford some microwave spectra. Microwave transitions between Λ doublets for several rotational states of the free radical OH have been observed and studied by Sanders, Schawlow, Dousmanis, and Townes [861], [971a].

Measurements on the ultraviolet spectrum of OH show that the rotational energy levels for this molecule can be fitted to an expression of the type [7-12] with $\gamma \approx 0$, $B_0 = 555,040$ Mc, and $A = -7.547 B_0$ for the ground vibrational state. Hence for relatively small values of J, the rotational frequency $2JB_0$ becomes appreciable with respect to the fine-structure energy A, and intermediate coupling occurs. For higher J, the molecule corresponds to Hund's case (b).

The Λ doubling measured from ultraviolet spectra of OH can be fitted rather well to (7-19) with $q_\Lambda = 1060$ Mc for the ground vibrational state and for values of N up to 15 as shown in Fig. 7-9. For higher values of N, the Λ-type doubling is somewhat better fitted by $q_\Lambda = 925$ Mc. This decrease in q_Λ is presumably due to centrifugal effects. It may be noted from Fig. 7-9 that, except for small values of N, the Λ splitting is approximately proportional to N^2, as is indicated in Table 7-2 for a $^2\Pi$ case (b).

Transitions between the Λ-type doublets of OH fall in the microwave region for small and medium values of N, and have been observed [861], [971a] by microwave techniques for $N = 2, 3, 4,$ and 5. The measured frequencies are fitted to expression (7-19) with an accuracy of about 40 Mc by the constants $q_\Lambda = 1159$ and $A'/B' = -6.073$. It should be noted that A'/B' is different from A/B, which was given above as -7.547.

In obtaining q_Λ and A'/B', small corrections to B and q_Λ in expression (7-19) have been made to allow for its variation with J due to centrifugal

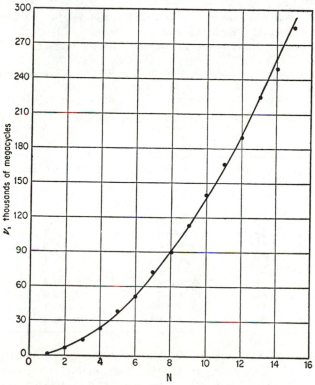

Fig. 7-9. Comparison of theory and experimental measurements of Λ-type doubling in OH. (*Experimental measurements from Dieke and Crosswhite* [266a].)

distortion. In addition, one may expect additional corrections to (7-19) of order B^3/ν_e^2, or about $\frac{1}{1000}$ as large as the measured splitting. These corrections are probably responsible for deviations noted between (7-19) and the experimental results for OH and for OD as given in Table 7-3.

Λ doubling in $O^{18}H$ and OD can be fairly well predicted by noting that B and B' are inversely proportional to the reduced mass and q_Λ is proportional to B^2. A comparison between experimental measurements on four Λ doublets for OD with expression (7-19) is shown in Table 7-3.

Again in this case, variation in B and q_Λ due to centrifugal distortion has been allowed for, as indicated in the title of Table 7-3.

In NO, B is much smaller and A larger than for OH, so that $A = 75B$, and NO is a rather good case (a) molecule. The Λ-type doubling of the $^2\Pi_{\frac{1}{2}}$ state has been accurately measured from the $\frac{3}{2} \leftarrow \frac{1}{2}$ rotational transition of NO [782b], [924], which gives the constant a of Table 7-2 a value of 355.2 Mc. From this value, the approximate magnitude of the constant b of Table 7-2 can be obtained, assuming the "pure-precession" relations of Table 7-2, as 0.13 Mc, and thus the Λ-type doubling of the $^2\Pi_{\frac{3}{2}}$ state estimated. Experimental measurement [671] of the doubling of the $^2\Pi_{\frac{3}{2}}$ state by microwave techniques gives the value $b = 0.28$ Mc.

TABLE 7-3. COMPARISON BETWEEN THEORETICAL AND EXPERIMENTAL
FREQUENCIES FOR Λ-TYPE DOUBLING IN $O^{16}D$
Theoretical results come from expression (7-19) with $A/B_0 = -14.147$,
$$\frac{A'}{B_0'} = -11.461, \ B = B' = B_0 - D_0N(N + 1), \ q_\Lambda = 327.32\left[1 - \frac{2D_0}{B_0}N(N + 1)\right]$$

(From Sanders, Dousmanis, and Townes [971a])

N	J	Measured frequency, Mc	Calculated frequency, Mc
$\Pi_{\frac{1}{2}}$ state			
3	$2\frac{1}{2}$	8,120.4	8,108.2
4	$3\frac{1}{2}$	9,587.9	9,582.1
5	$4\frac{1}{2}$	10,200.7	10,203.3
6	$5\frac{1}{2}$	9,922.8	9,931.4
$\Pi_{\frac{3}{2}}$ state			
5	$5\frac{1}{2}$	8,672.4	8,630.4
6	$6\frac{1}{2}$	12,918.0	12,882.8
7	$7\frac{1}{2}$	18,009.6	18,000.7
8	$8\frac{1}{2}$	23,907.1	23,949.5

Hyperfine structure appears prominently in the spectra of OH and NO. Such structure has been omitted in the discussion of rotational spectra of these molecules, and has been subtracted out of the experimental results to give the amount of Λ doubling. This hyperfine structure will be discussed in Chap. 8.

7-6. Nonlinear Molecules. Valence electrons in nonlinear molecules do not move in cylindrically symmetric fields, and hence no component of their orbital angular momentum is constant or "quantized." Rather, the electronic orbital momentum is part of the rotational momentum of the whole molecule. Any interaction between electron spin and orbit of the type $A\mathbf{L} \cdot \mathbf{S}$ comes about only through some slight uncoupling of L from the rotation of the molecule, and hence occurs as second- or higher-order perturbation effects. Henderson and Van Vleck [283] have shown that these have the form

$$
W = A \frac{\left[\alpha E + \beta \sum_K |a_{NK}|^2 K^2 + \gamma \right]}{N(N+1)} C +
$$

$$
\frac{A^2}{2} \left[\frac{\alpha' E + \beta' \sum_K |a_{NK}|^2 K^2}{N(N+1)} + \gamma' \right] \frac{\frac{3}{4}C(C+1) - S(S+1)N(N+1)}{(2N-1)(2N+3)} \tag{7-21}
$$

where $C = J(J+1) - S(S+1) - N(N+1)$

$\qquad E$ = energy of molecular rotation without electron spin effects

$\qquad a_{NK}$ = the coefficient in expression (4-17) which gives an expansion of any asymmetric-top wave function in terms of symmetric-top wave functions with quantum numbers K.

$\qquad \alpha, \beta, \gamma, \alpha', \beta'$ and γ' are all constants dependent on the structure of the molecule, and independent of angular-momentum quantum numbers.

It may be seen that the first term of (7-21) is a dipole-like interaction, and the second term a pseudo quadrupole interaction with the same dependence on angular-momentum quantum numbers as a genuine quadrupole interaction [cf. (5-45)]. In addition, the pseudo quadrupole term in (7-21) must be zero when S is less than 1, or for singlet and doublet states. For a symmetric-top molecule, $|a_{NK}|^2$ is unity for one particular value of K and otherwise zero, and since

$$
E = BN(N+1) + (A-B)K^2
$$

(7-21) reduces to

$$
W = \left[\frac{aK^2}{N(N+1)} + b \right] C
$$

$$
+ \left[\frac{a'K^2}{N(N+1)} + b' \right] \left[\frac{\frac{3}{4}C(C+1) - S(S+1)N(N+1)}{(2N-1)(2N+3)} \right] \tag{7-22}
$$

where $a, b, a',$ and b' are constants.

The only nonlinear molecules with unpaired electron spins which have so far been studied in the gaseous state are NO_2 and ClO_2. ClO_2 is somewhat asymmetric (asymmetry parameter $\kappa \approx 0.85$) but its spectrum observed in the optical region [144] fits expression (7-22) satisfactorily [283]. Some spectral lines of NO_2 have been observed in the microwave region, but their fine structure has not yet been well analyzed [497], [615].

CHAPTER 8

MAGNETIC HYPERFINE STRUCTURE IN MOLECULAR SPECTRA

8-1. Introduction. Although hyperfine structure due to nuclear magnetic dipole moments is not so prominent in molecular spectra as that involving nuclear electric quadrupole moments, it is by no means insignificant. In molecules with electronic angular momentum, magnetic hyperfine structure is comparable in size with magnetic hyperfine structure in atoms and is usually much larger than that due to quadrupole moments. It is only because this type of molecule is uncommon that magnetic hyperfine structure is not prominent in molecular spectra. For molecules in $^1\Sigma$ states the average of each component of the electronic angular momentum is so small that it is usually considered zero. Even for these molecules, however, there are weak interactions involving nuclear magnetic moments. These include interaction between magnetic moments of two nuclei in the same molecule (spin-spin interactions), interaction between a nuclear magnetic moment and the rather small magnetic fields produced by molecular rotation ($\mathbf{I} \cdot \mathbf{J}$ interactions), and magnetic polarization of a molecule by a nuclear magnetic moment (pseudo quadrupole interaction).

When a molecule has electronic angular momentum, hence is in some state other than $^1\Sigma$, the magnetic fields associated with this momentum interact strongly with the nuclear moments present in the molecule, giving a magnetic hyperfine structure comparable in size with that found in atoms (10^3 Mc would be typical). The interaction is due either to electronic orbital angular momentum, $\mathbf{L}$ or Λ, or to spin angular momentum, $\mathbf{S}$ or Σ.

In the case of orbital angular momentum, the interaction energy is given by (5-49) if the spin is assumed to be zero, *i.e.*, if we let $\mathbf{j} = 1$. Then (5-49) and (5-51) give the energy as

$$H_{Il} = \frac{2\mu_0\mu_I}{Ir^3} \mathbf{I} \cdot \mathbf{1} \tag{8-1}$$

where μ_0 = the Bohr magneton
μ_I = nuclear magnetic moment
$\mathbf{I}$ = nuclear spin

r = distance between the electron and the nucleus

$\mathbf{l}$ = orbital angular momentum of an electron with respect to the nucleus in units of $h/2\pi$ (the quantity $h/2\pi$ will frequently be written $\hbar$)

To calculate the energy resulting from (8-1) in the first-order perturbation approximation, we need to average this expression (since in a molecule neither $\mathbf{l}$ nor r is constant) and to sum over all the electrons which contribute to the total orbital angular momentum $\mathbf{L}$. In the simplest cases, the nucleus may be assumed to lie on the molecular axis. $\mathbf{L}$ has an average nonzero value only if the molecule is linear, so that the average value of $\mathbf{L}$ is just $\mathbf{k}\Lambda$, where $\mathbf{k}$ is a unit vector along the molecular axis and Λ is as usual the component of $\mathbf{L}$ along this direction. Then if only one electron contributes to the orbital angular momentum $\mathbf{L}$,

$$H_{IL} = \frac{2\mu_0\mu_I\Lambda}{I}\left(\frac{1}{r^3}\right)_{\text{av}}\mathbf{I}\cdot\mathbf{k} \qquad (8\text{-}2a)$$

If more electrons are involved,

$$H_{IL} = \frac{2\mu_0\mu_I}{I}\sum_n\left(\frac{1}{r_n^3}\right)_{\text{av}}\Lambda_n\mathbf{I}\cdot\mathbf{k} \qquad (8\text{-}2b)$$

where the sum is over each electron contributing to the hyperfine structure and Λ_n is the average projection of the orbital angular momentum of the nth electron on the axis.

Interaction between a nuclear magnetic moment and an electron spin moment is somewhat more complex. There is the classical interaction between two dipoles $\mathbf{\mu}_1$ and $\mathbf{\mu}_2$ of the form

$$W_{\mu_1\mu_2} = \frac{\mathbf{\mu}_1\cdot\mathbf{\mu}_2}{r^3} - \frac{3(\mathbf{\mu}_1\cdot\mathbf{r})(\mathbf{\mu}_2\cdot\mathbf{r})}{r^5} \qquad (8\text{-}3)$$

which for the nucleus $\left(\mathbf{\mu}_1 = \frac{\mu_I}{I}\mathbf{I}\right)$ and electron spin ($\mathbf{\mu}_2 = -2\mu_0\mathbf{S}$) becomes

$$(H_{IS})_1 = \frac{-2\mu_0\mu_I}{I}\left[\frac{\mathbf{I}\cdot\mathbf{S}}{r^3} - \frac{3(\mathbf{I}\cdot\mathbf{r})(\mathbf{S}\cdot\mathbf{r})}{r^5}\right] \qquad (8\text{-}4)$$

In addition, there is an interaction of the type found between a nuclear magnetic moment and an S electron in atoms, which is not given simply by (8-4). It can be written approximately as in the atomic case from (5-49) and (5-54) as

$$(H_{IS})_2 = \frac{16\pi}{3}\frac{\mu_0\mu_I}{I}\psi^2(0)\mathbf{I}\cdot\mathbf{S} \qquad (8\text{-}5)$$

Frosch and Foley [686] (cf. also [916a]) have discussed these magnetic interactions for a linear molecule, and obtain the sum of H_{IL}, $(H_{IS})_1$, and

$(H_{IS})_2$ in the common Hund's cases (a) and (b) as

$$H^1 = a\Lambda \mathbf{I} \cdot \mathbf{k} + b\mathbf{I} \cdot \mathbf{S} + c(\mathbf{I} \cdot \mathbf{k})(\mathbf{S} \cdot \mathbf{k}) \tag{8-6}$$

where

$$a = \frac{2\mu_0\mu_I}{I}\left(\frac{1}{r^3}\right)_{\mathrm{av}}$$

and to a good approximation

$$b = \frac{2\mu_0\mu_I}{I}\left[\frac{8\pi}{3}\psi^2(0) - \frac{3\cos^2\theta - 1}{2r^3}\right]_{\mathrm{av}}$$

$$c = \frac{3\mu_0\mu_I}{I}\left(\frac{3\cos^2\theta - 1}{r^3}\right)_{\mathrm{av}}$$

Expression (8-6) applies accurately only when Λ is a "good" quantum number. Here θ is the angle between the molecular axis and the radius r from the nucleus to the electron. Λ and S are assumed to be good quantum numbers, since second-order terms involving changes of Λ or S give extremely small contributions to the energy by comparison with the first-order terms of (8-6). Some of these second-order terms will be discussed below, however, in connection with magnetic effects in $^1\Sigma$ molecules, where the first-order terms given by (8-6) are zero.

Expression (8-6) applies to each electron in the molecule. Of course, most of the electrons occupy orbits in pairs with oppositely directed spins so that the second and third parts of (8-6) cancel out, and the orbits are usually filled so that the orbital angular momentum Λ of most of the electrons cancel. Expression (8-6) need then only be applied to each of the "unpaired" electrons whose angular momentum is not cancelled. The quantity a refers only to electrons with orbital angular momentum. A spherical distribution of an electron around the nucleus would make $\left(\dfrac{3\cos^2\theta - 1}{2}\right)_{\mathrm{av}}$ equal to zero, so that in this case the electron spin would interact with the nuclear magnetic moment only through the term in b proportional to $\psi^2(0)$. The probability $\psi^2(0)$ of the electron's being found at the nucleus is usually negligibly small for an electron in a p atomic orbit. For an S-type orbit, however, it is large enough so that $\dfrac{8\pi}{3}\psi^2(0)$ is much larger than $\left(\dfrac{3\cos^2\theta - 1}{2r^3}\right)_{\mathrm{av}}$ for a p orbit and $b \gg a$. Hence, whenever there is an appreciable amount of S character to the wave function of an unpaired electron, the hyperfine interaction which is proportional to $\psi^2(0)$ may be expected to dominate.

8-2. Coupling Schemes for Magnetic Hyperfine Structure. Evaluation of the energy resulting from the interaction (8-6) depends on the coupling scheme which is followed by the particular case of interest. In addition to the electron angular momentum which is coupled according to the usual Hund's coupling cases (*cf.* Chap. 7) to the molecular axis or to the momentum of end-over-end rotation of the molecule, we have also the nuclear spin momentum. The nuclear spin may be coupled with varying strength to the several molecular vectors, providing additional coupling possibilities. The commonly expected coupling schemes are shown in Fig. 8-1. They are classified according to Hund's scheme, with a subscript α indicating that the nuclear spin is most strongly coupled to the molecular axis [as is S in Hund's case (a)] and a subscript β indicating that the nuclear spin is not coupled to the molecular axis but to

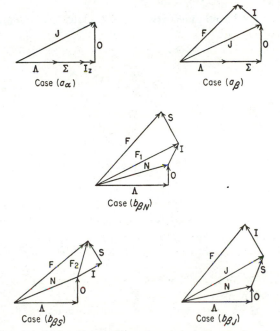

Case (a_α) Case (a_β)

Case $(b_{\beta N})$

Case $(b_{\beta S})$ Case $(b_{\beta J})$

FIG. 8-1. Molecular coupling schemes including nuclear spin.

some other vector [as in case (b)]. For Hund's case (a), one may expect the nuclear spin to be coupled either to the molecular axis [case (a_α)] or to J [case (a_β)]. However, for Hund's case (b), where the electron spin is not coupled to the molecular axis, it is very unlikely that the nuclear spin will be coupled to the molecular axis since the interaction of its small nuclear magnetic moment with the molecular fields should be considerably less than that between the electron moment and the molecular fields. Hence only case (b_β) is expected to occur.

In case (a), (8-6) becomes (omitting the negligible perturbation of Λ by hyperfine interactions)

$$H^1 = [a\Lambda + (b + c)\Sigma]\mathbf{I} \cdot \mathbf{k} \qquad (8\text{-}7)$$

since $\mathbf{S} \cdot \mathbf{k} = \Sigma$, and $\mathbf{I} \cdot \mathbf{S} = (\mathbf{I} \cdot \mathbf{k})(\mathbf{S} \cdot \mathbf{k})$ when $\mathbf{S}$ precesses about the molecular axis $\mathbf{k}$. The hyperfine energy for case (a_α) is the sum of a magnetic interaction given by (8-7) and a term dependent on the molecular moment of inertia. The entire effect of Ω_I, or $\mathbf{I} \cdot \mathbf{k}$, the projection of I on the axis, is

$$W = [a\Lambda + (b + c)\Sigma]\Omega_I - hB[\Omega_I + 2(\Lambda + \Sigma)]\Omega_I \qquad (8\text{-}8)$$

All parts of this expression are constant for a given electronic level except Ω_I, which has one of the values $I, I - 1, \ldots, -I$. The second term of (8-8) may be obtained from (7-2), which gives the rotational energy in case (a) as $W_J = hB[J(J + 1) - \Omega^2]$. But in case (a_α), Ω may include Ω_I, which therefore affects the rotational energy. In case (a_β), the vector model gives

$$\mathbf{I} \cdot \mathbf{k} = \frac{(\mathbf{I} \cdot \mathbf{J})(\mathbf{J} \cdot \mathbf{k})}{J(J + 1)}$$

or, since $\mathbf{J} \cdot \mathbf{k} = \Lambda + \Sigma = \Omega$,

$$W = [a\Lambda + (b + c)\Sigma] \frac{\Omega}{J(J + 1)} \mathbf{I} \cdot \mathbf{J} \qquad (8\text{-}9)$$

where

$$\mathbf{I} \cdot \mathbf{J} = \frac{F(F + 1) - J(J + 1) - I(I + 1)}{2}$$

In this case, the hyperfine structure decreases with increasing J, since $\mathbf{I}$ becomes more and more nearly perpendicular to the molecular axis.

When neither the electronic nor nuclear spin is coupled to the molecular axis, one encounters the further complication of three possible coupling schemes:

1. Case $(b_{\beta N})$: $\mathbf{N}$ and $\mathbf{I}$ are coupled to form $\mathbf{F}_1$, then $\mathbf{F}_1$ and $\mathbf{S}$ couple to form the total angular momentum $\mathbf{F}$.

2. Case $(b_{\beta S})$: $\mathbf{S}$ and $\mathbf{I}$ are coupled to form $\mathbf{F}_2$, then $\mathbf{F}_2$ and $\mathbf{N}$ couple to give $\mathbf{F}$.

3. Case $(b_{\beta J})$: $\mathbf{N}$ and $\mathbf{S}$ are coupled to form $\mathbf{J}$, then $\mathbf{J}$ and $\mathbf{I}$ couple to give $\mathbf{F}$.

It may be noted that a subscript has been added to the designation of the coupling scheme which indicates the vector to which $\mathbf{I}$ is coupled. These three cases are illustrated in Fig. 8-1. Case $(b_{\beta N})$ is not expected to occur commonly because the much larger magnetic moment associated with the electron spin should couple much more strongly to N than does the nuclear magnetic moment.

The energy resulting from the first two terms of the interaction (8-6) can be obtained from the vector model for each of the three above cases. However, the last term of (8-6) requires a more complicated procedure such as multiplication of the matrices for $\mathbf{I} \cdot \mathbf{k}$ and $\mathbf{S} \cdot \mathbf{k}$. Frosch and Foley [686] have given all matrix elements necessary to evaluate the energy from (8-6) using the coupling schemes $(b_{\beta S})$ and $(b_{\beta J})$. For the pure coupling case $(b_{\beta J})$, which is expected to be more common than $(b_{\beta S})$, the energies obtained are as follows:

When $S = \frac{1}{2}$:

$$W_{J=N+\frac{1}{2}} = \left[\frac{2a\Lambda^2}{(N+1)(2N+1)} + \frac{b}{2N+1} + \frac{c}{(2N+1)(2N+3)}\left(1 + \frac{2\Lambda^2}{N+1}\right) \right] \mathbf{I} \cdot \mathbf{J}$$

$$W_{J=N-\frac{1}{2}} = \left[\frac{2a\Lambda^2}{N(2N+1)} - \frac{b}{2N+1} + \frac{c}{(2N-1)(2N+1)}\left(1 - \frac{2\Lambda^2}{N}\right) \right] \mathbf{I} \cdot \mathbf{J}$$

(8-10)

where

$$\mathbf{I} \cdot \mathbf{J} = \frac{1}{2}[F(F+1) - J(J+1) - I(I+1)]$$

When $S = 1$:

$$W_{J=N+1} = \left[\frac{a\Lambda^2}{(N+1)^2} + \frac{b}{(N+1)} + \frac{c}{(N+1)(2N+3)}\left(1 + \frac{2\Lambda^2}{N+1}\right) \right] \mathbf{I} \cdot \mathbf{J}$$

$$W_{J=N} = \left[\frac{a\Lambda^2(N^2+N-1)}{N^2(N+1)^2} + \frac{b}{N(N+1)} + \frac{c}{N(N+1)}\left(1 - \frac{2\Lambda^2}{N(N+1)}\right) \right] \mathbf{I} \cdot \mathbf{J}$$ (8-11)

$$W_{J=N-1} = \left[\frac{a\Lambda^2}{N^2} - \frac{b}{N} + \frac{c}{N(2N-1)}\left(1 - \frac{2\Lambda^2}{N}\right) \right] \mathbf{I} \cdot \mathbf{J}$$

Because of the accuracy of microwave measurements, noticeable deviations from the pure coupling cases listed above are to be expected, and intermediate coupling cases must be considered. In addition, when large coupling constants occur, N may not be a "good" quantum number, and second-order effects involving excitation of other N states must be considered. Frosch and Foley [686] give most of the matrix elements needed for calculation of the energy in these cases.

8-3. Examples of Magnetic Hyperfine Structure in Molecules with Electronic Angular Momentum. The molecule NO is a rather good example of Hund's case (a), its ground electronic state being $^2\Pi$. The nuclear magnetic moment of nitrogen and the molecular fields produce a hyperfine structure which follows case (a_β). The $^2\Pi_{\frac{3}{2}}$ state shows a hyperfine structure which is given by expression (8-9) with

$$a\Lambda + (b + c)\Sigma = 74.1 \text{ Mc}$$

[899], [686]. Hyperfine structure of the $^2\Pi_{\frac{1}{2}}$ state involves some additional effects which are discussed in Sec. 8-6. The common isotope of

oxygen, O^{16}, has zero spin and hence produces no hyperfine structure in NO. The spectrum of the $^2\Pi_{\frac{3}{2}}$ state of NO was actually measured in a strong magnetic field [435], [899] and will be discussed in more detail in Chap. 11 in connection with Zeeman effects.

The ground state of the oxygen molecule O_2 is $^3\Sigma$, and hence a good Hund's case (b). The coupling between S and N is approximately 60,000 Mc, whereas the hyperfine coupling between S and I for O^{17} in $O^{16}O^{17}$ is only a few hundred megacycles, so that $O^{16}O^{17}$ represents a rather good case $(b_{\beta J})$. For this molecule, having $S = 1$ and $\Lambda = 0$, the energies given by (8-11) simplify to

$$W_{J=N+1} = \frac{\mathbf{I} \cdot \mathbf{J}}{N+1}\left(b + \frac{c}{2N+3}\right)$$

$$W_{J=N} = \frac{\mathbf{I} \cdot \mathbf{J}}{N(N+1)}\,(b + c) \tag{8-12}$$

$$W_{J=N-1} = \frac{\mathbf{I} \cdot \mathbf{J}}{N}\left(-b + \frac{c}{2N-1}\right)$$

where

$$\mathbf{I} \cdot \mathbf{J} = \frac{F(F+1) - I(I+1) - J(J+1)}{2}$$

The expressions (8-11) or (8-12) have been found to fit very satisfactorily the observed hyperfine splitting of the microwave spectrum of $O^{16}O^{17}$ with the values $I = \frac{5}{2}$, $b = -102$ Mc, and $c = 140$ Mc [841]. Effects of the quadrupole moment of O^{17} and second-order effects involving decoupling of S from N can be shown to be less than about 1 Mc, which was the accuracy of the measurements, and hence were neglected.

Magnetic hyperfine structure has also been observed in OH [861], [971a], where coupling intermediate between case (a) and case (b) occurs.

The way in which a, b, and c depend on the electronic wave function of the molecule, and the information about NO and O_2 which can be derived from them are discussed in Chap. 9.

8-4. Nonlinear Molecules. Electrons in molecules which are not linear have no constant component of orbital angular momentum, since the molecular electric fields interfere with each component of the electrons' orbital motion. Hence in first-order approximation the interaction (8-1) is zero because the average value of each component of L is zero. The electron spin, however, may still be nonzero in such a molecule and may give a sizable hyperfine structure through interactions of types (8-3) and (8-5). No detailed evaluation of the energies resulting from (8-3) and (8-5) has been made for nonlinear molecules. However, coupling of the case $(b_{\beta J})$ type should be expected. For a given rotational state of angular momentum N due to end-over-end rotation, and given value of $\mathbf{J} = \mathbf{N} + \mathbf{S}$, the energy will have the form

$$W = 2C(\mathbf{I} \cdot \mathbf{J}) = C[F(F+1) - I(I+1) - J(J+1)] \tag{8-13}$$

Two examples of hyperfine structure in asymmetric molecules with electronic angular momentum have so far been observed, NO_2 and ClO_2. Both these molecules have an odd number of electrons and a value $S = \frac{1}{2}$. Selection rules for the hyperfine structure ($\Delta F = 0, \pm 1$) and relative intensities of the components are the same as with fine structure or other forms of hyperfine structure (see Chap. 5). Except for very small values of J, the dominantly strong hyperfine components are those for which ΔF is the same as ΔJ. Assuming ($b_{\beta J}$) coupling, the frequencies of the strong hyperfine components ($\Delta F = \Delta J$) for a particular transition can be obtained from (8-13) as

$$\nu = \nu_0 + \frac{F - J}{h}[(C_2 - C_1)(F - J) + C_2(2J_2 + 1) - C_1(2J_1 + 1)]$$

(8-14)

where J_1 and J_2 = total angular momentum exclusive of nuclear spin in lower and upper states, respectively

C_1 and C_2 = hyperfine constants in (8-13) for lower and upper states, respectively

ν_0 = a constant given primarily by the frequency of the rotational transition without hyperfine structure

$F - J$ = difference between F and J in either upper or lower states, which may have the values $I, I - 1, \ldots, -I$

Expression (8-14) allows $2I + 1$ hyperfine components for each rotational transition and fine-structure component, i.e., for each transition for given values of N and J in initial and final states, since $F - J$ can have $2I + 1$ different values. This is provided $I < J$; otherwise there are $2J + 1$ components. The components may be approximately equally spaced if $C_2 - C_1$ is small compared with

$$C_2(2J_2 + 1) - C_1(2J_1 + 1)$$

or may tend to converge when the two terms are comparable in magnitude. Figure 8-2 shows part of the spectrum of ClO_2, which can be seen to consist of two groups of four lines corresponding to the $\frac{3}{2}$ spin of the two Cl isotopes (O^{16} has zero spin). One group of hyperfine components in this spectrum has approximately equal spacings, whereas spacings of the second group increase markedly from left to right.

It has been suggested [497], [615] that two nearby sets of three lines in the spectrum of NO_2 are the $J = 6_{06} \longleftrightarrow 5_{15}$ transition, one set with $J = N + \frac{1}{2}$, and the other with $J = N - \frac{1}{2}$. Certainly each set of three must correspond to the hyperfine lines due to N^{14} ($I = 1$) of some rotational transition. However, the ratios of the separations between these components has not yet been worked out in detail, and they do not seem to fit satisfactorily the approximate theoretical calculations which have been made. In addition, the Zeeman effect of these lines does not fit very well this assignment for the transitions [615].

8-5. Spin-spin Interaction between Nuclei. A nuclear magnetic moment may interact not only with an electron spin moment, but also with other nuclear magnetic moments which are present in a molecule. Expression (8-3) gives the interaction between two such dipole moments, from which the energy may be evaluated. This "spin-spin" interaction between nuclei is approximately 2000 times smaller than the hyperfine

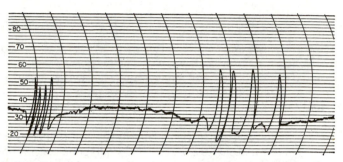

Fig. 8-2. Two lines of the asymmetric rotor ClO_2, each split into four components by magnetic hyperfine structure. (*From A. L. Schawlow and T. M. Sanders.*)

interactions in paramagnetic molecules because the magnetic moments associated with nuclei are this much smaller than those due to electrons. A rough magnitude for these effects can be obtained from μ_n^2/r^3, where μ_n is a nuclear magneton and r is the distance between two nuclei in a molecule. For a typical value of $r \approx 1.5$ A, μ_n^2/r^3 is only about 3 kc, which is so small that this "spin-spin" interaction is only rarely detected.

Because of their small size, nuclear "spin-spin" interactions will be of importance only for $^1\Sigma$ molecules, where the magnetic effect of electrons will also be very small. The vectors to be coupled are then the rotational angular momentum of the molecule J, and the spins of the two nuclei I_1 and I_2. If the nucleus having spin I_1 is more strongly coupled to J than is the other nucleus, then the coupling scheme for a $^1\Sigma$ symmetric top is similar to case $(b_{\beta J})$ discussed above, but with N replaced by J, Λ by K, S by I, and J replaced by $F_1 = I_1 + J$. This coupling scheme could occur when I_1 is coupled by electric quadrupole effects to J, and would be the only pure coupling scheme of much interest for two nuclei. The energies are given for two nuclei in symmetric-top molecules by simplified forms of (8-10) and (8-11) or (8-12), or in more general form by Gunther-Mohr, Townes, and Van Vleck [931].

For $I_1 = \frac{1}{2}$:

$$W_{F_1=J+\frac{1}{2}} = \frac{2J\mu_1\mu_2(3\cos^2\theta - 1)}{(2J+1)(2J+3)I_2r^3}\left[\frac{3K^2}{J(J+1)} - 1\right]\mathbf{I}_2 \cdot \mathbf{F}_1$$

$$W_{F_1=J-\frac{1}{2}} = \frac{-2(J+1)\mu_1\mu_2(3\cos^2\theta - 1)}{(2J-1)(2J+1)I_2r^3}\left[\frac{3K^2}{J(J+1)} - 1\right]\mathbf{I}_2 \cdot \mathbf{F}_1$$

(8-15a)

where r is the length of the vector between the two nuclei and θ is the angle between this vector and the molecular axis.

For $I_1 = 1$:

$$W_{F_1=J+1} = \frac{2J\mu_1\mu_2(3\cos^2\theta - 1)}{(J+1)(2J+3)I_2r^3}\left[\frac{3K^2}{J(J+1)} - 1\right]\mathbf{I}_2 \cdot \mathbf{F}_1$$

$$W_{F_1=J} = -\frac{2\mu_1\mu_2(3\cos^2\theta - 1)}{J(J+1)I_2r^3}\left[\frac{3K^2}{J(J+1)} - 1\right]\mathbf{I}_2 \cdot \mathbf{F}_1$$

$$W_{F_1=J-1} = -\frac{2(J+1)\mu_1\mu_2(3\cos^2\theta - 1)}{J(2J-1)I_2r^3}\left[\frac{3K^2}{J(J+1)} - 1\right]\mathbf{I}_2 \cdot \mathbf{F}_1$$

(8-15b)

A somewhat larger spin-spin effect, which occurs only when $K = 1$, will be discussed below in connection with the hyperfine structure of NH_3.

For intermediate coupling cases, the matrix elements given by Frosch and Foley [686] for the interaction of nuclear and electronic dipole moments may be used with a second nuclear moment replacing the electron spin S. The case of spin-spin interaction between two identical nuclei in a linear molecule, i.e., two protons in H_2 or two deuterons in D_2, has been discussed in some detail by Kellogg, Rabi, Ramsey, and Zacharias [91], [98]. However, such cases are not of great interest to microwave spectroscopy because linear molecules in which two identical nuclei are found would usually have no dipole moment.

8-6. Effect of Hyperfine Structure on Λ Doubling—Hyperfine Doubling. So far we have discussed only those types of hyperfine structure which would be identical for the two energy levels of a Λ-type doublet. However, for Π states certain additional hyperfine effects may occur which are different for the two Λ-doubled states.

Different hyperfine structure for two members of a Λ doublet arises only from electron spin–nuclear spin interactions. A general understanding of the phenomenon can be obtained from a consideration of the distribution of electron density in the two Λ-doubled states of a $^2\Pi$ case (b) molecule. The part of the electron wave function depending on angle ϕ of rotation about the symmetry axis has the form $e^{i\phi} \pm e^{-i\phi}$ for the two Λ-doublet states of a Π electron. The probability distributions for the electron are then proportional to $\cos^2\phi$ and $\sin^2\phi$ for the two states, giving large probabilities in the regions shown as shaded in Fig. 8-3. In Fig. 8-3a, corresponding to the lower Λ-doublet state with a $\sin^2 \phi$ distribution, the field of the electron at the nucleus is parallel to I (the electron magnetic moment is antiparallel to its spin S). In Fig. 8-3b, corresponding to the upper Λ-doublet state, the electron's field is directed oppositely to I. Hence the spin-spin interaction energy is quite different for the two cases. A reversal of I or S with respect to the

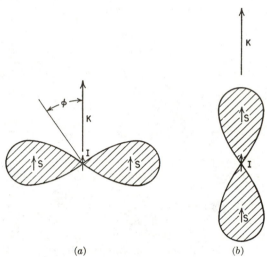

(a) $\qquad\qquad\qquad$ (b)

FIG. 8-3. Distribution of an unpaired electron in a $^2\Pi$ case (b) molecule for the two Λ-doublet states. The molecular axis is perpendicular to the page. Interaction energy between magnetic moments associated with the electron spin S and the nuclear spin I is different for the two cases. K is the angular momentum due to rotation of the molecule.

rotational angular momentum K of course reverses the sign of the hyperfine interaction energy for both Λ-doublet levels.

More formal reasoning shows that these effects are possible because the matrix elements of hyperfine interactions may connect states with a difference in Λ of ± 2 (and hence with differences in Ω [686]). The two Λ-doubled states actually involve equal mixtures of states with positive and negative values of Λ, in exact analogy with the l-doubled states discussed in Chap. 2 or the inversion states of Chap. 3. Hence in Hund's case (a) matrix elements connecting states with a difference in Ω of ± 1 or ± 2 give first-order (diagonal) effects for a $^2\Pi_{\frac{1}{2}}$ state but for no others. In case (b), first-order effects of such matrix elements are found for $^2\Pi$ or $^1\Pi$ states, but for no others.

The actual magnitude of the spin-spin interaction of the type described here, with case (a), case (b), or intermediate coupling, is given for a $^2\Pi$

state by [971a]

$$\Delta W = \pm \frac{d(X + 2 - A/B)}{4XJ(J + 1)} (J + \tfrac{1}{2}) \mathbf{I} \cdot \mathbf{J} \qquad (8\text{-}16)$$

where A = the fine-structure constant (energy = $A\mathbf{S} \cdot \mathbf{\Lambda}$)

B = the rotational constant

$X = \pm[(A/B)(A/B - 4) + 4(J + \tfrac{1}{2})^2]^{\frac{1}{2}}$

$d = 3\mu_0 \dfrac{\mu_I}{I} \left(\dfrac{\sin^2 \theta}{r^3}\right)_{\text{av}}$

r = distance from nucleus to electron

θ = angle between molecular axis and radius from nucleus to electron

The signs in (8-16) need explanation. When the constant A is positive (regular fine structure), then a positive sign for X is for the state which approaches $^2\Pi_{\frac{3}{2}}$ in case (a), and a negative sign for X for the state which approaches $^2\Pi_{\frac{1}{2}}$. When A is negative (inverted fine structure), a negative X gives the $^2\Pi_{\frac{3}{2}}$ state, and a positive X the $^2\Pi_{\frac{1}{2}}$ state. The upper (positive) sign in front of d in (8-16) applies to the upper Λ-doublet level of the state for which $J = N + \tfrac{1}{2}$ in the limiting case (b) or for both of the $^2\Pi_{\frac{3}{2}}$ and $^2\Pi_{\frac{1}{2}}$ states in case (a), and the lower sign to the lower state. The appropriate signs in front of d are the reverse in the state for which $J = N - \tfrac{1}{2}$ in the limiting case (b).

In case (a), $X = \pm A/B$, and (8-16) gives

$$\Delta W = \pm \frac{d(J + \tfrac{1}{2})}{2J(J + 1)} \mathbf{I} \cdot \mathbf{J}$$

for the $^2\Pi_{\frac{1}{2}}$ state and $\Delta W = 0$ for the $^2\Pi_{\frac{3}{2}}$ state. The upper sign in front of d applies to the upper Λ-doublet state. In case (b), (8-16) gives

$$\Delta W = \pm \frac{d(J - \tfrac{1}{2})}{4J(J + 1)} \mathbf{I} \cdot \mathbf{J} \qquad \text{for } J = N + \tfrac{1}{2}$$

and

$$\Delta W = \mp \frac{d(J + \tfrac{3}{2})}{4(J + 1)} \mathbf{I} \cdot \mathbf{J} \qquad \text{for } J = N - \tfrac{1}{2}$$

This type of hyperfine effect we shall call hyperfine doubling, since it can remove the degeneracy from or further split Λ-type doublets. For hyperfine doubling to occur, one of the interacting spins must lie off the axis of the molecule [$\theta > 0$ in the definition of d under expression (8-16)]. This doubling can also occur in a symmetric-top molecule in a $^1\Sigma$ state if there is interaction between a nuclear spin on the axis and one off the axis. An example is NH_3, which will be discussed more fully below. Here interaction between the N and H nuclear moments splits the otherwise degenerate levels $K = \pm 1$.

Hyperfine doubling has been found in OH [861], which is a case of

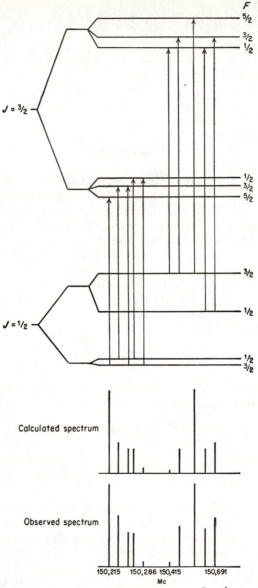

FIG. 8-4. Energy-level diagram and spectrum of the $J = \frac{3}{2} \leftarrow \frac{1}{2}$ rotational transition of the $^2\Pi_{\frac{1}{2}}$ state of $N^{14}O^{16}$. (*After Gallagher, Bedard, and Johnson* [924].)

intermediate coupling, and in the $^2\Pi_{\frac{1}{2}}$ state of NO [782b], [924], which is rather close to case (a). For OH, $d = 57$ Mc and for NO, $d = 112.6$ Mc. In both OH and NO, the hyperfine doubling is superimposed on a hyperfine structure of the form (8-6) which is identical for the two Λ doublets. Structures of the two lowest rotational states of NO are illustrated in Fig. 8-4. Here the hyperfine doubling is somewhat larger than other hyperfine effects and consequently has interchanged the order of the hyperfine levels in the two Λ-type doublets of each rotational state.

8-7. Electronic Angular Momentum in $^1\Sigma$ Molecules and Its Contributions to Molecular Energy. Although molecules in $^1\Sigma$ states are often said to have no electronic angular momentum, there is in fact a small electronic angular momentum when a molecule of this type is rotating. For example, consider the motion of closed shells of electrons associated with a nucleus. These move with the nucleus during rotation of the molecule and hence possess angular momentum. Their rotation produces magnetic fields which can give magnetic hyperfine structure. A more interesting and complex case is a valence electron which is not definitely associated with one nucleus. Valence electrons may only partly rotate with the molecule, but still usually produce the dominating magnetic fields at the nuclei. Their behavior also produces complications in the purely mechanical rotational energy of the molecule.

In order to examine the behavior of electrons in a rotating molecule, we shall start with the energy or Hamiltonian for electrons in a molecule with fixed nuclei, which is of the form

$$H_e = \frac{1}{2m} \sum_n [p_{nx}^2 + p_{ny}^2 + p_{nz}^2] + V$$

$$= \frac{1}{2m} \sum_n \sum_g p_{ng}^2 + V \tag{8-17}$$

where m is the electron mass, p_{ng} is the gth component of the momentum of the nth electron in Cartesian coordinates, and V is the potential energy of all the electrons. The energy of the electrons in the ground state may be found in principle by solving Schrödinger's equation with (8-17) as the Hamiltonian. Let this energy be W_{e0}.

If the molecule rotates, the Hamiltonian for the electrons is of the same form, but p_{ng} refers then to a Cartesian component of the generalized momentum of the nth electron referred to a set of axes fixed in the molecule and rotating with it. This momentum is not simply the mass times the velocity in the rotating coordinates, and must be found by the general method of differentiating the Lagrangian with respect to the velocity. The important point for our purposes, however, is that the Hamiltonian in rotating coordinates has the same form (8-17) as long as the coordinate system is Cartesian. The nuclei also acquire kinetic energy as a result of rotation, which may be written

$$H_n = \tfrac{1}{2} \sum_g \frac{O_g^2}{A_g} \qquad (8\text{-}18)$$

where A_g is the moment of inertia of the system of nuclei in the molecule about one of the principal axes, and O_g is their angular momentum about this axis. We shall for simplicity assume that the nuclei are rigidly fixed in the molecule and disregard possible vibrational energies. The component J_g of the total angular momentum about a principal axis of inertia is the sum of that due to the nucleus O_g and the small angular momentum L_g of the electrons, or

$$O_g = (J_g - L_g)\hbar \qquad (8\text{-}19)$$

where J_g and L_g are in units of $\hbar$.

Combining (8-17), (8-18), and (8-19), the Hamiltonian for the rotating system becomes

$$H = \tfrac{1}{2} \sum_g \frac{\hbar^2 J_g^2}{A_g} - \sum_g \frac{\hbar^2 J_g L_g}{A_g} + \tfrac{1}{2} \sum_g \frac{\hbar^2 L_g^2}{A_g} + \frac{1}{2m} \sum_n \sum_g p_{ng}^2 + V \qquad (8\text{-}20)$$

If the electrons are truly in a $^1\Sigma$ state, there is no electronic angular momentum and L_g is zero, so that the energy given by (8-20) is just the sum of the energy due to the last two terms, and that due to the first term. This first term gives the rotational energies W_R of a rigid rotor, but one without electrons since A_g is the moment of inertia of the nuclei only. The total energy in this approximation is then just

$$W_0 = W_R + W_{e0} \qquad (8\text{-}21)$$

It is only by allowing for perturbations of the ground $^1\Sigma$ electron states by the term $\sum_g (\hbar^2 J_g L_g / A_g)$ that the energy associated with the contribution of the electrons to the moment of inertia of the molecule can be obtained. This perturbation introduces a small probability for electronic angular momentum states other than $^1\Sigma$; hence it produces some electronic angular momentum which affects the rotational energy of the molecule. To a good approximation, the third term of (8-20), $\tfrac{1}{2} \sum_g (\hbar^2 L_g^2 / A_g)$, is a constant of the molecular rotation. Rotation will indeed produce perturbations of this term, but they are too small to be significant at present.

Since the perturbation $\sum_g (\hbar^2 J_g L_g / A_g)$ introduces electronic angular momentum, we must take into account the various other types of energy terms connected with electronic angular momentum in order to complete the Hamiltonian (8-20). One of these is the interaction between a nuclear magnetic moment and the magnetic fields produced by an electronic angular momentum. Such interactions are similar in principle to the magnetic hyperfine structure due to orbital motion of an electron in an atom, and may be written as in (5-49) after letting $\mathbf{j} = \mathbf{L}$,

$$a\mathbf{I} \cdot \mathbf{L}' = a \sum_g I_g L_g' \qquad (8\text{-}22)$$

where $\mathbf{L}'$ is the electronic orbital angular momentum about the nucleus of spin $\mathbf{I}$, and $a = (2\mu_I\mu_0/I)(1/r^3)_{\text{av}}$. However, in a molecule, $\mathbf{L}$ is not a constant of the motion, and hence both $\mathbf{L}'$ and $(1/r^3)_{\text{av}}$ must be averaged over the electronic wave function. It should be noted too that (8-22) contains $\mathbf{L}'$, the angular momentum about a particular nucleus, rather than $\mathbf{L}$, the angular momentum with respect to the center of mass of the molecule. $\mathbf{L}'$ and $\mathbf{L}$ are, of course, closely related.

Another common interaction associated with electronic angular momentum is the Zeeman effect due to an external applied field H. The interaction energy is given by

$$\mu_0\mathbf{H} \cdot \mathbf{L} = \mu_0 \sum_g H_g L_g \qquad (8\text{-}23)$$

where μ_0 is the Bohr magneton.

Although we shall not be concerned here with all possible types of magnetic interaction, still another interaction which we shall include is the direct interaction between the magnetic moment of the nucleus and the external field H. This is

$$-\mathbf{\mu}_I \cdot \mathbf{H} = -\sum_g \mu_g H_g \qquad (8\text{-}24)$$

where μ_I is the nuclear magnetic moment. Adding the three terms (8-22), (8-23), and (8-24) and omitting the constant term $\frac{1}{2}\sum_g \hbar^2(L_g^2/A_g)$ in (8-20), the Hamiltonian becomes

$$H = \frac{1}{2m} \sum_n \sum_g p_{ng}^2 + V + \tfrac{1}{2} \sum_g \frac{\hbar^2 J_g^2}{A_g}$$
$$+ \sum_g \left(\mu_0 H_g - \frac{\hbar^2 J_g}{A_g} \right) L_g + a\mathbf{I} \cdot \mathbf{L} - (\mathbf{\mu}_I \cdot \mathbf{H}) = H_0 + H' \quad (8\text{-}25)$$

Here H_0 represents the first three terms which are the main parts of the Hamiltonian. H' consists of the remaining small perturbing terms.

It may be seen from (8-25) that the effect of a magnetic field on electrons is just equivalent to that of molecular rotation since they both enter (8-25) only in the factor $\left(\mu_0 H_g - \dfrac{\hbar^2 J_g}{A_g} \right)$. Thus the effect of a magnetic field H_g is equivalent to a rotation $\dfrac{-\hbar^2 J_g}{\mu_0 A_g}$. Since $\hbar J_g = \omega A_g$, where ω is the angular velocity, the rate of angular rotation which produces an effect on the molecular electrons precisely the same as that of the field H_g is $\omega_g = \dfrac{eH_g}{2mc}$. This is Larmor's theorem, and is responsible for the direct connection $(1 - 47)$ between L uncoupling and the molecular g factor due to electrons.

The first-order correction to the energy given by H_0 is the average value of H' in the $^1\Sigma$ state of the molecule. The parts of H' proportional to L have an average value of zero, so that the only term of H' which gives an energy in first order is $-\mathbf{\mu}_I \cdot \mathbf{H}$. The second-order correction to the energy is given as usual by

$$W' = \sum_n \frac{|(0|H'|n)|^2}{W_0 - W_n} \tag{8-26}$$

where W_0 is the energy of the unperturbed state, and W_n the energy of the nth excited electronic state. $(0|H'|n)$ is the matrix element of H' between the unperturbed and an excited electronic state.

To illustrate the nature of the various terms in (8-26), let us assume that only one excited state is of importance so the sum over n may be omitted. Second-order terms in $\mathbf{\mu}_I \cdot \mathbf{H}$ will be assumed negligible, as they usually are.

$$
\begin{aligned}
W' = {} & \frac{|(0|a\mathbf{I} \cdot \mathbf{L}|n)|^2}{W_0 - W_n} + \sum_g \sum_{g'} \hbar^4 J_g J_{g'} \frac{(0|L_g|n)(n|L_{g'}|0)}{A_g A_{g'}(W_0 - W_n)} + \frac{|(0|\mu_0 \mathbf{H} \cdot \mathbf{L}|n)|^2}{W_0 - W_n} \\
& + \sum_{g'} \sum_g \mu_0 I_g H_{g'} \frac{(0|aL_g'|n)(n|L_{g'}|0) + (0|L_{g'}|n)(n|aL_g'|0)}{W_0 - W_n} \\
& - \sum_{g'} \sum_g \frac{\hbar^2 I_g J_{g'}}{A_{g'}} \frac{(0|aL_g'|n)(n|L_{g'}|0) + (0|L_{g'}|n)(n|aL_g'|0)}{W_0 - W_n} \\
& \qquad - 2 \sum_{g'} \sum_g \mu_0 \frac{\hbar^2 J_g H_{g'}}{A_g} \frac{(0|L_g|n)(n|L_{g'}|0)}{W_0 - W_n} \tag{8-27}
\end{aligned}
$$

Expression (8-27) appears rather formidable partly because it applies to the general asymmetric top and hence the sums over coordinate directions g and g' cannot be very much simplified. Its complexity is due also to the fact that each of the six terms of (8-27) corresponds to a different physical effect. These may be and usually are discussed separately.

The second term in (8-27) is the only one which does not involve magnetic effects. It is proportional to the square of the molecular angular momentum J, and represents the contribution of the electrons to the moment of inertia of the molecule or to the kinetic energy of rotation. Since $W_0 - W_n$ is negative, this term is negative and corresponds to a decrease in the energy of rotation. This is to be expected since the electrons should add to the moments of inertia A_g of the bare nuclei. It will be shown below that, at least for electrons closely bound to the nuclei, this term does in fact have the magnitude to be expected from the electronic contribution to the moment of inertia.

The first term in (8-27) is called the pseudo quadrupole effect [192] because it has the same form as a quadrupole interaction between I and the molecular electric fields. In a simple $^1\Sigma$ linear molecule, for example, L may be thought of as precessing around the molecular axis, so that $(\mathbf{I} \cdot \mathbf{L})^2$ becomes proportional to the square of the cosine of the angle between I and the molecular axis, which is the same dependence on this angle shown by an electric quadrupole interaction. Hence experimentally the effect of this term cannot simply be distinguished from that of a nuclear quadrupole interaction. Fortunately, however, the pseudo quadrupole interaction is almost always negligibly small, since from (8-27) it is proportional to the square of a magnetic hyperfine interaction divided by the separation between electronic levels. It is hence not usually larger than a few cycles per second.

The pseudo quadrupole effect may be regarded as a magnetic polarization of the molecule by the nuclear magnetic moment, and then an interaction between the resulting molecular magnetic field and the nuclear moment. In principle the interaction discussed by Hebb (cf. page 182) in ρ-type tripling is identical with this nuclear effect, but it involves fine-structure rather than hyperfine-structure interactions and is consequently enormously greater.

The third term of (8-27) is an energy associated with the magnetic susceptibility or polarization of the molecule by an external field (cf. [38], p. 227) and is not of great interest here. It has no dependence on molecular rotation and corresponds to the interaction between the field H and the molecular magnetic dipole induced by H.

The fourth term of (8-27),

$$\sum_{g'} \sum_{g} \mu_0 I_g H_{g'} \frac{(0|aL'_g|n)(n|L_{g'}|0) + (0|L_{g'}|n)(n|aL'_g|0)}{W_0 - W_n}$$

also does not depend on the molecular rotation. It corresponds to a magnetic polarization of the molecule by an external field H, or creation of electronic angular momentum L as a result of this field, and then an interaction between the nuclear magnetic moment and the magnetic field produced by the angular momentum L. These effects are usually unimportant to microwave spectroscopy of gases, since they are normally smaller than the direct interaction $\mathbf{\mu}_I \cdot \mathbf{H}$ between an external field H and a nuclear magnetic moment by a factor of 10^4. However, for nuclear magnetic resonances in solids or liquids such effects are measurable and are generally called "chemical effects," since they do not depend simply on the nucleus, but vary from one molecule to the next [569], [515].

The remaining two terms of (8-27),

$$-\sum_{g'}\sum_{g} \frac{\hbar^2 I_g J_{g'}}{A_{g'}} \frac{(0|aL_g'|n)(n|L_{g'}|0)(0|L_{g'}|n)(n|aL_{g'}'|0)}{W_0 - W_n}$$

and $-2\sum_{g'}\sum_{g} \frac{\mu_0\hbar^2 J_g H_{g'}}{A_{g'}} \frac{(0|L_g|n)(n|L_{g'}|0)}{W_0 - W_n}$, are both dependent on the rotational angular momentum J of the molecule and are interesting terms from the point of view of microwave spectroscopy. They may be thought of as due to an excitation of electronic orbital angular momentum L by the rotation J. This electronic angular momentum produces an internal molecular magnetic field which interacts with the nuclear magnetic moment, and also a molecular magnetic moment which interacts with the external field H. The latter term, linearly proportional to H and to J, is a Zeeman effect which will be discussed at length in Chap. 11. The term linearly proportional to I and J is part of the common variety of magnetic hyperfine structure in $^1\Sigma$ molecules which for simple linear molecules reduces to the form $C_I \mathbf{I} \cdot \mathbf{J}$, where C_I is a constant dependent on the molecule. This interaction is often called the "I dot J" interaction and will be discussed at length below.

8-8. Effect of Electronic Motion on Rotational Energy. We shall digress from magnetic effects briefly to examine parts of the energy (8-26) of the type illustrated by the second term of (8-27), *i.e.*,

$$W_2' = \sum_{g}\sum_{g'} \frac{\hbar^2 J_g J_{g'}}{A_g A_{g'}} \sum_{n} \frac{\hbar^2 (0|L_g|n)(n|L_{g'}|0)}{W_0 - W_n} \tag{8-28}$$

These terms give the effect of electrons on the rotational energy of the molecule, and illustrate some of the properties of the terms involved in magnetic hyperfine structure. For simplicity, consider a linear molecule and take the z direction along the molecular axis. Then $L_z = 0$, and the terms of (8-28) of the type

$$\sum_{n} \frac{\hbar^2}{W_0 - W_n} [(0|L_x|n)(n|L_y|0) + (0|L_y|n)(n|L_x|0)]$$

are also zero. The latter may be shown by a rotation of the molecule about the z axis by $\pi/2$. After the rotation, x becomes y, and y becomes $(-x)$. Since the directions x and y are equivalent, this term has changed sign but cannot have changed value, so that it must equal zero. The only nonzero terms of (8-28) are therefore

$$W_2' = \hbar^2 \frac{J_x^2}{A_x^2} \sum_{n} \frac{\hbar^2 |(0|L_x|n)|^2}{W_0 - W_n} + \hbar^2 \frac{J_y^2}{A_y^2} \sum_{n} \frac{\hbar^2 |(0|L_y|n)|^2}{W_0 - W_n}$$

or, since the x and y directions are equivalent,

$$W_2' = \hbar^2 \frac{J(J+1)}{A^2} \sum_n \frac{\hbar^2|(0|L_x|n)|^2}{W_0 - W_n} \quad \text{or}$$

$$4J(J+1)\hbar^2 B^2 \sum_n \frac{|(0|L_x|n)|^2}{W_0 - W_n} \quad (8\text{-}29)$$

where $J(J+1) = J_x^2 + J_y^2$ is the square of the total angular momentum, $A = A_x = A_y$, and $B = h/8\pi^2 A$ is the rotation constant.

The "slippage" of electrons with respect to the nuclei as the molecule rotates may be seen from (8-29). If the electron distribution and the fields were completely spherical about the center of mass, then $(0|L_x|n)$ would be zero. The fractional part of the valence electrons which is spherically distributed may be considered, as noted in Chap. 1, to slip with respect to the molecule and not contribute to the rotational energy.

The inner shells of electrons attached to a nucleus are not at all spherical with respect to the center of mass and may be more easily discussed with a coordinate system centered at this nucleus. Let τ be the distance from the center of mass to the nucleus. Then

$$\hbar L_x = \hbar L_x' - \tau p_y \quad (8\text{-}30)$$

where L_x' is the electronic angular momentum (in units of $\hbar$) of the spherical shell about the nucleus in question and p_y is the linear momentum in the y direction. But if we consider only a spherical closed shell of electrons in the spherical field of a nucleus, $(0|L_x'|n) = 0$ for any n, so that, using (8-30), for closed shells

$$\sum_n \frac{\hbar^2}{W_0 - W_n}|(0|L_x|n)|^2 = \tau^2 \sum_n \frac{|(0|p_y|n)|^2}{W_0 - W_n} \quad (8\text{-}31)$$

This can be simplified by the general identity* for any particle of mass m

$$\sum_n \frac{|(0|p_y|n)|^2}{W_0 - W_n} = \frac{-m}{2} \quad (8\text{-}32)$$

*This identity may be proved as follows:

$$p_y = m\frac{dy}{dt} = \frac{m}{i\hbar}(yH - Hy)$$

or

$$(0|p_y|n) = \frac{mi}{\hbar}(W_0 - W_n)(0|y|n)$$

(cf. [408], p. 139).
Hence

$$\sum_n \frac{(0|p_y|n)(n|p_y|0)}{W_0 - W_n} = \sum_n \frac{mi}{\hbar}\frac{(0|y|n)(n|p_y|0) - (0|p_y|n)(n|y|0)}{2} = \frac{-m}{2}$$

since $yp_y - p_y y = i\hbar$.

Hence (8-29) becomes for N electrons in spherical orbits around a nucleus which is a distance τ from the center of mass

$$W'_2 = \frac{-\hbar^2 J(J+1)Nm\tau^2}{2A^2} \tag{8-33}$$

This is a small decrease in the energy of rotation due to the presence of the electrons which should be added to the kinetic energy of the nuclei. Thus

$$W = W_0 + W'_2 = \frac{\hbar^2 J(J+1)}{2A} - \frac{\hbar^2 J(J+1)}{2A^2} Nm\tau^2 \approx \frac{\hbar^2 J(J+1)}{2(A+Nm\tau^2)} \tag{8-34}$$

Of course, (8-34) as written includes the effect of only those electrons "bound" to one particular nucleus, and all other closed electron shells should be similarly treated. This expression is sufficient to show, however, how the bound electrons add to the moment of inertia, and that, as stated in Chap. 1, the spherical closed electron shells in a molecule behave as if they follow the nuclei during molecular rotation, but remain fixed in orientation in space like the chairs on a rotating Ferris wheel.

For the valence shell of electrons, the sum in (8-29) cannot be so easily evaluated. One approximation which is often made involves the "pure-precession" hypothesis. In this case the electronic angular momentum L is assumed to have a fixed value and to precess around the molecular axis with a projection of zero on this axis for the ground state. The only low-lying electronic state for which $(0|L_x|n)$ is not zero is the Π state corresponding to a unit projection of L on the molecular axis. In this case $|(0|L_x|n)|^2 = L(L+1)/4$, so that (8-29) may be written

$$W'_2 = -J(J+1)B^2 \left[8\pi^2 \sum_s N_s m\tau_s^2 + \frac{L(L+1)\hbar^2}{(W_\Pi - W_\Sigma)} \right] \tag{8-35}$$

where the sum is over all "bound" electrons in closed shells around nuclei of distance τ_s from the center of gravity of the molecule, and J is the quantum number for the total angular momentum. The last term of (8-35) gives the contribution of valence electrons under the restrictive assumption of pure precession. $W_\Pi - W_\Sigma$ is the excitation energy of the lowest electronic state.

In nonlinear polyatomic molecules the full form (8-28) must be used to obtain the effect of electrons on the rotational kinetic energy. However, for almost all purposes sufficient accuracy may be obtained by assuming the electrons associated with each nucleus to be located at the nucleus and using moments of inertia which include these electrons in calculating rotational energy. Only for the valence electrons does this involve some small error, and usually only by very accurate measurements on diatomic molecules can such small effects be detected. In more complex molecules other inaccurately known effects tend to mask these errors.

8-9. Magnetic Hyperfine Interaction $(\mathbf{I} \cdot \mathbf{J})$ in $^1\Sigma$ Molecules. We shall examine now in more detail the term in (8-27) which is linearly proportional to I and J, and which represents a magnetic hyperfine interaction between a nucleus and electrons even in $^1\Sigma$ molecules. It is

$$W'_5 = -\sum_n \sum_{g'} \sum_g \frac{\hbar^2 I_g J_{g'}}{A_{g'}} \frac{(0|aL'_g|n)(n|L_{g'}|0) + (0|L_{g'}|n)(n|aL'_g|0)}{W_0 - W_n} \quad (8\text{-}36)$$

where the summation over all excited states n has been included rather than over only one state as in (8-27).

Now an angular velocity about the center of mass of the molecule is equivalent to the same angular velocity about any nucleus in the molecule plus an appropriate instantaneous linear velocity v. The magnetic fields produced at the nucleus by rotation about the two different points should differ only by the effect of the velocity v. At the nucleus the average electric field is zero (because there is no average force on the nucleus at equilibrium), so that the velocity v in fact produces no magnetic field at the nucleus [magnetic field $\approx (v/c) \times$ electric field]. Hence it may be expected that the angular momentum $L_{g'}$ about the center of mass can be replaced in (8-36) by the angular momentum $L'_{g'}$ about the nucleus in question without changing the magnetic energy given by this expression.* Furthermore, since the quantity a depends on $1/r^3$, but not on the angular position of the electron with respect to the nucleus, the part of the matrix element which depends on a may be factored out, that is, $(0|aL'_g|n) = a_{0n}(0|L'_g|n)$. a_{0n} may be taken as real so that $a_{0n} = a_{n0}$. Hence (8-36) becomes

$$W'_5 = -2 \sum_n \sum_{g'} \sum_g \frac{\hbar^2 I_g J_{g'}}{A_{g'}} a_{0n} \frac{(0|L'_g|n) \cdot (n|L_{g'}|0)}{W_0 - W_n} \quad (8\text{-}37)$$

The nature of expression (8-37) is perhaps most easily understood by applying it to the simple case of a diatomic molecule. It is evident that closed shells of electrons about the nucleus being considered do not contribute magnetic hyperfine structure, since for them $(0|L_g|n)$ is zero. Closed shells of electrons around other nuclei can be treated in a way similar to that used in reducing (8-28) for a diatomic molecule, and their contribution to (8-37) can be shown to be

$$-\frac{\mu_I \hbar \mathbf{I} \cdot \mathbf{J}}{IA} \sum_s \frac{e}{cr_s} = -4\pi \frac{\mu_I}{I} B\mathbf{I} \cdot \mathbf{J} \sum_s \frac{e}{cr_s}$$

Here e is the electron charge in esu (negative sign) and r_s is the distance of the center of the sth electron distribution from the nucleus of spin I

* For a more detailed discussion, see [686], Secs. 7 and 8.

and g factor g_I. Valence electrons which are not spherically distributed about any nucleus cannot be so simply treated and their contributions must be left in essentially the form (8-37). The charges on nuclei contribute in the same way as those of electrons in closed shells about the nuclei, as may be seen by formally considering their changes to be due to closed shells of positively charged electrons tightly bound to the nucleus. Therefore for the diatomic or linear molecule, the combined effect of electrons from (8-37) and the nuclear charges gives

$$W_5' + \text{effects of nuclear charges} = \mathbf{I} \cdot \mathbf{J} B \left[4h \sum_n \frac{a_{0n} |(0|L_x|n)|^2}{W_n - W_0} \right.$$
$$\left. - 4\pi \frac{\mu_I}{I} \sum_s \frac{q_s}{cr_s} \right] \quad (8\text{-}38)$$

where r_s is the distance between the nucleus for which (8-38) gives the magnetic hyperfine energy and any other nucleus in the molecule, and q_s is the net charge of the sth nucleus plus the electrons which are in closed shells about it. Assuming the pure precession approximation, (8-38) may be written

$$W_5' + \text{nuclear effects} = \mathbf{I} \cdot \mathbf{J} B \left[\frac{ha_{\Sigma\Pi}L(L+1)}{W_\Pi - W_\Sigma} - 4\pi \frac{\mu_I}{I} \sum_s \frac{q_s}{cr_s} \right] \quad (8\text{-}39)$$

The general form of the magnetic hyperfine interaction of a nucleus in a linear $^1\Sigma$ molecule (with no vibrational angular momentum) is, from (8-38)

$$W_{mag} = C_I \mathbf{I} \cdot \mathbf{J} \quad (8\text{-}40)$$

where the constant C_I is dependent on molecular parameters shown in (8-38). This type of interaction was first found in the hydrogen molecule by the use of molecular-beam techniques [91], and the constant C_I has now been evaluated for a number of molecules by both molecular beam and the usual microwave spectroscopic techniques. Table 8-1 shows the values of C_I for molecules so far measured, which vary from near 1 kc to somewhat greater than 100 kc. A few additional values of C_I have been reported for the alkali halides from measurements of line widths in molecular-beam magnetic resonance experiments. However, such measurements are subject to uncertainties in interpretation and hence these values have been omitted from Table 8-1. White [999] has shown that a number of interesting trends can be established from present measurements of C_I.

Examples of $\mathbf{I} \cdot \mathbf{J}$ *Interaction.* Usually μ_I for the nucleus is positive, and hence the first term of (8-38) due to the valence electrons is positive, while the second term due to nuclei and bound electrons is negative. In most cases C_I is positive. O^{17} and Se^{79} are the only nuclei with a negative μ_I for which C_I has been measured. It may be seen from Table 8-1 that, with the exception of hydrogen nuclei in H_2, these are precisely the only cases where C_I has been found to be negative. Hence except in the special cases of hydrogen the first term in (8-38), due to excitation of the valence electrons by rotation, always dominates.

TABLE 8-1. VALUES OF MAGNETIC HYPERFINE CONSTANTS C_I IN $^1\Sigma$ MOLECULES
(Magnetic hyperfine energy $= C_I \mathbf{I} \cdot \mathbf{J}$)

Molecule	Nucleus	C_I, kc	$\dfrac{C_I}{g_I B (1/r^3)_{\mathrm{av}}},$ 10^{-33} cm^3	Reference
Li^{6}F	F^{19}	37.3 ± 0.3	3.4	[753]
Li^{7}F	F^{19}	32.9 ± 0.1	3.3	[999]
KF	F^{19}	0 ± 10	0 ± 7	[463]
Rb^{85}F	F^{19}	11 ± 3	10	[476]
Rb^{87}F	F^{19}	14 ± 4	13	[476]
CsF	F^{19}	16 ± 2	14	[331]
Li^{7}F	Li^7	2.2 ± 0.6	90	[999]
DI	I^{127}	140	11	[999]
ClF	Cl^{35}	22 ± 3	54	[366] [999]
CS	S^{33}	19 ± 15	54	[999]
Tl^{205}Cl^{35}	Cl^{35}	1.4 ± 0.1	20	[676b]
Tl^{205}Cl^{35}	Tl^{205}	73 ± 2	110	[676b]
H$_2$	H	-113.904 ± 0.030	-40.8	[91] [716a] [807]
HD	H	-87.00 ± 0.85	-41.4	[91a]
HD	D	-12.6 ± 0.3	-39.0	[91a]
D$_2$	D	-8.445 ± 0.056	-39.1	[91a] [716a] [807]
OCS	O^{17}	-4.0 ± 1.5	28	[999]
OCS	S^{33}	2 ± 1	23	[999]
OCSe	Se^{79}	-3.2 ± 1.0	42	[999]
HCN	N^{14}	10 ± 4	34	[999]
DCN	N^{14}	8 ± 3	33	[999]
Cl^{35}CN^{14}	N^{14}	2.5 ± 0.8	62	[999]
Cl^{35}CN^{14}	Cl^{35}	3.0 ± 1.0	19	[999]
Cl^{35}CN^{15}	Cl^{35}	3.5 ± 0.6	24	[999]
Cl^{35}CN^{15} (in excited state $v_2 = 1$)	Cl^{35}	8 ± 5	52	[999]
NH$_3$	N^{14}	$6.1 \pm 0.2 +$ $(0.4 \pm 0.2) \dfrac{K^2}{J(J+1)}$	3.1	[999]
NH$_3$	H	See discussion in text		[999]

The proportionality of C_I to μ_I/I or to the gyromagnetic ratio

$$g_I = \frac{\mu_I}{I \mu_n}$$

and the rotational constant B is illustrated by comparing its value for H_2 and for D_2. The ratio of $g_I B$ for the two cases is 13.0, whereas the ratio of the value of C_I observed is 13.5. The small difference between these two values is due to the slightly different average internuclear distances for H_2 and D_2.

Since the dominant contribution to C_I in all molecules except H_2 comes from the valence electrons,

$$\frac{C_I}{g_I B} \text{ should be proportional to } \sum_n \frac{|(0|aL_x|n)|^2}{W_n - W_0}$$

To a rough approximation we may assume that only the first excited state is important in this summation and that a is proportional to the average value of $1/r^3$ for a valence p electron of the atom. Hence

$$\frac{C_I}{g_I B (1/r^3)_{av}} \text{ should be proportional to } \frac{|(0|L_x|1)|^2}{W_1 - W_0}$$

It may be seen from Table 8-1 that this quantity is of the same order of magnitude for most molecules.

For a given atom it may be assumed that the matrix elements $(0|aL_x|n)$ are approximately the same in a series of chemically similar molecules. The value of C_I for fluorine in LiF, RbF, and CsF represents such an example. The fact that $\frac{C_I}{g_I B (1/r^3)_{av}}$ for F increases regularly in progressing from LiF to the larger molecules RbF and CsF is evidently due to decreasing separations $W_n - W_0$ of the energy levels which can be expected for the larger molecules. Another similar series is provided by the values of $\frac{C_I}{g_I B (1/r^3)_{av}}$ for Cl in CH_3Cl, SiH_3Cl, and GeH_3Cl [999]. However, for these cases C_I is not yet very accurately known.

For O^{17} and S^{33} in the same molecule OCS, one would expect $\frac{C_I}{g_I B (1/r^3)_{av}}$ to be essentially the same, since O and S are very similar chemically, and hence their electronic surroundings are very similar. Table 8-1 shows that this expectation is correct. On the other hand, Tl and Cl in the same molecule TlCl as well as Cl and N in ClCN have very different values of $\frac{C_I}{g_I B (1/r^3)_{av}}$ since their electronic surroundings are quite dissimilar.

Certain types of molecular rotation may result in relatively little electronic angular momentum. Such is the case for the rotation of H_2, where the "slip effect" is quite important. Another case occurs in the bending mode of ClCN $(v_2 = 1)$. Here the bent ClCN molecule rotates very rapidly about the symmetry axis with a frequency roughly 100 times greater than the normal molecular rotational frequency. However, since the bending is slight, the rapid rotational motion does not excite much electronic angular momentum. This is shown by the fact that C_I in the bending mode is not much greater than C_I for the ground vibrational state of ClCN.

Relation between C_I and g_J. Even though magnetic hyperfine-structure interactions are complex and involve excited electronic states, the discussion above shows that reasonable correlations with certain molecular properties can be observed. Furthermore, the coefficients C_I are connected with other measurable quantities. If one low-lying excited electron state dominates in the sum $\sum_n \dfrac{|(0|L_x|n)|^2}{W_n - W_0}$, then C_I may be expected to be closely related to the molecular magnetic moment produced by electron excitation. For, from (8-38), (11-15), and (11-19),

$$C_I = \frac{g_J \mu_n a_{0n}}{\mu_0} = 2g_J g_I (\mu_n)^2 \left(\frac{1}{r^3}\right)_{av} \qquad (8\text{-}41)$$

Here g_J strictly includes only the contributions to g_J from the excited valence electrons. There should in addition be a connection between C_I and part of the rotational energy due to electrons as may be seen by comparing (8-29) or (8-35) with (8-39). OCS and OCSe are the only linear molecules in which both quantities C_I and g_J have already been measured so that a test of the relation (8-41) can be made. Assume that the electron which produces g_J spends half its time on the oxygen and half its time on the sulfur atom in OCS, so that $(1/r^3)_{av}$ is just one-half that for an atomic electron in each case. Then (8-41) predicts from the measured value $g_J = 0.025$ for OCS that C_I for O^{17} should be -2.3 kc and that for S^{33} C_I should be 1.5 kc. These values are very satisfactorily close to the measured values given in Table 8-1. Similarly, one may calculate the values of C_I for Se^{79} from relation (8-41) and the observed value of g_J for OCSe, which is $g_J = 0.019$. The result is $C_I = -1.4$ kc, again in reasonable agreement with the observed value of C_I for Se^{79} found in Table 8-1.

8-10. Magnetic Hyperfine Structure of Nonlinear Molecules in $^1\Sigma$ States. We now return to the general form (8-36) for any molecule. Since the hyperfine energy is a scalar and linearly proportional to the components of $\mathbf{I}$ and of $\mathbf{J}$, it must be of the form

$$W_{mag} = \mathbf{I} \cdot \mathbf{\alpha} \cdot \mathbf{J} = \sum_{gg'} \alpha_{gg'} I_g J_{g'} \qquad (8\text{-}42)$$

where $\mathbf{\alpha}$ is a dyadic, or $\alpha_{gg'}$ are components of a tensor which can be seen to be symmetric from inspection of (8-36). For a nucleus on the axis of a symmetric-top molecule, the principal axes of α must coincide with the principal axes of inertia, so that

$$W_{mag} = \alpha_{xx} I_x J_x + \alpha_{yy} I_y J_y + \alpha_{zz} I_z J_z \qquad (8\text{-}43)$$

Since from the molecular symmetry, $\alpha_{xx} = \alpha_{yy}$ (z is parallel to the molecular axis), (8-43) may be written

$$W_{mag} = \alpha_{xx}\mathbf{I} \cdot \mathbf{J} + (\alpha_{zz} - \alpha_{xx})I_z J_z \qquad (8\text{-}44)$$

Now $J_z = K$, and using the vector model $I_z = \mathbf{I} \cdot \mathbf{J}[K/J(J + 1)]$ so that for a nucleus on the axis of a symmetric top, the energy is of the form given by Henderson [283]

$$W_{mag} = \mathbf{I} \cdot \mathbf{J} \left[a + (b - a)\, \frac{K^2}{J(J + 1)} \right] \qquad (8\text{-}45)$$

For the general asymmetric rotor, (8-42) does not simplify.

N^{14} in NH_3 affords a particularly good test of the dependence of magnetic interaction on J and K, since inversion lines for this molecule with many different values of J and K appear in the microwave region. Accurate measurement [930] of the NH_3 hyperfine structure for values of J and K up to 6 shows that the relation (8-45) holds within the experimental error of about 5 per cent. The constants a and b are given in Table 8-1.

NH_3 also displays an interesting magnetic hyperfine structure associated with magnetic moments of the hydrogen nuclei. A rather complete treatment and examination of hyperfine effects in NH_3 has been given by Gunther-Mohr et al. [930], [931]. Still more refined measurements and additional theoretical discussion are given by Gordon [925]. We shall discuss the various effects involving the hydrogen nuclei one at a time for the sake of simplicity.

Consider first the rotational states for which $K \neq 1$ in order to avoid the complication of hyperfine doubling. Since the N^{14} spin I_N is coupled to J by a quadrupole interaction, I_N and J couple to form F_1, which then couples with the sum I_H of the three hydrogen spins to form the total angular momentum F. When K is a multiple of 3, the total spin I_H of the three hydrogens can only be $\frac{3}{2}$ because of the exclusion principle (see Sec. 3-4). When K is not a multiple of 3, I_H can only be $\frac{1}{2}$.

There is an interaction between the hydrogen magnetic moments and the nitrogen moment which is given by expressions (8-15b). A somewhat larger interaction occurs between the moments of the hydrogen nuclei themselves, which can change in magnitude when I_H changes orientation with respect to J or F_1. This interaction has the form [925]

$$W = \frac{3}{4}\, \frac{g_I^2 \mu_n^2}{r^3}\, (I_z^2 - \tfrac{5}{4}) \qquad (8\text{-}46)$$

where I_z is the component of I_H along the symmetry axis of the molecule and r is the distance between hydrogen nuclei. When K is not a multiple of 3, $I_H = \frac{1}{2}$ and $I_z^2 = \frac{1}{4}$ for all hyperfine states, so that (8-46) produces no variation of energy.   When K is a multiple of 3, however, $I = \frac{3}{2}$ and (8-46) is given [925] by

$$\Delta W_{I_H = \frac{3}{2}} = \frac{3}{4} \frac{g_I^2 \mu_n^2}{r^3} \left[\frac{3K^2}{J(J+1)} - 1 \right] \cdot \frac{4(\mathbf{I} \cdot \mathbf{F}_1)^2 + 2(\mathbf{I} \cdot \mathbf{F}_1) - 5F_1(F_1 + 1)}{2F_1(F_1 + 1)(2F_1 - 1)(2F_1 + 3)}$$
$$\cdot \frac{6(\mathbf{F}_1 \cdot \mathbf{J})^2 - 3(\mathbf{F}_1 \cdot \mathbf{J}) - 2F_1(F_1 + 1)J(J+1)}{(2J-1)(2J+3)} \quad (8\text{-}47)$$

The coefficient $\dfrac{3}{4} \dfrac{g_I^2 \mu_n^2}{r^3}$ in ΔW corresponds to 20.7 kc for hydrogen in NH_3, and hence spin-spin interactions of this type can be detected only by microwave spectroscopy of the highest resolution.

Rotation of the NH_3 molecule also creates a magnetic field which interacts with the moments of the hydrogen nuclei.   Since the hydrogens are not on the molecular axis, the complete dyadic form (8-42) of this $\mathbf{I} \cdot \mathbf{J}$ interaction must be considered.   From symmetry, one can immediately identify the principal axes of the dyadic for an individual hydrogen nucleus.   One is the symmetry axis of the molecule, which we shall call z; a second is the perpendicular direction from the z axis through the nucleus, which we shall call x, and the y direction is of course perpendicular to each of these.   Rotation of the molecule about the x, y, or z axis each produces a different magnetic field at the nucleus and hence a different $\mathbf{I} \cdot \mathbf{J}$ interaction energy.   If, with respect to these principal axes, the dyadic $\mathbf{\alpha}$ of Eq. (8-42) is taken to be $\alpha \mathbf{ii} + \beta \mathbf{jj} + \gamma \mathbf{kk}$, the energy of the three hydrogens due to molecular rotation when $K = 1$ is [931]

$$\Delta W_{\mathbf{I} \cdot \mathbf{J}} = \left[\alpha + \beta + \frac{(\gamma - \alpha - \beta)K^2}{J(J+1)} \right] \mathbf{I}_H \cdot \mathbf{J} \quad (8\text{-}48)$$

Here $\mathbf{I}_H \cdot \mathbf{J} = \dfrac{\mathbf{I}_H \cdot \mathbf{F}_1 \, \mathbf{F}_1 \cdot \mathbf{J}}{F_1(F_1 + 1)}$ can be evaluated by the vector model.

The hydrogen spin I can take on the usual orientations with respect to F_1 which are allowed by quantum mechanics, giving a total angular momentum $F = F_1 + I, F_1 + I - 1, \ldots, |F_1 - I|$.   The usual selection rules $\Delta F_1 = 0, \pm 1$ and $\Delta F = 0, \pm 1$ and intensity relations such as (6-6) apply.   The resulting spectrum is indicated in Fig. 8-5, and compared with experimental observations made in a high-resolution beam spectrometer [925].   It may be seen that the maximum splitting of the 3,3 inversion line due to energy of the type given by (8-48) is about 60 kc, and hence these effects are observable only with resolution higher than that of the usual microwave spectrometer.

When $K = 1$, hyperfine doubling can occur.   A small doubling is due to spin-spin interaction between the N and H nuclei, of the type discussed in Sec. 8.6.   Even though $I_H = \frac{1}{2}$, the interaction is just twice that given by (8-16), since three protons are involved rather than a single one [931]. Additional doubling occurs because of the variation of the $I_H \cdot J$ interaction with orientation. When $K = 1$, a distribution of hydrogen nuclear spin occurs which is similar to that of electron spin in Fig. 8-2. If I_H is fixed in orientation with respect to J, it may be seen from this figure and the inequivalence of magnetic effects in the x and y directions that the two otherwise degenerate K states do not have the same energy.

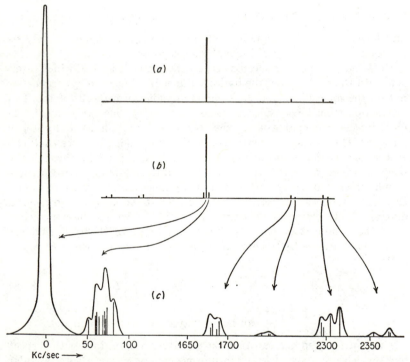

FIG. 8-5. Hyperfine structure of the NH_3 3,3 inversion spectrum. (a) Structure due to N^{14} quadrupole coupling alone. (b) Grosser features of structure due to N^{14} quadrupole coupling plus magnetic coupling of the three hydrogens. (c) Spectrum observed in a beam spectrometer, showing finer features of the individual lines of (b). (*After Gordon, Zeiger, and Townes* [925].)

A diagram of the energy levels of NH_3 for the 3,1 inversion transition and a particular value of F_1 is given in Fig. 8-6. Without hyperfine doubling, the stronger transitions would involve no change in magnetic energy (*cf.* Fig. 8-6b). Inversion transitions occur between different members of the two K-type levels, and hence the hyperfine doubling adds to the transition frequency for $F = \frac{7}{2}$, and subtracts from it for $F = \frac{5}{2}$, giving an appreciable doubling. The dotted levels of Fig. 8-6c are forbidden for $J = 3$ by the fact that hydrogen nuclei follow Fermi-Dirac

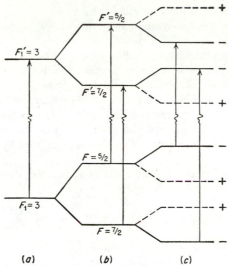

FIG. 8-6. Hyperfine structure of the 3,1 inversion transition of NH₃. Quantum numbers are $J = 3$, $K = 1$, $F_1 = |\mathbf{J} + \mathbf{I_N}| = 3$, $I_H = \frac{1}{2}$. (a) represents energies due to inversion and N^{14} quadrupole coupling only. (b) includes part of magnetic interaction with hydrogen moments which is the same for the two degenerate K states. (c) includes hyperfine doubling effects due to hydrogen moments. Only the stronger hyperfine transitions are shown. Dotted levels are forbidden by the exclusion principle.

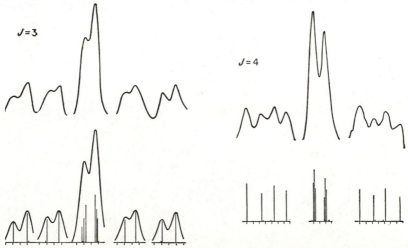

FIG. 8-7. Structure and hyperfine doubling of the $N^{14}H_3$ inversion spectrum for $J = 3$, $K = 1$, and $J = 4$, $K = 1$. Upper curves show experimental observations, and lower part of figure the theoretical expectations. Frequency increases from left to right with 60-kc intervals indicated. (*From Gunther-Mohr, White, Schawlow, Good, and Coles* [930].)

statistics (see Sec. 3-4 for a detailed explanation).   The levels indicated by a minus sign are antisymmetric with respect to interchange of two hydrogen nuclei; those labeled with a plus sign are symmetric with respect to such an interchange.   For $J = 4$, or any even value of J, the symmetry of the levels is reversed, so that the dotted levels of Fig. 8-6c are allowed, and the solid ones forbidden (cf. page 71).

Hyperfine structure of the 3,1 and 4,1 inversion lines of NH_3 is shown in Fig. 8-7, where the hyperfine doubling is quite prominent.   The alternation in relative intensity of the doublets in changing from $J = 3$ to $J = 4$ is due to the alternation in the levels permitted by statistics which was noted above.

The magnitude of the energy shift of each level due to this type of hyperfine doubling is given [931] by

$$\Delta W = \pm (\beta - \alpha)\mathbf{I}_H \cdot \mathbf{J} = \pm (\beta - \alpha) \frac{(\mathbf{F}_1 \cdot \mathbf{I})(\mathbf{F}_1 \cdot \mathbf{J})}{F_1(F_1 + 1)} \tag{8-49}$$

When the magnetic interaction is the same for the x and y directions, $\beta - \alpha = 0$, and this doubling disappears.

In the particular case of NH_3, $\beta - \alpha = -14.4$ kc.   From this value and the $\mathbf{I}_H \cdot \mathbf{J}$ interaction of the type given by (8-48) for hydrogen in inversion lines for which $K \neq 1$, one can obtain the values $\alpha = -1$ kc, $\beta = -16$ kc, and $\gamma = -19$ kc.   As for C_I in H_2, the $\mathbf{I} \cdot \mathbf{J}$ interaction constants here are negative, indicating a large amount of slippage of the valence electrons as the hydrogens rotate in NH_3.

CHAPTER 9

INTERPRETATION OF HYPERFINE COUPLING CONSTANTS IN TERMS OF MOLECULAR STRUCTURE AND NUCLEAR MOMENTS

Hyperfine structure in molecular spectra can be theoretically predicted with high accuracy if the rotational state of the molecule, the spin of the nucleus producing the hyperfine structure, and certain coupling coefficients between molecule and nucleus are known. The coupling coefficients are dependent on either the magnetic dipole or electric quadrupole moment of the nucleus and on various properties of the molecule. The purpose of this chapter is to examine in detail the dependence of these hyperfine coupling coefficients on molecular structure, and to show how they may be theoretically estimated or, when measured, how they may be interpreted to obtain information about molecular structure and nuclear moments. For most molecules hyperfine structure due to nuclear quadrupole moments is considerably more prominent than that due to magnetic moments. Hence we shall begin with a discussion of quadrupole coupling.

9-1. Introductory Remarks on Quadrupole Coupling. From measurement of hyperfine structure, the quadrupole coupling eqQ may be evaluated. The nuclear quadrupole moment Q is a property of the nucleus which depends on the state of the nucleus and may be considered fixed since nuclei will almost always be encountered in their ground state. The sign and magnitude of Q can be very roughly estimated for some nuclei ([326] and [513]), but present knowledge of nuclear structure allows only the crudest estimates of Q from nuclear theory. The quantity e is the proton charge, and q, or $(\partial^2 V/\partial z^2)_{\mathrm{av}}$, is a molecular property, depending on the distribution of charges in the molecule. It cannot be exactly evaluated, but Townes and Dailey ([239] and [422]) have shown how it can be evaluated accurately enough in certain molecules to allow a useful determination of Q from the molecular coupling constant eqQ. In some cases Q may also be approximately determined from optical atomic spectra and with more accuracy from radio-frequency atomic-beam spectroscopy. Once Q for a given nucleus is known, it can provide a very convenient probe to test the electronic distribution in various molecules, since the quadrupole coupling allows a very direct measurement of the second derivative of the potential at the nucleus due to molecular charges.

9-2. Quadrupole Coupling in Atoms. Before discussing in detail the relations between q and molecular structure and the evaluation of quadrupole coupling constants in molecules, we shall consider the simpler atomic case. Most of the electrons in atoms are arranged in groups of closed shells, corresponding to distributions of charge which are, on the average, spherical. These spherical shells produce zero average field at the nucleus and hence do not contribute to q (since from Chap. 5 charge density within the nucleus is neglected in obtaining q). In addition to spherical shells, the atom may contain one or more valence electrons which are not in closed shells. If we choose an atom with only one such valence electron, then from (5-43)

$$q_J = e \int \psi_{JJ}^* \frac{3 \cos^2 \theta - 1}{r^3} \psi_{JJ} \, d\tau \qquad (9\text{-}1)$$

If the electron is in a central field, then ψ may be separated into one factor depending on r, and another factor which is a spherical harmonic function of the angles. The angular part of the integral in (9-1) may be evaluated as $-2l/(2l + 3)$ so that

$$q_J = -\frac{2le}{2l + 3} \left(\frac{1}{r^3} \right)_{av} \qquad (9\text{-}2)$$

where l is the orbital angular momentum, e the electronic charge, and $(1/r^3)_{av}$ is the average inverse third power of the distance between nucleus and electron. Expression (9-2) neglects the electron spin, which for a real atom must be taken into account. However, in the parallel case of a molecule, which is of prime interest in this chapter, spin can correctly be neglected.

For a hydrogenlike wave function, $(1/r^3)_{av}$ is given by (5-13b). However, this quantity cannot usually be obtained with high accuracy because the radial wave functions for atoms are poorly known. Fortunately, other spectroscopically measurable quantities also depend on $(1/r^3)_{av}$, and from a measurement of these $(1/r^3)_{av}$ may often be obtained. Thus, using the fine-structure doublet separation $\Delta \nu$ and Eq. (5-13a)

$$q_J = q_{nll} = -\frac{2l - 1}{l + 1} q_{nl0} = -\frac{2le \, \Delta \nu}{Z_i R \alpha^2 a_0^3 (l + \frac{1}{2})(2l + 3)} \qquad (9\text{-}3)$$

where Z_i is the effective value of Z near the nucleus, which in heavy atoms is approximately the nuclear charge Z minus 4 for p electrons (cf. [895] for more precise values of Z_i). The quantities q_{nll} and q_{nl0} are introduced here for later reference. They are defined after Eq. (9-6).

In case the nuclear gyromagnetic ratio g_I and the resulting magnetic hyperfine structure are known, the uncertainty of an effective Z may be eliminated since the magnetic hyperfine structure depends directly on $(1/r^3)_{av}$ [cf. 5-51)]. Then

$$q_J = \left(\frac{\partial^2 V}{\partial z^2}\right)_{atom} = -\frac{aeMJ(J+1)}{g_I\mu_0^2 m(l+1)(2l+3)} \qquad (9\text{-}4)$$

where e, m = electronic charge and mass, respectively

$\quad\mu_0$ = Bohr magneton

$\quad M$ = proton mass

$\quad J$ = total angular momentum of the electron (l plus spin)

$\quad a$ = hyperfine structure parameter such that the energy of magnetic interaction between nucleus and electron is $a(\mathbf{I} \cdot \mathbf{J})$.

Values of Z_i and $(1/r^3)_{av}$ for a valence p electron of a number of atoms have been derived by Barnes and Smith [895] from data on atomic spectra. These and other values of $(1/r^3)_{av}$ obtained by interpolation are listed in Table 9-1. Table 9-1 neglects certain relativistic effects which are small for $Z < 65$, but which for the heaviest atoms may produce errors of 30 per cent in $(1/r^3)_{av}$ or in q [70].

TABLE 9-1. VALUES OF $(1/r^3)_{av}$ FOR VALENCE p ELECTRONS OF NEUTRAL ATOMS
Certain relativistic corrections are neglected. All values are in 10^{24} cm^{-3}. (After Barnes and Smith [895]. Value for N is from [916a].)

Li	0.26	Be	1.17	B	4.1	C	8.3	N	22.5	O	29	F	44
Na	1.65	Mg	5.2	Al	8.6	Si	15.6	P	24	S	34	Cl	48
K	3.0	Ca	7.6	Ga	24	Ge	39	As	51	Se	65	Br	92
Rb	5.7	Sr	13.5	In	39	Sn	76	Sb	88	Te	101	I	121
Cs	8.7	Ba	19.2	Tl	80	Pb	108	Bi	166				

Although expression (9-4) affords an exact way of obtaining q_J due to a single particle in a central field, in actual atomic cases it is in error by about 10 per cent. This is because the closed shells of electrons which have been assumed spherical are slightly polarized by the valence electron. As a result, electrons in the closed shell tend to move away from the position of the valence electron, producing a contribution to q_J at the nucleus which is of opposite sign to the part produced by the valence electron. Hence the closed shells may be regarded as shielding or screening the nucleus to some extent from fields of the valence electrons.

Accurate and detailed calculations of the magnitude of shielding are very difficult since they depend on radial wave functions of the electrons. However, Sternheimer, Foley, and others [539], [804], [979] have given approximate corrections for many atoms. For the ground states of atoms, the corrections to (9-3) or (9-4) correspond to approximately 10 per cent reduction in q_J. On the other hand, for an electron in an excited atomic state or for a charge as far away from the nucleus as an atomic radius, "antishielding" may occur, i.e., "shielding" may actually increase

q_J [923], [979]. The contribution of a charge one or two angstroms away to q at a nucleus may thus be increased by a factor of about 10 as a result primarily of polarization and distortion of the distribution of p electrons surrounding the nucleus.

Perturbation of the closed electron shells by a valence electron may affect magnetic hyperfine structure in atoms as well. This effect is, however, considerably smaller than the similar effect on quadrupole hyperfine structure [749].

In cases with two or more valence electrons in various atomic orbits, determination of their interaction with a nuclear quadrupole moment can be somewhat more complex. A few such cases, as well as corrections for relativistic effects, are discussed by Casimir [70].

9-3. Quadrupole Coupling in Molecules—General Considerations. For a nucleus in a molecule, contributions to q at the nucleus may come from the following sources:

1. Valence electrons of the nucleus or atom in question
2. Distortion of the closed shells of electrons around the nucleus
3. Charge distributions associated with adjacent atoms or ions, *i.e.*, charges essentially outside the radius of the atom

Contributions of type 3 might at first thought seem to be the only ones which differ from the atomic case. However, the wave functions or distribution of the valence electrons are very much affected by molecular bonds, and hence contributions of type 1 are much modified from the simpler atomic case.

In order to examine the contributions of valence electrons, we consider first an atom with one valence electron outside a closed shell. In the atom, this electron would be in some definite atomic state specified by the atomic wave function ψ_{nlm}. In the molecule, the wave function will be modified, perhaps radically changed, but it may be expressed as an expansion in terms of atomic wave functions:

$$\psi = \sum_{nlm} a_{nlm}\psi_{nlm} \qquad (9\text{-}5)$$

In some cases the larger terms in this expansion are fairly well known from molecular structure. A single bond between two atoms involves only terms with $m = 0$. If the bond is a covalent bond, then usually a good first approximation to the expansion (9-5) is to assume that the electron is entirely in the lowest-energy atomic states of the two bonded atoms, since the molecular bond would have to supply a considerable amount of energy to give large values to the coefficients of more highly excited states. Each atom supplies one electron to the bond, so that, although any one electron has a probability of $\frac{1}{2}$ of being found on a particular atom, there is on the average one electron in an orbit about each atom which is much the same as before the atoms were bonded in a molecule.

The bonding energy is provided by an overlap of the wave functions of the two atoms which gives an exchange energy. This exchange energy is a typical quantum-mechanical effect which depends on electrostatic interactions, and which increases as the wave functions of the two atoms coincide or overlap more completely (*cf.* [133]). Overlap of two $2p$ wave functions ($n = 2, l = 1$) of two atoms is shown in Fig. 9-1a. This overlap can be very much increased by allowing in the expansion (9-5) a small

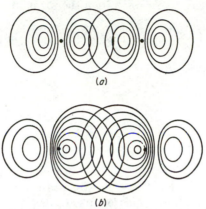

(a)

(b)

Fig. 9-1. Illustration of the increase of overlap or exchange integral with hybridization. (a) Overlap of two p wave functions; (b) overlap of two s-p hybrid wave functions. The lines represent contours of equal electron density for the atomic wave functions.

amount of $2s$ wave function, as shown in Fig. 9-1b. The s and p wave functions subtract on one side of an atom but add on the side of the bond, thus increasing the wave function in the region between the atoms substantially. The overlap and hence exchange energy of the bond increases until the bond has approximately 25 per cent s character ($a_{200}^2 \approx \frac{1}{4}$). That the exchange energy is a maximum does not necessarily mean that this is the lowest-energy condition for the bonds, since energy may be required to promote the atom from its ground state to obtain the mixed or hybrid bond wave function. Mixture or hybridization of a p wave function with a d wave function will achieve the same type of increase in the exchange energy, but hybridization of a p wave function with another p or an f wave function does not give the same advantage, since the two functions have the same symmetry and they either add or subtract on both sides of the atom. A second approximation to the correct expansion (9-5) can hence be obtained by adding a judicious amount of p atomic wave function if the atomic ground state corresponds to an s function, or of the lowest available s or d wave function if the atomic ground state is a p state. The "judicious amount" may be judged from a variety of helpful, though not too exact, methods (for a discussion see [913c]).

Of course, expression (9-5) must provide a probability of approximately $\frac{1}{2}$ for the electron's being found in an approximately atomic orbit of the adjacent atom to which it forms a bond. This results in the presence of a large number of atomic wave functions of rather large n and l, each, however, with rather small coefficients a_{nlm} by comparison with the lowest atomic energy states of the particular atom on which our attention is centered.

The quantity q may be obtained from the wave functions since

$$q = \left(\frac{\partial^2 V}{\partial z^2}\right)_{av} = e \int \psi^* \left(\frac{3 \cos^2 \theta - 1}{r^3}\right) \psi \, d\tau$$

Using the expression (9-5),

$$q = \sum_{nlm} |a_{nlm}|^2 q_{nlm,nlm} + \sum_{nlm,n'l'm'} a_{nlm} a^*_{n'l'm'} q_{nlm,n'l'm'} \tag{9-6}$$

where

$$q_{nlm,n'l'm'} = e \int \psi_{nlm} \left(\frac{3 \cos^2 \theta - 1}{r^3}\right) \psi^*_{n'l'm'} \, d\tau$$

and the second sum is over all nonidentical nlm and $n'l'm'$, since all terms for which $n = n'$, $l = l'$, and $m = m'$ are collected in the first sum. Since ψ_{nlm} is an atomic wave function, its angular variation is given by a spherical harmonic as in (5-2), and $q_{nlm,n'l'm'}$ is zero unless $m = m'$ and either $l = l' \neq 0$ or $l = l' \pm 2$. This conveniently eliminates a large number of terms from the second sum of (9-6).

The quantities $q_{nlm,nlm}$ of (9-6), which may be shortened to q_{nlm}, are simply the values of q for each of the atomic states, and are multiplied by $|a_{nlm}|^2$, the fractional importance of the respective atomic states in the molecular wave function. For an s state ($l = 0$), q_{n0m} is always zero, but for $l \neq 0$, q_{nlm} decreases very rapidly as n or l increase, because for these states the electron spends less time close to the nucleus. The dominating term in the first sum of (9-6) will usually be the state of lowest allowed n because not only will energy considerations make the amplitude a_{nlm} of this wave function large but, in addition, the q_{nlm} for this state is considerably larger than for the higher-energy states. This will be demonstrated for the most common case, where m, the projection of l on the axis, is zero.

The value of q_{nl0} for an atomic wave function differs from q_J discussed above, which has $m = l$, by the factor $-(l + 1)/(2l - 1)$. For hydrogenlike wave functions, evaluation of $(1/r^3)_{av}$ gives

$$q_{nl0} = \frac{2l(l + 1)e}{(2l - 1)(2l + 3)} \left(\frac{1}{r^3}\right)_{av} = \frac{4Z^3 e}{n^3 a_0^3 (2l - 1)(2l + 1)(2l + 3)} \tag{9-7}$$

Table 9-2 shows the relative magnitude of q_{nl0} for various states nl and compares the relative values computed for the hydrogenlike case from (9-7) with those obtained from the fine-structure splitting and Eq. (9-3). Because of dependence on n and screening which changes the effective value of Z, q_{nl0} for the $5p$ state of iodine is fourteen times larger than that for the $6p$ state of cesium. Similarly the value for the $6p$ state of Cs is considerably larger than that for any other state of Cs, the most marked differences occurring with a change of l and a consequent large change of screening. Although fine-structure measurements are less complete in F

TABLE 9-2. RELATIVE VALUES OF $q_{nl0} = \partial^2 V / \partial z^2_{n,l,m=0}$ FOR VARIOUS ATOMIC STATES

The very large effect of screening is shown by comparison of the third and fourth columns.

Electronic state	Atomic example for which fine structure is known	Values of q_{nl0} from fine structure, esu	Relative values of q_{nl0} assuming hydrogen wave functions and no screening [Eq. (9-7)]
			Relative to $5p$
$5p$	I	-45×10^{15}	1.00
$5d$	Cs	-0.31×10^{15}	0.14
$5f$	Cs		0.048
$6p$	Cs	-3.4×10^{15}	0.58
$6d$	Cs	-0.16×10^{15}	0.08
$6f$	Cs		0.028
$7p$	Cs	-1.1×10^{15}	0.36
$7d$	Cs	-0.09×10^{15}	0.05
			Relative to $2p$
$2p$	F	-21×10^{15}	1.00
$3p$	Na	-0.7×10^{15}	0.30
$4p$	Na	-0.2×10^{15}	0.12

and Na, Table 9-2 shows that, even in these light nuclei where screening effects are less important, q decreases very rapidly with increasing n or l.

Although most of the terms in the second sum of (9-6) are zero, the nonzero parts with $m = m'$ and $l = l' \neq 0$ or $l = l' \pm 2$ must be considered. The "cross-product integrals" $q_{nlm,n'l'm'}$ cannot be evaluated accurately, but they decrease rapidly with increasing n, l, n', or l'. In addition, they are usually not important because the coefficients $a_{nlm}a^*_{n'l'm'}$ are small. If, for example, a p-type function is the largest component of the molecular wave function, cross-product terms involving its amplitude a_{nlm} might be expected to be largest. The only terms of this type involve mixing this p state with other p and f states. But the discussion above of hybridization indicates that hybridization of a p function with either another p or f function should be small since it would not contribute very much to lowering the bond energy. The common types of hybridization, of a p function with either s or d, produce only zero cross terms involving a_{n1m} because then $l \neq l' \pm 2$. In some unusual cases, however, these cross terms could be an important part of q.

Contributions to q from neighboring atoms or ions (type 3) are much smaller than those of the valence electrons for many types of bonds. Assume, for example, that there is a neighboring ion with an average charge of one-half that of the electron and at a typical distance of 2.0 A from the nucleus of interest. It produces a value of $\partial^2 V / \partial z^2$, or q, at the nucleus of only 3×10^{13} esu.

A neighboring ion also distorts the electron distribution about the nucleus, including that of the closed shells. This distorted distribution in turn gives a contribution to q (type 2), which is the "antishielding" effect mentioned above (page 227). Calculations and experimental evidence indicate that such distortions may increase the contribution of a neighboring ion to q by a factor of about 10 [923]. Hence the value of q produced by an ion of the type mentioned above would be increased to 3×10^{14} esu. However, this is less than 2 per cent of the value due to a valence p electron in I or F, as shown in Table 9-2. Hence quadrupole effects due to neighboring ions or to distortion of surrounding spherical shells of electrons can be neglected in many cases.

Thus the lowest-energy atomic p wave function for the valence electrons may in most cases be considered to be the sole source of q. Contributions of the valence p electrons can be very simply obtained as $|a_{n1m}|^2 q_{n1m}$, where $|a_{n1m}|^2$ is the importance of the p wave function in an expansion of the type (9-5) and q_{n1m} is the value of q for an atomic state. However, there are cases for which the valence electrons produce only a very small contribution to q, and then some of the small and complicated terms discussed above must be taken into account. Some of these cases will be illustrated below.

9-4. Procedure for Calculating q in a Molecule. To evaluate q at a particular nucleus with the approximation discussed above, only the contributions of electrons in the valence shell need be considered, and these are taken to be of the form $\Sigma |a_{n1m}|^2 q_{n1m}$. The procedure will perhaps be made most clear by examples.

Consider first the value of $\partial^2 V / \partial z^2$ at the In nucleus in the diatomic molecule InCl, where the z direction is chosen along the molecular axis. In has three valence electrons, the configuration of the ground state of the atom being $5s^2 5p$. Cl has seven valence electrons, with a ground atomic state $3s^2 3p^5$. A first approximation to the structure of the InCl molecule might be that the two atoms are essentially in their ground atomic states and bound together by a covalent bond which uses the p atomic wave functions of each atom.   This would be a so-called p_σ bond, the σ meaning simply that $m = 0$.   The wave function then for the two electrons which form the bond may be written approximately

$$\psi = \frac{1}{\sqrt{2}} (\psi_{510})_{In} + \frac{1}{\sqrt{2}} (\psi_{310})_{Cl} \qquad (9\text{-}8)$$

where the subscripts In and Cl indicate the atomic wave functions of In and Cl, respectively.   If $(\psi_{310})_{Cl}$ is expanded in terms of indium atomic wave functions, it will involve a large number of smaller terms of highly excited states, which contribute essentially nothing to q at the In nucleus. However, the first term of (9-8) contributes for each bonding electron an amount $(1/\sqrt{2})^2 q_{510}$, or for the two bonding electrons an amount q_{510}. There are in addition two valence s electrons which are not involved in the bonding, but since for an s electron $q_{n00} = 0$, they contribute nothing. Hence a first (and very rough) approximation to the molecular structure suggests that there is one excess p electron around the In nucleus with $m = 0$, and that $q_{In} = q_{510}$.   The value of q_{510} may be calculated from the In atomic fine structure and expression (9-3), or more accurately from the magnetic hyperfine structure and expression (9-4).   In each case it should be noted that these expressions give $q_J = q_{n11}$, which equals $-\frac{1}{2}q_{n10}$.   In general the q's for different values of m are related by

$$q_{nlm} = q_{nl0} \left[1 - \frac{3m^2}{l(l + 1)} \right] \qquad (9\text{-}9)$$

Similarly there is at the Cl nucleus a contribution q_{310} from the bonding electrons, but in addition an amount $2q_{311}$ and $2q_{3,1,-1}$ from the four nonbonding valence electrons.   Now from (9-9) $q_{311} = q_{3,1,-1} = -\frac{1}{2}q_{310}$, so the total value of q_{Cl} is

$$q_{Cl} = q_{310} + 2q_{311} + 2q_{3,1,-1} = -q_{310} \qquad (9\text{-}10)$$

While In might be expected to have an excess of one p electron oriented along the axis, Cl has a defect of one p electron, and hence the negative sign in (9-10).

It may be noted that any covalent single-bonded In or Cl atom should have roughly the same values of q given above.   However, if these atoms are ionically bonded in a molecule, the situation is different.   Consider, for example, NaCl, which we shall first approximate as a completely ionic molecule $\overset{+}{\text{Na}}\overset{-}{\text{Cl}}$.   In this case the chlorine is surrounded by a closed spherical shell of electrons, the configuration of the ground state of $\overset{-}{\text{Cl}}$ being $3s^2 3p^6$, and hence in the approximation used here $q_{Cl} = 0$.   Similarly $\overset{+}{\text{Na}}$ has a closed-shell configuration of electrons and $q_{Na} = 0$.

Ionic Character and Hybridization. This introduces the question whether InCl is indeed covalently bonded as assumed or is more nearly ionic like NaCl. Actually it must be intermediate between these two extremes, and it is customary to describe such an intermediate bond as a mixture of covalent and ionic bonds, or as "resonating" between the two different types. If, for example, the ionic bond has a fractional importance x in the actual InCl structure and the pure covalent bond an importance $1 - x$, then each contributes amounts to q given by the product of its fractional importance times the value of q for the pure type of bond. Hence

$$q_{In} = (1 - x)q_{510} + (x)0 = (1 - x)q_{510}$$
$$q_{Cl} = -(1 - x)q_{310} + (x)0 = -(1 - x)q_{310}$$

(9-11)

A method of estimating the fractional importance, $1 - x$, of the covalent bond will be discussed below.

The quantity multiplying $-q_{510}$ or $-q_{310}$ in (9-11) is called the amount of unbalanced p electrons U_p oriented along the bond.* Thus for In in InCl, $U_p = -(1 - x)$ and for Cl, $U_p = 1 - x$. There may of course be a number of p valence electrons whose effects cancel, as in $\bar{Cl}$, $U_p = 0$. For any type of bond, the net effect of the valence p electrons may be expressed as the number of unbalanced p electrons U_p oriented along the bond. The quadrupole coupling constant is $-U_p$ times the coupling per p electron, or $-U_p eq_{n10}Q$.

Let us examine now the quadrupole coupling constants of Cl^{35} in a few typical molecules listed in Table 9-3. It is clear from the small coupling constant in NaCl that this molecule must be essentially ionic. TlCl must also be largely ionic, although not entirely so. The remaining molecules listed have bonds which are primarily covalent in nature. The value of eqQ for FCl is abnormally high because this molecule is partly ionic, but with the chlorine atom positive instead of negative since F is more electronegative (*i.e.*, tends more to attract electrons) than Cl. A $\overset{+}{Cl}$ ion has two p electrons missing from its valence shell, and the additional missing electron must have come from a $p_\sigma(m = 0)$ orbit if it is to migrate easily to the F atom. With two missing p_σ electrons, $\overset{+}{Cl}$ should have a value of $q = -2q_{310}^+$. The plus is put on q_{310} to indicate that q_{310} for $\overset{+}{Cl}$ may be slightly different from q_{310} for neutral Cl.

Most essentially covalently bonded Cl atoms have quadrupole coupling constants near -80 Mc, which is considerably lower than the value -109.6 Mc of $eq_{310}Q$ for atomic Cl. ICl is such an example, where other

---

* U_p might more logically be the quantity multiplying q_{510} rather than $-q_{510}$ but the above definition has the sanction of usage.

molecular information would indicate that no large amount of ionic bonding is present, yet the coupling is appreciably less than that expected for essentially atomic Cl. Hybridization of the bonding orbital for Cl

TABLE 9-3. QUADRUPOLE COUPLING CONSTANTS OF Cl^{35} IN A FEW TYPICAL MOLECULES

Molecule	eqQ
Cl (atomic)	-109.6 $(-eq_{310}Q)$
FCl	-146
ICl	-82.5
CH_3Cl	-74.8
ClCN	-83.3
TlCl	-15.8
NaCl	< 1

must be assumed for this and other similar cases. A decrease in eqQ is to be expected for hybridization of the $3p_\sigma$ orbit with some $3s$ wave function. For, if the bonding electrons use a small amount of s wave function, then the pairs of nonbonding electrons must use a corresponding amount of $3p_\sigma$ wave function. By this process part of the p_σ wave function becomes filled with two electrons rather than one, and the defect of electrons along the axis is reduced. Hybridization with a fraction y of s wave function gives for Cl a value $U_p = 1 - y$. If hybridization of the $3p$ orbit with a $3d$ wave function occurs, the value of the quadrupole coupling constant would have been increased rather than decreased, or $U_p = 1 + y$, where y is the fractional importance of the $3d$ wave function. This situation is quite different from hybridization with a $3s$ orbit, since the $3d$ orbit is unoccupied, while the $3s$ orbit contains two nonbonding electrons. In the case of In, where U_p is negative, s-hybridization increases the magnitude of the quadrupole coupling, giving $U_p = -1 - y$ for an amount y of s-hybridization.

Probably the quadrupole coupling for an atom in a molecule is also affected by the overlap of its valence wave function with that of the atom to which it is bonded [422], [973]. Unfortunately the precise effect of overlap on the quadrupole coupling constant is very difficult to evaluate theoretically. However, there seems to be good experimental evidence that overlap effects probably do not contribute much to the quadrupole coupling constants [913c], and hence we shall neglect them in subsequent considerations.

Once the amount and type of hybridization and the amount of ionic character of a bond is determined, then q may be rather quickly calculated. Unfortunately, however, the amount of hybridization and the ionic character cannot be separately determined from a measurement of eqQ alone, since the quadrupole effect cannot directly distinguish between the two. The per cent ionic character of a bond between two atoms

[x in (9-11)] can usually be approximately determined as a function of the electronegativity difference from the curve of Fig. 9-2. This curve has been obtained primarily from quadrupole coupling constants in diatomic molecules. After a small allowance for hybridization (described immediately below), these give values of the ionic character plotted in the figure.

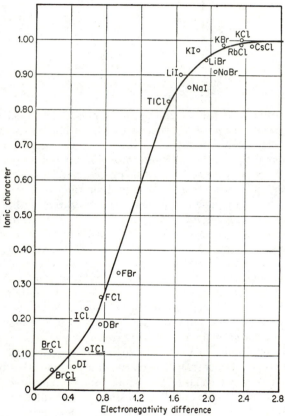

FIG. 9-2. Electronegativity difference. Ionic character of bonds vs. electronegativity difference between bonded atoms. (*After Dailey and Townes* [913c].)

Small deviations from the curve of Fig. 9-2 may occur for a variety of reasons, and hence no precise relation between ionic character and electronegativity difference can be established. The presence of three hydrogens near the carbon atom in CH_3Cl or CH_3I, for example, seems to decrease the effective electronegativity of C by several tenths of a unit. Internuclear distance and hybridization also affect the amount of ionic character. The normal electronegativities of various atoms are listed in Table 9-4.   Gordy [578] has suggested that ionic character is given by the linear relation: ionic character $= \frac{1}{2}$ (electronegativity difference).   This is a rough approximation to the solid curve of Fig. 9-2 but does not fit experimental observations quite so well [913c].

TABLE 9-4. ELECTRONEGATIVITIES AND COVALENT BOND RADII OF ATOMS
Values of electronegativities come from Huggins [814a] and Pauling [94] and are in arbitrary units. Differences in electronegativities may be considered approximately as electron volts. Atomic radii are those due to Pauling [94] and are given in angstrom units.

	H						
Electronegativity	2.2						
Single-bond radius	0.30						
	Li	Be	B	C	N	O	F
Electronegativity	1.0	1.5	2.0	2.6	3.0	3.5	3.9
Single-bond radius			0.88	0.771	0.70	0.66	0.72
Double-bond radius			0.76	0.665	0.60	0.55	0.62
Triple-bond radius			0.68	0.602	0.547	0.50	
	Na	Mg	Al	Si	P	S	Cl
Electronegativity	0.9	1.2	1.5	1.9	2.1	2.6	3.1
Single-bond radius				1.17	1.10	1.04	0.99
Double-bond radius				1.07	1.00	0.94	0.89
Triple-bond radius				1.00	0.93	0.87	
	K	Ca	Sc	Ge	As	Se	Br
Electronegativity	0.8	1.0	1.3	1.9	2.1	2.5	2.9
Single-bond radius				1.22	1.21	1.17	1.14
Double-bond radius				1.12	1.11	1.07	1.04
	Rb	Sr	Y	Sn	Sb	Te	I
Electronegativity	0.8	1.0	1.3	1.9	2.0	2.3	2.6
Single-bond radius				1.40	1.41	1.37	1.33
Double-bond radius				1.30	1.31	1.27	1.23
	Cs	Ba					
Electronegativity	0.7	0.9					

The amount of hybridization varies from bond to bond and no simple rules can give the hybridization exactly.   In the case of Cl, Br, and I, a reasonable approximation is 15 per cent s hybridization for all cases where these atoms are more electronegative than the atom to which they are bonded by 0.25 units, and otherwise no hybridization [913c].   If an atom is singly bonded to two or more other atoms, the angle between the bonds often gives a fairly reliable estimate of the amount of hybridization (cf. [133]).   If only s and p orbitals are involved in two or more equivalent bonds, then the amount of s hybridization is

$$x = \frac{\cos \theta}{\cos \theta - 1} \qquad (9\text{-}12)$$

where θ is the angle between two of the bonds. For example, the bond angles for $AsCl_3$ and NH_3 are $\theta = 98°25'$ and $\theta = 106°47'$, indicating s hybridizations of 13 and 18 per cent, respectively. However, expression (9-12) is not always an accurate indication of the hybridization; its use is especially misleading in some of the hydrides [903] which seem to involve more hybridization than is indicated by the bond angles. In the double or triple bonds of N, O, S, and probably those of other elements, the p_σ component of the bond appears to be hybridized by 10 to 25 per cent [903].

With the above rules for hybridization and ionic character, the quadrupole coupling constants of the halogens can be calculated in known molecules with an accuracy of a few per cent of $eq_{n10}Q$. For other atoms the approximations are not usually so accurate. For the alkalies, one can only say that the coupling should be rather small, since essentially no unbalanced p electrons are involved in bonding. For elements in the third to sixth columns of the periodic table, involving multiple bonds whose characteristics are not well known, the above approximations can usually give quadrupole coupling constants with an accuracy of about 25 per cent.

Multiple Bonds. In case there are many bonds to one atom in a molecule, as in the fifth-column elements mentioned above, the contributions to q of each bond and valence electron must be added. For example, a double bond, such as that for the end nitrogen in $\bar{N} = \overset{+}{N} = 0$, involves a p_σ orbit ($m = 0$) and a p_π orbit ($m = \pm 1$), so that the value of q if no hybridization occurs is $q = q_{\bar{2}10} + q_{\bar{2}11} + 2q_{\bar{2}11} = -\frac{1}{2}q_{\bar{2}10}$. If the p_σ bond is assumed to be 45 per cent s-hybridized, two of the nonbonding electrons must correspondingly have 45 per cent p wave function. The resulting contributions to q are as follows:

Nonbonding electrons in p_π orbit:	$2q_{\bar{2}11}$
Nonbonding electrons in s plus 45% p_σ orbit:	$0.90q_{\bar{2}10}$
Bonding electrons in p_π orbit:	$q_{\bar{2}11}$
Bonding electrons in p_σ plus 45% s orbit:	$0.55q_{\bar{2}10}$

This gives a net of $3q_{\bar{2}11} + 1.45q_{\bar{2}10} = -0.05q_{\bar{2}10}$.

A minus superscript was attached to q_{210} above to indicate that q_{210} should be evaluated for a negative ion $\bar{N}$ rather than for neutral N. Examination of the fine structure $\Delta\nu$ for several states of ionization of various atoms indicates that each stage of ionization modifies q by a factor $1 + \epsilon$ which is approximately 1.25; positive ionization increases q by pulling all electrons closer to the nucleus, and negative ionization decreases q. We shall hence assume $q^-_{210} = q_{210}/1.25$ and $q^+_{210} = 1.25q_{210}$. More precise values of ϵ can be found in Table 9-5. Thus for a structure like $\overline{N}$= with no hybridization, $q = -\frac{1}{2}(1/1.25)q_{210}$, or $U_p = 0.40$. If an atom in the first two rows of the periodic table has four covalent bonds, then all valence orbits are equally occupied by electrons, and $q = q_{210} + q_{211} + q_{2,1,-1} = 0$ $(U_p = 0)$. This is almost always the case

TABLE 9-5. VALUES OF ϵ FOR A NUMBER OF ELEMENTS

The coupling constant produced by a p electron is modified by a factor approximately $1 + \epsilon$ for each stage of ionization, being larger for positive ionization.

Be	0.90	B	0.50	C	0.45	N	0.30	O	0.25	F	0.20
Mg	0.70	Al	0.35	Si	0.30	P	0.20	S	0.20	Cl	0.15
Ca	0.60	Sc	0.30	Ge	0.25	As	0.15	Se	0.20	Br	0.15
Sr	0.60	Ga	0.20			Sb	0.15	Te	0.20	I	0.15

for C or Si, which are quadruply bonded. Nitrogen normally has three bonds and a quadrupole coupling constant near -4 Mc. However, in $\overline{N}=\overset{+}{N}=O$ and in CH_3—$N \equiv C$, the central nitrogens are quadruply bonded, and the observed coupling constants are very small as expected, being less than 0.3 Mc.

There are many cases of bonds which are partly single and partly double or triple, i.e., of resonance between single and multiple bond structures. Fortunately single, double, and triple bonds for a particular pair of atoms have distinctly different lengths, so that the internuclear distances may be used to determine the relative importance of single or multiple bonds between two atoms. If R_1 is the sum of the single-bond radii of two atoms, or the distance between them when there is a single bond, R_2 the double-bond distance, and R_3 the triple-bond distance, then the expected distance for an intermediate-type bond is (see [133] for a fuller discussion)

$$R = \frac{x_1R_1 + 3x_2R_2 + 6x_3R_3}{x_1 + 3x_2 + 6x_3} \tag{9-13}$$

where x_1 = fractional importance of the single-bond state

x_2 = fractional importance of the double-bond state

x_3 = fractional importance of the triple-bond state

Since $x_1 + x_2 + x_3 = 1$, if it is thought that only single and double bonds, or only double and triple bonds, are involved between two particular atoms, and if the internuclear distance R is known, then the fractional importance x of the two resonant bonding states can be determined. Standard atomic radii for single, double, and triple bonds for various atoms are listed in Table 9-4. One cannot expect high precision in determination of the amount of double- or triple-bond character from (9-13), since observed bond distances in molecules often differ by as much as 0.01 or 0.02 A from the values obtained by use of this expression and Table 9-4. In many of the heavier atoms the bond radii are still unknown.

A refinement in the calculation of internuclear distances is the Scho-

TABLE 9-6. THE NUMBER OF "UNBALANCED" p ELECTRONS, U_p, FOR VARIOUS BOND STRUCTURES

The fraction of s or d hybridization is indicated by the symbols s or d, respectively. It is assumed that the hybridizing s or d wave functions have the same principal quantum number as the p function. U_p is with respect to the axis of the bond or bonds unless otherwise stated. The quantity ϵ which appears here is given in Table 9-5.

Electron configuration of atom	Type of bond	Hybridization	U_p
s^2p^5 (like Cl)	Single covalent	s and d	$1 - s + d$
s^2p^6 (like Cl$^-$)	Single ionic	—	0
s^2p^4 (like Cl$^+$)	Single ionic	—	$2(1 + \epsilon)$
s^2p^4 (like O)	Double covalent	$\begin{cases} p_\sigma \ s \text{ and } d \\ p_\pi \text{ none} \end{cases}$	$\frac{1}{2} - s + d$
s^2p^3 (like N)	Triple covalent	$\begin{cases} p_\sigma \ s \text{ and } d \\ p_\pi \text{ none} \end{cases}$	$-s + d$
s^2p^4 (like O)	Two single covalent, each of ionic character i, with O positive	s	$\left(\frac{1}{2} + \frac{i}{2} + 2s \right)(1 + 3i\epsilon)$ (along direction bisecting bond angle) $(s - 1 - i)(1 + 2i\epsilon)$ (along direction perpendicular to plane of bonds)
	With O negative	s	$\dfrac{(\frac{1}{2} - 2s)(1 - i)}{1 + 2i\epsilon}$ (along direction bisecting bond angle) $\dfrac{(s - 1)(1 - i)}{1 + 2i\epsilon}$ (along direction perpendicular to plane of bonds)
s^2p^3 (like N)	Three single bonds, each of ionic character i, with N positive	s	$-3s(1 + i)(1 + 3i\epsilon)$
	With N negative	s	$-\dfrac{3s(1 - i)}{(1 + 3i\epsilon)}$
s^2p^2 (like C)	Four covalent bonds	Any $s - p$ hybridization	0
s^2p (like B)	Three bonds in a plane	s	1

maker-Stevenson rule [109], which states that ionic character of a bond shortens the bond by an amount

$$\Delta R = (-)0.09|x_1 - x_2| \qquad (9\text{-}14)$$

where $x_1 - x_2$ is the difference in electronegativity of the two bonded atoms. This shortening is particularly important for bonds to fluorine, but in most other cases it may be neglected, and in fact in some bonds which do not involve fluorine it does not seem to give correct results.

Examples and Tables of Quadrupole Coupling Constants. The number of unbalanced p electrons, U_p, for various important types of bonds is summarized in Table 9-6. U_p for structures which are intermediate between two or more varieties of bonds listed in this table may also be obtained from it by summing U_p times the fractional importance for each bond type.

The bond structure and expected values of U_p for a number of examples are indicated in Table 9-7. The percentage importance of the structures listed have been chosen from use of expressions (9-12) and (9-13) and consideration of molecular dipole moments and quadrupole couplings. The chosen combination of resonating structures gives values of quadrupole coupling (which may be obtained by multiplying the net unbalanced p electrons, U_p, by the negative of the quadrupole coupling per p electron given in Table 9-8) very close to those which are observed and at the same time is consistent with other known information about the molecules.

In rare cases the contribution to q of d or other orbitals may be taken into account in obtaining q. However, unless U_p is extremely small, contributions to q from p-type wave functions will be so much larger, as they are in all cases listed in Table 9-6, that the d orbital contribution may be considered to be zero.

Table 9-8 gives values for the quadrupole coupling constant of various isotopes produced by one p electron excess along the bond axis, *i.e.*, the value of eqQ when U_p, the "unbalance of p electrons," is unity. This might also be written $eq_{n10}Q$. These values are mostly obtained from observed coupling constants in a variety of molecules; however, some are obtained from measurements of atomic spectra. Table 9-8 also lists the best available values of q produced by one p electron excess along the bond axis, that is, q_{n10}. From eqQ and q, the nuclear quadrupole moments Q may be obtained, remembering that e is the charge of one proton. Best values of Q are given in Appendix VII.

9-5. Quadrupole Coupling in Asymmetric Molecules. The discussion so far has applied specifically to symmetric molecules, where one quadrupole coupling constant eqQ is sufficient to specify the energy of interaction between the nuclear quadrupole and the molecule. As mentioned in Chap. 6, for an asymmetric rotor two coupling constants are needed,

Quadrupole coupling constants may be obtained by multiplying U_p by $eq_{n10}Q$, the quadrupole coupling per p_σ electron.

Nucleus	Molecule	Partial structures	Comments	U_p for each structure	% importance	Net U_p
Cl	FCl	F—Cl	No hybridization	1.00	75	1.37
		$\bar{\text{F}}$ $\overset{+}{\text{Cl}}$		2.50	25	
	ICl	I—Cl	15 % s hybridization	0.85	85	0.72
		$\overset{+}{\text{I}}$ $\bar{\text{Cl}}$		0	15	
	TlCl	Tl—Cl	15 % s hybridization	0.85	18	0.15
		$\overset{+}{\text{Tl}}$ $\bar{\text{Cl}}$		0	82	
	SiH$_3$Cl	H$_3$Si—Cl	15 % hybridization	0.85	30	0.38
		$\text{H}_3\overset{+}{\text{Si}}\bar{\text{Cl}}$		0	40	
		$\text{H}_3\bar{\text{Si}}=\overset{+}{\text{Cl}}$	p_σ bond with 15 % s hybridization	0.40	30	
	C$_2$H$_3$Cl	(structure: CH$_2$=CHCl)	U_{pz} refers to z axis along C—Cl bond. y axis is in Cl—C—H plane and x axis is perpendicular. 15 % s hybridization	$U_{pz} = 0.85$ $U_{py} = -0.42$ $U_{px} = -0.42$	75	$U_{pz} = 0.66$ $U_{py} = -0.38$ $U_{px} = -0.28$
		(structure: $\bar{\text{C}}$—C with Cl$^+$)	p_π bond assumed perpendicular to Cl—C—H plane	$U_{pz} = 0.55$ $U_{py} = -1.16$ $U_{px} = 0.72$	5	
		(structure: C=$\overset{+}{\text{C}}$ with $\bar{\text{Cl}}$)		0	20	
S	OCS	$\bar{\text{O}}$—C≡$\overset{+}{\text{S}}$	No hybridization	0	14	0.51
		O=C=S		0.5	58	
		$\overset{+}{\text{O}}$≡C—$\bar{\text{S}}$		0.8	28	
		or $\bar{\text{O}}$—C≡$\overset{+}{\text{S}}$	25 % s hybridization of p_σ bond	−0.31	14	0.27
		O=C=S		0.25	58	
		$\overset{+}{\text{O}}$≡C—$\bar{\text{S}}$		0.60	28	
N	NH$_3$	$\bar{\text{N}}$—H (with H, H)	25 % hybridization	−0.40	100	−0.40
	N$_2$O	$\bar{\text{N}}$=$\overset{+}{\text{N}}$=O	45 % hybridization of p_σ bonds only	0.05	55	−0.17
		N≡$\overset{+}{\text{N}}$—$\bar{\text{O}}$	End nitrogen	−0.45	45	
		$\bar{\text{N}}$=$\overset{+}{\text{N}}$=O	Central nitrogen	0	55	0
		N≡$\overset{+}{\text{N}}$—$\bar{\text{O}}$		0	45	
As	AsH$_3$	As—H (with H, H)	10 % hybridization	−0.30	100	−0.30
	AsCl$_3$	$\overset{+}{\text{As}}$—Cl (with $\bar{\text{Cl}}$, Cl)	10 % hybridization	−0.25	50	−0.28
		As—Cl (with Cl, Cl)		−0.30	50	

$eQ \dfrac{\partial^2 V}{\partial z_m^2}$ and $eQ \left(\dfrac{\partial^2 V}{\partial x_m^2} - \dfrac{\partial^2 V}{\partial y_m^2}\right)$, where x_m, y_m, and z_m are directions of the principal axes of inertia.   For a symmetric molecule, if z is the axis of symmetry, $\dfrac{\partial^2 V}{\partial x_m^2} - \dfrac{\partial^2 V}{\partial y_m^2} = 0$.   Very often the electric field derivative

TABLE 9-8. QUADRUPOLE COUPLING CONSTANTS FOR VARIOUS NUCLEI DUE
TO ONE VALENCE p ELECTRON

A range of values is given for N^{14}, which is poorly known.

	Quadrupole coupling constant for one valence p electron with wave function oriented along bond $(eq_{n10}Q)$, in Mc	$\dfrac{\partial^2 V}{\partial z^2}$ for one valence p electron with $m = 0$ (q_{n10}), in -10^{15} esu
B^{10}	$-$ 10.9	1.6
B^{11}	$-$ 5.3	1.6
N^{14}	$-$ 10 to -24	8.6
O^{17}	3.3	11
Al^{27}	$-$. 37.5	3.3
S^{33}	55	13
S^{35}	$-$ 39	13
Cl^{35}	109.7	20
Cl^{36}	23.2	20
Cl^{37}	86.4	20
Ga^{69}	$-$ 125	7.5
Ga^{71}	$-$ 78.8	7.5
Ge^{73}	220	15
As^{75}	600	20
Se^{79}	-1400	25
Br^{79}	$-$ 769.8	34
Br^{81}	$-$ 643.1	34
In^{113}	$-$ 886.2	11
In^{115}	$-$ 899.1	11
Sb^{121}	2000	34
Sb^{123}	2500	34
I^{127}	2292.8	45
I^{129}	1688	45
Hg^{201}	-1000	23

tensor $\nabla\mathbf{E}$ is symmetric about some molecular bond.   For example, $\nabla\mathbf{E}$ at either Cl nucleus in CH_2Cl_2 is approximately symmetric about the C—Cl bond, and choosing this direction as an axis, eqQ may be calculated as above.   The quantities $eQ \dfrac{\partial^2 V}{\partial z_m^2}$ and $eQ \left(\dfrac{\partial^2 V}{\partial x_m^2} - \dfrac{\partial^2 V}{\partial y_m^2}\right)$ may be calculated from (6-23) by a rotation of coordinates to the principal axes of inertia.

There are some cases, however, for which the field is not symmetric about the bond axis [370], [459], as for Cl in the structure

where a double bond occurs. Taking the double bond to be a non-hybridized p_σ bond along the C—Cl or z_b direction plus a p_π bond perpendicular to the Cl—C—H plane or in the x_b direction, we obtain an excess of one p electron in a p_π orbit in the y_b direction, *i.e.*, in the Cl—C—H plane. Hence $\dfrac{\partial^2 V}{\partial y_b^2} = q_{310}(1 + \epsilon); \dfrac{\partial^2 V}{\partial x_b^2} = \dfrac{\partial^2 V}{\partial z_b^2} = -\tfrac{1}{2}q_{310}(1 + \epsilon)$. If this double-bonded structure is of importance x, and the structure

is of importance $1 - x$, then the sum of their contributions gives values

$$\frac{\partial^2 V}{\partial z_b^2} = \left[-1 + \frac{x}{2}(1 - \epsilon) \right] q_{310} \qquad \frac{\partial^2 V}{\partial x_b^2} = \left[\tfrac{1}{2} - x\left(1 + \frac{\epsilon}{2}\right) \right] q_{310}$$

$$\frac{\partial^2 V}{\partial y_b^2} = [\tfrac{1}{2} + x(\tfrac{1}{2} + \epsilon)] q_{310} \quad (9\text{-}15)$$

Values of $\dfrac{\partial^2 V}{\partial z_m}$ and $\dfrac{\partial^2 V}{\partial x_m^2} - \dfrac{\partial^2 V}{\partial y_m^2}$ for the principal axes of inertia may then be obtained from (9-15) by a rotation of coordinates in accordance with (6-23b). In this case, the transformation to principal axes of inertia is somewhat more complicated than (6-23a) because of the lack of axial symmetry of the field about the bond axis. Table 9-7 gives values of U_p for x_b, y_b, and z_b axes assuming 15 per cent s hybridization of the p_σ bond, and giving some importance to the structure

An interesting use of slight asymmetry occurs in the bending mode of molecules such as HCN, BrCN, and ICN [703], [754], [998a]. In the

ground state these molecules are linear and the electrostatic potential is axially symmetric. However, in the bending vibrational mode $v_2 = 1$, there is a slight asymmetry in the field due in part to the difference in direction between the principal axis of inertia and the C—N or I—C bonds, and in part to the fact that the dyadic ∇E is no longer symmetric about these bonds. Such effects afford a valuable measurement of the change in electronic structure of these molecules with bending.

9-6. Interpretation of Magnetic Hyperfine Coupling Constants. Magnetic hyperfine coupling constants are large and easily interpretable only for the rather rare molecules which have electronic angular momentum (*cf*. Chap. 8). For these molecules, however, magnetic hyperfine structure usually gives information about the electron distribution somewhat similar to that given by nuclear quadrupole effects.

It is primarily in diatomic or linear molecules that interpretation of magnetic coupling constants appears to be fruitful. For such cases, expressions (8-6) and (8-16) show that hyperfine effects depend primarily on four coupling constants:

$$a = \frac{2\mu_0\mu_I}{I}\left(\frac{1}{r^3}\right)_{\text{av}} \qquad b + \frac{c}{3} = \frac{16\pi}{3}\frac{\mu_0\mu_I}{I}\,\psi^2(0)$$

$$c = \frac{3\mu_0\mu_I}{I}\left(\frac{3\cos^2\theta - 1}{r^3}\right)_{\text{av}} \qquad d = \frac{3\mu_0\mu_I}{I}\left(\frac{\sin^2\theta}{r^3}\right)_{\text{av}} \tag{9-16}$$

The constant a contains $(1/r^3)_{\text{av}}$, which strictly is to be averaged over the electron or electrons which provide orbital angular momentum. The constants c and d involve similar averages, but over the electrons which provide spin angular momentum. Usually spin and orbital momentum involve precisely the same electrons, in which case the three constants a, c, and d are related by $c = 3(a - d)$. Since the averages in the expressions (9-16) are to be taken only over those electrons which contribute angular momentum, these magnetic coupling constants are more specific than the quadrupole coupling constant $eqQ = eQ\left(\dfrac{3\cos^2\theta - 1}{r^3}\right)_{\text{av}}$. The latter average must be taken over all electrons in the molecule.

A sufficiently complete examination of the magnetic hyperfine structure will yield numerical values for the constants a, b, c, and d given by (9-16) and hence for the four quantities $\left(\dfrac{3\cos^2\theta - 1}{r^3}\right)_{\text{av}}$, $\left(\dfrac{1}{r^3}\right)_{\text{av}}$, $\left(\dfrac{\sin^2\theta}{r^3}\right)_{\text{av}}$, and $\psi^2(0)$. The first may be evaluated or interpreted very much as the similar quantity q discussed above if it is remembered that only the electron or electrons which provide angular momentum are to be considered. The second quantity $(1/r^3)_{\text{av}}$ has also already been discussed in connection with quadrupole effects but provides additional information, a direct measurement of $(1/r^3)_{\text{av}}$, which is not given by the quadrupole coupling constant. This average applies only to an electron with orbital angular momentum, hence not to an s orbit for which $\left(\dfrac{3\cos^2\theta - 1}{r^3}\right)_{av}$ or $\left(\dfrac{\sin^2\theta}{r^3}\right)_{av}$ would also not apply. The fourth quantity, $\psi^2(0)$, is the square of the electron wave function at the nucleus, which is negligible except for an s orbit. This quantity therefore affords information which is never yielded by quadrupole effects, since they are never due to electrons in s orbits.

Since O_2 is in a $^3\Sigma$ state it has electronic spin momentum and provides an interesting example for examination of electron spin–nuclear spin interactions. The ordinary type of O_2 has no hyperfine structure because of the zero spin of O^{16}. However, $O^{16}O^{17}$ yields a rich hyperfine structure from which the constants b and c in (8-6) can be evaluated, and hence the quantities $\psi^2(0)$ and $\left(\dfrac{3\cos^2\theta - 1}{2r^3}\right)_{av}$ for the electrons with parallel spins [841], [842]. The value of $\left(\dfrac{3\cos^2\theta - 1}{2r^3}\right)_{av}$ agrees within about 10 per cent with calculations using the expected structure of O_2. The value of $\psi^2(0)$ corresponds to only 2.5 per cent s character for the electrons with parallel spins. Even this small amount of s character is not very well understood in terms of the molecular structure. In addition, it has a large effect on the observed hyperfine structure since magnetic hyperfine structure due to an s electron is usually enormously larger than that due to electrons in other types of orbits.

Other examples for which large magnetic hyperfine structure has been found are OH and NO, which are in $^2\Pi$ ground states. The coupling coefficients of OH have not yet been interpreted, but those for NO allow a simple and rewarding discussion [916a]. NO is a rather good Hund's case (a), so that the hyperfine energy is given by $a\Lambda + (b + c)\Sigma$ from (8-9) plus doubling effects in the $^2\Pi_{\frac{1}{2}}$ state from (8-16). Microwave measurements on the $^2\Pi_{\frac{3}{2}}$ state [899] give $a + (b + c)/2 = 74.1$ Mc, and those on the $^2\Pi_{\frac{1}{2}}$ state [782b], [924] give $a - (b + c)/2 = 92.2$ Mc and $d = 112.6$ Mc. If one assumes that the electron which contributes orbital angular momentum is identical with that which contributes spin momentum, then $c = -87.6$ Mc, and the following molecular constants may be derived.

$$\left(\frac{1}{r^3}\right)_{av} = 15 \times 10^{24}\,\text{cm}^{-3} \qquad (\sin^2\theta)_{av} = 0.90 \qquad \psi^2(0) = 0.85 \times 10^{24}\,\text{cm}^{-3}$$

The first two quantities may be compared with the values

$$\left(\frac{1}{r^3}\right)_{av} = 23 \times 10^{24}\,\text{cm}^{-3}$$

and $(\sin^2 \theta)_{av} = 0.80$ for a p_π electron in an atomic orbit about the nitrogen.   For an electron in an atomic orbit of the O atom, $(1/r^3)_{av}$, where r is the distance between the electron and N nucleus, is negligibly small.   Hence the hyperfine coupling constants provide good evidence that the electron with angular momentum is approximately 15/23, or 65 per cent of the time in a p_π orbit about the N atom.   The orbit is somewhat more confined to a plane through the N nucleus and perpendicular to the molecular axis than an atomic orbit would indicate, since $(\sin^2 \theta)_{av}$ is greater than its value for an atomic orbit.   The value of $\psi^2(0)$ represents only about 2.5 per cent probability for the unpaired electron to be in a $2s$ atomic nitrogen state.   As in the case of the O_2 molecule, only a small admixture of atomic s state is indicated, but this makes an important contribution to the hyperfine structure.

CHAPTER 10

STARK EFFECTS IN MOLECULAR SPECTRA

10-1. Introduction. Stark effects are the changes in the spectrum of a system which may be observed when the system is subjected to an electric field. The rotational spectrum of a molecule which has an electric dipole may be expected to be modified when the molecule is in an electric field, since the field exerts torques on the molecular dipole moment and thereby can change its rotational motion. These types of effects can be understood in a qualitative way from classical mechanics, although any detailed explanation of them requires a quantum-mechanical approach.

Consider first a rotating linear molecule with angular momentum perpendicular to the electric field. The field tends to twist the dipole and give it a faster rotation when the dipole is oriented in the direction of the field, and a slower rate of rotation when it is pointed oppositely to the field. The dipole is hence oriented away from the field more often than with it, so that, on the average, the dipole is directed oppositely to the field (contrary to what would be expected if there were no rotation). The fractional difference between the time the dipole points in the two directions can be shown to be proportional to the ratio of the energy of the dipole in the field to the rotational energy

$$f \propto \frac{\mu E}{\frac{1}{2} I \omega^2} \tag{10-1}$$

where μ is the dipole moment, E the electric field, I the moment of inertia, and ω the angular velocity. The change in energy due to the field is then $f \mu E$ or

$$\Delta W \propto \frac{(\mu E)^2}{h B J (J + 1)} \tag{10-2}$$

where the rotational energy $\frac{1}{2} I \omega^2$ has been written in terms of the rotational constant B and the rotational quantum number. There is also, of course, a change in the average rate of rotation of the molecule or its frequency.

If a linear molecule rotates with its angular momentum parallel or antiparallel to the electric field, then the rotating dipole is slightly twisted in the direction of the field, and the energy is decreased {an amount also proportional to $(\mu E)^2/[h B J (J + 1)]$}. It will be seen below that an average of the energy change over random orientations of a rotating molecule gives no net change in energy; the various positive and negative changes just cancel.

Symmetric-top molecules show a Stark effect of a rather different type, because their dipole moments may have components parallel to the angular momentum, and thus components which are fixed in direction rather than rotating. Thus for a symmetric top rotating about its symmetry axis, the dipole moment is in the direction of J and its energy in an electric field is $-\mu E \cos \theta$, where θ is the angle between J and the field E. The projection of J on a fixed direction, such as that established by the direction of E, is always an integer M, the "magnetic" quantum number. Hence the energy might be expected to be

$$-\mu E \cos \theta = -\mu E \left(\frac{M}{J} \right)$$

In the more general case of the angular momentum J and a component of the angular momentum K along the symmetry axis, the component of μ along the J direction is $\mu(K/J)$. Hence the energy might be expected to be

$$-\mu E \frac{K}{J} \cos \theta = - \frac{\mu E K M}{J^2}$$

Or, remembering that when the vector model is used J^2 must always be replaced by $J(J + 1)$,

$$\Delta W = - \frac{\mu E M K}{J(J + 1)} \tag{10-3}$$

This expression is correct but will be more rigorously derived below.

The change in energy, and hence in frequency of a symmetric-top molecule due to an electric field, is thus from (10-3) proportional to the first power of μE, whereas that for a linear molecule from (10-2) is proportional to the second power of μE and is much smaller because $\mu E/[BJ(J + 1)]$ is small (typically 0.01 to 0.001). These are often referred to as "first-order" and "second-order" Stark effects, respectively, names which might be attached to the power of μE which is involved, or to the order of perturbation approximation required to calculate these effects as will be seen below.

The first-order Stark effect characteristic of symmetric tops is more generally characteristic of a system with degenerate levels. It was shown in discussing nuclear moments (see page 132) that in the absence of an external field no system can have a dipole moment fixed in direction unless it is in a degenerate energy level. Symmetric tops can have a dipole moment of this type because of the degeneracy of the $+K$ and $-K$ levels, and this moment interacts with the electric field. Linear molecules have no degeneracy of this type, and a dipole moment must first be produced by "polarization" of the molecule by a field. It is interesting to note that, in the symmetric-top ammonia, the two levels which are ordinarily degenerate are split by the inversion frequency, and a first-order Stark effect does not occur. Classically the NH_3 dipole moment might be regarded as rapidly reversing in direction because of the inversion so that its average value in any direction is zero.

10-2. Quantum-mechanical Calculation of Stark Energy for Static Fields. The effect of an electric field on the molecular motion can be calculated quantum-mechanically by perturbation theory. The first-order perturbation is simply the average of the interaction energy over the quantum-mechanical state, or

$$\Delta W_1 = -\int \psi^* \mu E \cos \theta \, \psi \, dv \tag{10-4}$$

where θ is the angle between the molecular dipole μ and the field E. The expression (10-4) is just E times the z component of the dipole moment matrix element which is given in Table 4-4. For a symmetric-top wave function with rotational quantum numbers J, K, M this table gives

$$\Delta W_1 = -\mu E \phi_{JJ} \phi_{JKJK} (\phi_z)_{JMJM} = -\mu E \frac{MK}{J(J+1)} \tag{10-5}$$

which is identical with (10-3) and of course vanishes for a linear molecule when $K = 0$.

Transitions may occur with selection rules $\Delta J = \pm 1$, $\Delta K = 0$, and $\Delta M = 0$ or ± 1 (cf. Table 4-4). The observed frequencies are given for a transition $J + 1 \leftarrow J$ by $W_{J+1} - W_J$, so that, when $\Delta M = 0$,

$$\nu = 2B(J+1) + \frac{2MK\mu E}{J(J+1)(J+2)h} \tag{10-5a}$$

and when $\Delta M = \pm 1$,

$$\nu = 2B(J+1) + \frac{(2M \mp J)K\mu E}{J(J+1)(J+2)h} \tag{10-5b}$$

where M represents the "magnetic" quantum number for the initial or J state.

The next approximation, or second-order perturbation theory, takes into account the small changes in the molecular wave function due to the field, and the resulting energy may be written*

$$\Delta W_2 = \sum_{n'} \frac{|\mu_{nn'}|^2 E^2}{W_n - W_{n'}} \tag{10-6}$$

* The Stark effects included in (10-6) are due only to the molecular dipole moments. There are other very much smaller terms due to the polarization of electron wave functions within the molecule. Such effects are usually negligible but will be discussed later in this chapter.

where W_n is the energy of the undisturbed state and $W_{n'}$ is the energy of any other state unperturbed by the electric field. $\mu_{nn'}$ is the z component of the dipole moment matrix element between the two states indicated by quantum numbers n and n'. The two states n and n' are sometimes said to "interact" through the perturbation $\mu E \cos \theta$. It may be noted that two interacting states always repel each other, that is, if W_n is greater than $W_{n'}$, (10-6) shows that the presence of the state n' increases the energy of the state n when there is a field. Similarly it decreases the energy of n' by the same amount so that the levels become further separated. The net change in energy of the state must be obtained by summing all such repulsions as indicated in (10-6). Again the matrix elements can be obtained from Table 4-4. For a symmetric top, the matrix element is zero for all combinations of states except $J = J'$ or $J = J' \pm 1$, when $M = M'$, and $K = K'$ (since μ is always along the symmetry axis for a true symmetric top). The second-order energy ΔW_2 is hence affected only by the two neighboring states $J' = J + 1$ and $J' = J - 1$, and the sum of their effects is

$$\Delta W_2 = \frac{\mu^2 E^2}{2Bh} \left\{ \frac{(J^2 - K^2)(J^2 - M^2)}{J^3(2J - 1)(2J + 1)} - \frac{[(J + 1)^2 - K^2][(J + 1)^2 - M^2]}{(J + 1)^3(2J + 1)(2J + 3)} \right\} \quad (10\text{-}7)$$

This second-order energy is usually so much smaller than the first-order given by (10-5), that it is rather unimportant unless $K = 0$, so that the first-order energy is zero. For a linear molecule, or for any symmetric molecule with $K = 0$, ΔW_2 simplifies to

$$\Delta W_2 = \frac{\mu^2 E^2}{2hBJ(J + 1)} \frac{J(J + 1) - 3M^2}{(2J - 1)(2J + 3)} \quad (10\text{-}8)$$

However, in the special case where $J = 0$, (10-7) becomes

$$(\Delta W_2)_{J=0} = -\frac{\mu^2 E^2}{6hB} \quad (10\text{-}9)$$

The transition frequencies depend, of course, on the difference between Stark effects for the upper and lower levels of a transition. The resulting expression for the absorption of a linear molecule is given below by Eq. (10-25).

Second-order Stark energies are seen from (10-7) and (10-8) to be independent of the sign of M. Before Stark splitting there were $2J + 1$ different degenerate levels for each value of J corresponding to the different values of M. First-order Stark effect, when present, removes this degeneracy completely. Second-order Stark effects depend on M^2, so that the levels are separated into pairs of degenerate levels ($\pm M$) except for $M = 0$, which is nondegenerate.

Expression (10-8) shows that, for large J, for a molecule with angular momentum perpendicular to E ($M = 0$), the change in energy is positive and proportional to $\mu^2 E^2/hBJ^2$ as found above in expression (10-2). Similarly for large J and $M = J$, the energy is negative and proportional to $-\mu^2 E^2/hBJ^2$. For any particular value of J, other than $J = 0$, the average value of ΔW_2 from (10-8) is zero, since

$$3 \sum_{M=-J}^{M=J} M^2 = J(J + 1)(2J + 1) \tag{10-10}$$

so that the average value of $3M^2$ is just $J(J + 1)$.

Still higher-order perturbation terms may be included in the Stark energy. The fourth-order terms for the Stark energy of a linear molecule (third-order terms are zero) have been evaluated [20], [209], [375] but in most cases are very small. They are smaller than the second-order terms by somewhat less than the ratio of the second-order terms to the rotational frequency, or usually considerably less than 1 per cent. An exact expression for the energy W of a linear molecule in a strong electric field has also been written as the following continued fraction (W. E. Lamb, as quoted in [209]; cf. also [433]).

$$\frac{W}{hB} = M(M + 1)$$

$$-\cfrac{(\mu E/hB)^2 A_{MM}^2}{(M + 1)(M + 2) - \cfrac{W}{hB} - \cfrac{(\mu E/hB)^2 A_{M+1,M}^2}{(M + 2)(M + 3) - W/hB - \cdots}}$$

$$\tag{10-11}$$

where B, M, μ, and E have the meanings used above and

$$A_{xy}^2 = \frac{(x + 1)^2 - y^2}{(2x + 1)(2x + 3)}$$

Each of the many solutions to (10-11) for a given M corresponds to a different value of J. This continued fraction is not very convenient to use, but it has been evaluated for a number of conditions with small J [209].

Stark Effect for Two Nearby Levels. An important and interesting special case of Stark effects occurs when two "interacting" levels lie rather close together—considerably closer than the energy separation between either one and any third level. This occurs typically for slightly asymmetric rotors, and also for l-type doubled levels of linear molecules in excited states. (It may be remembered that linear molecules excited to degenerate vibration levels are formally very much like slightly asymmetric rotors.) For two such close levels the energy due to the electric field cannot very well be considered a small perturbation; so an exact solution is necessary. Assume the unperturbed wave functions for the two levels are ψ_1^0 and ψ_2^0. The wave functions after application of the field E may be written

$$\psi_1 = a(E)\psi_1^0 + b(E)\psi_2^0 \qquad \psi_2 = -b(E)\psi_1^0 + a(E)\psi_2^0 \qquad (10\text{-}12)$$

and the matrix element due to the perturbing interaction $-E\mu \cos \theta$ which connects the two states is

$$-E\mu_{12} = -E\mu \int \psi_1^0 \cos \theta \, \psi_2^0 \sin \theta \, d\theta \, d\phi \qquad (10\text{-}13)$$

This quantity is $-E$ times the dipole moment matrix element, which is proportional to the matrix element of the direction cosine.

This case is entirely parallel to the Fermi-resonance type of interaction between two neighboring states discussed in Chap. 2. The matrix element W_{12} of Chap. 2 corresponds to $-E\mu_{12}$, and the quantity δ is the energy difference between the unperturbed states.

$$\delta = W_1^0 - W_2^0 \qquad (10\text{-}14)$$

As in (2-22), the energy when a perturbing field is applied is given by

$$W = \frac{W_1^0 + W_2^0}{2} \pm \left[\left(\frac{W_1^0 - W_2^0}{2} \right)^2 + E^2\mu_{12}^2 \right]^{\frac{1}{2}} \qquad (10\text{-}15)$$

The values of a and b in (10-12) are, as in (2-24)

$$a = \left[\frac{\sqrt{\delta^2 + 4E^2\mu_{12}^2} + \delta}{2\sqrt{\delta^2 + 4E^2\mu_{12}^2}} \right]^{\frac{1}{2}}$$

$$b = \left[\frac{\sqrt{\delta^2 + 4E^2\mu_{12}^2} - \delta}{2\sqrt{\delta^2 + 4E^2\mu_{12}^2}} \right]^{\frac{1}{2}} \qquad (10\text{-}16)$$

Of first interest are the energies as given by (10-15). Let us assume $W_1^0 > W_2^0$. When the energy $E\mu_{12}$ is less than $(W_1^0 - W_2^0)/2$, (10-15) may be expanded

$$W = W_1^0 + \frac{E^2\mu_{12}^2}{W_1^0 - W_2^0} + \cdots \quad \text{or} \quad W_2^0 - \frac{E^2\mu_{12}^2}{W_1^0 - W_2^0} + \cdots \qquad (10\text{-}17)$$

This gives a Stark energy dependent on E^2 as is typical of a second-order perturbation, but perhaps a very large second-order effect because $W_1^0 - W_2^0$ may be rather small. When $|E\mu_{12}|$ becomes larger than $|W_1^0 - W_2^0|/2$, (10-15) may be expanded

$$W = \frac{W_1^0 + W_2^0}{2} \pm E\mu_{12} + \cdots \qquad (10\text{-}18)$$

In this approximation the Stark effect appears linearly dependent on E, as in a first-order perturbation. This is just the approximation which holds for a symmetric top with doubly degenerate levels, since in this case $W_1^0 - W_2^0 = 0$. Thus (10-15) shows a smooth transition from a "second-order" to a "first-order" type of Stark effect.

Usually this intermediate type of Stark effect for a pair of almost degenerate levels occurs for a slightly asymmetric top. In this case pairs of the symmetric-top energy levels (K-degenerate levels) are split by an amount given by (4-9) due to the asymmetry. Even with fairly large asymmetries, certain pairs of levels may be nearly degenerate. For the slightly asymmetric rotor, the dipole matrix elements are fairly accurately given by those for symmetric tops (Table 4-4), so that for these slightly split levels,

$$\mu_{12} = \frac{\mu M K}{J(J+1)} \qquad (10\text{-}19)$$

where K is the usual quantum number for the appropriate limiting symmetric top, and M is the projection of J in the direction of the field E. Linear molecules in excited bending vibrational states are very similar to slightly asymmetric molecules, and Stark effects on l-type doublets for these molecules are also described by (10-15) with matrix elements given by (10-19).

Stark Effects in Asymmetric Rotors. Stark effects in asymmetric rotors are usually "second-order," or proportional to E^2, since the energy levels are not degenerate, but not infrequently a pair of nearby levels give the type of Stark effect described above, and still other more special cases may occur. An extensive treatment has been given by Golden and Wilson [277]. For the usual case where near degeneracies do not occur, the Stark energy has the form (10-6), which involves the sum of a number of terms containing matrix elements between rotational states. The matrix elements take simple forms only when the rotor is approximately symmetric. However, the dipole matrix elements have been computed in connection with an evaluation of transition intensities, and certain sums $^x S_{J\tau,J'\tau'}^{\cdots}$ of these matrix elements have been tabulated [122].

For a dipole moment μ_x lying along a principal axis of inertia of the molecule, it is seen from (4-22) that

$$\sum_M \sum_{M'} |\mu_{J\tau M, J'\tau'M'}|^2 = (2J+1)\frac{|\mu_{ij}|^2}{3} = \frac{\mu_x^2}{3}\,{}^x S_{J\tau,J'\tau'}(\kappa) \qquad (10\text{-}20)$$

where $^x S_{J\tau,J'\tau'}(\kappa)$, or in other notation $^x S_{J_{kl}J'_{mn}}(\kappa)$ is tabulated in Appendix V as a function of the asymmetry parameter κ. The sum over M and M' in (10-20) may be undone by using Table 4-4 to give

$$|\mu_{J,\tau,M;J-1,\tau',M}|^2 = \mu_x^2\,\frac{J^2 - M^2}{J(4J^2-1)}\,{}^x S_{J\tau,J-1\tau'}$$

$$|\mu_{J,\tau,M;J,\tau',M}|^2 = \mu_x^2\,\frac{M^2}{J(J+1)(2J+1)}\,{}^x S_{J\tau,J\tau'} \qquad (10\text{-}21)$$

$$|\mu_{J,\tau,M;J+1,\tau',M}|^2 = \mu_x^2\,\frac{(J+1)^2 - M^2}{(J+1)(2J+1)(2J+3)}\,{}^x S_{J\tau,J+1\tau'}$$

Here M' has been set equal to M, since otherwise the matrix element is zero, or since the angular momentum M in the direction of the field E cannot change.

As in the case of line intensities, each component of the dipole moment may be treated separately, so that the entire Stark energy in the case of no degeneracy may be written:

$$W_{J_\tau M} = \sum_{x=a,b,c} \frac{\mu_x^2 E^2}{2J+1} {\sum_{\tau'}}' \left[\frac{J^2 - M^2}{J(2J-1)} \frac{{}^x S_{J\tau, J-1\tau'}}{W_{J_\tau}^0 - W_{J-1_{\tau'}}^0} \right.$$
$$\left. + \frac{M^2}{J(J+1)} \frac{{}^x S_{J\tau J\tau'}}{W_{J_\tau}^0 - W_{J_{\tau'}}^0} + \frac{(J+1)^2 - M^2}{(J+1)(2J+3)} \frac{{}^x S_{J\tau, J+1\tau'}}{W_{J_\tau}^0 - W_{J+1_{\tau'}}^0} \right] \quad (10\text{-}22)$$

where a, b, and c indicate the directions of the three principal axes of inertia and $W_{J_\tau}^0$ represents the unperturbed energy of the rotational state designated by J_τ. The summation Σ' is over all states except J_τ. This energy has the general form

$$W_{J_\tau M} = (A_{J_\tau} + B_{J_\tau} M^2) E^2 \quad (10\text{-}23)$$

Golden and Wilson [277] have tabulated quantities of the type A_{J_τ} and B_{J_τ} (differing, however, by certain factors) for all ranges of values of the rotational constants and for all levels with $J = 0$, 1, or 2. Unless high accuracy is needed, the Stark effect may be calculated for levels with J as large as 12 by inserting values of S given in Appendix V into (10-22).

Golden and Wilson [277] also consider a number of special cases of degeneracy, where no matrix elements occur between the degenerate levels. These result in Stark effects proportional to E^2 or some higher power of E.

10-3. Relative Intensities of Stark Components and Identification of Transitions from Their Stark Patterns. If M is the projection of J on the z axis or direction of a static electric field, then transitions produced by a microwave electric field in the z direction can occur only when

$$M' - M = \Delta M = 0$$

This may be understood classically from the fact that an electric field in the z direction can exert no torque about the z axis. If the microwave electric field is perpendicular to the static field, then $\Delta M = \pm 1$. From Table 4-4, dipole matrix elements may be obtained for each of these cases for symmetric tops. The dependence of these matrix elements on M is quite independent of the value of K. Since asymmetric-rotor wave functions are a combination of symmetric-rotor functions of the same J and M, but different K, the dependence of the matrix element on M is the same for asymmetric rotors as for symmetric ones. Hence the intensities of the various possible M transitions are proportional to the quantities in Table 10-1. This assumes that the Stark fields are small enough to be treated as small perturbations. In special cases (*cf.* page 269) this assumption is not justified.

The experimental arrangement most often used for applying a "Stark" field will be discussed below. In this arrangement, and in others most likely to be used for microwave spectroscopy, the static and microwave electric fields have essentially the same direction. Hence $\Delta M = 0$ tran-

TABLE 10-1. RELATIVE INTENSITIES OF STARK OR ZEEMAN COMPONENTS

J' in each case is the larger of the two values of J involved in the transition and M' the larger (more positive) M.

	$\Delta J = +1$	$\Delta J = 0$	$\Delta J = -1$
Static and microwave fields parallel: $\Delta M = 0$	$J'^2 - M^2$	M^2	$J'^2 - M^2$
Static and microwave fields perpendicular: $\Delta M = +1$	$(J' + M' - 1)(J' + M')$	$(J + M')(J - M' + 1)$	$(J' - M')(J' - M' + 1)$
$\Delta M = -1$	$(J' - M')(J' - M' + 1)$	$(J + M')(J - M' + 1)$	$(J' + M')(J' + M' - 1)$

sitions are by far the strongest and are ordinarily the only ones observed. In this case, from (10-23) the change in frequencies of the Stark components is given by

$$\Delta\nu = \frac{1}{h}[A_{J'\tau'} - A_{J\tau} + (B_{J'\tau'} - B_{J\tau})M^2]E^2 \quad \text{or}$$

$$\Delta\nu = (A + BM^2)E^2 \quad (10\text{-}24)$$

For the particular case of a linear molecule, the coefficients involved in (10-24) may be evaluated from (10-8) giving

$$\nu = 2B(J + 1) + \frac{\mu^2 E^2}{(J + 1)Bh^2} \frac{3M^2(8J^2 + 16J + 5) - 4J(J + 1)^2(J + 2)}{J(J + 2)(2J - 1)(2J + 1)(2J + 3)(2J + 5)} \quad (10\text{-}25)$$

For the special case $J = 0$,

$$\nu = 2B + \frac{4\mu^2 E^2}{15Bh^2}$$

If $\Delta M = 0$, the value of M must be no larger than the smallest J value involved in the transition. Hence a count of the number of components (different values of M^2) gives quite directly the value of the lowest J involved in the transition. However, in case $\Delta J = 0$, the intensity decreases rapidly with M^2. The component $M = 0$ is missing entirely, and other low values of M may be rather weak.

In case all Stark components are not visible, the relative spacings between components may be used to identify the largest value of M^2 and hence the lowest value of J for the transition. The spacing between successive Stark components increases with M according to (10-24) so that it is usually easy to distinguish the components of higher M from those of lower M. Furthermore, accurate measurements of the relative spacing for a particular value of E will usually allow a definite assignment of M^2 to each component.

Relative intensities of the Stark components usually allow an easy determination of the change in J involved. If $\Delta J = 0$, then from Table

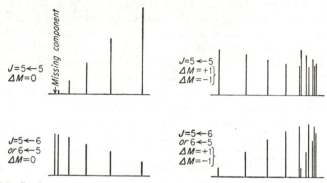

Fig. 10-1. Stark patterns for several types of transitions involving an energy level with $J = 5$. Relative spacings of components for $\Delta M = 0$ depend only on J and M, while those for $\Delta M = \pm 1$ depend to some extent on the particular energy levels involved.

10-1 the largest M values have the greatest intensity. If $\Delta J = \pm 1$, the intensity is proportional to $J'^2 - M^2$ so that the smallest M values have the largest intensity. In this latter case, too, no component has zero intensity, since J' is at least greater by one than M.

Examination of Stark patterns, then, allows a rather direct determination of the lowest J value involved in the transition and of whether or not J changes. Plots of the relative spacings and intensities for several typical cases of Stark splitting are given in Fig. 10-1.

The absolute magnitude of the Stark effect, *i.e.*, the values of the constants A and B in (10-24), may also at times be very helpful in identifying transitions in asymmetric tops. If A or B is large, for example, one of the energy levels involved in the transition may be expected to have a close neighbor which makes the energy denominator $W^0_{J_\tau} - W^0_{J'_{\tau'}}$ in (10-22) small. Often it is possible to evaluate $W^0_{J_\tau} - W^0_{J'_{\tau'}}$ approximately from the size of A or B. In other cases the magnitudes of A and B or their ratio may be used to distinguish between two or more possible identifications of a transition.

10-4. Stark Effect When Hyperfine Structure Is Present. When hyperfine structure is present due to a nucleus of spin I, the total angular momentum of a molecule is given by the quantum number $F = J + I$, $J + I - 1, \ldots, |J - I|$ rather than by J. The projection of the angular momentum on some chosen direction is $M_F = F, F - 1, \ldots,$ $-F$. Thus the number of Stark components depends on F rather than J, and it may be expected that energies of the Stark components will be rather different from the case with no hyperfine structure discussed above. It is convenient to discuss Stark effects under three types of conditions: weak field, strong field, and intermediate field.

In the weak-field case, the electric field is so small that the Stark energy is considerably less than the interaction between the nucleus and the molecule, *i.e.*, the hyperfine energy. In this case the molecular wave functions and the hyperfine structure are only very slightly perturbed by the electric field. Expressed classically, the precession of the molecule due to the Stark field is so slow and gentle that the interaction between the nucleus and molecule is very little disturbed. The molecular state is satisfactorily specified by the quantum numbers I, J, F, and M_F. M_J, the projection of J on the electric field, is not a good quantum number, or not a constant of the molecular motion. Each hyperfine line is then split by the Stark effect into various components according to the values of M_F, and this splitting is small compared with the hyperfine splitting.

In the strong-field case, the Stark energy is much larger than the hyperfine energy. The molecule is precessed so violently by the electric field that the nuclear orientation cannot follow the motion. I and J are said to be decoupled, and the hyperfine structure is radically changed. The quantum number F is no longer good, since the vector sum of I and J is no longer fixed. Appropriate quantum numbers for describing the molecular state are I, J, M_I, and M_J, where M_I is the projection of I on the direction of the field. The Stark energies are the same as when no hyperfine structure is present, and the hyperfine structure gives a splitting of each Stark level which is much smaller than the separation between Stark levels. It is possible to have an electric field so large that the Stark energy is large compared with the separation between rotational levels, so that J is no longer a good quantum number. We shall discuss here only the cases in which the Stark energy is large compared with the hyperfine energy, but small compared with the rotational energy, since these are the common conditions.

Intermediate-field cases, or intermediate-coupling cases, occur when the Stark and hyperfine energies are comparable. In these cases neither M_J nor M_F and F are good quantum numbers. The wave functions are combinations of those appropriate for the weak- or the strong-field cases, and calculation of wave functions and energy levels is generally rather

complex.   The Stark splitting is comparable with the hyperfine splitting, and relative intensities of the various components vary rapidly at
times with the strength of the field.   Wave functions, intensities, and
energies can be evaluated for these cases but they involve solving secular
equations which may be of high order if J and I are large.   An example
(although a rather specialized one) which has been worked out in some
detail is the Stark effect in the ammonia inversion spectrum discussed
by Jauch [212].

Weak Electric Fields.   In the weak-field case the hyperfine structure is
unperturbed, and the Stark splitting can be calculated by essentially the
same methods used for the case of no hyperfine structure.   Both first-
and second-order (linear and quadratic) Stark effects occur.   For a symmetric top the linear Stark effect may be calculated from the vector
model.   The Stark energy ΔW is $-\mu \cos \theta E$, where $\mu \cos \theta$ is the projection of the dipole moment on the field $\mathbf{E}$.   Now μ lies along the axis
of the molecule or the angular momentum K, which precesses around the
total rotational angular momentum $\mathbf{J}$, $\mathbf{J}$ precesses around $\mathbf{F}$, and $\mathbf{F}$ precesses around the direction of the field $\mathbf{E}$.   Averaged over time,

$$\cos \theta = \cos (\mathbf{KJ}) \cos (\mathbf{JF}) \cos (\mathbf{FE}) \qquad (10\text{-}26)$$

where $(\mathbf{KJ})$ represents the angular between the two vectors $\mathbf{K}$ and $\mathbf{J}$.
But $\cos (\mathbf{KJ}) = K/J$; $\cos (\mathbf{JF}) = (J^2 + F^2 - I^2)/2JF$;

$$\cos (\mathbf{FE}) = \frac{M_F}{F}$$

Hence

$$\Delta W = -\frac{\mu K M_F(J^2 + F^2 - I^2)E}{2J^2F^2}$$

Remembering that the square of any vector $\mathbf{J}$ must be replaced by
$J(J + 1)$ when vector model calculations are made,

$$\Delta W = -\frac{\mu K[J(J + 1) + F(F + 1) - I(I + 1)]M_F E}{2J(J + 1)F(F + 1)} \qquad (10\text{-}27)$$

It should be noted that (10-27) is proportional to K, so that no linear
Stark effect occurs when $K = 0$.   In addition, when $F = I + J = M_F$,
(10-27) becomes $\Delta W = -\mu EK/(J + 1)$, which is identical with what
would be obtained if $J = M_J$ and no hyperfine structure were present.
For all other cases, however, the Stark effect is modified by the presence
of hyperfine structure.   A more detailed derivation of (10-27) is given
by Low and Townes [391].

  Second-order Stark effect may be calculated in general by expanding
the wave function with hyperfine structure in terms of the rotational

wave functions which would occur if no hyperfine structure were present.

$$\psi(FMJ\tau I) = \sum_{M_J} C(FMJIM_J)U(J\tau M_J)\phi(IM_I) \qquad (10\text{-}28)$$

where $U(J\tau M_J)$ represents an asymmetric-top wave function with the indicated quantum numbers and $\phi(IM_I)$ is a wave function for the nuclear spin whose exact nature need not concern us. Only those functions occur for which $M_I + M_J = M$. The coefficients $C(FMJIM_J)$ are independent of the quantum number τ, and are evaluated in Condon and Shortley ([56], pp. 76–77). For the general asymmetric top without degeneracy and without hyperfine structure, second-order Stark effects were shown in (10-23) to be of the form

$$\Delta W_{J\tau} = (A_{J\tau} + B_{J\tau}M_J^2)E^2 \qquad (10\text{-}23)$$

Using the expansion (10-28), second-order Stark effects when hyperfine structure is present can be shown to be

$$\Delta W_{FJ\tau} = \sum_{M_J} |C(FMJIM_J)|^2(A_{J\tau} + B_{J\tau}M_J^2)E^2 \qquad (10\text{-}29)$$

which gives the net Stark effect as the sum of Stark effects for each value of M_J multiplied by the probability C^2 of the molecule's being in the state M_J. This expression (10-29) is valid only if the hyperfine energy is small compared with the separation between rotational energy levels, which is the usual case. The sum of all probabilities $\sum_{M_J} C(FMJIM_J)^2$ is unity, and from expressions given by Fano [269] the other sum

$$\sum_{M_J} |C(FMJ_IM_J)|^2M_J^2$$

which is necessary can be obtained. Then (10-29) becomes

$$\Delta W_{F,J\tau} = A_{J\tau}E^2$$
$$+ B_{J\tau}\left\{[3M^2 - F(F+1)]\frac{[3D(D-1) - 4F(F+1)J(J+1)]}{6F(F+1)(2F-1)(2F+3)} \right.$$
$$\left. + \frac{J(J+1)}{3}\right\}E^2 \qquad (10\text{-}30)$$

where $D = F(F+1) + J(J+1) - I(I+1)$. Thus the weak-field Stark effect for an asymmetric rotor with hyperfine structure can be expressed in terms of the Stark-effect coefficients A and B for the case with no hyperfine structure. Expression (10-30) can be applied to a linear molecule (or symmetric top with $K = 0$), for which it reduces to an expression given by Fano [269]

$$\Delta W_J =$$
$$- \frac{\mu^2 E^2}{Bh} \frac{[3M^2 - F(F+1)][3D(D-1) - 4F(F+1)J(J+1)]}{2J(J+1)(2J-1)(2J+3)2F(F+1)(2F-1)(2F+3)} \quad (10\text{-}31)$$

By using Eq. (10-22) to evaluate $A_{J\tau}$ and $B_{J\tau}$ in Eq. (10-30), the weak-field Stark effect when hyperfine structure is present may be expressed in terms of quantum numbers, rotational energies, and the line strengths tabulated in Appendix V (cf. [845]).

Strong Electric Fields. In the strong-field case the Stark energy is much larger than the hyperfine energy. The Stark splitting is identical with that obtained when no hyperfine structure is present as long as $|M_J| \neq 1$, and the hyperfine structure acts as a small perturbation. Hyperfine structure due to the nuclear magnetic dipole moment, for which the energy is of the form $W = a\mathbf{I} \cdot \mathbf{J}$, is given in this case by $W(\mu) = aM_IM_J$, where M_I takes on values $I, I - 1, \ldots, -I$ for each possible value of M_J. Usually this magnetic hyperfine structure is very small, so that the strong-field case almost always applies. Hyperfine energy due to a nuclear quadrupole moment is given in the strong-field case (and where $|M_J| \neq 1$) by

$$W(Q) = \frac{eqQ}{4I(2I-1)(2J-1)(2J+3)} \left[\frac{3K^2}{J(J+1)} - 1 \right]$$
$$[3M_I^2 - I(I+1)][3M_J^2 - J(J+1)] \quad (10\text{-}32)$$

In the strong-field case when $|M_J| = 1$ and $K = 0$, M_J is not necessarily a good quantum number because, regardless of how small the quadrupole coupling constant eqQ is, quadrupole effects may produce transitions between the degenerate levels $M_J = 1$ and $M_J = -1$. Quadrupole effects do not produce transitions between other pairs of degenerate levels, but only those for which $\Delta M_J = \pm 2$. The total angular momentum along the direction of the field must be constant, so that $M = M_J + M_I$ is a good quantum number, although M_J and M_I are not. When $M = I + 1$ or $M = I$, M_J can have values $+1$ only, so that then the state $M_J = -1$ does not occur, and energies are given simply by (10-32). In other cases, a quadratic secular equation must be solved [269].

Intermediate Electric Fields. The "intermediate-field" case occurs when Stark and hyperfine energies are comparable in magnitude. Usually the energy levels must be determined for these cases from the solutions of second- or higher-order equations. If J or I is small, this is not too difficult since the most complex equation which needs solution is of order $2J + 1$ or $2I + 1$, whichever is smaller. These equations are discussed by Fano [269] and Low and Townes [391]. The behavior of

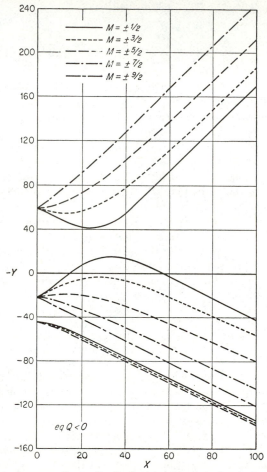

Fig. 10-2. Plot of energy due to Stark effect in a linear molecule for $J = 1$ when a nucleus of spin $I = \frac{7}{2}$ produces quadrupole hyperfine structure.

$$X = -\frac{21h^2}{2BeqQ}\mu^2E^2$$
$$Y = \frac{420W}{eqQ}$$

where W is the sum of Stark and hyperfine energy. (*After Fano* [269].)

hyperfine energy levels with all values of electric-field strength and for the particular case of a symmetric molecule with $K = 0$, $J = 1$, and $I = \frac{7}{2}$ is shown in Fig. 10-2. It may be seen from this figure that a particular component may vary with field in completely different ways for high- and low-field conditions. The component with $M = J + I$

always behaves simply because its secular equation is of first order, and Stark and hyperfine energies simply add without affecting each other. The plot of this level ($M = \frac{9}{2}$) is hence a straight line on Fig. 10-2. Figure 10-3 shows a comparison between actual measured hyperfine structure and theoretically calculated curves for an intermediate-field case.

Stark Effects in the Presence of Hyperfine Structure Due to Two Nuclei. In case two nuclei produce hyperfine structure, the Stark effect shows

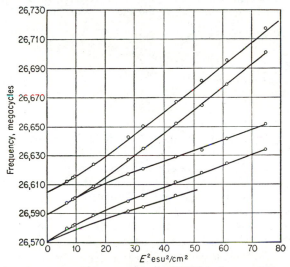

Fig. 10-3. Frequency of all components of CH_3Cl^{35} $J = 1 \leftarrow 0$ transition as a function of electric field strength. Curves are computed using $\mu = 1.869$ debye. (*From Shulman, Dailey, and Townes* [528].)

still more complications. The strong-field case is rather simple, for the hyperfine energy is just the sum of that which would be due to each nucleus individually. For example, if two nuclei have quadrupole moments Q_1 and Q_2, respectively, the hyperfine energy in the strong-field case is just $W(Q_1) + W(Q_2)$, where $W(Q)$ is given by (10-32). In weak or intermediate fields the situation is more complex. The case of two nuclei, one of which has quadrupole and magnetic dipole hyperfine structure, and the other magnetic dipole structure only, has been discussed [307].

Relative Intensities of Stark Components of Hyperfine Lines. Relative intensities of Stark components when hyperfine structure is present is analogous under weak- or strong-field conditions to what is obtained without hyperfine effects. For a very small field, the relative intensities of the Stark components of one hyperfine transition are given by when $\Delta F = 0$ and $\Delta M_F = 0$:

$$I \propto M_F^2 \tag{10-33}$$

when $\Delta F = \pm 1$ and $\Delta M_F = 0$:

$$I \propto F'^2 - M_F^2 \tag{10-34}$$

where F' represents the larger of the two F values involved in the transition. In the strong-field case, the relative intensities of each Stark component are the same as without hyperfine structure, or

when $\Delta J = 0$ and $\Delta M_J = 0$:

$$I \propto M_J^2 \tag{10-35}$$

when $\Delta J = \pm 1$ and $\Delta M_J = 0$:

$$I \propto J'^2 - M_J^2 \tag{10-36}$$

Each of these Stark components is broken up into hyperfine components corresponding to different M_I and which, for a given Stark component, all have the same intensity. For quadrupole hyperfine structure the components for $\pm M_I$ have the same energy, so that the observed hyperfine lines are all of the same intensity except that for $M_I = 0$, which is half the intensity of the other lines. Intensities of components under intermediate-field conditions may be estimated by interpolation between the two extreme cases of weak and strong fields. They may be calculated exactly by solving equations of the type encountered in obtaining exact energies for intermediate-field cases [391].

10-5. Determination of Molecular Dipole Moments. In addition to aiding in identification of transitions, and providing a good means of modulating microwave lines for detection, the Stark effect affords a very accurate and convenient means of measuring molecular dipole moments since its size depends on the product of the dipole moment and the electric field strength. For such a measurement, a fairly uniform known electric field is needed.

By far the most convenient method of obtaining Stark effects is to insert a conducting plate in the waveguide as shown in Fig. 10-4. This conducting plate cuts the microwave electric field perpendicularly and hence does not distort the microwave field appreciably or interfere with its propagation. The plate or septum may be mounted on narrow grooved insulating strips of polystyrene, teflon, or any other dielectric which is not too lossy at microwave frequencies. An electrical lead is usually brought out through a small hole in the center of the narrow face of the waveguide so that d-c or low-frequency voltages may be applied between the septum and the outside shell of the waveguide.

The electric field in a "Stark waveguide" may be seen in Fig. 10-5 to deviate considerably from uniformity near the edges of the septum. In addition, it is parallel to the direction of the microwave electric field only in the center of the waveguide. The presence of dielectric material of finite width modifies this field only slightly. Equipotential surfaces for a particular waveguide with dielectric material have been given by Sharbaugh [519]. Happily, it is the central part of the waveguide

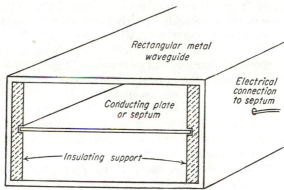

Fig. 10-4. Construction of waveguide for producing Stark effects. Central metal plate is parallel to broad face of waveguide and is insulated from it. Electrical connection to this plate is made by a lead through a small hole in the center of the narrow face of the waveguide.

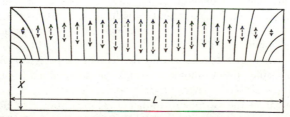

Fig. 10-5. Distribution of electric fields in "Stark waveguide." Dielectric supports of central septum are assumed to be of negligible thickness. D-c or low-frequency field direction is shown with solid lines, the density of lines being proportional to the intensity. The microwave electric field is everywhere in the same direction as indicated by the dotted lines whose length is proportional to the field intensity. Identical fields which are not shown exist in the lower half of the waveguide. (*From Shulman, Dailey, and Townes* [528].)

which is most important. The strength of the microwave electric field E is proportional to $\sin \pi y/L$, where L is the width of the guide and y the distance from one edge. The probability of a transition is proportional to E^2 or $\sin^2 \pi y/L$, which has a maximum in the center of the guide and is zero on the edges. Hence if gas pressure in the waveguide is large enough to prevent molecules from moving around appreciably during a transition (mean free path much less than L), most of the transitions occur near the center of the guide. Since also in the center of the waveguide the Stark and microwave electric fields are closely parallel, transitions of type $\Delta M = 0$ dominate. In fact, for the normal mode of microwave propagation (TE_{01}), transitions of the type $\Delta M = \pm 1$ are too weak to have been observed yet, although they must take place to some extent near the edges of the septum.

If the ratio of the guide width L to the spacing X between the septum and broad side of the waveguide is fairly large, then it can be assumed that only transitions of the type $\Delta M = 0$ occur and that molecular dipole moments can be calculated simply by assuming a uniform field equal to its value in the center of the waveguide. Shulman and Townes [530] have given approximate analytic expressions for the components of the Stark field and have calculated the effects of inhomogeneity in several waveguides. For a guide with the ratio $L/X = 6$, the measured dipole moment needs to be decreased by about 0.2 per cent because of field inhomogeneities, and for $L/X = 5$, the correction is approximately twice as large. Intensities of transitions $\Delta M = \pm 1$ for such waveguides are only a few tenths of 1 per cent as strong as $\Delta M = 0$ transitions. Although only small errors in the positions of the center of each Stark component are produced by the field inhomogeneities, the components are appreciably broadened. For a guide with $L/X = 6$, the breadth of a Stark component due to field inhomogeneities is roughly $\frac{1}{25}$ of the Stark displacement. Hence as the change of frequency due to Stark effect is increased, each component broadens out and becomes weaker. Because of this, individual Stark components cannot usually be followed more than a few hundred megacycles.

If the electric field E is known and the transition positively identified, then the molecular dipole moment μ may be determined by measuring the magnitude of the changes in frequency due to Stark effect. An appropriate one of the various formulas developed above for Stark energies is of course needed to connect the measured magnitude with the quantity μE. For symmetric molecules without hyperfine structure the case is particularly simple. When $K \neq 0$, (10-5a) may be used and the Stark effect is linearly proportional to E as long as the Stark displacement is not very large, so that second-order terms in E^2 are unimportant. It is usually advantageous to choose the most widely split Stark component, since it is more isolated and easily measured. However, other components may often be used. The Stark displacement may be measured for several different field strengths. If this is plotted against E, a straight line should be obtained whose slope gives the coefficient of E in (10-5a), or $2\mu KM/J(J + 1)(J + 2)h$. E is measured of course in electrostatic units, or volts per centimeter divided by 299.8. If $K = 0$, then the Stark displacement is proportional to E^2. It is again usually convenient to measure the component with largest Stark displacement and to plot this displacement against E^2. A straight line as shown in Fig. 10-6 should be obtained, whose slope is the coefficient of E^2 in (10-25), and hence gives the value of μ. In case hyperfine structure is present, similar measurements can be made, but interpretation of the Stark displacements from expressions (10-30), (10-32), or by calculation of intermediate-field energies is usually more complex. In addition, there are

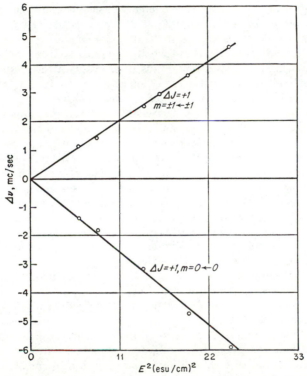

FIG. 10-6. Comparison between theory and experimental measurement of Stark effects in the $J = 2 \leftarrow 1$ transition of the linear molecule OCS. (*From Strandberg, Wentink, and Kyhl* [419].)

often many Stark components whose relative intensities and spacings vary with field strength in the intermediate-field region, so that care must be taken to select and identify one particular component.

The dipole moment of the ground vibrational state of OCS has been measured rather accurately in a number of places and can conveniently be used to calibrate any Stark apparatus for measuring dipole moments. Thus if the electrode spacing X is unknown so that field strength E cannot be accurately determined from the potential applied to the Stark waveguide, a measurement of the Stark effect in OCS and use of its known dipole moment will determine the proportionality constant between voltage or potential and the field strength. A weighted average of available values for the $O^{16}C^{12}S^{32}$ dipole moment in the ground state is $0.709 \pm 0.003 \times 10^{-18}$ esu, or 0.709 debye units.

Stark effects in asymmetric rotors may also be used to measure their dipole moments. In case the molecule has sufficient symmetry to establish the direction of the dipole moment as along one of the principal axes, calculation of the coefficient of μ^2E^2 may be made from (10-22) and μ fairly readily measured. If the direction of the dipole moment is not known, then Stark effects of several lines must be measured, and the several measured coefficients of E^2 used to evaluate the various components of μ along principal axes. This case is not common, but when it occurs, the direction of the dipole moment in the molecule may be established from the Stark measurements. Other techniques usually measure only the magnitude of the molecular dipole moment.

In addition to giving the direction of the dipole moment, determination of dipole moments by Stark effects has several other advantages. It can measure dipole moments as small as 0.1 debye with essentially the same accuracy as larger ones, or to about 0.2 per cent in waveguides of the type shown by Fig. 10-4. If a microwave spectrograph were designed with a very homogeneous field, dipole moments could undoubtedly be measured to an accuracy of 0.01 per cent or better. Another advantage of this type of dipole-moment determination over the classical technique of measuring dielectric constants is that dipole moments may be determined in rather impure gases, since a line of the particular molecule of interest may be singled out for measurement.

The dipole moment of molecules in a particular vibrational state may also be measured, rather than an average moment for all states as is done by other techniques. It has been found that the dipole moment of OCS in the first excited bending mode of vibration is 0.700 debye, or about 1.2 per cent less than the value 0.709 obtained for the ground state. This change in dipole moment depends not only on the bending or change in relative direction of the O—C and C—S bonds, but also on changes in the electronic wave functions involved in each bond due to the vibration. Since each different isotopic species of a molecule has a slightly different vibrational energy, it might be expected that the different isotopic molecules would have different dipole moments. However, a change in isotopic mass usually represents a smaller change in vibrational energy than a vibrational excitation. Hence this type of variation in dipole moment may be expected to be quite small, and not larger than a few tenths of 1 per cent except perhaps in the case of a hydrogen-deuterium substitution. In the case of OCS^{32} and OCS^{34}, for example, the difference in dipole moments seems to be less than 0.2 per cent.

10-6. Forbidden Lines and Change of Intensity Due to Stark Effect.
An electric field not only changes the rotational energies of a molecule
but also modifies the molecular wave functions and thereby affects the
dipole matrix elements and intensities of transitions. Let ψ_1 be the
wave function for one molecular energy level and ψ_2 that for another when
the electric field is zero. If $\psi_1(E)$ and $\psi_2(E)$ are the modified wave
functions after an electric field is applied and if E is not too large they
may be expanded in the form

$$\psi_1(E) = \psi_1 + \sum_n{}' C_{1n}\psi_n \qquad (10\text{-}37)$$

where

$$C_{1n} = \frac{\mu_{1n}E}{W_1 - W_n}$$

μ_{1n} is the dipole matrix element between the unperturbed state designated by
1 and that designated by n. W_1 and W_n are the energies of the two states.
The summation $\sum_n{}'$ is to be taken over all levels except level 1. If levels
are degenerate so that $W_1 = W_n$, as in the case of the K degeneracy of a
symmetric-top molecule, wave functions may be chosen to make $\mu_{1n} = 0$
so that the expression (10-37) has no divergent terms. $\psi_2(E)$ is similarly

$$\psi_2(E) = \psi_2 + \sum_n{}' C_{2n}\psi_n \qquad (10\text{-}38)$$

The transition probability between states 1 and 2 depends on the dipole
matrix element between them,

$$\mu_{12}(E) = \int \psi_1^*(E)\mu_z\psi_2(E)\, d\tau \qquad (10\text{-}39)$$

as in (1-59). Using (10-37) and (10-38),

$$\mu_{12}(E) = \mu_{12} + \sum_n{}' E\left(\frac{\mu_{1n}\mu_{n2}}{W_1 - W_n} + \frac{\mu_{2n}\mu_{n1}}{W_2 - W_n}\right) \qquad (10\text{-}40)$$

where μ_{12} is the matrix element for zero field. Expression (10-40) is
accurate only when the wave functions are not greatly changed by the
electric field, *i.e.*, when $\mu E/(W_1 - W_n) \ll 1$. E must be taken as positive, and the relative phases of the matrix elements are such that

$$|\mu_{12}(E)| = |\mu_{12}| + \sum_n{}' |\mu_{1n}||\mu_{2n}|E\left(\frac{1}{W_1 - W_n} + \frac{1}{W_2 - W_n}\right) \qquad (10\text{-}41)$$

Expression (10-41) allows an evaluation of the change in intensity of a
transition, since from (1-49) intensity is proportional to $|\mu|^2$. It should
be noticed that the fractional changes in intensity for the different Stark
components of a transition are not usually the same, since the matrix
elements μ_{1n}, etc., depend on the magnetic quantum number M.

If $|\mu_{12}|$ is not very small, (10-41) gives only a small fractional change in the dipole matrix element or in the intensity. Since intensities are generally not measured with an accuracy better than 5 per cent, a change in intensity of this type would usually be insignificant. Of course, Stark effects on energy levels are usually more obvious, since the transition frequencies are observed and measured to very high accuracy. When $|\mu_{12}|$ happens to be zero, a transition is said to be "forbidden" (at least for zero field). In this case the electric field may perturb the wave function enough that (10-41) gives an appreciable value to $|\mu_{12}(E)|$, and the transition can be observed when the electric field is present. Since the transition probability is zero when no field is present, such a change in intensity and the appearance of new lines, even though they are relatively weak, can be easily noticed. The intensities of these forbidden lines are, from (10-41), proportional to E^2.

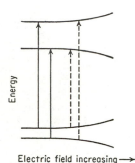

FIG. 10-7. Forbidden transitions between pairs of slightly split levels of a slightly asymmetric rotor. Solid lines represent permitted transitions; dotted lines forbidden transitions which occur when an electric field is applied.

When two energy levels lie rather close together, large perturbations may occur so that the expansion (10-37) and consequently (10-41) are no longer valid. This is the type of case discussed above (page 252). A particularly common example is the slightly split levels of a slightly asymmetric rotor, or the l-doubled levels of a linear molecule in an excited degenerate vibrational mode. This case is illustrated in Fig. 10-7. The perturbed wave functions are a mixture of the two unperturbed wave functions as in (10-12)

$$\begin{aligned} \psi &= a(E)\psi_1^0 + b(E)\psi_2^0 \\ \psi' &= -b'(E)\psi_1'^0 + a'(E)\psi_2'^0 \end{aligned} \tag{10-42}$$

where the unprimed quantities refer to the lowest level of Fig. 10-7, and the primed quantities to the uppermost. The coefficients a, b, or a' and b' are given by (10-16). The dipole matrix element $\int \psi \mu_z \psi' \, d\tau$ is proportional to $a'b - ab'$, which is zero for zero electric field, has a maximum near fields for which $\mu E\phi \approx W_{10} - W_{20}$, and then approaches zero again for large electric fields. The intensity of these forbidden transitions depends markedly on the magnetic quantum number M. When $M = 0$, for example, the matrix element ϕ between states 1 and 2 in slightly asymmetric rotors is zero, and forbidden transitions of the type $\Delta M = 0$ do not occur.

10-7. Polarization of Molecules by Electric Fields. So far interactions between an electric field and a molecular dipole moment of fixed value have been considered. In addition, the electric field can distort the distribution of electrons in the molecule or the relative positions of atoms, thus polarizing the molecule and changing its energy. These Stark effects are very much smaller than the type treated above but may not always be negligible. Stark energy due to these types of molecular polarization may be treated by second-order perturbation theory in much the same way as the larger effects discussed above. The result is [277]

$$\Delta W_p \equiv \frac{E^2}{2(2J+1)} \sum_{x=a,b,c} P_{xx} \sum_{\tau'} \left[\frac{J^2 - M^2}{J(2J-1)} {}^x S_{J\tau, J-1\tau'} + \frac{M^2}{J(J+1)} {}^x S_{J\tau J\tau'} \right.$$
$$\left. + \frac{(J+1)^2 - M^2}{(J+1)(2J+3)} {}^x S_{J\tau, J+1\tau'} \right] \quad (10\text{-}43)$$

This equation is analogous to (10-22), and the symbols have the same meaning. P_{xx} is the component of the polarizability tensor along one of the principal axes of inertia $x = a$, b, or c of the molecule, or simply the polarizability along one of these directions. Again, analogously to (10-22) the polarizability P_{xx} is the sum

$$P_{xx} = 2 \sum_n \frac{\mu_{0n}^2}{W_0 - W_n} \quad (10\text{-}44)$$

where μ_{0n}^2 is the dipole matrix element between the ground state of the molecule and any excited electronic or vibrational state, and $W_0 - W_n$ is their energy difference. The matrix-elements μ_{0n} are not much larger than 1 debye or 10^{-18} esu, and usually $W_0 - W_n$ is a few hundred wave numbers for vibrational states, or a few thousand wave numbers for excited electronic states. Hence the energy ΔW_p is in most cases less than $1/10,000$ as large as second-order Stark effects of the type given by (10-22) because rotational energy levels tend to be about 1000 times closer than vibrational energy levels and thus give energy denominators which are smaller by this factor. In addition, the vibrational matrix elements μ_{0n} are usually smaller than those for rotational transitions.

Rotational Transitions in Nonpolar Molecules. It may be possible, although very difficult, to induce rotational transitions in a nonpolar molecule by microwaves. If the molecule is nonpolar, it has zero electric dipole moment when unperturbed; but because of its polarization, a dipole moment may be induced by a large electric field and this induced dipole moment then used to produce rotational transitions [32].

As a simple typical case, consider a linear molecule such as CO_2 which has zero dipole moment as a result of its symmetry. Its polarizability may be assumed to be P along the axis and zero perpendicular to the axis. The change in energy due to an electric field E is then

$$\Delta W = -\tfrac{1}{2}E^2 P \cos^2 \theta \quad (10\text{-}45)$$

where θ is the angle between the molecular axis and the direction of E. Expression (10-45) may be taken as a perturbation in the Hamiltonian for the molecule which can cause transitions between two states having wave functions $\psi_1 e^{-iW_1 t/\hbar}$ and $\psi_2 e^{-iW_2 t/\hbar}$ in accordance with the size of the integral

$$\frac{P}{2} \int E^2 \cos^2 \theta \psi_1^* e^{iW_1 t/\hbar} \psi_2 e^{-iW_2 t/\hbar} \, d\tau \, dt \qquad (10\text{-}46)$$

Assume now that the field E is made of two parts, a static field E_s and a microwave field $E_m e^{i\omega t}$ which will be assumed for simplicity to be parallel. Then

$$E^2 = E_s^2 + 2E_s E_m e^{i\omega t} + E_m^2 e^{2i\omega t} \qquad (10\text{-}47)$$

If W_1 and W_2 are different, then the integration over time in (10-46) eliminates the constant term E_s^2 in (10-47). The second term of (10-47) gives a nonzero contribution to (10-46) if $\omega = (W_1 - W_2)/\hbar$ and if the matrix element

$$\mu_{12} = PE_s \int \psi_1^* \psi_2 \cos^2 \theta \, d\tau \qquad (10\text{-}48)$$

is not zero. Now ψ_1 and ψ_2 are known from Chap. 1 to involve Legendre polynomials of the form $P_J^M(\cos \theta)$. It can be shown from them that μ_{12} is zero unless the two states have values of J differing by 2, in which case

$$\mu_{J,J+2} = \frac{1}{2J+3} \left\{ \frac{[(J+2)^2 - M^2][(J+1)^2 - M^2]}{(2J+5)(2J+1)} \right\}^{\frac{1}{2}} PE_s \qquad (10\text{-}49)$$

From the above, it can be seen that an absorption occurs at a frequency

$$\nu = \frac{W_{J+2} - W_J}{h} = 2B(2J+3) \qquad (10\text{-}50)$$

with an intensity that can be obtained by substituting $\mu_{J,J+2}$ from (10-49) for the more usual dipole matrix element. The matrix element (10-49) is, however, very much smaller than a normal dipole matrix element of 10^{-18} esu. The polarizability P of a linear molecule could be calculated from an expression of the type (10-44) if the electronic wave functions were sufficiently known. More simply, P is usually given to a rough approximation by the cube of the molecular length. If this length is taken as 3 A, then E_s would have to be 300 esu, or about 10^5 volts/cm, in order for PE_s to be as large as 10^{-2} debye or 10^{-20} esu. Although it would be very useful to obtain rotational transitions of nonpolar molecules, these high field strengths which are necessary are not easily maintained in gases at low or moderate pressures.

The third term in (10-47) also produces transitions at frequencies given by

$$\nu = \frac{W_{J+2} - W_J}{h} = 2B(2J+3) \qquad (10\text{-}51)$$

The matrix element has the same form as (10-48), but with E_s replaced by E_m. Thus transitions in a nonpolar molecule can be produced simply by a microwave field of sufficiently high intensity. The required field strengths are again too high, however, for normal use.

10-8. Stark Effects in Rapidly Varying Fields—Nonresonant Case.
Stark effects are usually considered for static or essentially static fields, as has been done above. However, with the techniques of microwave spectroscopy, a number of interesting effects may be observed in rapidly varying fields. The normal or static type of Stark effect occurs in a slowly varying electric field, but when the frequency of variation becomes comparable with or rapid compared with the width of an absorption line, new effects occur. Still other new effects appear when the frequency of variation becomes comparable with or greater than the frequency difference between two energy levels between which electric-dipole transitions can occur.

We shall consider first the behavior of a molecule in an electric field varying sinusoidally at a frequency ν_0 which is considerably less than any transition frequency of the molecule, but which may be greater than the half width $\Delta\nu$ of an absorption line. The line width $\Delta\nu$ is $1/(2\pi\tau)$, where τ is the time between collisions. Hence $1/\Delta\nu$ is a measure of the "relaxation time" of the molecules in a gas, or the time required for any transient phenomenon to disappear. Therefore, if the frequency ν_0 of the varying electric field is considerably less than the half width $\Delta\nu$, the field may be considered essentially static at any time and the Stark effect calculated accordingly. If the field varies at a frequency $\nu_0 \ll \Delta\nu$ the frequency of an absorption line simply moves in synchronism with it. However, for $\nu_0 > \Delta\nu$ the molecular state cannot adjust rapidly enough to follow the varying field, and the Stark effect has a different character. This general type of effect was examined theoretically by Blockinzew [39] who remarked that it was too small to be observed. After the advent of microwave spectroscopy, however, Townes and Merritt [243] were able to demonstrate a variety of such effects.

Figures 10-8 and 10-9 illustrate the above behavior. They represent absorption by the $J = 2 \leftarrow 1$ transition of OCS. The gas was contained in an absorption cell made up of a short length in which there was zero electric field, and a larger length in which a field could be applied. Figure 10-8a shows a central absorption peak due to the portion of the gas which is not subjected to an electric field, a left-hand peak corresponding to absorption by molecules in a state $M = 0$, and a right-hand peak due to molecules in states $M = \pm 1$, both of which are displaced by a small amount due to an applied static field of approximately 650 volts/cm. In OCS, the change of frequency due to Stark effect is proportional to E^2, the square of the electric field, so that if the electric field varies slowly as $E = E_0 \cos 2\pi\nu_0 t$ the absorption frequency should vary at a

frequency of $2\nu_0$ between that of the undisplaced line and that corresponding to a static field E_0.   Such behavior is indicated by Fig. 10-8b, which
is the absorption by OCS in the same absorption cell as before, but with
a field varying at 1 kc, with a peak value E_0 essentially the same as the

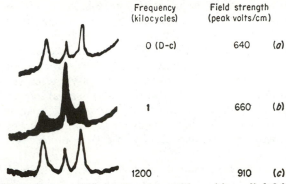

Frequency (kilocycles)	Field strength (peak volts/cm)	
0 (D-c)	640	(a)
1	660	(b)
1200	910	(c)

Fig. 10-8. Stark effect on OCS $J = 2 \leftarrow 1$ transition with applied fields of various frequencies.   (*From Townes and Merritt* [243].)

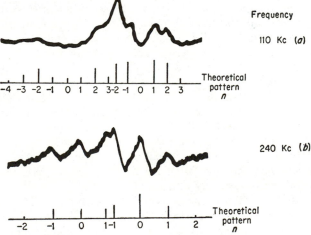

Fig. 10-9. Stark effect on OCS $J = 2 \leftarrow 1$ transition showing additional lines produced by fields of intermediate frequencies.   Peak field strengths of both fields are 640 volts/cm.   (*From Townes and Merritt* [243].)

previous static field.   The dark or fuzzy region in Fig. 10-8b corresponds
to the various possible values of absorption which may occur as the field
varies.   In Fig. 10-8c, the field is varied in the same way at 1200 kc, a
frequency much larger than the line half width of 100 kc, and also larger
than the frequency shift due to Stark effect of a static field of the same
magnitude.   The Stark pattern for this high-frequency field is seen to

have the same appearance as for the static field of Fig. 10-8a. Figure 10-9 shows other more complex Stark patterns associated with modulation frequencies ν_0 of intermediate values.

In order to understand the behavior of a molecular system in a varying field, we start with the wave equation including the time.

$$[H_0 + H'(t)]\psi = -\frac{\hbar}{i}\frac{\partial\psi}{\partial t} \tag{10-52}$$

where H_0 is the Hamiltonian operator without the varying electric field, and $H'(t)$ represents the small time-dependent perturbation produced by the field. Its precise form for a field $\mathbf{E}_0$ varying at a frequency ν_0 is

$$H'(t) = \mathbf{\mu} \cdot \mathbf{E}_0 \cos 2\pi\nu_0 t \tag{10-53}$$

To simplify the mathematics, let us assume that there are two nearby states which "interact" as the result of a dipole moment matrix element between them, and that no other states are sufficiently near in energy to be of importance in the Stark effect on these two. The unperturbed wave function for the lower state will be designated ψ_1, and that for the upper state ψ_2. The perturbed wave function for the lower state can then be written in the form

$$\psi'_1 = (a\psi_1 + b\psi_2) \exp\left[\frac{i}{\hbar}\int_0^t f(t)\,dt\right] \tag{10-54}$$

In this expression a is approximately unity, $-f(t)$ approximately the energy W_1 of the lower state, and b is small as long as the perturbation is small. Substituting (10-54) into the wave equation (10-52), letting $H_0\psi_1 = W_1\psi_1$, where W_1 is the energy for the unperturbed state, and $H_0\psi_2 = W_2\psi_2$,

$$a[W_1 + f(t) + H'(t)]\psi_1 + \frac{\hbar}{i}\dot{a}\psi_1 + b[W_2 + f(t) + H'(t)]\psi_2$$
$$+ \frac{\hbar}{i}\dot{b}\psi_2 = 0 \tag{10-55}$$

where $\dot{a}$ and $\dot{b}$ represent the time derivatives of a and b, respectively. If expression (10-55) is multiplied by ψ_1^* and integrated, then since ψ_1 and ψ_2 are orthogonal, one obtains

$$a[W_1 + H'_{11} + f(t)] + bH'_{12} + \frac{\hbar}{i}\dot{a} = 0 \tag{10-56}$$

where

$$H'_{11} = \int\psi_1^* H'(t)\psi_1\,d\tau \qquad \text{and} \qquad H'_{12} = \int\psi_1^* H'(t)\psi_2\,d\tau$$

similarly

$$b[W_2 + H'_{22} + f(t)] + aH'_{21} + \frac{\hbar}{i}\dot{b} = 0 \tag{10-57}$$

The solution of (10-56) and (10-57) may be obtained by successive approximations, assuming first that $(\hbar/i)\dot{a}$ and $(\hbar/i)\dot{b}$ are negligible. From (10-56), if $b = 0$,

$$f^{(1)}(t) = -W_1 - H'_{11} \qquad (10\text{-}58)$$

where $f^{(1)}(t)$ is the first-order approximation as in the case of a static field. By putting $f^{(1)}(t)$ from (10-58) into (10-57), one obtains

$$b = \frac{-H'_{21}}{W_2 + H'_{22} - W_1 - H'_{11}} \qquad (10\text{-}59)$$

And, combining (10-59) with (10-56), the second-order approximation to $f(t)$ may be obtained as

$$f^{(2)}(t) = -W_1 - H'_{11} + \frac{|H'_{12}|^2}{W_2 + H'_{22} - W_1 - H'_{11}} \qquad (10\text{-}60)$$

In order to examine the significance of the above solution (10-60) to the observed spectrum, we shall assume that $H'_{11} = H'_{22} = 0$, so that only a second-order Stark effect is present. Expression (10-60) may then be written

$$f^{(2)}(t) = -W_1 - \Delta W_1 \cos^2 2\pi\nu_0 t \qquad (10\text{-}61)$$

since $H'_{12} = \mu_{12} E_0 \cos 2\pi\nu_0 t$, where μ_{12} is the dipole moment matrix element. ΔW_1 is the change in energy which would be produced by the Stark effect of a static field E_0. Substituting (10-61) into (10-54) the wave function is then

$$\psi'_1 = (a\psi_1 + b\psi_2) \exp\left[-\frac{i}{\hbar}\left(W_1 t + \frac{\Delta W_1 t}{2} + \frac{\Delta W_1}{8\pi\nu_0}\sin 4\pi\nu_0 t \right)\right] \qquad (10\text{-}62)$$

Consider now a transition induced by a microwave field of frequency ν between ψ'_1 and a different energy level designated by a wave function ψ_3. The intensity of the transition will depend on the absolute magnitude of the integral

$$\iint \psi_3^* \mu \cos\theta \, \psi'_1 \, d\tau \, e^{-2\pi i\nu t} \, dt \approx$$

$$(10\text{-}63)$$

$$\mu_{31} \int \exp\left(i\int_0^t \left\{ \frac{1}{\hbar}[W_1 - W_3 + (\Delta W_1 - \Delta W_3)\cos^2 2\pi\nu_0 x] - 2\pi\nu \right\} dx \right) dt$$

where μ_{13} is the ordinary dipole moment matrix element between the two states. If the microwave frequency ν is constant, then the intensity of absorption depends on the square of the amplitude of the various Fourier components of

$$\int \psi_3^* \mu \cos\theta \, \psi'_1 \, d\tau = \mu_{31} \exp\left\{ \frac{i}{\hbar}\left[W_3 - W_1 + \frac{\Delta W_3 - \Delta W_1}{2} \right] t \right.$$

$$\left. + \frac{i}{\hbar}(\Delta W_3 - \Delta W_1)\frac{\sin 4\pi\nu_0 t}{8\pi\nu_0} \right\} \qquad (10\text{-}64)$$

$(W_1 - W_3)/h$ may be recognized as the frequency ν_{13} of the transition before application of a perturbing field and $(\Delta W_3 - \Delta W_1)/h$ as the change in frequency $\Delta\nu_{13}$ due to the Stark effect of a static field E_0. Hence the right-hand side of (10-64) can be written

$$\mu_{13} \exp\left[2\pi i\left(\nu_{13} + \frac{\Delta\nu_{13}}{2}\right)t + i\frac{\Delta\nu_{13}}{4\nu_0}\sin 4\pi\nu_0 t\right] \quad (10\text{-}65a)$$

or

$$\mu_{13} \exp\left[2\pi i \int_0^t (\nu_{13} + \Delta\nu_{13}\cos^2 2\pi\nu_0 t)\, dt\right] \quad (10\text{-}65b)$$

The part of (10-65a) involving $\sin 4\pi\nu_0 t$ can be expanded as a series of Bessel functions, giving

$$\int \psi_3'^* \mu \cos\theta\, \psi_1'\, d\tau = \mu_{13} e^{2\pi i\left(\nu_{13} + \frac{\Delta\nu_{13}}{2}\right)t} \sum_{n=-\infty}^{n=\infty} J_n\left(\frac{\Delta\nu_{13}}{4\nu_0}\right) e^{4\pi i n \nu_0 t} \quad (10\text{-}66)$$

The intensity of absorption of a particular frequency component of (10-65) is given by the square of its amplitude. Hence a transition of frequency $\nu_{13} + \dfrac{\Delta\nu_{13}}{2} + 2n\nu_0$ will appear with intensity

$$\text{Int}\left(\nu_{13} + \frac{\Delta\nu_{13}}{2} + 2n\nu_0\right) = IJ_n^2\left(\frac{\Delta\nu_{13}}{4\nu_0}\right) \quad (10\text{-}67)$$

where I is the intensity of the Stark component in a static field.

We can now discuss more completely the Stark effect with varying fields and compare expectations with observations shown in Figs. 10-8c and 10-9. The observed spectrum should be, according to (10-67), a series of lines differing in frequency by $2n\nu_0$ and centered at $\nu_{13} + \Delta\nu_{13}/2$. These equally spaced lines might be called "sidebands" corresponding to modulation of the molecular wave function, and are demonstrated in Figs. 10-9a and 10-9b where relative intensities are seen to agree well with the values predicted from (10-67). If the modulation frequency ν_0 is much larger than the static Stark effect ν_{13}, then all Bessel functions of (10-66) or (10-67) are quite small except $J_0(\nu_{13}/4\nu_0)$, which is approximately equal to unity. For such a case, all the intensity of the Stark component is concentrated in a single frequency $\nu_{13} + \Delta\nu_{13}/2$, displaced by an amount $\Delta\nu_{13}/2$ which is just the average of the Stark displacement to be expected from a slowly varying field of the same magnitude E_0. The molecule may be said in this case simply to average the Stark effect, since it cannot respond to the rapid variation. Figure 10-8c shows an observed spectrum under this condition. It should be noted that the Stark displacement appears to be identical with that obtained with a static field but that the peak field required is 910 volts/cm, which is larger than the static field of 640 volts/cm by $\sqrt{2}$.

In case first-order Stark effects occur, the phenomena observed with high-frequency modulation are very similar. However, in this case the average position of the Stark component is ν_{13}. The molecular frequency is modulated with a frequency ν_0 instead of the $2\nu_0$ which occurs with second-order Stark effect when the frequency depends on the square of the electric field. The observed frequencies are then $\nu_{13} + n\nu_0$ and the intensities are

$$\text{Int}(\nu_{13} + n\nu_0) = IJ_n^2 \left(\frac{\nu_{13}}{\nu_0}\right) \tag{10-68}$$

It is useful to observe from (10-63) that the absorption intensity depends only on the difference between the frequency of the absorbed microwaves and $\nu_{13} + \Delta\nu_{13} \cos^2 2\pi\nu_0 t$ which would be the frequency of the absorption line calculated on the basis of an essentially static field. Hence if the electric field is in fact constant and the microwave frequency ν is modulated, a similar breakup of the spectrum into "sidebands" can be expected.

In order for the multiple components to appear as indicated by (10-67) or (10-68), the line widths must of course be smaller than the separation between components. If the line width is larger than the separation [$2\nu_0$ or ν_0], then the intensity appears to be modulated, or the absorption line to move in frequency. When the line width is smaller than this modulation frequency, the line no longer appears to move, but to split up into its separate components. It is this effect which requires, as is discussed in Chap. 15, that the line width in a Stark modulation spectrometer be somewhat greater than the modulation frequency. A more detailed discussion of modulation effects of the above type, including modulation by a square wave instead of a sine wave, is given by Karplus [293].

It is appropriate now to examine the conditions under which $\dot{a}$ and $\dot{b}$ in Eqs. (10-56) and (10-57) can be properly omitted, as was done for the above solution, and the consequences of their inclusion. Using the value of b given by (10-59),

$$\frac{\hbar}{i}\dot{b} = \frac{h\nu_0 H'_{21} \sin 2\pi\nu_0 t}{i(W_2 + H'_{22} - W_1 - H'_{11})} \tag{10-69}$$

Hence $(\hbar/i)\dot{b}$ is comparable with the term aH'_{21} in (10-57) if

$$\frac{h\nu_0}{W_2 + H'_{22} - W_1 - H'_{11}} \approx 1$$

or if the frequency of modulation ν_0 becomes comparable with the resonance frequency between the two levels 1 and 2. The quantity $\dot{a}$ may be assumed to be zero, since any time variation of a may be taken as the part of the factor $\exp[-i/\hbar \int f(t)\, dt]$ in (10-54). In order to solve (10-57) to terms of order H' without neglecting $(\hbar/i)\dot{b}$, we may neglect H'_{22}, use

$f_1(t) = -W_1$ and assume b has the general form

$$b = A \cos 2\pi\nu_0 t + B \sin 2\pi\nu_0 t \qquad (10\text{-}70)$$

By substituting (10-70) into (10-57) and remembering that

$$H'_{12} = \mu_{12}E_0 \cos 2\pi\nu_0 t$$

the constants A and B can be determined. It will be seen below that the value of B is not important, so that after inserting the value of A into (10-70)

$$b = \frac{\mu_{12}E_0(W_1 - W_2)}{(W_1 - W_2)^2 - h^2\nu_0^2} \cos 2\pi\nu_0 t + B \sin 2\pi\nu_0 t \qquad (10\text{-}71)$$

From (10-71) and (10-56) an approximate value of $f(t)$ may be obtained which is good to terms in H'^2.

$$f_2(t) = -W_1 - \mu_{11}E_0 \cos 2\pi\nu_0 t - \frac{|\mu_{12}|^2 E_0^2 (W_1 - W_2) \cos^2 2\pi\nu_0 t}{(W_1 - W_2)^2 - h^2\nu_0^2}$$
$$- \mu_{12}E_0 B \sin 2\pi\nu_0 t \cos 2\pi\nu_0 t \qquad (10\text{-}72)$$

It has already been shown above that for small fields such that the Stark effect produces a frequency change much less than ν_0, the observed frequencies depend only on the average value of $f(t)$. Hence the first and third terms of (10-72) are the important ones, and the varying field produces an effective change in the energy level of

$$\Delta W = \frac{|\mu_{12}|^2 E_0^2 (W_1 - W_2)}{2[(W_1 - W_2)^2 - h^2\nu_0^2]} \qquad (10\text{-}73)$$

10-9. Stark Effects in Rapidly Varying Fields—Resonant Modulation. When $h\nu_0 \ll |W_1 - W_2|$, expression (10-73) reduces to the form

$$\frac{|\mu_{12}|^2 E_0^2}{2(W_1 - W_2)}$$

as was obtained previously for a rapidly oscillating field. As $h\nu_0$ approaches $|W_1 - W_2|$, however, the Stark effect from (10-73) increases in size, becomes infinite at the resonant frequency $\nu_0 = (1/h)|W_1 - W_2|$, and then reverses in sign for $h\nu_0 > |W_1 - W_2|$. This general type of behavior, and the reversal of sign of the Stark effect, has been observed by Autler and Townes [892c]. However, the Stark effect never becomes infinite at the resonance frequency, since then the approximations used above which treated the field as a small perturbation are no longer good, and (10-73) is incorrect. Near the resonance frequency

$$\nu_0 \approx \frac{1}{h}|W_1 - W_2|,$$

the wave equation must be solved by a still different mathematical technique described below.

The change in apparent energy of a system due to an oscillating electric field $E_0 \cos 2\pi\nu_0 t$ may be briefly summarized as follows:

Case 1. $\nu_0 \ll \Delta\nu$ (half width of energy level or line)

Stark effect can be calculated at any instant as if field were static.   Hence

$$\Delta W_1 = \mu_{11}E_0 \cos 2\pi\nu_0 t + \sum_n \frac{|\mu_{1n}|^2}{W_1 - W_n} E_0^2 \cos^2 2\pi\nu_0 t$$

$$= \Delta W_1^{(1)} \cos 2\pi\nu_0 t + \Delta W_1^{(2)} \cos^2 2\pi\nu_0 t \qquad (10\text{-}74)$$

Case 2. $\nu_0 \gg \Delta\nu$ and $\nu_0 \ll \dfrac{|W_1 - W_n|}{h}$ (transition frequency to any level connected by electric dipole matrix element)

A number of different component levels occur.   If first-order Stark effect is present $(\Delta W_1^{(1)} \neq 0)$, the changes in "energy" of these components are

$$\Delta W_1 = \pm m\nu_0 \qquad (10\text{-}75)$$

where m is an integer.   Intensity of each component is proportional to $[J_m(\Delta W_1^{(1)}/\nu_0)]^2$.   If $\Delta W_1^{(1)} = 0$, then only second-order Stark effects occur and the components are given by

$$\Delta W_1 = \pm 2m\nu_0 \qquad (10\text{-}76)$$

with intensities proportional to $[J_m(\Delta W_1^{(2)}/4\nu_0)]^2$.

Case 3. $\nu_0 \gg \Delta\nu$, $\nu_0 \gg \dfrac{|\Delta W_1^{(1)}| + |\Delta W_1^{(2)}|}{h}$ and $\nu_0 \neq \dfrac{W_1 - W_n}{h}$

$$\Delta W_1 = \sum_n \frac{|\mu_{1n}|^2 E_0^2 (W_1 - W_n)}{2[(W_1 - W_2)^2 - h^2\nu_0^2]} \qquad (10\text{-}77)$$

Case 4. $\nu_0 \approx \dfrac{W_1 - W_n}{h}$

This is the resonant case discussed below.   The level splits into two separated by $|\mu_{1n}E_0|/h$ if $\nu_0 = (W_1 - W_n)/h$.

If a molecule is initially in state 1, radiation of the resonant frequency will induce a transition to state 2, and then back to state 1, so that a regular oscillation between states 1 and 2 is produced.   The result is that the wave function is modulated at the frequency of this regular oscillation, and the observed spectral lines split into two components.

Assume that the upper level is state 2 and the frequency of the Stark field

$$\nu_0 = \frac{W_2 - W_1 + \epsilon}{h} \qquad (10\text{-}78)$$

where $\epsilon/(W_2 - W_1) \ll 1$, so that the above type of "resonant modulation" can occur. Then it can be shown that an appropriate wave function for the system can be written to a good approximation [892c].

$$\psi = \frac{\psi_2}{\sqrt{2}\,(\rho^2 + \beta^2)}\, e^{-i/\hbar(W_2 + \epsilon/2)t}[\rho(\rho - \beta e^{i\phi})e^{i\gamma t/\hbar} + \beta(\beta + \rho e^{i\phi})e^{-i\gamma t/\hbar}]$$

$$+ \frac{\psi_1}{\sqrt{2}\,(\rho^2 + \beta^2)}\, e^{-i/\hbar(W_1 - \epsilon/2)t}[-\beta(\rho - \beta^{i\phi})e^{i\gamma t/\hbar} + \rho(\beta + \rho e^{i\phi})e^{-i\gamma t/\hbar}]$$

$$(10\text{-}79)$$

where β is the amplitude of the electric field times the dipole moment matrix element μ_{12} between the two states, or

$$\beta = \mu_{12}E_0 \qquad \gamma = \tfrac{1}{2}\sqrt{|\beta|^2 + \epsilon^2} \qquad \rho = \sqrt{|\beta|^2 + \epsilon^2} - \epsilon \quad (10\text{-}80)$$

and where ϕ is an arbitrary phase angle whose value depends on the initial conditions at $t = 0$.

If a third state of energy W_3 is considered which is connected by a dipole transition to state 2 but not to state 1, then it may be seen from (10-79) that transition frequencies

$$\nu = \frac{1}{h}\,(W_2 - W_3 + \epsilon/2 \pm \gamma) \tag{10-81}$$

may be expected. The ratio of intensities of these two is given by the ratio of the square of the amplitude of the two terms of (10-79) which multiply ψ_2. After averaging over the arbitrary phase angle ϕ, this ratio is

$$R = \frac{\rho^2}{|\beta|^2} = \frac{|\beta|^2 - 2\epsilon\sqrt{|\beta|^2 + \epsilon^2} + 2\epsilon^2}{|\beta|^2} \tag{10-82}$$

Hence when ϵ is small and positive $[\nu_0 > (W_2 - W_1)/h]$, the transition frequency given by the plus sign in (10-81) is the weaker; when ϵ is small and negative $[\nu_0 < (W_2 - W_1)/h]$ this frequency is the stronger of the two. Precisely at resonance, $R = 1$ from (10-82) and the frequency difference between the two components of equal intensity is, from (10-81),

$$\Delta\nu = \frac{2\gamma}{h} = |\mu_{12}|\frac{E_0}{h} \tag{10-83}$$

Just at resonance, (10-79) shows that the wave function can be considered as oscillating back and forth between ψ_1 and ψ_2 because a proper choice of ϕ gives, from (10-79),

$$\psi = i\psi_1 e^{-iW_1 t/\hbar}\cos\left(\frac{\gamma t}{\hbar} + \frac{\pi}{4}\right) + \psi_2 e^{-iW_2 t/\hbar}\sin\left(\frac{\gamma t}{\hbar} + \frac{\pi}{4}\right) \tag{10-84}$$

Hence the frequency of oscillation between states is

$$\frac{\gamma}{h} = \frac{|\mu_{12}|E_0}{2h}$$

The type of splitting due to resonance modulation described above may be used to measure the separation between closely spaced levels [429]. Since the relative intensity of the two components is sensitively dependent on the deviation ϵ from resonant frequency $(W_2 - W_1)/h$, the resonant frequency can be rather accurately measured by varying ν_0 until the two intensities are equal. This technique is particularly useful when no microwave transitions to level 1 are observed, and some transition frequency $(W_2 - W_3)/h$ falls in the microwave region. Then if W_2

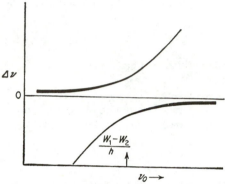

FIG. 10-10. Stark effect at frequencies ν_0 comparable to a resonance frequency $\frac{W_1 - W_2}{h}$. Width of lines correspond approximately to intensity of the Stark components. For low frequency, there is a single component. At resonance its frequency change $\Delta\nu$ is large, and a second component with an opposite frequency change has equal intensity. For $\nu_0 > \frac{W_1 - W_2}{h}$, the second component dominates in intensity.

is not very different from W_1, the energy separation $W_1 - W_2$ may be measured by resonant modulation.

For ϵ large and positive, R from (10-82) becomes zero, and for ϵ large and negative, $R \to \infty$, so that only one component has appreciable intensity in either case. This is in agreement with the nonresonant Stark effects discussed above and shows how the Stark effect can appear to change sign rather suddenly at resonance. In reality both components are always present, but on either side of resonance, different components predominate in intensity. This behavior is illustrated in Fig. 10-10.

Certain complications often occur in the use of resonant modulation. The above discussion assumes a simple nondegenerate transition so that $|\mu_{12}|$ has a unique value. In many cases, each microwave line is in fact a superposition of transitions involving various magnetic quantum numbers M. For each value of M the matrix element μ_{12} may be different, so that resonant modulation produces a number of pairs of components.

It should also be remembered that the wave function (10-79) is an approximation. Terms in various powers of $\beta/(W_1 - W_2)$ are neglected and these can be of importance for sufficiently large electric fields. A more complete solution has been obtained by Autler and Townes [892c]. They show that, for sufficiently strong electric fields, additional transition frequencies or "sidebands" occur at regular spacings equal to the modulation frequency ν_0. In addition, splitting of lines similar to that observed at the resonant frequency occurs at frequencies of $\dfrac{W_2 - W_1}{3h}$, $\dfrac{W_2 - W_1}{5h}$, $\dfrac{W_2 - W_1}{7h}$, $\cdot$ $\cdot$ $\cdot$ $\cdot$

ZEEMAN EFFECTS IN MOLECULAR SPECTRA

11-1. Introduction. Zeeman effects in molecular spectra bear about the same relation to Stark effects that magnetic hyperfine structure does to electric quadrupole hyperfine structure. That is, Stark effects are very much more prominent in the usual types of molecules, which are in $^1\Sigma$ states; Zeeman effects are relegated to small second-order effects or to the small moments of the nuclei. However, in the unusual molecules which have electronic angular momentum and hence are not in $^1\Sigma$ states, the Zeeman effect is quite large and comparable with Zeeman effects found in atomic spectra.

Perhaps the simplest cases to discuss and those giving the widest variety of effects are the cases of the rare molecules with electronic angular momenta; so they will be treated first. These molecules may also be classed as paramagnetic, since their paramagnetic response to a magnetic field occurs precisely because of the large Zeeman effect associated with electronic angular momentum. Each unit of angular momentum due to orbital motion of an electron produces one Bohr magneton, so that its magnetic moment is $\boldsymbol{\mu}_L = -\mu_0 \mathbf{L}$, where μ_0 is the Bohr magneton (taken as a positive quantity) and $\mathbf{L}$ is the orbital angular momentum in units of $\hbar$. Each unit of spin momentum produces slightly more than two Bohr magnetons, or $\boldsymbol{\mu}_s = -2.00229\mu_0 \mathbf{S}$. The energy of interaction between these magnetic dipoles and an external field is simply

$$\Delta W = -(\boldsymbol{\mu}_L \cdot \mathbf{H} + \boldsymbol{\mu}_s \cdot \mathbf{H}) \tag{11-1}$$

11-2. Zeeman Effect in Weak Fields for Molecules Having Electronic Angular Momentum. The energy (11-1) can easily be evaluated by vector-model calculations when $\mu_L H$ or $\mu_s H$ are considerably smaller than certain other molecular energies (weak-field case) and when some pure-coupling case applies. Consider, for example, Hund's coupling case (a) for a diatomic (or linear) molecule. $\mathbf{S}$ and $\mathbf{L}$ precess about the molecular axis, which precesses about $\mathbf{J}$ (*cf.* Chap. 7). When a magnetic field $\mathbf{H}$ is applied, $\mathbf{J}$ precesses about $\mathbf{H}$ with a projection M on the direction of H, where M has one of the values $J, J - 1, \ldots, -J$. Now, using the vector model, the average value of $\boldsymbol{\mu}_s \cdot \mathbf{H}$ is

$$\mu_s H[\cos(SH)]_{av} = \mu_s H[\cos(SA)]_{av}[\cos(AJ)]_{av}[\cos(JH)]_{av}$$

where **A** represents the molecular axis and $\cos (SA)$ the cosine of the angle between **S** and the axis. Thus,

$$\mathbf{y}_s \cdot \mathbf{H} = \frac{(\mathbf{y}_s \cdot \mathbf{k})(\mathbf{k} \cdot \mathbf{J})(\mathbf{J} \cdot \mathbf{H})}{J(J + 1)} \qquad (11\text{-}2)$$

where **k** is a unit vector along the molecular axis and J^2 in the denominator has been replaced by $J(J + 1)$ according to the usual vector-model rule. Now $\mathbf{y}_s \cdot \mathbf{k} \approx -2.002\mu_0$; $\mathbf{S} \cdot \mathbf{k} = -2.002\mu_0 \Sigma$; $\mathbf{k} \cdot \mathbf{J} = \Omega$, and

$$\mathbf{J} \cdot \mathbf{H} = MH$$

Hence

$$\mathbf{y}_s \cdot \mathbf{H} = \frac{-2.002\mu_0 \Sigma \Omega M H}{J(J + 1)} \qquad (11\text{-}3)$$

A similar calculation can be made for $\mathbf{y}_L \cdot \mathbf{H}$, so that the energy from (11-1) in Hund's case (a) is

$$\Delta W = \mu_0 \frac{(\Lambda + 2.002\Sigma)\Omega M H}{J(J + 1)} = \frac{(\Omega + 1.002\Sigma)\Omega M \mu_0 H}{J(J + 1)} \qquad (11\text{-}4)$$

Thus from (11-4) there are $2J + 1$ equally spaced Zeeman levels corresponding to the different possible values of M. The quantity $\mu_0 H/h$, which will occur frequently in this chapter, is conveniently expressed in megacycles, or $\mu_0/h = 1.39967 \pm 0.00005$ Mc/oersted. Hence, if the numerical coefficient of $\mu_0 H$ in (11-4) is approximately unity, a field of only 1 oersted (slightly larger than the earth's magnetic field) can produce Zeeman splittings as large as 1 Mc.

It is important to remember that Λ and Σ can be positive or negative in (11-4). Consider, for example, a $^2\Pi$ state. For $^2\Pi_{\frac{3}{2}}$, $\Lambda + 2.002\Sigma \approx 2$ (or -2), and Eq. (11-4) gives a large Zeeman effect, but for $^2\Pi_{\frac{1}{2}}$,

$$\Lambda + 2.002\Sigma = 0.001$$

so that (11-4) gives only a very small Zeeman effect. The only other important cases which give small Zeeman effects according to (11-4) are $^3\Delta_1$, where $\Lambda + 2.002\Sigma = 0.002$, and $^3\Pi_0$, where $\Omega = 0$.

For Hund's coupling case (b), a vector-model calculation can also be made. It gives the weak-field Zeeman energy as

$$\Delta W = \frac{1}{2J(J + 1)} \left\{ \frac{\Lambda^2[N(N + 1) + S(S + 1) - J(J + 1)]}{N(N + 1)} \right.$$
$$\left. + 2.002[J(J + 1) + S(S + 1) - N(N + 1)] \right\} M \mu_0 H \qquad (11\text{-}5)$$

For linear molecules with electronic spin in Σ' states ($\Lambda = 0$), or for nonlinear paramagnetic molecules where the electron orbital motion is quenched (Λ is undefined), the Zeeman effect given by (11-5) becomes

$$\Delta W = \frac{1.001}{J(J+1)} [J(J+1) + S(S+1) - N(N+1)]M\mu_0H \quad (11\text{-}6)$$

For the common case where $S = \frac{1}{2}$, (11-6) is simply

$$W_{J=N+\frac{1}{2}} = \frac{1.001}{J} M\mu_0H$$

$$W_{J=N-\frac{1}{2}} = -\frac{1.001}{J+1} M\mu_0H \quad (11\text{-}7)$$

In cases intermediate between (a) and (b), calculation of the weak-field Zeeman effect is more complex. This is discussed more fully below (page 289).

11-3. Characteristics of Zeeman Splitting of Spectral Lines. Although the expressions given above for magnetic energy do not cover all cases, they are typical enough to give a general picture of Zeeman splitting of spectral lines in magnetic fields which are not too large. It is customary to write the magnetic energy in terms of a molecular g factor, which is a pure number defined by stating that the magnetic energy is given by

$$\Delta W = -\mathbf{\mu} \cdot \mathbf{H} = -\mu_0 g_J \mathbf{J} \cdot \mathbf{H} = -g_J M\mu_0H \quad (11\text{-}8)$$

where μ_0 is the Bohr magneton (taken as a positive quantity). In molecules with electronic angular momentum, g_J is usually of the order of unity. However, for $^1\Sigma$ molecules and for the magnetic moments of nuclei, g_J is about 1000 times smaller. In these cases it is customary to write instead of (11-8)

$$\Delta W = -g_J M\mu_nH \quad (11\text{-}9)$$

where μ_n is the nuclear magneton, which is smaller than the Bohr magneton by the ratio of the electron-to-proton masses, or $\frac{1}{1836}$. Then g_J is again of the order of unity. It will usually be clear whether (11-8) or (11-9) is used to define g.

It is evident from some of the above expressions, e.g., (11-4), (11-5), or (11-7), that g_J may depend not only on the molecule but also on its rotational angular momentum J and other quantum numbers. There are, however, common cases where g does not vary with J (for example, in the $^1\Sigma$ molecule which will be discussed below), so that under some circumstances g may be considered a fixed constant of the molecule.

The energies given by (11-8) are proportional to the first power of M and of H, so that they correspond to "first-order" Stark effects seen in molecules with degenerate levels such as a symmetric top [cf. Eq. (10-5)]. However, Zeeman effects which are linear in the magnetic field do not require degeneracy as do "first-order" Stark effects. There are, in addition, second-order Zeeman effects proportional to H^2, but these are usually much smaller than those proportional to H, and hence have been omitted in the above formulas. A given energy level is, according to (11-8), split symmetrically about the zero field energy into $2J + 1$ equally spaced energy levels.

The selection rules for transitions involving Zeeman effects are identical with those when Stark effects are present. Thus, when the exciting microwave field is parallel to H, $\Delta M = 0$, and when it is perpendicular, $\Delta M = \pm 1$. Components of a line corresponding to $\Delta M = 0$ are sometimes called π components and those for $\Delta M = \pm 1$ are designated σ components. Relative intensities of these components for different values of J and M are similar to the Stark case and are given in Table 10-1.

A transition between upper and lower energy levels designated by J_1 and J_2, respectively, which has a frequency ν_0 in zero field will be split into a number of components by Zeeman effect with the following frequencies:

For $\Delta M = 0$ (π components),

$$\nu = \nu_0 + (g_{J_2} - g_{J_1})M\mu_0 H/h \tag{11-10}$$

For $\Delta M = M_2 - M_1 = \pm 1$ (σ components),

$$\nu = \nu_0 + [(g_{J_2} - g_{J_1})M_2 \pm g_{J_1}]\mu_0 H/h \tag{11-11}$$

where M_2 is the value of M in the lower state.

It may be seen that the average position of all the Zeeman components from either (11-10) or (11-11) is just ν_0, the position of the undisturbed line.   If intensities are allowed for, the "center of gravity" of all Zeeman components is also ν_0, since Zeeman patterns of this type are always symmetrical about ν_0 [cf. Fig. 11-1].   However, when the smaller Zeeman effects dependent on H^2 are allowed for, the "center of gravity" of a Zeeman split line need not be exactly equal to ν_0.

If the g factor g_J is constant, π components show no Zeeman effect from (11-10), while σ components from (11-11) are all superimposed on two frequencies only, $\nu_0 \pm g_J\mu_0 H/h$.   In most cases g_J does not change very much from the lower to the upper states, so that σ components show larger Zeeman effects than do the π components.   Thus for easiest observation of Zeeman effects in a rectangular waveguide, H should be either parallel to the length of the waveguide or parallel to its broadest faces, so that it is perpendicular to the electric field of the exciting microwave. If the g factor is not the same for the upper and lower molecular states, a more complex pattern is obtained, with $2J + 1$ π components or $2(2J + 1)$ σ components, where J is the smaller of J_1 and J_2.   General appearance of these spectra is shown in Fig. 11-1. This figure illustrates typical cases where $g_{J_1} - g_{J_2} = 0$ and where $g_{J_1} - g_{J_2}$ is not zero but is smaller than g_{J_1}. When H is neither parallel nor perpendicular to the exciting microwave field, both π and σ components appear.

It is possible to excite transitions of the type $\Delta M = M_2 - M_1 = +1$ without $\Delta M = -1$, or vice versa, by using circularly polarized microwaves. If the magnetic field is parallel to the direction of propagation of a circularly polarized microwave, then only one of the Zeeman σ components is excited, and it has all the intensity of the unsplit line. The

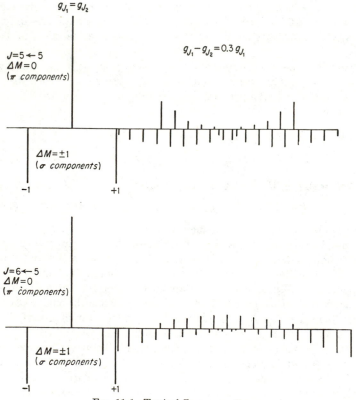

Fig. 11-1. Typical Zeeman patterns.

rotating electric component of the microwave field acts on the molecular dipole to give it some added rotational velocity in the same direction as the rotation of the field. This corresponds to $\Delta M = +1$, if the electric vector appears to rotate clockwise when viewed in the direction of the magnetic field, and $\Delta M = -1$ if the rotation is counterclockwise.

The value of using circular polarization to study the Zeeman effect is that it allows a determination of the sign of the molecular magnetic moment, or of g_J. This may be seen from (11-11), where frequencies for which the $+$ sign applies correspond to $\Delta M = +1$, and those with the $-$ sign to $\Delta M = -1$. Microwaves with circular polarization may be obtained in circular waveguide, or an approximately circularly polarized wave may be transmitted in a square waveguide. Even a small circular component of the microwave can give a noticeable difference in intensity of the two Zeeman σ components (cf. [705]).

11-4. Intermediate Coupling and Intermediate Fields. When a molecule follows no pure coupling scheme, or where the interaction between its magnetic moments and the external field becomes comparable with one of several other sources of molecular energy, the weak-field pure coupling case approximations used above may not be appropriate.

For example, the magnetic interaction $\mathbf{\mu}_s \cdot \mathbf{H}$ may be large enough to somewhat disturb the end-over-end rotational motion of the molecule, or to partially uncouple the spin S from the molecular axis or from N. Hill [14], [47] has obtained a complete and closed expression for the energy of doublet states intermediate between Hund's cases (a) and (b) when a magnetic field is applied. However, this expression is so involved and difficult to evaluate that it is not given here.

The Zeeman effect in Hund's coupling case (a) is essentially the same as the Stark effect in a symmetric molecule as long as the magnetic field is not sufficiently strong to uncouple S from the molecular axis. Hence for fields large enough to disturb the molecular rotation but not to uncouple S, expressions (10-5) and (10-7) apply to case (a) with E replaced by H, K by $\Lambda + \Sigma$, and μ by $\mu_0(\Lambda + 2.002\Sigma)$.

There is one striking difference between the Zeeman and Stark effects, however, when Λ doubling or inversion doubling occurs. An electric field produces matrix elements only between different l-doubled, Λ-doubled, or inversion-doubled levels, so that the behavior of the Stark effect depends on the amount of doubling (cf. page 252). Although matrix elements for a magnetic field have the same form and value when the doubled states are degenerate, they differ in that they do not connect the two doubled states (cf. [14], p. 1510). Hence, Λ-type or K-type doubled energy is simply added to the Zeeman energy, whereas its effect on Stark energy is more complex.

The effect of large magnetic fields in Hund's coupling case (b) may be obtained for doublet states from the results of Hill [14] or for a $^3\Sigma$ state from the recent work of Tinkham on O_2 [990]. For moderate fields second-order perturbations due to the magnetic field may be obtained in a straightforward way from the matrix elements which they give. These second-order effects involve a perturbation both of the rotational levels and of the precession of S about K, so that the results tend to be somewhat complicated.

11-5. Zeeman Effect with Hyperfine Structure. When hyperfine structure is present, the Zeeman effect is modified both by the introduction of a new angular momentum I, the spin of the nucleus producing the hyperfine structure, and also by the interaction between the external field and the nuclear magnetic moment. Almost always the nuclear spin I will couple to the molecular angular momentum J to produce the total angular momentum F, since the coupling between I and the molecule is weak relative to the coupling between electronic vectors and the molecule. This corresponds to coupling case (a_β) or (b_β) of Chap. 8.

If the magnetic field is too weak to disturb the coupling between any molecular vectors except perhaps that between $\mathbf{I}$ and $\mathbf{J}$, the molecule may be considered to have a magnetic moment $\mu_0 g_J J$ oriented along $\mathbf{J}$ and a moment $\mu_n g_I I$ oriented along $\mathbf{I}$. The magnetic energy is

$$\Delta W = -\mu_n g_I \mathbf{I} \cdot \mathbf{H} - \mu_0 g_J \mathbf{J} \cdot \mathbf{H} \qquad (11\text{-}12)$$

When $\mathbf{H}$ is so small that it does not disturb the coupling between $\mathbf{I}$ and $\mathbf{J}$, that is, when (11-12) is much smaller than the hyperfine energy, the energy (11-12) may be evaluated by the vector model. This is very similar to the case of first-order Stark effect with hyperfine structure [(10-26) and (10-27)]. The energy given by the vector model for weak fields is

$$\Delta W = \{ -\mu_n g_I [I(I+1) + F(F+1) - J(J+1)]$$
$$- \mu_0 g_J [J(J+1) + F(F+1) - I(I+1)] \} \frac{M_F H}{2F(F+1)} \qquad (11\text{-}13)$$

where M_F is the projection of the total angular momentum $\mathbf{F}$ on $\mathbf{H}$.

If the molecule is paramagnetic, then the first term of (11-13), which gives the interaction between $\mathbf{H}$ and the nuclear moment, is usually at least 1000 times smaller than the second term and is negligible. However, in $^1\Sigma$ molecules, $\mu_0 g_J$ is of the same order of magnitude as μ_n, and both terms of (11-13) are important.

When the Zeeman energy is not much smaller than the hyperfine energy, second-order perturbation theory must be used, or in larger fields a complete secular equation must be solved. The treatment is very similar to that for intermediate Stark effect in a symmetric top when $K \neq 0$ (cf. page 261). The case has also been discussed by Coester [444]. In magnetic fields strong enough to decouple $\mathbf{I}$ and $\mathbf{J}$, the Zeeman effect is just that found when no hyperfine structure is present. The hyperfine energy can be treated as a small perturbation, and is almost identical with that found in large Stark fields (cf. page 261). The only real difference is that the degeneracy between $M_J = 1$ and $M_J = -1$ is not present, so that the complications of transitions between these two levels produced by quadrupole effects do not occur.

11-6. Zeeman Effects in Ordinary Molecules ($^1\Sigma$ States). Most molecules are in $^1\Sigma$ states and hence have no electronic angular momentum. Their magnetic moments are proportional to the rotational angular momentum and are approximately the same as magnetic moments of nuclei, or $\frac{1}{1000}$ that of electrons. Hence, Zeeman effects in these molecules are quite small. Their moments would generally be neglected by comparison with electronic moments in paramagnetic molecules, but when the large electronic moments are not present, these small magnetic moments of $^1\Sigma$ molecules give noticeable effects.

Consider just the Zeeman effect for a molecule in a $^1\Sigma$ state with no hyperfine structure. The magnetic moment of the molecule comes partly from the rotation of the positively charged nuclei about the center of mass. This magnetic moment is usually canceled and reversed in sign by the electrons which provide a negative charge rotating with the nuclei. Closed electron shells about the nuclei can be simply thought of as moving with the nuclei, but the orientation of the shells remains fixed in space (slip effect, *cf.* page 213). The behavior of valence electrons is more complicated, since the amount of angular momentum these electrons acquire when the molecule rotates depends on details of their wave functions in the ground and excited states. It is expected that valence electrons will often have enough angular momentum to produce a magnetic moment larger than that due to the sum of the moments produced by the nuclei and bound electrons. Hence the observed sign of the magnetic moment of most molecules should be that produced by a negative charge rotating with the molecule.

The interaction between an external magnetic field and the electrons in a rotating molecule is given by terms similar to the last term in Eq. (8-27). That is,

$$\Delta W = -2 \sum_n \sum_{g'} \sum_g \mu_0 \hbar^2 \frac{J_g H_{g'}}{A_g} \frac{(0|L_g|n)(n|L_{g'}|0)}{W_0 - W_n} \tag{11-14}$$

For a linear molecule, (11-14) can be reduced, as was (8-36), to the form

$$\Delta W = -4\mu_0 Bh \mathbf{J} \cdot \mathbf{H} \sum_n \frac{|(0|L_x|n)|^2}{W_0 - W_n} \tag{11-15}$$

For N electrons in spherical orbits about a nucleus which is a distance τ from the center of mass of the molecule, (11-15) may be further reduced to a form similar to (8-33)

$$\Delta W = \frac{\mu_0}{A} N m \tau^2 \mathbf{J} \cdot \mathbf{H} \tag{11-16}$$

where m is the electron mass. Remembering from Chap. 8 that the spherical electron shells move with the nuclei, but slip so that they remain fixed in orientation, one can obtain (11-16) from the following simple classical calculation. The molecular angular momentum is $\hbar \mathbf{J}$, and the fraction of this momentum carried by the N electrons is $N m \tau^2/A$. Since the magnetic energy due to electron motion is $\Delta W = -\mathbf{\mu}_L \cdot \mathbf{H} = \mu_0 \mathbf{L} \cdot \mathbf{H}$, where now $\mathbf{L} = (N m \tau^2/A)\mathbf{J}$, expression (11-16) is easily obtained.

So far the magnetic moment due to motion of the nuclear charges has been neglected. These charges must give a magnetic energy essentially the same as that to be expected from electrons closely bound to the nucleus in a spherical shell but, of course, of opposite sign. If the nuclear charge is $+Ze$ and there are N electrons in closed shells around the nucleus, then the net charge due to the nucleus and electrons is $(Z - N)e = N_s e$. The net magnetic energy produced by nuclei and the spherical shells of electrons which may surround them is similar to (11-16), but with a modified charge.

$$\Delta W = - \sum_s \frac{\mu_0 N_s m \tau_s^2}{A} \mathbf{J} \cdot \mathbf{H} \tag{11-17}$$

where N_s is the net charge (nucleus and bound electrons) in units of the proton charge about the nucleus s, and τ_s is its distance from the center of mass.

The magnetic energy due to valence electrons which are not spherically distributed about nuclei must be obtained from an expression like (11-15). Hence

$$\Delta W = -2\mu_0 B \left[\sum_s \frac{N_s m \tau_s^2}{\hbar^2} + 2 \sum_n \frac{|(0|L_x|n)|^2}{W_0 - W_n} \right] \mathbf{J} \cdot \mathbf{H} \tag{11-18}$$

where B is the rotational constant $h/8\pi^2 A$, and L_x is the component of angular momentum of the valence electrons perpendicular to the molecular axis. It should be noted that (11-18) has the form

$$\Delta W = -\mu_n g_J \mathbf{J} \cdot \mathbf{H} = -\mu_n g M_J H \tag{11-19}$$

where g_J can be written g since it is independent of J. μ_n is the nuclear magneton, which is less than μ_0 by m/M, the ratio of the electron mass to that of the proton. The Zeeman splitting for such a molecule is hence given by the simple spectra for $g_{J_1} = g_{J_2}$ of Fig. 11-1.

If only one excited state of the valence electrons is important and pure precession is assumed, one can reduce (11-18) in analogy to (8-35) to

$$\Delta W = -2\mu_0 B h \left[\sum_s \frac{N_s m \tau_s^2}{\hbar^2} - \frac{L(L + 1)}{2(W_\Pi - W_\Sigma)} \right] M_J H \tag{11-20}$$

where $W_\Pi - W_\Sigma$ is the excitation energy of the lowest excited electronic state, and L the precessing angular momentum of the valence electrons.

Except for the cases where hydrogens rotate, such as in H_2 and NH_3, it seems likely that the second term of (11-18) or (11-20) dominates, and that hence molecular g factors are negative. That is, the magnetic moments of rotating molecules can be expected to have the sign given by rotating negative charges. This is true for OCS and OCSe, the only heavy molecules for which the sign of g has so far been measured.

The Zeeman effect for a general asymmetric rotor in a $^1\Sigma$ state is similar in principle to that of a linear molecule but, of course, more complicated. The magnetic moment due to electrons is still given by (11-14), and the moment produced by the motion of the nuclei can be calculated by semiclassical methods which are simple enough in principle. However, actual expressions for the molecular magnetic moments are rather complicated. The molecular magnetic moment along one of the principal axes of inertia of the molecule can always be written [682]

$$\mu_x = \mathfrak{M}_{xx}\Omega_x + \mathfrak{M}_{xy}\Omega_y + \mathfrak{M}_{xz}\Omega_z = \mathfrak{M}_{xx}\hbar\frac{J_x}{I_x} + \mathfrak{M}_{xy}\hbar\frac{J_y}{I_y} + \mathfrak{M}_{xz}\hbar\frac{J_z}{I_z} \quad (11\text{-}21)$$

where the $\mathfrak{M}$'s are components of a symmetric dyadic or tensor whose values depend only on the molecule, and I_x, I_y, I_z are the principal moments of inertia. The Ω's are components of angular velocity about the principal axes of inertia. Expressions similar to (11-21) can be written for μ_y and μ_z. The principal axes of the dyadic $\mathfrak{M}$ do not necessarily coincide with the principal axes of inertia. However, in molecules with symmetry it is often possible to simplify (11-21). For example, in H_2O, symmetry arguments show that the principal axes of inertia must coincide with those of the dyadic $\mathfrak{M}$, so that

$$\mu_x = \mathfrak{M}_{xx}\hbar\frac{J_x}{I_x} \qquad \mu_y = \mathfrak{M}_{yy}\hbar\frac{J_y}{I_y} \qquad \mu_z = \mathfrak{M}_{zz}\hbar\frac{J_z}{I_z}$$

The Zeeman energy is $-\mathbf{\mu}\cdot\mathbf{H} = -[\mu\cos(\mu H)]_{\text{av}}H$, where $\cos(\mu H)$ is the cosine of the angle between the net magnetic moment $\mathbf{\mu}$ and the field $\mathbf{H}$. Eshbach and Strandberg [682] have given matrix elements of $\mu\cos(\mu H)$ for symmetric-top wave functions, and from these,

$$[\mu\cos(\mu H)]_{\text{av}}$$

for an asymmetric top may be found using expansions of the type (4-17). In the case of a symmetric top, the average value of $\mu\cos(\mu H)$ is [592], [682]

$$[\mu\cos(\mu H)]_{\text{av}} = (JKM|\mu_z|JKM)$$
$$= \mu_n M\left[g_{xx} + (g_{zz} - g_{xx})\frac{K^2}{J(J+1)}\right] \quad (11\text{-}22)$$

where μ_n = the nuclear magneton
$g_{xx} = g_{yy} = \hbar\mathfrak{M}_{xx}/I_x\mu_n$
$g_{zz} = \hbar\mathfrak{M}_{zz}/I_z\mu_n$ (z is the symmetry axis)
M = projection of J on H

It should be noted that the Zeeman energy can also be written, as mentioned above, $\Delta W = -\mu_n g_J \mathbf{J}\cdot\mathbf{H}$, where $g_J = \dfrac{\mu_{\text{av}}}{\mu_n J}$ for the state $M = J$. For a linear molecule where $K = 0$, (11-22) gives the form (11-19).

The symmetric tops for which molecular g factors have been measured are the second group of molecules in Table 11-1. In NH_3 the general correctness of the form (11-22) has been checked experimentally for a variety of values of J and K [592], [682] and values $g_{xx} = +0.560$, $g_{zz} = +0.484$ obtained. The complication of hyperfine structure is present in NH_3, so that the Zeeman effect must be obtained from (11-13).

TABLE 11-1. MOLECULAR g FACTORS FOR MOLECULES IN $^1\Sigma$ STATES

$g_J = \mu/J\mu_n$, where μ is the molecular magnetic dipole moment, μ_n a nuclear magneton, and $\hbar J$ the angular momentum.

Molecule	g factors	Reference
H_2	0.88291 ± 0.00007	[695a]
N_2O	$\pm 0.086 \pm 0.004$	[592]
OCS	-0.025 ± 0.002	[682]
OCSe	-0.019 ± 0.002	[806a]
NH_3	$g_{zz} = 0.484 \pm 0.007$	[592] [682]
	$g_{xx} = g_{yy} = 0.560 \pm 0.007$	[682]
CH_3F	$g_{xx} = \pm 0.08$	[784]
CH_3CCH	$g_{zz} = \pm 0.30$	[784]
	$|g_{xx}| \ll |g_{zz}|$	
H_2O	$g_{aa} = 0.585$	[592] [742]
	$g_{bb} = 0.742$	[782]
	$g_{cc} = 0.666$	
H_2S	$g_J = \pm 0.24$ for 1_{01} and 1_{10} rotational states	[783]
O_2	$g_J = \pm 1.54 \pm 0.09$ for 1_{11} rotational state	[875]
	$g_J = 0.15 \pm 0.03$ for 2_{02} state	
SO_2	$g_J = \pm 0.084 \pm 0.010$ for 7_{26} and 8_{17} states	[592]
CH_3OH	$g_{zz} = \pm 0.078$	[592]
$KClFeCl_2$	$g_J = \pm 0.5$	[952a]
$KBrFeBr_2$	$g_J = \pm 0.25$	[952a]

Asymmetric rotors for which Zeeman effects have been studied are given in the last group of molecules of Table 11-1. The case of H_2O and HDO is particularly interesting since Zeeman effects on several lines have been measured, and since H_2O and HDO are similar electromagnetically. Schwarz [742] has related the g factors for these molecules and given the values listed in Table 11-1 for H_2O. g_{aa} indicates the g factor for the direction of smallest moment of inertia, I_a. SO_2 is, of course, rather similar to H_2O, but so far the values of g_J for SO_2 are known only for the $8_{17} \leftarrow 7_{26}$ transition, for which the g factor is given in the table. The compounds $KClFeCl_2$ and $KBrFeBr_2$ have particularly large values of g_J for such heavy molecules. They appear to be cases where some excited electronic level lies rather close to the ground level and hence makes large contributions to g_J of the type given by the second term of (11-18).

In H_2O the dyadic $\mathfrak{M}$ has principal axes which coincide with the principal axes of inertia as pointed out above. HDO is electrically similar to H_2O, and hence if the center of gravity of HDO were in the same place as that for H_2O, the principal axes and elements of the dyadic $\mathfrak{M}$ would be the same for the two molecules. However, in HDO the principal axes of inertia would no longer coincide with those for $\mathfrak{M}$. Actually the center of gravity of HDO is displaced from that of H_2O, which affects the dyadic $\mathfrak{M}$ because these molecules involve rotating electric dipoles.

Consider as a simple example the case of a positive and negative charge Ne separated by a distance x_0. The g factor is given, from (11-17), by

$$g = \sum_s \frac{N_s M x_s^2}{A}$$

where x_s is the distance of each charge from the center of gravity, M is the proton mass, and A is the moment of inertia. If now the mass of one of the charges is changed so that the center of gravity is shifted by an amount Δx and A is changed to A', the new g factor becomes

$$g' = \sum_s \frac{N_s M (x_s - \Delta x)^2}{A'} = \sum_s \frac{N_s M x_s^2}{A'} - \frac{2M \, \Delta x \sum_s N_s x_s}{A'} + \frac{M(\Delta x)^2}{A'} \sum_s N_s$$

Using the fact that $\sum_s N_s e x_s$ is the electric dipole moment D_x and $\sum_s N_s e$ is the total charge which is zero, we have then

$$g' = \frac{A}{A'} g - \frac{2M \, \Delta x D_x}{e A'}$$

Here the symbol D is used for electric dipole moment rather than the usual μ in order to avoid confusion with the magnetic moment μ. More generally, one can show that, if the center of gravity of a molecule is shifted with respect to the principal axes of $\mathfrak{M}$ by an amount x,y,z, components of the new dyadic $\mathfrak{M}'$ may be obtained from those of the old dyadic $\mathfrak{M}$ by the following relations [742]:

$$\mathfrak{M}'_{xx} = \mathfrak{M}_{xx} - \frac{2M\mu_n}{e\hbar} (y D_y + z D_z)$$

$$\mathfrak{M}'_{yy} = \mathfrak{M}_{yy} - \frac{2M\mu_n}{e\hbar} (x D_x + z D_z) \qquad (11\text{-}23)$$

$$\mathfrak{M}'_{zz} = \mathfrak{M}_{zz} - \frac{2M\mu_n}{e\hbar} (x D_x + y D_y)$$

where D_x, D_y, and D_z are components of the electric dipole moment along the principal axes of M. Thus the dyadic $\mathfrak{M}$, and hence the molecular g factors for HDO or D_2O, are derivable from those for H_2O, assuming the geometry and dipole moment of this molecule are known.

It is interesting to note that Eqs. (11-23) afford a means of determining the sign of molecular dipole moments, in contrast to Stark effects which can only determine dipole moment magnitudes. Thus if an isotopic substitution shifts the center of gravity of a molecule sufficiently so that the change in M due to the molecular dipole moment can be observed, the sign of this change will allow a determination of the sign of the terms involving components of the dipole moment.

11-7. Combined Zeeman-Stark Effects. Both electric and magnetic fields have been applied to molecules a number of times by microwave spectroscopists. However, in such cases the electric field has been used only as a means of sensitive detection by Stark modulation, and no experimental study of combined Zeeman-Stark splittings has been attempted. Rather complete theoretical description of the combined Zeeman-Stark effect has been worked out, however [444]. This work includes treatment of molecules with hyperfine structure due to one nucleus with quadrupole moment and various types of intermediate- and strong-field conditions.

If the magnetic and electric fields are parallel, then weak-field Zeeman and Stark effects are simply additive, since the molecular wave functions are the same for either field (projection M of J on either field is a good quantum number). If the Stark effect varies linearly with the electric field (first-order Stark effect), then each different value of M produces a different Stark component, and the application of a parallel magnetic field cannot further split this component but may change its frequency. If the Stark effect is proportional to the square of the electric field (second-order Stark effect), then positive and negative values of M coincide in the same component and may be further split by the magnetic field.

If the magnetic and electric fields are not parallel, then M is no longer a good quantum number and the frequencies of the components of the lines depend in a more complex way on the strengths of the two fields. In addition, the relative intensities of the components depend on the two field strengths. The reader is referred to Coester's work [444] for details of the crossed electric- and magnetic-field case, and for some of the intermediate-field cases.

11-8. Transitions between Zeeman Components. Transitions between the Zeeman components of an energy level, corresponding to $\Delta J = 0$, $\Delta M = \pm 1$, occur and give frequencies which increase more or less linearly with the magnetic field H. If a molecule has a magnetic moment as large as a Bohr magneton, then the Zeeman splittings can be so large that transitions between them fall in the microwave region. Since μ_0/h is about 1.4 Mc/oersted, a magnetic field of 15,000 oersteds would give a Zeeman splitting of 21,000 Mc if the molecular g factor is unity.

Beringer *et al.* have exploited transitions between Zeeman components in an interesting way to obtain microwave absorption spectra of several paramagnetic gases ([435], [443], [558]).   The gas is contained in a cavity between the pole pieces of an electromagnet.   The cavity is tuned to some convenient microwave frequency (*e.g.*, 24,000 Mc) and sensitive circuits are arranged to detect absorption of microwaves in the cavity. The magnetic field is then varied until some Zeeman component coincides with the cavity frequency, so that absorption due to the transition is

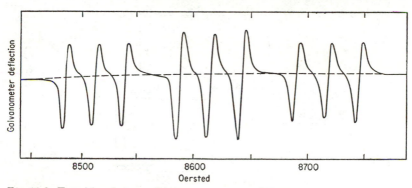

FIG. 11-2. Transitions between Zeeman components of $N^{14}O^{16}$ at a pressure of 1.0 mm Hg.   (*From Beringer and Castle* [435].)

detected.   This technique gives a plot of absorption at a fixed frequency due to Zeeman transitions vs. magnetic field instead of the usual rotational transitions vs. frequency.   A spectrum of this type for NO is shown in Fig. 11-2.   The spectrometer used for Fig. 11-2 presented the derivative of the absorption, which is the reason for the unusual shapes of the lines seen in this figure.

The molecule NO is in a $^2\Pi$ state, with the vectors coupled according to Hund's case (*a*).   The transitions seen in Fig. 11-2 are between Zeeman components of the $^2\Pi_{\frac{3}{2}}$ state with $J = \frac{3}{2}$, which is the state having the largest g factor.   All other states have g factors which are small enough to make transitions between their Zeeman components occur at a very much lower frequency in the magnetic fields used.   In the fields used, the molecule can be fairly accurately described as a case where the magnetic field is strong enough to uncouple the nuclear spin of N^{14} from the molecule (O^{16} has zero spin), but not sufficiently strong to uncouple **S** and **L**.   A diagram of the resulting energy levels is shown in Fig. 11-3.

The hyperfine structure indicated in Fig. 11-3 is in fact very much smaller than the separation between the main levels shown as stage (*b*). The latter separations can be obtained from (11-4) and are approximately

9,400 Mc. The magnetic hyperfine structure is, according to (8-7), $\Delta W_\mu = [a\Lambda + (b + c)\Sigma]\mathbf{I} \cdot \mathbf{k}$. From the vector model,

$$\mathbf{I} \cdot \mathbf{k} = \frac{M_I M_J \Omega}{J(J + 1)} = \tfrac{2}{5} M_I M_J$$

so that $\Delta W_\mu = A M_I M_J$. A is found experimentally to be 29.8 ± 0.3 Mc. The hyperfine energy due to quadrupole effects in this strong-field case

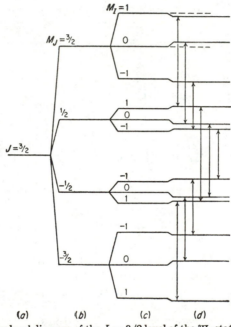

FIG. 11-3. Energy level diagram of the $J = 3/2$ level of the $^2\Pi_{\frac{3}{2}}$ state of $N^{14}O^{16}$. Stage (a) is in the absence of a magnetic field. Stage (b) shows the energy levels split by the presence of a magnetic field, but without hyperfine structure. Stage (c) shows the hyperfine structure due to the N^{14} nuclear magnetic moment. Stage (d) includes the energy due to the N^{14} electric quadrupole moment. (*After Beringer and Castle* [435].)

is given by (10-32), with a quadrupole coupling constant of -1.7 ± 0.5 Mc.

Each of the three groups of three lines shown in Fig. 11-2 corresponds to a particular M_J transition with hyperfine splitting. The above simple theory, where from (11-4)

$$\Delta W_H = \tfrac{2}{5}(\Lambda + 2.002\Sigma)M_J\mu_0 H$$

would make the centers of the three groups coincide. Their separation is due to small second-order effects proportional to H^2. More detailed discussion and more exact calculation of this spectrum have been given in several places ([468], [495], [686]).

Transitions between Zeeman levels in NO may occur because of interaction between the microwave field and either the molecular magnetic moment or its electric moment. The above discussion applies to either, except that transitions involving the electric moment involve Λ-type doubling. Allowed electric dipole transitions are similar to those allowed in a linear molecule where l-type doubling occurs, and these may be seen from expression (2-16) always to involve transitions between the two different states of the doublet. Hence the Zeeman transitions may be expected to be doublets separated by twice the Λ-type doubling if electric dipole transitions occur. Either magnetic or electric dipole transitions may be singled out by orienting the desired microwave field perpendicular to the fixed magnetic field and the other parallel to it. Beringer and Rawson were thus able to detect and measure the Λ doubling in NO as 1.7 Mc.

Transitions between Zeeman levels in O_2 have also been examined [558], [990]. Although the Zeeman effect on the ρ-type triplets of the O_2 spectrum is more complex than that for NO discussed above, and intermediate-field conditions occur, a fairly accurate theory for the observed transitions between Zeeman levels can be developed [74], [467], [990].

CHAPTER 12

THE AMMONIA SPECTRUM AND HINDERED MOTIONS

12-1. Introduction. Because of the intensity and richness of its spectrum, ammonia has played a major role in the development of microwave spectroscopy. It has provided a large number of easily observable lines on which to try both experimental techniques and theory. NH_3 also provides the simplest and most thoroughly worked out example of a class of spectra which will continue to occupy and puzzle microwave spectroscopists for many years—spectra involving hindered motions.

The most important hindered motions all involve the quantum-mechanical tunneling effect. That is, they are motions which cannot occur in classical mechanics because of energy considerations, but are allowed by the wave nature of quantum mechanics. For example, in the ground vibrational state of NH_3, the molecule does not have enough energy to allow the nitrogen to be found in the plane of the hydrogens because of the large potential-energy hump at this position, as indicated in Fig. 12-1. Its actual penetration of the plane of the hydrogens and rapid vibration from one side of it to the other is called the tunnel effect because, if the nitrogen cannot climb the potential hill, it must "tunnel" in order to get through. In principle, the NH_3 inversion is a vibrational motion. Although vibrations normally give frequencies in the infrared region, the NH_3 inversion is so much slowed down by the hindering potential that its frequency lies in the microwave region. The qualitative change from ordinary vibrational levels where there is no hump in the potential to vibrational levels occurring in pairs when the hindering hump occurs is discussed in Chap. 3. When the vibrational energy is so great that the NH_3 molecule has sufficient energy to invert classically, the vibrational levels no longer occur in close pairs but are similar to more normal equally spaced vibrational levels. However, they are still markedly influenced by the presence of the potential hump, and hence in this case of no "tunneling," one may still speak of some hindering of the motion.

Another type of hindered motion is the rotation of one part of a molecule with respect to the remainder. Thus the hydrogen attached to oxygen in CH_3OH (Fig. 12-2) has three possible positions of equal energy but must tunnel through a potential hump between them as indicated in Fig. 12-3. Hence the rotation of this hydrogen around the C—O bond is hindered. Similarly, the methyl group in CH_3CF_3 (Fig. 12-4) has three positions of equal energy separated by potential humps. Usually hindered motions do involve two or more positions of equal energy. If these positions are only minima, which are not equal in energy, many of

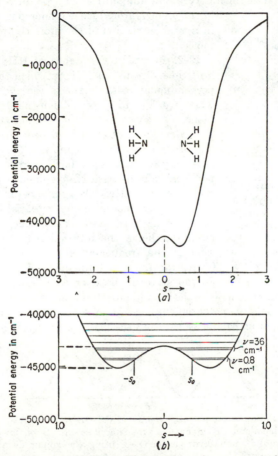

FIG. 12-1. Potential curve of NH_3. The variable s is a measure of the distance between the nitrogen and the plane of the hydrogens. (b) shows the lower part of the potential curve in more detail, and the energy levels.

the characteristic properties and interesting features of the hindered motions considered here do not occur.

When hindered motions have frequencies in or near the microwave region, the spectra are often quite difficult to interpret because the number of lines falling in the microwave region is greatly increased, and because there is no exact solution for the energy levels in these cases. In addition, the frequencies of these motions are so strongly dependent on the potential humps or barriers that usually no helpful estimates of the frequency to be expected in a molecule can be made. However, hindered motions are of considerable importance to a knowledge of molecular structure and to chemistry. The difficulties they produce will be frequently encountered since hindered motions occur in a sizable fraction of the more complex gaseous molecules.

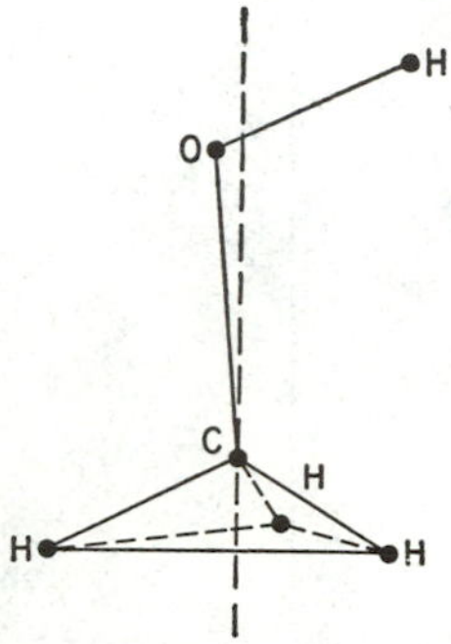

12-2. Inversion Frequency of NH_3. Consider now the vibration of NH_3 in which the N moves perpendicular to the plane of the hydrogens, and which would lead to inversion if the N crossed this plane. As a function of the distance s of the nitrogen from the plane of the hydrogens, the potential is approximately as shown in Fig. 12-1. The molecule may vibrate at a fairly rapid rate in the potential minima with N on one side of the hydrogen plane, and after a large number of such vibrations, it may penetrate the potential barrier and begin oscillations on the other side. The rapid vibrations lie in the infrared region (950 cm^{-1}), whereas the frequency of penetration back and forth is the inversion frequency which occurs for the ground state in the microwave range (0.8 cm^{-1}).

FIG. 12-2. Structure of methyl alcohol. The O—H bond may rotate about the axis with respect to the CH_3 group, but this rotation is hindered by mutual interaction.

The approximate energy of the ground vibrational state is indicated in Fig. 12-1, and equals the potential energy $V(s)$ at $s = \pm s_0$, so that classically the nitrogen atom cannot be closer to the plane of the hydrogens than s_0.

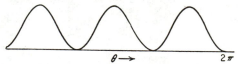

FIG. 12-3. Variation of the potential energy of CH_3OH with the relative rotation θ of the O—H bond and CH_3 group about the molecular axis. Probably the potential minima correspond to the hydrogen in O—H being a maximum distance from the nearest hydrogen of CH_3.

The inversion frequency has been calculated approximately by Dennison and Uhlenbeck [33] as

$$\nu = \frac{\nu_0}{\pi A^2} \tag{12-1}$$

where $A = \exp\left\{ \frac{2\pi}{h} \int_0^{s_0} [2\mu(V(s) - W)]^{\frac{1}{2}} \, ds \right\}$

ν_0 = frequency of vibration in one of the potential minima
μ = reduced mass for the vibrational motion
W = total vibrational energy

This expression was obtained by the Wentzel-Kramers-Brillouin (WKB) method which uses approximate wave functions in the various regions of the motion and matches them at the boundaries of these regions. The significance of (12-1) may perhaps better be made clear by the following approach.

Let the wave function corresponding to the lowest state of simple harmonic oscillation with the nitrogen on the left side of the hydrogens be u_L and that for the nitrogen on the right be u_R. The true molecular wave functions, having a definite energy, must be either symmetric or antisymmetric with respect to the inversion, so the wave functions for the two inversion levels of the lowest vibrational state must be the combinations

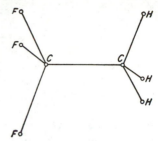

$$\psi_0 = (1/\sqrt{2})(u_L + u_R)$$
and $$\psi_1 = (1/\sqrt{2})(u_L - u_R),$$

as shown in Fig. 3-8. If now the N is definitely placed on the left side at a time $t = 0$, the appropriate wave function describing the system at that time (and not

FIG. 12-4. Structure of CH_3CF_3 where rotation of the CH_3 group with respect to the CF_3 group is hindered by mutual interaction.

having one definite energy) is $\psi = u_L = (1/\sqrt{2})(\psi_0 + \psi_1)$. If, as noted in Chap. 3, the time variation of the wave function is included, it becomes

$$\psi = \frac{1}{\sqrt{2}}[\psi_0 + \psi_1 e^{2\pi i \nu t}]e^{2\pi i W_0 t/h} \qquad (12\text{-}2)$$

where $h\nu$ is the energy difference between the two inversion states and W_0 is the energy of the lower state, which corresponds to ψ_0. After a time $t = 1/2\nu$, the wave function (12-2) becomes

$$\psi = \frac{1}{\sqrt{2}}[\psi_0 - \psi_1]e^{2\pi i W_0 t/h} = u_R e^{2\pi i W_0 t/h} \qquad (12\text{-}3)$$

so that the nitrogen has moved over to the right side. Only a short time after the nitrogen has been initially placed on the left, the part of (12-2) in brackets may be written by expanding the exponential

$$\psi = \left[\frac{1}{\sqrt{2}}(\psi_0 + \psi_1) + \psi_1 \frac{2\pi i \nu t}{\sqrt{2}} + \cdots\right]e^{2\pi i W_0 t/h}$$
$$= u_L + \pi i \nu t(u_L - u_R) + \cdots \qquad (12\text{-}4)$$

so that the amplitude of the wave function on the right has grown to $\pi \nu t$. This gives a measure of the rate of penetration of the potential barrier in terms of the frequency of inversion. This rate of penetration may also be calculated from the "tunnel effect."

Consider now the nitrogen striking the barrier at $s = -s_0$ from the left. Its wave function will partly penetrate the barrier and extend over into the right-hand potential minimum. The amount of penetration can be roughly evaluated by examining Schrödinger's equation in the classically forbidden region.

$$\frac{d^2\psi}{ds^2} - \frac{2\mu}{\hbar^2}(V - W)\psi = 0 \tag{12-5}$$

An approximate solution to (12-5) is

$$\psi = \exp\left\{-\frac{1}{\hbar}\int_{-s_0}^{s}[2\mu(V(s) - W)]^{\frac{1}{2}}\,ds\right\} \tag{12-6}$$

if V does not vary too rapidly with s. Since $V - W$ is positive in the classically forbidden region, this corresponds to an exponentially decreasing function which has unity value at the boundary where the nitrogen strikes (the increasing exponential which is also a solution is omitted, since it would give the N the largest probability of being found on the right). The amplitude of the wave function which penetrates the boundary is then

$$\psi_{s=s_0} = \exp\left\{-\frac{1}{\hbar}\int_{-s_0}^{s_0}[2\mu(V - W)]^{\frac{1}{2}}\,ds\right\}$$

$$= \exp\left\{-\frac{2}{\hbar}\int_{0}^{s_0}[2\mu(V - W)]^{\frac{1}{2}}\,ds\right\} = \frac{1}{A^2} \tag{12-7}$$

Now in a time t which is large enough for the nitrogen to oscillate on the left side many times, but small compared with the time required for inversion, the nitrogen will strike the left side of the potential barrier $\nu_0 t$ times, where ν_0 is the vibrational frequency. Each time it may be considered to partially penetrate and transmit an amplitude $1/A^2$ to the right side. The transmitted amplitudes add up to give a total amplitude after time t of $\nu_0 t/A^2$, so that the probability of penetration in time t is $(\nu_0 t/A^2)^2$. This is similar to, but differs slightly from, the usual expression for radioactive decay or other types of barrier penetration by tunneling ([305], p. 22), because the transmitted wave on the right-hand side of the barrier is trapped in an identical potential minimum and is added to by successive transmitted waves. Equating the amplitude $\nu_0 t/A^2$ to the amplitude for u_R in (12-4), we have

$$\pi\nu t = \frac{\nu_0 t}{A^2} \qquad \text{or} \qquad \nu = \frac{\nu_0}{\pi A^2} \tag{12-8}$$

which is the result obtained by Dennison and Uhlenbeck.

In the ground state of NH_3, $\nu/\nu_0 \approx 1200$, so that $A^2 \approx 400$ or e^6.

Because A^2 is large and varies exponentially with $\mu^{\frac{1}{2}}$ or $(V - W)^{\frac{1}{2}}$, changes in either of these quantities can very drastically affect the inversion frequency ν. For example, if the reduced mass μ is increased by a factor of 2, such as would be roughly done by changing from NH_3 to ND_3, ν decreases by $e^{6(\sqrt{2}-1)}$ or a factor of 11. The inversion frequency is correspondingly sensitive to the potential, and most molecules have such high potentials and heavy masses that the inversion frequencies are less than 1 cycle/sec. Many molecules invert so slowly that they have not succeeded in inverting during the few billion years' life of our planet.

NH_3 in the first excited level of this vibrational mode has a much higher inversion frequency because of the increased value of W. Dennison and Uhlenbeck [33] were able to assume a simple shape for the potential barrier and obtain values for the inversion frequencies in the ground and excited states which agreed roughly with experimental observations.

The exact form of the potential barrier assumed is not critical, since A depends only on an integral of this energy, and not on details of shape. However, Manning [61] found a potential function which has the general shape to be expected for NH_3, and for which the wave equation could be more easily solved and accurate values for the energies obtained. Manning's potential function is

$$\frac{V}{hc} = 66{,}551 \ \text{sech}^4 \frac{x}{2\rho} - 109{,}619 \ \text{sech}^2 \frac{x}{2\rho} \tag{12-9}$$

where V/hc = potential in units of cm^{-1}

x = a coordinate which is dependent on the distance of the nitrogen from the plane of the hydrogens

$\rho = 6.98 \times 10^{-8}/\mu^{\frac{1}{2}}$ where μ is the reduced mass in atomic mass units

This potential is zero for large s (or x), is symmetric about $s = 0$, and has a peak at $s = 0$, where the potential is $-43{,}068 \ cm^{-1}$. It has two minima where the potential is $-45{,}140 \ cm^{-1}$. The constants in (12-9) are also chosen to give close to the correct equilibrium configuration for the molecule, the vibration frequency, and the correct inversion frequency for one level.

If the distance between hydrogens is assumed to stay constant during the motion so that they move together as a rigid triangle, then x is taken as equal to s, the distance between the nitrogen and the plane of this triangle. If m is the mass of the hydrogens and M that of nitrogen, then the reduced mass for this case is simply $\mu = 3mM/(3m + M)$. However, probably a better approximation to the motion is to assume the N—H distance remains constant, and only the H—N—H angles change with vibration [110]. In this case x is taken as the distance along an arc moved by the hydrogens from the median plane, i.e., a plane through the nitrogen and perpendicular to the molecular axis. If the angle between this plane and the N—H bond is taken as α, the proper reduced mass can be shown to be $\mu = 3m(M + 3m \sin^2 \alpha)/(3m + M)$. This reduced mass varies only a small amount with α, and its value may be assumed to be that at the equilibrium angle of $\alpha_0 = 21°49'$. Different reduced masses in Manning's potential will give different equilibrium heights for the NH_3 pyramid. However, these heights do not deviate significantly from the observed value.

The values of inversion frequency for various isotopic species of ammonia calculated from Manning's potential are shown in Table 12-1

TABLE 12-1. INVERSION FREQUENCIES OF AMMONIA IN MEGACYCLES

The "first excited vibrational state" ($v_2 = 1$) corresponds to approximately 950 cm^{-1} excitation of the vibration against the plane of the hydrogens. The calculated values are given by Manning [61] and Newton and Thomas [306] except for the calculation of $N^{14}D_3$ with the Manning potential, which was done by A. Javan and J. Lotspeich.

	Ground state			First excited vibrational state	
	$N^{14}H_3$	$N^{14}D_3$	$N^{15}H_3$	$N^{14}H_3$	$N^{14}D_3$
Observed value, Mc.............	23,786	1600	22,705	1,095,000	117,000
Calculated from Manning potential.........................	25,000	1250	......	780,000	83,000
Calculated from Newton-Thomas potential....................	23,800	....	22,700	690,000	

and compared with experimental results. The potential constants have been chosen to fit some of the data, including the energy of the first vibrational level. The excited vibrational levels are fitted rather satisfactorily by this potential [61]. Although the inversion frequencies are given approximately correctly, it does not seem possible to obtain a really satisfactory fit of all the data with a potential of Manning's type. This will be seen again when fine structure of the inversion spectrum is discussed.

Newton and Thomas [306] used a potential of the form

$$\frac{V}{hc} = \left\{ \left[\frac{(0.377)^2 - s^2}{0.536 + s^2} \right]^2 - 1 \right\} \times 3.17 \times 10^4 \, cm^{-1} \qquad (12\text{-}10)$$

where s is measured in angstrom units. Here the hydrogens are considered to move as a rigid triangle. Values for some of the inversion frequencies calculated by an approximate method [306] from this potential are given in Table 12-1. Form (12-10) for the potential appears to give results which are comparable in accuracy with those from the potential (12-9).

12-3. Inversion of Other Symmetric Hydrides. A very simple form of potential has been found by Costain and Sutherland [678] to give fairly accurate values for the inversion frequency of NH_3, and has been used to estimate the inversion frequencies of PH_3 and AsH_3. This is a potential of the form

$$V = \tfrac{3}{2}k_r(\Delta r)^2 + \tfrac{3}{2}k_\delta(\Delta \delta)^2 \tag{12-11}$$

where Δr = change in the N—H bond length
$\Delta \delta$ = change in the H—N—H bond angle
The force constants k_r and k_δ, as well as the ratio of Δr to $\Delta \delta$, can be obtained by a normal coordinate treatment of the observed vibrational frequencies of NH_3. Their evaluation gives

$$V = 3.89 \times 10^4(\Delta \delta)^2 \qquad cm^{-1} \tag{12-12}$$

where δ is measured in radians. This gives a potential hill of 2077 cm^{-1} between the two minima in good agreement with Manning's value of 2072 cm^{-1}. In fact, the entire potential curve given by (12-12) is very close to that given by (12-9).

A similar evaluation of the potential hill from vibrational constants and molecular geometry for PH_3 and AsH_3 gives the following results [678].

For PH_3:

$V = 5.3 \times 10^4(\Delta \delta)^2$ cm^{-1} (12-13)
Height of potential hill above minimum = 6085 cm^{-1}
Inversion frequencies:
    Ground state = 0.14 Mc
    First excited state = 7.2 Mc

For AsH_3:

$V = 4.56 \times 10^4(\Delta \delta)^2$ cm^{-1} (12-14)
$\delta_0 = 0.585$ radian
Height of potential hill above minimum = 11,220 cm^{-1}
Inversion frequencies:
    Ground state = 1/2 cycle/year
    First excited state = 1 cycle/day

Thus the inversion splitting of PH_3 is probably large enough to give a splitting in microwave measurements of the rotational spectrum, whereas AsH_3 would take two years to go through a cycle of inversion and should hence have no observable splitting. These numbers illustrate the very rapid variation in frequency as the potential barrier height is changed.

12-4. Fine Structure of the Ammonia Inversion Spectrum Rotation-Vibration Interactions. Discussion of the inversion spectrum has so far neglected the rotational motion of the molecule. There is, of course, an interaction between rotation and vibration which in most molecules shows up as a series of closely spaced lines in the rotational spectrum, each being due to different vibrational states. In an inversion spectrum, a type of vibration, there is similarly a series of lines at different inversion frequencies, each corresponding to a different rotational state.

A good qualitative view of the effect of rotation on the inversion spectrum of NH_3 (or any other molecule) may be obtained from a classical discussion. The rotational frequencies of NH_3 are considerably higher than the inversion frequencies of the ground state, so that no peculiar resonant interactions occur, but rather the centrifugal forces due to rotation simply change the effective potential in which the molecule vibrates. Consider first a rotation of the molecule about the symmetry axis. The resulting centrifugal force tends to distort the molecule by flattening it or increasing the H—N—H angle. Hence as a result of the centrifugal force the hydrogens can somewhat more readily move past the nitrogen and the inversion frequency is increased. The centrifugal force is proportional to the square of the angular momentum about this axis, or to K^2, so the inversion frequency can be expected to be increased roughly by an amount bK^2, where b is some positive constant. Now consider a rotation about an axis perpendicular to the molecular axis. In this case the centrifugal force tends to elongate the NH_3 pyramid, or decrease the H—N—H angle. It is then more difficult for the hydrogens and nitrogen to move into the same plane, and the inversion frequency is decreased. Since the square of the angular momentum perpendicular to the symmetry axis is proportional to $J(J + 1) - K^2$, where J is the total angular momentum quantum number, this type of motion can be expected to decrease the inversion frequency roughly by an amount $a[J(J + 1) - K^2]$, where a is a positive constant. Thus the frequency should be of the form

$$\nu = \nu_0 - a[J(J + 1) - K^2] + bK^2 + \text{higher powers of } J \text{ and } K \quad (12\text{-}15)$$

A quantitative calculation of the effects of rotation can be made by considering in detail how the centrifugal forces act on the vibrational and inversion motion. They will affect the inversion frequency through the quantity A in expression (12-1). Allowing for rotation, A may be written [110]

$$A = \exp\left\{\frac{2\pi}{h} \int_0^{s_0 + \delta s} [2\mu(V + \delta V - W - \delta W)]^{\frac{1}{2}} \, ds\right\} \quad (12\text{-}16)$$

where δV and δE represent the respective changes in the potential and vibrational energy as a result of rotation. δs is the change due to value of s where the kinetic energy is zero. The change in effective potential V due to centrifugal forces equals simply the change in kinetic energy of rotation as the molecule vibrates, or

$$\delta V = \hbar^2[J(J+1) - K^2]\left(\frac{1}{2I_A} - \frac{1}{2I_A^0}\right) + \hbar^2 K^2\left(\frac{1}{2I_C} - \frac{1}{2I_C^0}\right) \quad (12\text{-}17)$$

where I_A and I_C are the moments of inertia of the molecule as a function of the vibrational coordinate, while I_A^0 and I_C^0 are the values for the molecule in the equilibrium condition. The change δW contributes only a small amount to A, and δs is quite negligible; so both these quantities will be neglected here. If (12-16) is expanded, assuming δV is small, and the new value of A is inserted into (12-1), the change in inversion frequency ν is

$$\delta\nu = -\frac{2\nu}{\hbar}\int_0^{s_0} \mu\, \delta V[2\mu(V-W)]^{-\frac{1}{2}}\, ds \quad (12\text{-}18)$$

Expression (12-18) must be numerically integrated, after inserting δV from (12-17) and allowing for the dependence of I_A and I_C on the parameter s. In addition, various possible paths for the motion of the hydrogens with respect to the nitrogen may be assumed. Sheng, Barker, and Dennison [110] choose a path intermediate between that obtained by considering the hydrogens as a rigid triangle and the path for the N—H

TABLE 12-2. VALUES OF THE FINE-STRUCTURE COEFFICIENTS FOR AMMONIA
$\nu = \nu_0 - a[J(J+1) - K^2] + bK^2 +$ higher powers in J and K

		Isotope				
		$N^{14}H_3$		$N^{15}H_3$	$N^{14}D_3$	
		Ground state	Excited state	Ground state	Ground state	Excited state
Experimental value, Mc	a	151.5	4860	141.9	7.16	
	b	59.9	1800	55.8	2.88	
Calculated value	a	102	4680	....	3.61	267
Manning potential, Mc	b	55.5	1980	....	2.60	187
Calculated value	a	180	....	186		
Newton-Thomas potential, Mc	b	27	....	6		

bond fixed in length during the inversion. The expression resulting from (12-18) is of the form (12-15) expected qualitatively. Calculated values of the coefficients a and b in (12-15) are compared in Table 12-2 with the experimentally determined constants.

Newton and Thomas [306] have calculated the fine structure of the inversion spectrum of $N^{14}H_3$ and $N^{15}H_3$ by introducing the centrifugal terms somewhat earlier in the calculation. These terms can be conveniently combined with their form of the potential (12-10) so that rotational effects are included in the calculation of inversion frequencies *ab initio*. Their values for the fine-structure constants are also given in Table 12-2.

Although the agreement between theory and experiment in Tables 12-1 and 12-2 is reasonably good, it is clear that none of the potentials assumed is capable of giving the behavior of the inversion spectrum for all isotopes and vibrational states to an accuracy better than about 10 per cent. Hadley and Dennison [157] have tried some modifications in the potentials assumed here, but with no marked improvement in the results. It appears that probably a treatment which is not restricted to one dimension as are all those so far used will be necessary in order to obtain much improvement in the theoretical results. The relative success of a one-dimensional treatment comes from the fact that there is one normal mode of vibration which is primarily involved in its inversion. However, other normal modes of vibration are of some importance in affecting inversion.

About 65 lines of the NH_3 inversion spectrum which have so far been measured are listed in Table 12-3 with their approximate intensities at room temperature. In computing relative intensities, it has been assumed that expression (13-61) is correct for the half widths of the lines. The extensive spectrum is often useful in calibrating wavemeters, checking the performance of spectrometers, and as standard frequency references in measuring accurately other spectra. Use of the NH_3 spectrum in testing has the disadvantage, however, that NH_3 tends to persist in the spectrometer absorption cell sometimes long after it is wanted.

The expression (12-15) is too much simplified to fit accurately all the lines of Table 12-3. A more complete expansion of the type indicated by (12-15), but including higher powers of $J(J + 1)$ and K^2 can be made to fit the experimental frequencies reasonably well. A number of expansions of this general type have been given [141], [155], [172], [236], [317], [410]. One example [317] containing five terms in J and K is

$$\nu = 23,787 - 151.3J(J + 1) + 211.0K^2 + 0.5503J^2(J + 1)^2$$
$$- 1.531J(J + 1)K^2 + 1.055K^4 \quad (12\text{-}19)$$

This fits the low J and K values best and gives deviations for some of the larger values of 25 to 50 Mc. The exponential dependence of (12-1) on the potential suggested to Costain that an exponential of various powers of $J(J + 1)$ and K^2 might fit the frequencies [564]. His expansion

$$\nu = 23,785.88 \exp\left[-6.36996 \times 10^{-3}J(J + 1) + 8.88986 \times 10^{-3}K^2\right.$$
$$+ 8.6922 \times 10^{-7}J^2(J + 1)^2 - 1.7845 \times 10^{-6}J(J + 1)K^2$$
$$\left.+ 5.3075 \times 10^{-7}K^4\right] \quad \text{Mc} \quad (12\text{-}20)$$

fits the lines in Table 12-3 with a mean deviation of 1.3 Mc and should hence predict other unmeasured lines with some accuracy.

There is a certain group of lines, those for which $K = 3$, which shows

TABLE 12-3. OBSERVED $N^{14}H_3$ INVERSION LINES

Rotational state		Frequency, Mc	Intensity, cm^{-1}
J	K		
9	5	16,798.3	8.7×10^{-6}
7	1	16,841.3	3.5×10^{-6}
7	2	17,291.6	1.0×10^{-5}
8	4	17,378.1	1.5×10^{-6}
7	3	18,017.6	4.3×10^{-6}
12	9	18,127.2	4.7×10^{-6}
11	8	18,162.6	4.8×10^{-6}
13	10	18,178.0	1.1×10^{-6}
10	7	18,285.6	9.4×10^{-6}
14	11	18,313.9	4.5×10^{-7}
6	1	18,391.6	4.2×10^{-6}
9	6	18,499.5	3.4×10^{-5}
15	12	18,535.1	3.6×10^{-7}
8	5	18,808.7	2.8×10^{-5}
16	13	18,842.9	6.6×10^{-8}
6	2	18,884.9	2.6×10^{-5}
7	4	19,218.52	4.0×10^{-5}
6	3	19,757.56	1.1×10^{-4}
5	1	19,838.4	1.8×10^{-5}
5	2	20,371.48	5.6×10^{-5}
8	6	20,719.20	1.0×10^{-4}
9	7	20,735.46	3.3×10^{-5}
7	5	20,804.80	7.4×10^{-5}
10	8	20,852.51	1.9×10^{-5}
6	4	20,994.62	9.9×10^{-5}
11	9	21,070.73	2.0×10^{-5}
4	1	21,134.37	4.0×10^{-5}
5	3	21,285.30	2.3×10^{-4}
12	10	21,391.55	5.2×10^{-6}
4	2	21,703.34	1.1×10^{-4}
13	11	21,818.1	6.0×10^{-7}
3	1	22,234.51	6.9×10^{-5}
14	12	22,355	2.2×10^{-6}
5	4	22,653.00	2.2×10^{-4}
4	3	22,688.24	4.4×10^{-4}
6	5	22,732.45	1.7×10^{-4}
3	2	22,834.10	2.0×10^{-4}
7	6	22,924.91	2.9×10^{-4}
15	13	23,004	4.8×10^{-7}
2	1	23,098.78	1.1×10^{-4}

TABLE 12-3. OBSERVED $N^{14}H_3$ INVERSION LINES (*Continued*)

Rotational state		Frequency, Mc	Intensity, cm⁻¹
J	K		
8	7	23,232.20	9.9×10^{-5}
9	8	23,657.46	6.5×10^{-5}
1	1	23,694.48	1.9×10^{-4}
2	2	23,722.61	3.2×10^{-4}
16	14	23,777.4	1.9×10^{-7}
3	3	23,870.11	7.9×10^{-4}
4	4	24,139.39	4.3×10^{-4}
10	9	24,205.25	7.8×10^{-5}
5	5	24,532.94	4.0×10^{-4}
17	15	24,680.1	1.1×10^{-7}
11	10	24,881.90	2.2×10^{-5}
6	6	25,056.04	6.9×10^{-4}
12	11	25,695.23	1.3×10^{-5}
7	7	25,715.14	2.7×10^{-4}
8	8	26,518.91	2.0×10^{-4}
13	12	26,655.00	1.3×10^{-5}
9	9	27,478.00	2.8×10^{-4}
14	13	27,772.52	3.0×10^{-6}
10	10	28,604.73	9.0×10^{-5}
15	14	29,061.14	1.4×10^{-6}
11	11	29,914.66	5.5×10^{-5}
12	12	31,424.97	6.2×10^{-5}
13	13	33,156.95	1.7×10^{-5}
14	14	35,134.44	8.7×10^{-6}
15	15	37,385.18	8.3×10^{-6}
16	16	39,941.54	1.9×10^{-6}

peculiar and systematic deviations from expressions (12-19) or (12-20) [198], [236]. The magnitude of this deviation increases rapidly with J and is alternately positive or negative according to whether J is odd or even, as may be seen from Table 12-4. The deviations for high J values are very much greater than the average error in expression (12-20) of 1.3 Mc, and have consequently not been averaged into this error.

H. H. Nielsen and Dennison [223] have shown that the deviations of lines for which $K = 3$ are due to still another rotation-vibration interaction of high order and hence small magnitude. It may be thought of as a splitting of the K degeneracy, *i.e.*, of the two levels corresponding to $K = \pm 3$. As explained in Chap. 3, the statistics and spin of the hydrogen nuclei make only one of these levels possible in NH_3, so one displaced line is seen rather than two separated lines. From the discussion in Chap. 3, it may be remembered that the type of wave function that is allowed by the nuclear properties depends on the oddness or evenness of J and on the inversion level. Hence in the lowest inversion level for odd J the higher-frequency K doublet occurs, whereas for even J only the lower frequency doublet appears. In the upper inversion level the situation is reversed. A transition, therefore, gives one line shifted to higher frequency for even J by an amount equal to the K doubling, and shifted to a lower frequency for odd J.

Symmetry considerations can show quite generally that no K-type doubling or further splitting due to rotation-vibration interactions can

TABLE 12-4. DEVIATIONS OF AMMONIA LINES WITH $K = 3$ FROM NORMAL
INVERSION FREQUENCIES AS GIVEN BY FORMULA (12-20)
Calculated shift for NH_3 is according to expression (12-21).

J	NH_3		ND_3 Calculated shifts [223], Mc
	Calculated shift, Mc	Measured shift [564], Mc	
3	− 0.25	− 0.21	±0.03
4	1.76	1.76	±0.24
5	− 7.06	− 7.03	±0.95
6	21.18	21.18	±2.85
7	−52.9	−52.39	±7.14

occur for levels where K is not a multiple of 3 (cf. [66]). In addition, although levels with $K = 6$ or 9 can, in principle, be split, the effect of vibration-rotation interaction in splitting these levels is very much smaller than their effect on the levels with $K = 3$. Even when $K = 3$, the disturbance of the levels by vibration-rotation interaction is a very small perturbation; Nielsen and Dennison [223] have shown that it is proportional to the fourth power of the ratio of rotational to vibrational energy: They obtain a value for the splitting of the two degenerate levels and hence for the shift in frequency of the observed lines

$$\Delta\nu = 3.50 \times 10^{-4}J(J + 1)[J(J + 1) - 2][J(J + 1) - 6] \quad \text{Mc} \quad (12\text{-}21)$$

The constant in (12-21) can be evaluated to about 10 per cent accuracy from known rotational and vibrational constants of the NH_3 molecule, but its precise value is picked to fit the data of Table 12-4 [564]. This table shows that expression (12-21) agrees quite well with the observed deviations of lines with $K = 3$.

In the case of ND_3, the nuclear spin of D is unity and both the K doublets are allowed. These have not yet been observed, but the calculated deviation of each member from normal frequency is given in Table 12-4.

12-5. Asymmetric Forms of Ammonia.   In addition to the symmetric forms of ammonia NH_3 and ND_3, the asymmetric rotors NH_2D and NHD_2 occur and have inversion spectra.   However, in these cases the inversion spectrum tends to be mixed up with rotational energies. Examination of the transitions allowed for NH_3 in Fig. 3-9 shows that rotational transitions occur only at the same time as an inversion (or vibrational) transition.   In addition, an inversion transition occurs only with a rotational transition, since upper and lower inversion levels have different rotational wave functions.   In the case of NH_3, the molecular symmetry gives the two different rotational wave functions the same rotation energy.   However, it can be shown to be always true that only transitions with a change of both rotational and inversion states occur, and in the cases of NH_2D and NHD_2 the rotational transition contributes to the frequency of the observed lines because of their asymmetry [662].

If the energy difference due to the rotational transition $J'_{\tau'} \leftarrow J_\tau$ is small, then a pair of lines is observed of frequency $\nu_i \pm \nu_{J_\tau J'_{\tau'}}$, where ν_i is the inversion frequency and $\nu_{J_\tau J'_{\tau'}}$ the rotational frequency.   If the inversion frequency is much smaller than that due to the rotational frequency, the observed spectrum has a pair of lines of frequencies $\nu_{J_\tau J'_{\tau'}} \pm \nu_i$.   In the extreme and very common case of an unobservably small inversion frequency, this last doublet becomes the single observed rotational frequency $\nu_{J_\tau J'_{\tau'}}$.   A number of lines of both the above types have been found in partially deuterated ammonia, and from these rotational constants and inversion frequencies have been obtained [662].   The rotational constants are consistent with the structural parameters of NH_3 usually assumed [130] after some allowance is made for centrifugal distortion.

The inversion frequencies of the asymmetric forms of ammonia cannot be treated as easily as those for NH_3 and ND_3.   In the vibration which inverts the asymmetric forms, the N does not move strictly perpendicularly to the plane of the hydrogens.   However, an approximate calculation of the inversion frequencies of these molecules, assuming the same motion and potential for the inversion as in the symmetric cases, gives surprisingly good agreement with the observed frequencies as shown in Table 12-5.   The inversion frequencies for the ground states are in very close agreement, although the excited state frequencies show a sizable discrepancy.

It is also more difficult to calculate the fine structure, or rotation-vibration interaction, in the asymmetric ammonias because both the vibrational and the rotational motions are rather complex.   Weiss and Strandberg [662] have obtained a reasonably good empirical approximation to the fine structure, however, by assuming it to be of the form

$$\nu = \nu_0 - a[J(J + 1) - (P_c^2)_{av}] + b(P_c^2)_{av} \qquad (12\text{-}22)$$

where $(P_c^2)_{av}$ is the average square of the angular momentum in units of $h/2\pi$ parallel to the principal axis of largest moment of inertia. This axis differs by only about $10°$ from the "symmetry" axis perpendicular to the plane of the hydrogens. $(P_c^2)_{av}$ is the analog of the quantity K^2 in expression (12-15) for the symmetric ammonias. The more intense microwave transitions in NH_2D and NHD_2 involve very little change in $(P_c^2)_{av}$ and no change in J, again in analogy with the symmetric cases where $\Delta K = 0$ and $\Delta J = 0$. Hence $(P_c^2)_{av}$ and J can be assumed the

TABLE 12-5. CONSTANTS OF THE INVERSION SPECTRUM OF NH_2D AND NHD_2
(From Weiss and Strandberg [662])

	NH$_2$D		NHD$_2$	
	Observed	Calculated	Observed	Calculated
Inversion frequency of ground state, Mc....................	12,182	12,100	5,111	5,160
Fine-structure constant of a	23.6		8.1	
ground state, Mc b	76.7		26	
Inversion frequency of first excited state, Mc..............	592,000	465,000	295,000	204,000

same for upper and lower states of the transition. Empirical values for a and b of expression (12-22) for NH_2D and NHD_2 are listed in Table 12-5.

The quantity $(P_c^2)_{av}$ is just $(\alpha_{zc}^2)_{av}J(J + 1)$, where α_{zc} is the cosine of the angle between the total angular momentum J and the molecular axis c of largest principal moment of inertia. $(\alpha_{zc}^2)_{av}$ can be evaluated as in Chap. 6 [cf., for example, expression (6-16)].

12-6. Hindered Torsional Motions in Symmetric Rotors. Another common type of hindered motion is the rotation of one part of a molecule with respect to the remainder, which, when hindering is large, becomes a torsional oscillation. An example is the relative rotation of CH_3 and CF_3 about the symmetry axis of the molecule $H_3C\text{—}CF_3$. Hindered torsional motion can also occur in an asymmetric rotor, as in CH_3OH, where the OH bond may rotate with respect to the CH_3 group.

Consider first a symmetric top such as H_3CCF_3 having an "internal" torsional motion of one end of the molecule with respect to the other end. In either one of the two extreme cases where the CH_3 rotates perfectly freely about the molecular axis, or where it interacts strongly with the CF_3 group so that it can scarcely rotate at all with respect to this group, the energy levels are relatively simple. We shall examine these two extreme cases first before proceeding to the more complex intermediate case where tunneling effects are prominent.

If the two parts of the molecule are bound together tightly, the energy may be obtained in a straightforward way as the sum of the rotational energy of a symmetric top and various molecular vibrational energies. We shall consider only the torsional vibration. The reduced moment of inertia for the CH_3 twisting with respect to CF_3 is

$$I_r = \frac{I_1 I_2}{I_1 + I_2} \text{ or } \frac{I_1 I_2}{I}$$

where I_1 is the moment of inertia for CH_3 about the symmetry axis, I_2 is that for CF_3, and $I = I_1 + I_2$. The potential energy has three minima corresponding to the three equivalent positions 120° apart of the CH_3 with respect to CF_3, and there is a very high potential hump between each minimum—so high that tunneling from one minimum to another is negligibly small. In one of the minima, the potential energy can then be written $V = \frac{1}{2} k \alpha^2$, where $\alpha = \chi_1 - \chi_2$ is the difference in the angular position χ_1 of CH_3 about the symmetry axis and the angle χ_2 of CF_3. The force, or rather torque, constant is k. The frequency of torsional oscillation is then

$$\omega = \frac{1}{2\pi} \sqrt{\frac{kI}{I_1 I_2}}$$

and the energy of torsional and rotational motions is simply

$$W = hB[J(J + 1) - K^2] + hCK^2 + h\omega(v + \tfrac{1}{2}) \qquad (12\text{-}23)$$

where B and C are the usual rotation constants ($C = h/8\pi^2 I$) and v is the vibrational quantum number for the torsional motion. Each torsional vibrational level is triply degenerate since there are three equivalent positions in which the oscillation may occur.

In the other extreme case, CH_3 and CF_3 can rotate freely about the molecular axis. Let the angular momentum about the axis for CH_3 be $m_1 \hbar$ and that of CF_3 be $m_2 \hbar$ so that the total angular momentum about the symmetry axis is $(m_1 + m_2)\hbar = K\hbar$. The usual quantum-mechanical conditions require that m_1, m_2, and K be integers. The energy of rotation is

$$W = hB[J(J + 1) - K^2] + \frac{(m_1 \hbar)^2}{2I_1} + \frac{(m_2 \hbar)^2}{2I_2}$$

or

$$W = hB[J(J + 1) - K^2] + hCK^2 + \frac{I\hbar^2}{2I_1 I_2}\left(m_1 - \frac{KI_1}{I}\right)^2 \qquad (12\text{-}24)$$

The energies due to deviations from these ideal cases, or in the intermediate case when the barrier is of moderate height, remain to be calculated. Connections between the energy levels for the two extreme cases are shown in Fig. 12-5.   It may be seen from this figure that some of the degenerate levels of the extreme cases split in the intermediate case of moderate potential barrier heights.   The amount of splitting must be obtained from a solution of the wave equation which, of course, gives (12-23) or (12-24) in the two extreme cases.

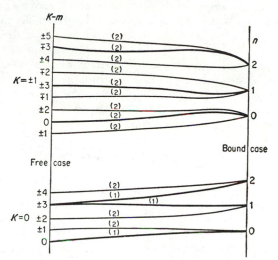

Fig. 12-5. Connections between energy levels with free internal rotation and those for torsional oscillation.   Three potential minima are assumed, as in the molecule CH_3CF_3.   Intermediate barrier heights, corresponding to neither very strong nor negligible interaction between the CH_3 and CF_3 groups, can be seen to split certain levels.   Numbers (1) and (2) indicate the number of levels of the same energy. (*After Koehler and Dennison* [99].)

The kinetic energy due to rotation is

$$T = \tfrac{1}{2}I_x\omega_x^2 + \tfrac{1}{2}I_y\omega_y^2 + \tfrac{1}{2}I_1\dot{\chi}_1^2 + \tfrac{1}{2}I_2\dot{\chi}_2^2 \qquad (12\text{-}25)$$

where I_x and I_y are the principal moments of inertia perpendicular to the axis and ω_x and ω_y, the angular velocities. I_1 and I_2 are the moments of inertia of the two parts of the molecule about which relative rotation occurs, and which is taken as the z direction with respect to the axis. χ_1 and χ_2 are angles specifying the rotation of these two parts about this axis.   If new variables are defined so that

$$\chi = \frac{I_1\chi_1 + I_2\chi_2}{I} \qquad \alpha = \chi_1 - \chi_2 \qquad (12\text{-}26)$$

Then the energy is, from (12-25)

$$W = \tfrac{1}{2}I_x\omega_x^2 + \tfrac{1}{2}I_y\omega_y^2 + \tfrac{1}{2}I\dot{\chi}^2 + \tfrac{1}{2}\frac{I_1I_2}{I}\dot{\alpha}^2 + \tfrac{1}{2}V_0(1 - \cos 3\alpha) \qquad (12\text{-}27a)$$

and the Hamiltonian is

$$H = \frac{1}{2}\frac{P_x^2}{I_x} + \frac{1}{2}\frac{P_y^2}{I_y} + \frac{1}{2}\frac{P_z^2}{I} + \frac{1}{2}\frac{I}{I_1 I_2} p_\alpha^2 + \frac{1}{2}V_0(1 - \cos 3\alpha) \quad (12\text{-}27b)$$

where the symbols involved have been defined above. It may be observed that α gives the angle of torsion within the molecule, and that $I\dot{\chi} = I_1\dot{\chi}_1 + I_2\dot{\chi}_2$ is the total angular momentum about the symmetry axis.

Here the potential energy for torsional motion is assumed to be $V = \frac{1}{2}V_0(1 - \cos 3\alpha)$, corresponding to three potential minima with intervening barriers of height V_0. The shape of the potential barrier will not, of course, be given perfectly by the $\cos 3\alpha$ variation but more generally should be written as a Fourier series

$$V = \sum_p (a_p \cos 3p\alpha + b_p \sin 3p\alpha)$$

However, the energy levels of the molecule will not depend strongly on minor details of the potential. In addition, potential curves which have been calculated on the basis of simple assumptions about the origin of the hindering forces are in fact fitted extremely well by a $\cos 3\alpha$ curve. There is some experimental evidence from the barrier in CH_3NO_2, to be discussed below, that the term proportional to $\cos 6\alpha$ in this Fourier expansion is not more than a few per cent as large as that proportional to $\cos 3\alpha$.

Properties of Quantum States with Hindered Torsional Motion. It may be seen that the first three terms of (12-27) have just the same form as Eq. (3-2), which gives the energy of rotation for a rigid top. In this case, $I_x = I_y$ so that the top is symmetric. Consequently the wave function describing the rotation and torsion of the molecule has the form [36], [99]

$$\psi = \frac{1}{2\pi} e^{iK\chi}e^{iM\phi}\Theta_{JKM}(\theta)\mathfrak{M}(\alpha) \quad (12\text{-}28)$$

where θ and ϕ are the usual Eulerian angles and $e^{iK\chi}e^{iM\phi}\Theta(\theta)$ is identical with expression (3-12a), the wave function for a rigid symmetric top $\mathfrak{M}$ satisfies the equation

$$\frac{I\hbar^2}{2I_1 I_2}\frac{d^2\mathfrak{M}}{d\alpha^2} + \left[W_\alpha - \frac{V_0}{2}(1 - \cos 3\alpha) \right]\mathfrak{M} = 0 \quad (12\text{-}29)$$

The energy is

$$W = W_R + W_\alpha = hB[J(J + 1) - K^2] + hCK^2 + W_\alpha \quad (12\text{-}30)$$

where W_R is the rotational energy of the molecule considered as a rigid symmetric top and W_α the torsional energy. B and C are the usual rotational constants with

$$C = \frac{h}{8\pi^2(I_1 + I_2)}$$

Hence, the only new problem is to solve (12-29), a form which is equivalent to Mathieu's equation, and hence to obtain W_α. Characteristics of the functions $\mathfrak{M}(\alpha)$ have been discussed by Koehler and Dennison [99], whose treatment we shall follow.

Solutions of equation (12-29) are known from Floquet's theorem (see [65] and *Tables Relating to Mathieu Functions*, Columbia University Press, New York, 1951) to be of the type

$$\mathfrak{M}(\alpha) = e^{i\sigma\alpha}F(\alpha) \tag{12-31}$$

where $F(\alpha)$ is periodic with period $2\pi/3$, that is, $F(\alpha)$ may be expanded in the form

$$F(\alpha) = \sum_p a_p e^{3ip\alpha} \tag{12-32}$$

where p is any integer. The constant σ must be real in order to make the wave function finite everywhere. The other conditions imposed on the wave function by the physical situation it must describe are that it must be unchanged when either end of the molecule is rotated any number of complete rotations, that is, when

$$\chi_1 \rightarrow \chi_1 + 2\pi n_1 \quad \text{and} \quad \chi_2 \rightarrow \chi_2 + 2\pi n_2 \tag{12-33}$$

Making these substitutions into (12-28) and (12-31),

$$e^{iK\chi}e^{iM\phi}\Theta(\theta)e^{i\sigma\alpha}F(\alpha) = e^{iK\chi}e^{iM\phi}\Theta(\theta)e^{i\sigma\alpha}F(\alpha)e^{2\pi i\left[\frac{K(n_1I_1+n_2I_2)}{I}+\sigma(n_1-n_2)\right]}$$

so that

$$\frac{K(n_1I_1 + n_2I_2)}{I} + \sigma(n_1 - n_2) = p \tag{12-34}$$

where p is an integer.

Equation (12-34) can hold only if K is an integer (as would be expected since $K\hbar$ is the total angular momentum around the symmetry axis) and if

$$\sigma = s - \frac{KI_1}{I} \tag{12-35a}$$

where s is any integer.

There are only three types of solutions, which may be obtained with values of $s = 0$, 1, or 2; that is, with

$$\sigma = -\frac{KI_1}{I}, \quad 1 - \frac{KI_1}{I}, \quad \text{or} \quad 2 - \frac{KI_1}{I} \tag{12-35b}$$

For, if $\sigma = 3p + s - KI_1/I$, where p is an integer, the exponent $3p$ may be taken as part of $F(\alpha)$ in (12-31), so that the solution is equivalent to that with $\sigma = s - KI_1/I$. (These three types of solutions have in previous treatments often been designated by a quantum number τ, with $\tau = 1, 3$, and 2, corresponding to $s = 0, 1$, and 2, respectively.) The solutions (12-31) are also periodic in K with a period $3I/I_1$, for if $K = 3I/I_1$, $\sigma = s - 3$, which is equivalent to the solution with $\sigma = s$.

Actual solutions $\mathfrak{M}(\alpha)$ must be obtained by using the expanded form (12-32) of $F(\alpha)$, substituting into the Eq. (12-29), and evaluating the coefficients a_p (or from tabulations). This also gives the various allowed values of the "internal" energy W_α associated with the angle α. There are an infinite number of allowed values of W_α, but for reasonably high barriers the lower values are grouped in threes near the energies to be expected from the bound case with vibrational quantum numbers $v = 0$, $1, 2, 3, \ldots$. The calculated values of these energies for a particular case where

$$V_0 = 770 \text{ cm}^{-1} \qquad \frac{Ih}{8\pi^2 I_1 I_2 c} = 24.8 \text{ cm}^{-1} \qquad \frac{I_1}{I} = 0.21$$

are plotted in Fig. 12-6. The lower energy levels correspond essentially to vibrational levels which are slightly split into three levels. The higher levels (*e.g.*, with vibrational quantum number $v = 2$ or 3) are still partly grouped by vibrational quantum numbers but are far enough above the hindering potential to correspond fairly closely to the levels of a free rotor which are plotted in the same figure for comparison. The internal energy of the free rotor is, from (12-24),

$$W_x = \frac{I\hbar^2}{2I_1 I_2} \left(m_1 - \frac{KI_1}{I} \right)^2$$

For a fixed value of m_1 and varying K, this gives a parabola with a minimum energy at $K = Im_1/I_1$. The various parabolas in Fig. 12-6 correspond to different integral values of m_1.

Figure 12-6 also illustrates the rapid decrease in the frequency of tunneling as the energy sinks farther below the top of the potential barrier. Both sets of levels $v = 1$ and $v = 0$ are below the top of the barrier, and the spread in frequency in each set corresponds to the tunneling frequency. For $v = 0$, the tunneling frequency is very low, and it would be decreased further, of course, if the potential barrier were assumed to be higher.

The three different types of levels, with $s = 0, 1$, or 2, are indicated by solid, short-dashed, or long-dashed curves, respectively, as are their counterparts in the free-rotor levels. The effect of a hindering potential is to separate groups of the free-rotor levels at the points where levels of like types cross in Fig. 12-6. Each type of level may be seen to be periodic in K with a period as shown above of $3I/I_1$, or approximately 14.

There are many molecules having hindered torsional motions with more or fewer equilibrium positions than the three which apply to our present example. The ethylene molecule H_2CCH_2 has only two equilibrium positions, while F_3CSF_5 has twelve equilibrium positions of the CF_3 group with respect to the four off-axis fluorines bonded in the SF_5

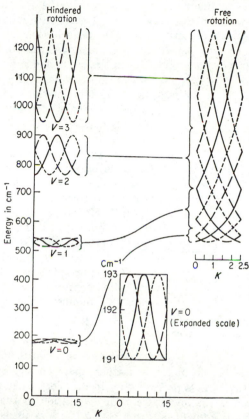

FIG. 12-6. Comparison of energy levels of torsional motion of a molecule of the type H_3CCF_3 with free rotation and with a hindering potential. The hindering potential is assumed to have a height $V_0 = 770$ cm^{-1}, $\dfrac{Ih}{8\pi^2 I_1 I_2 c} = 24.8$ cm^{-1}, and $\dfrac{I_1}{I} = 0.21$. (*Adapted from Koehler and Dennison* [99].)

group (Fig. 12-7). The general structure of the energy levels is similar to that of the case discussed above, but the number of levels is, of course, changed. If there are n equilibrium positions the potential may be written

$$V = \frac{V_0}{2}(1 - \cos n\alpha) \tag{12-36}$$

There are then n levels in each vibrational group, which repeat as K increases with a period of nI/I_1.

A truly symmetric rotor has a dipole moment only along the symmetry axis; hence a microwave field is not able to change the angular momentum around the symmetry axis of any part of the symmetric molecule with hindered rotation discussed here. It can only induce transitions in the total angular momentum J, so that the selection rules are $\Delta J = 0$, ± 1, $\Delta K = \Delta m_1 = \Delta v = 0$. The frequencies observed are therefore just $\nu = 2BJ$, and one might suppose the above discussion of internal energies which do not give spectra has been fruitless. This is not the case, however, because the same energy levels are good approximations to those of the slightly asymmetric rotor with hindered torsional motion, where dipole transitions between these levels do occur. In addition, the effects of the torsional motions on the rotational constant B are often seen.

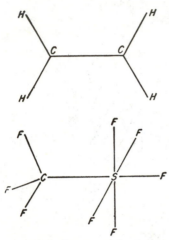

FIG. 12-7. Structure of ethylene with two potential minima and CSF_8 with 12 minima.

12-7. Heights of Hindering Barriers. In the simplest case of strongly hindered torsion, where the levels are essentially those of a harmonic oscillator, their effect on B is expressible in terms of a rotation-vibration constant α and the vibrational quantum number v, that is,

$$B = B_e - \alpha(v + \tfrac{1}{2})$$

One can be sure that the effect of torsional vibrational will be to push the two interfering parts of the molecule farther apart on the average, which will almost always give a positive sign to α. Very frequently also, the torsional motion is the lowest-frequency vibration in the molecule, so that one can identify the lines with $v = 1, 2, \ldots$ as the strongest set of excited vibrational lines in the rotational spectrum with a positive value of α. A measurement of the relative intensities of the ground state rotational line and one or more of the excited states at a given temperature allows a determination of the torsional frequency ω, since the intensities are proportional to the Boltzmann factor $e^{-h\omega/kT}$. A knowledge of the torsional frequency and the approximate moments of inertia involved in turn allows a determination of the barrier height, since approximately

$$\omega = \frac{1}{2\pi} \sqrt{\frac{kI}{I_1 I_2}} \qquad (12\text{-}37)$$

where k is the torque constant at the position of the potential minimum,

If the potential has n minima, and is of the form (12-36), then

$$k = \left(\frac{d^2V}{d\alpha^2}\right)_{\alpha=0} = \frac{n^2V_0}{2}$$

and from (12-37),

$$V_0 = \frac{8\pi^2\omega^2}{n^2} \frac{I_1I_2}{I} \tag{12-38}$$

The barrier heights for a number of molecules determined by measurement of intensity ratios and use of (12-38) are listed in Table 12-6 with values of α for the torsional vibration. Table 12-6 also lists some barrier

TABLE 12-6. BARRIER HEIGHT AND ROTATION-VIBRATION CONSTANT α FOR
HINDERED MOLECULAR VIBRATIONS

All barrier heights listed have been determined by microwave methods except for those in [130].

Molecule	Barrier height, cm^{-1}	Rotation-vibrational constant α, Mc./sec.	Reference
CH_3NO_2	2.10	. . . . .	[986]
CF_3SF_5	220	0.05	[713]
CH_3CCl_3	950	. . . . .	[130]
CH_3SiF_3	410	4.2	[618] [641]
CH_3OH	375	. . . . .	[561] [817]
CH_3SH	400	. . . . .	[978a] [949a]
CH_3SiH_3	558	30	[604]
CH_3SnH_3	. . . .	10.8	[603]
CH_3CF_3	1200	. . . . .	[618]
CH_3CH_3	960	. . . . .	[130]
$(CH_3)_2O$	1000	. . . . .	[130]
CH_3CHF_2	1200	. . . . .	[978]
CH_3NH_2	685	. . . . .	[976]
H_2O_2	113	. . . . .	[960]

heights which have been more precisely determined from spectroscopic observations, and some obtained by thermodynamic techniques. Usually the intensity ratios have not been determined more accurately than ± 10 per cent, so that there are sizable uncertainties in V_0. However, this method is one of the best available for barrier-height determinations, since the torsional frequencies do not often give detectable effects in either vibrational or Raman spectra (cf. [130], p. 496), and a determination from thermodynamic measurements, which has sometimes been used, is often tedious and inaccurate (cf. [130], p. 520).

Of course (12-38) assumes the particular form (12-36) for the potential. However, as pointed out above, calculations of potential shapes with reasonable assumptions about the interactions give curves very close in shape to (12-36), so that actual potential curves are not likely to differ greatly from it. Expression (12-38) also assumes that the excited vibrational states whose relative intensities are measured lie sufficiently close to the bottom of the potential for a harmonic-oscillator approximation to apply. If this is not true, then the intensity ratios of successive excited states will not be constant, nor will α be constant. In one or two cases so far measured, such deviations seem to have been observed.

The separation between the three levels in an individual vibrational state may be sufficient to produce slightly different values of the rotation-vibration constant α. For example, Lide and Coles [604] report that in excited torsional vibrational states the $J = 1 \leftarrow 0$ transition of CH_3SiH_3 splits into two lines representing slightly different values of α, and that the $J = 1 \leftarrow 0$, $K = 1$ transitions split into three components, corresponding to expectations from Fig. 12-6 (see also [948a]).

12-8. Hindered Torsional Motions in Asymmetric Rotors. Consider now the case where one of the two parts of the molecule with hindered torsional motion is not a symmetric rotor. Let the hindered motion involve rotation of the part which is a rigid symmetric rotor about its axis and with respect to a second part of the molecule which comprises a rigid asymmetric rotor. Also let the asymmetric rotor have a plane of symmetry which includes the axis of the symmetric rotor. Such a model is of course not completely general, but it includes a wide variety of interesting cases such as CH_3OH, CH_3NO_2, CH_3NH_2, and CF_3CH_2Cl. Our treatment will follow closely that of Hecht and Dennison [934a], which in turn relies heavily on a discussion by Burkhard and Dennison [561].

The kinetic energy due to rotation of such a molecule is

$$T = \tfrac{1}{2}I_x\omega_x^2 + \tfrac{1}{2}I_y\omega_y^2 + \tfrac{1}{2}I_1\dot{\chi}_1^2 + \tfrac{1}{2}I_2\dot{\chi}_2^2 - D\omega_y\dot{\chi}_1 \qquad (12\text{-}39)$$

where, as in Eq. (12-25), I_x is the principal moment of inertia of the entire molecule about an axis perpendicular to the plane of symmetry and to the axis of hindered motion which is the z direction, and I_y is the moment of inertia about an axis perpendicular to x and z. The corresponding angular velocities are ω_x and ω_y. I_1 and I_2 are the moments of inertia of the two parts of the molecule about the axis or z direction, χ_1 is the angle specifying the rotation of the asymmetric part about the axis and χ_2 is the similar angle for the symmetric part. Since I_1 is not necessarily a principal moment of inertia of the asymmetric rotor, the term $-D\omega_y\dot{\chi}_1$ is needed in (12-39), where D is the product of inertia

$$D = \sum_i m_i y_i z_i \qquad (12\text{-}40)$$

Here m_i is the mass of an atom of the asymmetric rotor and y_i, z_i the coordinates of this mass with respect to the center of gravity of the entire molecule. No products of inertia involving the x axis occur in (12-39), because the yz plane is taken to be the plane of symmetry of the molecule so that these products of inertia are zero.

Suitable rotations of axes and introduction of the variable $\alpha = \chi_1 - \chi_2$ as in (12-26) transform [934a] the Hamiltonian derived from (12-39) to

$$
\left.
\begin{aligned}
H = &\left[\frac{1}{4I_x} + \frac{I_y}{4(I_y^2 + D^2)} \right](P_{x'}^2 + P_{y'}^2) \\
&+ \frac{1}{2}\left[\frac{I_y + I}{I_y I - D^2} - \frac{I_y}{I_y^2 + D^2} \right] P_{z'}^2 \\
&+ \frac{(I_y I - D^2)}{2I_2(I_y I_1 - D^2)}\, p_\alpha^2 + V(\alpha)
\end{aligned}
\right\} \text{(I)}
$$

$$
\left.
\begin{aligned}
&+ \frac{1}{4}\left(\frac{1}{I_x} - \frac{I_y}{I_y^2 + D^2} \right)\left[(P_{x'}^2 - P_{y'}^2)\cos 2\left(\frac{I_2}{I} \right)^{*}\alpha \right.\\
&\qquad\qquad \left. + (P_{x'}P_{y'} + P_{y'}P_{x'})\sin 2\left(\frac{I_2}{I} \right)^{*}\alpha \right] \\
&+ \frac{D}{2(I_y^2 + D^2)}\left[(P_{y'}P_{z'} + P_{z'}P_{y'})\cos\left(\frac{I_2}{I} \right)^{*}\alpha \right.\\
&\qquad\qquad \left. - (P_{x'}P_{z'} + P_{z'}P_{x'})\sin\left(\frac{I_2}{I} \right)^{*}\alpha \right]
\end{aligned}
\right\} \text{(II)}
$$

(12-41)

where $P_{x'}$, $P_{y'}$, and $P_{z'}$ are components of the total angular momentum operator, p_α is the momentum which is canonically conjugate to α or $\chi_1 - \chi_2$, $I = I_1 + I_2$, and $\left(\dfrac{I_2}{I} \right)^{*} = \dfrac{I_2 \sqrt{I_y^2 + D^2}}{I_y I - D^2}$. The potential $V(\alpha)$ is assumed to have the form $V(\alpha) = (V_0/2)(1 - \cos n\alpha)$, where n is an integer.

The part of the Hamiltonian indicated by (I) is identical in form with (12-27b) for the completely symmetric hindered rotor and hence has solutions and energy values of the same type. In solutions of (12-41), however, the following quantities which occur in the symmetric-rotor cases will be replaced by their equivalents:

$$
\frac{I}{I_1 I_2} \text{ is replaced by } \left(\frac{I}{I_1 I_2} \right)^{*} = \frac{I_y I - D^2}{I_2(I_1 I_y - D^2)} \tag{12-42a}
$$

$$
\frac{I_2}{I} \text{ by } \left(\frac{I_2}{I} \right)^{*} = \frac{I_2 \sqrt{I_y^2 + D^2}}{I_y I - D^2} \tag{12-42b}
$$

and

$$
\frac{I_1}{I} \text{ by } \left(\frac{I_1}{I} \right)^{*} = 1 - \frac{I_2 \sqrt{I_y^2 + D^2}}{I_y I - D^2} \tag{12-42c}
$$

Solutions of part (I) are

$$
\psi_{JKMvs} = \frac{1}{2\pi}\, e^{iKx'} e^{iM\phi'} \Theta_{JKM}(\theta') e^{i\sigma\alpha} F_{Kvs}(\alpha) \tag{12-43}
$$

where χ', ϕ', and θ' are Eulerian angles for the system corresponding to the transformed axes x', y', and z'. $F_{Kvs}(\alpha)$ is a function of the form (12-32) appropriate to a particular value of K, a particular torsional-vibrational quantum number v, and a particular value of s. The quantum number s is defined as $\sigma + K(I_1/I)^*$ in analogy with (12-35a). In the common case when the potential has three minima, $s = 0$, 1, or 2 [cf. expression (12-35b)]. The energies are given by

$$H^{J,Kvs}_{JKvs} = hB[J(J+1) - K^2] + hCK^2 + W^{Kvs}_\alpha \qquad (12\text{-}44)$$

where $B = \dfrac{h}{16\pi^2}\left(\dfrac{1}{I_x} + \dfrac{I_y}{I_y^2 + L^2}\right)$

$$C = \dfrac{h}{8\pi^2}\left(\dfrac{I_y + I}{I_y I - D^2} - \dfrac{I_y}{I_y^2 + D^2}\right)$$

W^{Kvs}_α = the torsional or internal energy for the particular state designated by the quantum numbers K, v, and s.

The terms of the Hamiltonian (12-41) designated as (II) may be treated as perturbations of the solutions (12-43) for the symmetric case. Part (II) has no matrix elements diagonal in J, K, v, s, and the off-diagonal elements are

$$H^{J,K\pm1,v,s'}_{JKvs} = \dfrac{h^2 D(2K \pm 1)\,\sqrt{(J \mp K)(J \pm K + 1)}}{16\pi^2(I_y^2 + D^2)}$$
$$\int_0^{2\pi} e^{i(s'-s\mp1)\alpha}F^*_{Kvs}(\alpha)F_{K\pm1,v,s'}(\alpha)\,d\alpha \qquad (12\text{-}45)$$

$$H^{J,K\pm2,v,s'}_{JKvs} = \dfrac{-h^2}{32\pi^2}\left(\dfrac{1}{I_x} - \dfrac{I_y}{I_y^2 + D^2}\right)$$
$$\sqrt{(J \mp K)(J \mp K - 1)(J \pm K + 1)(J \pm K + 2)}$$
$$\int_0^{2\pi} e^{i(s'-s\mp2)\alpha}F^*_{Kvs}(\alpha)F_{K\pm2,v,s'}(\alpha)\,d\alpha \qquad (12\text{-}46)$$

When the potential has three minima, $F_{Kvs}(\alpha)$ from (12-32) is a sum of terms of the type $a_p e^{3ip\alpha}$, where p is an integer. Hence the integral in (12-45) is different from zero only when $s' - s \mp 1 = 3p$, and that in (12-46) is nonzero only when $s' - s \mp 2 = 3p$. Therefore, for $K \rightarrow K \pm 1$, $s' - s = \pm 1$ or ∓ 2, and for $K \rightarrow K \pm 2$, $s' - s = \pm 2$ or ∓ 1 (cf. [561]). These rules may also be expressed as $\Delta s = \Delta K \pm 3p$, where p is an integer.

In principle, energies W of all levels can be found from the secular equation of the determinantal form

$$|H^{JK'v's'}_{\cdot Kvs} - W\delta^{K'v's'}_{Kvs}| = 0 \qquad (12\text{-}47)$$

where the matrix elements $H^{JK'v's'}_{JKvs}$ are given by (12-44), (12-45), and (12-46). The symbol $\delta^{K'v's'}_{Kvs}$ equals unity when $K' = K$, $v' = v$, and $s' = s$, but otherwise it is zero. For such a calculation, however, the functions F_{Kvs} must be obtained and the integrals indicated in (12-45) and (12-46) carried out. This procedure is not practical except in individual cases, or when special conditions make simplifying approximations appropriate. However, before considering such approximations, we shall discuss the general nature of the solutions of the secular equation.

When the hindering potential has three minima

$$[V(\alpha) = \tfrac{1}{2}V_0(1 - \cos 3\alpha)]$$

the levels for which $K = s \pm 3p \pm 1$, where p is an integer, are always doubly degenerate. This degeneracy corresponds in the symmetric-rotor case to the degeneracy of the $\pm K$ states but is not removed by

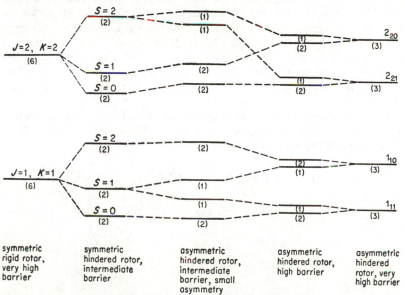

Fig. 12-8. Behavior of energy levels of a hindered rotor with three potential minima and with various asymmetries and barrier heights. Numbers in parentheses under energy levels represent the multiplicity of the level. (*After Ivash and Dennison* [817].)

asymmetry. However, for the states having $K = s \pm 3p$, asymmetry does remove the K degeneracy and splits the two levels by an amount which increases with asymmetry and decreases rapidly with increasing values of K. The general behavior of these levels for various barrier heights and asymmetries is illustrated in Fig. 12-8.

In the limit of a very high barrier, the molecule becomes a rigid asymmetric rotor. The energies W_α^{Kvs} become simply the energies of torsional oscillation in a single potential minimum, and the integrals in the matrix elements (12-45) and (12-46) both equal unity.

Whether or not a barrier is "high" or "low" depends on the ratio of its height V_0 to the kinetic energy associated with torsional momentum, or hence to the ratio

$$V' = \frac{V_0}{\frac{\hbar^2}{2}\left(\frac{I}{I_1 I_2}\right)^*} \tag{12-48}$$

where

$$\left(\frac{I}{I_1 I_2}\right)^* = \frac{I_y I - D^2}{I_2(I_1 I_y - D^2)}$$

For a barrier with $V' > 200$, the molecule may be treated as a rigid rotor for the lowest torsional-vibrational state. If $V' \leq 100$ the lowest torsional-vibrational state will be appreciably split, and the rigid-rotor approximation is not accurate. When $V' \approx 50$, the lowest vibrational state is split by many megacycles and the third vibrational state lies near the top of the barrier. If V' is less than unity, the barrier is low enough for something like free rotation to exist in the lower torsional states.

The High-barrier Case. Good approximations for the energies when $V' \geq 50$ have been worked out by Hecht and Dennison [934a]. They obtain for the torsional or internal energy

$$W_\alpha^{Kv0} = W^v - \tfrac{2}{3}\Delta_v \cos \frac{2\pi K}{3}\left(\frac{I_1}{I}\right)^*$$

$$W_\alpha^{Kv1} = W^v - \tfrac{2}{3}\Delta_v \cos \frac{2\pi}{3}\left[K\left(\frac{I_1}{I}\right)^* - 1\right] \tag{12-49}$$

$$W_\alpha^{Kv2} = W^v - \tfrac{2}{3}\Delta_v \cos \frac{2\pi}{3}\left[K\left(\frac{I_1}{I}\right)^* + 1\right]$$

where

$$W^v = \frac{\hbar^2}{\frac{2(I_1 I_2)^*}{I}}\left\{3\sqrt{V'}\,(v+\tfrac{1}{2}) - \tfrac{9}{8}[(v+\tfrac{1}{2})^2 + \tfrac{1}{4}]\right.$$

$$\left. - \frac{27}{64}\frac{1}{\sqrt{V'}}\,[(v+\tfrac{1}{2})^3 + \tfrac{3}{4}(v+\tfrac{1}{2})] + \cdots \right. \tag{12-50}$$

and, for the ground vibrational state $(v = 0)$, Δ_0 is given to within a few per cent accuracy by

$$\Delta_0 = 7.05(V')^{\frac{3}{4}} \exp\left(-1.379\sqrt{V'}\right) \tag{12-51}$$

A general expression for Δ_v by use of the *WKB* approximation has also been obtained [934a]. The matrix elements (12-45) and (12-46) can be evaluated for the ground vibrational state of the high-barrier case from the following values of the integrals which are involved:

$$\int_0^{2\pi} F_{K0s}(\alpha)e^{i(s'-s\mp2)\alpha}F_{K\pm2,0,s'}(\alpha)\,d\alpha = 1 - \frac{2}{3\sqrt{V'}}\left[1 - \left(\frac{I_1}{I}\right)^*\right]^2$$
$$- \frac{4}{9V'}\left\{\frac{9}{8}\left[1 - \left(\frac{I_1}{I}\right)^*\right]^2 - \frac{1}{2}\left[1 - \left(\frac{I_1}{I}\right)^*\right]^4\right\} + \cdots \quad (12\text{-}52a)$$

$$\int_0^{2\pi} F_{K0s}(\alpha)e^{i(s'-s\mp1)\alpha}F_{K\pm1,0,s'}(\alpha)\,d\alpha = 1 - \frac{1}{6\sqrt{V'}}\left[1 - \left(\frac{I_1}{I}\right)^*\right]^2$$
$$- \frac{1}{9V'}\left\{\frac{9}{8}\left[1 - \left(\frac{I_1}{I}\right)^*\right]^2 - \frac{1}{8}\left[1 - \left(\frac{I_1}{I}\right)^*\right]^4\right\} + \cdots \quad (12\text{-}52b)$$

These integrals are approximately independent of the value of s. Hecht and Dennison [934a] have given explicit expressions for the energies in the high-barrier case for states with $J = 1$ and $J = 2$.

Barriers of Intermediate Height. When the barrier is intermediate or low, that is, when V_0 is not much greater than $h^2I_1I_2/2I$, the formulation and matrix elements discussed above may still be used. However, the approach of Burkhard and Dennison [561] may be advantageous in actual calculation of energy levels. They do not make the transformation of Eq. (12-39) required to eliminate the cross term $D\omega_y\dot{\chi}_1$ but rather divide the Hamiltonian into two parts H_0 and H', where

$$H = H_0 + H' = \frac{(I_x + I_y)I_1 - D^2}{4I_x(I_yI_1 - D^2)}(P_x^2 + P_y^2) + \frac{1}{2(I_1 + I_2)}P_z^2$$
$$+ \frac{I_1 + I_2}{2I_1I_2}p_\alpha^2 + V(\alpha) + H' \quad (12\text{-}53)$$

H_0 in this expression is identical in form with Eq. (12-27b) for the completely symmetric hindered rotor with the two equal moments of inertia replaced by I_B, where

$$I_B = \frac{2I_x(I_yI_1 - D^2)}{(I_x + I_y)I_1 - D^2} \quad (12\text{-}54)$$

Wave functions identical with Eq. (12-28) are therefore appropriate and matrix elements of $H = H_0 + H'$ for these functions can be evaluated. They are

$$H_{JKvs}^{JKvs} = \frac{h^2}{8\pi^2I_B}[J(J+1) - K^2] + \frac{h^2K^2}{8\pi^2(I_1 + I_2)}$$
$$- \frac{h^2D^2}{8\pi^2I_1(I_yI_1 - D^2)}\int_0^{2\pi}e^{-is\alpha}F_{Kvs}^{*(\alpha)}\frac{d^2}{d\alpha^2}\left(e^{is\alpha}F_{Kvs}^{(\alpha)}\right)\,d\alpha \quad (12\text{-}55)$$

$$H_{JKvs}^{JKv's} = -\frac{h^2D^2}{8\pi^2I_1(I_yI_1 - D^2)}\int_0^{2\pi}e^{-is\alpha}F_{Kvs}^{*(\alpha)}\frac{d^2}{d\alpha^2}\left(e^{is\alpha}F_{Kv's}^{(\alpha)}\right)\,d\alpha \quad (12\text{-}56)$$

$$H^{J,K\pm1,v's'}_{JKvs} = \frac{h^2 D \sqrt{(J \mp K)(J \pm K + 1)}}{8\pi^2(I_y I_1 - D^2)} \int_0^{2\pi} \left[\pm \frac{e^{-i(s\pm1)\alpha}}{i} F^{*(\alpha)}_{Kvs} \frac{d}{d\alpha} \right.$$

$$\left. (e^{is'\alpha} F_{K\pm1,v's'}) - \frac{e^{i(s'-s\mp1)\alpha}}{2} F^{*(\alpha)}_{Kvs} F^{(\alpha)}_{K\pm1,v',s'} \right] d\alpha \quad (12\text{-}57)$$

$$H^{J,K\pm2,v',s'}_{JKvs} = h^2 \cdot \frac{(I_x I_1 - I_y I_1 + D^2)}{32\pi^2 I_x(I_y I_1 - D^2)} \sqrt{(J \mp K)(J \mp K - 1)}$$

$$\times \sqrt{(J \pm K + 1)(J \pm K + 2)} \int_0^{2\pi} e^{i(s'-s\mp2)\alpha} F^{*(\alpha)}_{JKvs} F_{J,K\pm2,v',s'} \, d\alpha \quad (12\text{-}58)$$

As for Eqs. (12-45) and (12-46), these matrix elements are nonzero only when $\Delta S = \Delta K \pm 3p$, where p is an integer. Energy levels may be determined from them by solution of a secular equation of the form (12-47).

The previous treatment, which eliminated the cross term $D\omega_y \dot{\chi}_1$ and gave (12-41), is of advantage in avoiding matrix elements between states of different vibrational quantum number v. These are especially troublesome only in the high-barrier case, where their elimination affords considerable simplification.

The Low-barrier Case. When the barrier is very low, that is, when $V_0 \leq \dfrac{\hbar^2}{2(I/I_1 I_2)^*}$, the most appropriate starting point for approximations is free internal rotation. If V_0 is assumed to equal zero, the part of the wave function for the internal motion, which is equivalent to (12-29), becomes

$$\frac{\hbar^2}{2}\left(\frac{I}{I_1 I_2}\right)^* \frac{d^2\mathfrak{M}}{d\alpha^2} + W_\alpha \mathfrak{M} = 0$$

Hence

$$\mathfrak{M}(\alpha) = \frac{1}{\sqrt{2\pi}} e^{\pm i[m_1 - K(I_1/I)^*]\alpha} \quad (12\text{-}59)$$

and

$$W^{m_1}_\alpha = \frac{\hbar^2}{2}\left(\frac{I}{I_1 I_2}\right)^*\left[m_1 - K\left(\frac{I_1}{I}\right)^*\right]^2 \quad (12\text{-}60)$$

To satisfy boundary conditions, as in (12-35a), m_1 must be an integer. $F_{Kvs}(x)$ is just $1/\sqrt{2\pi}\, e^{3ip\alpha}$, where $s + 3p = m_1$. It may be seen from (12-24) that m_1 is just the angular momentum in units of $\hbar$ of the asymmetric part of the molecule about the z' axis. From the similarity of the two parts of the molecule, $m_1 - K(I_1/I)^*$ in (12-59) and (12-60) might have been replaced by $m_2 - K(I_2/I)^*$, where $m_2\hbar$ is the angular momentum of the symmetric part and $m_1 + m_2 = K$. The quantum numbers v and s may be replaced by the number m_1 or m_2. The matrix elements (12-45) and (12-46) then become

$$H_{JKm_1}^{J,K\pm1,m_1\pm1} = H_{JKm_2}^{J,K\pm1,m_2} = \frac{h^2 D(2K \pm 1) \sqrt{(J \mp K)(J \pm K + 1)}}{16\pi^2 (I_y^2 + D^2)} \quad (12\text{-}61)$$

$$H_{JKm_1}^{J,K\pm2,m_1\pm2} = H_{JKm_2}^{J,K\pm2,m_2} = \frac{-h^2}{32\pi^2}\left(\frac{1}{I_x} - \frac{I_y}{I_y^2 + D^2}\right)$$

$$\sqrt{(J \mp K)(J \mp K - 1)(J \pm K + 1)(J \pm K + 2)} \quad (12\text{-}62)$$

If the barrier is low, but not zero, then it may be considered a small perturbation. The largest effects of such a perturbation are to add a constant $V_0/2$ to W_α^m, and to split pairs of levels which are close together and for which m differs by n, where n is the number of potential minima. This splitting is due to the existence of off-diagonal matrix elements of

$$V(\alpha) = \frac{V_0}{2}(1 - \cos n\alpha)$$

which are

$$V_{JKm}^{J,K,m\pm n} = -\frac{V_0}{4} \quad (12\text{-}63)$$

Matrix elements such as (12-63) are most important when there is a near degeneracy of the two levels which they connect.

Other Cases. A number of other useful approximations can be obtained for various special cases. Wilson, Lin, and Lide [1,000a] have discussed the case of a hindered asymmetric rotor with the product of inertia D equal to zero. They use a similar but somewhat different approach from that adopted above and give useful approximations for cases of small asymmetry and of high and low barriers.

Burkhard [782a] has formulated the rather general case of two asymmetric parts of a molecule with a potential hindering their relative motion, and with the sole limitation that the center of mass of one of these parts must lie on the axis of hindered rotation. He has written out the wave equation for this case and the matrix elements needed for a solution. Burkhard has also written down the wave equation and matrix elements for the still more general case of hindered rotation of two asymmetric parts with the center of gravity of neither part on the axis of rotation [908a]. However, this case is too unwieldy to be very useful unless simplifying approximations apply.

12-9. Selection Rules. For hindered rotors, the usual selection rule $\Delta J = 0, \pm 1$ for the total angular momentum applies, as well as

$$\Delta M = 0, \pm 1$$

for the projection of J on a space-fixed axis. One other general rule can be stated for transitions of a hindered rotor of which one part is a symmetric top. It was pointed out in discussion of the matrix elements (12-45) and (12-46) that if the potential has three minima these matrix elements connect only states for which

$$\Delta K = \Delta s \pm 3p$$

where p is an integer. This rule breaks up the states into three types which may be divided as follows:

$$K = s \pm 3p \qquad K = s \pm 3p + 1 \qquad K = s \pm 3p - 1 \qquad (12\text{-}64)$$

For a potential with n minima, the states may similarly be divided into n groups. There are also no electric dipole transitions between the different types of states specified by (12-64), so that one may write the selection rule $\Delta K = \Delta s \pm 3p$, or more generally

$$\Delta K = \Delta s \pm np \qquad (12\text{-}65)$$

where p is an integer and n is the number of potential minima. It should be noted that, if the rotor is asymmetric, it will involve a sum of symmetric-rotor states of various values of K so that K is not a well-defined quantity. However, from (12-64) a given state involves values of $K - s$ which differ only by $\pm 3p$, so that the selection rule (12-65) is meaningful.

For the general asymmetric rotor, all transitions are allowed which satisfy (12-65) and for which $\Delta J = 0$ or ± 1 and $\Delta M = 0$ or ± 1. However, unless the rotor is very asymmetric, the stronger transitions are given by the somewhat more restrictive selection rules of a symmetric rotor, which will now be discussed.

The electric dipole moment of a strictly symmetric rotor will always be parallel to the molecular axis. However, since we are also concerned with nearly symmetric tops, it will be assumed that the dipole moment may have a component perpendicular to the axis as well as one parallel to the axis. Selection rules are different for the two types of dipole moments, and intensities are in each case proportional to the square

TABLE 12-7. SELECTION RULES FOR SYMMETRIC OR NEARLY SYMMETRIC ROTORS WITH HINDERED ROTATION

The rules $\Delta J = 0, \pm 1$ and $\Delta M = 0, \pm 1$ apply to all transitions. $m_1\hbar$ is the angular momentum about the molecular axis of that part of the molecule which may have a dipole moment perpendicular to the axis.

	Very high barrier	Free rotation (zero barrier)	Intermediate barrier
Dipole moment parallel to molecular axis	$\Delta K = 0$ $\Delta s = 0$ $\Delta v = 0$	$\Delta K = 0$ $\Delta m_1 = \Delta m_2 = 0$	$\Delta K = 0$ $\Delta s = 0$ $\Delta v = 0$
Dipole moment perpendicular to molecular axis	$\Delta K = \pm 1$ $\Delta v = 0, \pm 1$	$\Delta K = \pm 1$ $\Delta m_1 = \pm 1$ $\Delta m_2 = 0$	$\Delta K = \pm 1$ $\Delta K = \Delta s \pm np$ $\Delta v = $ anything

of the component of the dipole moment involved. The selection rules are given in Table 12-7. The very-high-barrier case with the dipole moment parallel to the axis corresponds to the normal rigid symmetric rotor. In the case of free rotation with a dipole perpendicular to the axis, the angular momentum about the axis $m_1\hbar$ of only the asymmetric part of the molecule changes, since it alone would allow a perpendicular component of the dipole moment.

12-10. Examples of Hindered Torsional Motion in Asymmetric Rotors. The case of hindered torsional motion which has received by far the most complete and detailed analysis is CH_3OH. This is the molecule to which Koehler and Dennison [99] applied the symmetric-rotor approximation. However, little quantitative progress could be made until the advent of microwave spectroscopy allowed measurement of the CH_3OH spectrum under high resolution. Burkhard and Dennison [561] first worked out the CH_3OH hindered motion in detail, and quantitatively fitted the rather extensive microwave spectrum measured by Hughes, Good, and Coles [590] and by others. From this work the structure of CH_3OH, the components of the dipole moment perpendicular and parallel to the CH_3 axis, and the barrier height listed in Table 12-6 was obtained. A similar but still more complete discussion of CH_3OH has been given by Ivash and Dennison [817].

The CH_3OH lines of primary interest here make up an intense series beginning near 25,000 Mc, extending to about 31,000 Mc, and then returning toward lower frequencies. Thirty members of the series have been found. Stark effects of these lines show that the series corresponds to $\Delta J = 0$ transitions, that for the first line of the series $J = 2$, and that others involve successively higher values of J. The frequencies of the first dozen members of this series (for the common isotopic species of CH_3OH) are fairly accurately given [561] by

$$\nu = 24{,}948.13 - 2.9656J(J+1) + 0.11258J^2(J+1)^2 - 0.4094$$
$$\times 10^{-4}J^3(J+1)^3 - 0.3168 \times 10^{-6}J^4(J+1)^4 \qquad \text{Mc} \qquad (12\text{-}66)$$

Burkhard and Dennison [561] showed that the only explanation of the origin of this series consistent with reasonable parameters for the CH_3OH molecule is a transition of the type $v = 0, \Delta J = 0, K = 2 \leftarrow 1, s = 0 \leftarrow 2$. The increase of rotational energy due to the transition $K = 2 \leftarrow 1$ is approximately 10 cm^{-1}, which is almost canceled by the decrease in internal energy of about 9 cm^{-1} due to the transition $s = 0 \leftarrow 2$. The difference between these two quantities depends to some extent on J, which allows the series of lines for different values of J. The lowest value of J for such a series equals, of course, the maximum value of K, which is 2. This is in agreement with the results of Stark-effect measurements.

The value of $V' = \dfrac{V_0}{\hbar^2 I/2I_1 I_2}$ in CH_3OH is approximately 13, so that the barrier is of intermediate height, and neither the high-barrier nor low-barrier approximations discussed above are accurate. Burkhard and Dennison [561] have, however, worked out approximate expressions for the torsional energy.

An interesting example of a low barrier is afforded by CH_3NO_2, which has been studied by Tannenbaum, Johnson, Myers, and Gwinn [986]. CH_3NO_2 has a potential with 6 minima and a height of 2.1 cm^{-1}. The quantity $\hbar^2 I/2I_1 I_2$ is approximately 5.8 cm^{-1}, so that $V' = 0.36$, and the low-barrier approximation is appropriate. Table 12-8 gives observed components of the $J = 2 \leftarrow 1$ transition of this molecule and calculated frequencies assuming either a zero barrier or a barrier of height $V_0 = 2.10$

TABLE 12-8. STRUCTURE OF THE $J = 2 \leftarrow 1$ TRANSITION OF CH_3NO_2, A MOLECULE WITH HINDERED TORSION AND A LOW BARRIER
Frequencies are calculated assuming $B + C = 16{,}419.3 - 0.32m_2{}^2$ Mc,
$$B - C = 4{,}666.0 \text{ Mc},$$
$h/8\pi^2 I_1 = 13{,}277.5$ Mc, and $h/8\pi^2 I_2 = 160{,}000$ Mc. (From Tannenbaum, Johnson, Myers, and Gwinn [986].)

K	m_1	m_2	Calculated frequencies, Mc		Measured frequencies, Mc
			$V_0 = 0$	$V_0 = 2.10 \text{ cm}^{-1}$	
0	0	0	30,010.7	30,011.5	30,035.6
± 1	± 2	∓ 1	32,033.4	32,034.1	32,034.1
± 1	0	± 1	33,642.5	33,643.5	33,643.5
0	∓ 2	± 2	32,959.8	32,959.8	32,959.2
± 1	∓ 2	± 3	33,174.4	33,474.6	33,476.5
∓ 1	± 2	∓ 3	33,174.4	31,676.2	31,677.3
± 1	± 4	∓ 3	32,491.6	32,191.4	32,189.7
∓ 1	∓ 4	± 3	32,491.6	33,989.8	33,988.5
0	∓ 4	± 4	32,856.6	32,856.6	32,859.5

cm^{-1}. Since NO_2 has no dipole moment perpendicular to the molecular axis, the torsional energy does not appear very directly in the transition frequencies, and hence the free-rotation approximation is fairly accurate. However, the potential does strongly affect the energy levels with $m_2 = \pm 3 = \pm n/2$. These levels are split by an interaction of the type indicated by (12-63).

In CH_3NO_2, the zero spin and Bose-Einstein statistics of O^{16} require that only levels with even m_1 are permitted. For, if two oxygens in the NO_2 group are interchanged, this is equivalent to a rotation about the axis of $180°$, so that the wave function is changed by $e^{im_1\pi}$. Since the O^{16} spins are zero and only a symmetric-spin wave function can be formed, the spatial partial of the wave function must also be symmetric. This requires that $e^{im_1\pi} = 1$, or that m_1 is an even integer.

The magnitude of V_0 gives some interesting qualitative information about the nature of barriers in molecules with three potential minima. Consider the simplified molecule CH_3NO, which would have three potential minima, with a potential of the general form

$$V = \sum_p V_p \cos 3p\alpha \qquad (12\text{-}67)$$

Presumably the height of the barrier would be approximately the same as found in most molecules of this geometry, that is, V_1 equals a few hundred wave numbers. If a second oxygen 180° away from the first is added to the molecule to produce CH_3NO_2, the potential is

$$V = \sum_p V_p[\cos 3p\alpha + \cos 3p(\alpha + \pi)]$$
$$= \sum_p 2V_{2p} \cos 6p\alpha \qquad (12\text{-}68)$$

Since experimentally $2V_2 = 2.1$ cm^{-1}, we have a good indication that V_2/V_1 is only a few parts in 100; therefore, higher terms in a series such as (12-67) are probably not important. In view of this it is very surprising, however, that the 12-minima potential of CF_3SF_5 is as large as is given in Table 12-6.

CHAPTER 13

SHAPES AND WIDTHS OF SPECTRAL LINES

A truly isolated, undisturbed, and stationary molecular system would have the attractive feature of definite and fixed energy levels, but various types of unavoidable disturbances do in fact vary the energy levels, giving a width to spectral lines and varying their average or center frequencies. The sources of spectral line broadening which need to be considered are:

1. Natural line breadth
2. Doppler effect
3. Pressure broadening, *i.e.*, disturbances due to interactions between molecules
4. Saturation broadening
5. Collisions between molecules and the walls of a containing vessel.

13-1. Natural Line Breadth. The natural line breadth may be interpreted classically as due to radiation damping, or quantum-mechanically as a disturbance of the molecule by zero-point vibration of electromagnetic fields which are always present in free space. For a transition of frequency ν from an excited state to the ground state of the system, zero-point electromagnetic fields give an absorption line a half width at half maximum intensity of

$$\Delta\nu = \frac{32\pi^3\nu^3}{3hc^3}\,|\mu|^2 \qquad \text{cycles/sec} \qquad (13\text{-}1)$$

where μ is the quantum-mechanical matrix element of the dipole moment —usually of the order of 1 debye unit, or 10^{-18} esu. For radiation of 1 cm wavelength $\Delta\nu$ is, from Eq. (13-1), approximately 10^{-7} cycle/sec. For radio frequencies and ordinary temperatures, thermal radiation consists of stronger electromagnetic fields than the usual zero-point fields, since each mode of vibration of the field has a mean energy kT rather than $\frac{1}{2}h\nu$. This increases the value of $\Delta\nu$ by a factor $2kT/h\nu$, or approximately 400 for room temperature, giving a "natural" width of 4×10^{-5} cycle/sec. This width is quite negligible in comparison with that caused by other types of broadening.

The natural line breadth is often regarded as an unchangeable effect of disturbance of the system by electromagnetic fields, which are uni-

formly present in all space.   However, in the radio-frequency range the
zero-point electromagnetic fields need not be uniform because cavities or
circuits may be as small as one wavelength.   Thus in a cavity with
perfectly reflecting walls only certain resonant frequencies can occur, and
a particular frequency necessary to cause a particular microwave transition may not occur.   In this case no "spontaneous" emission of this
frequency can occur, and the natural line breadth is zero.   Similarly
natural line breadths may be increased by the presence of resonant circuits
which increase the local strength of zero-point electromagnetic vibrations.

Radiation broadening is of importance for microwave spectroscopy
when transitions between levels of excited electronic states are observed.
Then the line width is large because it is proportional to ν_0^3, where ν_0 is
now the frequency of the transition to the ground electronic state.   For
instance, the $2p\ ^2P_{\frac{3}{2}}$ state of hydrogen, discussed in connection with the
Lamb-Retherford experiment in Chap. 5, has a natural half width of
50 Mc.

13-2. Doppler Effect.   The Doppler effect occurs when a molecule is
moving parallel to the direction of propagation of the radiation being
absorbed, and gives a frequency shift of $\pm \nu(v/v_p)$, where ν is the resonant
frequency without Doppler shift, v the molecular velocity, and v_p the
velocity of phase propagation of the radiation.   Although under some
conditions v_p may be larger than c (*e.g.*, for propagation in a waveguide
near cutoff), usually $v_p \approx c$, and the fractional frequency shifts are
simply v/c.   The probability that a molecule in a gas at temperature T
has a velocity v in a particular direction is proportional to $e^{-mv^2/2kT}$, where
m is the molecular mass.   Hence the line intensity as a function of change
ϵ from the resonance frequency is $e^{-\frac{mc^2}{2kT}\left(\frac{\epsilon}{\nu}\right)^2}$.   The line is consequently
symmetric and has a half width at half maximum of

$$\Delta\nu = \frac{\nu}{c}\sqrt{\frac{2kT}{m}\ln 2} = \frac{\nu}{c}\sqrt{2kN_0 \ln 2}\sqrt{\frac{T}{M}} = 3.581 \times 10^{-7}\sqrt{\frac{T}{M}}\,\nu \quad (13\text{-}2)$$

where M is the molecular weight and N_0 is Avogadro's number.   For
an ammonia molecule at room temperature, $\Delta\nu/\nu = 1.5 \times 10^{-6}$.   Doppler effect can be decreased to some extent by use of heavier molecules
and lower temperatures, but a decrease in line width of more than a
factor of 2 can hardly be expected because at low temperatures molecules
have not sufficient vapor pressure ($\sim 10^{-2}$ mm Hg) to absorb radiation.
A great decrease in Doppler width is obtained in some optical spectroscopy experiments by observing an atomic beam at right angles to its
direction of motion [238].   This method is not so easy to apply or so
popular in microwave spectroscopy, but it has been used in two different
types of microwave spectrometers [709], [925], [982].   Newell and Dicke
[622] have also developed a technique of selecting absorption only from

molecules in a certain narrow velocity range and thus decreasing the effect of Doppler broadening on the line width by a factor of 10 or more. Although these techniques sacrifice sensitivity to eliminate Doppler effect and obtain narrow lines, they should be useful for resolving very closely spaced lines in strong microwave spectra.

13-3. Pressure Broadening. The most important source of broadening in many microwave experiments is pressure broadening. It is also the most interesting because it provides information about how molecules behave in intermolecular collisions and hence about molecular force fields. This broadening arises from collisions between molecules.

The spectral distribution of a molecular oscillation of finite lifetime was first considered by Lorentz [2]. For an oscillator whose amplitude decreases exponentially with time ($a = a_0 e^{-t/\tau}$), the radiation distribution corresponds to the well-known resonance-type curve with a half width in frequency of $1/2\pi\tau$. Exactly the same result is obtained for a group of oscillators, each of which oscillates with a constant amplitude but is abruptly stopped after time t, where the number oscillating for time t is given by $n_t = n_0 e^{-t/\tau}$. The theory assumes that after a collision, when the oscillation is stopped, it starts again with a phase having no relationship to the phase before collision; i.e., "strong" collisions are assumed. When applied to the case of rotating molecules, this assumption is equivalent to assuming that the orientation of the molecules after collision is random.

A fairly complete qualitative description of pressure broadening in the microwave and radio-frequency region can be obtained with the simple assumption that collisions are very brief, but so strong that the behavior of the molecule after collision has no particular relationship to that before collision. We shall first explore the consequences of collisions of this type to obtain a general description of pressure broadening, and then return to examine in more detail what happens during a collision and the relation between pressure broadening and intermolecular forces.

Debye ([12], Chap. 5) considered the case of the fixed dipole with no rotation or translation energy. After each collision the dipole is assumed to be not completely random in orientation, but oriented with respect to the electric field present at the moment in accordance with the Boltzmann distribution $\exp(-\mathbf{E} \cdot \mathbf{\mu}/kT)$, where $\mathbf{E}$ is the electric field strength existing at the time, $\mathbf{\mu}$ is the dipole moment, and k and T are the Boltzmann constant and the absolute temperature, respectively. If the field has oscillated many times before the molecule makes another collision, the dipole has no special orientation with respect to the field at the time of this next collision. During the next collision, however, it is again oriented with respect to the field present, again absorbing a small quantity of energy from the field during this orientation process. Such a process is repeated many times and thus absorbs energy although there

is no characteristic resonance peak.   Debye calculated the theoretical
expression for this type of absorption (cf. Van Vleck and Weisskopf
[136] for some discussion and generalization).   In this case, the absorption per unit length is

$$\gamma = \frac{\omega}{c} \frac{4\pi N \mu^2}{3kT} \frac{\omega \tau}{1 + \omega^2 \tau^2} \qquad (13\text{-}3)$$

where ω is the angular frequency ($2\pi\nu$) of the incident radiation, τ is
the mean lifetime between collisions, N is the number of molecules per
cubic centimeter, μ is the dipole moment of a molecule, c is the velocity
of light, k is the Boltzmann constant, and T is the absolute temperature.
    The Van Vleck–Weisskopf Line Shape.   The Debye and Lorentz
theories have been synthesized by Van Vleck and Weisskopf ([136]; see
also H. Fröhlich [154] for another derivation).   Assuming that the molecule undergoes a violent collision the phase of its oscillation after such a
collision will not be greatly dependent on its phase at the start of the
collision.   In this case there must be thermodynamic equilibrium
between the molecule and the existing electric field immediately after
each collision somewhat like the assumed equilibrium distribution in the
orientation of the fixed dipole mentioned above.   Using this assumption
rather than Lorentz's assumption that the phase after a collision is
arbitrary, an expression similar to Lorentz's formula may be obtained
which is consistent with the Debye case.
    Since the rotation of a molecule can always be resolved into two
perpendicular vibrations, it is sufficient to consider a linear vibrator.
To determine the absorption and the dielectric constant associated with a
vibrating charge in a classical way, one needs only to solve the equation
of motion of the oscillator in the field subject to the right boundary
conditions.   The equation is of the type

$$\ddot{x} + \omega_0^2 x = \frac{eE}{m} \cos \omega t \qquad (13\text{-}4)$$

where $\omega_0 = 2\pi$ times the natural molecular frequency
        $\omega = 2\pi$ times the frequency of oscillation of the field E
Before solving Eq. (13-4) we shall show how the absorption and dielectric
constant may be obtained from the solution for x.   The dielectric constant is defined as usual

$$K = \frac{D}{E} = 1 + 4\pi \frac{P}{E} \qquad (13\text{-}5)$$

where P is the polarization per unit volume.   After x has been averaged
over all molecules, it will be of the form

$$\bar{x} = aE \cos \omega t + bE \sin \omega t \qquad (13\text{-}6)$$

The real part of the polarization is $P = naEe \cos \omega t$, so that

$$K = 1 + 4\pi nae \qquad (13\text{-}6a)$$

where n is the number of such oscillators per unit volume.

To obtain the fractional absorption of power per unit distance, consider a cube of unit volume with the radiation traveling through it perpendicular to one of the faces. The radiation absorbed during a time T will be

$$n \int_0^T e\dot{x}E \cos \omega t \, dt$$

The total radiation energy passing into the cube will be $c(E^2/8\pi)T$, where c is the velocity of light, so that the fractional absorption per unit distance or the absorption coefficient will be

$$\frac{neE \int_0^T \dot{x} \cos \omega t \, dt}{c(E^2/8\pi)T}$$

Remembering again the form of $\bar{x}$ and integrating over a long period of time the absorption coefficient is seen to be

$$\gamma = \frac{4\pi neb\omega}{c} \qquad (13\text{-}7)$$

We return to a solution of the equation of motion. For ease of solution we use a complex quantity $e^{i\omega t}$ for $\cos \omega t$ in the equation of motion so that x will be the real part of the solution, which is of the form

$$\frac{eEe^{i\omega t}}{m(\omega_0^2 - \omega^2)} + c_1 e^{i\omega_0 t} + c_2 e^{-i\omega_0 t} \qquad (13\text{-}8)$$

where c_1 and c_2 depend on the initial values of x and $\dot{x}$. The average initial value of x and $\dot{x}$ can be found from the energy

$$II = \frac{m}{2}(\dot{x})^2 + \frac{m}{2}(\omega_0 x)^2 - exE \cos \omega t \qquad (13\text{-}9)$$

thus

$$\bar{x}_0 = \frac{\int xe^{-H/kt} \, dx \, d\dot{x}}{\int e^{-H/kt} \, dx \, d\dot{x}} = \frac{Ee \cos \omega t}{m\omega_0^2} \qquad (13\text{-}10)$$

and similarly

$$\bar{\dot{x}}_0 = \dot{\bar{x}}_0 = 0 \qquad (13\text{-}11)$$

For some time of interest t the constants c_1 and c_2 will be functions of t_1, the time of last collision of the molecule. We must average then over all values of t_1. The distribution of collisions in time can be written according to kinetic theory

$$n(t_1) = \frac{1}{\tau} e^{-(t-t_1)/\tau} \, dt_1 \qquad (13\text{-}12)$$

where τ is the mean time between collisions, where $n(t_1)$ is the probability that a molecule made its last collision between time t_1 and $t_1 + dt_1$. Our expression for x must be averaged over this distribution. After performing the average one can find the in-phase and the quadrature terms of the real part of $\bar{x}$ to be

$$a = \frac{e}{m(\omega_0^2 - \omega^2)}$$
$$\left\{ 1 - \frac{\omega}{2\omega_0^2\tau^2} \left[\frac{\omega_0 + \omega}{(1/\tau)^2 + (\omega_0 - \omega)^2} + \frac{\omega - \omega_0}{(1/\tau)^2 + (\omega_0 + \omega)^2} \right] \right\} \quad (13\text{-}13)$$

$$b = \frac{e\omega}{2m\omega_0^2\tau} \left\{ \frac{1}{(1/\tau)^2 + (\omega_0 - \omega)^2} + \frac{1}{(1/\tau)^2 + (\omega_0 + \omega)^2} \right\} \quad (13\text{-}14)$$

Correspondingly, from Eq. (13-6a) the dielectric constant is

$$K = 1 + \frac{ne^2}{\pi m(\nu_0^2 - \nu^2)} \left\{ 1 - \frac{\nu}{2\nu_0^2(2\pi\tau)^2} \left[\frac{\nu_0 + \nu}{(1/2\pi\tau)^2 + (\nu_0 - \nu)^2} \right. \right.$$
$$\left. \left. + \frac{\nu - \nu_0}{(1/2\pi\tau)^2 + (\nu_0 + \nu)^2} \right] \right\} \quad (13\text{-}15)$$

and the absorption coefficient from Eq. (13-7) is

$$\gamma = \frac{ne^2\nu^2}{mc\nu_0^2} \left[\frac{1/2\pi\tau}{(\nu - \nu_0)^2 + (1/2\pi\tau)^2} + \frac{1/2\pi\tau}{(\nu + \nu_0)^2 + (1/2\pi\tau)^2} \right] \quad \text{cm}^{-1} \quad (13\text{-}16)$$

This is the complete expression for a classical oscillator. They must be modified to some extent, however, for the quantum-mechanical case. In such equations it is found that e^2/m in a classical expression corresponds to $(8\pi^2/3h)|\mu_{ij}|^2\nu_0$ in the corresponding quantum-mechanical expression. This may be seen by comparing Eq. (19) on p. 38 with Eq. (19) on p. 180 of Heitler's *The Quantum Theory of Radiation* [936]. Here μ_{ij} is the matrix element of the dipole moment, or may be called the dipole moment for the transition from state i to state j. These substitutions give the proper transition to the quantum-mechanical expression. It should be noted that $|\mu_{ij}|^2$ is an average of the square of the matrix element for a transition from the lower state i to the upper state j. It is defined, as in Eq. (1-76), by

$$|\mu_{ij}|^2 = \sum_{M'} |\mu_x(JMJ'M')|^2 + |\mu_y(JMJ'M')|^2 + |\mu_z(JMJ'M')|^2$$

We have assumed so far that oscillators are absorbing energy only and not emitting energy. From the quantum-mechanical viewpoint there must be oscillators in the upper state of the transition as well as those in the lower state, and it can be shown that the electromagnetic field induces the oscillators in the upper state to emit with the same probability that the oscillators in the lower state absorb. Our net absorption is then proportional to the difference in the number of oscillators in the upper and lower state, which is

$$\Delta n = (1 - e^{-h\nu_0/kT})n \tag{13-17}$$

where n is the number of molecules per unit volume in the lower state. In the radio-frequency region $h\nu_0 \ll kT$, so that this can be well approximated by

$$\Delta n = \frac{h\nu_0}{kT} n \tag{13-18}$$

In addition to these two states there may be many other molecular states which are occupied by molecules of our material. We may represent by f the fraction of the total which is in the lower of the two states of interest, so that n in the above equations may be replaced by Nf, where N is the total number of molecules per unit volume. Making this substitution we obtain the final expression

$$\gamma = \frac{8\pi^2 Nf}{3ckT} |\mu_{ij}|^2 \nu^2$$
$$\left[\frac{1/2\pi\tau}{(\nu - \nu_0)^2 + (1/2\pi\tau)^2} + \frac{1/2\pi\tau}{(\nu + \nu_0)^2 + (1/2\pi\tau)^2} \right] \quad \text{cm}^{-1} \tag{13-19}$$

This can be shown to reduce to the Debye case when ν_0 is 0, for then γ becomes

$$\gamma = \frac{8\pi Nf\mu^2}{3ckT} \frac{\omega^2\tau}{1 + \omega^2\tau^2} \quad \text{cm}^{-1} \tag{13-20}$$

In that case also $f = \frac{1}{2}$ and

$$\gamma = \frac{4\pi N\mu^2}{3ckT} \frac{\omega^2\tau}{1 + \omega^2\tau^2} \quad \text{cm}^{-1} \tag{13-21}$$

which is identical with expression (13-3) derived directly from the Debye theory.

At low pressures (*i.e.*, where $1/2\pi\tau \ll \nu_0$), the first term in Eq. (13-19) is the predominant one. The intensity at the line center is then proportional to N, where N is the number of molecules per unit volume and τ is the mean time between molecular collisions. But the time between collisions is inversely proportional to the pressure, so that γ is independent of pressure. Thus the intensity at the peak of a microwave line is independent of pressure over a wide range of pressures. Moreover the line width is proportional to $1/2\pi\tau$, and therefore to the pressure.

The first term of Eq. (13-19) was given in Chap. 1 without proof as Eq. (1-49). It is the most commonly used expression because it does fit the observed microwave line intensities and widths well at low and medium pressures. Karplus and Schwinger [297] have given a quantum-mechanical derivation of the Van Vleck–Weisskopf formula for the shape and intensity of a microwave spectral line. The assumptions are the same, *i.e.*, collisions so strong that the Boltzmann energy distribution is restored after each collision, and that the duration of a collision is short enough that the field does not change appreciably during the collision. It is not necessary that $\Delta \nu$ be less than ν, and the theory would appear to apply even at quite high pressures.

Van Vleck and Margenau [424] have also examined the Van Vleck–Weisskopf theory. They evaluated separately the work done between collisions and that done impulsively by sudden changes of position of the molecule in the electric field at collisions.

The dielectric constant K and absorption coefficient γ are both given for a particular line shape by (13-15) and (13-16). There are general relations between the dielectric constant and absorption coefficient for any system [8a], [135a], of which these equations are a special case. The general expressions are known as the Kramers-Kronig relations, and may be written

$$K(\omega) - 1 = \frac{2c}{\pi} \int_0^\infty \frac{\omega' \gamma(\omega') d\omega'}{\omega[\omega'^2 - \omega^2]}$$

$$\gamma(\omega) = \frac{-2\omega^2}{\pi c} \int_0^\infty \frac{[K(\omega') - 1]d\omega'}{\omega'^2 - \omega^2}$$

13-4. Absolute or Integrated Line Intensity. The Van Vleck–Weisskopf equation (13-19) in the region of ν_0 can be approximated

$$\gamma = \frac{8\pi^2 Nf}{3ckT} |\mu_{ij}|^2 \nu^2 \frac{\Delta \nu}{(\nu - \nu_0)^2 + (\Delta \nu)^2} \tag{13-22}$$

if $\Delta \nu \ll \nu_0$. This is the form of a typical resonance absorption of half width $\Delta \nu = 1/2\pi\tau$. If one integrates over the absorption line, assuming $\Delta \nu \ll \nu_0$, the integral $\int \gamma \, d\nu$ becomes

$$\frac{8\pi^3 Nf}{3ckT} |\mu_{ij}|^2 \nu_0^2 \tag{13-23}$$

which is often called the absolute or integrated line intensity.

The approximation $\Delta \nu \ll \nu_0$ is good in the infrared and optical regions, but not always good in the microwave region. In fact, sometimes $\Delta \nu > \nu_0$ at high pressures. Absolute or integrated line intensity may be more appropriately defined as

$$\int_0^\infty \frac{\gamma}{\nu^2} \, d\nu$$

It may be noted that the first term of the expression in brackets in Eq. (13-19) corresponds to a resonant absorption at frequency ν_0, and the second term to a resonant absorption at frequency $-\nu_0$. Integrating the sum of the two terms from 0 to ∞ is equivalent to integrating either one from $-\infty$ to $+\infty$. Hence

$$\int_0^\infty \frac{\gamma}{\nu^2}\, d\nu = \frac{8\pi^2 Nf}{3ckT}\, |\mu_{ij}|^2 \int_{-\infty}^\infty \frac{1/2\pi\tau}{x^2 + (1/2\pi\tau)^2}\, dx = \frac{8\pi^3 Nf}{3ckT}\, |\mu_{ij}|^2 \qquad (13\text{-}24)$$

This is independent of τ and hence of the line width $\Delta\nu$. The invariance of the absolute intensity with perturbations such as collisions is connected with the principle of "spectroscopic stability." This principle is usually applied to a spectroscopic line split by Zeeman, Stark, or other effects into a fine structure. It states that the sum of intensities of all fine-structure components of a line is equal to the intensity of the unsplit line which would occur if the cause of fine structure were removed, or if the fine structure were not resolved.

13-5. Comparison of the Van Vleck–Weisskopf Line Shape with Experiment. The following general features of collision or pressure broadening have been demonstrated from microwave measurements and are predicted from the Van Vleck–Weisskopf theoretical shape [Eq. (13-19)].

1. The half width $\Delta\nu$ is proportional to pressure over a wide range of low pressures.

2. The peak absorption intensity is independent of pressure over a wide range of low pressures.

3. The apparent resonant frequency ν_0 is constant over a wide range of low pressures.

4. At low pressure the line shape is fitted very accurately by a simple resonant expression.

5. At moderate pressures (1 atm) the absorption-line shape is very asymmetric and given qualitatively by Eq. (13-19).

6. At high frequencies ($\nu \gg \nu_0$) the absorption is constant and has the value

$$\gamma_\infty = \frac{8\pi Nf}{3ckT\tau}\, |\mu_{ij}|^2 \qquad (13\text{-}25)$$

The theoretical line shapes for several values of $\Delta\nu$ are shown in Fig. 13-1. It should be noted that properties 1 to 4 are characteristic of impact theories in general, and only 5 and 6 distinguish the Van Vleck–Weisskopf formulation.

The observed shape of a line in the inversion spectrum of ammonia is compared at pressures near 1 mm Hg with a Lorentz resonance line shape in Fig. 13-2. The Van Vleck–Weisskopf line shape reduces to this Lorentz shape at these low pressures, since $\Delta\nu \ll \nu_0$.

At a pressure of 0.27 mm, the fit is good. When the pressure is raised to 0.83 mm, both the frequency and the intensity of the peak absorption remain unchanged, as predicted, and the line shape fits the theory, except on the low-frequency side where there is overlapping with the edge of a neighboring line [172]. Bleaney and Penrose [180] have shown that $\Delta\nu/p$

is a constant from 0.5 mm up to 10 cm pressure while others have shown that this holds down to about 10^{-3} mm Hg. Below this pressure other causes of broadening become important.

A more complete test of theoretical line shapes and discrimination between the Lorentz shape and its modification by Van Vleck and Weisskopf must be made at higher pressures.   Only when the lines are so broad

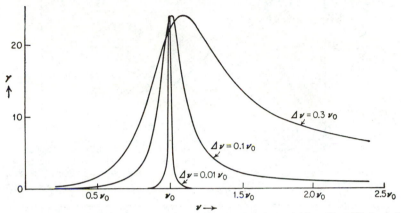

FIG. 13-1. Theoretical shape of pressure-broadened line. (*After Van Vleck and Weisskopf* [136].)

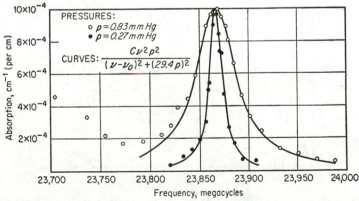

FIG. 13-2. Effect of pressure broadening on the NH_3 3,3 absorption line. (*From Townes* [172].)

that $\Delta \nu$ is comparable with ν_0 do the effects of the ν^2 factor and the "negative frequency resonance" term in (13-19) become apparent.   The shape of a microwave line between 15,000 and 35,000 Mc for water vapor in air is compared in Fig. 13-3 with theoretical line shapes.   Here $\Delta \nu \approx \nu_0$, and it is evident that the Van Vleck–Weisskopf line shape is different from the Lorentz theory and more nearly correct. However, there is a deviation from the Van Vleck–Weisskopf line shape on the high-frequency side of the water line.

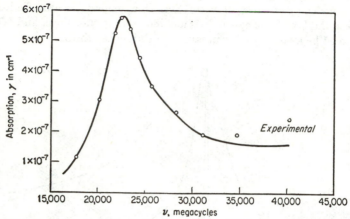

FIG. 13-3. Absorption by water vapor in air (10 g of H_2O per cubic meter). (*From Becker and Autler* [137].)

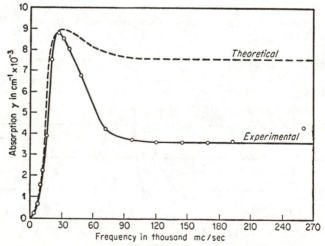

FIG. 13-4. Absorption in NH_3 at 1 atm pressure. (*From Bleaney and Loubser* [439] *and Nethercot, Klein, Loubser, and Townes* [734].)

The microwave spectrum of ammonia at a pressure of 1 atm is shown in Fig. 13-4 [439], [734]. Since at this pressure the width of any one line is considerably greater than the over-all spacing of the individual lines, the spectrum should fit a Van Vleck–Weisskopf curve fairly well with a single value of ν_0 and of $\Delta\nu = 1/2\pi\tau$. Qualitatively the spectrum does have the predicted shape. In particular the flat "tail," or region of constant absorption at high frequencies, is observed. However, a close comparison shows many deviations from the Van Vleck–Weisskopf line shape. If the line shape is fitted near the peak of the line, the theoretically expected absorption in the flat region should be considerably larger than the observed value. Birnbaum and Maryott [780] have shown that at pressures of 5 to 30 cm of Hg, the low-frequency tail of the ammonia absorption is about 40 per cent larger than that given by the Van Vleck–Weisskopf shape. Too large an observed intensity is, of course, necessary on the low-frequency side of the line to compensate for too small an intensity on the high-frequency side if the integrated intensity (13-24) is to remain constant. Furthermore, the magnitude of $\Delta\nu$ used for the theoretical fit in Fig. 13-4 is only 14,000 Mc, whereas from measurements of line widths at lower pressures, and assuming $\Delta\nu/p$ is constant, it would be expected to be 22,000 Mc. The fact that $\Delta\nu/p$ is less at atmospheric pressure than at low pressure indicates that collisions of more than two molecules at a time have become important, a circumstance not allowed for by Van Vleck and Weisskopf.

Still another complication is that the best fits of the experimental absorption curves to the Van Vleck–Weisskopf shape require a decrease in ν_0 with pressures as high as 1 atm. At 2 atm pressure and higher, ν_0 must be taken as zero [255], [321], [338]. For ND_3, which has a lower inversion frequency than NH_3, the shift in ν_0 has been found [779] to occur at a proportionally lower pressure. This is in accord with the theory of Margenau [394] and Anderson [341]. They show that when the average energy of interaction between molecules is comparable with the inversion energy, the wave functions are sufficiently disturbed to shift the microwave absorption to lower frequencies. Such effects should occur for all gases of sufficient density, since the microwave absorption must approach that given by Debye (13-20) for the liquid state. The density at which such shifts occur can be expected to depend on the strength of the intermolecular interactions and to be approximately proportional to the frequency of the microwave transition.

13-6. Pressure Broadening and Intermolecular Forces. Even if a Lorentz-type theory were developed which fits the line shapes exactly, it would still not be completely satisfactory because the quantity $\Delta\nu$ or $\Delta\nu/p$ is taken as an empirical parameter. A more complete theory should evaluate $\Delta\nu$ in terms of intermolecular forces or some known molecular properties. One might expect that the time between collisions, τ, and hence $\Delta\nu$ could be obtained from classical kinetic theory and measurements of collision diameters from viscosity or from van der Waals' equation of state. However, observed line widths are in most cases greater than those obtained by such methods, so that collision diameters for broadening of microwave lines are sometimes several times larger than diameters calculated from kinetic theory (see table of values in [473]). This may be expected since kinetic-theory diameters are determined by the requirement that molecules come close enough for intermolecular forces to cause a transfer of most of the kinetic energy kT. Microwave lines on the other hand are disturbed by more distant collisions which transfer considerably less energy since $h\nu \ll kT$.

A complete treatment of pressure broadening and evaluation of $\Delta\nu$ is so complex that a variety of different types of treatments and approximations have been developed to account for the wide range of phenomena encountered. Some approximations are suitable in the optical and infrared regions, others in the microwave region.

One is forced into making simplifying assumptions and approximations. However, no one approximate theory can be used for the wide range of phenomena encountered. Many of the earlier theories use approximations which are appropriate for optical and infrared frequencies, but not for the very different microwave frequencies. In addition some theories are nicely applicable to some types of intermolecular forces but must give way to other approximations for other types of forces.

All intermolecular forces are usually called van der Waals forces. This name includes a number of short-range interactions [93] which depend in various ways on the relative angles of orientation of the two interacting molecules and the distance r between them. The most important types are listed in Table 13-1. Note that each different type of force depends on the distance in a different way except types 4, 5, and 6 which are variations on the same basic type of interaction between a dipole and induced dipole. When the longer-range forces (dependent on lower inverse powers of r) are present they are generally more important than the shorter-range forces in producing collisions simply because of the larger radius of effectiveness. For this reason it is often possible to attribute the major responsibility for pressure broadening to one or two types of interaction rather than all types which are known to be present. One of the reasons for studying pressure broadening is the information about the occurrence and relative importance of the various types of van der Waals forces which may be obtained.

13-7. Comparison of Methods of Treating Pressure Broadening. The pressure-broadening problem can be formulated in general terms without approximations. In arriving at useful numerical results, however, simplifying approximations must always be introduced. Jablonski [130a], [161] has pointed out that an entire volume of gas may be considered as one system with bands of energy levels between which transitions occur. Interactions between molecules are just part of the Hamiltonian of the complete system whose energy levels need to be determined. The mathematical difficulty of obtaining energy levels and transition intensities of the entire collection of molecules treated as one system is so great,

TABLE 13-1. TYPES OF VAN DER WAALS FORCES IMPORTANT TO
PRESSURE BROADENING

Type of interaction	Variation of potential with r	Discussion
1. Dipole-dipole.............	r^{-3}	Interaction between two dipoles fixed in orientation, or "resonant," so that they may vary synchronously
2. Quadrupole-dipole....:......	r^{-4}	Dipole and quadrupole fixed in orientation
3. Quadrupole-quadrupole....	r^{-5}	Interaction between two quadrupoles
4. Keesom alignment.........	r^{-6}	Two dipoles not fixed in orientation. One dipole induces an alignment of the other and hence a second-order dipole-dipole interaction
5. Dipole–induced dipole......	r^{-6}	Molecular dipole which perturbs electronic wave function of second molecule thus inducing and interacting with a dipole moment. Same as type 4 but induced dipole due to perturbation of electron state
6. London dispersion........	r^{-6}	Electrons in molecule or atom inducing electronic dipole moment. Same as type 4 but both dipoles due to electronic motions rather than being fixed molecular dipoles
7. Quadrupole–induced dipole.	r^{-7}	Dipole moment of first molecule induces dipole moment in second molecule which reacts back on quadrupole moment of first molecule
8. Exchange forces..........	Exponential or very high power of $1/r$	Strong, usually repulsive, forces due to direct interaction of electronic distributions in two molecules

however, that his approach has been used only in approximations which give essentially the same results obtained by the Kuhn-London statistical theory discussed below.

All other treatments of pressure broadening can be classified as collision or statistical theories. Collision theories assume that, during most of the time a molecule is sufficiently far from other molecules, it may be considered free. Occasionally it comes close enough to one or more other molecules for the intermolecular fields to perturb its energy levels appreciably. After collision the molecule may be in the same state as before the encounter, with only a change in phase of its wave function, or a transition to another state may have been induced by the collision. Both types of collision contribute to pressure broadening. Collision theories usually assume that the radiation takes place only while the molecule is undisturbed by intermolecular interactions, the collisions being so brief and infrequent that they serve only to interrupt or change the phase of the normal process of radiation. Statistical theories, on the other hand, consider that molecules are always under the influence of intermolecular interactions, even though these may be weak, and that the frequency radiated depends on the amount of interaction occurring during radiation. The intensity emitted at a particular frequency ν depends simply on the probability that a molecule is perturbed by other molecules just the correct amount to make its frequency ν. Statistical theories hence always involve finding the probabilities of molecules being within a range R and $R + \Delta R$ apart with various possible angles of orientation and hence of their levels being perturbed by various amounts. Collision theories, on the other hand, require calculation of the probability of various types of collisions, the changes in molecular states which occur during these collisions, and a Fourier analysis of the molecular radiation which has intermittent disturbances due to collisions.

Either the collision or statistical approach can give fairly complete and accurate theories if they are developed sufficiently far. However, collision theories usually neglect radiation during a collision and hence become poor approximations at pressures of a few atmospheres where molecules are always close together and collisions are frequent. Statistical theories cannot very well take into account the changes of interactions with time due to molecular motion and hence are good only when molecular velocities are so low that the rate of change of intermolecular interactions may be neglected. We shall see below that this limitation prevents the statistical approach from being very accurate in the microwave range, although it can be used as a guide and rough approximation.

Let us examine now the effect of collisions on microwave radiation, assuming that pressures are low enough so that only collisions between two molecules are important. Using classical language, assume the first molecule is oscillating or rotating so that it radiates a frequency of interest and that it collides with a second molecule. During the collision its oscillation is somewhat modified in frequency because of the interaction, but if the oscillation continues without loss in energy, the molecule emerges from the collision oscillating as before, but with a change in phase due to the changes in frequency during the collision. Such a collision is called adiabatic since no energy has been lost from the oscillation of interest during collision.   The radiated wave is as intense as before the collision, but new frequencies have been introduced because the change in phase due to the collision can be represented as a Fourier distribution of frequencies.   Quantum-mechanically, this change of frequency during collision corresponds to a change in energy separation between the ground and excited states due to intermolecular interactions, and the change in phase is a change in relative phases of the ground and excited-state wave functions.   The molecule could also appreciably change its energy of oscillation during collision, *i.e.*, make a transition from the excited to the ground state as a result of the collision or vice versa.   Such a collision is diabatic.*   In this case the radiation emitted, if any, after collision will have no particular relation to that emitted before the disturbance.

In order for a change in phase as large as 1 radian to occur during an adiabatic collision,

$$2\pi\epsilon t \geq 1 \qquad \text{or} \qquad \epsilon \geq \frac{1}{2\pi t} \qquad (13\text{-}26)$$

where ϵ is an average change in frequency during the collision and t is the duration of the collision.   A rough measure of ϵ from the energy of the interaction W is

$$\epsilon = \frac{W}{h} \qquad (13\text{-}27)$$

The time of collision is given approximately by

$$t = \frac{R}{v} \qquad (13\text{-}28)$$

where R is the distance between molecules required to produce an appreciable interaction, and v is the thermal velocity of the molecule.   Since R is a few angstroms and v near 10^5 cm/sec, t is approximately 10^{-13} sec. Hence $1/2\pi t$ is greater than any microwave frequency ν, and from (13-26), (13-27), and (13-28),

$$W > h\nu \qquad (13\text{-}29)$$

This shows that the energy of interaction required to give an appreciable change in phase during collision is greater than enough to cause a transition between the ground and excited states.   Furthermore this energy fluctuates in a time short compared with the period of oscillation $1/\nu$, and hence the fluctuation has frequency components which can easily produce transitions.   Another requirement for transitions to occur is that the kinetic energy of the molecules be sufficiently high to provide the

* Adiabatic means without a transfer of energy.   To describe the case where energy is transferred we use diabatic rather than the more usual but clumsy double negative nonadiabatic.

necessary energy, or

$$kT > h\nu \tag{13-30}$$

This condition is, of course, fulfilled for microwave frequencies. For these reasons, almost all collisions which are effective in broadening a microwave line are also strong enough to leave the molecule more or less randomly in the upper or lower state. That this is true can be shown from experimental measurements of saturation discussed below, as well as from comparison of observed line shapes with those predicted by various theories.

When energy levels are so far separated that transitions occur in the infrared or optical region, the situation is very different. There $1/t$ is smaller than the radiated frequency ν', and hence for a change of phase of approximately 1 radian

$$W < h\nu' \tag{13-31}$$

In addition the interactions fluctuate too slowly to excite frequencies as high as those of interest, and the kinetic energy is usually insufficient to cause these higher-frequency transitions ($kT < h\nu'$). Thus, in contrast to the microwave region, adiabatic collisions are the common type of importance to pressure broadening of optical or infrared lines. It is for this reason that most of the early theories of pressure broadening, which were developed for the optical and infrared region, are not very good when applied to pressure broadening of microwave lines. Furthermore, such theories can be expected to be inadequate whenever fine structure is resolved which is so small that diabatic collisions producing transitions between the fine structure levels commonly occur.

Since statistical theories of pressure broadening have not been developed to take into account the variations of intermolecular interactions with time, they suffer the same difficulty as collision theories which allow only adiabatic collisions. In fact the two can be shown to be equivalent for slow collisions or for frequencies far removed from the line center [104], [472], [494]. However, these types of approximations are in some cases simpler than a theory of diabatic collisions, which should be more accurate for microwave lines, and hence are still of value.

The statistical method was introduced by Kuhn and London [50], [51] in a very simple form. Let the transition frequency be ν_0 and its change due to intermolecular interactions be of the form

$$\nu - \nu_0 = \frac{B}{r^n} \tag{13-32}$$

where B and n are constants and r is the distance between two molecules. If only two molecules are considered, the probability that the intermolecular distance will lie between r and $r + dr$ is

$$dP = Ar^2 \, dr \tag{13-33}$$

where A is a constant. Then the fraction of the radiation intensity with frequency between ν and $\nu + d\nu$ is, from (13-32) and (13-33),

$$I = \frac{AB^{3/n}}{n(\nu - \nu_0)^{(n+3)/n}} \tag{13-34}$$

This simple form of the statistical theory gives an infinite relative intensity at $\nu = \nu_0$ because there is an infinite probability of two molecules being separated by an infinite distance. Finite molecular densities must be taken into account to remove this divergence. In addition, expression (13-34) gives a shift of frequency only to one side of the resonance frequency ν_0, and hence a very asymmetric line. Such asymmetric lines are observed in some optical spectra, but this type of asymmetry never occurs in the microwave region because of the prominence of diabatic collisions. However, the tail of a microwave line can be approximately fitted by an expression of the type (13-34). The tail is sometimes fitted to the tail of a resonance-shaped curve, and from this a half width of the resonance can be obtained.

Margenau has given a much more sophisticated statistical theory of the pressure broadening of the ammonia inversion spectrum [393] in case collisions between more than two molecules are negligible. The energy of interaction of the dipole moments of two symmetric tops is [82]

$$V(JKJ'K'\lambda) = \frac{\mu^2}{r^3} \frac{KK'}{J(J+1)J'(J'+1)} \, \epsilon_\lambda \tag{13-35}$$

where J, K and J', K' are the quantum numbers of the two molecules, each of which has a dipole moment μ. λ is an index replacing the individual M's of the molecules, which are not good quantum numbers during close approaches because the separate angular momenta about the intermolecular axis are not conserved. ϵ_λ is a numerical factor for each λ state. The frequency of the line absorbed by one of the molecules (J,K) is then

$$\nu = \nu_0 + \frac{B_{\lambda\lambda'}}{r^3} \tag{13-36}$$

where ν_0 is the unperturbed resonance frequency and

$$B_{\lambda\lambda'} = \frac{\mu^2}{h} \frac{KK'}{J(J+1)J'(J'+1)} \, (\epsilon_\lambda - \epsilon_{\lambda'}) \tag{13-37}$$

In passing from one inversion state to the other the dipole changes its orientation so that the system of two molecules changes from state λ to state λ'. By averaging over all possible types of collisions, or hence over all λ, λ', K', and J', and using the same basic statistical approach discussed above, Margenau obtains

$$\frac{\Delta\nu}{p} = 33.9 \left[\frac{K^2}{J(J+1)} \right]^{\frac{1}{2}} \quad \text{Mc/mm Hg at } 20°\text{C} \tag{13-38}$$

where $\Delta\nu$ is the half width of the line at half maximum intensity and p is the pressure in millimeters of mercury.   This theory gives line widths of the right order of magnitude and varying in roughly the right way with K and J, as shown by the comparison with experimental line widths in Table 13-3 (page 362).   However, a systematic deviation between the theoretical and experimental results can be seen in this table.

Collision theories for the optical and infrared region have been given by Lindholm [131] and Foley [153].   They consider the phase shifts produced during collisions, and assume that diabatic collisions are unimportant, so that their approximations are not good for pressure broadening of microwave lines, although they apply well in the higher-frequency infrared and optical regions.   This type of theory considers the approach of a perturbing molecule during a collision to change the energies of both ground and excited states of the emitting molecule, the difference between their changes being $W(t)$.   This changes the frequency of the emitting molecule by $W(t)/h$, and if the change in frequency persists for a time dt, there is a phase change of $[2\pi W(t)/h]\, dt$ over what would have occurred for normal oscillation without a collision.   The total phase shift due to the collision may be obtained by integrating over the duration of the collision as

$$P_1 = 2\pi \int_{-\infty}^{\infty} \frac{W(t)\, dt}{h} \tag{13-39}$$

The straight-line paths of the colliding molecules are assumed to be unaffected by the collision.

P_1 is a function of the impact parameter b (distance of closest approach), being large when b is small.   If b is so small that the phase shift is very large, then the phase after collision has no very close connection with that before collision, and the collision can be considered strong, producing a complete and arbitrary interruption of the emitted wave train. Weisskopf, who originated this type of calculation [44], assumes that a phase shift larger than 1 radian is equivalent to a complete interruption of the radiation, and obtains an approximate collision diameter as the value of b for this particular phase shift, considering a collision to occur only for b less than this value.

The strong collisions, giving large phase shifts, produce a symmetrical line broadening of the Lorentz type with no shift in the central frequency. By taking into account the phase shifts produced by all types of collisions, including the weaker ones, and making a Fourier analysis of the resulting wave trains, Foley and Lindholm showed, however, that there is often an appreciable shift of the central line frequency due to phase shifts during collisions.   Thus if $W(t)/h$ corresponds to a decrease in frequency during a distant collision, the line will be both broadened and shifted slightly to lower frequencies by the distant or weak collisions.

These distant collisions are much less important than closer collisions if the potential drops off very rapidly with r, that is, when $V = B/r^n$, if n is large. The phase-shift approximation therefore gives an absorption line of the form

$$\gamma = \frac{A}{(\nu - \nu_0 \pm a\,\Delta\nu)^2 + (\Delta\nu)^2} \qquad (13\text{-}40)$$

where the change in the center of the line $a\,\Delta\nu$ is proportional to the line width and dependent on the force law as shown in Table 13-2. When n is 3, no frequency shift is observed because only "resonant" type interactions give $n = 3$ (cf. Table 13-1), and they give a symmetrical splitting of energy levels. Although this type of frequency shift is often observed in infrared or optical spectra, no such frequency shifts have yet been found in the microwave region. They are in most cases certainly less than $0.05\,\Delta\nu$. This is because adiabatic collisions are of little importance in broadening microwave lines. However, some collisions of this type occur, and they undoubtedly produce small frequency shifts which may be found with refined techniques.

TABLE 13-2. RATIO a OF SHIFT IN FREQUENCY TO LINE BREADTH $\Delta\nu$ ON PHASE-SHIFT THEORY

Potential of interaction of two molecules is assumed to be of the form $V = B/r^n$

n	3	4	5	6	7	∞
a	0	0.866	0.500	0.363	0.289	0

13-8. Impact Theory—Anderson's Treatment. Anderson has given a more complete treatment of pressure broadening of the collision type which allows adequately for diabatic collisions, *i.e.*, those which induce transitions [342], [343]. Where the computations involved are not too complex, Anderson's theory may be very satisfactorily applied to pressure broadening in the microwave region. In some cases this approach, combined with experimental measurements, can be used to determine the magnitude of certain intermolecular interactions. However, there are always minor contributions to collision effects which involve computations that are prohibitively complex, and in many cases even the major sources of pressure broadening still involve such difficulties.

Anderson makes some assumptions usual to collision theories:

1. Colliding molecules follow definite classical paths. In very close collisions this is a poor assumption, but the errors introduced are unimportant since any path giving a close collision involves complete interruption of the radiation, and details of the path are unnecessary. Collisions near the limit of the effective collision radius are the ones which must be accurately treated, and the quantum-mechanical wave packet for each molecule may in almost all cases be considered sufficiently small compared with this distance to make the classical path a very good approximation.

2. The duration of a collision is small compared with the time between collisions. This is always true at sufficiently low pressures, and is true for most molecules if the pressure is below about 1 atm.

A line shape similar to that of Van Vleck and Weisskopf is obtained, but with the possibility of a shift in the central frequency,

$$\gamma = \frac{8\pi^2 Nf}{3ckT} |\mu_{ij}|^2 \nu^2 \left[\frac{\Delta\nu}{(\nu - \nu_0 - a\,\Delta\nu)^2 + (\Delta\nu)^2} \right.$$
$$\left. + \frac{\Delta\nu}{(\nu + \nu_0 + a\,\Delta\nu)^2 + (\Delta\nu)^2} \right] \quad (13\text{-}41)$$

However, the frequency change $a\,\Delta\nu$ is not so prominent as in the phase-shift theory or (13-40), and is usually negligible for microwave lines.

The number of collisions per second is given by $Nv\sigma_2$, where σ_2 is an effective cross section, v the molecular velocity, and N the number of molecules per unit volume. Hence

$$\Delta\nu = \frac{Nv\sigma_2}{2\pi} \quad (13\text{-}42)$$

Similarly, the frequency shift is

$$a\,\Delta\nu = \frac{Nv}{2\pi}\sigma_1 \quad (13\text{-}43)$$

where σ_1 is the effective cross section for frequency shifts. These cross sections may be written

$$\sigma = \int_0^\infty 2\pi b S(b)\,db \quad (13\text{-}44)$$

where b is the impact parameter, or distance of closest approach of the molecules, so that $2\pi b\,db$ is proportional to the probability of a collision with impact parameter b, and $S(b)$ is a weight factor which indicates whether or not a collision of this type is effective in disturbing the molecular radiation. $S(b)$ is unity when b is small since every such collision is effective in broadening the spectrum and for larger b the most important parts of S are given by

$$S_2(b) = \frac{1}{2} \sum_{l,M} \left[\frac{|(iM|P|l)|^2}{2J_i + 1} + \frac{|(fM|P|l)|^2}{2J_f + 1} \right] \quad (13\text{-}45)$$

where i,M represent all the quantum numbers of the initial state, and f,M those of the final state of the radiating transition. J_i and J_f are the angular momenta of initial and final states. l represents the quantum numbers of any state to which transitions are induced by the perturbing intermolecular interaction. The matrix elements $(i,M|P|l)$ are

$$(a|P|b) = \frac{2\pi}{h} \int_{-\infty}^\infty [a|V_1(t)|b] \exp(2\pi i\nu_{ab}t)\,dt \quad (13\text{-}46)$$

where $V_1(t)$ is the perturbing interaction, and ν_{ab} is the frequency of a transition between states a and b. The sum over M divided by $2J + 1$ involved in (13-45) is simply an average over the $2J + 1$ different possible values of the magnetic quantum numbers M.

From (13-45), $S_2(b)$ may be interpreted as the probability that the collision at a distance b produces a transition, averaged over the ground and excited states, and over the various possible orientations of the angular momentum. This is because $|(iM|P|l)|^2$ is just the quantity which gives the transition probability (*cf.* [153]). It has been suggested [536] that a simplifying assumption might be applied to Anderson's theory by considering an effective collision to take place for any b small enough

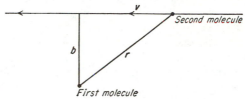

FIG. 13-5. Geometry of molecular collision.

to make this probability (S_2) larger than $\frac{1}{4}$. This gives approximately the same result as an integration of (13-44) over all collision parameters. However, for force laws of the form $1/r^n$ this approximation gives little simplification.

The expression for frequency shift similar to (13-45) is

$$S_1 = \sum_M \left[\frac{(iM|P|iM)}{2J_i + 1} - \frac{(fM|P|fM)}{2J_f + 1} \right] \tag{13-47}$$

which is zero for most of the common interactions because the matrix elements in the sum are zero.

The relative positions of the two colliding molecules and the path over which the time integral in (13-46) is made is indicated by Fig. 13-5. The path of the second molecule with respect to the first is assumed to be a straight line since only the closer collisions which always interrupt the radiation will involve appreciable curvature of the path. Then

$$r(t)^2 = b^2 + v^2t^2 \tag{13-48}$$

For an interaction energy which varies as $1/r^n$ with the distance between molecules, $[a|V_1(t)|b]$ is the form $K/[r(t)]^n$. Letting $x = vt/b$ and $k = (2\pi b/v)\nu_{ab}$, (13-46) becomes

$$(a|P|b) = \frac{2\pi K}{hb^{n-1}v} \int_{-\infty}^{\infty} \frac{e^{ikx}\, dx}{(1 + x^2)^{n/2}} \tag{13-49}$$

When $k \approx 0$, the collision might be called "fast," and the P matrix element tends to be large if the interaction energy K/b^n is large. This value does not decrease much for values of k up to 1, but then falls off rapidly, being very small for k as large as 4 or 5.

The parameter k is roughly the ratio of the time b/v required for the collision to be completed to the time $1/\omega$ required for the radiation to change phase. For a collision parameter b as large as 10 A, and v a typical thermal velocity of 10^5 cm/sec, $b/v \approx 10^{-12}$ sec. This is considerably smaller than $1/\omega$ for 1 cm wavelength radiation, which is $1/\omega = 3 \times 10^{-11}$.

One may see from (13-49) two criteria for a transition to occur during collision: first the interaction energy must be large, and secondly the collision must be "fast," *i.e.*, the time variation of the interaction must involve frequencies as high as the transition frequency ν_{ab}. For microwave transitions, k is almost always less than unity, so that diabatic collisions occur, and the exponential in (13-49) may be ignored. However, for optical frequencies the exponential could not be ignored, and the integral in (13-49) would take on a very different character.

Letting $k = 0$, S_2 is of the form

$$S_2 = \frac{A}{b^{2n-2}} \tag{13-50}$$

where b is the impact parameter and A is a quantity depending on various properties of the molecules and the quantum states involved. With some approximations which introduce errors of 10 per cent or less, the cross section and half width $\Delta\nu$ may then be found from (13-42), (13-44), and (13-50) as

$$\sigma = \pi \left(\frac{n-1}{n-2}\right) A^{1/(n-1)} \qquad \Delta\nu = \frac{Nv}{2}\left(\frac{n-1}{n-2}\right) A^{1/(n-1)} \tag{13-51}$$

The problem is then reduced to finding A, or evaluating the sum in (13-45). Since a particular molecule of interest may collide with molecules in a variety of states, each having a different effective cross section, $\Delta\nu_a$ must be obtained for each different type of molecule with which collisions may occur, and weighed according to the fractional abundance f_a of each type. Then

$$\sigma = \sum_a f_a \sigma_a \qquad \Delta\nu = \sum_a f_a \, \Delta\nu_a \tag{13-52}$$

In some cases the sum in (13-45) is very difficult to obtain, especially when both colliding molecules may make a variety of transitions. It has been worked out, however, for the important cases of dipole-dipole, dipole–induced dipole, and quadrupole–induced dipole interactions (types 1, 5, and 7 of Table 13-1).

The interaction energy of two dipole moments $\mathbf{\mu}_1$ and $\mathbf{\mu}_2$ separated by a distance r is

$$V_1 = \left[\mathbf{\mu}_1 \cdot \mathbf{\mu}_2 - \frac{3(\mathbf{\mu}_1 \cdot \mathbf{r})(\mathbf{\mu}_2 \cdot \mathbf{r})}{r^2} \right] r^{-3} \qquad (13\text{-}53)$$

For two interacting symmetric-top molecules, this energy is of the first order and proportional to $1/r^3$. This is because a symmetric-top molecule has a component of the dipole moment along the angular momentum J, which is fixed in orientation. For a linear molecule, no component of the dipole moment is fixed in direction, and the energy (13-53) averages to zero in the first order unless two interacting molecules happen to be rotating at essentially the same rate. Even two molecules like ammonia, which invert at microwave frequencies and thus reverse the direction of their dipole moments, give a first-order dipole-dipole interaction. Averaging over a long period of time would give no first-order interaction because unless the two ammonia molecules are in the same rotational state they invert at different frequencies and would change relative orientation with time. However, over the short duration of a collision, they may be considered to invert synchronously and remain with the same relative orientation. The matrix elements needed to evaluate these first-order interactions produced by dipole-dipole interactions [V_1 given by (13-53)] turn out to be just those given by Table (4-4) for dipole radiation. After the time integration indicated by (13-46) is done, the matrix elements inserted and summed according to (13-45), one obtains [342], [343]

$$S_2(b) = \frac{8}{9} \frac{\mu_1^2 \mu_2^2}{b^4 v^2 (h/2\pi)^2} \frac{K_1^2 K_2^2}{J_1(J_1 + 1)J_2(J_2 + 1)} \qquad (13\text{-}54)$$

where K_1, J_1, K_2, J_2 are the usual angular-momentum quantum numbers for molecules 1 and 2, respectively, which are involved in the collision. v is the relative velocity of the two molecules.

Letting the molecule with index 1 be the one emitting microwave radiation of interest, and summing over all other colliding molecules, from (13-51), (13-52), and (13-54),

$$\Delta\nu = \frac{4\pi \sqrt{2} \, N|\mu_1 K_1|}{3h \sqrt{J_1(J_1 + 1)}} \sum_{\mu_2, J_2, K_2} \frac{|\mu_2 K_2|}{\sqrt{J_2(J_2 + 1)}} f_{J_2 K_2 \mu_2} \qquad (13\text{-}55)$$

where $f_{J_2 K_2 \mu_2}$ is the fraction of molecules having dipole moment μ_2 and quantum numbers K_2, J_2 and N is the number of molecules per unit volume. The cross section σ may be easily obtained from $\Delta\nu$ by (13-42) if some average velocity v is assumed. Sometimes an effective diameter b_e is calculated for collisions, which is given by

$$\sigma = \pi b_e^2 \qquad (13\text{-}56)$$

Another case where the dipole moments of two molecules interact strongly, or in first-order perturbation approximation, is the case of "rotational resonance." Two molecules in rotational resonance may be considered classically to be rotating at the same rate so that the interaction between their dipole moments when averaged over time is not zero. Quantum-mechanically, rotational resonance requires two identical molecules with angular momenta differing by one unit ($J_1 = J_2 \pm 1$). Using the interaction (13-53), this gives [342], [343], for collisions with $J_2 = J_1 - 1$,

$$S_2(b) = \frac{8}{9} \frac{\mu_1^4}{b^4 v^2 (h/2\pi)^2} \frac{(J_1^2 - K_1^2)(J_1^2 - K_2^2)}{J_1^2 (2J_1 + 1)(2J_1 - 1)} \tag{13-57}$$

$$\Delta\nu = \frac{4\pi \sqrt{2}}{3h} \frac{N\mu_1^2}{J_1} \left[\frac{(J_1^2 - K_1^2)}{(2J_1)^2 - 1} \right]^{\frac{1}{2}} \sum_{K_2} (J_1^2 - K_2^2)^{\frac{1}{2}} f_{J_1-1, K_2} \tag{13-58}$$

When $J_2 = J_1 + 1$, the number J_1 in (13-58) need only be replaced by $J_1 + 1$. The notation is the same as in Eqs. (13-54) and (13-55).

"Rotational-resonance" interactions can occur with either symmetric tops or linear molecules ($K = 0$) and tend to be large when the dipole moments μ_1 and μ_2 are large and the fraction f of molecules in the proper state is not too small. However, for most rotational lines of molecules falling in the microwave region, $f \approx h\nu/kT$, or $\frac{1}{200}$ [cf. (1-56)], and the rotational-resonance effects are usually small. The case of ammonia is somewhat unusual since its microwave lines show large effects due to rotational resonance. This is because the ammonia lines in the microwave region are due to inversion rather than rotational transitions, and the molecule is so light that for some rotational states f is as large as $\frac{1}{16}$. Other molecules do show a quasi resonance, because collisions are so short that the frequency cannot be established accurately. If a collision lasts 10^{-12} sec, all levels within about 10^{11} cycles are near enough to behave as resonant.

Other types of interactions which have been treated by Anderson's method involve a dipole moment induced by the distortion of the distribution of electrons on one molecule, which we shall label 1, by the molecular dipole moment of a second molecule labeled 2. The induced dipole moment on molecule 1 may interact back on the dipole moment of molecule 2, with an energy proportional to $1/r^6$ (type 5 of Table 13-1). Because of certain symmetry relations, this type of interaction usually contributes a negligible amount to pressure broadening [342], [343], [428].

The induced dipole moment of molecule 1 of the last paragraph may also interact with the electric quadrupole moment of molecule 2 (type 7, Table 13-1). If molecule 2 is symmetric, the interaction energy is then

$$V_1 = -\frac{6\alpha_1 \mu_2 Q_2}{r^7} \cos^3 \theta \tag{13-59}$$

where α_1 = polarizability of molecule 1

μ_2, Q_2 = dipole and quadrupole moments, respectively, of molecule 2

r = radius between the two molecules

θ = angle between r and the axis of molecule 2

Anderson has obtained [428]

$$S_2 = \frac{3}{b^{12}}\left(\frac{15\pi^2\alpha_1\mu_2 Q_2}{64vh}\right)^2 \frac{K^2}{J(J+1)}\left\{1 + \frac{22}{5}\frac{K^2}{J(J+1)} + \frac{121}{21}\frac{K^4}{J^2(J+1)^2}\right.$$
$$\left. - \frac{K^2}{3J^2(J+1)^2}\left[\frac{22}{5} + \frac{121}{7}\frac{K^2}{J(J+1)}\right] + \frac{121}{63}\frac{K^4}{J^4(J+1)^4}\right\} \quad (13\text{-}60)$$

which is of the form given by (13-50)

$$S_2 = \frac{A}{b^{12}}$$

so that from (13-51)

$$\sigma = 1.2\pi A^{\frac{1}{5}}$$

Leslie [602] has given a treatment of pressure broadening similar to that of Anderson but has eliminated the assumption of a classical path and used a Boltzmann distribution of velocities throughout rather than assuming an average molecular velocity. Although such a treatment is naturally more complicated because of elimination of these two simplifying assumptions, Leslie has obtained results similar to those discussed above.

13-9. Comparison of Theories with Experiment. After the above lengthy discussion of theoretical evaluation of line widths, the reader might expect extensive confirming comparisons between theory and experiment. Unfortunately, successful comparisons are rather limited, and many attempts to fit particular calculations to observations have been inconclusive. One of the reasons for this is that there are few cases in which only one of the wide variety of possible interactions predominates, and even the rough magnitudes of some of the molecular constants required by theory are unknown. In addition, there has been much conflicting experimental data. The one outstanding exception to these uncertainties is dipole-dipole broadening in NH_3, which is so far the only microwave case for which extensive quantitative agreement between theory and experiment seems to have been achieved.

The NH_3 $J = 3$, $K = 3$ inversion line in pure NH_3 at a pressure of 1 mm Hg and 0°C has a measured half width at half maximum of 30 ± 1 Mc.* The half width calculated from expression (13-54) is 31 Mc for the

---

* This value is an average of the results of Bleaney and Penrose [179] and of Townes [172]. Both the directly measured line widths and those calculated from the measured intensities were averaged. A dipole moment of 1.468 debye units [562] was used. The reported widths were reduced to 0°C assuming the width is inversely proportional to temperature.

same conditions. Rotational-resonance terms (13-57) are zero for this line because $J = K$. The agreement between theory and experiment is well within the accuracy of the theoretical approximations and the experimental uncertainties.

The variation of the NH_3 line-breadth parameter $\Delta\nu$ with J and K is shown in Table 13-3 and compared with predictions of several types of theories. It may be seen that Anderson's approach, using broadening

TABLE 13-3. COMPARISON BETWEEN EXPERIMENTAL AND THEORETICAL LINE
WIDTHS FOR AMMONIA SELF-BROADENING
Experimental values apply to a temperature of 20°C.

Line		Line breadth, Mc/mm Hg					
J	K	Experimental, Bleaney and Penrose [179]	Anderson [342]	Anderson (neglecting rotational resonance) [342]	Bleaney and Penrose [257]	Margenau [393]	Mizushima [619]
2	1	16	16	13	16	14	12
3	1	14	14	9	13	10	9
3	2	19	20	18	20	20	16
3	3	27	27	27	27	29	23
4	4	27	28	28	28	30	24
5	1	11	11	6	10	6	6
5	2	16	15	11	15	12	11
5	3	20	20	17	20	19	16
5	5	28	29	29	28	31	24
6	3	19	17	14	19	15	13
6	4	22	21	19	22	21	17
6	6	28	29	29	28	31	24
7	5	25	22	21	23	23	18
7	6	23	26	25	26	27	22
8	7	25	26	26	26	28	22
10	9	25	27	27	27	29	23
11	9	17	25	25	25	27	21

of type (13-54) and allowing for rotational resonance (13-57), gives rather good agreement with experimental results. Omitting the rotational-resonance type of interaction (column 5 of Table 13-3) gives poor agreement when $K < J$. The values in Table 13-3 are measured by Bleaney and Penrose [257], and all the results of Anderson's theory have been multiplied by a constant factor to give exact agreement between this theory and the experimental value for the 3,3 line.

The closest fit to the experimental data is a formula given by Bleaney and Penrose (column 6, Table 13-3)

$$\Delta\nu = 30 \left[\frac{K^2}{J(J+1)} \right]^{\frac{1}{3}} \qquad (13\text{-}61)$$

where $\Delta\nu$ is the half width in megacycles for a pressure of 1 mm Hg. The form of this expression is derived on the assumption that a collision occurs when two molecules approach closely enough for the interaction of their dipoles to reach some critical value W. However, since any effect of the perturbation depends on its duration, the criterion for a collision should be that the product of the interaction energy W and the duration of the collision reaches some critical value. With this modification, Bleaney and Penrose's approach becomes similar to the statistical theory and gives an expression of the form $\Delta\nu \propto K/\sqrt{J(J+1)}$. This is the type of variation found by Margenau from a statistical type of theory (column 7, Table 13-3). It does not fit so well as (13-61). Bleaney and Penrose's formula must be regarded essentially as empirical since their assumption in deriving it is quite artificial. Moreover, as may be seen from the discussion below, it gives the wrong temperature dependence of line width. Mizushima's calculation of line breadths (Table 13-3) is based on an adiabatic type of collision theory. The similarity between the results of Bleaney and Penrose, Mizushima, Margenau, and Anderson shows that approximate agreement with experiment can be obtained under a variety of assumptions. However, the line widths calculated by Anderson are clearly more accurate than those given by Margenau or Mizushima in the cases where rotational resonance is of importance.

Broadening of ammonia lines by foreign gases is treated in exactly the same manner as self-broadening if the perturbing molecule is a symmetric top with a reasonably large dipole moment (e.g., CH_3Cl or $CHCl_3$) and similarly good agreement is obtained ([536], Table 1). Line breadths and effective collision diameters for pressure broadening of NH_3 by a number of gases are given in Table 13-4. These collision diameters are in many cases larger than the collision diameters given by kinetic theory, which are listed in Table 13-4 for comparison. This is because for pressure broadening the long-range and weak interactions can be important, whereas in processes of significance to kinetic theory the shorter-range stronger interactions are necessary. In the cases where the collision diameter for broadening is large, one can expect to obtain a relatively simple theory of the collision since the many short-range forces can be neglected. In the case of self-broadening in NH_3, for example, the collision diameter is more than three times as large as the kinetic theory diameter, or the cross section about ten times as large. The importance of long-range dipole-dipole forces in the case of NH_3 self-broadening is the reason it can be treated so successfully with a theory that neglects other types of interactions.

Smith and Howard [536] have shown that the large effective collision diameters between NH_3 and N_2, CO_2, COS, and CS_2 are very probably due to interaction between the NH_3 dipole moment and the molecular quadrupole moments of these molecules [536].   This type of interaction has not yet been treated accurately, but if broadening of the NH_3

TABLE 13-4. COLLISION DIAMETERS AND LINE-WIDTH PARAMETERS FOR
BROADENING OF THE NH_3 3,3 LINE BY VARIOUS GASES*

$\Delta\nu$ is the half width at half maximum intensity for a small amount of NH_3 in 1 mm Hg pressure of the colliding molecule.   Collision diameters computed from kinetic theory are included for comparison.

Colliding molecule	Dipole moment, debye units, or 10^{-18} esu	$\Delta\nu$, Mc	Effective collision diameter b_e, angstroms	Kinetic theory collision diameter b, angstroms
NH_3	1.47	27	13.8	4.43
He	0	1.3	2.4	3.31
A	0	1.7	3.7	4.04
H_2	0	3.0	3.1	3.59
N_2	0	3.8	5.5	4.09
O_2	0	2.3	4.3	4.02
CO_2	0	6.8	7.6	4.46
COS	0.720	6.5	7.6	
CS_2	0	6.5	7.7	
HCN	2.96	13	10.0	
ClCN	2.80	16	11.9	
CH_3Cl	1.87	15	11.3	5.14
CH_2Cl_2	1.59	12	10.3	
$CHCl_3$	0.95	20	13.7	
CCl_4	0	5.5	7.2	
SO_2	1.7	12	10.4	

* W. V. Smith and R. Howard [536].   Where other measurements have been published [658], [627] the results have been averaged with those of Smith and Howard. The numbers presumably all apply to room temperature.

3,3 line is due to the quadrupole moment Q of a foreign molecule, then approximately

$$|Q| = 5.3 \times 10^5 b_e^3 \left(\frac{M_1 + M_2}{M_1 M_2} \right)^{\frac{1}{2}}$$ (13-62a)

(cf. [536]).   Here M_1 and M_2 are the molecular weights of the colliding molecules.   The quadrupole moment of a symmetric molecule is defined as

$$Q = \int \rho(3z^2 - r^2) \, dv$$ (13-62b)

where ρ is the charge density at a point z,r in the molecule.   Coordinates are measured from the center of mass and z is in the direction of the molecular-symmetry axis. This definition is that commonly used in molecular theory but differs by a factor of the electron charge from that used for nuclear quadrupole moments.

Table 13-5 gives the molecular quadrupole moments for some molecules which have been measured by their effectiveness in broadening the ammonia 3,3 line [536], [586].

The values in the table are those measured directly for the normal rotating molecule. On the other hand, theoretical calculations [using (13-62)] of molecular quadrupole moments for a known or postulated molecular structure assume that the molecule is at rest. The effective quadrupole moment for a classically rotating linear molecule, averaged over the rotation, is one-half of that for the stationary molecule. This

TABLE 13-5. MOLECULAR QUADRUPOLE MOMENTS

Q' is the effective quadrupole moment for rotating molecules. For a linear molecule, $Q' = Q/2$, where Q is defined by (13-62b).

Molecule	$b \times 10^8$, cm, kinetic theory	$b \times 10^8$, cm, from NH_3 3,3 line broadening	Molecular Q', 10^{-26} esu
N_2	4.09	6.0	1.5
O_2	4.02	4.18	<0.55
NO	3.90	5.64	1.4
CO	3.96	5.97	1.6
CO_2	4.46	7.59	3.1
COS	....	7.56	2.9
CS_2	....	7.72	3.1
N_2O	4.35	9.1	4.4
HCN	....	10.0	7.7
ClCN	....	11.9	11.5
C_2H_2	....	8.79	5.3
C_2H_4	4.79	6.67	2.3
C_2H_6	4.86	5.64	<1.3

classical average is usually adequate in pressure-broadening studies, since the most abundant states of the perturbing molecules are usually those with high J values for which the classical approximation is good. Sometimes, particularly in molecular theory, the molecular quadrupole moment is defined as half the value given here.

Estimates of molecular quadrupole moments using what is known about molecular structures and bonding are consistent with the measured values [582].

For collisions between NH_3 and H_2, He, or A, interactions of the quadrupole–induced dipole type discussed above [cf. (13-60)] may be of importance. The rather large dipole moment of NH_3 may induce a dipole moment in a He or A atom, and this induced dipole moment reacts back on the quadrupole moment of the NH_3 molecule. It is perhaps surprising that such an interaction is more important than the reaction of the induced dipole back on the NH_3 dipole. This is because the symmetry of the dipole–induced dipole interaction is such that it produces no inversion transitions and hence a small disturbance of the inversion spectrum [428]. The quadrupole–induced dipole interaction has such a symmetry that it can cause inversion transitions and is therefore of more importance. Unfortunately the quadrupole moment of NH_3 is not accurately known. Estimates of this quadrupole moment and application of expressions (13-59) and (13-60) show that the collision diameters for broadening of the NH_3 3,3 line by the gases H_2, He, A, and O are of approximately the magnitude to be expected from this type of interaction [428]. However, these collision diameters are not larger than the kinetic theory diameters (cf. Table 13-4), so other short-range interactions such as type 8 of Table 13-1 may also be of importance.

13-10. Self-broadening of Linear Molecules. Self-broadening of linear molecules in the infrared region has received considerable attention, and dipole-dipole interactions of the Keesom alignment type have been rather completely worked out for adiabatic collisions, which predominate in the infrared region. Rotational-resonance-type interactions are also of importance for the infrared spectra of these molecules and have been fairly completely treated [93], [153].

For microwave transitions adiabatic collision approximations are not very satisfactory. It might be thought that rotational resonance is unimportant because at room temperature the most abundant molecular states have high J values ($J \approx 30$), whereas only molecules with low J values give microwave transitions. Hence an encounter between a molecule absorbing microwaves and another molecule differing by one unit in J is a rare event and contributes little to the line breadth. However, the short duration of collisions between molecules (10^{-12} sec) makes the requirements for rotational resonance much less stringent. That is, during a collision lasting τ sec, the energy levels are uncertain by about $h/2\pi\tau \approx 5$ cm^{-1}. Thus all levels within this distance are effectively in resonance with the transition being studied. For OCS, where $B = 0.2$ cm^{-1}, the first five J levels are effective in rotational-resonance broadening of the lowest rotational level. The number of molecules with energies near enough to cause resonance broadening increases as J increases.

These qualitative expectations are confirmed by the measurements of Johnson and Slager [706] on OCS line widths. It is found that line widths increase with increasing J, which is to be expected as the number of molecules within the resonance interval is greater for those with higher values of J. The relative increase with J is greater at lower temperatures where the peak of the population distribution is shifted toward lower J.

By considering interactions between a given rotational state and all other rotational states which may be considered resonant, Smith, Lackner, and Volkov [to be published] have calculated collision diameters for OCS and BrCN in good agreement with experimental observations. Thus for the $J = 3 \leftarrow 2$ transition of BrCN, the theoretical collision diameter is 19.3 A while the experimental value is 19.0 A. Recent measurements on OCS by R. S. Anderson [to be published] give similarly good agreement.

No complete treatment of broadening of microwave lines by either dipole-quadrupole or Keesom alignment forces is available. However, Keesom alignment forces have been approximately treated by adiabatic theories [153], [303], [398], [499], [619].

Keesom alignment interactions are difficult to evaluate from Anderson's theory because of the particular matrix elements involved. While the mean of the diagonal elements of the square of the interaction matrix is desired, some idea of the magnitude of these effects may be obtained from the ordinary linear average of the energy which is more easily computed [93]. It is

$$(W_{dd})_{av} = \frac{2\mu^4}{3Br^6} \frac{J_1(J_1 + 1) + J_2(J_2 + 1)}{(J_1 + J_2)(J_1 + J_2 + 2)(J_1 - J_2 - 1)(J_1 - J_2 + 1)} \tag{13-63}$$

where μ is the permanent dipole moment, B the rotational constant, J_1 and J_2 the rotational quantum numbers of the two colliding molecules, and r the distance between them. The most important collisions involve large quantum numbers J_2 for the colliding molecules, or $J_2 \gg J_1$. For this case

$$(W_{dd})_{av} = \frac{\mu^4}{3r^6 E_{J_2}}$$

where E_{J_2} is the rotational energy of the colliding molecule, which may be approximated as kT, or

$$(W_{dd})_{av} = \frac{\mu^4}{3kTr^6} \tag{13-64}$$

An average value for the dipole-quadrupole interaction is [536]

$$(W_{dq})_{av} = \frac{1}{3} \frac{\mu Q_e}{r^4} \tag{13-65}$$

In many typical collisions the energies of interaction W_{dd} and W_{dQ} are comparable in magnitude. However, for small dipole moments and distant collisions the longer-range dipole-quadrupole interactions may predominate. Thus for the collision diameter of OCS, $b_e = 7.6$ A and $Q = 0.6 \times 10^{-16}$ cm^2, the dipole moment μ must be as large as 2 debyes for W_{dd} to equal W_{dq}. Since the dipole moment of OCS is only 0.7 debye, the dipole-quadrupole interactions may be expected to dominate. For dipole moments as large as 2 or 3 debyes, the dipole-dipole Keesom alignment interactions usually are most important. While dipole moments are readily measurable, quadrupole moments have so far only been estimated roughly from the molecular structure [382] or deduced from ammonia pressure broadening. Smith and Howard [536] have estimated the quadrupole moments of several molecules in this way from their effectiveness in broadening the NH_3 3,3 line and have used the resulting values to calculate dipole-quadrupole self-broadening in these gases. They obtained fair agreement with experiment.

13-11. Oxygen Line Breadths. The oxygen spectrum is unusual in that it arises from a molecule having a zero electric dipole moment. It might therefore be expected to have unusually narrow lines, and this is indeed the case. Because of its importance to atmospheric transmission, breadths of oxygen lines in pure oxygen and in air have been studied since the early days of microwave spectroscopy [139], [244], [301], [416], [576]. Half widths of the microwave oxygen lines are about 2 Mc per mm of Hg and are approximately independent of the rotational quantum numbers [667], [669], [770]. This, plus theoretical estimates of the magnitudes of various interactions, indicate that the width of these lines is mainly caused by the same short-range repulsive forces responsible for kinetic energy transfer [770]. Even though they decrease very rapidly with distance, they are probably large enough to cause rotational transitions (which involve little energy) at a distance $1\frac{1}{2}$ times the kinetic theory radius, and so to cause the observed line widths.

13-12. Temperature Dependence of Line Widths. An elementary model of the collision process might be two colliding "hard" molecules of definite boundaries, so that the collision cross section is independent of the velocity of collision. Then since the number of molecules at a given pressure is inversely proportional to the temperature, and the velocity is proportional to the square root of the temperature,

$$\Delta \nu \propto nv\sigma \propto \frac{1}{\sqrt{T}}$$

In contrast to expectations from this model, experimentally the ammonia line width of fixed pressure is approximately proportional to $1/T$ [536] so the collision cross section cannot be independent of velocity. Some such variation is to be expected from a more refined collision theory. A slow molecule spends more time near the radiator and so causes more disturbance than one which passes quickly.

The dependence of the collision cross section on velocity for an intermolecular potential proportional to $1/r^n$ may be readily obtained. For such a potential, a measure of the disturbance is the matrix element $(a|P|b)$ whose square is the probability of a collision-induced transition.

From (13-49),

$$(a|P|b) = \frac{\text{const}}{b^{n-1}v} \tag{13-66}$$

For a collision just strong enough to be effective in line broadening or to give a critical value of $(a|P|b)$, the collision parameter b is the effective collision diameter b_e. From (13-66)

$$b_e^{n-1} \propto \frac{1}{v}$$

Hence the line-breadth parameter is

$$\Delta \nu \propto n v b_e^2 \propto n v^{1-2/(n-1)}$$

Remembering that the average velocity is proportional to $T^{\frac{1}{2}}$ and n to T^{-1},

$$\Delta \nu \propto T^{-(n+1)/2(n-1)} \tag{13-67}$$

For $n = 3$ (the ammonia case), $\Delta \nu \propto T^{-1}$; for $n = 6$, $\Delta \nu = T^{-0.7}$. If T is increased by a factor of 2, $\Delta \nu$ would change to 0.5 or 0.62, respectively, of its original value. Thus the temperature dependence of line width, when accurately measured, may give information about the force law between molecules. However, if it were to be so used, very precise line widths would be needed.

Furthermore, the measured collision diameter varies with temperature whenever resonance collisions between molecules are important because the distribution of molecules among the rotational or other quantum states changes with temperature. For the $J = 2 \leftarrow 1$ transition of OCS the line-width parameter $\Delta \nu$ varies as $T^{-0.9}$ [921], so that, from Eq. (13-67), the collision diameter is approximately proportional to $T^{-0.2}$. About half of this temperature variation of collision diameter has been shown to be due to the change in first-order (resonant) dipole interactions [921].

13-13. Effect of Temperature on Intensities. Temperature enters into the peak intensity of a microwave line [Eq. (19)] through $\Delta \nu$, kT, N, and f, where f is the fraction of molecules in the lower state of the transition and N is the number of molecules per cubic centimeter. From the above discussion, at a given pressure $\Delta \nu$ varies with temperature as $T^{-\frac{1}{2}}$ for a very short-range force law, and as T^{-1} for the longer-range force laws. Assuming $\Delta \nu \propto T^{-1}$, Eq. (1-77) shows that for a diatomic or linear molecule the variation of intensity with T is

$$\gamma_{\max} \propto \frac{1}{T^2} \tag{13-68}$$

For a symmetric rotor, from (3-52),

$$\gamma_{\max} \propto \frac{1}{T^{\frac{5}{2}}}$$

In these expressions most of the variation with temperature is due to the change in the Boltzmann distribution which populates lower rotational states more fully at lower temperatures. That portion of the variation which is caused by changes in population of the vibrational state has not been taken into account. Although population of vibrational states is also affected by temperature, its variation can be neglected in the important case where the vibrational frequency is so high that almost all molecules are in the ground vibrational state.

For linear and symmetric rotor molecules in the ground vibrational state, intensity is always increased by operating at as low a temperature as possible, consistent with a reasonable vapor pressure. For microwave transitions between high-energy rotational levels of asymmetric tops, or those involving excited vibrational states, a decrease in temperature may decrease the intensity or leave it relatively unchanged.

13-14. High Pressures. Pressures may be called high from the point of view of line breadths when collisions involving more than two molecules become frequent enough to be important. In such cases the line width is no longer proportional to pressure, since the number of effective collisions undergone by one molecule is not simply proportional to the density of molecules. "High" pressure may occur as low as $\frac{1}{2}$ atm for molecules with large collision diameters. For example, at 1 atm the average distance between molecules is approximately 30 A, which is only twice the effective collision diameter for NH_3. "Low" pressure conditions may still apply at 1 atm to other gases such as O_2 which have very small collision diameters. Collision theories which assume two-body impacts only and a long interval of time between impacts are, of course, not strictly applicable to high-pressure conditions. Statistical types of theories may be more appropriate since the radiating molecule is almost continually perturbed. However, as yet no quantitative theory of line broadening at high pressures is available.

The ammonia spectrum has been studied experimentally at pressures up to a few atmospheres by several observers [779], [257], [439], [321], [734]. It is found that the spectrum is fitted near the peak by a Van Vleck–Weisskopf shape but that ν_0 may be assumed to decrease with increasing pressure. For pressures equal to or greater than 2 atm $\Delta\nu$ increases less than linearly with pressure and the best fit is obtained with $\nu_0 = 0$. This shift in ν_0 is proportional to the square of the pressure rather than its first power as might be expected from phase-shift theories. Diabatic collisions, of course, predict no shift at all. The observed shift in frequency has been explained qualitatively by Anderson [341] and by Margenau [393]. They point out that, with the small intermolecular distances which are common at high pressures, the perturbations are large enough to cause a change in molecular wave functions and hence in selection rules. In a strong field the molecule is characterized by either of the wave functions ψ_{+K} or ψ_{-K} rather than by their weak-field combinations $\psi = \psi_{+K} \pm \psi_{-K}$. In other words, the inversion is eliminated and the nitrogen is held on one side of the hydrogens by the electric field. The only allowed transitions are changes in orientation or of one unit in magnetic quantum number. These transitions involve much smaller frequencies than inversion. At 1 atm pressure, where this effect begins to appear experimentally, the interaction between dipole moments is greater than the inversion splitting so that the strong-field approximation holds for about 50 per cent of the time. The importance of this effect and interpretation of the high-pressure line shape has, however, been questioned [504] on the basis of infrared measurements.

13-15. Saturation Effects. The well-known Lambert's law states that each layer of material of equal thickness absorbs an equal fraction of radiation which traverses it. From Lambert's law is derived the exponential decrease in intensity

$$I = I_0 e^{-\gamma x} \tag{13-69}$$

where γ is constant. In optical spectra departures from Lambert's law are usually associated with polychromatic radiation, individual components of which are absorbed at different rates. In the microwave region, however, Lambert's law breaks down even for monochromatic radiation because of saturation effects. The intensity of radiation can be made so large that absorbing molecules of a gas cannot get rid of the absorbed energy rapidly enough and γ becomes dependent on I.

Consider a molecular ground state in which there are n_0 molecules per unit volume and an excited state containing n_1 molecules per unit volume, with a microwave transition possible between the two states. From the derivation of Eq. (13-19), it can be seen that

$$\gamma = \frac{8\pi^2}{3ch}(n_0 - n_1)|\mu_{01}|^2 \frac{\nu(1/2\pi\tau)}{(\nu - \nu_0)^2 + (1/2\pi\tau)^2} \tag{13-70}$$

where the "negative frequency resonance" term of Eq. (13-19) has been omitted since it is quite unimportant under the low-pressure conditions which show saturation effects.

If essentially no microwave radiation is present, then collisions maintain an equilibrium between n_0 and n_1 such that

$$n_1 = n_0 e^{-h\nu/kT} \tag{13-71}$$

If $1/t_{01}$ and $1/t_{10}$ are the probabilities per second that a molecule is transferred from state 0 to state 1 or from state 1 to state 0, respectively, by collision, then for equilibrium with no radiation

$$\frac{n_0}{t_{01}} = \frac{n_1}{t_{10}} \tag{13-72}$$

If radiation of intensity I (in units of quanta per second per unit cross section of area) is absorbed with an absorption coefficient γ, then a new equilibrium condition must be set up

$$\frac{n_0}{t_{01}} + I\gamma = \frac{n_1}{t_{10}} \tag{13-73}$$

From Eqs. (13-71) and (13-72), $1/t_{01} = (1 - h\nu/kT)/t_{10}$, assuming $h\nu/kT$ is small, so that Eq. (13-73) becomes

$$(n_0 - n_1) = n_0 \frac{h\nu}{kT} + I\gamma t_{10} \tag{13-74}$$

Furthermore, when no radiation is present but $n_0 - n_1$ is disturbed from its equilibrium value $(n_0 - n_1)_{\text{equ}}$,

$$\frac{d(n_0 - n_1)}{dt} = \frac{-2}{t_{10}} [(n_0 - n_1) - (n_0 - n_1)_{\text{equ}}]$$

so that $t = t_{10}/2$ gives the rate of approach to equilibrium.

Combining Eqs. (13-74) and (13-70), and letting $t = t_{10}/2$

$$\gamma = \frac{8\pi^2}{3ch} \left(n_0 \frac{h\nu}{kT} - 2I\gamma t \right) |\mu_{01}|^2 \nu \frac{1/2\pi\tau}{(\nu - \nu_0)^2 + (1/2\pi\tau)^2} \tag{13-75}$$

or

$$\gamma = \gamma_0 \left(1 - \frac{2I\gamma kTt}{n_0 h\nu} \right) \tag{13-76}$$

where γ_0 is the absorption coefficient when I is very small. When I is very large then $\gamma \ll \gamma_0$ and the quantity in the parentheses in Eq. (13-76) approaches zero; so that for large radiation flux

$$\gamma \approx \frac{n_0 k\nu}{2IkTt} \tag{13-77}$$

This equation written

$$\gamma I \approx \frac{n_0}{2t} \frac{h\nu}{kT}$$

is simply equivalent to saying that the total number of quanta absorbed per unit time must equal the rate at which the energy of these quanta can be transformed into kinetic energy by collision.

Solving Eq. (13-75) for γ gives

$$\gamma = \frac{8\pi^2 n_0}{3ckT} |\mu_{01}|^2 \nu^2 \frac{1/2\pi\tau}{(\nu - \nu_0)^2 + (1/2\pi\tau)^2 + \dfrac{16\pi^2 t}{3ch} |\mu_{01}|^2 \nu I (1/2\pi\tau)} \tag{13-78}$$

Thus, as a result of saturation, γ is decreased at all frequencies. The most noticeable effects occur, of course, near the peak absorption $\nu = \nu_0$. The line shape is altered only in that the maximum intensity is decreased by the factor

$$\frac{1.}{1 + \dfrac{16\pi^2|\mu_{01}|^2\nu I t}{3ch} 2\pi\tau}$$

and the half width increased by the factor

$$\sqrt{1 + \frac{16\pi^2|\mu_{01}|^2\nu I t 2\pi\tau}{3ch}} \tag{13-79}$$

Essentially the same result is given by a more sophisticated quantum-mechanical treatment [297], [322].

At high pressures, saturation is generally unobservable. It becomes noticeable when

$$\frac{16\pi^2|\mu_{01}|^2\nu I t 2\pi\tau}{3ch} \approx 1$$

Since both t and τ are inversely proportional to the number of collisions per second, the radiation intensity I at which saturation occurs is proportional to the pressure squared. Experimentally, saturation is often noticed when the power flux is as high as one milliwatt per square centimeter and the pressure low enough to give line widths less than 1 Mc [172]. It is not uncommon for saturation effects to set a lower limit on the width of microwave lines. In order to obtain very narrow lines, the radiation intensity must be kept low.

In considering saturation, each Zeeman (or Stark) component of a line should be treated individually. Although individual components all have the same frequency, their matrix elements differ (*cf.* page 23) so that some components saturate more readily than others [295]. However, saturation effects may be described approximately by taking some average matrix element $|\mu_{10}|^2$ for the entire transition.

Measurement of saturation effects allows determination of $1/t$, the rate at which the molecular distribution approaches equilibrium. In ammonia, it has been shown that $t \approx \tau$ [172], [256], [294], [295], so that each collision which is effective in producing line broadening is also effective in restoring an equilibrium distribution to the colliding molecules. This is the result to be expected from the discussion on pages 351 and 352.

Saturation effects may also be used to obtain an estimate of the matrix element $|\mu_{01}|^2$ in certain cases. The intensity of a line is influenced both by the dipole matrix elements $|\mu_{01}|^2$ and by the number of molecules in the particular ground state from which transitions occur. Saturation effects may be seen from (13-78) or (13-79) not to depend on the number of molecules normally in the ground state, but they do depend on the matrix element. Thus the relative ease with which two different lines can be saturated affords some information about the relative sizes of their dipole matrix elements.

13-16. Broadening by Collisions with Walls. If a gas molecule strikes a wall of the cavity or waveguide in which it is contained, the process of absorption is interrupted.

A good approximate treatment of the broadening effects of wall collision can be obtained by assuming that the line shape due to wall collisions is the same as that produced by intermolecular collisions, *i.e.*

$$\gamma = \frac{\gamma_{max}(\Delta\nu)^2}{(\nu - \nu_0)^2 + (\Delta\nu)^2} \tag{13-80}$$

Here the line-breadth parameter is given by $\Delta\nu = 1/2\pi\tau$, where τ is the mean time between collisions. τ may be evaluated from kinetic theory.* The number of molecules hitting the surface of total area A in each second is

$$n = \tfrac{1}{4}N\bar{v}A \tag{13-81}$$

where N is the number of molecules per cubic centimeter of gas, $\bar{v}$ is the average molecular velocity $= 4(RT/2\pi M)^{\frac{1}{2}}$, R is the gas constant, T is the absolute temperature, and M is the molecular weight. Hence the number of collisions per second is

$$n = NA \left(\frac{RT}{2\pi M}\right)^{\frac{1}{2}} \tag{13-82}$$

But the total number of molecules in the absorption cell is NV, where V is the volume of the cell. Thus the average time between collisions for any molecule is

$$\tau = \frac{NV}{NA(RT/2\pi M)^{\frac{1}{2}}} = \frac{V}{A}\left(\frac{2\pi M}{RT}\right)^{\frac{1}{2}} \tag{13-83}$$

so that

$$\Delta\nu = \frac{A}{V}\left(\frac{RT}{8\pi^3 M}\right)^{\frac{1}{2}} \tag{13-84}$$

At 300°K,

$$\Delta\nu = 1.00 \times 10^4 \frac{A}{V} M^{-\frac{1}{2}} \tag{13-85}$$

As an example, for ammonia in a cell with a spacing of 4 mm between a Stark plate and the opposite face, and all other dimensions considerably larger, $\Delta\nu = 12$ kc.

A more rigorous treatment of wall broadening has been given by Danos and Geschwind [788]. The line shape is shown to be close to the Lorentz shape and the line width is about 10 per cent greater than that given by the elementary treatment above.

If both Doppler and collision widths are appreciable, the total line-

---

* See, for instance, M. Knudsen, *Kinetic Theory of Gases*, Methuen & Co., Ltd., London, and John Wiley & Sons, Inc., New York, 1950.

width parameter is given very nearly by*

$$\Delta \nu \approx [(\Delta \nu_{Doppler})^2 + (\Delta \nu_{collision})^2]^{\frac{1}{2}} \qquad (13\text{-}86)$$

Wall-collision broadening is ordinarily much less than pressure broadening. Where necessary for high-resolution spectroscopy it can always be reduced as much as desired by using a sufficiently large absorption cell.

13-17. Microwave Absorption in Nonpolar Gases. A nonpolar gas ordinarily does not absorb microwaves. However, if the gas molecules are sufficiently polarizable, some dipole moment may exist during collisions. At high pressures, the molecules are in collision for a large part of the time, so that an appreciable absorption occurs. Such a pressure-dependent absorption has been found in CO_2 by Birnbaum, Maryott, and Wacker [905]. The observed absorption was approximately proportional to the square of the pressure and reached 2.3×10^{-5} at 3.3 cm wavelength for a pressure of 45 atm and a temperature of 25°C.

* See numerical tabulation in M. Born, *Optik*, pp. 486, 431*ff.*, Springer-Verlag OHG, Berlin, 1933.

CHAPTER 14

MICROWAVE CIRCUIT ELEMENTS AND TECHNIQUES

14-1. Introduction. Electromagnetic Fields and Waves. Although microwaves were produced by Hertz in the very earliest demonstration of electromagnetic waves, for many years waves of these short lengths were not much used. The interesting features of wavelengths comparable with the dimensions of laboratory apparatus (1 to 1000 mm) were recognized but could not easily be studied because of the lack of suitable generators.

With the development of klystron and magnetron oscillators and of waveguide techniques near the beginning of World War II, large-scale military applications of radar became possible. The attendant intensive development program saw the introduction or development of most of the microwave techniques currently used in spectroscopy. The status of the art at the end of the war is described most completely in the 28 volumes of the Massachusetts Institute of Technology Radiation Laboratory Series (McGraw-Hill Book Company, Inc., New York, 1947). A number of shorter treatments of microwave theory and practice on various levels have appeared since the war. Many of these are listed at the end of the present chapter.

Because the basic microwave techniques are described in detail elsewhere, they will be only outlined here except for those peculiar to microwave spectroscopy. References will for the most part be to books where fairly complete treatments are available rather than to original papers.

Devices with dimensions comparable with the radiation wavelength cannot usually be treated by the lumped circuit-element approximation used in ordinary electrical-circuit theory. On the other hand, microwave devices are seldom large enough in comparison with the wavelength for the approximations of geometrical optics to be employed. Both limiting cases do occasionally occur, but most microwave apparatus has to be discussed in terms of electromagnetic wave theory.

The behavior of the entire electromagnetic spectrum may be obtained from a study of Maxwell's equations combined with additions from quantum mechanics which are necessary to understand interactions between radiation and matter. In Gaussian units (for comparison with mks units see, for instance, [537]), Maxwell's equations are

$$\nabla \times \mathbf{H} = \frac{4\pi \mathbf{i}}{c} + \frac{1}{c} \frac{\partial \mathbf{D}}{\partial t} \tag{14-1}$$

$$\nabla \times \mathbf{E} = -\frac{1}{c} \frac{\partial \mathbf{B}}{\partial t} \tag{14-2}$$

$$\nabla \cdot \mathbf{B} = 0 \tag{14-3}$$

$$\nabla \cdot \mathbf{D} = 4\pi \rho \tag{14-4}$$

where $\mathbf{H}$ = magnetic field
$\quad\;\; \mathbf{B}$ = magnetic induction
$\quad\;\; \mathbf{E}$ = electric field
$\quad\;\; \mathbf{D}$ = electric displacement
$\quad\;\; \mathbf{i}$ = electric current
$\quad\;\; \rho$ = electric charge density
$\quad\;\; \nabla$ = the operator $\mathbf{a}(\partial/\partial x) + \mathbf{b}(\partial/\partial y) + \mathbf{c}(\partial/\partial z)$

$\mathbf{a}$, $\mathbf{b}$, and $\mathbf{c}$ are unit vectors along the x, y, and z axes, respectively. To these equations must be added relationships which have to do with the properties of matter:

$$\mathbf{E} = \tau \mathbf{i} \tag{14-5}$$

$$\mathbf{B} = \mu \mathbf{H} \tag{14-6}$$

$$\mathbf{D} = \epsilon \mathbf{E} \tag{14-7}$$

where τ = resistivity
$\quad\;\; \mu$ = permeability and
$\quad\;\; \epsilon$ = dielectric constant

From these seven equations the following important relations may be derived:

Energy density of an electric field = $\epsilon \mathbf{E}^2/8\pi$
Energy density of a magnetic field = $\mu \mathbf{H}^2/8\pi$

Energy flow from a volume bounded by a surface $S = \dfrac{c}{4\pi} \displaystyle\int_S \frac{\mathbf{E} \times \mathbf{B}}{\mu} \, ds$

The quantity $\dfrac{c\mathbf{E} \times \mathbf{B}}{4\pi\mu}$ is called Poynting's vector and usually can be regarded as the energy flux vector per unit surface, although it can be rigorously used only with integrations over a closed surface. For non-isotropic, ferromagnetic, or ferroelectric material, the relations (14-5), (14-6), and (14-7) are replaced by more complicated expressions. In addition, when one proceeds to a detailed study of interactions between microwaves and matter, a number of interesting cases appear in which τ, μ, and ϵ may no longer be usefully regarded as constants. However, in developing the general characteristics of microwave propagation, we shall accept τ, μ, and ϵ as constant properties of bulk matter.

If there are no free charges in a material ($\rho = 0$), by taking the curl of Eq. (14-1) or Eq. (14-2) and using Eqs. (14-3), (14-4), (14-5), (14-6), and

(14-7), one obtains the wave equations

$$\nabla^2 \mathbf{E} = \frac{4\pi\mu}{c^2\tau} \frac{\partial \mathbf{E}}{\partial t} + \frac{\mu\epsilon}{c^2} \frac{\partial^2 \mathbf{E}}{\partial t^2} \tag{14-8}$$

$$\nabla^2 \mathbf{B} = \frac{4\pi\mu}{c^2\tau} \frac{\partial \mathbf{B}}{\partial t} + \frac{\mu\epsilon}{c^2} \frac{\partial^2 \mathbf{B}}{\partial t^2} \tag{14-9}$$

A solution of Eq. (14-8) for the electric field $\mathbf{E}$ is a satisfactory solution of Maxwell's equations if it is paired with the proper solution of Eq. (14-9) for the magnetic field, if it satisfies Eq. (14-4), and if the appropriate boundary conditions are satisfied. The proper solution for $\mathbf{B}$ may be obtained from $\mathbf{E}$ by using Eq. (14-2).

Focusing attention on solutions of Eqs. (14-8) or (14-9) which involve a wave of definite frequency $\nu = \omega/2\pi$ traveling along the z axis, we substitute the form $E = E_0 e^{j(\omega t - \gamma z)}$ into Eq. (14-8), where $j = \sqrt{-1}$. The wave equation (14-8) is satisfied if

$$\gamma = \pm \sqrt{\frac{\mu\epsilon}{c^2} \omega^2 - \frac{4\pi\mu}{c^2\tau} j\omega} \tag{14-10}$$

This solution is given in the complex notation commonly employed in electrical theory. The actual physical quantities are the real parts of these expressions.

If the propagation constant γ is positive and real, the wave progresses in the positive z direction without damping or diminution. If γ is negative, the direction of propagation is in the negative z direction. If $\gamma = j\alpha + \beta$ is complex, then $E = E_0 e^{j(\omega t - \beta z) - \alpha z}$ represents an exponentially increasing or exponentially decreasing wave. To prevent violation of the principle of conservation of energy, the wave must usually decrease as it progresses, so that α and β must have opposite signs. An exceptional case occurs inside the traveling-wave amplifier or oscillator tube where energy is supplied to the wave as it advances.

If the wave is in a good dielectric material (resistivity τ very large), the propagation constant $\gamma = \pm (\omega/c) \sqrt{\mu\epsilon}$ is real, giving an undamped wave in a medium with index of refraction $\sqrt{\mu\epsilon}$.

If the wave is propagated in a good conductor (resistivity τ small), then usually $4\pi/\tau \gg \epsilon\omega$, and the first term under the radical of Eq. (14-10) can be neglected so that $\gamma = \pm \sqrt{\frac{2\pi\mu\omega}{c^2\tau}} (1 - j)$. Such a propagation constant corresponds to a very strongly damped wave since E decreases by $1/e$ in a distance

$$\delta = \sqrt{\frac{c^2\tau}{2\pi\mu\omega}} \tag{14-11}$$

Because of the large damping, such a wave is usually appreciable only at the surface of a conductor, and δ is called the skin depth since it gives the amount of penetration of the wave into the material. For a good conductor and microwave frequencies $\delta = 10^{-3}$ cm or less.

14-2. Waveguides. We shall now proceed to show by means of these relations that electromagnetic waves of suitable length are propagated in a hollow rectangular metal pipe or "waveguide." Figure 14-1 shows the coordinate system. In these coordinates the boundary conditions are

$$E_x = 0 \qquad \text{at } y = 0 \text{ and at } y = b \qquad (14\text{-}12)$$
$$E_y = 0 \qquad \text{at } x = 0 \text{ and at } x = a \qquad (14\text{-}13)$$

of the walls are perfect conductors.

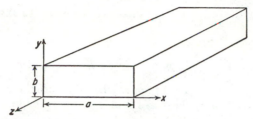

Fɪɢ. 14-1. Rectangular hollow metal pipe waveguide.

For a perfect dielectric (infinite resistivity) inside the waveguide $\tau = \infty$ and the wave equation (14-8) becomes

$$\nabla^2 \mathbf{E} = \frac{\mu\epsilon}{c^2} \frac{\partial^2 \mathbf{E}}{\partial t^2} \qquad (14\text{-}14)$$

Assume a solution of this wave equation, subject to the boundary conditions, of the form of a traveling wave with no longitudinal electric-field component.

$$E_x = -E_0 \frac{k_y}{k_x} \cos k_x x \, \sin k_y y \, e^{+i(\omega t - \gamma z)}$$
$$E_y = E_0 \sin k_x x \, \cos k_y y \, e^{+i(\omega t - \gamma z)} \qquad (14\text{-}15)$$
$$E_z = 0$$

These satisfy the wave equation provided that

$$k_x^2 + k_y^2 + \gamma^2 = \frac{\mu\epsilon\omega^2}{c^2} = \left(\frac{2\pi}{\lambda_0}\right)^2 \qquad (14\text{-}16)$$

where λ_0 is the wavelength in the free dielectric without walls.

From the boundary conditions at $x = a$ and at $y = b$

$$k_x = \frac{m\pi}{a} \qquad k_y = \frac{n\pi}{b} \qquad (14\text{-}17)$$

where m and n are integers.   Then

$$\gamma = 2\pi \sqrt{\frac{1}{\lambda_0^2} - \left(\frac{m}{2a}\right)^2 - \left(\frac{n}{2b}\right)^2} \tag{14-18}$$

For waves whose λ_0 is sufficiently small, γ is real and propagation occurs.   Longer waves are not propagated through the guide, for γ is imaginary when γ_0 is greater than a critical "cutoff" value designated λ_{0c}.   The cutoff wavelength is given by

$$\frac{1}{\lambda_{0c}^2} = \frac{m^2}{(2a)^2} + \frac{n^2}{(2b)^2} \tag{14-19}$$

The length of the wave in the guide λ_g is related to the propagation constant by

$$\lambda_g = \frac{2\pi}{\gamma} = \frac{\lambda_0}{\sqrt{1 - (m\lambda_0/2a)^2 - (n\lambda_0/2b)^2}} = \frac{\lambda_0}{\sqrt{1 - (\lambda_0/\lambda_{0c})^2}} \tag{14-20}$$

λ_g approaches infinity when λ_0 approaches the cutoff wavelength.   For shorter wavelengths, λ_g decreases and eventually approaches λ_0 when both are much smaller than λ_{0c}.

It is interesting at this point to compare the behavior of low-frequency electromagnetic waves and of light in a hollow tube with that of microwaves.   A low-frequency wave requires two insulated conductors for propagation.   It cannot go down a single-conductor waveguide because its wavelength is beyond cutoff.   Microwaves can be propagated down a waveguide, but the propagation is usually strongly affected by the boundary conditions required by the walls; i.e., the microwaves are somewhat constrained in being made to "fit into" a waveguide.   For optical wavelengths, however, the walls have relatively little effect since the fields vary rapidly with distance, and boundary conditions in a tube of ordinary dimensions are easily met.   A tube of 1 cm dimension is almost equivalent to free space for light waves.

As mentioned above, Eq. (14-20) shows that, for λ_0 just slightly smaller than the cutoff wavelength, γ is very small and hence the apparent wavelength measured along the z axis is very large.   Since the phase velocity equals ω/γ, the phase velocity may become very large and is always greater than $c/\sqrt{\mu\epsilon}$, the velocity of light in the free medium.   Of course, the group velocity must obey the principles of relativity and cannot exceed $c/\sqrt{\mu\epsilon}$.   The group velocity may be shown in the usual way to be $v_g = (c/\sqrt{\mu\epsilon})(\lambda_0/\lambda_g)$.   The group velocity is always less than $c/\sqrt{\mu\epsilon}$, approaching this value when $\lambda_0 \ll a$, or for very short wavelengths.

The modes which have been discussed are those for which $E_z = 0$, that is, the electric field is purely transverse, and are designated TE$_{mn}$ modes (transverse electric).   Equation (14-15) gives the components of **E**, and the components of **H** may be obtained from Eq. (14-15) and Eq. (14-2) as

$$H_x = - \frac{c\gamma}{\mu\omega} E_0 \sin k_x x \cos k_y y \; e^{j(\omega t - \gamma z)}$$

$$H_y = - \frac{c\gamma k_y}{\mu\omega k_x} E_0 \cos k_x x \sin k_y y \; e^{j(\omega t - \gamma z)} \qquad (14\text{-}21)$$

$$H_z = \frac{cj(k_x^2 + k_y^2)}{\mu\omega k_x} E_0 \cos k_x x \cos k_y y \; e^{j(\omega t - \gamma z)}$$

where $k_x = m\pi/a$, $k_y = n\pi/b$, $k_x^2 + k_y^2 + \gamma^2 = \mu\epsilon\omega^2/c^2$, and m and n are integers.

For any given rectangular waveguide having $a > b$, the lowest frequency, or longest λ_0, mode of the TE type which can be propagated is the one for which $m = 1$, $n = 0$. It is known as the TE_{10} mode, or the dominant mode, and is almost the only one used for microwave transmission and spectroscopy.

The complete set of field equations for the dominant mode may be written

$$E_x = 0 \qquad\qquad\qquad H_x = - \frac{c\gamma}{\mu\omega} E_0 \sin k_x x \; e^{j(\omega t - \gamma z)}$$

$$E_y = E_0 \sin k_x x \; e^{j(\omega t - \gamma z)} \qquad H_y = 0 \qquad\qquad (14\text{-}22)$$

$$E_z = 0 \qquad\qquad\qquad H_z = \frac{cjk_x}{\mu\omega} E_0 \cos k_x x \; e^{j(\omega t - \gamma z)}$$

Except near cutoff, H_z is considerably smaller than H_x, and the wave may be visualized as an electric and a magnetic vector both approximately perpendicular to the direction of propagation and traveling down the waveguide in phase. Actually H_z is by no means negligible.

An instructive viewpoint is provided by realizing that the fields in the dominant mode given above are the same as those due to two plane waves reflected successively from the walls of the waveguide so that each zigzags its way down the guide in the z direction. The velocity with which power is propagated down the waveguide is slowed up an amount appropriate to the zigzag motion of the hypothetical plane waves.

Figure 14-2 shows the instantaneous electric field and wall currents in a rectangular waveguide carrying a TE_{10} wave. Magnetic fields, which can also be obtained from the wave equations, are shown by broken lines. Wall currents are perpendicular at every point to the magnetic field.

Other modes satisfying the wave equation and boundary conditions exist, including some with a purely transverse magnetic field. The latter are designated TM waves, with the particular mode indicated by subscripts analogous to those used for TE modes. Their cutoff wavelength is also given by Eq. (14-20), but m and n must be greater than zero, so that no TM mode has a cutoff wavelength as long as that of the dominant TE mode.

Circular-cylinder waveguides also can propagate microwaves in various modes governed by the boundary condition that the electric field at the wall must be perpendicular to the wall. Both TE and TM modes exist and are designated by those letters with added subscripts. The first subscript is the number of maxima of the field in a 180° angle measured

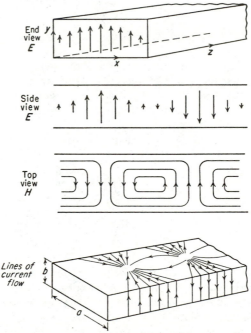

Fig. 14-2. Electric and magnetic fields and wall currents in a rectangular waveguide (TE$_{1,0}$ mode). (*After Pollard and Sturtevant* [308].)

in a plane perpendicular to the axis of the cylinder. The second subscript is the number of maxima between the center and the wall of the waveguide. The fields of some of the circular modes are shown in Fig. 14-3, along with their cutoff wavelengths. As in the rectangular waveguide, λ_g is related to λ_0 and λ_{0c} by Eq. (14-20), *i.e.*,

$$\lambda_g = \frac{\lambda_0}{\sqrt{1 - (\lambda_0/\lambda_{0c})^2}}$$

While each of these modes has uses, most microwave spectrographs employ the TE$_{01}$ mode in a rectangular guide. This mode has the advantage that, if the size of the pipe relative to the wavelength is properly chosen, no other mode can be propagated. In addition, no current flows across the center of the broad face, so that a longitudinal slot may be cut there without disturbing the fields. Finally, the electric field is always perpendicular to the broad face, so that a metal plate parallel to that face may be put inside the guide without disturbing the fields.

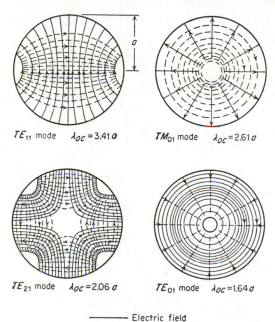

FIG. 14-3. Fields in circular waveguides.

14-3. Attenuation. Waveguide walls made of copper, brass, or silver approximate perfect conductors well enough that the field distributions discussed above are good approximations. However, since currents flow in the walls it is evident that some loss of power will accompany propagation of the wave. These losses give a small imaginary contribution to the propagation constant γ which may be fairly accurately determined by assuming that the fields in the waveguide are those appropriate to the perfectly conducting waveguide and computing the ohmic losses due to current flow.

The boundary conditions at the surface of a perfect conductor require that H must be parallel to the surface. Considering any small area of a waveguide wall over which the oscillating magnetic field has essentially constant amplitude, the solution of Eq. (14-9) inside the conducting wall (analogous to the solution discussed for E in conducting material) is

$$H_y = H_{0y}e^{j(\omega t - z/\delta) - z/\delta} \tag{14-23}$$

where $\delta = \sqrt{c^2\tau/2\pi\mu\omega}$ is the skin depth.   For copper $\delta = 3.8 \times 10^{-5} \sqrt{\lambda}$ cm, where δ and λ are in centimeters.   Axes have been oriented so that the origin is at the surface, y is in the direction of **H** over the small element of area having **H** essentially constant, and the positive z axis extends perpendicularly into the wall.

From Eq. (14-1), since $\partial \mathbf{D}/\partial t \approx 0$ in the conductor,

$$i = \frac{c}{4\pi} \nabla \times \mathbf{H} \qquad \text{or} \qquad i_x = \frac{c}{4\pi\delta} (1 + j)H_{0y}e^{j(\omega t - z/\delta) - z/\delta} \qquad (14\text{-}24)$$

The power loss due to ohmic resistance per unit volume is $(i_R^2)_{\text{av}}\tau$ or $\frac{1}{2}ii^*\tau$, where i_R is the real part of Eq. (14-24).   Integrating over z from 0 to ∞, the power loss per unit area is obtained as

$$\text{Power loss/area} = \int_0^\infty \frac{ii^*}{2} \tau \, dz = \frac{c^2\tau}{32\pi^2\delta} H_{0y}^2 = \frac{\mu\nu\delta}{8} H_{0y}^2 \qquad (14\text{-}25)$$

where ν is the frequency.   The last expression for power loss has a simple interpretation.   The total energy per unit area built up in the magnetic field inside the metal on every half cycle is approximately $(\mu H_{0y}^2/8\pi)\delta$ since $\mu H_{0y}^2/8\pi$ is the energy density at the metal surface and δ is approximately the depth of penetration of the field.   If all this energy associated with the magnetic field in the metal were dissipated in the metal on every half cycle, then the rate of power loss per unit area would be

$$2\nu \frac{\mu H_{0y}^2}{8\pi} \delta$$

which is approximately the same as Eq. (14-25).   Thus the energy of the electromagnetic field penetrating the metal may be considered lost on every half cycle.

If the behavior of some field component of a microwave is described by $E = E_0 e^{j(\omega t - \beta z) - \frac{\alpha}{2}z}$, where α and β are real, the constant $\frac{\alpha}{2}$ is sometimes called the attenuation constant.   Its unit is nepers per unit length. More commonly the quantity α is called the attenuation constant, since the power or energy associated with the wave depends on the square of field strength, or as $P = P_0 e^{-\alpha z}$.   The attenuation constant α for power is given in units of inverse length, such as cm^{-1}.   It can be calculated by computing the power flow through a waveguide and the power loss per unit length.   Thus

$$\alpha = \frac{\displaystyle\int_{S_1} \frac{\mu\nu\delta}{8} H_{S_1}^2 \, dS_1}{\displaystyle\int_{S_2} \frac{c}{4\pi\mu} \mathbf{E} \times \mathbf{B} \cdot d\mathbf{S}_2} \qquad (14\text{-}26)$$

where H_{s_1} is the component of $\mathbf{H}$ at the surface of the waveguide, S_1 is the surface of the waveguide walls of unit length, S_2 is the cross-sectional surface of the waveguide (perpendicular to the direction of power flow).

Evaluation of α for the TE_{10} mode gives, for a silver waveguide at wavelengths well below the cutoff wavelength

$$\alpha = \frac{3.384 \times 10^{-4}}{a^{\frac{3}{2}}} \left\{ \frac{(a/2b)(2a/\lambda_0)^{\frac{3}{2}} + (2a/\lambda_0)^{-\frac{1}{2}}}{[(2a/\lambda_0)^2 - 1]^{\frac{1}{2}}} \right\} \qquad \text{cm}^{-1} \quad (14\text{-}27)$$

where a is the long dimension and b is the short dimension of the waveguide, both being measured in centimeters. λ_0 is the free-space wavelength (in centimeters) of the radiation. To obtain the power loss in decibels per centimeter, the α given by Eq. (14-27) should be multiplied by 4.343.

For metals other than silver, the attenuation is similar to that given by (14-27), but is proportional to the square root of the resistivity. For copper waveguide, the attenuation given by Eq. (14-27) is to be multiplied by 1.05; for gold the factor is 1.26 and for brass it is 2.08.

At the cutoff frequency, the attenuation given by this expression is infinite, and for slightly higher frequencies the attenuation falls rapidly. The large attenuation near the cutoff wavelength is to be expected from the equivalence of the wave to two plane waves zigzagging across the waveguide. Near cutoff the plane waves are reflected back and forth between the walls many times while moving down the waveguide one wavelength. There is consequently opportunity for considerable loss of power. The attenuation passes through a minimum at higher frequencies and then rises approximately proportionally to the frequency. For typical "K-band" copper waveguide, having inside dimensions 0.420 by 0.170 in., Eq. (14-25) gives for $\lambda_0 = 1.25$ cm, $\alpha = 1.1 \times 10^{-3}$ cm^{-1}, or 0.48 db/meter. The actual measured attenuation is likely to be between 20 and 100 per cent greater because of surface imperfections ([312], pp. 117, 191). Similar expressions may be given for other modes and other types of waveguides, most of which give attenuation of the same order of magnitude, and with analogous frequency dependence. An exceptional case is the TE_{01} mode in a circular waveguide for which the attenuation decreases without limit as the frequency increases. If, in going to higher frequencies, the size of the waveguide is changed in proportion to the wavelength, the attenuation increases as $a^{-\frac{3}{2}}$, or $\lambda_0^{-\frac{3}{2}}$. If the size of guide is not changed, the attenuation increases for smaller wavelength only as $\lambda_0^{-\frac{1}{2}}$. A waveguide larger than that required to avoid cutoff is often used in the millimeter-wavelength region because of its lower attenuation. For wavelengths beyond cutoff ($\lambda_0 > \lambda_{0c}$), the attenuation is given by $\alpha = 54.6[1/\lambda_{0c}^2 - 1/\lambda_0^2]^{\frac{1}{2}}$ db/unit length and the impedance is reactive. Waveguides beyond cutoff are sometimes used as standards of attenuation.

14-4. Reflections in Waveguides. Reflection occurs whenever a wave in a guide encounters any irregularity. The irregularity may be a change in guide size, a bend or twist, or an obstacle. A treatment of reflection at irregularities is very complicated for the general waveguide case because the obstacle is likely to introduce higher modes into the reflected and transmitted waves. Very great simplifications are possible, however, when only one mode can be propagated at the wavelength considered. Then a reflected wave can be propagated only in the same mode as the incident wave, as in the case of a low-frequency transmission line. The analogy between a single-mode waveguide and a transmission line is very far-reaching and is quite useful because of the familiarity of transmission-line theory and behavior.

A transmission line for electrical impulses usually consists of two parallel conductors with current flowing along each conductor. Current also flows between the two conductors either because of current paths purposely introduced or because of the inevitable interconductor capacitance and leakage resistance. Let the series impedance of the line per unit length be T, and the shunt impedance per unit length be S. Then (cf. [117], Chap. 1) the instantaneous current at any point distant z along the line is given by

$$i = i_0 \exp j\omega t \pm \sqrt{T/S}\, z \qquad (14\text{-}28)$$

where $i_0 e^{j\omega t}$ represents the current at $z = 0$. The potential difference V between the conductors at any point is obtained by multiplying their shunt impedance per unit length by the current flowing through this impedance. The latter is $-di/dz$ since the shunt current is that which is lost from the main line. Therefore

$$V = \mp \sqrt{TS}\, i \qquad (14\text{-}29)$$

The characteristic impedance of the transmission line is defined as $Z = V/i$. If the wave travels toward positive z ($i = i_0 e^{j\omega t - \sqrt{T/S}z}$), $Z = \sqrt{TS}$, and if the wave is traveling toward negative z the impedance appears to be $Z = -\sqrt{TS}$. Actually the impedance is usually taken as positive for both directions and $V = Zi$ for the wave advancing in the z direction, $V = -Zi$ for the reversed wave.

If the transmission line is not lossy, then T is usually purely inductive and may be written $T = j\omega L$, while S is purely capacitive and is $S = 1/j\omega C$. Hence $Z = \sqrt{L/C}$ and

$$i = i_0 e^{j(\omega t \pm \gamma z)} \qquad (14\text{-}30)$$

where $\gamma = \sqrt{LC}$ is the propagation constant. Expressed in this form, it is clear that propagation of electromagnetic waves along a transmission line has much in common with the propagation of microwaves in a waveguide or in free space.

Instead of propagation on an infinitely long uniform transmission line, consider now the behavior of a current wave at a junction between two transmission lines of impedance Z_0 and Z_1 as indicated in Fig. 14-4. At the junction itself the current and voltage can be computed from the point of view of either transmission line, which must of course give equivalent results. Thus the current is either the algebraic sum of an incident current wave i_i and the reflected wave i_r or equals the transmitted wave i_t. Hence when $z = 0$,

$$i_i - i_r = i_t \qquad (14\text{-}31)$$

and

$$i_i Z_0 + i_r Z_0 = i_t Z_1 \qquad (14\text{-}32)$$

where the subscripts i, r, and t denote, respectively, the incident, reflected, and transmitted currents. It is important to remember that in Eq. (14-32) for a wave traveling to the right in Fig. 14-4, $V = iZ$, but for one traveling to the left, $V = -iZ$. Solving Eqs. (14-31) and (14-32) for i_t and i_r when $z = 0$,

Fig. 14-4. Currents at a junction of transmission lines of impedance Z_0 and Z_1.

$$i_t = \frac{2Z_0}{Z_1 + Z_0}\, i_i \qquad (14\text{-}33)$$

$$i_r = \frac{Z_1 - Z_0}{Z_1 + Z_0}\, i_i \qquad (14\text{-}34)$$

At any point z to the left of the junction a new impedance may be defined as the ratio of voltage to current

$$Z = \frac{V}{i} = \frac{i_i Z_0 + i_r Z_0}{i_i - i_r} = \frac{Z_0 \left(e^{-i\gamma z} + \dfrac{Z_1 - Z_0}{Z_1 + Z_0} e^{+i\gamma z} \right)}{e^{-i\gamma z} - \dfrac{Z_1 - Z_0}{Z_1 + Z_0} e^{+i\gamma z}}$$

$$= Z_0 \left[\frac{-\sinh j\gamma z + (Z_1/Z_0)\cosh j\gamma z}{\cosh j\gamma z - (Z_1/Z_0)\sinh j\gamma z} \right] \qquad (14\text{-}35)$$

It is to be noted that, for points to the left of $z = 0$, z is a negative quantity. If $Z_1 = Z_0$, from Eq. (14-34) it is seen that there is no reflected wave. This condition is usually desirable because it gives a maximum power transfer and because it avoids standing waves on the transmission line.

If the transmission line to the right of $z = 0$ consists simply of a short circuit ($Z_1 = 0$) and γ is real, then from Eq. (14-35), $Z = -Z_0 j \tan \gamma z$, so that the impedance changes from zero to infinity when z varies from 0 to $-\pi/2\gamma$. Thus the short circuit at $z = 0$ produces what appears to be an open-circuited line at $z = -\pi/2\gamma$ (or $-\lambda/4$), and the impedance varies periodically with z between the extremes of zero and infinity. If, however, there is some loss in the transmission line to the left of $z = 0$,

γ is partly imaginary ($\gamma = j\alpha/2 + \beta$), and some distance from the junction $z = 0$ the periodic impedance variation decreases in magnitude and Z approaches a constant value Z_0.

The impedance concept is particularly useful for transmission lines because Eq. (14-35) permits calculation of the effects of a discontinuity at any point on the line, provided that the propagation constant γ is known. That is, any discontinuity can be expressed as an equivalent impedance at that place, or by Eq. (14-35) at any other place on the line.

For microwave transmission in waveguides these impedance concepts and formulas are still useful. Since there is only one conductor and since different current densities flow over the various waveguide surfaces, both voltage and current are difficult to define. Although no unique definition of impedance can be made, it is customary to define the impedance as the ratio of electric to magnetic field strengths (E/H) at some point in the guide. It is not strictly necessary to define the characteristic impedance. Z_0 can be taken as an arbitrary constant, and other impedances, such as those due to obstacles in the guide, can be expressed in terms of Z_0 because Eq. (14-35) involves only ratios Z/Z_0 and Z_1/Z_0. Discontinuity in the guide propagation due to a sudden change of dielectric properties is easily interpreted as a sudden change in impedance, and the reflected-wave characteristics are given by Eq. (14-34).

However, if the waveguide suddenly changes dimensions, the reflected wave is not so easily obtained since it depends on the particular geometry at the discontinuity. Near the discontinuity the fields are complicated and can be represented as a combination of higher-order propagation modes with the fundamental modes. However, if the higher-order modes cannot be propagated, a wavelength or two from the discontinuity on the incident side one finds only an incident and a reflected wave in the dominant mode. As in a transmission line, these waves can be expressed in terms of an effective impedance at the position of the discontinuity which replaces the discontinuity for the calculation of the fields in the waveguide. Near the discontinuity higher modes are present and this approach cannot be used. It makes difficulty also if two discontinuities occur within a half wavelength or so, for then the higher-order waves from one discontinuity can reach the other before being completely attenuated. If several modes can be propagated, a different discontinuity impedance and propagation constant must be given for each mode; so the concept loses much of its usefulness. Moreover, a reflection can introduce a mode which was not present in the original wave.

In many cases it is desirable to minimize reflections at the junctions of transmission lines by making the terminating impedance for the first line equal to its characteristic impedance. This may be done at low frequencies by a simple transformer involving mutual inductance between two coils. A quarter wavelength of transmission line of characteristic impedance Z' between the two lines of characteristic impedances Z_0 and Z_1 can also act as a matching transformer. Thus if the junction between Z' and Z_1 is at $z = 0$ and Z' extends one-quarter wavelength in the negative z direction the impedance of the combination is, from Eq. (14-35), $Z = (Z')^2/Z_1$. If it is desired to make Z equal Z_0 so that the transmission lines may be joined with no reflections, then $Z' = \sqrt{Z_1 Z_0}$. Reflection can be eliminated by a number of other types of "transformers." Generally it is necessary to have two independent variables such as, in this case, the length and characteristic impedance of the auxiliary transmission line.

In many cases another approach based on simple ideas of wave interference and diffraction can be used to investigate microwave behavior at a discontinuity. For example, the quarter-wave transformer discussed above may be viewed as setting up two reflected waves. These waves are of equal magnitude because the ratios of characteristic impedance at the two reflecting discontinuities are equal. They are just opposite in phase and hence cancel because the path length for one is one-half wavelength (two quarter wavelengths) longer than for the other. The principle is the same as that used for making nonreflecting glass, where a quarter wavelength of surface material with index of refraction intermediate between air and glass produces two canceling reflections, or "matches impedances."

In many cases it is desirable to have a transformer produce little reflection over a wide range of frequencies. The quarter-wave transformer and many other impedance matching devices depend somewhat critically on the ratio of certain distance parameters to the wavelength and hence are not useful over a very wide frequency range. However, any change in waveguide dimensions or impedance which is so gradual that only a small fractional change occurs over one wavelength affords a good transformer from one impedance to another and shows approximately the same characteristics over a wide frequency range. A gradual change of impedance such as a long taper from one dimension of waveguide to another can easily be seen to produce very little reflection of power because for any reflecting point in the taper an essentially similar reflecting point can be found one-quarter wavelength farther down the taper and the reflected waves from these two will cancel.

Still another important method of reducing reflection is by means of attenuation. Equation (14-35) shows that the impedance approaches the characteristic impedance as z decreases if there is attenuation in the transmission line. This is because any reflected wave decreases as a result of the attenuation. A short section of highly attenuating waveguide will give the same effect. Such a section is customarily inserted between a klystron oscillator and a waveguide to minimize the effect of reflected waves in the guide on the operation of the klystron.

14-5. Cavity Resonators. A hollow space enclosed by metallic walls can support electromagnetic waves of some particular wavelengths in modes which fit the boundary conditions. Consider, for example, a section of rectangular waveguide which is closed at both ends by flat conducting plates (Fig. 14-5). The boundary conditions at the side walls are satisfied for any TE_{mn} waves provided only that their wavelength is short enough for propagation. If the length of the waveguide is l and the wavelength in the guide is λ_g for a particular TE_{mn} mode, the end boundary conditions require that E_x and E_y vanish at $z = 0$ and $z = l$. From Eq. (14-15) this will be satisfied if $e^{j(\omega t - \gamma l)} = \pm e^{j\omega t}$, or $\gamma l = \pi p$, where p is an integer. Since $\gamma = 2\pi/\lambda_g$,

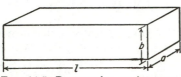

FIG. 14-5. Rectangular cavity resonator.

$$p = \frac{2l}{\lambda_g} \qquad (14\text{-}36)$$

This relation is just what would be expected for a wave of length λ_g which is reflected with a 180° change of phase at the end of the guide so that a standing wave is produced. For nodes at $z = 0$ and $z = l$, the guide must contain an integral number of half wavelengths.

From Eqs. (14-36) and (14-20), the values of λ_0 for which the cavity is resonant may be obtained.

$$l = \frac{p\lambda_g}{2} = \frac{p\lambda_0}{2 \sqrt{1 - (m\lambda_0/2a)^2 - (n\lambda_0/2b)^2}} \qquad (14\text{-}37)$$

so that

$$\left(\frac{m\lambda_0}{2a}\right)^2 + \left(\frac{n\lambda_0}{2b}\right)^2 + \left(\frac{p\lambda_0}{2l}\right)^2 = 1 \qquad (14\text{-}38)$$

From this relation, the wavelength λ_0 at which the cavity is resonant may be determined if its dimensions are known.

Rectangular cavities are not much used as resonators although a section of waveguide with partial reflections at each end behaves somewhat like a resonator.

Circular cylindrical cavities are more often used because they can easily be accurately machined and are suitable for wavelength measurements. As for a rectangular cavity, the resonance condition is that the length of the cavity be an integral multiple of $\lambda_g/2$. For either a TE or TM mode in a circular waveguide, the free-space resonant wavelength is given by

$$\lambda_0 = \frac{2}{\sqrt{(2x_{mn}/\pi D)^2 + (p/L)^2}}$$

where D = the diameter and L the length of the cavity

m = the number of half-period variations of E_r with respect to θ (of H_r for TM modes)

n = the number of half-period variations of E_θ with respect to r (of H_θ for TM modes)

p = the number of half-period variations of E_r with respect to z (of H_r for TM modes)

x_{mn} = the nth root of $J_m'(x) = 0$ for the TE modes or of $J_m(x) = 0$ for the TM modes; the J's are Bessel functions

Table 14-1 gives the values of some of the roots. For modes with $p = 0$ the field does not vary in the z direction and so the resonant wavelength does not depend on the length of the cavity, but only on its diameter. This can occur only for TM modes. For wavemeter applications, the TE_{01p} mode is commonly used.

TABLE 14-1. ROOTS OF $J_m(x)$ AND $J_m'(x)$

TE mode	x_{mn}	TM mode	x_{mn}
11p	1.841	01p	2.405
21p	3.054	11p	3.832
01p	3.832	21p	5.136
31p	4.201	02p	5.520
41p	5.318	31p	6.380
12p	5.332	12p	7.016

A useful figure of merit for a resonator is the Q, defined as

$$Q = 2\pi \frac{\text{electromagnetic energy}}{\text{energy lost per cycle}} = -2\pi W \frac{\nu \, dt}{dW} \qquad (14\text{-}39)$$

From this definition it follows that the energy in a freely oscillating resonator decays according to the equation

$$W = W_0 e^{-\omega t/Q} \qquad (14\text{-}40)$$

where W is the energy remaining after time t and W_0 is the initial energy at $t = 0$. Q is also a measure of the sharpness of resonance, for the width in frequency $\Delta\nu$ of the resonance curve between points at which the response is one-half that at maximum is

$$\frac{\Delta\nu}{\nu_0} = \frac{1}{Q} \qquad (14\text{-}41)$$

where ν_0 is the resonant frequency (*cf.* [121], p. 137).

Q may be calculated from Eq. (14-39) by a procedure resembling that used to calculate attenuation (*cf.* [486], pp. 73–77). For instance, for a rectangular box with $l = a$, for the TE_{101} mode (**E** parallel to the b

dimension)

$$Q = 0.71 \frac{\lambda_0}{\delta} \frac{b}{a + 2b} \qquad (14\text{-}42)$$

where δ is the skin depth and λ_0 the wavelength. Since for copper $\delta = 3.8 \times 10^{-5} \sqrt{\lambda_0}$ when δ and λ_0 are in centimeters, a cubic resonator in this mode at $\lambda = 1$ cm would have a Q value of 6200. This Q is much higher than those obtainable with coils and condensers at frequencies where the latter are appropriate, but still higher Q's can be obtained with other shapes or larger dimensions ([221], Chaps. 5 and 6). In general the Q is roughly proportional to the volume divided by the surface area because stored energy depends on the volume while losses occur only at the walls. An exceptional case is the TE_{01p} mode of a circular cylinder at high frequencies. For this mode the side-wall losses decrease continuously with increasing frequency, and it may be used to obtain very high Q's.

14-6. Coupling of Cavities to Waveguide. In practice a cavity must be coupled to a microwave system. Then, in addition to the losses at the walls, there will be some energy delivered from the cavity to the system, so that the energy loss is greater, and the Q less, than for a completely enclosed cavity. As before, $Q_E = 2\pi \times$ (electromagnetic energy in the cavity)/(energy lost per cycle), where the denominator is the total energy lost and the subscript E denotes the "external" Q with a coupling connection. This may be expressed in terms of the "internal" unloaded Q_U, involving only wall losses, and the "radiation" Q_R involving only losses from the cavity to the external circuit, for

$$\frac{1}{Q_E} = \frac{1}{Q_U} + \frac{1}{Q_R} \qquad (14\text{-}43)$$

The resonant frequency of the cavity may also be affected by the connection to the microwave circuit if the latter is reactive. However, the coupling is often not large enough to affect greatly either the Q or the resonant frequency ([221], Chap. 5), although in some applications these effects must be considered.

Coupling to a cavity is usually accomplished by one or two holes in the cavity wall. At lower frequencies (3000 Mc or less), coupling loops entering through a suitable hole provide a direct connection to a coaxial line. Usually the coupling is designed to excite preferentially some particular mode by making the electric or magnetic lines of force at the coupling region coincide with those of the desired mode. For instance, the TE_{01p} mode in a cylindrical cavity may be excited by two diametrically opposite holes in the end wall [178]. These holes coincide with holes in the narrow face of the waveguide approximately $\lambda_g/2$ apart. The only field existing at the narrow edge of a waveguide carrying the dominant mode is a longitudinal magnetic field. Through the coupling holes this field is transmitted to the cavity with opposite phases for the two holes, and so matches the TE_{01p} mode, but not other possible modes except the TE_{02p} mode. The latter may be avoided in a transmission-type resonator by taking the output from a coupling hole in the curved wall of the cylinder at a 45° angle to the input holes. Then an output is obtained only if the cavity is at or very near resonance for the desired mode.

The TE_{01p} mode has the advantage for wavemeter applications of a high Q, so that resonances are sharp. Moreover, there are no radial currents at the edges of the end plates, so that the latter need not make good contact with the walls. This is particularly useful in wavemeters where one end plate must be movable for tuning.

Another mode which is often used for wavemeters is the TE_{11p} mode. This is the lowest-frequency mode for a given cavity, and so the diameter of the cavity may be chosen to permit only this mode over an appreciable band of wavelengths (around 25 per cent—cf. Table 14-1). However, this mode is polarized in that it has a plane of symmetry containing the axis of the cylinder (cf. Fig. 14-3) and can be resolved into similar modes in any pair of perpendicular directions. If the cylinder is slightly elliptical, these modes have different propagation constants and so give slightly different resonant frequencies. A horseshoe-shaped strap across the coupling hole perpendicular to the electric vector in the waveguide produces a large effective eccentricity so that these resonances are far apart and cross coupling is prevented. Only the mode with polarization parallel to the electric vector in the waveguide is excited.

Good effective contact between the walls and the plunger is obtained by a quarter-wave "choke." That is, the plunger is one-half wavelength long, and in the back is cut a groove one-quarter wavelength deep. The wavelength involved here is the free-space wavelength since either in the groove or in the space between the plunger and the walls waves travel in the coaxial TE_{11} mode for which λ_g is very close to λ. The effect of the groove is to produce a high impedance at the edge of the plunger for waves flowing toward the back of the plunger and so to confine the radiation to the region in front of the plunger. Any radiation which does reach the region behind the plunger and might cause spurious resonances is further reduced by some suitable absorbing material such as "poly-iron" ([221], p. 723).

Resonant cavities may be used as either transmission or reaction wavemeters. For use as a transmission-type wavemeter, the cavity of Fig. 14-6 is modified to use two coupling holes, diametrically opposed on the curved wall $\lambda_g/4$ from the base. Then the transmission is almost zero until near resonance it rises to a large value. A typical wavemeter might transmit 25 per cent of the incident power at resonance.

Alternatively, the wavemeter may be coupled to a hole in the side of the waveguide. At the end of the waveguide is a crystal detector which matches the waveguide well enough to absorb most of the power reaching it. In the guide directly opposite the wavemeter is a post or "iris" diaphragm which is designed to cancel out as nearly as possible the reflections produced by the wavemeter when not in resonance. At resonance the wavemeter absorbs some power and reflects more, so that the power reaching the detector is reduced. Sometimes the detector is also coupled to the side of the waveguide before the wavemeter and detects a change in the reflected power. Depending on the relative phases of the wavemeter reflection and other reflections in the system, the crystal current may either increase or decrease at resonance. In either

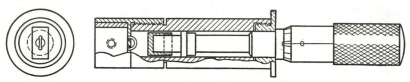

FIG. 14-6. A typical wavemeter for the 1.25-cm region (MIT Radiation Laboratory type TFK-2). Barrel diameter (inside) = 0.3750 ± 0.0002 in. Coupling hole diameter = 0.111 in.

case the detector used with a reaction wavemeter may also be used to monitor the output of the microwave oscillator, for a signal is obtained as long as the wavemeter is not at resonance.

A loaded Q of the order of 3500 may be expected from a wavemeter in the 1-cm region, such as that shown in Fig. 14-6. As the plunger is moved by the micrometer screw two or three resonances separated by $\lambda_g/2$ are reached, and from λ_g the wavelength λ is readily calculated. In practice, if many wavelength readings are to be made, a conversion table from λ_g to wavelength or frequency is worth the trouble required to make it.

14-7. Directional Couplers. It is often desirable to monitor the power delivered by a microwave generator to a waveguide system. For this purpose a directional coupler is very useful, for it permits a sampling of the power from the oscillator without any of the reflected waves from the system reaching the monitor detector.

The easiest type of directional coupler to understand is analogous to directional antennas. Two holes approximately $\lambda_g/4$ apart are cut in the narrow edge of a waveguide which is to carry waves in the dominant mode. The holes match similar holes in the narrow edge of another waveguide (Fig. 14-7). Then a wave traveling from left to right in waveguide (1) will reach hole B one-quarter of a period later than hole A. A wave in the same direction starting at the same time and passing through hole A will reach hole B at the same time and the two will reinforce to produce a combined wave in waveguide (2) in the same direction.

On the other hand, a wave going through (1) to hole B and then back to A will arrive at A one-half wavelength behind a wave which passes directly through A and so be canceled by the wave from A. Thus a wave traveling from left to right in waveguide (1) will produce only a wave in the same direction in waveguide (2). A reflected wave traveling from right to left in (1) will produce only a wave traveling from right to left in (2), which will be absorbed by a suitable termination of the left end of (2). A wedge-shaped strip of lossy dielectric can be made to absorb this wave without serious reflection.

A detector at the right end of (2) then receives only waves from the oscillator, and none of the reflected power. It is thus suitable for monitoring the oscillator output power. A cavity wavemeter connected at

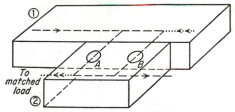

Fig. 14-7. Two-hole directional coupler.

this point is assured of a reliable power input regardless of reflections and in turn causes a minimum disturbance to the rest of the microwave circuit. For wavemeter connection a directional coupler which transmits from 1 to 10 per cent of the incident power into (2) is useful ("a 20-db" or "a 10-db" coupler).

Other types of directional couplers have been used which are merely variants of that described. For instance, the coupling holes may be in the broad face of the waveguide as long as they are far enough from the center line or wide enough to provide sufficient radiation. It should be mentioned that for either type the holes are usually large enough so that they cannot be treated exactly as points, and so their optimum spacing may not be exactly $\lambda_g/4$. In any case the spacing is not excessively critical so that the coupler is useful over a considerable wavelength band. However, better directivity (forward wave to back wave amplitude ratio) over a wide wavelength band can be obtained by using more than two coupling holes, with neighboring holes $\lambda_g/4$ apart [222]. The diameters of the holes are chosen so that α_r, the amplitude of the wave originating at the rth hole, is given by

$$\alpha_r = \alpha_0 \frac{n!}{r!(n-r)!} \tag{14-44}$$

where $n + 1$ is the number of holes. That is, the amplitudes are proportional to the binomial-expansion coefficients. Then the amplitude of the reflected wave is proportional to $\cos^n (2\pi d/\lambda_g)$, where d is the distance between successive holes. The reflected wave is zero for $\lambda_g = 4d$ regardless of the number of holes, but the larger n is, the more slowly the reflected wave increases when λ_g deviates from $4d$. Many other varieties of directional coupler have been used, and several are described in [221], Chap. 14.

Another waveguide device related to the directional coupler is the "magic T" shown in Fig. 14-8. Its operation may be understood from the fact that the plane bisecting arms C and D is a plane of symmetry.

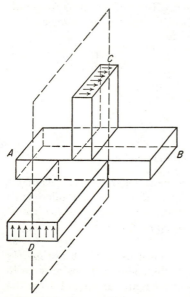

Thus a wave entering through branch D in the principal mode has its electric vector symmetric about that plane. However, a wave which can be propagated in C must have an electric vector which is antisymmetric with regard to the symmetry plane. Thus a wave entering from either C or D will excite waves in A and B, either symmetrically or antisymmetrically, but waves will not pass directly from C to D. Similarly waves from A or B can excite waves in C or D, but the amplitude in C will be proportional to the difference, and in D to the sum, of the amplitudes of the waves entering through C and D.

In a typical application as a directional coupler, the wave from the oscillator enters through A, and the waveguide system is connected to C.

Fig. 14-8. Waveguide "magic T."

D has a power-monitoring crystal detector or wavemeter, and B has a matched load such as a tapered lossy plastic strip designed to prevent reflections. Power is then transmitted to C and D, but reflected power from C cannot enter D. A monitor in D measures only the power delivered by the oscillator to the load, regardless of reflections.

Another important application of the magic T is to balanced systems analogous to bridge circuits in ordinary electrical measurements. Use is made of the fact that arm C gives a signal depending on the difference of the signals in A and B. It can then be used to show departures from balance in arms A and B.

To prevent reflections by the magic T itself, it is usually necessary to match it to the waveguides by "irises" or diaphragms in the waveguide positioned so as to reflect a wave of phase opposite to, and amplitude equal to, the reflection from the magic T.

14-8. Attenuators. Most waveguide systems contain at least one attenuator. Sometimes an attenuator is used to reduce power input to the system, *e.g.*, to prevent power saturation in a spectroscope absorption cell. Equally often, the attenuator is inserted principally to prevent reflections in the waveguide system from reaching the oscillator and affecting its operation.

It is always desirable, and sometimes very important, that standing waves be kept low to avoid variations in power at the detector as the frequency is varied. Attenuation in the system helps to reduce standing waves since reflected waves must pass an attenuator twice in each back-and-forth trip through the waveguide.

Most commonly attenuation is obtained by inserting a strip of lossy material such as bakelite coated with carbon (with surface resistance of a few hundred ohms per square) into the guide through a longitudinal slit in the center of the broad face. The strip is a few wave-

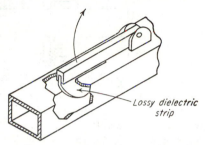

Lossy dielectric strip

FIG. 14-9. Waveguide attenuator.

lengths long and is tapered to prevent reflections from the ends. Often it is pivoted at one end, and the attenuation is controlled by lowering the other end into the guide (see Fig. 14-9).

A fixed attenuator can be constructed in many ways. One of the simplest is to attach a strip of lossy dielectric to the narrow wall of the waveguide. In order that the attenuator will not itself introduce reflections its ends should be tapered. With carefully made attenuators of this type, it is possible to reduce the variation in transmitted power due to reflections to as low as about 1 per cent.

When the attenuation must be known absolutely, a short section of waveguide which is too small to propagate the wavelength produced by the oscillator may be used. The attenuation can then be calculated from the dimensions of the guide and the free-space wavelength (*cf.* page 385).

14-9. Joints in Waveguide Systems. Waveguide sections and devices are commonly joined by attaching suitable flanges and bolting them together. The flanges need only be large enough so that they can be conveniently bolted, but they must be soldered carefully flush with the ends of the waveguide. Especially if the guide is to transmit wavelengths short enough so that higher modes could be propagated, it is important to obtain good alignment of the two guides and locating pins may be provided for this purpose.

In a flat-flange joint, the longitudinal current components are carried by the capacitance between the flanges even if the contact between them is poor. However, there is a small voltage drop across the joint, and if the guide is to carry high power, as from a magnetron, arcing may occur at the joint, or there may be a small leakage of power. For these applications "choke" flanges are useful. A choke-flange joint is shown in Fig. 14-10. It differs from a flat-flange joint in that a groove λ/4 deep is cut in one flange. The inner and outer walls of this groove act as conductors of a quarter-wave coaxial line, short-circuited at one end and

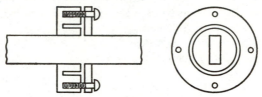

FIG. 14-10. Choke-flange waveguide joint.

thus having a very large impedance at the open end. This large impedance makes difficult a radial flow of power between the flanges. If, moreover, the distance from the groove to the waveguide is roughly another quarter wavelength, then the impedance at the waveguide wall is very small. This permits the wall currents to flow without large potential drops.

Since the dimensions of a choke flange depend on the wavelength, it is frequency-sensitive and so not well suited to equipment required to operate over large frequency ranges.

If leakage from flanges must be minimized, the completed joint may be wrapped in steel wool. If this is not done, waves radiated from the joint may be reflected from objects in the room and reenter the waveguide to cause a small change in the signal which varies as people move in the room. This precaution is usually needed only for spectrographs which do not employ modulation because they are sensitive to any change in microwave power at the detector, no matter how slowly the change occurs.

14-10. Waveguide Windows. It is often necessary to confine a gas to one section of a waveguide system. Windows for this purpose must be vacuumtight and yet reflect or absorb very little microwave power. A suitable window may be constructed of 0.001-in.-thick mica sheet sealed by a flat gasket stamped from rubber or polyethylene sheet a few mils thick and sandwiched between flat flanges. "Dental-dam" rubber obtained from dental supply houses is suitable for gaskets. A light coating of stopcock grease is used on the gasket.

For higher-temperature operation or for spectroscopy on certain reactive compounds, the rubber gaskets may be replaced by solder, lead, or gold, and thin quartz windows have been used. A design for mica windows with lead gaskets has been described [637]. If the flanges are made sufficiently sturdy and the area of contact with the mica sufficiently small so that considerable pressure can be applied to the mica window, a good vacuum seal can be obtained without the use of a gasket. Mica windows may also be sealed permanently to metals. Such windows have been used in klystron tubes and might be applied to spectroscopy.

14-11. Plungers. Often, as in a cavity wavemeter or in a matching section, a waveguide must be terminated by a movable short-circuiting plunger. If the mode is such that currents flow across the gap between the plunger and the walls, care must be taken to provide a dependable low-impedance path for them. Sometimes, particularly at lower microwave frequencies, spring contacting fingers are sufficient. A good effective contact may be obtained by using the principle of the choke flange,

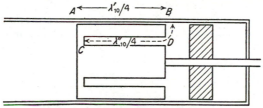

FIG. 14-11. Waveguide choke plunger.

as in Fig. 14-11, which shows a type of plunger often used in wavemeters. The inner and outer walls of the slot CD act as the conductors of a quarter-wave coaxial line short-circuited at C. Thus at D there is a high impedance which is transformed by the quarter-wave section BA to give a very low effective impedance at A. The lengths λ'_{10} and λ''_{10} are not free-space wavelengths and are not equal because the coaxial lines involved are not excited in the principal mode but in the TE_{10} mode for the appropriate coaxial waveguide ([312], Chap. 8). This type of device is of course designed for a particular wavelength but is fairly satisfactory over a wavelength range of 5 to 10 per cent.

14-12. Other Types of Guided Waves. It is possible to confine microwave radiation to a restricted region without requiring that region to be part of a hollow-tube waveguide. For instance, a horn radiator can be used to beam radiation toward a similarly directional receiving horn. A glass or quartz absorption cell between the horns could have the advantage of bringing no metal in contact with the sample, which would be helpful for corrosive materials.

If the radiation from the horns is polarized, thin metal plates can be placed in the path of the beam with their planes perpendicular to the electric vector without disturbing the fields. A low-frequency electric field can then be applied between the plates for Stark modulation.

When the horns are arranged to confine the radiation to the space between the plates, the plates constitute a parallel-plate waveguide. Alternatively the ordinary rectangular waveguide in the dominant mode may be imagined to have its width increased indefinitely. The electric field remains perpendicular to the broad faces and has a maximum value in the center, decreasing toward the now distant walls. If the extreme edges are then cut off at a point where the field is sufficiently small, the remaining portion behaves very much as the infinite-parallel-plate guide. The coupling horns serve to direct the waves from ordinary waveguides to the central region of the flat plate guide.

The use of parallel-plate waveguides for microwave spectroscopy has been suggested by Gordy [278] and investigated by Baird, Fristrom, and Sirvetz [430]. It has the advantage that small spacings can be used, and a large uniform d-c field can be applied between the plates. Moreover, the insulating spacers may be kept entirely outside the region occupied by the microwave energy so that their dielectric-loss properties are unimportant. For this reason, parallel-plate waveguides would appear to be

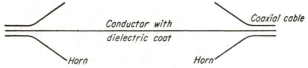

Fig. 14-12. Surface wave transmission line.

useful for operation at high temperatures where the insulators used to support Stark-effect electrodes inside ordinary waveguides give trouble. However, unless very large horns are used, the radiation tends to spread and produce rather bad frequency-sensitive reflections from obstacles outside the absorption cell.

Some of the advantages of the free-space waveguide cell, without the need for extremely large horns to provide high directivity, might be gained by using a single-conductor surface wave transmission line [575a]. This consists of a single wire with a thin coating of dielectric. The wave is launched and received by coaxial cables having their outer conductors flared to form horns (Fig. 14-12).

14-13. Microwave Applications of Ferrites. The name "ferrite" is applied to a group of materials which have high magnetic permeability and low electrical conductivity. Because of their low conductivity, microwaves can propagate through them without excessive loss. When a magnetic field is applied to a ferrite, the unpaired electron spins, which produce the high permeability, precess. A broad resonance of this precession is obtained at microwave frequencies in fields of a few thousand oersteds.

For fields lower than that needed to give resonance, a large Faraday rotation of the plane of polarization is obtained.   Figure 14-13 shows one way in which this rotation can be applied to provide variable attenuation.

The incoming wave is plane-polarized, and in a circular waveguide section the ferrite rotates the plane of polarization.   If the plane of polarization is rotated 45° in one direction, the wave is transmitted by the output waveguide, but if it is rotated 45° in the opposite direction, the output waveguide cannot transmit it.   Thus a control over the amplitude of the transmitted wave is obtained by varying the current in the solenoid and with it the degree of rotation.

If the magnetic field is adjusted to give 45° rotation for maximum transmission, then a reflected wave coming in the opposite direction is rotated a further 45° in the same sense.   Thus the reflected wave emerges with its plane of polarization perpendicular to the short face of the input waveguide and so is not transmitted.   The device therefore is a one-way

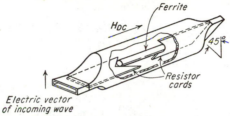

FIG. 14-13. A variable attenuator and one-way transmission device (isolator) using a ferrite.   (*After Rowen* [860].)

transmitter and can be used, for instance, to isolate a klystron from a long waveguide section in which reflections can occur.

Other useful devices employing ferrites can be constructed and some of them will undoubtedly find applications in microwave spectroscopy [696a], [860].

14-14. Microwave Generators.   More than any other one thing work in the microwave region is characterized by the electronic sources used for microwave generation.   Heat sources used in the infrared region hardly produce usable amounts of energy for wavelengths beyond $\frac{1}{2}$ mm. The typical microwave sources provide milliwatts of power in a frequency band less than 1 Mc and sometimes less than $\frac{1}{100}$ Mc.   The temperature required of a hot body to produce a similar radiation would be approximately 10^{14}°C.

Electronic generators have the advantage over heat sources that their radiation is entirely confined to a small portion of the spectrum.   Hence even a small amount of power from an electronic generator may represent a very high effective temperature and be easily detectable.   Moreover, this nearly monochromatic radiation makes possible the study of absorption spectra by tuning the source rather than by making a selection from the radiation of a broad source, as by a prism or grating.

However, microwaves lie just in the region of the high-frequency limit for satisfactory operation of a normal electronic tube. This is because the electrons take a time which is an appreciable part of a cycle to pass between the electrodes. At microwave frequencies the field may actually alternate many times during the electron transit time unless the electrode spacing is extremely small. The time average of the field acting on the electron is then zero regardless of the field strength. Thus the grid in a conventional triode cannot control at a microwave frequency the flow of the electrons through the tube. Triodes have been made to operate at 10,000 Mc by the use of extremely small grid-to-cathode spacing (less than the thickness of the oxide coating on most cathodes). However, with such close spacing interelectrode capacitance is a serious difficulty.

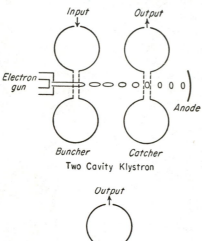

Two Cavity Klystron

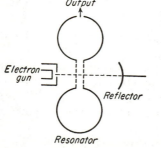

Reflex Klystron

FIG. 14-14. Klystron oscillators.

14-15. Klystrons. Most microwave generators allow for and make use of the electron transit time. The klystron [281] has two resonant cavities through which an electron beam passes. A radio-frequency field in the first cavity bunches the electrons into groups which pass through the second cavity and induce radio-frequency fields in it. More exactly, the first cavity accelerates some electrons slightly and decelerates others, depending on what portion of the radio-frequency cycle they traverse. After a distance of some millimeters the fast electrons have caught up with the slower ones ahead and maximum bunching occurs. It is at this point that the second resonator is placed, for farther along the beam the accelerated electrons have passed the slow ones and the electrons are again debunched. The two-cavity and reflex klystrons are illustrated in Fig. 14-14.

If some of the radio-frequency energy from the second resonator is fed back to the buncher resonator in the right phase, as by a coaxial transmission line of suitable length, the klystron becomes an oscillator. The frequency of oscillation is determined primarily by the size of the cavities, i.e., by their resonant frequency. It can also be influenced slightly by changing the electron velocity, but if the frequency is changed too much from the center of the cavity resonance by changing the electron velocity, oscillations are not maintained.

The second cavity may be eliminated and the klystron considerably simplified if the electron beam is reversed by a suitable reflector electrode, negatively charged. After reflection, the beam retraverses the cavity, delivering to it more energy than originally received from the cavity if the beam velocity and reflector distance are right.

"Reflex" klystrons of this type have been by far the most commonly used microwave oscillators for spectroscopy. They are available commercially for almost all frequencies from below 3000 Mc to about 60,000 Mc (Sperry Gyroscope Co., Great Neck, N.Y.; Raytheon, Manufacturing Co., Waltham, Mass.; Varian Brothers, San Carlos, Calif.). Individual types of reflex klystrons require accelerating voltages in the range from 300 to 3000 volts. The reflector is operated at a negative voltage which is usually between zero and 300 volts. It is necessary to stabilize the power supplies, and this is especially true for the reflector voltage. Most general spectroscopic work requires a stability such that possible short period changes in voltage (such as ripple from the alternating current which has not been perfectly filtered after rectification) can cause a frequency change not greater than $\frac{1}{50}$ Mc. This means that for the type 2K33 the accelerating voltage must be held within 0.1 volt and the reflector voltage within 0.01 volt. In addition, for this type of stability it is often helpful to operate the klystron cathode heater on direct current either from batteries or from a rectifier. Typical klystron power supplies suitable for general-purpose microwave spectroscopy use a series regulator tube controlled by reference to a voltage-regulator tube with a single stage of amplification ([221], Sec. 2-13, and [519]). An additional stage of d-c amplification in the control circuit may be needed for higher stability, particularly with 2K50 tubes.

The 2K33 family of tubes (Raytheon) has been most widely used because tubes are available for the entire wavelength region from 18,000 to 60,000 Mc with the same wiring connections and similar power requirements for all the members of the group. They have a variety of type numbers and each type covers a frequency range of only about 10 per cent. The tubes are reflex klystrons with a cavity partly inside and partly outside the vacuum. The cavity involves a bellows arrangement so that it can be compressed and thereby tuned. While the nominal range of the type 2K33 is 22,000 to 25,000 Mc, individual models sometimes may be found to oscillate anywhere in the range from 18,000 to 28,000 Mc. If a tube is not on hand for a particular desired frequency, it is possible to extend the range of another tube by loosening the locking screws which limit the range of the cavity adjustment (*cf.* [519]). Too large a change of this type may, however, damage the tube.

Another useful klystron type for the same region as the 2K33 is the

2K50. This is a smaller tube mechanically and somewhat more difficult to build. However, the completed tube is very convenient to use and requires only 300 volts on the accelerator and −150 volts on the reflector. Tuning is accomplished by changing the size of cavity thermally. One of the cavity supports is made the anode of an auxiliary triode. When this triode plate current is increased, this support is heated by electron bombardment and expands enough to tune the klystron cavity. The tube may be tuned in 2 or 3 sec over about 2000 Mc by changing the tuner grid current. The tuner grid has a sensitivity of about 120 Mc/volt [225] so that its voltage must be accurately regulated. This is not very difficult in practice, because the voltage is low and little current is drawn. Batteries may be used if a suitable power supply is not available. Moreover the tuner response is slow enough that the effect of power-supply ripple is greatly reduced. The 2K50 reflector regulation requirements are similar to those for the 2K33 family. This tube is now being manufactured by Bendix Red Bank, Eatontown, N.J.

The Varian klystrons (*e.g.*, types X-12 and X-13) are available only for frequencies below 20,000 Mc. However, they give somewhat more power than the preceding types and are therefore especially useful for driving crystal harmonic generators.

If it is desired to sweep a 2K33 slowly through a frequency band wider than can be obtained by reflector voltage variation, a suitably geared motor can be attached to the screw which varies the cavity size. A friction clutch must be included in the drive to allow slippage at the end of the tuning range and thus avoid damage to the tube. In thermally tuned klystrons, a wide frequency sweep is obtained by deriving the tuner grid voltage from a motor-driven potentiometer, preferably of the 10-turn type (*e.g.*, Helipot potentiometer).

The sources of noise and extraneous modulation in the output of klystrons are varied, and their noise spectra are poorly known at frequencies near the oscillator frequency itself. However, the noise generated by a klystron oscillator is not usually so large as that produced by the crystal detector. Occasionally an individual tube will produce exceptionally large noise, particularly near the end of a mode. In addition, klystron noise is larger than that generated by a bolometer detector and is the factor limiting the ultimate sensitivity of spectrographs using thermal detectors.

In case klystron noise has a harmful effect on a microwave measurement, it can usually be decreased by one of a number of techniques. Beyond 50 or 100 Mc from the center frequency it is usually negligible, so that noise from a local oscillator can be eliminated by using a sufficiently high intermediate frequency. Local oscillator noise can also be avoided by the use of a balanced mixer. Noise at low frequencies (*i.e.*, close to the frequency of klystron oscillation) may be eliminated by dividing the emitted microwave into two parts, one part transversing, for example, an absorption cell before detection and the other part being separately detected.   If the two detected signals are compared, variations due to oscillator noise may be eliminated and distinguished from effects of absorption in the cell (see [172]).   A microwave bridge also tends to balance out noise.

14-16. Magnetrons.   Multicavity magnetrons have been made to produce wavelengths as short as 2.6 mm.   However, they have not been much used for spectroscopy because of the difficulty in tuning them [262]. Moreover, magnetrons are often rather noisy, and their large power output is seldom necessary.

Magnetrons have been used for spectroscopy in the millimeter region and may find further application for that purpose (see Chap. 15).

14-17. Traveling-wave and Backward-wave Tubes.   A more recent type of tube which is not yet commercially available for the regions of most interest for spectroscopy is the traveling-wave oscillator.   Several varieties have been constructed experimentally and they have been operated at wavelengths as short as 6 mm [511], [617].   In these tubes an electron beam is surrounded by a structure through which a wave travels more slowly than the electrons.   At low frequencies (hundreds of megacycles) the wave may be guided by a helix; at higher frequencies a corrugated waveguide has been used.   With suitable wave and electron velocities and appropriate coupling between them, the electron beam imparts energy to the wave, thereby increasing its amplitude.   An amplification of 18 db has been obtained over a 3 per cent bandwidth near 6 mm wavelength, and such a tube can be used as an oscillator.

The "backward-wave" tube [830] is related to the traveling-wave tube. It can be constructed to give good amplification or oscillation at millimeter wavelengths and is tunable over a wide range of frequencies.   In this device the electron beam passes along a corrugated waveguide to a collecting anode.   An electromagnetic wave sent in from near the collecting anode emerges near the cathode and hence the wave is backward from that of a traveling-wave tube.   Along its path it interacts with the electron beam at regular intervals, determined by the waveguide corrugations.   The beam is thereby bunched and on traveling toward the collector interacts further with the electromagnetic wave.   This provides a feedback which can be positive if rates of progress of the waves and hence phases are correct.   Since the feedback phase and therefore the frequency of maximum amplification is determined by the electron-beam velocity, the frequency can be controlled by varying electrode voltages. One model has oscillated at wavelengths from 6 to 7.5 mm and has been used as an amplifier with a gain of 20 db.   The power output of the oscillator is of the order of 10 mw.

Table 14-2 summarizes the microwave generators which have been

TABLE 14-2. CHARACTERISTICS OF MICROWAVE SOURCES

Source	Wavelength range	Average power output	Comments
Hot body	All wavelengths	$kT \, \Delta\nu$ (into transmission line) as long as $h\nu < kT$. For bandwidth $\Delta\nu = 10^6$ cycles/sec $kT \, \Delta\nu \lesssim 5 \times 10^{-14}$ watts	Of use only in exceptional cases for microwave work because of low power
Spark discharge*	∞–0.2 mm	$10^{-5} - 10^{-8}$ watt for bandwidth $\Delta\nu \approx 10^8$ cycles/sec	Low power and not monochromatic, but for short wavelengths one of the few available sources
Triode electronic tubes	∞ –3 cm	10–0.5 watts	Convenient for longer wavelengths
Klystrons	50 cm–5 mm	100–10^{-3} watts	Very convenient and tunable over about 10% range for high-frequency types
Klystrons plus crystal multipliers (harmonics)*	50 cm–0.6 mm	10^{-2}–10^{-9} watts	One of most convenient sources below 1 cm wavelength, and best source for spectroscopy below 4 mm
Magnetrons (fundamental)	50 cm–3 mm	100–1 watts	High power. Often pulsed with very high peak power. Usually tunable only over small range
Magnetron harmonics*	3 cm–1 mm	10^{-1}–10^{-9} watt	Good monochromatic source for wavelengths below 2.5 mm, but usually tunable only over small range
Traveling-wave and backward wave tubes	1 m–6 mm	100–10^{-3} watts	Expected performance similar to klystrons but not yet commercially available for short wavelengths. Tunable over wide range

* Discussed more fully in Chap. 15.

used or suggested for microwave spectroscopy. The higher powers quoted are representative of performance at the lower frequency end of the range of each type. Tubes for higher frequencies are necessarily smaller and so have reduced power, dissipation, and usually lower efficiency.

14-18. Detectors. Crystal rectifiers are used almost exclusively for detectors of microwave power in spectroscopy, although thermal detectors have been applied in some special cases. The crystal detector [325] consists of a fine wire in contact with a block of semiconducting material (most often silicon but sometimes germanium). The contact resistance is greater in one direction that in the reverse, and the current-voltage characteristic is very nonlinear near the origin so that rectification occurs when an alternating voltage is applied. Because the contact is a fine point, contact capacitance is small and the rectifier can be used up to extremely high frequencies. Nevertheless, in the millimeter-wavelength range the stray capacitance shunting the contact becomes a limiting factor.

The sensitivity of a crystal is determined by its forward and backward low-current impedances. However, the usable sensitivity is limited by crystal noise. Both these factors vary greatly between individual crystals, but a typical good type 1N26 crystal matched to a waveguide gives an output current of about 1 ma/mw. The internal output impedance is near 200 ohms when 1 mw of power is received, and considerably higher for lower received powers. The impedances which govern the performance at microwave frequencies differ from those at low frequencies by the effects of shunt capacitance and series inductance, so that low-frequency measurements are only a rough guide to the performance of a crystal at microwave frequencies. The noise power generated in a crystal may be divided into two parts. The first is the thermal agitation, or "Johnson" noise, $kT \Delta \nu$, where k is Boltzmann's constant = 1.380×10^{-23} joule per degree, T the absolute temperature, and $\Delta \nu$ the output frequency bandwidth. The second part of the noise is not strongly temperature-dependent but is approximately proportional to the square of the crystal current and inversely proportional to the output frequency. Thus the noise power of a crystal is given by

$$ P = \left(kT + \frac{CI^2}{\nu} \right) \Delta \nu \tag{14-45} $$

where C = a constant
I = the d-c crystal current, amp
ν = output frequency

At room temperature (20°C), $kT = 4.04 \times 10^{-21}$ watt/cycle/sec. For a reasonably good K-band crystal C is about 10^{-7} ohms. Thus for an output frequency of 30 Mc, which is typical of superheterodyne detection, the second term of (14-45) is less than the first for normal crystal currents of 1 ma or less. For direct or "video" detection, however, lower output frequencies are used and the second term normally predominates. For $\nu = 6000$ cycles/sec, I must be reduced to a few microamperes to keep excess noise due to the second term less than "thermal" noise. Such low crystal currents have distinct disadvantages.

Since the current-voltage characteristic of a crystal is represented fairly well by $I = KV^2$, where K is some constant, the output power is approximately proportional to the square of the input power, that is, $P_{out} \propto I^2 \propto V^4 \propto P_{in}^2$. Hence, for small variations ΔP_{in} of the input power such as would be produced by gas absorption, the change in output signal is

$$\Delta P_{out} \propto P_{in}\Delta P_{in} \qquad (14\text{-}46)$$

The conversion gain may be defined as $\Delta P_{out}/\Delta P_{in}$, which is a measure of the efficiency of the rectification process. It is usually considerably less than unity, and from (14-46) may be seen to decrease with decreasing power level. Hence usually the crystal current should be high enough to give good conversion and to make crystal noise predominate over amplifier noise. Since crystal noise power is proportional to the square of the current or input power from (14-45), and detected signal for a given fractional absorption is also proportional to the square of the power from (14-46), the precise value of crystal current does not affect the sensitivity or signal-to-noise ratio as long as the crystal current is large enough to make crystal noise predominate over other noise sources. Crystal currents as low as a few microamperes may hence still allow good sensitivity if exceptionally noise-free amplifiers are used. For very large crystal currents (greater than about 0.5 ma), the conversion gain no longer increases with current, which sets an upper limit to the desired crystal current.

The noise which is proportional to crystal current may be regarded as originating in a variable resistance of the rectifier. If, instead of a direct current through the rectifier, the microwaves are completely modulated with frequency ν_0 so that the rectified crystal current varies at frequency ν_0, the noise of this type becomes $\dfrac{CI^2}{\nu - \nu_0} \Delta \nu$ rather than $\dfrac{CI^2}{\nu} \Delta \nu$ as given in Eq. (14-45). For this reason amplitude modulation of the source of microwave power at a high frequency does not obviate the large crystal-produced noise in a narrow band about the frequency of modulation ν_0.

Figure 14-15 shows two typical crystal mounts for 1N26 crystals ([309], p. 171). The mount in Fig. 14-15a is tunable for optimum impedance match at a particular frequency, while the other is broadly resonant over the K-band region. In the latter, especially, some adjustment may be obtained at the ends of the frequency band by moving the crystal slightly in or out of its holder.

Crystals are also used for harmonic generation, being well suited for these purposes because their nonlinear resistance characteristic persists to high frequencies, and for superheterodyne mixers. When superheterodyne detection is used, the beat oscillator should give adequate power (approximately 1 milliwatt) to the crystal to give good conversion

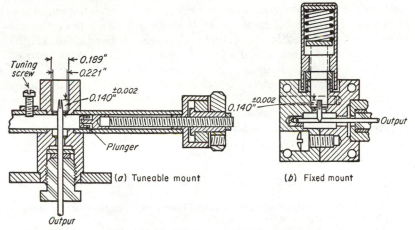

FIG. 14-15. Microwave crystal mounts.

gain [cf. (19-46)]. Its frequency should also differ from the signal frequency by an amount sufficiently large to prevent difficulty from crystal noise of the type indicated by the second term of (19-45) (cf. Sec. 15-7).

Welded contact germanium crystals have been used and are very sensitive and stable. However, they tend to have higher noise than good silicon crystals ([325], Chap. 13). When used as mixers, they may have a conversion gain greater than unity, giving an intermediate-frequency output power greater than the signal power, the local oscillator providing the difference. In the longer wavelength range, these crystals are very good for harmonic generation but unfortunately are no longer being manufactured.

Bolometers or barretters can also provide useful detection of microwaves if the power level is not too low. A bolometer consists of a small length of fine metal wire which is heated by the presence of a microwave signal, with a resulting change in resistance. The change in resistance can be detected as a voltage signal by passing a small current through the bolometer. Bolometers and their mounts are manufactured by the Sperry Gyroscope Company, Great Neck, N.Y., and by the Polytechnic Research and Development Co., Brooklyn, N.Y., with approximately the following characteristics: resistance = 200 ohms; change of resistance = 8 ohms/milliwatt; maximum power before burnout = 15 milliwatts; time constant = 300 μsec. Units are made for microwave frequencies near 3 cm and near 1 cm which give satisfactory reception over variations in wavelength of ± 10 per cent or more. If 5 ma of current is passed through the bolometer, the above change of resistance gives a voltage change of 0.04 volts/milliwatt of microwave power.

One of the most important advantages of a bolometer is that, in contrast with crystal detectors, it gives very low noise at low (audio) frequencies. Thus the bolometer noise is not much larger than thermal noise, which is about 2×10^{-9} volts for a 1 cycle/sec bandwidth and the bolometer described above. If an amplifier is used with a noise figure of 100, this means that microwave power changes of about 5×10^{-10} watts can be detected. Bolometers will not generally respond well to changes in microwave power more rapid than about 1000 cycles/sec (time constant $\approx 10^{-4}$ sec), but for slower modulation frequencies they can be very convenient and sensitive. Additional discussion of their design and use can be found in reference [221].

References: Books on Microwaves

Barlow, H. M., and A. L. Cullen, *Microwave Measurements*, Constable & Co., Ltd., London, 1950.

Bronwell, Arthur B., and Robert E. Beam, *Theory and Application of Microwaves*, McGraw-Hill Book Company, Inc., New York, 1947.

Lamont, H. R. L., *Wave Guides*, 3d ed., Methuen & Co., Ltd., London, and John Wiley & Sons, Inc., New York, 1950.

MIT Radiation Laboratory Series, Vols. 1 to 28, members of the staff of the Massachusetts Institute of Technology Radiation Laboratory, McGraw-Hill Book Company, Inc., New York, 1947.

Principles of Radar, 3d ed., members of the staff of the Radar School, Massachusetts Institute of Technology, McGraw-Hill Book Company, Inc., New York, 1952.

Moreno, T., *Microwave Transmission Design Data*, McGraw-Hill Book Company, Inc., New York, 1948.

Pollard, Ernest C., and Julian M. Sturtevant, *Microwaves and Radar Electronics*, John Wiley & Sons, Inc., New York, and Chapman & Hall, Ltd., London, 1948.

Reich, H. J., P. F. Ordung, H. L. Kraus, and J. G. Skalnick, *Microwave Theory and Techniques*, D. Van Nostrand Company, Inc., New York, 1953.

Sarbacher, R. I., and W. A. Edson, *Hyper and Ultra-high Frequency Engineering*, John Wiley & Sons, Inc., New York, and Chapman & Hall, Ltd., London, 1943.

Skilling, H. H., *Fundamentals of Electric Waves*, 2d ed., John Wiley & Sons, Inc., New York, and Chapman & Hall, Ltd., London, 1948.

Slater, J. C., *Microwave Electronics*, D. Van Nostrand Company, Inc., New York, 1950.

Slater, J. C., *Microwave Transmission*, McGraw-Hill Book Company, Inc., New York, 1942.

Wind, M., and H. Rapaport, *Handbook of Microwave Measurements*, Polytechnic Institute of Brooklyn Microwave Research Institute, Brooklyn, 1954.

CHAPTER 15

MICROWAVE SPECTROGRAPHS

15-1. General Principles and Ultimate Sensitivity. Microwave absorption in gases is usually detected by passing microwave radiation from an oscillator through a long waveguide cell containing the gas, and measuring the amount of transmission as a function of the oscillator frequency. If the gas pressure is low enough so that the lines are not more than a megacycle or so wide, this is conveniently done by sweeping the oscillator frequency periodically and putting a voltage corresponding

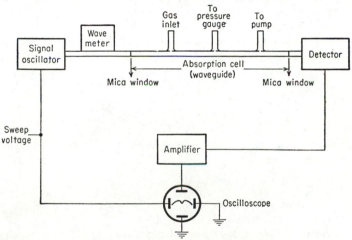

Fig. 15-1. A simple microwave spectrograph.

to this sweep on the horizontal axis of an oscilloscope, and a voltage corresponding to the transmitted power on the vertical axis. Such a system, shown in Fig. 15-1, thus plots the spectrum on the cathode-ray tube. The oscillator is swept over a region only 10 to 50 Mc wide, the center of its frequency being varied by hand to search for absorption lines or to move from one line to another. If, as is customary, the oscillator is a reflex klystron, a sweep of the order of 30 Mc is obtainable electrically by varying the repeller potential.

The type of spectrometer shown in Fig. 15-1 is simple and convenient if the absorption lines are strong, *i.e.*, produce changes in power as large as 0.1 per cent. However, many lines give very weak absorption, so that, even with a cell length of several meters, changes in power due to gas absorption of 1 part in a million or less would have to be observed on the oscilloscope. Interfering variations in power can easily be produced by the variation in oscillator output as a function of frequency, or by variation in transmission due to reflections and standing waves in the waveguide. These difficulties generally limit a spectrometer of this type to detection of lines with absorption coefficients greater than 10^{-6} cm^{-1}, or producing a power change of 0.1 per cent in a 10-meter absorption cell. Another type of interfering fluctuation is "thermal noise." This is electromagnetic radiation in the waveguide and detector produced by thermal agitation of their electrons and is the only interfering variation of a fundamental nature. Other sources of variation can presumably be eliminated by sufficiently careful design and construction of a spectrometer, so that it is thermal noise which limits the ultimate sensitivity attainable. This limit on the minimum detectable absorption coefficient for a "perfect" spectrometer, *i.e.*, a spectrometer whose sensitivity is limited only by thermal noise, is discussed below.

If power P_0 is introduced into a waveguide it will be attenuated by losses in the waveguide walls and also by possible gas absorption. These two sources of attenuation are designated by attenuation coefficients α_0 and α_{gas}, respectively. The power after traversing a length L of the waveguide will therefore be

$$P = P_0 e^{-(\alpha_0 + \alpha_{\text{gas}})L} \tag{15-1}$$

If the gas is removed, the power is $P = P_0 e^{-\alpha_0 L}$, so that the change in power due to the gas is

$$\Delta P = P_0 e^{-\alpha_0 L}(1 - e^{-\alpha_{\text{gas}}L}) \approx \alpha_{\text{gas}} L P_0 e^{-\alpha_0 L} \tag{15-2}$$

where the exponential has been expanded because we shall be interested in very weak absorptions such that $\alpha_{\text{gas}} L \ll 1$.

To find the minimum detectable absorption, we need also to calculate the fluctuations in power caused by thermal agitation. Consider a long waveguide of length L which, for simplicity, supports only one mode of propagation for frequencies near ν. The waveguide will be assumed to be lossless and to be shorted at both ends. This guide is in reality a cavity and its resonant modes are given by $n = 2L/\lambda_g$, where λ_g is the wavelength of radiation in the guide and n is an integer. Since λ_g is not very different from the wavelength λ in a vacuum (in most practical cases $\lambda_g \approx 1.2\lambda$), we shall write $n = 2L/\lambda$ and neglect the complications of modified wavelength and velocities in the waveguide. Then $n = 2L\nu/c$. As long as $kT \gg h\nu$, the classical equipartition law holds, and the average energy in each mode of oscillation is kT. Since the number of modes per frequency interval is $dn/d\nu = 2L/c$, the energy per frequency interval is

$$\frac{dW}{d\nu} = \frac{2LkT}{c} \quad \text{or} \quad \Delta W = \frac{2LkT}{c} \Delta\nu \qquad (15\text{-}3)$$

Each standing-wave mode can be resolved into two equal traveling waves moving in opposite directions. Each wave contains half the energy, so that the total energy per second moving down the guide in one direction is one-half the energy density per unit length times the velocity c, or

$$\frac{c \, \Delta W}{2L} = kT \, \Delta\nu \qquad (15\text{-}4)$$

This is the thermal power in the frequency interval $\Delta\nu$ which travels in either direction down the waveguide. It is independent of the length of the waveguide and so holds equally well for an infinite waveguide or for its practical equivalent, a finite lossy waveguide or one terminated by matching impedances. For instance, the thermal power flowing from a waveguide into a matched crystal detector is given by (15-4). In addition the crystal must radiate an equal amount of power into the waveguide if the entire system is in thermal equilibrium.

This thermal energy $kT \, \Delta\nu$ produces a fluctuating signal at the crystal detector. It might, at first thought, appear that a signal due to gas absorption would be detectable as long as the change in power produced at the crystal was somewhat larger than the thermal power $kT \, \Delta\nu$. However, the thermal power is associated with a field strength in the waveguide which combines with the field strength due to the purposely transmitted microwave to produce apparent changes in power at the crystal considerably greater than $kT \, \Delta\nu$.

For convenience, let the transmitted microwave power from the signal oscillator be represented by a voltage defined so that $V^2 = 2ZP$, where P is the power and Z is the guide impedance. V is the voltage amplitude of the wave emitted by the oscillator. The similar voltage due to thermal radiation will then be $(\Delta V)^2 = 4ZkT \, \Delta\nu$, where an additional factor of 2 is required because there is a wave of thermal radiation in each direction along the guide which contributes to voltage fluctuations at the detector. The net power flow in the guide (assuming no gas absorption) is

$$\frac{(V \pm \Delta V)^2}{2Z} = P \pm 2 \sqrt{2kTP \, \Delta\nu} + 2kT \, \Delta\nu \qquad (15\text{-}5)$$

Equation (15-5) shows that the thermal radiation produces changes of power of magnitude approximately $\sqrt{PkT \, \Delta\nu}$, which is considerably larger than the thermal power itself, since usually $P \approx 10^{12}kT \, \Delta\nu$. For a signal to be noticeable, the power change involved must be approximately as large as or larger than $\sqrt{PkT \, \Delta\nu}$. Then using the change in power due to gas absorption given by Eq. (15-2), the smallest detectable absorption coefficient for a waveguide of length L is given when

$$\alpha_{\text{gas}} L P_0 e^{-\alpha_0 L} \approx 4 \sqrt{2 P_0 e^{-\alpha_0 L} k T \, \Delta \nu} \qquad (15\text{-}6)$$

or

$$\alpha_{\text{gas}} = \frac{4}{L} \sqrt{\frac{2kT \, \Delta \nu}{P_0 e^{-\alpha_0 L}}} \qquad (15\text{-}7)$$

This expression is a minimum for a certain optimum waveguide length

$$L_{op} = \frac{2}{\alpha_0} \qquad (15\text{-}8)$$

Then

$$\alpha_{\text{gas}}(\text{min}) = 2 e \alpha_0 \sqrt{\frac{2kT \, \Delta \nu}{P_0}} \qquad (15\text{-}9)$$

for the optimum waveguide length $2/\alpha_0$. The same result follows from a more rigorous treatment ([328]; *cf.* also [445], [278]).

Instead of absorption, it is possible to use dispersion, or variation in dielectric constant, to detect a gas resonance. The ultimate sensitivity attainable can be shown to be essentially the same as by detection of absorption [653].

The optimum guide length $L = 2/\alpha_0$ is generally between 5 and 30 meters since $\alpha_0 \approx 10^{-3}$ cm^{-1}. If the power used in the waveguide is 1 mw, and the bandwidth of the detecting circuits is 30 cycles, the minimum detectable absorption coefficient would be $\alpha_{\text{gas}}(\text{min}) \approx 10^{-10}$ cm^{-1}. Actual spectrometers have not reached this ideal sensitivity, the best being approximately 30 times less sensitive for bandwidths of 30 cycles as assumed here.

The actual sensitivity of microwave spectrometers is usually limited by one of the following instead of by fundamental thermal fluctuations:

1. Random fluctuations of power due to
   a. Noise in excess of thermal noise as a result of current flowing in the detecting crystal
   b. "Oscillator noise," or variations in oscillator output
   c. Noise in excess of thermal noise in the amplifying circuits
2. Changes in power which vary systematically with oscillator frequency as a result of
   a. Variation of oscillator power as a function of frequency
   b. Variation with frequency of transmission of the waveguide, absorption cell, and detecting crystal
3. "Conversion loss" in the crystal detector, *i.e.*, the loss in signal power when a microwave signal is converted into a lower-frequency signal by the detector.

The first and third difficulties are usually allowed for by introducing a noise figure N, which is the factor by which the ratio of noise to signal power coming through the amplifying circuits has been changed by troubles of types 1 and 3. Thus for $N = 1$, a signal power of $kT \Delta \nu$ would appear as large as noise, but for $N = 20$, the signal power would have to be 20 times larger to equal noise power. Hence (15-9) becomes for a nonideal system

$$\alpha_{gas}(\min) = 2e\alpha_0 \sqrt{\frac{2kT \Delta \nu N}{P}} \qquad (15\text{-}10)$$

The noise figure N can easily vary with conditions and usually depends to some extent on the power P and the frequency. In typical good microwave spectrographs N is as large as 10^3, as may be surmised from the discrepancy of a factor of 30 between $\alpha_{gas}(\min)$ observed and the ideal value. The fact that the crystal noise and hence the noise figure usually decreases with decreasing current means that optimum sensitivity is obtained with waveguides somewhat longer than that given by (15-8). However, from (15-7) the sensitivity varies only slowly with variations in lengths, which are near optimum, so that the precise value of the optimum length is not critical.

Usually the worst source of random noise is the detecting crystal, for which the noise at low frequencies is often much greater than thermal (*cf.* Chap. 14). Crystals vary by orders of magnitude; it is well worth while to select crystals for low noise. Oscillator-tube noise is usually less than crystal noise, but some klystrons have noisy frequency regions, particularly near the edge of a mode. Oscillator noise is also less serious than crystal noise because if necessary it can always be eliminated by bridge or compensating systems.

The most troublesome cause of systematic variation with frequency is usually reflections or standing waves in the absorption cell which vary the transmission from oscillator to crystal. As the signal oscillator frequency is varied, this produces a pattern of peaks and valleys in the power received by the crystal. If the waveguide cell is of length L and the offending reflections occur as usual near the two ends of the cell, the peaks and valleys repeat every time the oscillator frequency is varied by $c/2L$. If the cell is long, these reflection effects produce narrow peaks which closely resemble absorption lines. They may be reduced by introducing an attenuator between the two sources of reflections. The attenuator itself may, however, introduce new reflections. Carefully tapered attenuators can reduce power variations of this type to $\frac{1}{10}$ per cent, but even so reflections make it difficult to detect a total gas absorption much less than this amount. Thus the simple microwave spectrograph of Fig. 15-1 is suitable only for the stronger absorption lines.

Electrical filters can help discriminate between absorption lines and noise or reflections. In a waveguide which is not too long, reflections give a slower variation of power with frequency than does a line, so that a high-pass filter tends to suppress reflections with respect to the line. Since noise occurs with all frequencies, much of the noise can be eliminated by filtering out all frequencies higher than those essential to pass the absorption line. Often noise can be best eliminated by sweeping very slowly over a small region about the absorption line and rejecting all frequencies higher than about ten times the sweep frequency, which is all that is necessary to reproduce the line. If the sweep frequency is ν_0, then this represents a bandwidth of $10\nu_0$, and if the noise power per frequency interval $P/\Delta\nu$ is constant as in thermal noise, ν_0 should be made as small as possible to minimize the bandwidth $\Delta\nu$. On the other hand, if crystal noise dominates as is often the case, the noise power depends on ν. From Eq. (14-45) the noise power per frequency interval is $P/\Delta\nu = C/\nu$, where C is a constant. The total noise power is the integral

$$ P = \int_{\nu_1}^{\nu_2} \frac{CI^2}{\nu}\, d\nu = CI^2 \log \frac{\nu_2}{\nu_1} $$

The lower limit ν_1 of frequency passed should be not more than two or three times the sweep frequency ν_0, and the highest frequency ν_2 not much lower than $10\nu_0$. Hence $P = CI^2 \log (\nu_2/\nu_1) \sim CI^2 \log 5$, and is independent of the sweep rate. This independence of noise power on sweep frequency (with appropriate filtering) holds only under conditions where crystal noise predominates. If any other sources of noise are important, a slow sweep rate is usually advantageous.

It might be thought that modulating the klystron power at a high frequency (*i.e.*, 100 kc) and amplification of the 100-kc signal on the crystal would allow discrimination against this crystal noise since crystal noise has a maximum at low frequency from (14-45). Unfortunately such a system gives no particular advantage, however, because the crystal noise acts as if it were produced by a variable resistance. Hence the maximum noise always occurs at the frequency of the current flowing through the crystal. Modulating the amplitude of the microwave power at 100 kc simply produces a maximum of the objectionable noise at 100 kc, and no improvement in signal-to-noise ratio (*cf.* Sec. 14-18).

15-2. Source Modulation. Considerable improvement in signal-to-noise ratio can be obtained by a small frequency modulation of the klystron oscillator at, say, a frequency of the order of 100 kc in addition to the slower sweep frequency [278], [286]. This modulation can be achieved easily by adding to the klystron slow sweep a 100-kc sine-wave or square-wave sweep voltage. If at a particular instant the klystron is just at the peak of an absorption line and a voltage pulse is applied to the repeller, the klystron moves off the line and the absorption disappears so that more power reaches the crystal. With a 100-kc modulation the line appears and disappears 100,000 times a second. If a line is present, the 100-kc frequency modulation is converted into amplitude variations at the crystal, which may be amplified by a tuned amplifier. If the amplifier is tuned to 100 kc, only noise components near that frequency are amplified and these are much smaller than the low-frequency components. Note that this frequency modulation does not produce a large modulation of the crystal current, as would amplitude modulation of the klystron source.

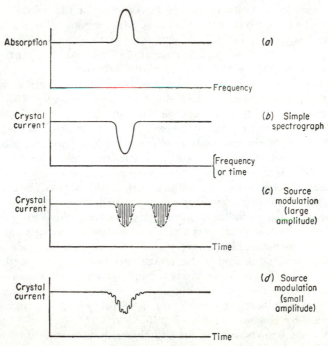

Fig. 15-2. Absorption and corresponding waveforms, one with a simple spectrograph and one using frequency modulation of the microwave source in addition to a slow frequency sweep.

Sharp variations in output power with frequency, as, for instance, those due to certain types of reflections, may also produce 100-kc components and such reflections have often been mistaken for lines. Transmitted power variations with frequency due to reflections are usually less rapid than those due to absorption lines and hence may often be distinguished from them. However, variations due to reflections are a serious limitation on the ultimate sensitivity of this type of spectrometer.

Figure 15-2 shows the crystal-current waveform for the simple spectrograph and for the frequency-modulated, or "double-modulated," spectrograph. In Fig. 15-2c, the modulation of the instantaneous klystron frequency is assumed to be greater than the line width. The line first appears as one tip of the instantaneous frequency excursion strikes the absorption. However, the two appearances of the line are at opposite phases of the modulation frequency and if necessary they can be distinguished by a phase-sensitive detector.

If the modulation is small compared with the line width, the modulation-frequency component of the crystal current is proportional to the slope of the absorption curve. The line may be said to be differentiated, and the two modulation-frequency peaks occur at the points of largest slope, while the output at the peak absorption is zero. For a line with a Lorentz shape, as given by the first term of the Van Vleck–Weisskopf equation (13-19), these peaks occur at $\Delta\nu/\sqrt{3}$ from the center of the absorption line, where $\Delta\nu$ is the line-width parameter.

Square-wave modulation, which has been assumed, gives the most faithful reproduction of the line shape (or its derivative) [465]. However, sine-wave modulation may be used for simplicity [202], [248] with little loss of sensitivity, but with some distortion of the line shape.

15-3. Stark Modulation. If an electric field is applied to a polar molecule its absorption frequencies are shifted because of the Stark effect (Chap. 10). If at some instant the klystron frequency coincides with the peak of an absorption line, when the electric field is applied the absorption decreases. Thus high-frequency modulation can be obtained by subjecting the gas to a periodically interrupted electric field [210], [396]. In addition to the reduction in crystal and tube noise at the high modulation frequency, the method has the very great advantage of being almost completely insensitive to systematic power variations due to anything but spectral lines. Reflections and klystron power variations associated with the low-frequency sweep usually produce very small interfering signals at the high modulation frequency. This useful type of spectrometer was introduced by Hughes and Wilson [210].

Square-wave modulation is almost universally used with Stark spectrographs, with one tip of the square wave being at zero field. Then during this half of the cycle, the absorption pattern is that of the undisplaced line, while during the other half cycle the Stark spectrum occurs. With a phase-sensitive detector, the oscillograph pattern shows both the absorption line and the Stark pattern with the latter inverted. The Stark pattern may then be used to identify the transition or to measure the molecular dipole moment.

Figure 15-3 shows a cross section of a waveguide cell suitable for Stark modulation and a block diagram of the instrument. Stark spectrographs have been described by several authors [419], [396], [519]. The cell is usually a section of waveguide about 3 meters long, as the optimum length is reduced by the extra attenuation of the Stark plate and its supporting insulation. It may be constructed from ordinary waveguide, with the addition of a central flat plate parallel to the broad faces of the guide and so perpendicular to the microwave electric field. The plate is supported by strips of good insulating material such as polystyrene or teflon in which guiding grooves are milled. Connection to the Stark electrode is made by a wire through a hermetic seal in the side wall, which may terminate in a screw threaded into the plate. An alternative support using mica strips rather than plastic is suitable for higher temperatures [637].

Most often the waveguide size chosen is that appropriate to the wavelength band being used, but larger sizes are sometimes desirable to reduce

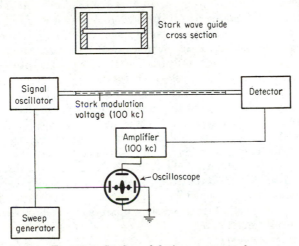

FIG. 15-3. Stark-modulation spectrograph.

the microwave energy density and so avoid saturation. However, larger sizes require larger applied voltages for a given Stark field, and this may be difficult to obtain. A large enough modulating field strength should be used so that all the Stark components are displaced more than the line width, if possible. Otherwise some of the Stark components (inverted by the phase-sensitive detector) will overlap the main line and subtract from it. A few volts per centimeter is usually sufficient for a line with a low J value having a first-order Stark effect. If only a second-order Stark effect occurs several hundred to several thousand volts per centimeter may be needed.

The requirements for the square-wave generator are made more severe by the capacitance between the central plate and the rest of the guide which may be as much as 1000 $\mu\mu f$. For this reason the square-wave generator must have a low output impedance and be capable of supplying large output currents. One such generator, constructed for the Columbia Radiation Laboratory by S. Geschwind, takes a sine wave from a

100-kc oscillator and clips it in several stages of pentode limiting amplifiers to produce a square voltage wave. The necessary low output impedance is obtained from a final cathode follower stage of two 829B or 3E29 tubes in parallel. Provision is made so that an adjustable direct voltage may be added to the square wave to ensure that one end of the square wave corresponds to zero voltage between the Stark plate and the waveguide. For lines which have only small second-order Stark effect, it is sometimes useful to adjust the d-c voltage so that the low-voltage half of the square wave is several hundred volts above zero. Since a second-order Stark effect is proportional to the square of the field, a given change in voltage produces a larger modulation at high field strengths, and so it may be useful to start from a nonzero field.

Another suitable type of square-wave generator [373], [585], [519] uses two sets of parallel output tubes. One set, connected to the high voltage, charges the capacitance of the Stark electrode while the other set discharges it on alternate half cycles. The two groups of output tubes are each triggered at the right time by a blocking oscillator controlled by a sine-wave input voltage of the same frequency as the desired square wave. If the square wave is to be very near zero on one half-cycle, an additional clamping diode is needed. Thus the plates of one group of triggered output tubes may be at a high positive voltage and the cathodes of the other group at about minus 40 volts. The diode clamping tubes are connected to ground so that the voltage cannot actually decrease to negative values, but is stopped sharply near zero. This type of circuit can provide a better and higher voltage square wave at high frequencies than the limiting amplifier, but the blocking oscillator must be readjusted if the frequency is changed.

15-4. Modulation-frequency Signal Amplifiers. The modulation-frequency amplifiers need to have sufficient gain and sufficiently low noise that crystal noise is the only limiting factor in the spectrograph's sensitivity. Since the signal is lowest in the first amplifier stage, that is the most critical one from the standpoint of noise. In fact, the signal voltage at the input may be as low as the crystal noise which, for a 30-cycle bandwidth and a few hundred microamperes of crystal current, is of the order of 10^{-9} volt. A considerable increase in signal voltage can be obtained by using a series-resonant input circuit tuned to the modulation frequency, as shown in Fig. 15-4. The device may be considered as a matching network from the low crystal impedance (300 to 5000 ohms) to the high input impedance of the amplifier. Since the matching network output impedance is high, the cable connecting it to the amplifier must be short and of low capacitance. It is important that the matching network be well shielded. The circuit of Fig. 15-4 also includes a separate low-pass filter to permit measurement of the direct crystal current. If, to avoid saturation of the spectral line, it is necessary to operate at low power levels so that the crystal noise and signal are both small, the resonant input circuit can be followed by a cascade amplifier using a low-noise tube such as the 12AY7 [460]. With this combination the amplifier noise is only slightly greater than the thermal noise of a resistor equal in value to the crystal impedance, *i.e.*, its noise figure is not much more than unity.

It is always easy to test whether or not the amplifier is sufficiently noise-free. If it is, then a decrease in crystal current should produce a decrease in noise voltage through the amplifier or on the oscilloscope which is approximately proportional to the change in current. If such a

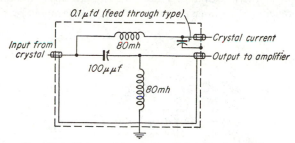

Fig. 15-4. Resonant input circuit for signal amplifier.

decrease occurs, the noise must originate either in the microwave oscillator or in the crystal, and for normal oscillators it will be crystal noise.

Various types of modulation-frequency amplifiers have been used. It is possible to use a commercial broad-band preamplifier followed by a tunable low-frequency radio receiver. Since the receiver does not need to be easily tunable and the preamplifier does not require a broad band, a specially constructed amplifier can be somewhat simpler. Many other amplifier combinations are possible, and some of them are described in the references [419], [396], [519], [460].

The bandwidth used in the receiver should, to reduce noise, be as low as possible. If it is too small, the line will not be faithfully reproduced unless the sweep rate is reduced. Moreover, a too narrow receiver band may lead to errors in frequency measurement [460]. The necessary ratio of bandwidth to sweep rate for good portrayal of a line cannot be specified uniquely, but is often of the order of 20. It is not difficult to have a sweep rate lower than 1 per second on a long-persistence oscilloscope. For use with a recorder on the receiver output, the klystron tuner may be driven slowly by a motor so as to cause the line to be traversed even more slowly. Bandwidths as low as 1 cycle/sec may then be usable. Since in this case the noise power is proportional to bandwidth, a narrow band and slow sweep are highly desirable for good sensitivity. Such a narrow effective bandwidth is most readily attained by means of a phase-sensitive detector (or "lock-in amplifier," as it is sometimes called).

Figure 15-5 is the circuit diagram of a simple type of phase-sensitive detector. The signal is applied to the control grid of a pentode amplifier. From the modulation generator a large voltage (10 to 100 volts depending on the tube used) is applied, through a phase shifter, to the suppressor grid. During one-half of the modulation cycle, the tube functions as a class A amplifier. During alternate half cycles, the lock-in voltage is sufficiently negative to cut off the plate current and with it the amplification.

In the absence of a signal, the plate current consists of a series of pulses once each cycle of the modulation frequency. If a signal of the same frequency and phase is present, the pulses are larger, while if it has opposite phase the pulses are smaller. The plate load resistance R and by-pass capacitor C produce an output voltage determined by the average

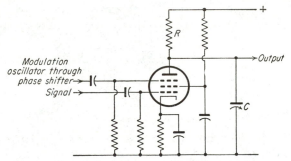

FIG. 15-5. A phase-sensitive detector.

size of these pulses averaged over a time determined by the time constant CR. Thus a signal of the modulation frequency will either increase or decrease this averaged output voltage, depending on its phase. A signal of any other frequency will have no fixed phase relative to the lock-in voltage and so will produce no effect on the average plate current or output voltage.

Interfering signals or noise at frequencies near the modulation frequency will produce fluctuations in output voltage if the time constant is not sufficiently large. The fluctuations might appear to be transient signals of the modulation frequency encountered during a sweep. It will be seen that the longer the time constant CR, the nearer an interfering signal must be in frequency to produce a signal, and so the narrower the effective bandwidth.

Figure 15-6 shows the signal waveforms at different places in a Stark spectrograph using a tuned amplifier and lock-in detector. Phases are indicated by "off" or "on" referring to the portions of the modulation cycle in which the Stark voltage, is, or is not, zero, respectively.

Both the line and its Stark components appear in the final output voltage. Since they occur at opposite phases, the Stark component is inverted relative to the main line. The line shown in Fig. 15-6 is assumed to have only one Stark component, but commonly there are several.

To use a recording milliameter with a lock-in detector of the type shown, the average plate current in the absence of a signal must be balanced out by an auxiliary supply. The balancing current may be obtained by means of a potentiometer from the plate supply.

For greater stability, the signal and balancing voltages may be obtained from matched amplifiers; *i.e.*, the balancing voltage may be supplied

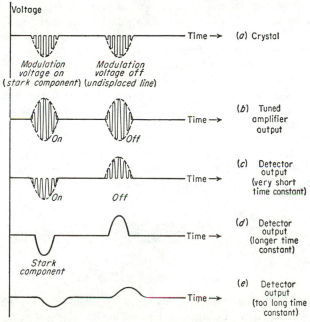

FIG. 15-6. Signal waveforms in a Stark-modulation spectrograph using a phase-sensitive detector.

by a second phase detector whose phase is such that its plate current charges in the opposite direction to the first one when a signal is received [396]. Inverse feedback may be applied to the amplifiers to improve the linearity and stability [737].

One of the primary advantages of a lock-in amplifier is that it provides an easy method for obtaining a very narrow bandwidth about any desired frequency. If a lock-in amplifier is used as part of an amplifying system, the effective bandwidth, *i.e.*, the band from which noise is received, is just half the reciprocal of the time constant, or $\Delta\nu_2 = 1/2CR$. (This assumes that $\Delta\nu_2$ is the smallest bandwidth of the system, which is almost always the case.) If a lock-in is not used, the signal would typically be passed through an amplifier of bandwidth $\Delta\nu_1$, then rectified by a conventional detector followed by a filter of time constant CR or bandwidth $\Delta\nu_2$. The effective bandwidth may then be considerably greater than $\Delta\nu_2$. This is because pairs of noise components throughout the whole amplifier bandwidth $\Delta\nu_1$ are mixed in the detector to produce frequencies which lie within the band $\Delta\nu_2$. The effective bandwidth of this system is $\sqrt{\Delta\nu_1\,\Delta\nu_2}$, so that the noise power may be considerably greater than that obtained with a lock-in amplifier.

If the detector time constant is too long relative to the time for a sweep, the output voltage cannot change fast enough to reproduce the line shape accurately. The line then appears to be broadened and reduced in height. Moreover, its peak is delayed and so comes at a point in the sweep corresponding to a different microwave frequency.

If a long time constant (*i.e.*, narrow bandwidth) must be used to reduce noise, the sweep rate must be proportionately lowered. The ultimate limit to the degree of improvement in signal-to-noise ratio which can be thus achieved is set by fluctuations in the klystron output frequency. Further improvement requires increased oscillator stability, which can be attained by use of an external stabilizing cavity resonator or by a controlling quartz-crystal oscillator with frequency multipliers.

15-5. Zeeman Modulation Spectrographs. For paramagnetic molecules such as NO, O_2, NO_2, ClO_2, and free radicals, a magnetic modulation

FIG. 15-7. Zeeman modulation cell.

analogue of the Stark spectrograph can be constructed. The high-frequency magnetic field is most easily applied by a solenoid surrounding the waveguide.

Unless the modulation frequency is quite low, the guide must be slotted longitudinally to reduce eddy currents; such a slot can be put in the center of the broad face without disturbing the microwave fields. A glass tube surrounding the waveguide serves as a support for the coil and keeps the sample gas in the guide region (Fig. 15-7). At modulation frequencies less than 1000 cycles/sec, it is often possible to operate without the slot and then the glass envelope is no longer needed.

Since a square current wave through an inductance is hard to generate at high frequencies, sine-wave modulation may be used with a direct current added so that one extreme of the sine wave occurs at approximately zero field. The coil can then be part of a series-resonant circuit if large currents are desired.

15-6. Choice of Modulation Frequency for Spectrographs. Although crystal noise is reduced as the modulation frequency is raised [Eq. (14-45)], the apparent line width is increased (Chap. 10). The line breadth due to a square-wave modulation is somewhat greater than the modulation frequency, so that for the fine structures commonly encountered in microwave spectroscopy, modulation frequencies much greater than 100 kc have seldom been found desirable. Indeed, for particularly small line spacings even considerably lower frequencies may be desirable, even though the crystal noise is increased.

Whatever the modulation frequency, the amplifier should if possible have a low enough noise figure so that it contributes less noise than the crystal. As long as this is so, decreasing the microwave power level decreases the signal and the noise proportionately. Thus the signal-to-noise ratio is nearly independent of power. Very low microwave power levels (a few microwatts) are sometimes required in high resolution spectrographs to avoid saturation broadening, and then special precautions for amplifier input circuits may be needed. If such a low level must be used that most noise comes from the amplifier, the noise is independent of the signal. In that case, for any given microwave absorption, the signal is proportional to the power level, and so the sensitivity is also proportional to the power.

15-7. Superheterodyne Detection. At low microwave power levels, crystals are inefficient detectors, as their output voltage is approximately proportional to the square of the radio-frequency amplitude. It is then advantageous to use superheterodyne detection to produce a signal as large as possible relative to noise in the following amplifiers. Moreover, if the intermediate frequency is reasonably high (*e.g.*, 30 to 60 Mc) crystal noise within the intermediate-frequency amplifier bandwidth is reduced practically to the thermal noise.

Superheterodyne detection requires an auxiliary microwave oscillator which is kept at a constant frequency difference from the signal oscillator. This local oscillator may be made to follow the signal oscillator by a discriminator and automatic frequency-control system ([308], Sec. 8.3; [573]). Then as the signal oscillator is tuned, the frequency difference between it and the local oscillator is always such that the mixer output is at the desired intermediate frequency.

The need for an extra oscillator and automatic frequency control makes the superheterodyne more complicated than the simple detector. It is not usually possible to make the local oscillator follow automatically over more than a limited region. Moreover, the local oscillator may be an additional source of noise, although this noise can always be eliminated by the use of a balanced mixer [309].

15-8. Bridge Spectrographs. It has been seen that superheterodyne detection can be employed as an alternative to modulation for reducing crystal noise. However, the carrier is enormously larger than the useful absorption signal. It persists at the output of the mixer and is amplified along with the signal. Thus saturation of the amplifier by the carrier is likely to occur before the signal has reached a suitable level. In addition, small fluctuations in power supplies or amplifier characteristics introduce noise into the carrier which is easily confused with the slow variation in the carrier level produced by gas absorption.

To reduce the carrier relative to the useful signal, a balanced bridge may be used. The bridge is shown schematically in Fig. 15-8. The microwave power is split by the first magic T into parts which travel

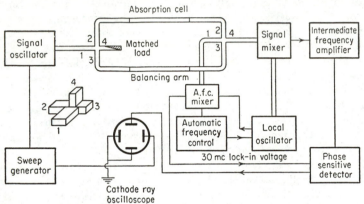

Fig. 15-8. Microwave bridge spectrograph. Many details, such as attenuators and phase shifters, have been omitted for simplicity.

through two waveguide arms. One arm is the absorption cell; the other is made as nearly identical to it as possible, although provision must be made for a final balance of phase and attenuation. A second T combines the waveguide outputs to give a voltage at the mixer which is the difference between those transmitted by the two arms. This difference signal is combined with the local oscillator in the signal mixer to produce the intermediate-frequency signal, which is then amplified by a fixed-tuned intermediate-frequency amplifier.

A second mixer combines the output of the second magic T's summation arm with the local oscillator to produce a second intermediate-frequency voltage. This signal is used as a lock-in reference voltage and also is applied to a discriminator to control the local oscillator's frequency. Finally, the output of the phase-sensitive detector is displayed on an oscilloscope or recorder.

It has been mentioned that the bridge permits the use of superheterodyne detection without overloading the intermediate-frequency amplifiers. It has the additional advantage that any signal-oscillator noise appears equally in both bridge arms and so is balanced out. Moreover, with careful construction, reflections in the bridge arms tend to cancel although reflections outside the bridge proper do not cancel.

The bridge is more complicated to build and adjust than a modulation spectrograph. Its balance is usually good over only a small region because the two arms are never completely identical. Thus searching for unknown lines with a bridge spectrograph is very difficult. However, it is very effective in high-resolution studies of known lines, for which the absence of complications due to modulation and the ability to work with very small power levels to avoid saturation are important.

15-9. High-resolution Spectrometers. For some studies it is important to have as high resolving power as possible even at the expense of additional complications and reduced flexibility. At the same time it is desirable to sacrifice no more sensitivity than necessary. The principal sources of line breadth which need to be overcome or at least reduced are (*cf*. Chap. 13):

1. Collisions with other molecules (pressure broadening)
2. Collisions with the waveguide walls
3. Modulation broadening
4. Power saturation
5. Doppler broadening
6. Oscillator frequency fluctuations

Of these, items 1, 2, 4, and 5 are discussed in Chap. 13, while item **3** is described in Chap. 10.

Pressure breadth can be made as small as desired by using sufficiently low pressures. As long as the pressure width is predominant, reducing the pressure only narrows the line without decreasing the peak intensity [*cf*. Eq. (13-19)]. However, if the width is appreciably greater than that due to pressure, the peak of the observed line is proportional to the integrated intensity and so does decrease with further reduction of pressure. Thus in a high-resolution spectrograph the pressure is chosen just low enough so that it is no longer the limiting factor. Usually the Doppler width is the most difficult to reduce, so that the pressure is adjusted to give a width less than the expected Doppler width.

With such low pressures, collisions with the walls may actually be more frequent than collisions with other molecules, and so a suitably large waveguide or cavity must be chosen. Collisions with the walls are also made somewhat less frequent if molecular velocities are reduced by cooling.

As shown in Chap. 10, the use of modulation broadens the line. The width contributed by a square-wave modulation is somewhat greater than the modulation frequency because of the harmonics present in any square wave. However, the line is unable to follow modulation rates greater than the pressure width, and so in practice a width of about two or three times the modulation frequency is produced. Thus for high-resolution spectroscopy the modulation frequency needs to be kept two or three times less than the pressure width, which in turn is a little less than the expected Doppler width. A few kilocycles is low enough for most gases unless special means are used to reduce the Doppler broadening. The use of low modulation frequencies implies relatively high crystal noise unless proportionately low microwave powers are used, since the noise power is approximately proportional to the square of the crystal current divided by the modulation frequency [Eq. (14-45)].

To avoid power saturation the energy density in the waveguide must be kept low. From Chap. 13, a power density of a few microwatts per square centimeter may produce a breadth comparable with Doppler width for some gases. With a given total power, the density is less when the waveguide is considerably larger than needed for propagation. However, a guide having a cross section more than a factor of 10 or so greater than ordinary K-band waveguide is likely to be inconveniently large and heavy. Thus even with oversize waveguide small total power is necessary to prevent saturation broadening. With small power, super-heterodyne detection, preferably with a balanced waveguide bridge, is helpful in obtaining a good signal-to-noise ratio.

When narrow spectral lines are to be studied, the microwave oscillator frequency must be very stable. High-resolution spectroscopy usually involves measurement of frequencies to better than 1 part in a million, with correspondingly high requirements for oscillator stability. Not only must the oscillator power supplies be well regulated, but the tube must be protected against temperature variation and vibration. Air currents are the commonest source of rapid temperature fluctuations. Their effects can be greatly reduced by immersing the tube in an oil bath or by putting it in good thermal contact with a metal block, preferably of copper. With this thermal regulation and protection from vibration, a klystron can be made to stay within 1 Mc over a period of some minutes.

Better stability may be obtained by using an external cavity to control the frequency (see Chap. 17). A well-constructed cavity with temperature compensation can be used to hold frequency within 1 kc. It is also possible to stabilize a klystron by comparing its frequency with harmonics of a good crystal oscillator [cf. Chap. 17]. A crystal oscillator may be used to stabilize frequency even more directly by using special frequency-multiplier klystrons. They have been used at the National Bureau of Standards to give useful amounts of microwave power as harmonics of the crystal oscillator frequency [724].

15-10. Some High-resolution Spectrometers. A Stark modulation frequency of 6 kc has been used at Massachusetts Institute of Technology for high-resolution spectroscopy [419], [546]. "X-band" waveguides (about 1 by $1\frac{1}{2}$ in.) reduced collisions with walls and energy density in the waveguide ([710], but *cf.* also [788]). Both superheterodyne detection and direct crystal detection with a low-noise audio-frequency amplifier have been used with this spectrometer.

While the bridge spectrograph is more complex and less suited to searching than modulation spectrographs, its advantages for high-resolution spectroscopy with good sensitivity have been confirmed by the instrument at Columbia University [573], [688]. With this instrument, line widths not much greater than the Doppler breadth can be obtained for most molecules (*i.e.*, as narrow as 50 kc).

In its original form, the bridge operates without modulation. Its sensitivity is then limited by frequency-dependent reflections and vibration which make it difficult to maintain an accurate balance. Lines having an absorption α as low as 2×10^{-8} cm^{-1} have been detected. More recently the bridge has been modified by the addition of a very low frequency (about 1000 cycles) Stark modulation [930]. This has eliminated the effects of reflections and increased the usable sensitivity by about a factor of 10.

Any one gas molecule in a spectrograph has a definite velocity in the direction of propagation of the microwave signal. Its microwave absorption is displaced in frequency by the Doppler effect but not broadened. However, an actual gas contains molecules with all velocities so that the over-all absorption is a broadened average. This Doppler width can be reduced somewhat by cooling the gas since from (13-2) it is proportional to $\sqrt{T}$. The degree of improvement obtainable in this way is, however, limited by condensation of most molecular gases at a fairly high absolute temperature.

Any desired reduction in Doppler width can in principle be obtained by somehow selecting a group of molecules with only a small spread in velocity. There is necessarily a corresponding reduction in intensity of absorption because of the decreased number of effective molecules in the sample and the lower microwave power necessary to avoid saturation.

Molecular Beams for Microwave Spectroscopy. One method of accomplishing the above type of selection which has been applied in optical spectroscopy is the method of molecular-beam absorption [238]. The molecules are confined by collimating slits to a narrow beam through which the radiation passes transversely. Then there are no molecules having more than a very small component of velocity in the direction of propagation and so the line width is reduced in proportion to the degree of collimation. Doppler widths can rather easily be reduced by a factor of 10 this way [869a], [925], [982].

Johnson and Strandberg have constructed a microwave molecular-beam spectrograph [709]. A plane wave from a linear array antenna was passed transversely through a molecular beam. After passing through the molecular beam, the wave was reflected back into the antenna. A magic T separated the reflected wave from the incident power. Stark modulation of 660 cycles was provided by electrodes on opposite sides of the beam. A line width (total width at half maximum) for the ammonia 3,3 line of 40 kc was obtained, whereas the Doppler breadth was 70 kc. Considerably narrower lines were obtained by a later beam spectrometer [982]. However, the power needed to avoid saturation and the density of

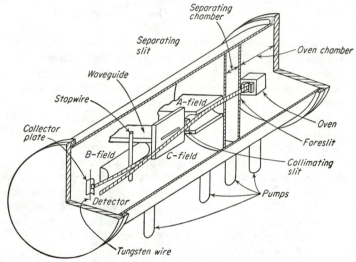

Fig. 15-9. Molecular-beam-resonance apparatus. (*From Lee, Fabricand, Carlson, and Rabi* [835].)

molecules obtainable in the beam are so low that only the strongest microwave lines could be observed.

Molecular beams can be, and most often are, used in spectroscopy in arrangements such that the resonance absorption of radiation is detected by its effect on the absorbing molecule rather than on the radiation. This type of technique has been extensively developed, particularly at frequencies somewhat lower than those in the microwave region, and has given a large amount of valuable information [163], [969a]. We shall not attempt any complete summary of this type of spectroscopy but shall rather try to indicate briefly the general principles involved and how this technique compares with more usual forms of microwave spectroscopy.

Figure 15-9 shows the arrangement of a typical molecular-beam deflection and resonance experiment. The molecules to be studied are evaporated in an oven and emerge through a narrow slit into a good vacuum. The beam is defined by one or more slits and then passes through two deflecting fields marked A and B in the diagram. These are nonuniform electric (or magnetic) fields, the gradient of which acts on the electric (or magnetic) dipole moment to provide a force deflecting the molecule.

A molecule in an electric or magnetic field has a potential energy which is given by the Stark or Zeeman effects (*cf*. Chaps. 10 and 11) and which depends on the field strength and on the particular quantum state. Since the potential energy is dependent on the field strength, a varying or inhomogeneous static field exerts a force on the molecule and hence may deflect its path.

If the A and B fields are equal and oppositely directed their deflections can cancel each other, and the molecules can reach a detector placed in line with the beam's original direction. A stop wire in front of the detector excludes those molecules in states which are not appreciably deflected by the fields.

In the space between the two fields there is a region in which a uniform radio-frequency field and a steady field are applied. If the radio frequency is that corresponding to the energy difference between two molecular states, some of the molecules will make an induced transition in the C-field region. These molecules will have different quantum numbers and hence a different energy and accelerating force in the B-field region. Thus the molecules which have undergone a transition do not receive equal and opposite deflections in the A and B fields and so do not reach the detector. Alternatively, the fields can be arranged so that only the molecules which have undergone the transition reach the detector.

In either case, the resonance is indicated by a change in the number of molecules reaching the detector. For alkali metals and some others with low ionization potentials, the detector can be a hot tungsten wire, from which the incident atoms evaporate as ions [163]. A more complicated, but more generally applicable, detector introduced by Lew and Wessel [886a] uses a transverse electron beam to ionize the molecules. For each quantum absorbed, the path of a molecule in the beam is changed, so that one molecule more (or less, depending on the experimental arrangement) reaches the detector. This exchange of a quantum of radiation for a molecule becomes increasingly advantageous as the frequency is lowered, for the energy in the quantum is proportional to the frequency. At higher frequencies the advantage in sensitivity of this technique is not so marked, and the detection of quanta emitted or absorbed rather then deflected molecules becomes relatively more profitable.

The beams used in these resonance experiments are so narrow and undirectional that the Doppler effect can be eliminated to a large extent. The chief limitation on resolving power is usually the transit time through the C field. This time is ordinarily about 10^{-4} sec, and leads to a line

width of a few kilocycles.   At microwave frequencies the line width will
be much less than that given by ordinary Doppler effect only if the microwave field in the C region is approximately constant in amplitude and
phase over a distance along the beam path which is greater than the
wavelength in free space.   This can be achieved by passing the beam
parallel to the broad face of a very broad rectangular waveguide in the
TE_{01} mode, or parallel to the axis of a waveguide near cutoff so that the
wavelength in guide has been increased.   The line width can also be
reduced by using two radio-frequency fields separated by some distance
[857], or by selecting the slowest molecules which therefore spend the
longest time in the C region.

The resolving power of molecular-beam resonance experiments can
be greater than that of most absorption microwave spectroscopy.   However, the detected signals are often very weak so that the location and
measurement of a spectrum is usually a slow process.   The method is
therefore particularly adapted to high-resolution work and might well
be used to give detailed information about lines which have been located
by absorption spectroscopy.   In addition, in common with all methods
employing beams, it does not require that the molecule studied be particularly stable, or that it have appreciable vapor pressure at ordinary
temperatures.

A spectroscopic device which borrows from both molecular-beam
deflection techniques and microwave absorption spectroscopy has been
described by Gordon, Zeiger, and Townes [925].   This device can be
used as a high-resolution spectrometer, or as a very stable microwave
oscillator and frequency standard.   The name "maser," an acronym for
"microwave amplification by stimulated emission of radiation," has been
given to this general type of device.   A schematic indicating the general
operation of the molecular beam maser is shown in Fig. 15-10.

A gas (NH_3 in the illustration) issues from a number of small holes in
a chamber at a pressure of about 1 mm Hg into a region where the
vacuum is sufficiently low that a beam is formed.   The beam enters a
focusing region where inhomogeneous fields are arranged to deflect molecules in an upper state toward the axis and those in the lower state away
from the axis.   Thus a beam of molecules which are largely in an excited
state is made to enter a cavity which is tuned to the resonant frequency
W/h. W is of course the energy difference between the ground and
excited states being considered.   If a small amount of microwave power
of frequency W/h is introduced into the cavity, it will react strongly
enough with the molecules to make them give up their energy, which
then increases the flow of power from the cavity into the output waveguide.   The increased power occurs only if the microwave introduced is
very near the resonant frequency, so that its occurrence indicates the
presence of a molecular resonance.   Therefore, as the frequency of the

microwave input power is varied, one may pass over one or more resonances and obtain a spectrum of the molecules.

The field in the focusing region comes from a potential of the approximate form [cf. [572a])

$$V = V_0 + axy$$

where V_0 and a are constants. The equipotentials are hyperbolas which, for the case of an electrostatic potential, are approximated by the inner

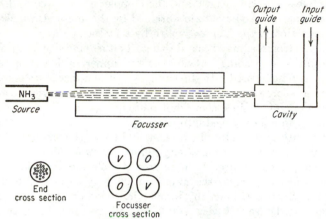

FIG. 15-10. Molecular-beam-emission spectrometer and oscillator (maser). (*From Gordon, Zeiger, and Townes* [925].)

surfaces of the four focusing electrodes. The x component of the electric field is $-\dfrac{\partial V}{\partial x} = -ay$ and the y component is similarly equal to $-ax$. Hence the magnitude of the electric field is

$$|E| = |a| \sqrt{x^2 + y^2} = |a|r$$

where r is the distance from the axis. Therefore, if the Stark effect is such that the energy increases with increasing magnitude of E (as for the excited inversion level of NH_3) the molecule is accelerated toward the axis with a force dependent only on r and not on its angular position with respect to the electrodes. If the molecule does not have too high a velocity perpendicular to the axis, it may be regarded as trapped in a potential well with a minimum on the axis. The molecule is brought back to its original displacement from the axis or focused after a distance of travel which depends on its axial velocity, the strength of the field E, the molecular dipole moment, and the quantum state involved. Molecules with a Stark energy which decreases with increasing field (as for the ground inversion level of NH_3) are deflected away from the axis and hence would not in most cases enter the cavity. A number of other forms of fields may be used for similarly directing molecules in selected states into a cavity.

This type of device allows high resolution because the molecules whose transitions are detected are traveling more or less in one direction along the axis of the apparatus and Doppler effect can therefore be reduced. If the cavity is of such a diameter that it is very near cutoff for the microwave frequency, the wavelength in the cavity is considerably greater than the free-space wavelength, and the Doppler effect is reduced. NH_3 lines as narrow as 7 kc have been obtained by this method, which represents a reduction of the Doppler width by a factor of 10.

Such a spectrometer may be used to detect either emission or absorption of microwaves, depending on whether the upper or lower state of a transition is best focused into the cavity. If the upper state is focused, emission of microwave energy is detected, and the device acts as a microwave amplifier, since somewhat more energy may be emitted from the cavity than is introduced into it.

The intensity of the induced emission, and hence the amount of amplification, increases with an increasing number of molecules in the beam. If the beam gives a sufficiently large flow of molecules and the Q of the cavity is sufficiently high, the amplification may become infinite and radiation is emitted without any input microwave energy. Under such conditions the device is a very stable microwave oscillator with power being supplied from the molecular excitation, and with a frequency primarily determined by the molecular resonance.

The signal obtained by this device from the strongest ammonia lines is approximately 10^{-9} watt, which is as much as a few thousand times noise in a well-designed spectrometer system. Its output power as an oscillator, which is the same 10^{-9} watt, is not large, but it is large enough to serve as a frequency standard (see Sec. 17-7).

Stark-wave Spectrograph. Newell and Dicke [622] have found a method of selecting only those molecules in a gas within a small range of velocities in the direction of microwave propagation. A special electric field is used which is periodic in the direction of microwave propagation with the wavelength, $\lambda/2$, where λ is the microwave length. It is equivalent then to a forward and a backward Stark field traveling wave each with velocity $\Omega\lambda/2$ where Ω is the Stark wave frequency. The Stark modulation provides a regular variation in phase of the reflected wave so that those molecules moving with either Stark wave reflect energy coherently in the backward direction. Others produce reflections with random phases which are much weaker. Widths as low as 7 to 10 kc have been achieved in this way for the ammonia 3,3 line. The theoretically obtainable sensitivity of this device is less than that of other spectrometers by approximately the square of the ratio of line width to Doppler width, so that only rather strong microwave lines can be observed. However, in some cases the higher resolution may be extremely valuable. Fig. 15-11 is a schematic of the apparatus.

15-11. Cavity Spectrographs. Microwave absorption can be detected by its effect on the resonance of a cavity. When an absorbing

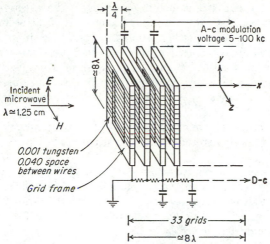

FIG. 15-11. Stark-wave spectrograph cell. (*From Newell and Dicke* [622].)

gas is present, the losses in the resonator are thereby increased, so that its quality factor Q is decreased. This change is shown either by a decrease in the relative amplitude of a wave transmitted through the cavity, or by a change in the wave reflected by the cavity into a waveguide coupled to it.

If the absorption line is narrower than the cavity resonance, it may be displayed by sweeping a microwave oscillator connected to the cavity through the resonance curve. Figure 15-12 shows the pattern observed with a detector and oscilloscope arranged to display either the transmitted or the reflected power. The absorption is obtained as the difference between the curves with and without the absorber. Very wide lines, as in gases at high pressures, can be studied by tuning the cavity

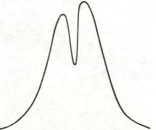

FIG. 15-12. Oscilloscope pattern with resonant cavity spectrograph and sharp absorbing line.

to a number of frequencies in the line's width. At each point the change in Q is observed as a change in relative height or width of the cavity resonance when the gas is introduced, and from it the absorption is deduced [179]. Of course, if the effective absorption at a particular frequency can be removed by a suitable electric or magnetic field as in a Stark spectrograph, this can be used as an alternative to actual removal of the absorber from the cavity.

Whether used for narrow or broad lines, the cavity may be considered as a short section of waveguide in which microwave radiation is reflected back and forth many times before emerging. The effective number of reflections or the effective path in wavelengths is of the order of Q. The radiation finally emerges either from the input hole, in which case the cavity is used in reflection, or from another hole after transmission through the cavity. Although the external microwave circuits differ considerably in the two cases, there is little essential difference in the operation of the cavity. We shall consider further a reflection cavity coupled by a single hole.

As in Chap. 14, the quality factor Q of the resonator is defined as $2\pi \times$ (average energy stored)/(energy lost per cycle) $= 1/\delta$, where δ is a loss factor.

When the cavity is coupled to a waveguide, δ is increased an amount δ_1 by the energy loss through the coupling hole. The reflection factor Γ_0 in the waveguide for an empty cavity, defined as the ratio of the complex amplitudes of the incident and reflected waves, and for frequencies ν near the resonant frequency ν_0, is [291]

$$\Gamma_0 = \frac{2\delta_1}{\delta_1 + \delta_0 + 2j[(\nu - \nu_0)/\nu_0]} - 1 \qquad (15\text{-}11)$$

where δ_0 is the energy loss due to the cavity walls and $j = (-1)^{\frac{1}{2}}$. When a gas with complex dielectric constant $\epsilon = \epsilon' - j\epsilon''$ is introduced (where usually $\epsilon'' \ll \epsilon' \approx 1$), the resonant frequency is shifted to $\nu_0/\epsilon^{\frac{1}{2}} \approx \nu_0/\epsilon'^{\frac{1}{2}}$ and δ_0 is increased to $\delta_0 + \epsilon''$, so that the reflection factor becomes

$$\Gamma_g = \frac{2\delta_1}{\delta_1 + \delta_0 + \epsilon'' + 2j[(\epsilon'^{\frac{1}{2}}\nu - \nu_0)/\nu_0]} - 1 \qquad (15\text{-}12)$$

The fractional change in voltage amplitude of the reflected wave at resonance $\nu = \nu_0$ or $\nu = (\epsilon')^{\frac{1}{2}}\nu$ is then

$$\frac{\Delta V}{V_0} = (\Gamma_0 - \Gamma_g)_{\text{res}} = \left[\frac{2\delta_1}{(\delta_1 + \delta_0)^2}\right]\epsilon'' \qquad (15\text{-}13)$$

For a given δ_0, $\Delta V/V_0$, is maximum when $\delta_1 = \delta_0$. If then

$$\delta_1 + \delta_0 = \delta = \frac{1}{Q}$$

$$(\Delta V)_{\max} = Q\epsilon'' V_0 \qquad (15\text{-}14)$$

This voltage change may be expressed in terms of the free-space attenuation α of the gas by using the relation

$$\alpha = \frac{2\pi}{\lambda}\frac{\epsilon''}{\epsilon'^{\frac{1}{2}}} \approx \frac{2\pi}{\lambda}\epsilon'' \qquad (15\text{-}15)$$

where λ is the free-space wavelength.   Then

$$(\Delta V)_{\max} = \frac{Q\lambda}{2\pi} V_0 \alpha \tag{15-16}$$

The minimum detectable absorption is obtained by setting $(\Delta V)_{\max}$ equal to the thermal rms noise voltage $(4kTN \, \Delta\nu \, Z_0)^{\frac{1}{2}}$ so that

$$\alpha_{\min} = \left(\frac{4kTN \, \Delta\nu}{P_0}\right)^{\frac{1}{2}} \frac{2\pi}{Q\lambda} \tag{15-17}$$

where k = Boltzmann's constant
$\quad T$ = absolute temperature
$\quad N$ = noise figure, or factor by which the noise exceeds thermal noise
$\quad \Delta f$ = frequency band width of the amplifiers
$\quad P_0$ = power reflected from the cavity

For a cavity made of waveguide with attenuation α_0 per unit length, the factor $2\pi/Q\lambda$ becomes approximately α_0, so that (15-17) is closely equivalent to (15-10), and the limit of sensitivity for a cavity spectrometer is much the same as that for one using a waveguide (10^{-9} to 10^{-8} in typical cases).

The factor $Q\lambda/2\pi$ in Eq. (15-16) is the equivalent absorption-path length in free space.   Since Q can be quite large, very long effective path lengths can be obtained in a small space.   This property is particularly advantageous for experiments on Zeeman effects in ordinary molecules, where the necessary large magnetic fields can only be obtained over a small volume.   However, spectrometers using small cavities tend to have much more difficulty with saturation than do those using waveguide absorption cells.   Because of the smaller volume in which absorption takes place, field strengths are higher in the cavities and each molecule must absorb more energy.

For high sensitivity Q should be as large as possible.   If the absorption line is to be displayed within the width of the cavity resonance, the usable value of Q is limited by the need for a resonance wide enough to include the entire line and perhaps its fine structure.   Thus Q must be several times less than $\nu/\Delta\nu$ where ν is the frequency of the line, and $\Delta\nu$ its half width at half maximum.   It is convenient to use a tunable cavity large enough so that resonances can be obtained in several modes with different Q's to suit individual lines.

As with the Stark modulation spectrograph, the bandwidth of the amplifiers following the crystal detector must be great enough to reproduce the line shape but small enough to give low noise.   If the absorption line is swept out in a time $1/t$, then for accurate reproduction of the line shape the bandwidth in cycles per second must be as large as about $20/t$.

The microwave circuits used in cavity spectrographs resemble those used for cavity wavemeters ([221], pp. 308–318). When the cavity is used by transmission an arrangement similar to that of Fig. 15-13 is needed, and the instrument then resembles the simple waveguide absorption spectrograph of Fig. 15-1. This type of instrument without the sweep and oscillograph can be used for studying the pressure-broadening of strong lines [179].

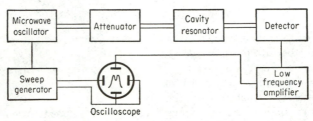

FIG. 15-13. Simple microwave cavity transmission spectrograph.

The simple cavity spectrograph is, like its waveguide prototype, limited in sensitivity by crystal noise. The same remedies of source modulation, field modulation, and superheterodyne detection are effective when they can be applied. Stark modulation is not so easily obtained as in a waveguide without disturbing the microwave fields, because the fields of the cavity modes usually employed are more complicated than those involved in waveguide transmission. For paramagnetic gases, weak-field Zeeman modulation by an external solenoid or short coils is satisfactory [642],

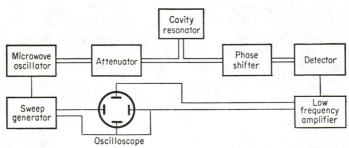

FIG. 15-14. Simple microwave cavity reflection spectrograph. The detector and resonator may be interchanged.

[643], [435], [558], although eddy currents in the walls usually limit the modulation to relatively low frequencies.

In reflection spectrographs the cavity may be placed on either the side or the end of a waveguide. When placed on the side as in Fig. 15-14, a resonance in the cavity changes an effective impedance in parallel with the guide and so affects the power reaching a detector at the end. Depending on the phase of the cavity reflection relative to the standing-wave pattern in the guide, the detector power may increase or decrease at resonance; *i.e.*, the change in impedance of the cavity at resonance may either partially cancel reflections already existing in the waveguide or may add to them. At certain positions along the guide the resonance may have very small effect, but such difficulties may be avoided if necessary by the use of a phase-shifting adjustment between the cavity and the detector.

With only low modulation frequencies available and power limited by the need for avoiding saturation, superheterodyne detection is helpful in obtaining a good signal-to-noise ratio. It is then necessary that most of the carrier be balanced out before reaching the second detector, in

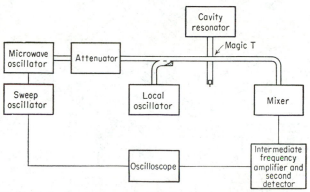

Fig. 15-15. Reflection cavity spectrograph with superheterodyne detection.

order that the intermediate-frequency amplifiers will not be overloaded. The balancing can be accomplished by a waveguide bridge. Partial balancing can be obtained with a magic T having a movable plunger in one arm and the cavity in another, as in Fig. 15-15 [291].

15-12. Large Untuned Cavity. Since the Q factor of a cavity is the ratio of energy stored to energy lost in the cavity walls, it is very large for a large cavity which has a high volume-to-area ratio. Nevertheless, such a cavity does not usually show a sharp resonance because many modes are excited simultaneously. Therefore its Q cannot be measured by the shape of the resonance curve. Q can be obtained, however, either from the decay time constant when the cavity is excited by a pulse, or by a measurement of the energy density in the cavity for a given exciting power.

Both methods involve sampling the energy in the cavity at a sufficient number of points to get a good average energy density. The sampling is carried out by a large number of detectors [137], such as strings of fine-wire bolometers [338] arranged at random in the cavity (Fig. 15-16). The thermal detectors need to have a sufficiently short time constant that they will respond to changes in cavity energy density as the oscillator is tuned, or to modulation in oscillator power if that is used.

To ensure that many modes are excited and good average energy densities are measured, a metal-bladed fan, or "mixer," may be rotated in front of the input coupling horn. The incoming microwaves are reflected from the fan in varying directions as it rotates. Only when a large number of different modes are excited simultaneously is the response of the detectors proportional to Q [165].

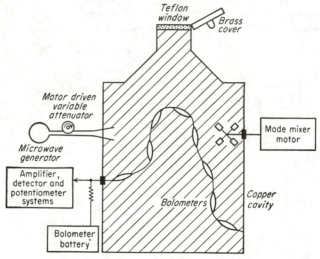

Fig. 15-16. Untuned cavity spectrograph.

This Q can be considered as being controlled by losses from three sources, to each of which we may attribute a separate Q.

1. Q_C: Losses in the walls and fittings of the cavity and through the coupling holes

2. Q_G: Losses in absorbing gas

3. Q_A: Losses through an aperture which can be opened in the cavity

Then the total Q is given by

$$\frac{1}{Q} = \frac{1}{Q_C} + \frac{1}{Q_G} + \frac{1}{Q_A} \tag{15-18}$$

Q_G can be deduced by comparing expressions for the decay of a microwave pulse in the cavity viewed as a lossy cavity and as absorption of a microwave traveling through a lossy medium.

$$\left. \begin{aligned} W &= W_0 e^{-\omega t/Q_G} \\ &= W_0 e^{\alpha x} = W_0 e^{-\alpha ct} \end{aligned} \right\} \tag{15-19}$$

where c is the velocity of light, t is the time, α is the absorption coefficient in units of reciprocal length, and x is the path length in the vapor. Then

$$\frac{1}{Q_G} = \frac{\lambda\alpha}{2\pi} \tag{15-20}$$

which demonstrates that the absorption in the cavity is equivalent to that of a free-space path length of $Q\lambda/2\pi$.

The loss through the aperture is expressed by Q_A which, if its area is large enough to avoid diffraction effects but small in comparison with the wall area, is given by [165]

$$Q_A = \frac{8\pi V}{\lambda A} \tag{15-21}$$

where V is the volume of the box and A is the area of the hole. Thus

$$\frac{1}{Q} = \frac{1}{Q_C} + \frac{\lambda\alpha}{2\pi} + \frac{\lambda A}{8\pi V} \tag{15-22}$$

The response of the detectors is proportional to Q, with a constant of proportionality determined by the input power and the detector sensitivity. Let E_1 be the detector output voltage with the cavity empty and the aperture closed, E_2 be the output voltage with the aperture open, and E_1' and E_2' be the corresponding quantities after the absorbing sample has been admitted to the cavity. Then from the above equations

$$Q_C = \left(\frac{E_1 - E_2}{E_2}\right)\frac{8\pi V}{\lambda A} \tag{15-23}$$

and

$$\alpha = \frac{A}{4V}\left(\frac{E_2'}{E_1' - E_2'} - \frac{E_2}{E_1 - E_2}\right) \tag{15-24}$$

If the absorption lines are not too broad, E_1 and E_2 can be measured without removing the sample by using a nearby frequency where the absorption is small.

The large cavity has not been much used because it is slow and cumbersome for general work. However, it is particularly suitable for measurements of absolute intensities and line widths.

15-13. Spectrographs for Measurements of Zeeman Effect. The type of spectrograph needed for Zeeman studies depends largely on the sensitivity of the particular molecule's frequencies to a magnetic field, *i.e.*, on the magnitude of the Zeeman effect. Paramagnetic molecules like O_2, NO_2, and ClO_2 which exhibit a large Zeeman effect can be conveniently studied in magnetic fields of only a few oersteds. Stark modulation [615] can be used for most paramagnetic molecules, with an additional small magnetic field provided by a long single-layer solenoid enclosing the waveguide. For these same molecules, Zeeman modulation is an alternative method of sensitive detection, while for O_2 it is the only method of modulation because the molecule has no electric dipole [441]. For high-field studies of paramagnetic molecules and for nonparamagnetic molecules which require large magnetic fields, a cavity is advantageous because of its small volume [435], [291].

For paramagnetic molecules, the cavity need not be tunable. Since the Zeeman effect is so large, individual Zeeman components can be brought to the desired frequency by varying the magnetic field [435], [558]. This gives a rather different view of the spectrum, however, than that obtained by varying the frequency of observation until an absorption line is found, and then applying a weak magnetic field to produce a small Zeeman splitting.

A moderately strong magnetic field can be obtained economically over the length of an ordinary waveguide absorption cell either by coiling the cell [371] or by using special pole pieces, long and narrow like the waveguide [683]. With a magnetic field transverse to the direction of propagation, it may still be either parallel or perpendicular to the microwave electric vector. In the former case, π components ($\Delta M = 0$) are observed, while the second arrangement gives the σ components

$$(\Delta M = \pm 1)$$

A longitudinal field, as from a solenoid, is always perpendicular to the microwave electric vector if the principal mode is being used, and so gives σ components.

The signs of the nuclear and molecular magnetic moments cannot be determined as long as linearly polarized microwaves are used, because the Zeeman patterns are symmetrical; $i.e.$, the Zeeman pattern consists of pairs of lines equally displaced in frequency from the zero field position, and of equal intensity. In order to determine the sign, circularly polarized microwave fields in a square or circular guide may be used since circularly polarized radiation carries angular momentum [681], [683]. $\Delta M = +1$ transitions occur when the microwave electric vector rotates in a clockwise direction when viewed by a person looking in the direction of the longitudinal magnetic field. If the direction either of rotation or of the magnetic field is reversed, $\Delta M = -1$ transitions occur.

The circular polarization can be obtained by a suitable length of waveguide having rhombic (or elliptical) cross section, connected to square waveguide by tapered sections. Such a rhombic waveguide can support modes in which the electric vector near the center is parallel to either the long or the short diagonal. These modes have different phase velocities and so different guide wavelengths, λ_g. If the length of the rhombic guide is chosen so that one wave lags 90° in phase behind the other, the emergent wave is circularly polarized. Adjustment can be made by squeezing the polarizing section and thus varying the phase difference between the two components of the wave.

If Stark effect is used to obtain high sensitivity in an absorption cell supporting a circularly polarized microwave, the usual flat septum must be avoided since it will disturb the microwave field distribution. The Stark electrode may be a wire or a rod of circular cross section along the axis of the waveguide. This arrangement gives a nonuniform Stark field, but the nonuniformity may be useful in smearing out the Stark components so that they do not interfere with observation of the Zeeman components. Further smearing can be obtained by use of a trapezoidal or similarly shaped Stark voltage rather than a square wave.

The hyperfine-structure transitions in atomic spectra of alkali metal vapors resemble paramagnetic molecules in their large Zeeman effect. In these cases interactions between the valence s electron and the nucleus are so large that the transitions between individual hyperfine components lie in the microwave range. Cesium [405] and sodium [642], [643] transitions have been observed by microwave absorption in resonant cavities. The alkali metal vapors were contained in glass or quartz liners within the cavity. External coils provided magnetic fields of a few oersteds for Zeeman modulation.

Roberts, Beers, and Hill [405] detected the dispersion on the edges of a line rather than the absorption at its center. The imaginary part of the permeability, which is maximum at the point of greatest change of absorption, slightly modifies the cavity's resonant frequency. This cavity controlled the frequency of an oscillator, and so varying the magnetic field caused a frequency modulation which was detected.

15-14. Spectrometers for High and Low Temperatures. Low temperatures are often used to increase the population of lower vibrational and rotational energy levels. The intensities of most microwave lines are thereby increased. Lines involving excited vibrational states can be distinguished because their intensities do not increase so much or may even decrease on cooling.

Few polar molecules have an appreciable vapor pressure at very low temperatures, so that it is seldom possible to observe their spectra at temperatures lower than those obtainable with dry ice. A rectangular trough of sheet copper, thermally insulated and supported in a wooden box, serves to hold either dry ice alone or a dry ice–acetone mixture. For the lowest temperatures such a trough can be used to contain liquid air, although the liquid air evaporates rapidly.

High-temperature spectroscopy is much more difficult, chiefly because the materials used to support a Stark electrode become poor insulators at high temperatures (cf. review in [728]). Waveguide windows tend to be lossy at high temperatures, and difficulties are experienced in sealing them to the waveguide.

For moderate temperatures (150 to 250°C), the designs used differ from ordinary Stark spectrographs only in details. The polystyrene or teflon insulation of the Stark electrode can be replaced by mica fins crimped into slots on the electrode edges [637] or by grooved quartz strips [728]. Rather than rubber gaskets at the mica windows, lead gaskets [637] or copper gaskets [728] may be used. High temperatures are useful for outgassing the absorption cell and remaining traces of molecules previously introduced into the cell. If the instrument is used for spectroscopy at high temperatures rather than just for outgassing, the sample may condense out at lower temperatures, and if so it must be introduced from a heated container through heated valves and connections.

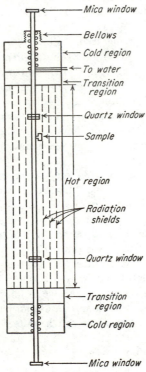

FIG. 15-17. Microwave spectrograph for high temperatures.

For the high-temperature range (250 to 1000°C), which is necessary to obtain sufficient vapor pressure of many interesting diatomic molecules, more radical changes are needed to avoid the serious insulation difficulties and prevent rapid oxidation of metal surfaces. Several approaches have been used. One method reduces the amount of insulation required to position the Stark plate by supporting the entire waveguide vertically [980a]. The waveguide is contained inside an evacuated cylinder with radiation shields (Fig. 15-17). Since the container is evacuated, the waveguide seals need only have a reasonably slow leak rate for a pressure differential less than $\frac{1}{10}$ mm Hg rather than for the entire pressure of the atmosphere. Quartz or ceramic windows with copper gaskets are used at the end of the hot waveguide. Outer mica windows at the points where the waveguide enters and leaves the vacuum jacket are at room temperature and so present no unusual difficulties.

The waveguide cell is made of nickel and gold-plated on the inside to resist corrosion by the gases being studied and to reduce the attenuation of the microwaves. Small ceramic spacers serve to center the Stark electrode. Modulation voltage is applied through a tantalum wire insulated by a lava bushing in the waveguide wall. Tantalum wire is used for heating the cell, and extra heaters near the ends assure that the windows are hotter than the rest of the system and so prevent condensation on the windows.

A molecular beam spectrograph has been designed to study substances which require high temperatures for their evaporation. Only the oven need be at a high temperature, and Stark modulation electrodes can be outside the beam. Alternatively, mechanical modulation of the beam by a cooled shutter may be used [728]. The low modulation frequency attainable in this way necessitates a bolometer or superheterodyne detector for good signal-to-noise ratio. The method avoids the problems of windows and electrode insulation. Moreover, the beam can be used to reduce Doppler broadening, since the path of the radiation is transverse to the beam. However, the amount of absorption obtainable with a molecular beam is so small that only the strongest lines can give detectable signals.

A third type of high-temperature spectrograph uses a thin quartz liner for the waveguide [728]. The guide is split and the Stark voltage is applied between the two halves. Although this gives a nonuniform field so that Stark components are smeared, it should be adequate for detection of the lines. The guide assembly is inserted in a heated iron pipe and fed by horns from an external waveguide.

15-15. Spectrographs for Intensity and Line-shape Measurements. Line widths and relative intensities of nearby lines can be measured on a Stark modulation spectrograph if sufficient care is taken to minimize reflections in the waveguide system. When the microwave oscillator frequency is changed, any reflections present cause changes in the amount of power reaching the crystal. With Stark modulation, these changes do not usually cause spurious lines, but the apparent intensity of a real line changes with the power level at the crystal (cf. [968a]).

If standing waves are present in a section of waveguide because of reflections at its ends, it acts as a resonator. At those frequencies for which it is resonant, the energy density is large, and the attenuation caused by absorption lines is consequently large. Alternatively it may be said that the reflections produce a long effective path length because the radiation is reflected back and forth several times. When the input frequency is changed slightly, the section becomes antiresonant so that the energy density, and with it the attenuation produced by the line, becomes small.

Consider as an example a waveguide absorption cell of length L with small equal reflections occurring at its two ends with a fraction r of power reflected at either single reflection. The power transmitted through the cell may be written

$$P = \frac{P_0 e^{-\alpha L}}{[1 - re^{-(\alpha + 4\pi j/\lambda_g)L}]^2} \tag{15-25}$$

where P_0 is a constant proportional to, but not exactly equal to, input power; α is the attenuation coefficient per unit length in the waveguide; and λ_g is the radiation wavelength in the waveguide. Expression (15-25) shows the variation in power transmitted through the guide to the detector with variation in frequency (or hence λ_g). The ratio of maximum to minimum power (or of detected crystal current, I, since power is approximately proportional to I) is, from (15-25),

$$\frac{P_{\max}}{P_{\min}} = \frac{I_{\max}}{I_{\min}} = \left(\frac{1 + re^{-\alpha L}}{1 - re^{-\alpha L}}\right)^2 \qquad (15\text{-}26)$$

Clearly from (15-26) attenuation tends to reduce the fractional variation in crystal current, since $\dfrac{P_{\max}}{P_{\min}} \to 1$ when αL becomes large.

In order to determine the effect on transmitted power of a small change in the absorption coefficient α such as might be introduced by a gas absorption, (15-25) may be differentiated with respect to α. This gives

$$\Delta P = -P_0 e^{-\alpha L} \frac{[1 + re^{-(\alpha + 4\pi j/\lambda_g)L}]}{[1 - re^{-(\alpha + 4\pi j/\lambda_g)L}]^3} \alpha_{\text{gas}} L \qquad (15\text{-}27)$$

where α_{gas} is the absorption coefficient of the gas. For a given value of gas absorption α_{gas}, (15-27) shows the rapid change of absorbed power ΔP with frequency or λ_g. The ratio of maximum to minimum values of ΔP for a given α is, from (15-27) and (15-26),

$$\frac{\Delta P_{\max}}{\Delta P_{\min}} = \left(\frac{1 + re^{-\alpha L}}{1 - re^{-\alpha L}}\right)^4 = \left(\frac{I_{\max}}{I_{\min}}\right)^2 \qquad (15\text{-}28)$$

Hence the variations in apparent absorption coefficient due to reflections are considerably greater than the variations in crystal current. In some cases, approximate corrections for reflections might be possible by use of (15-27).

For low reflection, the Stark electrode must be tapered at each end. Individual sections of waveguide should have nearly identical cross sections and be carefully aligned where they meet. Windows should be thin, and of good dielectric material.

In addition to keeping reflections low, if intensities are to be measured accurately, the crystal current must not be large enough to cause nonlinearity of the microwave-power–crystal-current relation. Usually a few hundred microamperes are permissible if the load resistance for the crystal is not too large. Finally, circuits which amplify the signal must be linear over the range of signals encountered (*cf.* [901]).

15-16. Gas Handling for Microwave Spectrographs. Most microwave spectrographs use gases at pressures of from 10^{-3} to 1 mm mercury. For ordinary gases, the sample is kept in a glass bulb, with glass connections to the waveguide and vacuum pumps. Glass stopcocks, lubricated with some vacuum grease, regulate the flow of gas into the waveguide. If two stopcocks in series are used between the sample holder and the waveguide cell, gas may be admitted first into the space between the stopcocks. The sample tube is then shut off and the gas in the small space between the stopcocks is allowed to expand into the waveguide. This arrangement is convenient because it admits only the small amounts of gas usually needed to fill the absorption cell. These small amounts of gas are often difficult to control by a single stopcock opening directly into the sample bulb. Pressure can be measured by any of the ordinary vacuum gauges (S. Dushman, *Scientific Foundations of High Vacuum Technique*, John Wiley & Sons, Inc., New York, 1950; R. T. Sanderson, *Vacuum Manipulation of Volatile Compounds*, John Wiley & Sons, Inc., New York, 1948).

Some gases react with the glass, stopcock grease, metal waveguide, mica windows, or plastic insulation and require careful choice of the materials in contact with the gas. Teflon (polymerized tetrafluoroethylene) is inert to most chemicals and can be used for insulating the Stark septum and leads. Polytrifluorochloroethylene (Kel-F) tubing is often useful in handling samples of fluoride compounds. Copper equipment is suitable for reactive fluorides and can be used for the entire gas-handling system.

Chlorinated and fluorinated greases are available for stopcocks. While they are generally more expensive and poorer lubricants than ordinary stopcock greases, they are much more resistant to chemical attack. A thin coating of such a grease can be used to protect windows. When highly reactive or unstable compounds are to be used in the absorption cell, the observable effect of decomposition is minimized by allowing a sample of gas to remain in the system for a few hours or even days in order that initial surface reaction may go to completion. The sample may be replenished occasionally or, if necessary, a continuous-flow method at low pressure may be used. For substances with sufficiently high vapor pressures, reaction and decomposition may be reduced by cooling the entire system [521], [447].

Some substances, such as ammonia and water, are strongly absorbed on the walls of the waveguide and are then evolved very slowly. It may take several weeks at room temperature before the ammonia or water lines disappear from a sensitive spectrometer. Heating the absorption cell to about 100°C usually outgases it sufficiently in a few hours. Differential absorption may also change the composition of a gas mixture, giving rise to serious errors in experiments on line broadening by foreign gases or in quantitative analysis.

15-17. Spectrometers for Free Radicals. Free radicals are usually extremely unstable or reactive and often exist as separate molecules for only a thousandth of a second or less (*cf.* E. W. R. Steacey, *Atomic and Free Radical Reactions*, Reinhold Publishing Corporation, New York, 1946; W. A. Waters, *Chemistry of Free Radicals*, Oxford University Press, Oxford, 1946; F. O. Rice and K. K. Rice, *The Aliphatic Free Radicals*, Johns Hopkins Press, Baltimore, 1935). Nevertheless some molecules with odd numbers of electrons are stable (*e.g.*, NO, NO₂, ClO₂), and so radicals with intermediate stability and lifetimes should occur and give detectable microwave spectra. Lifetimes of these chemically active substances are often limited by the presence of other materials which combine with the radicals or catalyze their recombination.

Attempts to obtain microwave spectra of free radicals have been hampered by the paucity of knowledge about them and by the lack of suitable tests for the presence of radicals or for the elimination of interfering substances from the system (*cf.* review by J. Mays [728]). The microwave spectrum of the OH radical, arising from transitions between members of a Λ doublet, has been observed by Sanders, Schawlow, Dousmanis, and Townes [861], [971] using a system in which radicals

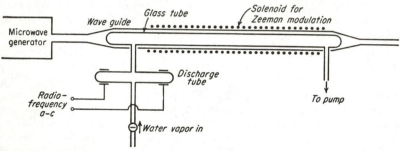

FIG. 15-18. Microwave spectrograph used to observe the spectrum of the free radical OH.

come in contact with no metal (Fig. 15-18). Radicals are produced by dissociation of water vapor in a radio-frequency discharge tube with external electrodes, and are then pumped through a straight tube of low-loss glass. This tube is the lining of a waveguide of circular cross section which is tapered at its ends to an ordinary rectangular waveguide. Zeeman modulation is applied by superimposed direct and 100-kc currents through a solenoid around the waveguide. The waveguide is slotted longitudinally to reduce eddy currents and allow the magnetic field to penetrate it. This method of modulation is particularly suitable for radicals which have large magnetic moments arising from an unpaired electron. Absorptions caused by substances other than radicals are not modulated and so are not observed.

15-18. Microwave Radiometers. A type of spectrograph introduced by Dicke [148] is especially suitable for microwave spectroscopy of astronomical sources. In radio astronomy neither the original source nor the absorber is under the control of the observer, so that neither can be modulated. Furthermore, the radiation from astronomical sources is usually spread over some range of frequencies rather than being essentially monochromatic as is a microwave oscillator. The instrument

designed by Dicke may be called a radiometer, since it detects the noise
power radiated from an extended source and determines the apparent
temperature of the source at microwave frequencies.

In the radiometer, which is shown in Fig. 15-19, a movable absorber is
placed in the waveguide between the antenna and the detector. The
absorber is moved in and out of the guide at a rate of 30 cycles/sec.
When it is in the guide, the incoming signal is replaced by thermal
radiation from the absorber at room temperature. Thus small variations
between the effective temperature of the astronomical radiator and of the
spectrograph show up as variations in the noise power received by the

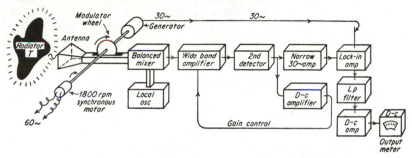

Fig. 15-19. Microwave radiometer. (*From Dicke* [148].)

detector during the two portions of the cycle. A relatively large intermediate-frequency bandwidth (8 Mc) is used so that all the thermal noise
radiated by the source within a band of that size on either side of the
oscillator frequency is available as signal to be amplified and detected.

The noise-power levels in the two positions (absorber in and absorber
out of the guide) are compared sensitively by a lock-in detector which
has a time constant large enough to average over a relatively long period.
In this way fluctuations in the thermal noise power and in amplifier
gain are minimized.

As in other spectrographs dealing with low signal powers, superheterodyne detection is used. A balanced mixer ensures that local oscillator
power is not radiated into the waveguide where it might be reflected
differently from the absorber than from the open waveguide.

With this instrument a change in source temperature somewhat less
than 1°C is measurable, corresponding to a noise power of 10^{-16} watt. It
should be noted that the effective temperature measured with microwaves
often differs from that of the optical region because of selective absorption
or emission.

The change in temperature which can be detected by such a system is
given [148] approximately by

$$\Delta T = NT \left(\frac{\Delta \nu_2}{\Delta \nu_1}\right)^{\frac{1}{2}}$$

where T is the temperature of the receiver (room temperature) and N is the receiver's noise figure (that is, NT is the apparent temperature of the receiver input circuit judged by the residual noise when no signal is present) $\Delta\nu_1$ is the bandwidth of the receiver system before the second detector of Fig. 15-19, and $\Delta\nu_2$ is the bandwidth of the low-pass filter following lock-in.

Microwave radiation from hydrogen atoms in interstellar space is spread over a frequency range of only a few tens of kilocycles. Hence for its detection a receiver bandwidth of a few tens of kilocycles is used, and the received frequency is shifted back and forth periodically at the lock-in frequency. This shifting of the frequency replaces the variable absorber of Fig. 15-19 and allows an accurate comparison between the amount of noise power radiated at the hydrogen resonance and that at other frequencies.

CHAPTER 16

MILLIMETER WAVES

16-1. Introduction. Conventional vacuum tubes and circuits do not operate well for wavelengths as short as the microwave region, because they use lumped circuit constants which require that the wavelength be considerably larger than the size of the tube or circuit element. Klystrons, magnetrons, and other microwave devices which do work successfully in the microwave region usually require only that the wavelength be comparable with the dimensions of the tube, and hence they are appropriate for microwaves as short as 1 cm. For wavelengths below about 4 mm, even these devices no longer work well because they cannot be satisfactorily scaled down to a sufficiently small size.

Short wavelengths are of course emitted by hot bodies, and this source of radiation is commonly used in infrared spectroscopy. However, the intensity of radiation in a given bandwidth or range of frequencies decreases at longer infrared wavelengths as $1/\lambda^2$ and is so small beyond wavelengths of a few tenths of 1 mm that it can be used for spectroscopy only with great difficulty. For example, one square centimeter of a black body at 2500°K radiates only 5×10^{-8} watt within a 1 per cent range of frequencies centered at 1 mm wavelength. If one wishes a bandwidth or resolution of 1 Mc at this wavelength, the power is reduced to 10^{-11} watt, which is too small to be useful at present—especially since heterodyne detection cannot be used without a more powerful local oscillator of the same frequency.

Thus neither electronic oscillators nor infrared sources provide much usable radiation in the wavelength range between a few tenths and a few millimeters. In approaching this range from either side, spectroscopy becomes increasingly difficult. However, techniques are now available which permit some rewarding spectroscopy for wavelengths as short as 1 mm, and these will be discussed. Spectroscopy in the longer millimeter region is not so difficult, but its techniques are sufficiently different from those in the centimeter range to warrant treatment in this chapter.

16-2. Spark Oscillators for Millimeter Waves. Since thermal sources give so little power in the millimeter and submillimeter region, stronger sources of broad-band radiation were sought many years before continuous-wave generators even approached the centimeter band. Spark

451

oscillators were used by Lebedew [1] to generate 0.6-mm waves in 1895, and by Nicholls and Tear in 1922 to reach a wavelength of 0.22 mm [3], [4].

A variant of the spark oscillator is the mass radiator, exploited first by Glagolewa-Arkadiewa [5], in which the sparks pass between many small metal particles. A pair of nearby particles with a spark passing between them acts as a dipole radiator of short wavelength. But since the particle separation is not fixed, a wide range of wavelengths is generated. Usually, the metal particles are immersed in an oil bath which flows or is carried past high-voltage electrodes [5], [13], [49], [92], [6], [128]. Wavelengths from 0.1 to 50 millimeters have been generated in this manner and used, with a diffraction grating, for low-resolution spectroscopy. One serious difficulty with this source, besides the continuous spectrum and low intensity, is the large voltage pulse as the spark breaks down the main gap. Even with careful shielding, this pulse tends to limit the usable amplifier sensitivity.

A variant of the mass radiator is obtained by dropping charged mercury droplets into a mercury pool of the opposite charge [503a]. Small powers in the 2- to 10-mm range have been produced by this method.

16-3. Vacuum-tube Generators. While vacuum-tube oscillators for the millimeter region are more difficult to construct than incoherent sources, their advantages are great. Since all the radiation is confined to a narrow band, the spectrometer need not have high resolving power and so can be relatively simple and efficient. In favorable cases, almost all the radiated energy lies in one narrow band.

One method of obtaining millimeter wave tubes is to scale down those types used at centimeter wavelengths. Klystrons, traveling-wave tubes, and backward-wave oscillators have been built to produce wavelengths near 4 mm, while pulsed magnetrons have reached 2.5 mm (*cf.* review in [510]). Of these, the rather new backward-wave oscillator appears most promising as a flexible high-frequency source for spectroscopy. The first three types of tubes are in practice inefficient and are necessarily small, so that neither conduction nor radiation is sufficient to dissipate heat for more than very low power operation. Furthermore circuit losses increase approximately as the square of the frequency. Magnetrons, by virtue of their pulsed operation, are able to operate with moderate efficiency and power output in the millimeter region. Figure 16-1 shows how the heat transfer limits possible power output. Since tubes are usually of the order of the wavelength λ in dimensions, the upper line on the graph of Fig. 16-1 is a natural upper limit to the amount of heat which can be dissipated by conduction. Actual tubes, as shown by the shaded area, come fairly close to this limit in the centimeter region. The power radiated by a hot body is very much less, especially when only those wavelengths in a small band are selected.

If the efficiency of vacuum-tube generators is not very small, they can produce powers enormously greater than hot bodies in the millimeter region. Efficiencies are usually somewhat greater for pulsed tubes than for continuous types; this accounts for some of the spread of the output powers shown for tubes of Fig. 16-1.

When tubes are made to work at shorter wavelengths by scaling down dimensions, the power output will be unaffected provided the current and losses are unchanged. But scaling down the tube reduces the cathode area in proportion to the square of the linear dimensions, so

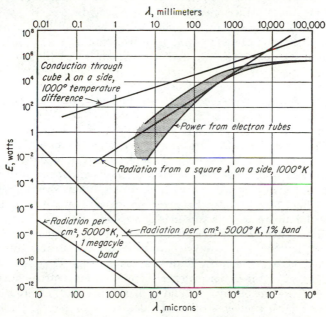

FIG. 16-1. Power limits of thermal sources, and thermal limitations on vacuum tubes. (*From J. R. Pierce* [510].)

that attainable beam current will fall unless means are found for greatly increasing emission-current density. Higher emission currents are accompanied by increased space-charge effects which modify the operation of the tube. Very high emission-current densities can be achieved for a short period, if the tube is allowed to rest before the next pulse. Thus a pulsed magnetron, with plate potential applied for about one out of every thousand microseconds and having intensive back bombardment of the cathode by some of the electrons, attains a much higher emission density than ordinary tubes. The high plate potential during the pulse tends to overcome the effects of space-charge concentration near the cathode.

Finally, fabrication tolerances become exceedingly critical when centimeter-wave klystrons are scaled to operate in the millimeter-wave range. For these reasons it is desirable to have tube types whose dimensions are all large in comparison with the wavelength. Klystrons, traveling-wave or backward-wave tubes, and magnetrons are an advance over conventional triodes for the microwave region in that they have sizes not very much less than the wavelength generated. Neither they nor any other type yet invented have critical dimensions which are greater than the output wavelength. Nevertheless, all these types have been successfully scaled down to operate in the millimeter region, and in spite of the difficulties, further progress may be expected.

Klystrons are commercially available to produce wavelengths as short as 5 mm (Raytheon Manufacturing Co., Waltham, Mass.) with a power output of a few milliwatts. Similar reflex klystrons for the longer millimeter wavelengths have been constructed in England (cf. review in [388]). However, millimeter-wave klystrons tend to be noisier and less stable than their centimeter-wave counterparts, and the power emitted sometimes varies rapidly over the range of tuning.

Traveling-wave generators of several types can be constructed for the millimeter region. One traveling-wave tube has been operated in the 6- to 8-mm region [617] as both an amplifier and an oscillator. Backward-wave oscillators have also been made in the 5-mm region with an output power of about 10 mw and with electronic tuning over a frequency range of about 30 per cent. However, no traveling-wave or backward-wave oscillators are as yet commercially available.

Pulsed magnetrons have been built to oscillate at wavelengths as short as 2.5 mm, but the shorter-wavelength tubes are not tunable (Bernstein, M. J., and others, Columbia Radiation Laboratory Progress Reports, 1947–1954). For the 6-mm region a few pulsed magnetrons have been built which are tunable over a band of 1 or 2 per cent. Because these magnetrons have simpler structures than, for example, klystrons and because of the advantages of pulsed operation discussed above, efficiencies of the order of 5 per cent are achieved. This efficiency permits fairly high output powers, and the millimeter-wave magnetrons have peak pulse outputs of around 25 kw.

16-4. Harmonics from Vacuum Tubes. Since several kinds of tubes are available which give power at a wavelength of about 1 cm, their overtones have been investigated as sources of millimeter waves. Magnetrons generate so much power that even relatively weak overtones are detectable. Klein, Loubser, Nethercot, and Townes [716] have detected harmonics as high as the tenth (1.25-mm wavelength) from K-band tubes and the third (1.1 mm) from a 3.3-mm tube. Peak powers of a few hundred microwatts were obtained at the shortest wavelengths. In the apparatus used, a side arm and phaser produced an adjustable load by which the harmonic output could be maximized (Fig. 16-2). The filter which selected the high harmonics was a section of waveguide too small to permit the longer wavelengths to pass. The magnetron source was also used in conjunction with an echelette grating similar in principle to those used for infrared spectroscopy. This reflection grating was operated in free space and diffracted energy from the transmitter waveguide to the section containing the detector crystal. With this apparatus it was found that not all higher frequencies emitted by magnetrons are harmonics, *i.e.*, integral multiples of the fundamental frequency.

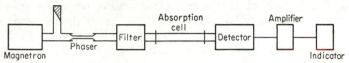

FIG. 16-2. Apparatus for isolating magnetron harmonics. (*From Klein, Loubser, Nethercot, and Townes* [716].)

These nonharmonic frequencies are, however, weaker than the true harmonics, so that they would not ordinarily lead to confusion. Individual tubes, even of the same type, vary considerably in their harmonic and overtone content. The use of magnetron harmonics for spectroscopy is limited by the difficulty of obtaining magnetrons which are tunable over a sufficient frequency range.

Klystrons for the 1-cm region give considerable amounts of power and are reasonably stable. Their output occasionally contains an appreciable power at some particular harmonic frequency, but the harmonic output is not sufficiently certain or controllable to be very useful.

Harmonics of microwave oscillators can also be produced by nonlinear circuit elements. Before discussing harmonic generation in a silicon crystal, which is the most convenient variety of nonlinear element, we shall consider detection of millimeter waves. Detection involves a similar, but somewhat simpler, application of silicon crystals.

16-5. Detection of Millimeter Waves. Techniques adapted from those used in both the infrared and the centimeter-wave region can be applied to detect millimeter waves. In the infrared region, thermal detectors predominate; for centimeter waves thermal detectors are sometimes used, but crystal detectors are usually best and are most widely used.

In the early work on mass radiators, various kinds of thermal detectors were employed. Cooley and Rohrbaugh [128] used a bismuth-antimony thermopile which had all junctions covered by a 2-mm-thick coating composed of equal parts lampblack and tellurium-plated cork dust. The mixture was equivalent to many absorbers whose size and separation were of the order of the wavelengths being received. The thermopile was used over a range of 0.2 to 2.2 mm and had a sensitivity such that, after amplification, a flux of about 3×10^{-8} watt/cm^2 of exposed thermopile surface could be detected.

A more sensitive thermal detector is the Golay cell [195], [196]. The Golay detector contains an air space between two films, one of which is an absorber of radiation. The other film is a very light flexible mirror. When radiation falls on the absorber, it heats the film and hence the air in the cell which then expands and slightly deflects the mirror. Motion of the mirror affects the amount of light falling on a photocell, and the resultant photocurrent is proportional to the radiation intensity.

When the cell is used as a detector of millimeter waves, the end of the waveguide is pointed at the cell [716]. Some resonance effects are observed in this region because of the finite size of the cell aperture, so that it is helpful to optimize the cell response by adjusting its distance from the waveguide. The radiation is chopped at 10 cycles/sec by a rotating semicircular absorber passing into the waveguide through a slot. When used with a circuit having a time constant of 5 sec, one cell of this type had a sensitivity of about 5×10^{-11} watt. The same sensitivity can be expected for shorter wavelengths, and so the Golay cell may be useful for submillimeter waves.

In the longer millimeter range the packaged cartridge-type crystals designed for 1.25-cm (1N26) or 8-mm (1N53) wavelength are convenient as detectors or harmonic generators. One design of detector for cartridge-type crystals due to Beringer [139] introduces the signal from the waveguide into the crystal through a short coaxial line. Similar but simplified designs such as that shown in Fig. 16-3 have been extensively and successfully used for spectroscopic work in the millimeter range by Gordy and coworkers [413], [457], [694]. Such detectors, using commercial crystals, are most successful at wavelengths above 4 mm, but they have detected radiation from harmonic generators of similar design (see Fig. 16-3) at wavelengths as short as 2 mm. Their use at such short wavelengths is quite difficult because very careful selection of crystals is required and each crystal operates over only a narrow range of frequencies.

For most spectroscopic purposes, it is best to use components which are "broad-band" so that they operate satisfactorily over a wide range of frequencies without complicated and critical tuning. Detectors and other circuit components which are simple and broad-band are particularly desirable in the highest-frequency ranges. This is partly because at the shortest wavelengths any critical dimensions become so difficult to control that they are best avoided. Furthermore, present spectroscopy below a few millimeters is still in a rather primitive state and is limited enough by other difficulties that an additional restriction of detectors to a narrow frequency range is very objectionable.

Although detectors using pieces of silicon mounted individually in a

waveguide without a surrounding cartridge are somewhat time-consum-
ing to construct and less rugged than the commercial 1N26 and 1N53
crystals, they give enormously better performance at the shortest wavelengths. Problems of matching the microwave power into the crystal
are very much reduced and the detector is usable over a much broader
frequency range. Detectors of this type, illustrated in Fig. 16-4, were

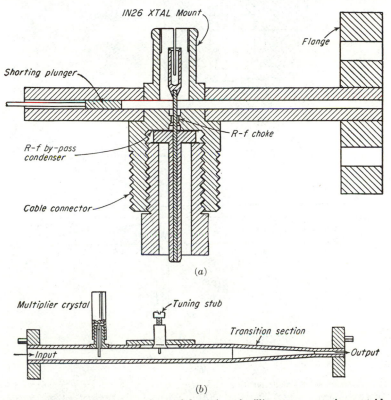

(a)

(b)

FIG. 16-3. Apparatus for generation and detection of millimeter waves using cartridge-
type crystals. (a) Detector. (b) Simple harmonic generator. (From C. M. Johnson, reproduced in [694].)

shown by Klein, Loubser, Nethercot, and Townes [716] to be fairly
sensitive over a wide range of frequencies and down to wavelengths at
least as short as 1 mm. Near 1 mm, these detectors were only about one
hundred times less sensitive than the best crystals at 1 cm wavelength.
Johnson, Slager, and D. D. King [944] made a systematic comparison of
several detectors and harmonic generators for the range 6 to 2 mm wavelength. They tested cartridge-type (1N26, 1N31, and 1N53) crystals,
cartridge-type crystals with part of the walls of the cartridge cut away,

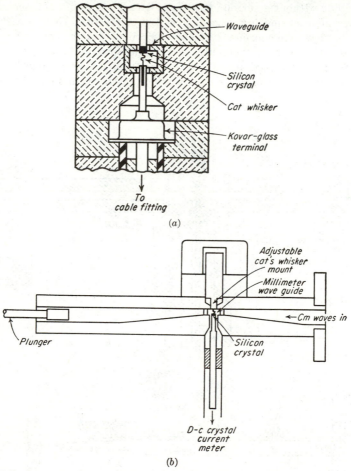

FIG. 16-4. Apparatus for generation and detection of millimeter waves using open-guide crystal mounts. (a) Detector. (*From Klein, Loubser, Nethercot, and Townes* [716].) (b) Harmonic generator. Waveguide for high harmonics crosses waveguide for fundamental at crystal, and is perpendicular to the page. (*Design due to E. Richter* [733].)

and "open-guide" mounts such as that of Fig. 16-4. They found that the last type were distinctly superior as detectors, especially at the highest frequencies.

16-6. Semi-conducting Crystal Harmonic Generators. Semiconducting crystal harmonic generators receive fundamental microwave power from an electronic oscillator (usually a klystron), and because of the nonlinear response of the semiconductor to the microwave field, they produce harmonics of the fundamental. Good harmonic generators involve efficient matching of the fundamental microwave power into a crystal and appropriate nonlinear characteristics. They also require efficient radiation of the generated harmonics into a waveguide. Since they involve matching of the crystal to a waveguide at both the fundamental and harmonic frequencies they are somewhat more complex than are detectors. However, the principles involved in detectors and generators are very similar, and hence the designs used for the two functions in any one experiment usually involve the same types of crystal mounts as indicated by Figs. 16-3, 16-4, and 16-5. The generator design shown in Fig. 16-3 is quite convenient at the longer wavelengths but tends to be narrow-band and weak in output at the shortest wavelengths. The open-guide crystal mount such as that shown in Fig. 16-4 has strong advantages in being broad-band [733], and at the higher frequencies gives appreciably greater output [944] than the type shown in Fig. 16-3.

W. C. King and Gordy [823], [946] have recently made notable progress in pushing the limit of operation of crystal harmonic generators down to 1 mm or slightly below. The generators and detectors which they use are of the open-guide type and are illustrated in Fig. 16-5. The shortest wavelength which has so far been obtained by this type of harmonic generation is 0.77 mm [910] but it seems reasonable to expect that detectable amounts of power may eventually be obtained at wavelengths of a few tenths of 1 mm.

There is as yet very little definite or quantitative information about the best techniques for crystal harmonic generation of wavelengths near 1 millimeter. However, we shall attempt to summarize what information and suggestions are available.

For the shortest wavelengths it is quite important that the crystal be in an open and simple mount (see Figs. 16-4 and 16-5) with simple tuning adjustments. Several different mounts of this type have worked with some success [733], [944], [946], but details of an optimum design are not clear. It should be noted that each of these generators produces high harmonics in a variety of waveguide modes, so that transmission and matching problems cannot be treated so simply or successfully as in a waveguide with only one mode of propagation.

No study has yet been made of the best crystal material. King and Gordy [823], [946] have used pieces of silicon about 1 mm in diameter broken from the silicon slabs of 1N26 crystals. They indicate that small crystals are considerably better than large ones, but experimental evidence for this is not clear. 1N23 crystals contain cylinders of silicon of about 1 mm diameter which have also been successfully used. Best types of etching, polishing, or doping of the crystal surface are unknown. However, it is important to keep the crystals rather dry. High humidity will make at least some crystal contacts deteriorate rapidly.

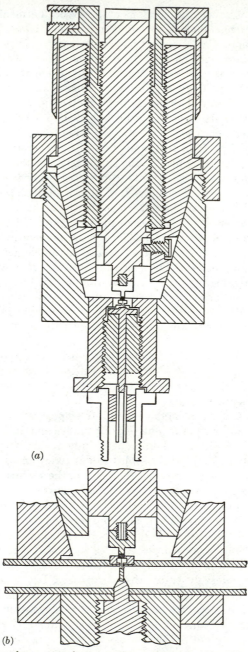

(a)

(b)

FIG. 16-5. Design of apparatus for generation and detection of millimeter waves using open-guide mount. In each case the waveguide for millimeter waves is perpendicular to the page. (a) Detector. (b) Harmonic generator. (*From King and Gordy* [946].)

Attention must be paid to obtaining a very small contact point on the crystal. A large contact surface provides a capacitance which shunts the desired nonlinear conduction characteristics of the contact and the impedance of the smallest contacts obtainable are still objectionably low at wavelengths near 1 mm. "Cat's whiskers," the fine tungsten wires which make contact, are usually about 2 mils (0.005 cm) in diameter. These may be sharply pointed by electrolytic etching [325]. Fine points may be blunted by pressure on the silicon surface, so that carefully controlled contact pressure is desirable, and repeated contacts may produce objectionable blunting. For accurate pressure control, a differential screw drive as shown in Fig. 16-5 is useful. Usually nearly optimum performance is obtained when the cat's whisker first contacts the silicon. Slightly increasing the pressure may worsen the performance, but there is often a somewhat higher pressure at which the contact properties pass through another optimum approximately as good as the first.

For high harmonics it is very important to drive the harmonic generator with a large amount of fundamental power. The harmonic output usually increases rapidly with increasing fundamental power until it reaches a "saturation" point where it stays approximately constant with further increases in fundamental power. At least for the first four harmonics of K band, each successive harmonic requires larger amounts of fundamental power to reach the saturation point.

Very probably the production of very high harmonics depends on having a microwave voltage at the crystal contact which is so large that the majority of the nonlinear crystal response is traversed by the varying voltage of the fundamental in a time comparable with the period of the harmonic. For the highest harmonics, fundamental power at least as large as 100 mw should be used. Crystal contacts can usually stand a few hundred milliwatts without damage due to "burn-out."

It might be expected that shorter wavelengths could be obtained by using fundamental microwave power of wavelength shorter than 1 cm. However, the higher-frequency klystrons emit appreciably less power than do those for K band, and the harmonic power at very short wavelengths which can be obtained from them is consequently no greater than that obtained from harmonics of the rather powerful 2K33 tubes.

Good harmonic generator and detector design should produce nearly 1 mw of power or 1 ma of detected current at the second harmonic of K band. Detected signals from harmonics up to the fifth or sixth will decrease by about a factor of 10 per harmonic. However, for harmonics above the seventh or eighth, the decrease in signal per harmonic is only a factor of 3 or 4 if the fundamental power is sufficiently large.

Harmonic generation and detection of wavelengths shorter than about 1.5 mm require at present a considerable amount of painstaking trial-and-error work in varying conditions and adjustments. The harmonic obtained may be identified by the use of short sections of waveguide which are beyond cutoff for the lower harmonics and thus act as high-pass filters. King and Gordy [946] recommend that adjustments be made at a frequency for which the fundamental or one of the lower harmonics coincides with the rotational absorption line of a linear or symmetric molecule. This means that higher harmonics will coincide closely (but not exactly, because of centrifugal stretching) with higher rotational lines of the same molecule. As the frequency of the klystron providing fundamental power is then varied, absorption lines corresponding to successive harmonics appear—generally with decreasing amplitude

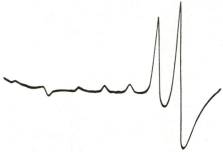

Fig. 16-6. The 8th, 10th, 12th, 16th, 18th and 20th (from right to left) rotational lines of OCS obtained with the 4th to 10th harmonics of 1.23 cm klystron. Frequency of fundamental increases from left to right. Harmonic frequencies range from 97,000 Mc to 243,000 Mc (3.08-mm to 1.23-mm wavelength). (*From King and Gordy* [946].)

as shown in Fig. 16-6. The number of harmonics present may then be judged by the number of absorption lines present, and a particular harmonic optimized by observing the magnitude of its absorption. For use of this technique, it is important to find a molecule with rotational lines very near the frequency which is desired, since harmonic power will not remain optimized when the klystron frequency is varied.

16-7. Propagation of Millimeter Waves. A conventional rectangular waveguide which propagates microwaves only in the dominant (TE_{10}) mode is usable for millimeter waves, although attenuation increases with frequency [*cf.* Eq. (14-27)] and, at the shortest wavelengths, attenuation may be prohibitively large. Oversize waveguide (for example, K-band size) may be used if it is matched to the generator, or to a small guide used to filter out low frequencies, by a suitably gradual taper. However, such a guide will transmit any higher modes generated at discontinuities, and so special care must be taken to avoid irregularities. For instance, flanged joints must be carefully assembled to ensure accurate alignment of the waveguide sections. Many generators of millimeter waves emit the waves into a number of different modes in an oversized waveguide. In such cases a complex of modes is already present, but careful elimination of irregularities at flanges is still important to reduce reflections and losses.

Attenuators used for millimeter waves resemble those for longer wavelengths, although they are reduced in dimensions. Carbon-coated tapered strips of mica inserted through an axial slot in the broad waveguide face are satisfactory.

Sometimes techniques based on those of optical spectroscopy may be used to advantage. For instance, reasonably small horns will produce a fairly narrow beam in free space or through an absorbing gas. This beam may also be reflected by a diffraction grating which permits rough wavelength measurements. For high efficiency, the grating can be of the echelette type, in which the rulings are shaped to throw as much of the diffracted radiation as possible into one order. One typical echelette grating, designed to operate around 1.6-mm wavelength, had eighty $\frac{1}{8}$-in. grooves milled into a flat metal surface. Focusing of the microwaves was effected by two spherical mirrors as shown in Fig. 16-7. An echelette grating is most efficient only for one particular wavelength, although the maximum

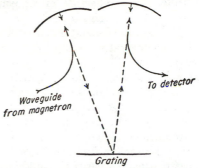

Fig. 16-7. Diffraction grating for measuring millimeter wavelengths. (*From Klein, Loubser, Nethercot, and Townes* [716].)

is fairly broad. A more elaborate design suitable for somewhat longer wavelengths has been constructed of semicircular rods which can be rotated to suit the wavelength [260].

16-8. Frequency Measurement. The methods employed for the measurement of frequency in the millimeter range are not essentially different from those used for longer microwaves (see Chap. 17). However, the relative importance and usefulness of the various techniques of frequency measurement are changed by the change in wavelength.

Cavity resonator wavemeters can be used for the longer millimeter waves. As the wavelength is shortened, construction tolerances become more critical. In addition, Q decreases for most modes as the wavelength decreases. Some improvement in Q may be obtained by using a larger cavity in a higher mode, such as one of the circular electric modes which have no currents across the gap between the end plunger and the wall. However, in this case mode ambiguity can occur, and so it is necessary to make sure that the resonances observed really correspond to the desired mode. One good way to do this is to use the wavemeter to check the frequency of a known spectral line. Another method of checking the oscillator frequency roughly, and so of verifying the wave-meter mode, is to use a low-Q wavemeter.

In the millimeter region broad-band interferometers based on similar optical types can serve this purpose. Among these are the Fabry-Perot [786], [770a] and Michelson types [214], [265], [385], [404]. These inter-ferometers resemble their optical counterparts in most respects, although some of them use magic T's instead of partly reflecting mirrors as beam splitters. They can be refined to give moderately good accuracy but even in quite simple form can discriminate between successive harmonics of an oscillator, or between different wavemeter modes. A diffraction grating, either reflection or transmission, can also be used for rough frequency measurement.

If a harmonic generator is used, it is often more convenient to measure the fundamental wavelength by a wavemeter connected before the gen-erator. The order of the harmonic can be identified by some rough frequency measurement as described above. Or the successive rotation lines of a simple molecule may be used to identify the frequency as indicated in the discussion on harmonic generation. A heterodyne method has also been used to discriminate between harmonics [457]. Two harmonic generators driven by separate klystrons are used, one of which is swept in frequency synchronously with an oscilloscope sweep. Beat frequencies between the klystron harmonics are passed through a tuned intermediate-frequency amplifier and applied to the oscilloscope. Different harmonics appear at different places on the oscilloscope screen and so may be identified. This is because the difference frequency between harmonic pairs is proportional to the harmonic number.

For precision frequency measurements, the unknown frequency may be compared with harmonics of a standard crystal oscillator (*cf.* Chap. 17). The comparison may be made at the klystron fundamental fre-quency where there is more than sufficient power to beat with the fre-quency standard.

16-9. Absorption Spectrographs for the Millimeter Region. In view of the difficulty in making satisfactory signal sources and detectors, it is fortunate that absorption lines in the millimeter region tend to be stronger than those at lower frequencies. This is partly because of the (fre-quency)^2 factor in Eq. (13-19), and partly because the higher-frequency transitions usually belong to states of higher rotational quantum number J, which are more populated at ordinary temperatures. For linear molecules, the intensity is proportional to J^3 or ν^3.

The line intensities are sometimes high enough so that straight absorp-tion, without modulation, may be used, and absorption cells of a few feet or even a few inches may be sufficiently long. Stark modulation is more difficult to apply than at lower frequencies, because the Stark effect is proportional to $1/J^3$. Besides, Stark modulation cells often have high attenuation at millimeter wavelengths. Frequency modulation of the source ("double modulation," see sec. 15-2) is a useful alternative in cases where Stark modulation cannot be applied.

Otherwise, millimeter-wave spectrographs generally resemble those for longer wavelengths and considerations similar to those of Chap. 15 are applicable.

CHAPTER 17

FREQUENCY MEASUREMENT AND CONTROL

Many of the lines observed in microwave spectroscopy are so narrow as to warrant extremely high precision in the measurement of their frequency. For instance, if a line is 100 kc wide, its center can be located to a tenth of its width, or 10 kc, without much difficulty, and in some cases the line center can be still more accurately determined. And if the line is near 25,000 Mc, 10 kc represents an accuracy of a part in 2.5 million. Precision of this order is rare in physical measurements and requires very good standards of frequency. In fact, refined techniques permit locating the line center to an accuracy at least as great as that of the very best earlier standards of frequency or time. Thus microwave spectral lines may themselves be used as frequency standards, making possible "atomic clocks."

The methods used for measuring microwave line frequencies and the converse problem of using the lines to control electronic oscillators or clocks will be discussed in this chapter.

17-1. Wavemeters. For rough frequency measurements, as when a new line is found, cavity wavemeters are extremely useful (*cf*. Chap. 14). For instance, a 1.25-cm-band wavemeter of the type shown in Fig. 14-6 has an unloaded Q of 8000 to 10,000, which is reduced to near 5000 by loading. With this instrument settings can be made on an oscillator frequency to about 1 Mc, or 1 part in 25,000. Even this accuracy, which requires very careful wavemeter construction, is much poorer than the narrowness of the spectral lines permits. Hence some better measuring device such as a quartz-crystal-controlled frequency standard is needed for accurate measurements.

A crystal-controlled frequency standard is usually designed to give a series of harmonic frequencies separated by perhaps 30 Mc. In this case a cavity wavemeter is needed to distinguish between the different harmonics, and its accuracy need only be as good as about 10 Mc. This accuracy (near 1 part in 3000) can be attained with ordinarily good machining tolerances and without compensation for temperature or atmospheric conditions. It is probably representative of the performance of the wavemeters used in most microwave spectroscopy.

When the wavemeter is used with a spectrograph, a circuit like that of

Fig. 17-1 is convenient.   The wavemeter absorption pip is centered on the mode oscilloscope, which otherwise displays the variation of klystron output during its sweep.   If previously a microwave line has been brought to the center of the spectrum oscilloscope by tuning the microwave oscillator, its frequency is thus measured by the wavemeter.

While the second oscilloscope is convenient because it permits viewing the klystron mode at the same time as the line, it can be eliminated. Then both the wavemeter and spectrograph signals are applied to the same oscilloscope.   The pattern seen then is the sum of the sharp microwave resonance and the broader cavity resonance, which can be brought into coincidence with it.   Probably the most satisfactory arrangement would be to use a double-beam oscilloscope if one is available.

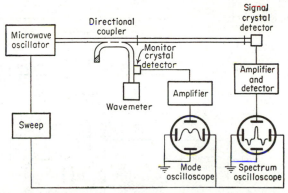

FIG. 17-1.   Microwave spectrograph with absorption wavemeter.

To attain an accuracy appreciably greater than about 1 part in 3000 with a wavemeter alone, it must be constructed so that the plunger displacement can be measured to a thousandth of a millimeter or better. Compensation or correction for temperature changes is needed ([221], pp. 384–392; [288]).   For a cavity constructed entirely of one material, the temperature coefficient of frequency is approximately the same as the coefficient of linear expansion, $i.e.$, about 2×10^{-5} per degree centigrade for brass.

The wavemeter of course directly measures wavelength, and conversion to frequency requires a knowledge of the dielectric constant or refractive index of air.   This varies with temperature and humidity.   The refractive index of air for the ranges encountered in the laboratory is given by the empirical equation [453], [675]

$$(N_{t,p} - 1)10^6 = \frac{103.49p_1}{T} + \frac{177.4p_2}{T} + \frac{96.0}{T}\left(1 + \frac{5208}{T}\right)p_3 \quad (17\text{-}1)$$

where p_1, p_2, and p_3 are the respective partial pressures of dry air, carbon dioxide, and water vapor in mm. Hg and $T = 273 + t$ is the absolute temperature, t being the temperature in degrees centigrade. From this it follows that changes in temperature and humidity can shift the apparent frequency by as much as 0.02 per cent.

If the waveguide system to which the cavity is connected has standing waves, another error can be introduced. The standing waves correspond to a reactance such that the total circuit reactance will be a minimum at a frequency slightly different from the resonance frequency of the cavity alone. Errors of the order of 0.02 per cent may be produced in this way·

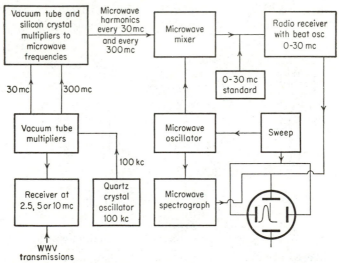

FIG. 17-2. Quartz-crystal-controlled microwave frequency standard.

While these sources of error can be reduced, even the best wavemeters are not good enough by themselves for really precise measurements of microwave spectral-line frequencies.

17-2. Quartz-crystal-controlled Frequency Standards. As noted above, for accurate measurements the spectral-line frequency is compared with harmonics of a high quality quartz-crystal-controlled oscillator. This in turn is standardized by comparison with the radio transmissions of a standard frequency like those of station WWV in Washington. Ultimately these transmitted frequencies are regulated by checking them against astronomical observations.

A commonly used type of frequency standard is shown schematically in Fig. 17-2. The multiplier chain produces harmonics in the microwave band whose frequencies are accurate multiples of 30 Mc. These are combined with the microwave oscillator output in a silicon or germanium crystal mixer to produce a beat frequency between 0 and 30 Mc. The beat is observed at the output of the radio receiver only when it nearly coincides with the frequency to which the receiver is tuned. Therefore, as the klystron is swept over a range of a few megacycles to display an absorption line on the spectrograph, a sharp pulse appears as the difference frequency passes that to which the receiver is tuned. This pulse is displayed on the same oscilloscope or recorder as the line, and by tuning the receiver the pulse can be made to coincide with the line center. Then the receiver frequency is read from its calibrated dial or measured by a low-frequency (0 to 30 Mc) tunable standard. The latter need be of only moderate precision since it measures the difference between the 30-Mc harmonics and the line frequency. This difference is a small fraction of the whole frequency.

Beats are obtained whether the klystron frequency is above or below the harmonic or marker frequency. They may be distinguished by changing the receiver frequency a small amount in a known direction and noting the direction in which the frequency marker moves.

Occasionally a microwave line will lie so close to one of the 30-Mc harmonics that, when the klystron is adjusted into coincidence with the line, the beat frequency is below the receiver's tuning range. It is then necessary to use the beat with an adjacent harmonic, which will be just greater than 30 Mc.

In setting on a line, it must be noted that the line and the marker pulse or "pip" have very different waveforms and that they pass through separate and dissimilar amplifiers. They will each suffer a time delay during amplification, and the amounts of these delays will be unequal. Therefore if the true line frequency is to be determined accurately, a measurement should be made with the sweep going in the direction of increasing frequency, and again with the direction reversed (cf. [445]). The average of these two settings gives the true line frequency. The error may also be minimized by sweeping at a rate slow compared with the response time of the amplifying and "pip"-forming circuits.

The quartz crystal which controls an accurate frequency standard should be of a type such as the GT cut having a small temperature coefficient of frequency and furthermore should be operated at a fixed temperature if very high accuracy is desired. As the frequency generated by a crystal-controlled oscillator is affected slightly by the tube and other circuit elements, the circuit employed should be designed to minimize the effect of such variations. In the circuits commonly employed for frequency standards, the quartz crystal is placed in one arm of a bridge which controls the oscillator feedback [86a], [626a]. The same arm contains a small adjustable reactance which is used to adjust the frequency to coincide exactly with a primary standard such as station WWV. Because conditions for the reception of WWV are not always good, and the received signal sometimes exhibits fading or frequency flutter [523], the oscillator ought to be stable enough that only occasional checking is required. To this end the oscillator should be protected from load variations, being coupled to output circuits through one, or possibly even several, successive cathode followers. This precaution is especially important if the crystal oscillator output is to be available for use as a general laboratory standard, so that different test circuits may be connected to it.

Recently spherically contoured polished quartz plates of very high stability have been developed. They can be used to control directly oscillators in the 5 to 10-Mc region, thereby eliminating several frequency-multiplier stages in the standard. Their Q is so high (several million) that fairly simple oscillator circuits, such as that in Fig. 17-3, may be used [760a], [882].

Frequency multipliers for use in the early stages of microwave standards are not essentially different from those used in radio transmitters (see, for instance, [121], pp. 458–462; [221], pp. 365–374). Harmonics are generated by a nonlinear amplifier, which has its output circuit tuned to the desired harmonic. Usually the harmonic generator is a class C amplifier with the grid biased to, or slightly beyond, cutoff, so that plate current flows only in bursts which are rich in harmonics. For best stability, no grid current should flow during any part of the cycle. Push-push amplifiers give only even harmonics, while a push-pull operation gives only odd harmonics. Some typical circuits are given in Fig. 17-4.

A higher order of multiplication is needed to produce microwave frequencies especially if one starts from an oscillator in the region of 100 kc. Therefore, care must be taken to avoid introducing undesired frequencies in the early stages, which will produce a rich spectrum in the microwave region, so that the desired harmonics are hard to identify. Harmonic generators, being highly nonlinear amplifiers, will mix undesired frequencies, such as 60 cycles from power supplies, to produce side frequencies, each of which has its own overtones. Excellent filtering is needed in the power supplies, which can best be achieved by the use of a series-tube regulated power supply. If successive multiplier stages are too tightly coupled, their tuning will be broadened sufficiently to permit them to pass undesired harmonics. Loosely coupled series-tuned link coupling between stages helps to select the desired frequency.

In some microwave-frequency standards conventional tubes are used up to a frequency of some hundreds of megacycles, and then followed by a silicon crystal harmonic generator. In one design, for example, the final vacuum-tube stage is a pair of 2C40's tripling from 270 to 810 Mc. Others use special multiplier klystrons up to 3000 Mc, or even higher, before the final crystal harmonic generator. Such klystrons give frequency multiplication by factors up to about 12 and yield much larger

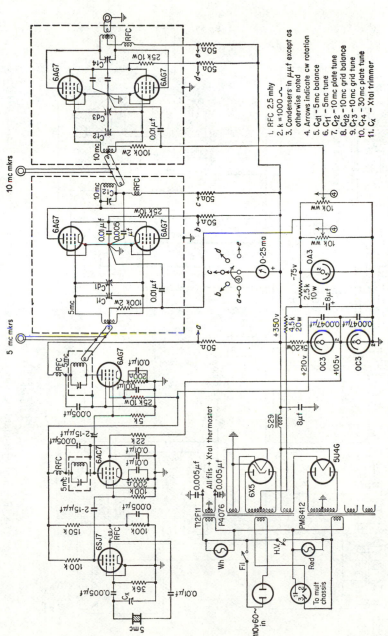

FIG. 17-3. Frequency standard—oscillator chassis schematic diagram. (*From L. C. Hedrick* [809].)

471

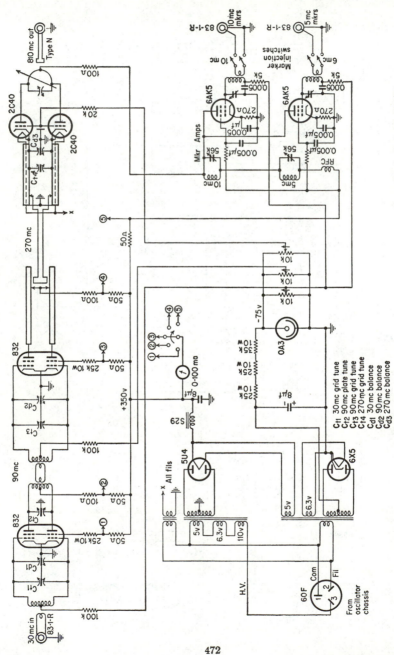

Fig. 17-4. Frequency standard—multiplier chassis, schematic diagram. *(From L. C. Hedrick [809].)*

472

power than can be obtained in the corresponding harmonic from a crystal harmonic generator. Somewhere in or just before the final stage, provision is made for mixing in frequencies such as 270, 30, or 10 Mc to give markers more closely spaced than the overtones of the last multiplier.

The crystal harmonic generator is similar in principle to those employed for generating millimeter waves (sec. 16-6). However, the lower input frequency usually requires coaxial cable connections rather than waveguide.

Some frequency standards generate a variable microwave frequency which is adjustable to the line frequency, rather than a series of fixed harmonics. To get an accurately known variable frequency, the output of a fixed crystal oscillator or one of its harmonics is combined with a low-frequency oscillator covering a small range. Then the sum frequency, which is selected by a filter, is known to the same absolute accuracy as the variable oscillator but to a much higher percentage accuracy ([221], p. 365).

There are many variations possible in the particular combination of circuits used to measure frequencies. However, those above appear at present to be the most convenient and widely used.

17-3. Measurement of Frequency Differences. Sometimes, as for hyperfine-structure measurements of microwave lines, only frequency differences are needed rather than accurate absolute frequencies. While these differences can be obtained readily by measuring the two components with an absolute frequency standard, the use of a standard can sometimes be avoided by measuring differences directly. This can be done by electronically frequency-modulating the klystron with an alternating repeller voltage whose frequency is adjustable over the range of separations to be measured. Then the klystron output contains the center frequency and sidebands separated from it by the modulation frequency. To make the measurement, the klystron is set on one component of the hyperfine structure and a sideband is simultaneously adjusted to coincide with the other component. The repeller modulation frequency is then the separation of the components, and since this is usually of the order of a few megacycles it can be easily measured with an accuracy comparable with the line width [145], [317]. This technique is useful only if the hyperfine structure is very simple so that there is no confusion between the several overlapping patterns which are produced. It is to be noted that the separation in megacycles of the line images is equal to the modulation frequency even for harmonics of the klystron frequency.

Another somewhat more complex, but more generally useful, system involves stabilizing a microwave oscillator on some fixed frequency near the lines to be measured. Its output is then mixed with the output of a low-frequency oscillator to provide frequency markers with separations which are accurately known, even though the absolute frequencies are not known.   Methods for stabilizing a microwave oscillator are discussed in the following section.

17-4. Frequency Stabilization of Microwave Oscillators.   The frequency generated by a microwave oscillator can be stabilized by comparing it with an external standard.   Since the standard can be relatively free from the thermal, electrical, and mechanical disturbances to which the klystron is sensitive, a considerable improvement in stability can be achieved.   Among the standards which may be used as references for stabilization are resonant cavities, microwave spectral lines, and harmonics of quartz-crystal-controlled oscillators.   These will be discussed separately.

Stabilization relative to another oscillator may be achieved by combining the outputs in a mixer and applying the difference frequency to the

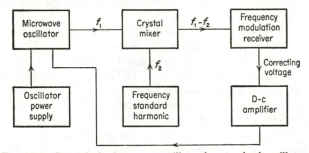

Fɪɢ. 17-5.   Control of microwave oscillator by standard oscillator.

input of a conventional frequency-modulation receiver.   From the receiver's discriminator a voltage is obtained which is zero when the microwave oscillators differ by just the frequency to which the receiver is tuned, and which changes sign on either side of the null point.   Any discriminator output voltage can then be amplified by a d-c amplifier and applied to the klystron reflector or tuner electrode in such a direction as to counteract the frequency changes which produced it.   (See Fig. 17-5.)   The microwave oscillator can be made to follow the frequency-modulation receiver tuning over a range of some megacycles depending chiefly on the electronic tuning range of the tube and spurious competing signals picked up by the receiver.   This type of stabilization is usually quite convenient for microwave spectroscopy, since a frequency standard is often readily available and can be used as the oscillator with respect to which the klystron is stabilized.

Sometimes the microwave oscillator is a reflex klystron in which the frequency-controlling electrode (the·reflector) is operated at a high negative potential relative to ground.   Then it may be convenient to couple it to the amplified discriminator signal through a magnetic control tube, such as the 2B23 (General Electric) [567]. The signal from the d-c amplifier controls the current through a field coil, and this in turn varies the current through a resistor in series with the klystron reflector. The reflector voltage is thereby controlled without a direct connection between the high-voltage electrode and the d-c amplifier.

17-5. Control of Frequency by a Resonant Cavity. Since cavity resonators can be constructed with high Q and good stability, they can be used for controlling the frequency of a microwave oscillator. Some improvement in klystron stability can sometimes be made even by such a simple means as using a stabilizing voltage derived from a cavity wavemeter. If the oscillator is at a frequency which falls on one side of the sharp wavemeter response, a shift in frequency toward the wavemeter peak increases the detector crystal voltage, while a shift in the

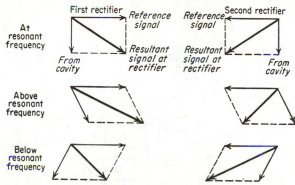

Fig. 17-6. Phase and amplitude relationships in microwave discriminator.

opposite direction decreases the voltage. After d-c amplification, this signal is applied to a controlling electrode of the oscillator. Such a method suffers from being sensitive to changes in oscillator amplitude, which can result from changes in loading. It is also sometimes a disadvantage that the frequency at which the oscillator is stabilized is not the cavity resonance center frequency.

The disadvantages of the simple method can be overcome by using a microwave discriminator analogous to those used with lumped circuit elements at lower frequencies. The discriminator uses two rectifiers in a circuit in which their outputs just balance each other at resonance. Changes in oscillator amplitude affect both outputs equally and so do not displace the balance frequency. On one side of resonance the first crystal rectifier produces a larger signal than does the second, and on the other side of resonance the situation is reversed.

This result may be obtained by combining the wave reflected from the cavity with waves which are, at resonance, respectively, 90° behind and 90° ahead of it in phase (Fig. 17-6). As the frequency changes on either side of resonance, the phase of the reflected wave shifts rapidly in a direction dependent on the sense of the frequency change. The resultant amplitudes at the two crystals change accordingly, so that the difference between their rectified outputs indicates the magnitude and sign of any departure from resonance.

A microwave discriminator of this type is shown as part of a Pound frequency-controlling circuit [227], [221], in Fig. 17-7. The cavity and a short circuit are at opposite ends of the coplanar arms (numbered 1 and 2 here) of a magic T (T_1 in the diagram) but are positioned so that their effective distances from the median plane differ by $\lambda_g/8$ near the cavity resonance frequency. Thus waves reflected from the cavity

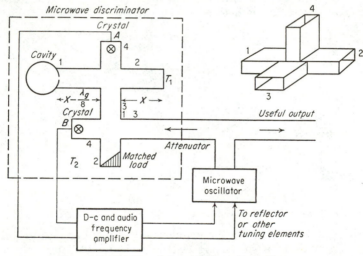

Fig. 17-7. The Pound d-c stabilizer—block diagram.

and the short differ by 90° in phase on reaching the median plane. Here the reflected waves combine to send a sum wave into arm 3 and a difference wave into arm 4, thus providing the two crystal detectors, A and B, respectively, with signals in which the cavity wave lags or leads the other wave by 90° in phase at resonance. On either side of resonance the cavity wave undergoes a rapid change of phase, producing the discriminator action described above. The second magic T (T_2) directs only the reflected signal to crystal B, while sending the incoming klystron wave first to T_1. Crystal B actually receives only half as much power as crystal A, but this can be compensated by a balancing attenuator at the amplifier input.

For this circuit, the rate of change of discriminator voltage with frequency near resonance is

$$\frac{dV}{d\nu} = DP_0 \frac{Q_0}{\nu_0} \frac{4\alpha}{(1 + \alpha)^2} \tag{17-2}$$

where D = detector sensitivity in volts per unit power input

P_0 = microwave power applied to the discriminator

Q_0 = unloaded Q of the cavity resonator

ν_0 = resonant frequency of the cavity

α = δ_1/δ_0, where δ_0 = decrement of the unloaded cavity and δ_1
= change in decrement when the cavity is coupled to a
matched waveguide

For optimum sensitivity α should be 1.

The d-c stabilizer is relatively simple and effective but it is limited by crystal noise, which is always largest at low frequencies. Moreover d-c amplifiers are seldom completely free from drift. Even so, d-c stabilizers have been used to hold two 9000-Mc oscillators within a few kilocycles over a period of hours.

The disadvantages of the d-c stabilizer are largely overcome by a circuit in which the signal amplification occurs at a high frequency ([227]; [221], pp. 58–78; [333]). The discriminator produces an intermediate-frequency voltage whose amplitude is proportional to the amount of the departure of the oscillator frequency from the cavity resonance and whose phase depends on the direction of departure. The phase-sensitive detector, similar to those described in Chap. 15, derives from this a d-c voltage whose magnitude and sign measure the departure of the oscillator from cavity resonance.

The error voltage derived from a microwave discriminator of either a d-c or an intermediate-frequency type can be applied to a servomechanism which tunes the microwave oscillator. Servomechanisms cannot respond as quickly as can an electronic system to rapid (audio-frequency) fluctuations in frequency, but are effective in counteracting slow drifts [228]. The electronic and servomechanical systems can be combined to control both the rapid changes in frequency and slow drifts [687].

17-6. Stabilization of Microwave Oscillators by Absorption Lines. Since a waveguide cell filled with a suitable gas shows a sharp absorption peak at the resonant frequency of the gas, it can be used in place of a resonant cavity to control the frequency of a microwave oscillator. Moreover the position of the line's center is nearly independent of temperature and pressure (Chap. 13), so that it can be made to serve as an absolute standard of frequency. An oscillator controlled by a spectral line can then be used to drive a clock through a chain of frequency dividers [492], [724].

If the microwave line is to be used as a primary frequency standard or "atomic clock" (a time standard dependent on a frequency of an approximately isolated nuclear, atomic, or molecular system), the associated circuits should hold the oscillator as accurately as possible to the center of the line. Since deviations from the center of a line are detected by the change in amplitude of a wave transmitted through a cell containing the gas, the minimum observable change in frequency is closely related to the minimum observable absorption. For example, in Chap. 15 it is shown that the minimum detectable absorption, limited by thermal noise, is of the order of 10^{-10} cm^{-1} (for a 30-cycle noise bandwidth). Since the ammonia 3,3 line has a peak absorption roughly 10^7 times stronger than this, it ought to be possible to locate the line center to about (1/10^7) of its width. Since this width, as limited by Doppler effect at low pressures, is around 100 kc, *i.e.*, approximately $1/(2 \times 10^5)$ of the line frequency, it seems reasonable that the ammonia line might ultimately be used to control microwave frequencies to about 1 part in 5×10^{13}, or even better with a narrower noise bandwidth. Although the method used in making this estimate is rather rough, the estimate of the ultimate stabilization limit agrees well with the more accurate treatment below (*cf.* also [653]). Actual atomic clocks constructed so far are limited by less fundamental difficulties to a much lower precision.

It has been shown previously that $\Delta\gamma$, the smallest detectable change in absorption coefficient, is [Eq. 15-9]

$$(\Delta\gamma)_{\min} = 2e(\alpha_0 + \gamma)\left(\frac{2kT\,\Delta f}{P_0}\right)^{\frac{1}{2}} \tag{17-3}$$

where γ = the absorption coefficient of the gas, cm^{-1}
$\quad \alpha_0$ = the loss coefficient of the waveguide, cm^{-1}
$\quad e$ = the base of natural logarithms
$\quad k$ = Boltzmann's constant
$\quad T$ = the absolute temperature
$\quad P_0$ = the microwave power reaching the detector
$\quad \Delta f$ = bandwidth of the detecting system

It is assumed that the optimum waveguide length, $l = 2/(\alpha_0 + \gamma)$, is used and that no sources of noise other than thermal are present.

The absorption coefficient of a narrow microwave line may be written approximately as [*cf.* Eq. (13-22)]

$$\gamma = \frac{\gamma_{\max}(\Delta\nu)^2}{(\nu - \nu_0)^2 + (\Delta\nu)^2} \tag{17-4}$$

where $\gamma_{\max}$ = the absorption at the peak of the line
$\quad \nu$ = the microwave frequency
$\quad \Delta\nu$ = the half width of the line at the half maximum
$\quad \nu_0$ = the frequency at the peak of the line

Some of the devices used at present for stabilizing on a spectral line set on the peak of the absorption. In that case a change $\Delta\gamma$ in γ occurs if the frequency ν differs from ν_0 by an amount ϵ where

$$\epsilon = \Delta\nu\left(\frac{\Delta\gamma}{2\gamma_{\max}}\right)^{\frac{1}{2}} \tag{17-5}$$

Considerably better performance can be obtained by using the steep slopes of the absorption lines, since then a small change in frequency corresponds to a greater change $\Delta\gamma$. For a distance from resonance $\nu - \nu_0 = a\,\Delta\nu$,

$$\frac{\epsilon}{\nu} = \frac{\Delta\gamma(a^2 + 1)^2}{2a\gamma_{max}}\frac{\Delta\nu}{\nu} \tag{17-6}$$

Using Eq. (17-3), the smallest detectable fractional change in frequency is

$$\frac{\epsilon}{\nu} = \frac{e(\alpha_0 + \gamma)(a^2 + 1)^2}{a\gamma_{max}}\frac{\Delta\nu}{\nu}\left(\frac{2kT\,\Delta f}{P_0}\right)^{\frac{1}{2}} \tag{17-7}$$

For γ_{max} considerably smaller than α_0, the smallest value of ϵ/ν, or the steepest part of the resonance curve, is obtained when $a = \pm(\frac{1}{3})^{\frac{1}{2}}$, that is, $\nu = \nu_0 \pm (\frac{1}{3})^{\frac{1}{2}}\,\Delta\nu$. Best performance for larger values of γ_{max} is obtained somewhat farther from the resonance frequency. For the other extreme $\gamma \gg \alpha_0$, optimum values of a are ± 1, and $\dfrac{(\alpha + \gamma)(a^2 + 1)^2}{a\gamma_{max}} = 2$. This case may be assumed, since for strong absorbers such as NH_3, it is possible to have $\gamma_{max} > \alpha_0$. Thus the minimum detectable frequency change given by Eq. (17-7) becomes

$$\frac{\epsilon}{\nu} = 2e\frac{\Delta\nu}{\nu}\left(\frac{2kT\,\Delta f}{P_0}\right)^{\frac{1}{2}} \tag{17-8}$$

The above expression would indicate that the error in frequency stabilization could be made arbitrarily small by making $\Delta\nu$ or T small or ν or P_0 large. However, T cannot be reduced indefinitely without reducing the gas pressure to too low a value. Usually an upper limit to the power P_0 is set by saturation effects. The line width $\Delta\nu$ cannot be reduced below the limit set by Doppler effect except by methods which reduce the number of molecules available for absorption and so reduce γ. Thus the only method of obtaining arbitrarily small error is to decrease the bandwidth Δf.

If a particular absorption line is considered, an optimum power flux can be determined. As shown in Chap. 13, half width is obtained when saturation occurs which, when inserted in place of the $\Delta\nu$ occurring in (17-8), gives

$$\frac{\epsilon}{\nu} = 2e\frac{\Delta\nu}{\nu}\left(1 + \frac{8\pi^2|\mu|^2\nu It}{3ch\,\Delta\nu}\right)^{\frac{1}{2}}\left(\frac{2kT\,\Delta f}{P_0}\right)^{\frac{1}{2}} \tag{17-9}$$

where $\mu = $ the dipole moment matrix element of the transition
$c = $ the velocity of light
$h = $ Planck's constant
$t \approx \dfrac{1}{2\pi\,\Delta\nu}$

Since I, the number of quanta per second per unit cross-sectional area, is directly proportional to P_0, this expression approaches a limiting minimum value for large P_0 and differs from the minimum value only by a factor of $\sqrt{2}$ when saturation first becomes noticeable, $i.e.$, when

$$I \approx \frac{3ch(\Delta\nu)^2}{4\pi|\mu|^2\nu} \tag{17-10}$$

If the cross-sectional area of the waveguide is A, then $P_0 = Ah\nu I$, and for power P_0 large enough to give saturation

$$\frac{\epsilon}{\nu} = \frac{8e|\mu|}{h}\left(\frac{\pi kT\,\Delta f}{3cA\nu^2}\right)^{\frac{1}{2}} \tag{17-11}$$

This expression very significantly does not contain the line width $\Delta\nu$. The advantage in sharpness of the line which is obtained by decreasing $\Delta\nu$ is just counteracted by the loss in sensitivity due to power saturation of the absorption. Thus there is no strong reason for attempting to attain lines so narrow that their width is determined by Doppler effects. In practice a very wide line is not desirable because the maximum power available P_0 and the power at which efficient detectors operate is usually near 1 mw. In addition nonfundamental variations of circuit response with frequency are very troublesome, and hence a sharp line is desirable to minimize their effects. However, Eq. (17-11) does show that, so far as the limitations of thermal noise are concerned, the line width is unimportant and can be adjusted to any convenient value. Since saturation of most spectral lines occurs with line widths of about 1 Mc and a power flux of a few milliwatts per square centimeter, pressures such that the line is somewhat narrower than 1 Mc would be a convenient choice. Of course the radiation density or the field strength is not uniform throughout the waveguide cross section or throughout the length of the guide because of attenuation. However, a reasonably accurate approximation for ϵ/ν is obtained by considering the "average" saturation condition at the end of the guide where the radiation is introduced.

If it is assumed that, to avoid higher modes of propagation, the waveguide cross-sectional dimensions are of the order of the waveguide λ, then $A \approx (c/\nu)^2$, and

$$\frac{\epsilon}{\nu} = \frac{8e|\mu|}{hc}\left(\frac{\pi kT\,\Delta f}{3c}\right)^{\frac{1}{2}} \tag{17-12}$$

For a numerical evaluation of (17-12), the specific case of the strongest ammonia line, the 3,3 line at 23,870 Mc is of interest. Taking $T = 200°K$ and $\Delta f = 1$ cycle/sec, one obtains, since $\mu = 1.4 \times 10^{-18}$ esu,

$$\frac{\epsilon}{\nu} = 1.5 \times 10^{-13} \tag{17-13}$$

for the accuracy limit imposed by thermal noise.

If the peak of the line is used in the same way for stabilization rather than the points of maximum slope the ultimate accuracy may be decreased by a factor of more than a thousand [724]. In this case, as when certain less efficient detectors are used, or noise is present which is greater than fundamental thermal noise, the attainable accuracy is no longer independent of the line-width parameter.

Use of Dispersion. The above discussion has considered using only the absorption of a molecular resonance in order to stabilize an atomic clock. The dispersion, or reactive part of the resonance, is equally usable. Figure 17-8 shows the behavior of the anomalous dispersion or variation in the dielectric constant near a resonant absorption. Near an absorption line the dielectric constant can be written [*cf.* Eq. (13-15)]

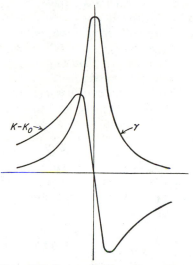

FIG. 17-8. Loss and dispersion near a spectral line.

$$K = K_0 + \frac{\lambda(\nu_0 - \nu)}{\Delta\nu}\frac{\gamma}{2\pi}$$

$$= K_0 + \frac{\lambda(\nu_0 - \nu)\gamma_{max}\,\Delta\nu}{(\nu - \nu_0)^2 + (\Delta\nu)^2} \tag{17-14}$$

where K_0 is the dielectric constant some distance on either side of the line, λ is the wavelength, and other quantities are as defined for expression (17-4). For a gas at low pressure, K_0 may be taken as 1 and K is not very different from 1, so that the index of refraction n is

$$n = K^{\frac{1}{2}} = 1 + \frac{\lambda(\nu_0 - \nu)\gamma}{4\pi\,\Delta\nu} \tag{17-15}$$

By methods shown in reference [653], one finds that the signal power derivable from this dispersion is a maximum for a length of waveguide $l = 2/(\alpha + \gamma_{max})$, and has the value

$$\Delta P = P_0 \left(\frac{\gamma_{max}e^{-1}}{\gamma_{max} + \alpha}\frac{\epsilon}{\Delta\nu}\right)^2 \tag{17-16}$$

where P_0 is the initial power introduced into the waveguide. If this power is equal to the thermal noise $2kT\,\Delta f$, then the fractional frequency error is

$$\frac{\epsilon}{\nu} = \frac{(\alpha + \gamma_{max})e}{\gamma_{max}}\frac{\Delta\nu}{\nu}\left(\frac{2kT\,\Delta f}{P_0}\right)^{\frac{1}{2}} \tag{17-17}$$

Thus use of dispersion gives not only essentially the same optimum length for the microwave path in gas, but also within a small factor the same fractional frequency error, as is shown by comparing (17-7) and (17-17).

17-7. The Molecular-beam Maser.   One of the most promising atomic frequency standards is the molecular-beam oscillator shown in Fig. 15-10. This device can serve as a microwave amplifier or it can oscillate at a frequency determined primarily by the molecular resonance.   It has the strong advantage of obtaining microwave oscillations directly from the molecules rather than requiring stabilization of an electronic oscillator on a molecular resonance.

Consider the amplification of noise by a *maser*—a cavity with wall and load losses given by $1/Q$, and containing a dielectric material with dielectric constant $\epsilon = \epsilon' - i\epsilon'' = 1 + 4\pi\chi' + i(4\pi\chi'')$.   Here χ' is the in-phase component of the polarizability per unit volume and χ'' is the component which is 90° out of phase with the polarizing electric field. χ' and χ'' depend, of course, on the properties of the beam of molecules and, if amplification rather than power loss occurs, χ'' must be positive. The power which is generated at various frequencies in the cavity is given [925] by

$$P \, d\nu = \frac{(4\pi\chi''kT/Q)d\nu}{[(1/2Q) - 2\pi\chi'']^2 + [(\nu - \nu_c')/\nu_c']^2} \tag{17-18}$$

where ν_c' is the resonance frequency of the cavity as modified by the presence of the molecules, or

$$\nu_c' = \nu_c(1 + 4\pi\chi')^{-\frac{1}{2}} \approx \nu_c(1 - 2\pi\chi') \tag{17-19}$$

since $\chi' \ll 1$.   Here ν_c is the resonance frequency when no molecules are present in the cavity.   As in (15-12), the imaginary part of the dielectric constant or out-of-phase component of the polarization simply modifies the apparent loss factor $1/Q$ of the cavity.

The maximum power generation occurs, from (17-18), when $\nu_c = \nu_c'$, and is large, corresponding to an oscillation, only if $4\pi\chi'' \approx 1/Q$.   Near the center of the molecular resonance, χ' may be written approximately as

$$\chi' = \frac{\chi_0''(\nu - \nu_0)}{\Delta\nu} \tag{17-20}$$

where χ_0'' is the value of χ'' at the resonance frequency ν_0 of the molecular line, and $\Delta\nu$ is the half width of the line at half maximum intensity. Hence the oscillation occurs at

$$\nu = \nu_c' = \nu_0 + (\nu_c - \nu_0)\left(1 + \frac{2\pi\chi''\nu_c}{\Delta\nu}\right)^{-1} \approx \nu_0 + \frac{Q}{Q_L}(\nu_c - \nu_0) \tag{17-21}$$

Here $\nu_c/2 \, \Delta\nu$ has been set equal Q_L, which is an effective Q for the molecular line, and the condition of oscillation $4\pi\chi'' \approx 1/Q$ has been used, which makes $2\pi\chi''\nu_c/\Delta\nu \gg 1$.

Expression (17-21) shows that the oscillator frequency will be primarily determined by the molecular frequency ν_0, but that if the cavity frequency ν_c is not equal to ν_0, the frequency will be "pulled" by approximately $Q/Q_L(\nu_c - \nu_0)$. For a typical case, $Q/Q_L \approx \frac{1}{1000}$, so that any variation δ in cavity frequency will vary the oscillator frequency by about $\delta/1000$. Therefore, if the cavity frequency is constant to one part in 10^8, the oscillator should be constant to about one part in 10^{11}.

The total power, which is given by integrating over-all frequencies, should equal the power P_B delivered by the molecular beam. From this one can evaluate $1/2Q - 2\pi\chi''$ which, although very small, is not exactly zero. Substituting the value of $1/2Q - 2\pi\chi''$ into (17-18), it takes the form

$$P\,d\nu = \frac{4kT(\Delta\nu)^2\,d\nu}{[4\pi kT(\Delta\nu)^2/P_B]^2 + [\nu - \nu_0 - (Q/Q_L)(\nu_c - \nu_0)]^2} \quad (17\text{-}22)$$

This expression for the spectrum of the oscillator is not strictly correct because, as in any oscillator, nonlinearities affect the frequency distribution. However, it gives an approximately correct value for the half width of the oscillator output, which is

$$\Delta\nu_{\text{osc}} = \frac{4\pi kT(\Delta\nu)^2}{P_B} \quad (17\text{-}23)$$

For a typical case, $P_B \approx 10^{-10}$ watts and $\Delta\nu \approx 2000$ cycles/sec, so that $\Delta\nu_{\text{osc}} \approx 4 \times 10^{-3}$ cycles/sec. Experimental observations have shown that $\Delta\nu_{\text{osc}} < 10^{-1}$ cycle/sec, and it seems probable that $\Delta\nu_{\text{osc}}$ is actually near the theoretical value, which is $2/10^{13}$ of the frequency ν_0. This is by far the most monochromatic radiation which has yet been produced.

17-8. Realization of Atomic Frequency and Time Standards. No microwave frequency standard yet constructed approaches in accuracy the limits imposed by thermal noise. Difficulties involved in setting accurately and automatically on the center of a line to within a very small fraction of its width are great, even though in principle they can be overcome.

Almost all atomic frequency standards which have been based on a gas absorption line have used the ammonia 3,3 line. This line is very suitable because it is so strong ($\gamma_{\text{max}} = 8 \times 10^{-4}$ cm^{-1}). Its only likely competitor is a line in the oxygen spectrum near 5-mm wavelength.

The oxygen molecule has a smaller dipole moment matrix element than ammonia since its electric dipole moment is zero and only the magnetic dipole moment exists to produce transitions. Thus the lines are not easily broadened or saturated. It has not been used because until recently oscillators giving sufficient power in the 5-mm region to use these lines effectively were not available.   Moreover, the oxygen lines exhibit a large Zeeman-effect broadening and so need to be shielded from fluctuating magnetic fields.   Fortunately, the "first order" Zeeman effect is symmetrical and does not produce a displacement of the line center.

Even though one uses a single preferred line as a microwave standard, it can be used to control oscillators at other frequencies.   These frequencies may be derived by using frequency multipliers and dividers, or by mixing in lower-frequency oscillators to generate sum or difference frequencies.   The low-frequency oscillators need not be known to so great a percentage accuracy as the primary standard.

Several different methods have been employed to stabilize an oscillator relative to the ammonia line.   They have succeeded in stabilizing to a

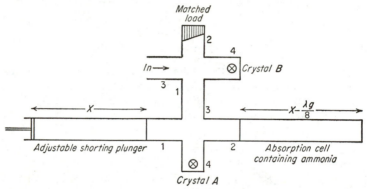

Fig. 17-9. Microwave discriminator using gas-absorption line in place of resonant cavity.

small fraction of the line's width.   Nevertheless even the best is quite far from the theoretical limit [Eq. (13)].

The Pound microwave discriminator ([221], pp. 58–78) has been modified so that the reference cavity resonator is replaced by a short-circuited waveguide cell containing the gas under study [235], [310].   Figure 17-9 shows the modified discriminator.   The discriminator output was applied through a d-c amplifier having a gain of 2000 to the frequency-controlling electrode of a 2K50 klystron oscillator.   By this means drift was reduced by a factor of 1000 relative to that of an unstabilized tube.

Another method of locating the center of a spectral line is to sweep an oscillator across the line and detect the peak or the points of maximum slope by the ordinary methods of microwave spectroscopy [724], [287], [470], [600], [719], [329], [454].   For example, the microwave oscillator may be swept over a narrow range of frequencies (less than a line width) and the crystal detector current observed with a phase-sensitive detector. Output of the phase-sensitive detector passes through zero and reverses at the peak of the line. A voltage derived from the detector is used to control the center frequency of this or another microwave oscillator which is compared with it. If the microwave oscillator is quite stable even when uncontrolled, relatively small and infrequent corrections will be needed. Then a more sensitive control system with a narrower band pass and a higher gain will be usable, resulting in a better final stability than if the oscillator is unstable. Quartz-crystal oscillators with harmonic generators can have excellent short-period stability and thus are advantageous [724].

These methods depend on the variation with frequency of the microwave amplitude at the detector. As noted in Chap. 15, reflections in the waveguide, and variations in klystron power, can cause changes of amplitude with frequency even without the presence of a line. These will limit the accuracy of any atomic clock, so that in practice it may be an advantage to have a narrow line even at some sacrifice of intensity. As with spectrographs, the effects of reflections and certain other disturbances can be minimized by using Stark modulation [329], [975], and some improvement in accuracy is possible by this means. Shimoda [975] has been able to stabilize an oscillator to an accuracy of about one part in 10^9 by the use of a combination of Stark, Zeeman, and source-frequency modulation.

Other natural frequency standards which promise accuracies comparable with microwave absorption are provided by molecular beams and by nuclear quadrupole resonances in solids. The molecular-beam lines are very sharp. Q's of 30 million have already been obtained with a cesium atomic beam [724], and these may afford even better "atomic clocks" than do absorption lines in gases. Nuclear quadrupole resonances give Q's comparable with the microwave lines. At ordinary temperatures the resonances are not intense and their frequency depends strongly on temperature, but in the liquid helium range they are much less sensitive to temperature changes and may afford accurate frequency standards.

The molecular-beam maser, discussed in sec. 17-7, appears to provide one of the simplest and most accurate frequency standards. Two such devices have been compared, using cavities which were not thermostated, and are shown to agree within one part in 10^{10} over a time of about one-half hour. Presumably, thermostating of the cavities and reasonably careful design should allow this type of accuracy for an indefinite time. The oscillations of each such device were extremely monochromatic, as indicated in sec. 17-7, so that they could easily be compared over short periods of time to a few parts in 10^{12}.

CHAPTER 18

THE USE OF MICROWAVE SPECTROSCOPY FOR CHEMICAL ANALYSIS

The well-known varieties of spectroscopy have been so widely and successfully used for chemical analysis that the reader has undoubtedly already wondered whether or not microwave spectroscopy can also be successfully applied in this way. Although microwave spectroscopy appears to be well suited for certain varieties of analytical work, actual applications of this type have so far been very limited. Several articles [361], [538], [589], [700] discussing analytic uses of microwaves have, however, appeared. It is the purpose of this chapter to examine the possible uses and limitations of microwave spectroscopy for gas analysis, and to give the reader basic information needed for such application.

18-1. Microwave Spectroscopy for Analysis. Spectroscopy in each of the major regions of the spectrum has its own characteristic techniques, and as a result of these and the nature of the spectra observed, it has particular advantages and disadvantages. The advantages of microwave spectroscopy of gases spring primarily from its high resolution, the use of very low pressure gases, and from the electronic techniques which are characteristic of spectroscopy in this region.

The high resolution and consequent accuracy of measurement mean that lines of two different substances—regardless of how small their difference—are generally well separated and easily differentiated. For example, the spectra of different isotopic species of the same molecule are almost always easily resolved and identified. Microwave spectra under high resolution are usually so highly specific that a measurement of one line will suffice to identify the molecule to which it belongs.

High resolution makes it possible in addition to isolate and identify the lines of a large number of substances in one gaseous mixture. Thus between 20,000 and 30,000 Mc there is room for about 40,000 microwave absorption lines (assuming a width of $\frac{1}{4}$ Mc for each line). If each of 100 substances in a gas mixture has 20 lines in this region, there is less than one chance in 10^6 that more than one-third of the lines of some one of the substances will be overlapped by other lines. Thus, if the microwave lines are sufficiently intense, each of a very large number of substances can be identified in a gas mixture. The situation is very different in the infrared region, where the rotation-vibration bands of one substance often give lines separated by less than the resolving power of infrared spectroscopy, so that the spectrum of each substance becomes a series of continuous bands, and interference between the spectra of two or more substances is often very troublesome.

The lines of even very small impurities in a gas mixture would usually not be masked by lines of the more abundant components. For some substances which have relatively strong lines, such as H_2O or NH_3, fractional abundances as small as 1 part in 10^5 or 10^6 can be detected by microwave spectroscopy. This is comparable with the minimum abundances which can be detected in the most favorable cases by other types of spectroscopy.

In addition to high resolution, another natural advantage of microwave spectroscopy is the rather small amount of gas required to detect absorption. Typically, a microwave absorption cell would have a volume of a few hundred cubic centimeters and be filled to a pressure of about 10^{-2} mm Hg. This corresponds to only about 10^{-7} mole of gas, or a few micrograms of material. If the gas has very strong lines, then an amount 10^5 times smaller than this, or about 10^{-12} mole, is sufficient for detection. The small quantities of gas needed thus make analysis by microwave spectroscopy practical for microchemical work. In addition, microwave spectroscopy of a sample does not destroy it as might analysis by optical spectroscopy or by a mass spectrometer.

Although the electronic equipment needed for microwave spectroscopy may seem unfortunate to those who look for simplicity, the fact that microwave spectroscopy is based on electronic techniques offers certain advantages. The presence of absorption is indicated by an electrical voltage, which can hence be easily used to operate automatic controls or recording instruments. In addition, the electronic detecting circuits can be made to act very rapidly for rapid control or recording. Thus absorption by a strong line can be detected and translated into a voltage in times as short as 1 millisecond. It must be pointed out that, for the most sensitive detection of small amounts of material or of weak lines, a few seconds are required. Even this is very rapid, however, compared with many other types of analyses.

The natural limitations of analysis by microwave spectroscopy are mainly of two general types. First, one must usually deal with a dipolar gas. There are indeed certain types of characteristic microwave absorptions in liquids or solids, such as paramagnetic absorptions. However, these can be used only in rather limited and special ways for analysis, and analysis by microwave spectroscopy is practically restricted to gases. The substance to be analyzed need not have a vapor pressure more than about 10^{-3} mm Hg at some obtainable temperature, such as a few hundred degrees centigrade, so that many substances ordinarily thought of as liquids or solids can be used.   However, restriction of analysis to gases does rule out direct study of a wide variety of interesting materials. The requirement that the molecule have a dipole moment eliminates a further number of molecules, *e.g.*, CO_2, N_2, benzene, and others, from detection and determination by microwave absorption.

A second variety of disadvantage inherent in analytic applications of microwaves springs, as do some of its strongest advantages, from the very high resolution and specificity of this technique.   The microwave spectra of molecules depend on all the minute details of molecular structure; the slightest variation in structure may radically change a microwave spectrum.

This sensitivity of microwave spectra to very small changes within the molecule tends to prevent successful work with molecules having a large number of atoms.   Thus a molecule with 25 atoms has approximately 70 modes of internal vibration, many of which may be excited at ordinary temperatures to split each rotational line into as many different frequencies.   If each rotational line is so split into multiplets, each component is so weak that detection may be difficult, and it is hence doubtful that microwave spectroscopy will be very successful with molecules having more than about 25 atoms.

The great sensitivity of microwave spectra to details of molecular structure prevents the existence of any effect in the microwave range which is comparable with the characteristic vibrations of certain atomic groups or bonds which are prominent in infrared spectra.   In some special cases hyperfine structure in microwave spectra can be used to identify atoms which are responsible for it in a way somewhat similar to the use of characteristic vibrational frequencies in the infrared region.   However, this use of hyperfine structure has only limited application.   The absence of spectra which are characteristic of certain groups within the molecule is a disadvantage only to those interested in studying new microwave spectra, and not to analysis of gases whose spectra are already known.

Microwave spectroscopy is characteristically done at very low pressures.   There may be some occasions where pressures as high as a few centimeters of Hg or more might be used.   One such case in which higher pressures could be useful is the case mentioned above where there are too many rotation-vibration lines.   Higher pressure would amalgamate these lines and thus produce a stronger absorption.   At these pressures, however, the resolution would be rather poor, so that probably not more than one or two different components of a gas mixture could be identified from the microwave spectrum.

18-2. Qualitative Analysis.   Qualitative analysis by microwave spectroscopy is normally very much easier than quantitative measurement and hence will be discussed first.   Identification of mixtures of gases for which microwave spectra are known is in fact very simple and direct, once a sensitive microwave spectrometer is available.

Microwave spectra can be tabulated as a list of frequencies rather than as series of curves as are infrared spectra, since normally individual lines are resolved and measured. A table of known microwave lines of gases has been prepared by Kisliuk and Townes [714] under the auspices of the National Bureau of Standards. The latest edition of this table was published in 1952 and included approximately 1800 lines of 92 different substances which represented the microwave spectra known by 1950. This table probably will be revised and brought up to date from time to time by the National Bureau of Standards.

It has already been pointed out above that the high resolution and accuracy obtainable with microwave spectroscopy results in very little possibility of overlap or confusion of lines. This is illustrated by the fact that there are only about 10 cases among the 1800 lines listed by Kisliuk and Townes [714] where two lines of different substances are closer than 0.25 Mc, which is approximately the resolving power of an ordinary spectrometer. For each substance listed, it is possible to find lines which are more than 0.5 Mc away from any known lines of other substances. The measurement of a single line to an accuracy of about 0.1 Mc would hence suffice for positive identification of any one of this group of 92 substances.

As shown in Chap. 17, an accuracy in measurement of 0.1 Mc is easily achieved if a frequency standard based on a quartz-crystal oscillator is available. The considerably simpler type of frequency meter, a tunable resonant cavity, can measure frequencies to an accuracy of a few megacycles and hence would suffice in most cases to positively identify a substance by the presence of a single line. If identification were still in doubt, more than one line might be measured, or the Stark effect of the line might be examined. As discussed in Chap. 10, microwave transitions have characteristic Stark effects which differ in the number of Stark components, the spacing of these components, and their intensities. In the rare cases where frequency measurement does not suffice for identification of a line, the characteristic Stark effect of the line may be useful for identification.

A microwave spectrometer need not operate over all microwave frequencies, since most substances have a number of lines throughout the microwave region. A relatively limited frequency range can be chosen for a spectrometer which would allow analysis of a large fraction of gases suitable for microwave spectroscopy. Most microwave spectroscopy has been done in the K-band region (near 25,000 Mc) since this is the highest frequency for which components have been readily available. Suppose a spectrometer made for qualitative analysis operates in this region. The frequency range over which it must be operable to obtain absorption lines for a given fraction of the substances listed in Appendix VI as having microwave spectra can be judged to some extent from Fig. 18-1. This figure gives the fraction of molecules with presently known microwave spectra which have at least one line within a frequency interval $\pm\Delta f$ from 25,000 Mc. From Fig. 18-1, it may be seen that a spectrometer operating over a range 20,000 to 30,000 Mc would be able to detect lines of 90 per cent of the molecules whose microwave spectra are known. The molecules which could not be detected in this range are certain rather light linear and symmetric-top molecules for which spectra occur only at higher frequencies. There is nothing unusual about the region

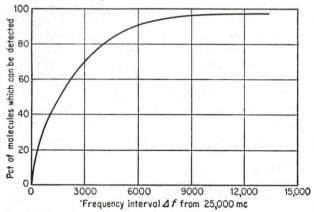

FIG. 18-1. Percentage of molecules with known microwave spectra having lines within a given frequency internal Δf from 25,000 Mc. Thus, within the range 23,000 to 27,000 Mc ($\Delta f = 2000$), 57 per cent of the molecules with known microwave spectra have lines which may be used for spectroscopic identification.

near 25,000 Mc; so that a similar fraction of molecules can be expected to have spectra in almost an arbitrary spectral region of 10,000 Mc width in the microwave range.

If a spectrometer is required only to check the abundance of a number of possible known components of a gas mixture, then it can be adjusted at certain known frequencies to detect the required lines. This type of operation may be the most common one in analytic uses of microwave spectrometers. However, in some cases it may be useful to search over a wide range of frequencies in order to examine the composition of a completely unknown gas, or in order to find lines in a known gas for which the microwave spectrum has not yet been examined. In these cases consideration must be given to the time required for such a search.

A spectrometer sensitive enough to detect a line with an absorption coefficient as small as 10^{-9} cm^{-1} must have such a narrow bandwidth that it takes as long as about 2 sec to sweep over an individual line, or a distance of 0.5 Mc.   Hence a search for lines with this sensitivity over a 1000-Mc range requires approximately 1 hr, and searching times can be long enough to be inconvenient.   If the sensitivity required is less, searching can be much faster.   Thus search over 1000 Mc to detect an absorption as large as 4×10^{-9} cm^{-1} would require in principle only about 5 min since sensitivity is proportional to the square root of the bandwidth [*cf*. (15-9)].

The fractional abundance of a given gas in a mixture required for detection of its microwave lines can be determined approximately from the intensity of its lines and the sensitivity of the spectrometer used. The intensity of a line due to a gas of fractional abundance x is given roughly by x times the intensity of the line for the pure gas (this relation

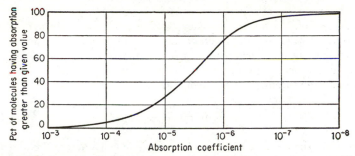

FIG. 18-2. Maximum intensities of known microwave spectra in the range 20,000 to 30,000 Mc.   The curve gives the percentage of those molecules having spectra between 20,000 and 30,000 Mc for which at least one line in this region has an absorption coefficient greater than the values listed along the horizontal axis.

is not precise, as will be shown below in the discussion of quantitative analysis).   Thus the intensity of the strongest line of NH_3 is 8×10^{-4} cm^{-1}, so that a spectrometer sensitive enough to detect an absorption coefficient of 10^{-9} cm^{-1} would be able to detect NH_3 in a gas mixture with a fractional abundance as low as about 10^{-6}.   More commonly, stronger microwave lines of molecules have absorption coefficients near 10^{-5} or 10^{-6} cm^{-1}, so that such a spectrometer could detect abundances as low as about 10^{-4} or 10^{-3}.

If a spectrometer operates in the frequency range 20,000 to 30,000 Mc, Fig. 18-2 indicates the sensitivity needed for detection of a certain fraction of the molecules which have microwave spectra (or for detection of a given fractional abundance of these molecules in a gas mixture). This figure shows the fraction of molecules listed in the tables of Kisliuk and Townes [714] which have a line with intensity greater than a given amount and falling in the frequency range 20,000 to 30,000 Mc.   Since intensities of microwave lines increase with frequency approximately as ν^2 or ν^3 [*cf*. (1-49) and (1-78)], typical intensities in other frequency ranges can be roughly obtained by multiplying those of Fig. 18-2 by the square of the frequency ratio.

It should be noted that microwave spectroscopy done so far has been somewhat concentrated on the simpler molecules. These tend to have fewer and stronger lines than the more complex molecules. Hence one may guess that, as work on additional molecules proceeds, a larger fraction of molecules will fall in a given range in Fig. 18-1, and line intensities will on the average be somewhat less than indicated in Fig. 18-2.

18-3. Quantitative Analysis. The most obvious type of quantitative analysis by microwave spectroscopy involves comparison of the strength of absorption lines due to an unknown mixture of gases with that in a known mixture. However, before discussing this technique, we shall examine other methods and some of the theory on which quantitative analysis by microwave methods depends.

It is possible, although not easy, to use microwave absorption lines for a quantitative determination of the abundance of a gas in an absolute way without using known samples for comparison. For this, the integrated intensity of the line must be measured. As shown in (13-23) the integrated intensity of a reasonably narrow line is

$$\int_0^\infty \gamma \, d\nu = \frac{8\pi^3 Nf}{3ckT} |\mu_{ij}|^2 \nu^2 \tag{18-1}$$

where N = the number of molecules per cubic centimeter, T is the absolute temperature, ν the frequency, f and $|\mu_{ij}|^2$ are determinable from properties of the molecule involved, and c and k are known constants. Hence if the absorption coefficient γ could be accurately measured and the integrated intensity obtained, the density N of molecules of a given type could be found.

At the low pressures normally used in microwave spectroscopy, absorption lines are close to a Lorentz shape as given by (13-22), and the integrated intensity is

$$\int_0^\infty \gamma \, d\nu = \gamma_{max} \Delta\nu = \frac{8\pi^3 Nf}{3ckT} |\mu_{ij}|^2 \nu^2 \tag{18-2}$$

where γ_{max} is the absorption coefficient at the peak intensity, and $\Delta\nu$ the half width at half maximum. Hence a measurement of the peak absorption coefficient and the line width at half maximum absorption is sufficient to give the density of molecules N from (18-2).

The absolute value of the absorption coefficient γ_{max} is rather difficult to measure, although it can with care be measured to an accuracy of about 5 per cent (*cf.* pages 445–446). It is usually sufficient and much easier, however, to measure γ_{max} relative to some absorption line of a known γ_{max} by comparing peak responses of a spectrometer to the two lines in question. The half width $\Delta\nu$ may be measured to an accuracy of about 5 per cent by setting frequency markers at points on the absorption line corresponding to half intensity. This method for obtaining the density of molecules is a rather cumbersome method, however, and is usually limited to an accuracy of about 5 per cent. Furthermore, the density N of molecules per cubic centimeter is usually not so much desired as the percentage abundance, which can be obtained from N only with an additional measurement of the pressure.

The simplest quantity to measure, and one which can be measured most accurately, is the maximum signal produced in a spectrometer by an absorption line, which is nearly proportional to γ_{max} for most spectrometers. If a spectrometer is adjusted for a line at a particular frequency and all conditions are maintained constant, it will be found that the response or deflection of the spectrometer due to the line can be measured reproducibly to an accuracy of 1 per cent or better, assuming the signal due to the line is sufficiently large compared with noise.

Fortunately the deflection, which is proportional to γ_{max}, is not only the easiest quantity to measure, but also it is rather directly related to the fractional abundance of the gas responsible for the absorption. From (18-2), the fractional abundance is proportional to $(\gamma_{max} \, \Delta\nu)/p$, where p is the pressure. However, γ_{max} is independent of pressure in the range of pressure normally used for spectroscopy (*cf.* Chap. 13), and $\Delta\nu/p$ is also independent of pressure, since the line width is proportional to p. Thus it appears that γ_{max} and hence the maximum spectrometer response is directly proportional to the fractional abundance of the gas responsible for the microwave absorption.

For some purposes, γ_{max} can in fact be taken as proportional to fractional abundance, and the maximum spectrometer response affords a ready measurement of concentration of a component of a gas mixture. However, accurate measurements of fractional abundance require one to allow for the fact that the line-width parameter $\Delta\nu$ is somewhat dependent on the composition of the gas and that the fractional abundance is strictly proportional not to γ_{max} but to $\gamma_{max} \, \Delta\nu/p$.

From (13-42), the number of collisions per second of a molecule of type 1 with molecules of type 2 in a gas mixture is

$$\frac{1}{\tau_{12}} = N_2 v_{12} \sigma_{12} \qquad (18\text{-}3)$$

where N_2 is the density of molecules of type 2, v_{12} an average relative velocity for the two molecular types, and σ_{12} the cross section between them. The number of collisions per second for molecules of type 1 in a mixture of two components only is therefore

$$\frac{1}{\tau_{11}} + \frac{1}{\tau_{12}} = N_1 v_{11} \sigma_{11} + N_2 v_{12} \sigma_{12} \qquad (18\text{-}4)$$

and the half width of an absorption line is [cf. (13-19)]

$$\Delta\nu_1 = \frac{1}{2\pi}\left(\frac{1}{\tau_{11}} + \frac{1}{\tau_{12}}\right) = \frac{1}{2\pi}(Nx_1v_{11}\sigma_{11} + Nx_2v_{12}\sigma_{12}) \qquad (18\text{-}5)$$

where x_1 and x_2 are the fractional abundance of each molecular species and N is the total molecular density. Expression (18-5) may also be written

$$\Delta\nu_1 = x_1\,\Delta\nu_{11} + x_2\,\Delta\nu_{12} \qquad (18\text{-}6)$$

where $\Delta\nu_{12}$ would be the half width of the absorption line of molecule 1 if it were in an almost pure sample of molecule 2. The fractional abundance of the first component of the gas is then proportional to

$$x_1 \propto \frac{\gamma_{\max}\,\Delta\nu_1}{p} = \gamma_{\max}\frac{N}{2\pi p}(x_1v_{11}\sigma_{11} + x_2v_{12}\sigma_{12}) \qquad (18\text{-}7)$$

Since $N/2\pi p$ is a universal constant at a given temperature, the last part of expression (18-7) depends only on the fractional abundances x_1 and x_2 and not on the pressure. It may also be written

$$x_1 \propto \gamma_{\max}\left(x_1\frac{\Delta\nu_{11}}{p} + x_2\frac{\Delta\nu_{12}}{p}\right) \qquad (18\text{-}8)$$

where $\Delta\nu_{11}/p$ and $\Delta\nu_{12}/p$ are independent of pressure. For a mixture of more than two components, (18-8) can be easily generalized so that the fractional abundance of the ith component of a mixture of n different gases is proportional to

$$x_i \propto \gamma_{\max}\sum_{j=1}^{n} x_j\frac{\Delta\nu_{ij}}{p} \qquad (18\text{-}9)$$

If component 2 were just as effective in broadening a microwave line as component 1, then $\Delta\nu_{11}/p = \Delta\nu_{12}/p$ and (18-8) could be reduced to

$$x_1 \propto \gamma_{\max}\frac{\Delta\nu_{11}}{p}(x_1 + x_2) = \gamma_{\max}\frac{\Delta\nu_{11}}{p} \qquad (18\text{-}10)$$

and the fractional abundance x_1 would be strictly proportional to $\gamma_{\max}$ regardless of the amount of dilution. If $\Delta\nu_{11}/p \neq \Delta\nu_{12}/p$, this strict proportionality is not true, as is shown in extreme cases by Fig. 18-3.

Usually for a mixture of more than two gases, the various values of line-width parameters $\Delta\nu_{ij}/p$ will not be known, and one of the following procedures must be adopted:

1. Measure $\gamma_{\max}$, $\Delta\nu$, and p.

2. Measure only $\gamma_{\max}$ and accept the approximate results given by assuming $\gamma_{\max}$ proportional to the fractional abundance.

3. Dilute the unknown mixture with a large amount of known gas for which the necessary information on line-width parameters $\Delta\nu_{ij}/p$ is known.

4. Compare γ_{max} for the unknown sample with values for known mixtures which are similar to it in composition.

The procedure indicated by item 3 above is useful when the absorption lines being examined are strong so that they can still be measured if a few per cent of the sample is mixed with a known gas (*e.g.*, N_2). If most of the molecules in a sample are of one type (*e.g.*, N_2), then the broadening produced by them is the only important source of broadening, and only one of the values of $\Delta\nu_{ij}$ need be known to measure each component of the gas. This may be seen from Eq. (18-8) if x_2 becomes very much larger than x_1, or more generally from Eq. (18-9).

The comparison method with use of standard mixtures indicated by item 4 above appears to be the simplest and most reliable technique if a large number of measurements are to be made on mixtures whose composition is approximately known. Such a measurement involves only a comparison of the peak response of a spectrograph to the absorption lines of two different samples of gas. Good accuracy depends only on having stable reproducible conditions in the spectrometer and an adequate

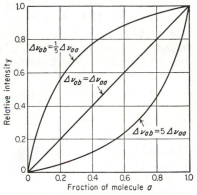

FIG. 18-3. Peak intensity of a line due to molecule *a* which is in a mixture of gases *a* and *b*. The peak intensity varies linearly with concentration only if the broadening effect of each molecule is the same. The curved lines indicate deviations from linearity which are as large as will ordinarily occur. (*After Hughes* [700].)

signal-to-noise ratio for the observed line. This type of measurement can fairly easily be made to an accuracy of a few per cent, and higher accuracy appears feasible if considerable care is taken.

In case a Stark spectrometer is used with Stark components and undisplaced absorption lines giving deflections in the opposite directions (*cf.* Chap. 15), the total deflection difference between the strongest Stark component and the undisplaced line may be used as a measure of intensity. This eliminates some uncertainties which might otherwise exist in the exact position of the base line where zero absorption occurs.

The comparison method has been tested by Southern, Morgan, Keilholtz, and Smith [650] for isotopic mixtures of NH_3 and $ClCN$. For NH_3, the N^{15}/N^{14} ratio was determined to an accuracy of about 3 per cent with the N^{15} concentration in the range 0.38 to 4.5 per cent; and for $ClCN$, the ratio C^{13}/C^{12} was determined to an accuracy of 2 per cent with C^{13} abundances between 1.1 and 10 per cent. Weber and Laidler [660] have studied exchange in NH_3 by similar analysis of isotopic mixtures.

Unless very special spectrometers are built which allow absolute measurements of intensities, accurate quantitative work always requires some type of comparison method on absorption lines which are identical or very similar in frequency, width, and other characteristics. The inevitable presence of reflections in a waveguide absorption cell produces variations in the field strength with the absorption cell, and hence variations in the spectrometer's response as the frequency is varied. As shown in Chap. 15, if reflections produce a 10 per cent variation in power transmitted through the absorption cell as a function of frequency, they can produce a 20 per cent variation in the response of the spectrometer to a given absorption coefficient γ_{max} [cf. Eq. (15-28)]. Nevertheless, if care is taken, the relative intensities of lines of different frequencies can be measured to an accuracy of about 5 per cent. For lines of the same frequency, reflections are of course constant and cause no errors in measurement of relative intensities.

Spectrometers using Stark modulation are the most convenient type of spectrometer for high sensitivity. However, they introduce additional uncertainties in the measurement of γ_{max} unless comparison is made between two very similar lines.

A Stark modulation spectrometer detects lines by the periodic application of an electric field which shifts the frequency of the line because of the Stark effect (cf. sec. 15-3). In an electric field the line usually breaks up into a number of Stark components, with widely different amounts of frequency change for a given field strength. If a square wave of electric field is applied to the gas which is sufficiently large to shift all Stark components well away (more than the line width) from the absorption which occurs without an electric field, then the full intensity of the absorption line appears. For smaller fields, however, some of the Stark components may not be moved appreciably from the absorption line, and the apparent intensity as determined by the spectrometer response is correspondingly less. Partial splitting of the absorption line by Stark effect also distorts the line shape and gives errors in measurement of line width if width measurements are attempted. In addition, if the square-wave voltage is not nearly zero during half its cycle of variation, then the line is partly split at all times, and the line is broadened so that γ_{max} is reduced. If a sine wave is used rather than a square wave, these difficulties are particularly acute, since the sine wave cannot usually have enough amplitude at maximum to move all Stark components far enough from the undisplaced line without at the same time broadening the undisplaced line because it is not sufficiently close to zero during the other half of the cycle. If a line is not completely split by Stark effect, then a variation in pressure, which produces a variation of line width, will also affect the apparent value of γ_{max}. As the width increases the line is less completely split and γ_{max} decreases. For the above reasons, it is important in most cases to compare only similar lines under similar conditions if quantitative work with a Stark spectrometer is attempted.

18-4. Special Equipment and Techniques for Spectroscopic Analysis. Equipment for some types of analytic work with microwave spectroscopy need not be very special. Almost any sensitive spectrometer and moderately accurate frequency-measuring device should suffice for qualitative analysis and for rough quantitative work. Techniques and apparatus specifically for analytic work have not yet been well developed. However, certain considerations and problems of technique involved in analysis are well enough known to be discussed here.

A Stark modulation spectrometer probably provides the simplest means for obtaining good sensitivity, although it has some handicaps for quantitative work. If quantitative work is planned, care should be taken to ensure that the signal amplifiers are linear (*i.e.*, give an output strictly proportional to signal), that reflections are minimized, and that the modulation is produced by a well-formed and large square wave of electric field. Other precautions, which are common to other types of spectrometers, are to see that the rate of sweep over a line is sufficiently slow that its shape (and therefore height) is not distorted, and that the power is sufficiently low so that the absorption lines do not suffer from power saturation. Reflections in any spectrometer can usually be decreased by inserting well-tapered attenuators at either end of the absorption cell. In the Stark spectrometer, it is particularly important that the septum for producing the Stark field and its insulating supports inside the absorption cell be well tapered at the ends.

An absorption cell without a Stark septum is almost always freer of reflections than is a Stark cell. In addition, a simple spectrometer without Stark modulation has none of the complications of incomplete modulation of the line mentioned above. However, this type of spectrometer is usually suitable only for work with the most intense lines, because of the greater difficulty in obtaining good sensitivity.

In many cases a spectrometer may be used to observe the intensity of some particular line over a long period of time. In such circumstances, a cavity can conveniently be used for an absorption cell. Special care must be exercised, however, to see that power saturation of the absorption line does not occur when a cavity is used, since the power level in a cavity is likely to be higher than that in a waveguide cell.

There are many ways of measuring the magnitude of response of a spectrometer to an absorption line. Some have been tried and described [700], [901], but none seems to have particularly strong or unique advantages. The method chosen in an individual case should probably depend on details of the desired functions of the apparatus and on the available equipment.

Absorption of gases on surfaces can be a serious problem in the use of microwave spectroscopy for analysis. Since very small samples of gas are used, a relatively small amount of material absorbed by waveguide surfaces or outgased from them can appreciably change the pressure and composition of a sample being analyzed. It is common experience, for example, that, if ammonia is let into an absorption cell, its lines may persist in a microwave spectrometer even after many days of pumping to evacuate the cell. Cells for Stark modulation are worse from the point of view of absorption than others, since they have the additional surfaces of septum and insulators. A cavity would minimize the ratio of surface to volume and thus decrease difficulties due to surface absorption. In the usual Stark cell absorption is a serious problem to analytic work, since certain components of the gas sample may be selectively absorbed, and foreign gas from previous samples may be released by the cell walls.

There are a number of ways of dealing with absorption.

1. Cell design should minimize absorption. It has been found, for example, that stainless steel absorbs gases far less than silver or copper, and stainless steel may be used for a cell. Teflon is also less absorbent than polystyrene and hence better suited for insulators of a Stark septum.

2. Several batches of gas may be let into the cell, or the gas may be flowed through the cell until no further exchange of gas with the walls occurs.

3. Chemical "getters" may be used to destroy gases on the walls which produce spurious lines. Thus ClF_3 let into a waveguide will destroy water absorbed on the surfaces, and $BrCN$ will help eliminate NH_3.

4. The cell may be heated up to outgas its walls or may be operated at an elevated temperature to minimize absorption. A temperature of $100°C$ seems to be fairly effective in outgassing, and $200°C$ would probably be adequate for most purposes.

A somewhat specialized technique to which microwave spectroscopy seems well adapted is the continuous monitoring of a mixture of gases. Gas may be flowed at very low pressure through the absorption cell of a spectrometer and the output of the spectrometer used to register or control the concentration of a given component. Changes in concentration would be very easily and sensitively detected, since the spectrometer itself would be under fixed adjustments and conditions. Furthermore, these changes could be quite rapidly recorded or controlled by the output of the spectrometer. In order to eliminate the effects on spectrometer output of possible fluctuations in klystron output powers, a bridge system can be used, with one arm of the bridge containing the absorption cell through which gas is being flowed, and the other arm involving only fixed components.

APPENDIX I

Intensities of Hyperfine Structure Components and Energies Due to Nuclear Quadrupole Interactions

Intensities of individual lines are given as a percentage of the intensity of the entire transition (sum of all hyperfine components). They are obtained from expressions (6-6). Values for angular momentum J between 0 and 10 and for the nuclear spin I equal to $\frac{1}{2}$, 1, $\frac{3}{2}$, 2, $\frac{5}{2}$, 3, $\frac{7}{2}$, 4, $\frac{9}{2}$, or $\frac{11}{2}$ are included. The quantity F is the magnitude of $\mathbf{I} + \mathbf{J}$. These tables may also be used for relative intensities of atomic fine-structure transitions if I, J, and F are replaced by S, L, and J, respectively. The tabulation then corresponds to formulas (5-17) and (5-18). For $J > 10$, almost all the intensity of a transition is concentrated in components such that $\Delta F = \Delta J$, that is, with $J + 1 \leftarrow J$ and $F + 1 \leftarrow F$ or $J \leftarrow J$ and $F \leftarrow F$. Approximate intensities of transitions for large J are as follows:

For transitions $J + 1 \leftarrow J$, $F + 1 \leftarrow F$ Intensity is proportional to F
                             $F \leftarrow F$ Intensity $\sim 1/2J^2$ of entire intensity
                            $F - 1 \leftarrow F$ Intensity $\sim 1/10J^4$ of entire intensity
For $J \leftarrow J$, $F \leftarrow F$ Intensity is proportional to F
            $F \pm 1 \leftarrow F$ Intensity $\sim 1/2J^2$ of entire intensity

The table also gives Casimir's function

$$f(I,J,F) = \frac{\frac{3}{4}C(C + 1) - I(I + 1)J(J + 1)}{2I(2I - 1)(2J - 1)(2J + 3)}$$

for rotational angular momentum J between 0 and 10, for the nuclear spin $I = 1$, $\frac{3}{2}$, 2, $\frac{5}{2}$, 3, $\frac{7}{2}$, 4, $\frac{9}{2}$, or $\frac{11}{2}$ and all possible values of F. Here

$$C = F(F + 1) - I(I + 1) - J(J + 1)$$

The interaction energy due to a nuclear quadrupole moment coupled to a linear molecule is given by multiplying the appropriate value of $f(I,J,F)$ in this table by $-eqQ$. For a symmetric-top molecule the proper multiplying factor is $eqQ\left[\dfrac{3K^2}{J(J + 1)} - 1\right]$ [cf. Eq. (6-4)]. When $J > 10$, the stronger hyperfine components of a transition (where $\Delta F = \Delta J$) involve a change in quadrupole energy which is a very small fraction of the quadrupole coupling constant, being in almost all cases less than $eqQ/4J^2$. For the other, very much weaker, components, the change in energy is larger and can be approximated by

$$\Delta W_Q \,(\Delta F = \Delta J \pm 1) = \mp \frac{3[2(F - J) + 1]}{8I(2I - 1)}\left[\frac{3K^2}{J(J + 1)} - 1\right]eqQ$$
$$+ \text{ terms of order } eqQ/2J$$

499

$$I = \tfrac{1}{2}$$

J	F	$f(I,J,F)$	$J+1 \leftarrow J$			$J \leftarrow J$	
			$F+1 \leftarrow F$	$F \leftarrow F$	$F-1 \leftarrow F$	$F \leftarrow F+1$ or $F+1 \leftarrow F$	$F \leftarrow F$
0	1/2	0	66.7	33.3			
1	3/2	0	60.0	6.66			55.6
	1/2	0	33.4			11.1	22.2
2	5/2	0	57.1	2.86			56.0
	3/2	0	40.0			4.00	36.0
3	7/2	0	55.6	1.59			55.1
	5/2	0	42.9			2.04	40.8
4	9/2	0	54.5	1.01			54.3
	7/2	0	44.4			1.23	43.2
5	11/2	0	53.8	0.699			53.7
	9/2	0	45.5			0.826	44.6
6	13/2	0	53.3	0.513			53.3
	11/2	0	46.2			0.592	45.6
7	15/2	0	52.9	0.392			52.9
	13/2	0	46.7			0.444	46.2
8	17/2	0	52.6	0.310			52.6
	15/2	0	47.1			0.346	46.7
9	19/2	0	52.4	0.251			52.4
	17/2	0	47.4			0.277	47.1
10	21/2	0	52.2	0.207			52.2
	19/2	0	47.6			0.227	47.4

$$I = 1$$

J	F	$f(I,J,F)$	$J + 1 \leftarrow J$			$J \leftarrow J$	
			$F + 1 \leftarrow F$	$F \leftarrow F$	$F - 1 \leftarrow F$	$F \leftarrow F + 1$ or $F + 1 \leftarrow F$	$F \leftarrow F$
0	1	0	55.5	33.3	11.1		
1	2	0.050000	46.6	8.33	0.556		41.7
	1	−0.250000	25.0	8.33		13.9	8.33
	0	0.500000	11.1			11.1	0
2	3	0.071429	42.9	3.70	0.106		41.5
	2	−0.250000	29.6	3.70		5.19	23.1
	1	0.250000	20.0			5.00	15.0
3	4	0.083333	40.7	2.08	0.033		40.2
	3	−0.250000	31.3	2.08		2.68	28.0
	2	0.200000	23.8			2.65	21.2
4	5	0.090909	39.4	1.33	0.013		39.1
	4	−0.250000	32.0	1.33		1.63	30.1
	3	0.178571	25.9			1.62	24.3
5	6	0.096154	38.5	0.926	0.006		38.3
	5	−0.250000	32.4	0.926		1.09	31.1
	4	0.166667	27.3			1.09	26.2
6	7	0.100000	37.8	0.680	0.003		37.7
	6	−0.250000	32.7	0.680		0.785	31.8
	5	0.159091	28.2			0.783	27.4
7	8	0.102941	37.3	0.521	0.002		37.2
	7	−0.250000	32.8	0.521		0.590	32.2
	6	0.153846	28.9			0.590	28.3
8	9	0.105263	36.8	0.412	0.001		36.8
	8	−0.250000	32.9	0.412		0.460	32.4
	7	0.150000	29.4			0.460	29.0
9	10	0.107143	36.5	0.333	0.001		36.5
	9	−0.250000	33.0	0.333		0.368	32.6
	8	0.147059	29.8			0.368	29.5
10	11	0.108696	36.2	0.275	0.001		36.2
	10	−0.250000	33.1	0.275		0.302	32.7
	9	0.144737	30.2			0.302	29.9

$$I = \tfrac{3}{2}$$

J	F	$f(I,J,F)$	$J+1 \leftarrow J$			$J \leftarrow J$	
			$F+1 \leftarrow F$	$F \leftarrow F$	$F-1 \leftarrow F$	$F \leftarrow F+1$ or $F+1 \leftarrow F$	$F \leftarrow F$
0	3/2	0	50.0	33.3	16.7		
1	5/2	0.050000	40.0	9.0	1.00		35.0
	3/2	−0.200000	21.0	10.7	1.67	15.0	4.44
	1/2	0.250000	8.33	8.33		13.9	2.78
2	7/2	0.071429	35.7	4.08	0.204		34.3
	5/2	−0.178571	24.5	5.22	0.286	5.71	17.3
	3/2	0	16.0	4.00		7.00	8.00
	1/2	0.250000	10.0			5.00	5.00
3	9/2	0.083333	33.3	2.31	0.067		32.7
	7/2	−0.166667	25.5	3.02	0.085	2.98	21.8
	5/2	−0.050000	19.1	2.30		3.83	14.7
	3/2	0.200000	14.3			2.86	11.4
4	11/2	0.090909	31.8	1.49	0.028		31.5
	9/2	−0.159091	25.8	1.96	0.034	1.82	23.6
	7/2	−0.071429	20.7	1.48		2.38	18.1
	5/2	0.178571	16.7			1.79	14.9
5	13/2	0.096154	30.8	1.04	0.013		30.6
	11/2	−0.153846	25.9	1.37	0.016	1.22	24.4
	9/2	−0.083333	21.7	1.03		1.61	19.9
	7/2	0.166667	18.2			1.21	17.0
6	15/2	0.100000	30.0	0.762	0.007		29.9
	13/2	−0.150000	25.9	1.01	0.008	0.879	24.9
	11/2	−0.090909	22.3	0.761		1.16	21.0
	9/2	0.159091	19.2			0.874	18.4
7	17/2	0.102941	29.4	0.584	0.004		29.3
	15/2	−0.147059	25.9	0.775	0.005	0.662	25.1
	13/2	−0.096154	22.8	0.583		0.877	21.8
	11/2	0.153846	20.0			0.659	19.3
8	19/2	0.105263	28.9	0.462	0.003		28.9
	17/2	−0.144737	25.9	0.613	0.003	0.516	25.3
	15/2	−0.100000	23.1	0.461		0.685	22.3
	13/2	0.150000	20.6			0.515	20.1
9	21/2	0.107143	28.6	0.374	0.002		28.5
	19/2	−0.142857	25.8	0.497	0.002	0.414	25.4
	17/2	−0.102941	23.3	0.374		0.549	22.7
	15/2	−0.147059	21.1			0.413	20.6

$$I = \tfrac{3}{2} \ (Continued)$$

J	F	f(I,J,F)	$J+1 \leftarrow J$			$J \leftarrow J$	
			$F+1 \leftarrow F$	$F \leftarrow F$	$F-1 \leftarrow F$	$F \leftarrow F+1$ or $F+1 \leftarrow F$	$F \leftarrow F$
10	23/2	0.108696	28.3	0.309	0.001		28.2
	21/2	−0.141304	25.8	0.411	0.001	0.339	25.4
	19/2	−0.105263	23.5	0.309		0.450	23.0
	17/2	0.144737	21.4			0.338	21.1

$$I = 2$$

J	F	f(I,J,F)	$J+1 \leftarrow J$			$J \leftarrow J$	
			$F+1 \leftarrow F$	$F \leftarrow F$	$F-1 \leftarrow F$	$F \leftarrow F+1$ or $F+1 \leftarrow F$	$F \leftarrow F$
0	2	0	46.7	33.3	20.0		
1	3	0.050000	36.0	9.33	1.33		31.1
	2	−0.175000	18.7	11.7	3.00	15.6	2.78
	1	0.175000	7.00	9.00	4.00	15.0	5.00
2	4	0.071429	31.4	4.29	0.286		
	3	−0.142857	21.4	6.00	0.571	6.00	30.0
	2	−0.053571	13.7	5.71	0.571	8.00	14.0
	1	0.125000	8.00	4.00		7.00	5.00
	0	0.250000	4.00			4.00	1.00
3	5	0.083333	28.9	2.44	0.095		28.3
	4	−0.125000	22.0	3.54	0.179	3.14	18.1
	3	−0.091667	16.4	3.47	0.159	4.46	11.3
	2	0.050000	11.9	2.38		4.28	7.14
	1	0.200000	8.57			2.86	5.71
4	6	0.090909	27.3	1.58	0.040		27.0
	5	−0.113636	22.1	2.31	0.073	1.93	19.7
	4	−0.105519	17.6	2.29	0.061	2.80	14.5
	3	0.017857	14.0	1.56		2.75	11.0
	2	0.178571	11.0			1.85	9.26
5	7	0.096154	26.1	1.10	0.020		26.0
	6	−0.105769	22.0	1.62	0.035	1.30	20.4
	5	−0.112179	18.4	1.62	0.028	1.91	16.2
	4	0	15.3	1.09		1.89	13.2
	3	0.166667	12.7			1.27	11.5
6	8	0.100000	25.3	0.810	0.011		25.2
	7	−0.100000	21.9	1.20	0.019	0.934	20.8
	6	−0.115909	18.8	1.20	0.015	1.38	17.2
	5	−0.011364	16.1	0.806		1.37	14.6
	4	0.159091	13.8			0.923	12.9
7	9	0.102941	24.7	0.621	0.007		24.6
	8	−0.095588	21.7	0.924	0.011	0.704	20.9
	7	−0.118212	19.1	0.922	0.008	1.04	17.9
	6	−0.019231	16.7	0.619		1.04	15.6
	5	0.153846	14.7			0.698	14.0

$I = 2$ (*Continued*)

J	F	$f(I,J,F)$	$J + 1 \leftarrow J$			$J \leftarrow J$	
			$F + 1 \leftarrow F$	$F \leftarrow F$	$F - 1 \leftarrow F$	$F \leftarrow F + 1$ or $F + 1 \leftarrow F$	$F \leftarrow F$
8	10	0.105263	24.2	0.491	0.004		24.2
	9	−0.092105	21.6	0.732	0.007	0.549	21.0
	8	−0.119737	19.3	0.731	0.005	0.817	18.4
	7	−0.025000	17.2	0.490		0.815	16.3
	6	0.150000	15.3			0.546	14.7
9	11	0.107143	23.8	0.398	0.003		23.8
	10	−0.089286	21.5	0.594	0.005	0.440	21.0
	9	−0.120798	19.4	0.594	0.003	0.656	18.7
	8	−0.029412	17.5	0.398		0.655	16.8
	7	0.147059	15.8			0.439	15.4
10	12	0.108696	23.5	0.329	0.002		23.4
	11	−0.086957	21.4	0.492	0.003	0.361	21.0
	10	−0.121568	19.5	0.492	0.002	0.538	18.9
	9	−0.032895	17.8	0.329		0.538	17.2
	8	0.144737	16.2			0.360	15.8

$$I = \tfrac{5}{2}$$

J	F	$f(I,J,F)$	$F + 1 \leftarrow F$	$F \leftarrow F$	$F - 1 \leftarrow F$	$F \leftarrow F + 1$ or $F + 1 \leftarrow F$	$F \leftarrow F$
0	5/2	0	44.4	33.3	22.2		
1	7/2	0.050000	33.3	9.52	1.59		28.6
	5/2	−0.160000	17.1	12.2	4.00	15.9	1.90
	3/2	0.140000	6.22	9.33	6.67	15.6	6.67
2	9/2	0.071429	28.6	4.41	0.353		27.2
	7/2	−0.121428	19.4	6.45	0.816	6.17	11.9
	5/2	−0.071429	12.2	6.61	1.14	8.57	3.43
	3/2	0.071429	6.86	5.42	1.06	8.00	0.148
	1/2	0.200000	2.96	3.70		5.19	1.48
3	11/2	0.083333	25.9	2.53	0.120		25.3
	9/2	−0.100000	19.7	3.85	0.265	3.25	15.7
	7/2	−0.100000	14.6	4.16	0.340	4.85	9.10
	5/2	−0.006667	10.4	3.63	0.265	5.10	4.90
	3/2	0.110000	7.14	2.38		4.29	2.59
	1/2	0.200000	4.76			2.65	2.12
4	13/2	0.090909	24.2	1.63	0.052		23.9
	11/2	−0.086364	19.6	2.53	0.110	1.99	17.2
	9/2	−0.107792	15.6	2.79	0.135	3.06	12.1
	7/2	−0.037662	12.3	2.46	0.096	3.33	8.57
	5/2	0.071429	9.52	1.59		2.91	6.35
	3/2	0.178571	7.41			1.85	5.56

$$I = \tfrac{5}{2} \ (Continued)$$

J	F	$f(I,J,F)$	$J+1 \leftarrow J$			$J \leftarrow J$	
			$F+1 \leftarrow F$	$F \leftarrow F$	$F-1 \leftarrow F$	$F \leftarrow F+1$ or $F+1 \leftarrow F$	$F \leftarrow F$
5	15/2	0.096154	23.1	1.13	0.026		22.9
	13/2	−0.076923	19.4	1.79	0.054	1.35	17.8
	11/2	−0.110256	16.1	1.98	0.064	2.11	13.8
	9/2	−0.053846	13.4	1.76	0.043	2.31	10.8
	7/2	0.050000	11.0	1.12		2.04	8.77
	5/2	0.166667	9.09			1.31	7.79
6	17/2	0.100000	22.2	0.840	0.014		22.1
	15/2	−0.070000	19.2	1.32	0.029	0.970	18.0
	13/2	−0.110909	16.4	1.48	0.034	1.53	14.7
	11/2	−0.063636	14.1	1.31	0.022	1.69	12.2
	9/2	0.036364	12.0	0.832		1.50	10.4
	7/2	0.159091	10.3			0.950	9.31
7	19/2	0.102941	21.6	0.645	0.009		21.5
	17/2	−0.064705	19.0	1.02	0.017	0.731	18.1
	15/2	−0.110859	16.6	1.14	0.020	1.15	15.3
	13/2	−0.070136	14.5	1.01	0.013	1.29	13.1
	11/2	0.026923	12.7	0.641		1.14	11.5
	9/2	0.153846	11.1			0.722	10.4
8	21/2	0.105263	21.1	0.511	0.005		21.0
	19/2	−0.062526	18.8	0.809	0.011	0.571	18.1
	17/2	−0.110526	16.7	0.906	0.012	0.903	15.7
	15/2	−0.074737	14.9	0.806	0.008	1.01	13.8
	13/2	0.020000	13.2	0.508		0.897	12.3
	11/2	0.150000	11.8			0.566	11.2
9	23/2	0.107143	20.6	0.414	0.004		20.6
	21/2	−0.057143	18.6	0.658	0.007	0.458	18.1
	19/2	−0.110084	16.8	0.737	0.008	0.726	16.0
	17/2	−0.078151	15.1	0.656	0.005	0.813	14.3
	15/2	0.014706	13.6	0.413		0.722	12.9
	13/2	0.147059	12.3			0.455	11.8
10	25/2	0.108696	20.3	0.343	0.003		20.3
	23/2	−0.054348	18.5	0.545	0.005	0.375	18.1
	21/2	−0.109610	16.8	0.611	0.005	0.596	16.2
	19/2	−0.080778	15.3	0.543	0.003	0.668	14.6
	17/2	0.010526	13.9	0.342		0.594	13.3
	15/2	0.144737	12.7			0.373	12.3

$$I = 3$$

J	F	$f(I,J,F)$	$J+1 \leftarrow J$			$J \leftarrow J$	
			$F+1 \leftarrow F$	$F \leftarrow F$	$F-1 \leftarrow F$	$F \leftarrow F+1$ or $F+1 \leftarrow F$	$F \leftarrow F$
0	3	0	42.9	33.3	23.8		
1	4	0.050000	31.4	9.64	1.79		26.8
	3	−0.150000	16.1	12.5	4.76	16.1	1.39
	2	0.120000	5.71	9.52	8.57	15.9	7.94
2	5	0.071429	26.5	4.49	0.408		25.1
	4	−0.107142	18.0	6.73	1.02	6.29	10.5
	3	−0.078571	11.2	7.14	1.63	8.93	2.50
	2	0.042857	6.12	6.12	2.04	8.57	0.00
	1	0.171429	2.45	4.08	2.04	5.71	2.86
3	6	0.083333	23.8	2.58	0.142		23.2
	5	−0.083333	18.1	4.05	0.340	3.32	14.0
	4	−0.100000	13.3	4.59	0.510	5.10	7.65
	3	−0.033333	9.35	4.37	0.567	5.61	3.57
	2	0.063333	6.24	3.54	0.425	5.10	1.28
	1	0.150000	3.83	2.30		3.83	0.255
	0	0.200000	2.05			2.04	0
4	7	0.090909	22.1	1.67	0.062		21.8
	6	−0.068182	17.8	2.68	0.144	2.04	15.4
	5	−0.103246	14.1	3.11	0.208	3.24	10.5
	4	−0.061039	11.0	3.04	0.216	3.71	7.00
	3	0.019481	8.44	2.52	0.144	3.57	4.63
	2	0.107143	6.35	1.59		2.91	3.24
	1	0.178571	4.76			1.79	2.98
5	8	0.096154	20.9	1.17	0.031		20.7
	7	−0.057692	17.5	1.90	0.071	1.38	15.9
	6	−0.102564	14.6	2.23	0.100	2.23	12.1
	5	−0.074571	12.0	2.20	0.100	2.60	9.14
	4	−0.003846	9.80	1.84	0.062	2.55	7.03
	3	0.083333	7.95	1.14		2.11	5.68
	2	0.166667	6.49			1.30	5.19
6	9	0.100000	20.0	0.862	0.017		19.9
	8	−0.050000	17.2	1.41	0.039	0.994	16.1
	7	−0.100000	14.8	1.67	0.054	1.62	13.0
	6	−0.081818	12.6	1.65	0.052	1.91	10.5
	5	−0.018182	10.7	1.38	0.031	1.88	8.63
	4	0.068182	9.04	0.848		1.57	7.36
	3	0.159091	7.69			0.962	6.73

$I = 3$ (*Continued*)

J	F	$f(I,J,F)$	$J + 1 \leftarrow J$			$J \leftarrow J$	
			$F + 1 \leftarrow F$	$F \leftarrow F$	$F - 1 \leftarrow F$	$F \leftarrow F + 1$ or $F + 1 \leftarrow F$	$F \leftarrow F$
7	10	0.102941	19.3	0.662	0.011		19.3
	9	−0.044118	17.0	1.09	0.023	0.750	16.1
	8	−0.099095	14.9	1.29	0.032	1.23	13.5
	7	−0.086425	13.0	1.28	0.030	1.45	11.4
	6	−0.027828	11.3	1.07	0.018	1.44	9.73
	5	0.057692	9.82	0.655		1.20	8.54
	4	0.153846	8.57			0.735	7.84
8	11	0.105263	18.8	0.524	0.007		18.7
	10	−0.039474	16.8	0.863	0.015	0.586	16.1
	9	−0.097368	14.9	1.03	0.020	0.963	13.9
	8	−0.089474	13.2	1.02	0.018	1.14	12.0
	7	−0.034737	11.7	0.856	0.011	1.14	10.5
	6	0.050000	10.4	0.520		0.950	9.40
	5	0.150000	9.24			0.578	8.67
9	12	0.107143	18.4	0.425	0.004		18.3
	11	−0.035714	16.6	0.702	0.010	0.470	16.1
	10	−0.095798	14.9	0.837	0.013	0.775	14.1
	9	−0.091597	13.4	0.834	0.012	0.922	12.4
	8	−0.039916	12.1	0.697	0.007	0.919	11.1
	7	0.044118	10.9	0.423		0.768	10.0
	6	0.147059	9.77			0.465	9.31
10	13	0.108696	18.0	0.352	0.003		18.0
	12	−0.032609	16.4	0.582	0.007	0.385	16.0
	11	−0.094394	14.9	0.694	0.009	0.636	14.3
	10	−0.093135	13.6	0.693	0.008	0.759	12.8
	9	−0.043936	12.3	0.579	0.004	0.757	11.4
	8	0.039474	11.2	0.350		0.632	10.5
	7	0.144737	10.2			0.383	9.82

$$I = \tfrac{7}{2}$$

J	F	$f(I,J,F)$	$J + 1 \leftarrow J$			$J \leftarrow J$	
			$F + 1 \leftarrow F$	$F \leftarrow F$	$F - 1 \leftarrow F$	$F \leftarrow F + 1$ or $F + 1 \leftarrow F$	$F \leftarrow F$
0	7/2	0	41.7	33.3	25.0		
1	9/2	0.050000	30.0	9.72	1.94		25.5
	7/2	−0.142857	15.3	12.7	5.36	16.2	1.06
	5/2	0.107143	5.36	9.64	10.0	16.1	8.93
2	11/2	0.071429	25.0	4.55	0.455		23.6
	9/2	−0.096939	16.9	6.93	1.19	6.36	9.47
	7/2	−0.081633	10.5	7.48	2.04	9.17	1.90
	5/2	0.025510	5.61	6.53	2.86	8.93	0.071
	3/2	0.153061	2.14	4.29	3.57	6.00	4.00
3	13/2	0.083333	22.2	2.62	0.160		21.6
	11/2	−0.071429	16.8	4.20	0.406	3.37	12.8
	9/2	−0.097619	12.3	4.89	0.661	5.28	6.62
	7/2	−0.047619	8.60	4.84	0.850	5.95	2.72
	5/2	0.035714	5.61	4.21	0.893	5.61	0.638
	3/2	0.119048	3.27	3.17	0.694	4.46	0
	1/2	0.178571	1.49	2.08		2.68	0.893
4	15/2	0.090909	20.5	1.70	0.071		20.1
	13/2	−0.055195	16.5	2.78	0.175	2.07	14.0
	11/2	−0.097403	13.1	3.34	0.275	3.37	9.32
	9/2	−0.071892	10.1	3.43	0.337	3.98	5.90
	7/2	−0.009276	7.66	3.13	0.325	4.01	3.53
	5/2	0.065399	5.63	2.49	0.212	3.57	2.02
	3/2	0.132653	4.00	1.56		2.75	1.19
	1/2	0.178571	2.78			1.62	1.16
5	17/2	0.096154	19.2	1.19	0.036		19.1
	15/2	−0.043956	16.1	1.97	0.087	1.40	14.5
	13/2	−0.094322	13.4	2.40	0.134	2.32	10.8
	11/2	−0.082418	11.0	2.50	0.159	2.80	7.95
	9/2	−0.032051	8.90	2.32	0.146	2.89	5.82
	7/2	0.036630	7.14	1.86	0.087	2.65	4.33
	5/2	0.107143	5.68	1.14		2.11	3.44
	3/2	0.166667	4.55			1.27	3.27
6	19/2	0.100000	18.3	0.877	0.020		18.2
	17/2	−0.035174	15.8	1.47	0.048	1.01	14.6
	15/2	−0.090909	13.5	1.80	0.073	1.69	11.6
	13/2	−0.087662	11.5	1.89	0.085	2.06	9.25
	11/2	−0.045455	9.70	1.76	0.075	2.16	7.38
	9/2	0.018831	8.15	1.42	0.043	2.00	6.01
	7/2	0.090909	6.84	0.855		1.60	5.13
	5/2	0.159091	5.77			0.962	4.81

$$I = \tfrac{7}{2} \ (Continued)$$

J	F	f(I,J,F)	$J+1 \leftarrow J$			$J \leftarrow J$	
			$F+1 \leftarrow F$	$F \leftarrow F$	$F-1 \leftarrow F$	$F \leftarrow F+1$ or $F+1 \leftarrow F$	$F \leftarrow F$
7	21/2	0.102941	17.6	0.674	0.012		17.6
	19/2	−0.029412	15.5	1.14	0.029	0.764	14.6
	17/2	−0.087750	13.6	1.40	0.043	1.28	12.1
	15/2	−0.090498	11.8	1.48	0.049	1.58	10.1
	13/2	−0.054137	10.2	1.38	0.042	1.66	8.46
	11/2	0.006787	8.87	1.11	0.023	1.55	7.21
	9/2	0.079670	7.67	0.663		1.24	6.35
	7/2	0.153846	6.67			0.741	5.93
8	23/2	0.105263	17.1	0.534	0.008		17.1
	21/2	−0.024436	15.3	0.903	0.018	0.600	14.6
	19/2	−0.084962	13.6	1.12	0.027	1.01	12.5
	17/2	−0.092105	12.0	1.18	0.030	1.24	10.7
	15/2	−0.060150	10.6	1.10	0.026	1.31	9.23
	13/2	−0.001880	9.39	0.889	0.014	1.23	8.08
	11/2	0.071429	8.29	0.528		0.985	7.25
	9/2	0.150000	7.35			0.585	6.77
9	25/2	0.107143	16.7	0.433	0.005		16.6
	23/2	−0.020408	15.0	0.734	0.012	0.479	14.5
	21/2	−0.082533	13.5	0.910	0.018	0.810	12.7
	19/2	−0.093037	12.2	0.965	0.020	1.00	11.1
	17/2	−0.064526	10.9	0.903	0.017	1.06	9.79
	15/2	−0.008403	9.79	0.727	0.009	0.993	8.73
	13/2	0.065126	8.78	0.430		0.798	7.94
	11/2	0.147059	7.89			0.472	7.42
10	27/2	0.018696	16.3	0.359	0.004		16.3
	25/2	−0.017081	14.9	0.609	0.008	0.392	14.4
	23/2	−0.080418	13.5	0.756	0.012	0.666	12.8
	21/2	−0.093576	12.3	0.802	0.013	0.826	11.4
	19/2	−0.067833	11.1	0.752	0.011	0.876	10.2
	17/2	−0.013485	10.1	0.604	0.006	0.820	9.24
	15/2	0.060150	9.17	0.357		0.659	8.48
	13/2	0.144737	8.33			0.389	7.94

$$I = 4$$

J	F	$f(I,J,F)$	$J + 1 \leftarrow J$			$J \leftarrow J$	
			$F + 1 \leftarrow F$	$F \leftarrow F$	$F - 1 \leftarrow F$	$F \leftarrow F + 1$ or $F + 1 \leftarrow F$	$F \leftarrow F$
0	4	0	40.7	33.3	25.9		
1	5	0.050000	28.9	9.78	2.07		24.4
	4	−0.137500	14.7	12.8	5.83	16.3	0.833
	3	0.098214	5.09	9.72	11.1	16.2	9.72
2	6	0.071429	23.8	4.59	0.494		22.5
	5	−0.089286	16.0	7.06	1.33	6.42	8.69
	4	−0.082908	9.90	7.71	2.38	9.33	1.50
	3	0.014031	5.24	6.79	3.53	9.17	0.216
	2	0.140306	1.94	4.41	4.76	6.17	4.94
3	7	0.083333	21.0	2.65	0.176		20.4
	6	−0.062500	15.9	4.30	0.463	3.40	11.8
	5	−0.094643	11.6	5.09	0.794	5.40	5.87
	4	−0.055952	8.02	5.16	1.10	6.19	2.14
	3	0.017857	5.16	4.63	1.32	5.95	0.309
	2	0.098214	2.91	3.64	1.39	4.85	0.110
	1	0.163690	1.21	2.31	1.23	2.98	1.79
4	8	0.090909	19.2	1.72	0.079		18.9
	7	−0.045455	15.5	2.86	0.202	2.10	13.0
	6	−0.091721	12.2	3.50	0.337	3.46	8.43
	5	−0.077110	9.43	3.70	0.449	4.17	5.09
	4	−0.026670	7.07	3.54	0.505	4.32	2.78
	3	0.038729	5.11	3.06	0.471	4.01	1.30
	2	0.102389	3.50	2.36	0.314	3.33	0.463
	1	0.151786	2.22	1.48		2.38	0.093
	0	0.178571	1.23			1.23	0
5	9	0.096154	17.9	1.20	0.040		17.8
	8	−0.033654	15.0	2.03	0.102	1.42	13.4
	7	−0.086767	12.5	2.53	0.167	2.39	9.82
	6	−0.085165	10.2	2.73	0.216	2.94	7.05
	5	−0.048077	8.22	2.66	0.233	3.14	4.94
	4	0.008013	6.53	2.36	0.204	3.03	3.41
	3	0.069368	5.10	1.85	0.121	2.65	2.38
	2	0.125000	3.93	1.12		2.04	1.80
	1	0.166667	3.03			1.21	1.82

$I = 4$ (*Continued*)

J	F	$f(I,J,F)$	$J + 1 \leftarrow J$			$J \leftarrow J$	
			$F + 1 \leftarrow F$	$F \leftarrow F$	$F - 1 \leftarrow F$	$F \leftarrow F + 1$ or $F + 1 \leftarrow F$	$F \leftarrow F$
6	10	0.100000	17.0	0.889	0.023		16.9
	9	−0.025000	14.7	1.52	0.057	1.03	13.5
	8	−0.081981	12.5	1.90	0.092	1.74	10.6
	7	−0.088474	10.6	2.07	0.116	2.17	8.29
	6	−0.060065	8.95	2.04	0.122	2.35	6.45
	5	−0.010389	7.47	1.83	0.013	2.31	5.04
	4	0.048864	6.20	1.44	0.057	2.05	4.04
	3	0.107955	5.13	0.855		1.60	3.43
	2	0.159091	4.27			0.950	3.32
7	11	0.102941	16.3	0.683	0.014		16.3
	10	−0.018382	14.3	1.17	0.034	0.774	13.5
	9	−0.077771	12.5	1.48	0.054	1.32	11.1
	8	−0.090417	10.9	1.62	0.068	1.67	9.11
	7	−0.067469	9.44	1.60	0.070	1.82	7.50
	6	−0.022503	8.13	1.44	0.057	1.80	6.23
	5	0.034947	6.99	1.13	0.031	1.61	5.28
	4	0.096154	6.00	0.667		1.26	4.67
	3	0.153846	5.19			0.741	4.44
8	12	0.105263	15.8	0.541	0.009		15.7
	11	−0.013158	14.1	0.932	0.022	0.605	13.4
	10	−0.074154	12.5	1.18	0.034	1.04	11.4
	9	−0.090132	11.1	1.30	0.042	1.32	9.66
	8	−0.072368	9.78	1.29	0.043	1.44	8.24
	7	−0.031015	8.61	1.16	0.034	1.43	7.09
	6	0.024906	7.57	0.911	0.018	1.28	6.21
	5	0.087500	6.66	0.533		1.01	5.60
	4	0.150000	5.88			0.588	5.29
9	13	0.107143	15.3	0.440	0.006		15.3
	12	−0.008929	13.8	9.759	0.015	0.486	13.3
	11	−0.071053	12.5	0.965	0.023	0.838	11.5
	10	−0.090036	11.2	1.06	0.028	1.06	10.0
	9	−0.075780	10.0	1.06	0.028	1.17	8.78
	8	−0.037290	8.97	0.953	0.022	1.16	7.73
	7	0.017332	8.01	0.747	0.011	1.05	6.91
	6	0.080882	7.17	0.434		0.819	6.31
	5	0.147059	6.43			0.477	5.96

$I = 4$ (*Continued*)

J	F	$f(I,J,F)$	$J + 1 \leftarrow J$			$J \leftarrow J$	
			$F + 1 \leftarrow F$	$F \leftarrow F$	$F - 1 \leftarrow F$	$F \leftarrow F + 1$ or $F + 1 \leftarrow F$	$F \leftarrow F$
10	14	0.108696	15.0	0.364	0.004		14.9
	13	−0.005435	13.6	0.630	0.010	0.399	13.2
	12	−0.068384	12.4	0.802	0.016	0.689	11.7
	11	−0.089715	11.3	0.885	0.019	0.877	10.3
	10	−0.078253	10.2	0.882	0.019	0.966	9.18
	9	−0.042089	9.24	0.795	0.015	0.962	8.22
	8	0.001142	8.37	0.622	0.007	0.866	7.45
	7	0.075658	7.58	0.361		0.678	6.87
	6	0.144737	6.88			0.393	6.49

$I = \frac{9}{2}$

J	F	$f(I,J,F)$	$F + 1 \leftarrow F$	$F \leftarrow F$	$F - 1 \leftarrow F$	$F \leftarrow F+1$ or $F+1 \leftarrow F$	$F \leftarrow F$
0	9/2	0	40.0	33.3	26.7		
1	11/2	0.050000	28.0	9.82	2.18		23.6
	9/2	−0.133333	14.2	12.9	6.22	16.4	0.673
	7/2	0.091667	4.89	9.78	12.0	16.3	10.4
2	13/2	0.071429	22.9	4.62	0.527		21.5
	11/2	−0.083333	15.4	7.16	1.45	6.46	8.08
	9/2	−0.083333	9.45	7.88	2.67	9.45	1.21
	7/2	0.005952	4.95	6.97	4.08	9.33	0.381
	5/2	0.130952	1.79	4.49	5.71	6.29	5.71
3	15/2	0.083333	20.0	2.67	0.190		19.4
	13/2	−0.055556	15.1	4.38	0.513	3.43	11.1
	11/2	−0.091667	11.0	5.24	0.909	5.50	5.29
	9/2	−0.061111	7.58	5.39	1.32	6.36	1.73
	7/2	0.005556	4.81	4.91	1.70	6.19	0.136
	5/2	0.083333	2.65	3.92	2.00	5.10	0.326
	3/2	0.152777	1.05	2.44	2.22	3.14	2.57
4	17/2	0.090909	18.2	1.73	0.086		17.9
	15/2	−0.037879	14.6	2.92	0.226	2.12	12.1
	13/2	−0.086580	11.5	3.62	0.392	3.53	7.72
	11/2	−0.079545	8.88	3.91	0.551	4.31	4.48
	9/2	−0.037879	6.61	3.83	0.673	4.55	2.25
	7/2	0.020563	4.71	3.45	0.727	4.32	0.854
	5/2	0.081169	3.15	2.84	0.679	3.71	0.152
	3/2	0.132576	1.89	2.07	0.485	2.80	0.015
	1/2	0.166667	0.889	1.33		1.63	0.593

$$I = \tfrac{9}{2} \ (Continued)$$

J	F	f(I,J,F)	J + 1 ← J			J ← J	
			F + 1 ← F	F ← F	F − 1 ← F	F ← F + 1 or F + 1 ← F	F ← F
5	19/2	0.096154	16.9	1.21	0.044		16.7
	17/2	−0.025641	14.2	2.08	0.115	1.44	12.5
	15/2	−0.080128	11.7	2.63	0.196	2.44	9.05
	13/2	−0.085470	9.57	2.89	0.269	3.06	6.34
	11/2	−0.057692	7.68	2.91	0.318	3.33	4.28
	9/2	−0.010684	6.05	2.71	0.326	3.31	2.76
	7/2	0.043803	4.66	2.33	0.280	3.03	1.70
	5/2	0.096154	3.50	1.79	0.168	2.55	1.02
	3/2	0.138889	2.55	1.09		1.89	0.655
	1/2	0.166667	1.82			1.09	0.727
6	21/2	0.100000	16.0	0.898	0.025		15.9
	19/2	−0.016667	13.8	1.55	0.065	1.04	12.6
	17/2	−0.074242	11.8	1.98	0.109	1.78	9.80
	15/2	−0.087121	9.95	2.21	0.147	2.26	7.54
	13/2	−0.068182	8.35	2.25	0.169	2.51	5.73
	11/2	−0.028788	6.94	2.12	0.168	2.54	4.32
	9/2	0.021212	5.71	1.85	0.137	2.38	3.26
	7/2	0.073485	4.65	1.43	0.075	2.05	2.53
	5/2	0.121212	3.77	0.848		1.57	2.12
	3/2	0.159091	3.08			0.923	2.15
7	23/2	0.102941	15.3	0.691	0.015		15.2
	21/2	−0.009804	13.4	1.20	0.039	0.783	12.5
	19/2	−0.069193	11.7	1.54	0.065	1.36	10.2
	17/2	−0.087104	10.2	1.73	0.087	1.74	8.32
	15/2	−0.074284	8.79	1.78	0.098	1.94	6.74
	13/2	−0.040347	7.55	1.69	0.095	1.98	5.47
	11/2	0.006222	6.45	1.48	0.075	1.88	4.48
	9/2	0.058069	5.49	1.14	0.039	1.64	3.77
	7/2	0.108974	4.67	0.667		1.26	3.34
	5/2	0.153846	4.00			0.735	3.27
8	25/2	0.105263	14.7	0.547	0.010		14.7
	23/2	−0.004386	13.1	0.956	0.025	0.612	12.4
	21/2	−0.064912	11.7	1.23	0.041	1.07	10.5
	19/2	−0.086404	10.3	1.39	0.054	1.37	8.85
	17/2	−0.078070	9.09	1.43	0.061	1.54	7.46
	15/2	−0.048246	7.99	1.37	0.058	1.59	6.32
	13/2	−0.004386	6.99	1.20	0.044	1.51	5.41
	11/2	0.046930	6.11	0.924	0.023	1.32	4.73
	9/2	0.100000	5.35	0.535		1.02	4.28
	7/2	0.150000	4.71			0.588	4.12

$$I = \tfrac{9}{2} \ (Continued)$$

J	F	$f(I,J,F)$	$J+1 \leftarrow J$			$J \leftarrow J$	
			$F+1 \leftarrow F$	$F \leftarrow F$	$F-1 \leftarrow F$	$F \leftarrow F+1$ or $F+1 \leftarrow F$	$F \leftarrow F$
9	27/2	0.107143	14.3	0.444	0.007		14.2
	25/2	0	12.9	0.778	0.017	0.491	12.3
	23/2	−0.061275	11.6	1.01	0.027	0.859	10.7
	21/2	−0.085434	10.4	1.14	0.036	1.11	9.22
	19/2	−0.080532	9.31	1.18	0.040	1.25	7.98
	17/2	−0.053922	8.31	1.13	0.037	1.29	6.95
	15/2	−0.012255	7.41	0.987	0.028	1.23	6.11
	13/2	0.038515	6.59	0.760	0.014	1.08	5.46
	11/2	0.093137	5.88	0.437		0.831	5.01
	9/2	0.147059	5.26			0.478	4.78
10	29/2	0.108696	13.9	0.368	0.005		13.9
	27/2	−0.003623	12.7	0.646	0.012	0.403	12.2
	25/2	−0.058162	11.5	0.839	0.019	0.707	10.8
	23/2	−0.084382	10.5	0.950	0.025	0.916	9.48
	21/2	−0.082189	9.47	0.984	0.027	1.04	8.37
	19/2	−0.058162	8.56	0.942	0.025	1.07	7.43
	17/2	−0.018307	7.73	0.826	0.019	1.03	6.65
	15/2	0.031941	6.98	0.635	0.009	0.898	6.03
	13/2	0.087719	6.30	0.364		0.690	5.58
	11/2	0.144737	5.71			0.396	5.32

$$I = \tfrac{11}{2}$$

J	F	$f(I,J,F)$	$F+1 \leftarrow F$	$F \leftarrow F$	$F-1 \leftarrow F$	$F \leftarrow F+1$ or $F+1 \leftarrow F$	$F \leftarrow F$
0	11/2	0	38.9	33.3	27.8		
1	13/2	0.05000	26.7	9.87	2.35		22.4
	11/2	−0.12727	13.5	13.1	6.82	16.5	0.466
	9/2	−0.08273	4.60	9.85	13.3	16.4	11.4
2	15/2	0.07143	21.4	4.66	0.582		20.1
	13/2	−0.07468	14.4	7.29	1.65	6.52	7.20
	11/2	−0.83117	8.79	8.09	3.12	9.62	0.839
	9/2	−0.00455	4.55	7.18	4.94	9.55	0.701
	7/2	0.11818	1.60	4.59	7.14	6.42	6.91
3	17/2	0.08333	18.5	2.70	0.214		18.0
	15/2	−0.04545	14.0	4.48	0.595	3.47	9.96
	13/2	−0.08636	10.1	5.45	1.10	5.62	4.45
	11/2	−0.06667	6.92	5.68	1.68	6.59	1.20
	9/2	−0.01000	4.33	5.26	2.31	6.49	0.010
	7/2	0.06364	2.31	4.23	2.98	5.40	0.806
	5/2	0.13788	0.860	2.58	3.70	3.32	3.83

$$I = \tfrac{11}{2} \ (Continued)$$

J	F	$f(I,J,F)$	$J+1 \leftarrow J$			$J \leftarrow J$	
			$F+1 \leftarrow F$	$F \leftarrow F$	$F-1 \leftarrow F$	$F \leftarrow F+1$ or $F+1 \leftarrow F$	$F \leftarrow F$
4	19/2	0.09091	16.7	1.75	0.098		16.4
	17/2	−0.02686	13.4	3.00	0.267	2.14	10.9
	15/2	−0.07804	10.5	3.79	0.485	3.62	6.69
	13/2	−0.08070	8.06	4.18	0.725	4.50	3.62
	11/2	−0.05077	5.94	4.20	0.964	4.84	1.55
	9/2	−0.00207	4.16	3.92	1.18	4.71	0.378
	7/2	0.05372	2.69	3.37	1.35	4.17	0
	5/2	0.10702	1.52	2.59	1.45	3.24	0.389
	3/2	0.15041	0.613	1.58	1.52	1.93	1.78
5	21/2	0.09615	15.4	1.23	0.051		15.2
	19/2	−0.01399	12.9	2.14	0.138	1.46	11.2
	17/2	−0.06935	10.6	2.76	0.247	2.51	7.92
	15/2	−0.08322	8.64	3.12	0.363	3.21	5.33
	13/2	−0.06748	6.89	3.25	0.471	3.58	3.35
	11/2	−0.03263	5.37	3.17	0.556	3.67	1.91
	9/2	0.01224	4.05	2.92	0.604	3.51	0.921
	7/2	0.05944	2.94	2.53	0.599	3.14	0.321
	5/2	0.10268	2.00	2.02	0.524	2.60	0.039
	3/2	0.13706	1.22	1.45	0.356	1.91	0.027
	1/2	0.15909	0.589	0.926		1.09	0.421
6	23/2	0.10000	14.4	0.911	0.029		14.3
	21/2	−0.00455	12.4	1.60	0.079	1.05	11.2
	19/2	−0.06182	10.6	2.09	0.139	1.84	8.60
	17/2	−0.08223	8.94	2.40	0.201	2.39	6.44
	15/2	−0.07521	7.47	2.53	0.256	2.72	4.69
	13/2	−0.04917	6.15	2.52	0.295	2.85	3.32
	11/2	−0.01157	5.00	2.38	0.311	2.81	2.26
	9/2	0.03116	3.99	2.13	0.293	2.62	1.48
	7/2	0.07355	3.11	1.78	0.235	2.31	0.938
	5/2	0.11116	2.37	1.34	0.134	1.88	0.589
	3/2	0.14050	1.76	0.806		1.37	0.407
	1/2	0.15909	1.28			0.783	0.499
7	25/2	0.10294	13.7	0.701	0.018		13.7
	23/2	0.00267	12.0	1.24	0.048	0.794	11.1
	21/2	−0.05551	10.5	1.63	0.084	1.40	8.98
	19/2	−0.08013	9.10	1.89	0.120	1.84	7.16
	17/2	−0.07896	7.83	2.02	0.151	2.11	5.64
	15/2	−0.05903	6.69	2.03	0.172	2.25	4.39
	13/2	−0.02664	5.66	1.94	0.176	2.25	3.39
	11/2	0.01267	4.75	1.76	0.160	2.14	2.61
	9/2	0.05407	3.95	1.48	0.123	1.92	2.03
	7/2	0.09350	3.26	1.12	0.064	1.61	1.63
	5/2	0.12762	2.68	0.655		1.20	1.43
	3/2	0.15385	2.22			0.698	1.52

$$I = \tfrac{11}{2} \ (Continued)$$

J	F	$f(I,J,F)$	$J + 1 \leftarrow J$			$J \leftarrow J$	
			$F + 1 \leftarrow F$	$F \leftarrow F$	$F - 1 \leftarrow F$	$F \leftarrow F + 1$ or $F + 1 \leftarrow F$	$F \leftarrow F$
8	27/2	0.10526	13.2	0.556	0.012		13.1
	25/2	0.00837	11.7	0.989	0.031	0.621	11.0
	23/2	−0.05024	10.4	1.31	0.054	1.10	9.21
	21/2	−0.07775	9.19	1.52	0.076	1.45	7.64
	19/2	−0.08077	8.07	1.63	0.095	1.69	6.31
	17/2	−0.06531	7.06	1.66	0.106	1.81	5.19
	15/2	−0.03684	6.14	1.59	0.107	1.82	4.27
	13/2	0.01297	5.32	1.45	0.095	1.75	3.53
	11/2	0.04019	4.58	1.23	0.070	1.58	2.96
	9/2	0.08072	3.94	0.927	0.035	1.34	2.56
	7/2	0.11818	3.39	0.533		1.01	2.34
	5/2	0.15000	2.94			0.578	2.36
9	29/2	0.10714	12.7	0.452	0.008		12.7
	27/2	0.01299	11.5	0.807	0.021	0.499	10.9
	25/2	−0.04580	10.3	1.07	0.036	0.890	9.33
	23/2	−0.07540	9.23	1.25	0.051	1.18	7.97
	21/2	−0.08155	8.24	1.35	0.063	1.37	6.80
	19/2	−0.06952	7.33	1.37	0.069	1.48	5.79
	17/2	−0.04412	6.50	1.32	0.069	1.50	4.95
	15/2	−0.00970	5.75	1.21	0.060	1.44	4.26
	13/2	0.02983	5.07	1.02	0.043	1.32	3.71
	11/2	0.07105	4.47	0.771	0.021	1.11	3.31
	9/2	0.11096	3.95	0.439		0.837	3.07
	7/2	0.14706	3.51			0.477	3.03
10	31/2	0.10870	12.3	0.374	0.006		12.3
	29/2	0.01680	11.2	0.670	0.015	0.410	10.8
	27/2	−0.04202	10.2	0.892	0.025	0.733	9.40
	25/2	−0.07320	9.24	1.04	0.035	0.974	8.21
	23/2	−0.08178	8.35	1.13	0.043	1.14	7.16
	21/2	−0.07245	7.53	1.15	0.047	1.23	6.25
	19/2	−0.04951	6.78	1.11	0.046	1.25	5.47
	17/2	−0.01690	6.08	1.02	0.040	1.21	4.83
	15/2	0.02182	5.46	0.864	0.028	1.10	4.31
	13/2	0.06348	4.89	0.649	0.013	0.935	3.92
	11/2	0.10526	4.40	0.366		0.702	3.66
	9/2	0.14474	3.97			0.397	3.57

Second-order Energies Due to Nuclear Quadrupole Interactions in Linear Molecules and Symmetric Tops

Entries in this table should be multiplied by $\dfrac{(eqQ)^2}{B_0} \times 10^{-3}$ and added to the first-order energy of nuclear quadrupole interaction obtained from use of Appendix I [cf. Eq. (6-9)]. B_0 is the rotational constant. Values for $I = \frac{3}{2}, \frac{5}{2},$ and $\frac{7}{2}$ only are given since these spins are most commonly associated with large quadrupole coupling constants where second-order corrections are important. (This table was computed by N. M. McDermott.)

$$I = \tfrac{3}{2}$$

J	F	$K = 0$	$K = 1$	$K = 2$	$K = 3$
0	3/2	−10.417			
1	5/2	−6.000	−9.469		
	3/2	−2.250	−10.875		
	1/2	0	−11.719		
2	7/2	−4.100	−5.649	−7.289	
	5/2	−2.187	2.456	−3.887	
	3/2	10.417	5.208	−10.417	
	1/2	0	11.719	0	
3	9/2	−3.086	−3.852	−5.382	−5.376
	7/2	−1.929	0.217	3.472	−1.712
	5/2	6.000	3.447	−2.604	−7.324
	3/2	2.250	5.667	10.417	0
4	11/2	−2.465	−2.892	−3.909	−4.729
	9/2	−1.690	−0.594	1.857	3.155
	7/2	4.100	2.821	−0.461	−4.078
	5/2	2.187	3.566	6.492	7.324
5	13/2	−2.048	−2.308	−2.980	−3.738
	11/2	−1.494	−0.868	0.716	2.378
	9/2	3.086	2.372	0.444	−2.042
	7/2	1.929	2.611	4.278	5.790
6	15/2	−1.750	−1.920	−2.378	−2.970
	13/2	−1.333	−0.945	0.097	1.422
	11/2	2.465	2.029	0.818	−0.877
	9/2	1.690	2.075	3.080	4.263

$I = \frac{3}{2}$ (Continued)

J	F	K = 0	K = 1	K = 2	K = 3
7	17/2	−1.527	−1.644	−1.967	−2.417
	15/2	−1.202	−0.945	−0.233	0.756
	13/2	2.048	1.764	0.958	−0.223
	11/2	1.494	1.731	2.376	3.227
8	19/2	−1.353	−1.437	−1.673	−2.016
	17/2	−1.093	−0.915	−0.410	0.327
	15/2	1.750	1.554	0.993	0.147
	13/2	1.333	1.490	1.925	2.538
9	21/2	−1.215	−1.262	−1.440	−1.706
	19/2	−1.002	−0.873	−0.503	0.054
	17/2	1.527	1.386	0.981	0.357
	15/2	1.202	1.311	1.617	2.067
10	23/2	−1.102	−1.149	−1.285	−1.492
	21/2	−0.925	−0.828	−0.550	−0.122
	19/2	1.353	1.253	0.948	0.476
	17/2	1.093	1.172	1.396	1.733

$I = \frac{5}{2}$

J	F	K = 0	K = 1	K = 2	K = 3
0	5/2	−5.833			
1	7/2	−2.893	−6.417		
	5/2	−2.083	−3.639		
	3/2	−0.540	−8.235		
2	9/2	−1.804	−3.140	−4.844	
	7/2	−1.804	2.686	−0.826	
	5/2	4.948	0.066	−3.985	
	3/2	−0.197	5.699	−5.112	
	1/2	0	−0.833	−2.083	
3	11/2	−1.277	−1.911	−3.242	−3.557
	9/2	−1.477	0.301	3.140	−0.442
	7/2	2.000	0.786	−1.667	−1.797
	5/2	1.807	1.750	0.764	−3.593
	3/2	0.540	1.744	2.681	−2.788
	1/2	0	0.833	2.083	0
4	13/2	−0.975	−1.319	−2.160	−2.923
	11/2	−1.229	−0.404	1.453	2.495
	9/2	0.976	0.613	−0.280	−1.114
	7/2	1.508	1.201	0.330	−0.960
	5/2	0.886	1.200	1.735	1.266
	3/2	0.197	0.792	2.091	2.637

$$I = \tfrac{5}{2} \ (Continued)$$

J	F	$K = 0$	$K = 1$	$K = 2$	$K = 3$
5	15/2	−0.783	−0.989	−1.529	−2.169
	13/2	−1.045	−0.607	0.503	1.671
	11/2	0.523	0.399	0.071	−0.341
	9/2	1.184	0.964	0.356	−0.482
	7/2	0.893	0.966	1.093	1.004
	5/2	0.276	0.623	1.486	2.327
6	17/2	−0.652	−0.784	−1.146	−1.626
	15/2	−0.905	−0.648	0.041	0.915
	13/2	0.290	0.247	0.126	−0.045
	11/2	0.947	0.802	0.396	−0.184
	9/2	0.828	0.832	0.820	0.723
	7/2	0.296	0.508	1.069	1.753
7	19/2	−0.557	−0.647	−0.898	−1.255
	17/2	−0.797	−0.634	−0.185	0.438
	15/2	0.158	0.145	0.106	0.046
	13/2	0.777	0.680	0.405	−0.002
	11/2	0.754	0.738	0.685	0.575
	9/2	0.293	0.431	0.807	1.315
8	21/2	−0.485	−0.549	−0.730	−0.998
	19/2	−0.711	−0.603	−0.296	0.151
	17/2	0.079	0.077	0.071	0.061
	15/2	0.652	0.585	0.393	0.102
	13/2	0.685	0.665	0.604	0.497
	11/2	0.282	0.376	0.637	1.012
9	23/2	−0.429	−0.476	−0.612	−0.815
	21/2	−0.642	−0.566	−0.348	−0.021
	19/2	0.029	0.031	0.039	0.050
	17/2	0.558	0.511	0.373	0.160
	15/2	0.625	0.606	0.548	0.452
	13/2	0.268	0.334	0.523	0.802
10	25/2	−0.384	−0.420	−0.523	−0.681
	23/2	−0.584	−0.529	−0.369	−0.125
	21/2	−0.004	0.000	0.013	0.033
	19/2	0.486	0.451	0.349	0.190
	17/2	0.573	0.556	0.505	0.421
	15/2	0.253	0.302	0.442	0.654

$$I = \tfrac{7}{2}$$

J	F	$K = 0$	$K = 1$	$K = 2$	$K = 3$
0	7/2	−4.464			
1	9/2	−2.020	−5.030		
	7/2	−1.804	−2.159		
	5/2	−0.656	−6.416		
2	11/2	−1.184	−2.369	−3.890	
	9/2	−1.455	2.462	−0.637	
	7/2	3.483	−0.530	−2.079	
	5/2	−0.409	2.095	−9.028	
	3/2	−0.084	−1.209	−2.882	
3	13/2	−0.802	−1.349	−2.514	−2.861
	11/2	−1.127	0.301	2.532	−0.595
	9/2	1.090	0.429	−0.852	−0.649
	7/2	1.272	0.937	−0.122	−2.062
	5/2	0.454	3.167	6.709	−2.720
	3/2	−0.039	0.642	2.346	−2.047
	1/2	0.000	−0.149	−0.478	−0.628
4	15/2	−0.592	−0.884	−1.601	−2.271
	13/2	−0.902	−0.281	1.101	1.803
	11/2	0.359	0.269	−0.016	−0.198
	9/2	0.913	0.633	−0.070	−0.845
	7/2	0.730	0.737	0.624	−0.016
	5/2	0.342	0.613	1.101	0.838
	3/2	0.084	0.370	2.320	0.973
	1/2	0.000	0.149	0.478	0.628
5	17/2	−0.463	−0.635	−1.088	−1.632
	15/2	−0.745	−0.432	0.354	1.157
	13/2	0.075	0.095	0.141	0.171
	11/2	0.611	0.454	0.087	−0.391
	9/2	0.664	0.589	0.363	−0.018
	7/2	0.450	0.520	0.658	0.644
	5/2	0.202	0.361	0.735	1.018
	3/2	0.039	0.196	0.587	1.074
6	19/2	−0.377	−0.486	−0.786	−1.187
	17/2	−0.630	−0.455	0.013	0.596
	15/2	−0.051	−0.014	0.085	0.207
	13/2	0.420	0.346	0.142	−0.137
	11/2	0.561	0.494	0.302	0.001
	9/2	0.454	0.462	0.468	0.418
	7/2	0.252	0.333	0.538	0.754
	5/2	0.067	0.181	0.483	0.854

$$I = \tfrac{7}{2} \ (Continued)$$

J	F	$K = 0$	$K = 1$	$K = 2$	$K = 3$
7	21/2	−0.317	−0.391	−0.597	−0.891
	19/2	−0.544	−0.438	−0.144	0.258
	17/2	−0.111	−0.076	0.020	0.152
	15/2	0.297	0.257	0.145	−0.018
	13/2	0.471	0.421	0.278	0.061
	11/2	0.427	0.417	0.382	0.310
	9/2	0.266	0.309	0.422	0.558
	7/2	0.081	0.162	0.382	0.680
8	23/2	−0.272	−0.324	−0.472	−0.690
	21/2	−0.478	−0.409	−0.216	0.065
	19/2	−0.140	−0.112	−0.030	0.089
	17/2	0.215	0.193	0.129	0.033
	15/2	0.397	0.361	0.256	0.096
	13/2	0.393	0.379	0.335	0.260
	11/2	0.265	0.288	0.351	0.433
	9/2	0.088	0.145	0.306	0.537
9	25/2	−0.238	−0.276	−0.385	−0.550
	23/2	−0.426	−0.379	−0.246	−0.046
	21/2	−0.153	−0.130	−0.065	0.036
	19/2	0.159	0.146	0.108	0.050
	17/2	0.340	0.313	0.236	0.117
	15/2	0.360	0.346	0.303	0.234
	13/2	0.256	0.269	0.305	0.354
	11/2	0.090	0.132	0.252	0.430
10	27/2	−0.211	−0.239	−0.322	−0.449
	25/2	−0.383	−0.350	−0.255	−0.110
	23/2	−0.158	−0.140	−0.086	−0.004
	21/2	0.119	0.111	0.088	0.053
	19/2	0.294	0.275	0.217	0.127
	17/2	0.330	0.317	0.280	0.219
	15/2	0.246	0.254	0.274	0.303
	13/2	0.090	0.121	0.212	0.351

Coefficients for Energy Levels of a Slightly Asymmetric Top

Rotational energy is given by

$$w = K^2 + C_1 b + C_2 b^2 + C_3 b^3 + C_4 b^4 + C_5 b^5 + \cdots$$

For a prolate top, energy $= W = \dfrac{B + C}{2} J(J + 1) + \left(A - \dfrac{B + C}{2} \right) w$

$$b = b_p = \frac{C - B}{2A - B - C}$$

For an oblate top, energy $= W = \dfrac{A + B}{2} J(J + 1) + \left(C - \dfrac{A + B}{2} \right) w$

$$b = b_0 = \frac{A - B}{2C - B - A}$$

Where the first few constants K, C_1, C_2 ... are identical for pairs of degenerate levels, they are usually listed for only the first of the two levels. (C_1, C_2, and C_3 were computed by J. F. Lotspeich; C_4 and C_5 by J. Kraitchman and N. Solimene.)

Level designation		K^2	C_1	C_2	C_3	C_4	C_5
Prolate top	Oblate top						
$0_{0,0}$	$0_{0,0}$	0	0	0	0	0	0
$1_{1,0}$	$1_{0,1}$	1	-1	0	0	0	0
$1_{1,1}$	$1_{1,1}$	1	1	0	0	0	0
$1_{0,1}$	$1_{1,0}$	0	0	0	0	0	0
$2_{2,0}$	$2_{0,2}$	4	0	3	0	-2.25	0
$2_{2,1}$	$2_{1,2}$	4	0	0	0	0	0
$2_{1,1}$	$2_{1,1}$	1	-3	0	0	0	0
$2_{1,2}$	$2_{2,1}$	1	3	0	0	0	0
$2_{0,2}$	$2_{2,0}$	0	0	-3	0	2.25	0
$3_{3,0}$	$3_{0,3}$	9	0	1.875	-1.40625	0.615234	0.197754
$3_{3,1}$	$3_{1,3}$	9	0	1.875	1.40625	0.615234	-0.197754
$3_{2,1}$	$3_{1,2}$	4	0	15	0	-56.25	0
$3_{2,2}$	$3_{2,2}$	4	0	0	0	0	0
$3_{1,2}$	$3_{2,1}$	1	-6	-1.875	1.40625	-0.615234	-0.197754
$3_{1,3}$	$3_{3,1}$	1	6	-1.875	-1.40625	-0.615234	0.197754
$3_{0,3}$	$3_{3,0}$	0	0	-15	0	56.25	0

Level designation		K^2	C_1	C_2	C_3	C_4	C_5
Prolate top	Oblate top						
$4_{4,0}$	$4_{0,4}$	16	0	2.33333	0	1.73380	0
$4_{4,1}$	$4_{1,4}$					-0.453704	0
$4_{3,1}$	$4_{1,3}$	9	0	7.875	-9.84375	4.55273	13.6890
$4_{3,2}$	$4_{2,3}$	9	0	7.875	9.84375	4.55273	-13.6890
$4_{2,2}$	$4_{2,2}$	4	0	42.6667	0	-488.296	0
$4_{2,3}$	$4_{3,2}$	4	0	-2.33333	0	0.453704	0
$4_{1,3}$	$4_{3,1}$	1	-10	-7.875	9.84375	-4.55273	-13.6890
$4_{1,4}$	$4_{4,1}$	1	10	-7.875	-9.84375	-4.55273	13.6890
$4_{0,4}$	$4_{4,0}$	0	0	-45	0	486.563	0
$5_{5,0}$	$5_{0,5}$	25	0	2.8125	0	0.736084	-0.769043
$5_{5,1}$	$5_{1,5}$					0.736084	0.769043
$5_{4,1}$	$5_{1,4}$	16	0	9	0	12.9375	0
$5_{4,2}$	$5_{2,4}$					-6.75	0
$5_{3,2}$	$5_{2,3}$	9	0	18.1875	-39.375	22.8890	150.886
$5_{3,3}$	$5_{3,3}$	9	0	18.1875	39.375	22.8890	-150.886
$5_{2,3}$	$5_{3,2}$	4	0	96	0	-2592	0
$5_{2,4}$	$5_{4,2}$	4	0	-9	0	6.75	0
$5_{1,4}$	$5_{4,1}$	1	-15	-21	39.375	-23.625	-150.117
$5_{1,5}$	$5_{5,1}$	1	15	-21	-39.375	-23.625	150.117
$5_{0,5}$	$5_{5,0}$	0	0	-105	0	2579.06	0
$6_{6,0}$	$6_{0,6}$	36	0	3.3.	0	.847688	0
$6_{6,1}$	$6_{1,6}$					.847688	0
$6_{5,1}$	$6_{1,5}$	25	0	10.3125	0	3.02124	-8.45947
$6_{5,2}$	$6_{2,5}$					3.02124	8.45947
$6_{4,2}$	$6_{2,4}$	16	0	19.2	0	59.2695	0
$6_{4,3}$	$6_{3,4}$					-39.168	0
$6_{3,3}$	$6_{3,3}$	9	0	34.6875	-118.125	92.6038	950.999
$6_{3,4}$	$6_{4,3}$	9	0	34.6875	118.125	92.6038	-950.999
$6_{2,4}$	$6_{4,2}$	4	0	187.5	0	-10199.2	0
$6_{2,5}$	$6_{5,2}$	4	0	-22.5	0	38.3203	0
$6_{1,5}$	$6_{5,1}$	1	-21	-45	118.125	-95.625	-942.539
$6_{1,6}$	$6_{6,1}$	1	21	-45	-118.125	-95.625	942.539
$6_{0,6}$	$6_{6,0}$	0	0	-210	0	10139.06	0
$7_{7,0}$	$7_{0,7}$	49	0	3.79166	0	0.965032	0
$7_{7,1}$	$7_{1,7}$						
$7_{6,1}$	$7_{1,6}$	36	0	11.7	0	3.21019	0
$7_{6,2}$	$7_{2,6}$						
$7_{5,2}$	$7_{2,5}$	25	0	20.95834	0	7.77481	-50.7568
$7_{5,3}$	$7_{3,5}$					7.77481	50.7568
$7_{4,3}$	$7_{3,4}$	16	0	34.13334	0	210.599	0
$7_{4,4}$	$7_{4,4}$					-150.338	0
$7_{3,4}$	$7_{4,3}$	9	0	59.625	-295.313	308.985	4355.86

Prolate top	Oblate top	K^2	C_1	C_2	C_3	C_4	C_5
$7_{3,5}$	$7_{5,3}$	9	0	59.625	295.313	308.985	-4355.86
$7_{2,5}$	$7_{5,2}$	4	0	332.16667	0	-32686.4	0
$7_{2,6}$	$7_{6,2}$	4	0	-45.83333	0	147.128	0
$7_{1,6}$	$7_{6,1}$	1	-28	-84.375	295.313	-317.725	-4305.10
$7_{1,7}$	$7_{7,1}$	1	28	-84.375	-295.313	-317.725	4305.10
$7_{0,7}$	$7_{7,0}$	0	0	-378	0	32472.6	0
$8_{8,0}$	$8_{0,8}$	64	0	4.28572	0	1.08509	0
$8_{8,1}$	$8_{1,8}$						
$8_{7,1}$	$8_{1,7}$	49	0	13.125	0	3.48633	0
$8_{7,2}$	$8_{2,7}$						
$8_{6,2}$	$8_{2,6}$	36	0	23.01428	0	7.29259	0
$8_{6,3}$	$8_{3,6}$						
$8_{5,3}$	$8_{3,5}$	25	0	35.625	0	18.6035	-219.946
$8_{5,4}$	$8_{4,5}$					18.6035	219.946
$8_{4,4}$	$8_{4,4}$	16	0	55.2	0	624.552	0
$8_{4,5}$	$8_{5,4}$					-458.261	0
$8_{3,5}$	$8_{5,3}$	9	0	95.625	-649.688	882.510	16079.8
$8_{3,6}$	$8_{6,3}$	9	0	95.625	649.688	882.510	-16079.8
$8_{2,6}$	$8_{6,2}$	4	0	547.5	0	-90112.6	0
$8_{2,7}$	$8_{7,2}$	4	0	-82.5	0	449.883	0
$8_{1,7}$	$8_{7,1}$	1	-36	-144.375	649.688	-904.600	-15859.8
$8_{1,8}$	$8_{8,1}$	1	36	-144.375	-649.688	-904.600	15859.8
$8_{0,8}$	$8_{8,0}$	0	0	-630	0	89479.7	0
$9_{9,0}$	$9_{0,9}$	81	0	4.78125	0	1.20665	0
$9_{9,1}$	$9_{1,9}$						
$9_{8,1}$	$9_{1,8}$	64	0	14.5714	0	3.80084	0
$9_{8,2}$	$9_{2,8}$						
$9_{7,2}$	$9_{2,7}$	49	0	25.2187	0	7.36478	0
$9_{7,3}$	$9_{3,7}$						
$9_{6,3}$	$9_{3,6}$	36	0	37.9286	0	15.0663	0
$9_{6,4}$	$9_{4,6}$						
$9_{5,4}$	$9_{4,5}$	25	0	55.3125	0	43.0826	-769.812
$9_{5,5}$	$9_{5,5}$					43.0826	769.812
$9_{4,5}$	$9_{5,4}$	16	0	84	0	1617	0
$9_{4,6}$	$9_{6,4}$					-1198.31	0
$9_{3,6}$	$9_{6,3}$	9	0	145.688	-1299.38	2229.47	50660.4
$9_{3,7}$	$9_{7,3}$	9	0	145.688	1299.38	2229.47	-50660.4
$9_{2,7}$	$9_{7,2}$	4	0	853.5	0	-221323	0
$9_{2,8}$	$9_{8,2}$	4	0	-136.5	0	1179.45	0
$9_{1,8}$	$9_{8,1}$	1	-45	-231	1299.38	-2281.13	-49890.6
$9_{1,9}$	$9_{9,1}$	1	45	-231	-1299.38	-2281.13	49890.6
$9_{0,9}$	$9_{9,0}$	0	0	-990	0	219687	0

| Level designation | | K^2 | C_1 | C_2 | C_3 | C_4 | C_5 |
Prolate top	Oblate top						
$10_{10,0}$	$10_{0,10}$	100	0	5.27778	0	1.32912	0
$10_{10,1}$	$10_{1,10}$						
$10_{9,1}$	$10_{1,9}$	81	0	16.0313	0	4.13529	0
$10_{9,2}$	$10_{2,9}$						
$10_{8,2}$	$10_{2,8}$	64	0	27.5079	0	7.66709	0
$10_{8,3}$	$10_{3,8}$						
$10_{7,3}$	$10_{3,7}$	49	0	40.6354	0	13.8541	0
$10_{7,4}$	$10_{4,7}$						
$10_{6,4}$	$10_{4,6}$	36	0	57.2143	0	30.3788	0
$10_{6,5}$	$10_{5,6}$						
$10_{5,5}$	$10_{5,5}$	25	0	81.1458	0	95.5982	−2309.44
$10_{5,6}$	$10_{6,5}$					95.5982	2309.44
$10_{4,6}$	$10_{6,4}$	16	0	122.333	0	3767.88	0
$10_{4,7}$	$10_{7,4}$					−2801.12	0
$10_{3,7}$	$10_{7,3}$	9	0	213.188	−2413.13	5107.54	141215
$10_{3,8}$	$10_{8,3}$	9	0	213.188	2413.13	5107.54	−141215
$10_{2,8}$	$10_{8,2}$	4	0	1272.67	0	−495992	0
$10_{2,9}$	$10_{9,2}$	4	0	−212.333	0	2761.81	0
$10_{1,9}$	$10_{9,1}$	1	−55	−351	2413.13	−5221.13	−138906
$10_{1,10}$	$10_{10,1}$	1	55	−351	−2413.13	−5221.13	138906
$10_{0,10}$	$10_{10,0}$	0	0	−1485	0	492185	0
$11_{11,0}$	$11_{0,11}$	121	0	5.775	0		
$11_{11,1}$	$11_{1,11}$						
$11_{10,1}$	$11_{1,10}$	100	0	17.5	0		
$11_{10,2}$	$11_{2,10}$						
$11_{9,2}$	$11_{2,9}$	81	0	29.85	0		
$11_{9,3}$	$11_{3,9}$						
$11_{8,3}$	$11_{3,8}$	64	0	43.5714	0		
$11_{8,4}$	$11_{4,8}$						
$11_{7,4}$	$11_{4,7}$	49	0	60	0		
$11_{7,5}$	$11_{5,7}$						
$11_{6,5}$	$11_{5,6}$	36	0	82.1786	0		
$11_{6,6}$	$11_{6,6}$						
$11_{5,6}$	$11_{6,5}$	25	0	114.375	0		
$11_{5,7}$	$11_{7,5}$						
$11_{4,7}$	$11_{7,4}$	16	0	171.750	0		
$11_{4,8}$	$11_{8,4}$						
$11_{3,8}$	$11_{8,3}$	9	0	301.875	−4222.97		
$11_{3,9}$	$11_{9,3}$	9	0	301.875	4222.97		
$11_{2,9}$	$11_{9,2}$	4	0	1830	0		
$11_{2,10}$	$11_{10,2}$	4	0	−315	0		
$11_{1,10}$	$11_{10,1}$	1	−66	−511.875	4222.97		
$11_{1,11}$	$11_{11,1}$	1	66	−511.875	−4222.97		
$11_{0,11}$	$11_{11,0}$	0	0	−2145	0		

| Level designation | | K^2 | C_1 | C_2 | C_3 | C_4 | C_5 |
Prolate top	Oblate top						
$12_{12,0}$	$12_{0,12}$	144	0	6.27272	0		
$12_{12,1}$	$12_{1,12}$						
$12_{11,1}$	$12_{1,11}$	121	0	18.975	0		
$12_{11,2}$	$12_{2,11}$						
$12_{10,2}$	$12_{2,10}$	100	0	32.2273	0		
$12_{10,3}$	$12_{3,10}$						
$12_{9,3}$	$12_{3,9}$	81	0	46.650	0		
$12_{9,4}$	$12_{4,9}$						
$12_{8,4}$	$12_{4,8}$	64	0	63.2857	0		
$12_{8,5}$	$12_{5,8}$						
$12_{7,5}$	$12_{5,7}$	49	0	84	0		
$12_{7,6}$	$12_{6,7}$						
$12_{6,6}$	$12_{6,6}$	36	0	112.414	0		
$12_{6,7}$	$12_{7,6}$						
$12_{5,7}$	$12_{7,5}$	25	0	156.375	0		
$12_{5,8}$	$12_{8,5}$						
$12_{4,8}$	$12_{8,4}$	16	0	235.8	0		
$12_{4,9}$	$12_{9,4}$						
$12_{3,9}$	$12_{9,3}$	9	0	415.875	−7038.28		
$12_{3,10}$	$12_{9,3}$	9	0	415.875	7038.28		
$12_{2,10}$	$12_{10,3}$	4	0	2553	0		
$12_{2,11}$	$12_{11,2}$	4	0	−450	0		
$12_{1,11}$	$12_{11,1}$	1	−78	−721.875	7038.28		
$12_{1,12}$	$12_{12,1}$	1	78	−721.875	−7038.28		
$12_{0,12}$	$12_{12,0}$	0	0	−3003	0		

Energy Levels of a Rigid Rotor

Energy (in cycles/sec) $= W/h = \frac{1}{2}(A + C)J(J + 1) + \frac{1}{2}(A - C)E_\tau$.

E_τ is tabulated as a function of the rotational level $J_{K_{-1}K_1}$ (or J_τ) and of the asymmetry parameter $\kappa = \dfrac{2B - A - C}{A - C}$.

Values for positive κ only are tabulated, since those for negative κ can be obtained from the relation $E_\tau(\kappa) = -E_{-\tau}(-\kappa)$. For further explanation see Chap. 4.

This table was reproduced with the permission of the Ballistics Research Laboratories, Aberdeen, Md., from *Ballistics Research Laboratories Report* No. 878 (September, 1953), by T. E. Turner, B. L. Hicks, and G. Reitwiesner. It was prepared for reproduction by S. Poley with the aid of an IBM card-controlled typewriter at the Watson Scientific Computing Laboratory.

Tables of E_τ for J up to 40 and values of $\kappa = 0, 0.1, 0.2, 0.3, \ldots 1.0$ are given by G. Erlandsson, *Arkiv för Fysik*, to be published.

$3_{2,2}$ / 3_0	$3_{2,1}$ / 3_1	$3_{3,1}$ / 3_2	$3_{3,0}$ / 3_3	$2_{0,2}$ / 2_{-2}	$2_{1,2}$ / 2_{-1}	$2_{1,1}$ / 2_0	$2_{2,1}$ / 2_1	$2_{2,0}$ / 2_2	$1_{0,1}$ / 1_{-1}	$1_{1,1}$ / 1_0	$1_{1,0}$ / 1_1	$0_{0,0}$ / 0_0	J_{K_{-1},K_1} / J_r / κ
0.0000000	1.8989794	7.7459666	7.8989794	-3.4641016	-3.0000000	0.0000000	3.0000000	3.4641016	-1.0000000	0.0000000	1.0000000	0	.00
0.0400000	1.9735759	7.7659925	7.9245871	-3.4441593	-2.9900000	0.0400000	3.0100000	3.4841593	-0.9900000	0.0000000	1.0100000	0	.01
0.0800000	2.0483734	7.7860699	7.9504020	-3.4243325	-2.9800000	0.0800000	3.0200000	3.5043325	-0.9800000	0.0000000	1.0200000	0	.02
0.1200000	2.1233690	7.8061990	7.9764272	-3.4046211	-2.9700000	0.1200000	3.0300000	3.5246211	-0.9700000	0.0000000	1.0300000	0	.03
0.1600000	2.1985597	7.8263797	8.0026659	-3.3850252	-2.9600000	0.1600000	3.0400000	3.5450252	-0.9600000	0.0000000	1.0400000	0	.04
0.2000000	2.2739426	7.8466121	8.0291212	-3.3655446	-2.9500000	0.2000000	3.0500000	3.5655446	-0.9500000	0.0000000	1.0500000	0	.05
0.2400000	2.3495148	7.8668961	8.0557964	-3.3461794	-2.9400000	0.2400000	3.0600000	3.5861794	-0.9400000	0.0000000	1.0600000	0	.06
0.2800000	2.4252733	7.8872317	8.0826947	-3.3269294	-2.9300000	0.2800000	3.0700000	3.6069294	-0.9300000	0.0000000	1.0700000	0	.07
0.3200000	2.5012155	7.9076189	8.1098195	-3.3077946	-2.9200000	0.3200000	3.0800000	3.6277946	-0.9200000	0.0000000	1.0800000	0	.08
0.3600000	2.5773384	7.9280578	8.1371739	-3.2887749	-2.9100000	0.3600000	3.0900000	3.6487749	-0.9100000	0.0000000	1.0900000	0	.09
0.4000000	2.6536394	7.9485482	8.1647615	-3.2698703	-2.9000000	0.4000000	3.1000000	3.6698703	-0.9000000	0.0000000	1.1000000	0	.10
0.4400000	2.7301158	7.9690902	8.1925854	-3.2510800	-2.8900000	0.4400000	3.1100000	3.6910805	-0.8900000	0.0000000	1.1100000	0	.11
0.4800000	2.8067648	7.9896838	8.2206493	-3.2324055	-2.8800000	0.4800000	3.1200000	3.7124055	-0.8800000	0.0000000	1.1200000	0	.12
0.5200000	2.8835838	8.0103290	8.2489564	-3.2138465	-2.8700000	0.5200000	3.1300000	3.7338451	-0.8700000	0.0000000	1.1300000	0	.13
0.5600000	2.9605703	8.0310257	8.2775102	-3.1953992	-2.8600000	0.5600000	3.1400000	3.7553992	-0.8600000	0.0000000	1.1400000	0	.14
0.6000000	3.0377216	8.0517739	8.3063142	-3.1770677	-2.8500000	0.6000000	3.1500000	3.7770677	-0.8500000	0.0000000	1.1500000	0	.15
0.6400000	3.1150352	8.0725737	8.3353720	-3.1588503	-2.8400000	0.6400000	3.1600000	3.7988503	-0.8400000	0.0000000	1.1600000	0	.16
0.6800000	3.1925087	8.0934250	8.3646871	-3.1407470	-2.8300000	0.6800000	3.1700000	3.8207470	-0.8300000	0.0000000	1.1700000	0	.17
0.7200000	3.2701396	8.1143278	8.3942630	-3.1227575	-2.8200000	0.7200000	3.1800000	3.8427575	-0.8200000	0.0000000	1.1800000	0	.18
0.7600000	3.3479255	8.1352820	8.4241032	-3.1048816	-2.8100000	0.7600000	3.1900000	3.8648816	-0.8100000	0.0000000	1.1900000	0	.19
0.8000000	3.4258639	8.1562877	8.4542114	-3.0871191	-2.8000000	0.8000000	3.2000000	3.8871191	-0.8000000	0.0000000	1.2000000	0	.20
0.8400000	3.5039526	8.1773449	8.4845913	-3.0694698	-2.7900000	0.8400000	3.2100000	3.9094698	-0.7900000	0.0000000	1.2100000	0	.21
0.8800000	3.5821893	8.1984534	8.5151803	-3.0519335	-2.7800000	0.8800000	3.2200000	3.9319335	-0.7800000	0.0000000	1.2200000	0	.22
0.9200000	3.6605716	8.2196133	8.5461133	-3.0345099	-2.7700000	0.9200000	3.2300000	3.9545099	-0.7700000	0.0000000	1.2300000	0	.23
0.9600000	3.7390973	8.2408246	8.5773964	-3.0171988	-2.7600000	0.9600000	3.2400000	3.9771988	-0.7600000	0.0000000	1.2400000	0	.24
1.0000000	3.8177643	8.2620873	8.6088989	-3.0000000	-2.7500000	1.0000000	3.2500000	4.0000000	-0.7500000	0.0000000	1.2500000	0	.25
1.0400000	3.8965703	8.2834013	8.6406911	-2.9829130	-2.7400000	1.0400000	3.2600000	4.0229130	-0.7400000	0.0000000	1.2600000	0	.26
1.0800000	3.9755133	8.3047665	8.6727768	-2.9659378	-2.7300000	1.0800000	3.2700000	4.0459378	-0.7300000	0.0000000	1.2700000	0	.27
1.1200000	4.0545910	8.3261831	8.7051596	-2.9490739	-2.7200000	1.1200000	3.2800000	4.0690739	-0.7200000	0.0000000	1.2800000	0	.28
1.1600000	4.1338015	8.3476508	8.7378432	-2.9323211	-2.7100000	1.1600000	3.2900000	4.0923211	-0.7100000	0.0000000	1.2900000	0	.29
1.2000000	4.2131427	8.3691698	8.7708313	-2.9156791	-2.7000000	1.2000000	3.3000000	4.1156791	-0.7000000	0.0000000	1.3000000	0	.30
1.2400000	4.2926126	8.3907399	8.8041376	-2.8991476	-2.6900000	1.2400000	3.3100000	4.1391476	-0.6900000	0.0000000	1.3100000	0	.31
1.2800000	4.3722092	8.4123612	8.8377552	-2.8827262	-2.6800000	1.2800000	3.3200000	4.1627262	-0.6800000	0.0000000	1.3200000	0	.32
1.3200000	4.4519307	8.4340337	8.8716584	-2.8664146	-2.6700000	1.3200000	3.3300000	4.1864146	-0.6700000	0.0000000	1.3300000	0	.33
1.3600000	4.5317750	8.4557571	8.9059006	-2.8502124	-2.6600000	1.3600000	3.3400000	4.2102124	-0.6600000	0.0000000	1.3400000	0	.34
1.4000000	4.6117403	8.4775317	8.9404653	-2.8341194	-2.6500000	1.4000000	3.3500000	4.2341194	-0.6500000	0.0000000	1.3500000	0	.35
1.4400000	4.6918248	8.4993572	8.9753562	-2.8181351	-2.6400000	1.4400000	3.3600000	4.2581351	-0.6400000	0.0000000	1.3600000	0	.36
1.4800000	4.7720266	8.5212338	9.0105768	-2.8022591	-2.6300000	1.4800000	3.3700000	4.2822591	-0.6300000	0.0000000	1.3700000	0	.37
1.5200000	4.8523440	8.5431613	9.0461307	-2.7864912	-2.6200000	1.5200000	3.3800000	4.3064912	-0.6200000	0.0000000	1.3800000	0	.38
1.5600000	4.9327752	8.5651396	9.0820212	-2.7708308	-2.6100000	1.5600000	3.3900000	4.3308308	-0.6100000	0.0000000	1.3900000	0	.39
1.6000000	5.0133185	8.5871689	9.1182520	-2.7552777	-2.6000000	1.6000000	3.4000000	4.3552777	-0.6000000	0.0000000	1.4000000	0	.40
1.6400000	5.0939725	8.6092490	9.1548264	-2.7398314	-2.5900000	1.6400000	3.4100000	4.3798314	-0.5900000	0.0000000	1.4100000	0	.41
1.6800000	5.1747345	8.6313798	9.1917477	-2.7244915	-2.5800000	1.6800000	3.4200000	4.4044915	-0.5800000	0.0000000	1.4200000	0	.42
1.7200000	5.2556039	8.6535614	9.2290194	-2.7092576	-2.5700000	1.7200000	3.4300000	4.4292576	-0.5700000	0.0000000	1.4300000	0	.43
1.7600000	5.3365788	8.6757937	9.2666448	-2.6941292	-2.5600000	1.7600000	3.4400000	4.4541292	-0.5600000	0.0000000	1.4400000	0	.44
1.8000000	5.4176575	8.6980766	9.3046269	-2.6791060	-2.5500000	1.8000000	3.4500000	4.4791060	-0.5500000	0.0000000	1.4500000	0	.45
1.8400000	5.4988386	8.7204102	9.3429692	-2.6641874	-2.5400000	1.8400000	3.4600000	4.5041874	-0.5400000	0.0000000	1.4600000	0	.46
1.8800000	5.5801203	8.7427943	9.3816745	-2.6493732	-2.5300000	1.8800000	3.4700000	4.5293732	-0.5300000	0.0000000	1.4700000	0	.47
1.9200000	5.6615014	8.7652290	9.4207461	-2.6346627	-2.5200000	1.9200000	3.4800000	4.5546627	-0.5200000	0.0000000	1.4800000	0	.48
1.9600000	5.7429802	8.7877141	9.4601870	-2.6200555	-2.5100000	1.9600000	3.4900000	4.5800555	-0.5100000	0.0000000	1.4900000	0	.49
2.0000000	5.8245553	8.8102496	9.5000000	-2.6055512	-2.5000000	2.0000000	3.5000000	4.6055512	-0.5000000	0.0000000	1.5000000	0	.50

$J_{K_{-1}K_1}$ / J_τ / κ	$0_{0,0}$ / 0_0	$1_{1,0}$ / 1_1	$1_{1,1}$ / 1_0	$1_{0,1}$ / 1_{-1}	$2_{2,0}$ / 2_2	$2_{2,1}$ / 2_1	$2_{1,1}$ / 2_0	$2_{1,2}$ / 2_{-1}	$2_{0,2}$ / 2_{-2}	$3_{3,0}$ / 3_3	$3_{3,1}$ / 3_2	$3_{2,1}$ / 3_1	$3_{2,2}$ / 3_0
.50	0	1.5000000	0.0000000	-0.5000000	4.6055512	3.5000000	2.0000000	-2.5000000	-2.6055512	9.5000000	8.8102496	5.8245553	2.0000000
.51	0	1.5100000	0.0000000	-0.4900000	4.6311494	3.5100000	2.0400000	-2.4900000	-2.5911494	9.5401879	8.8328355	5.9062252	2.0400000
.52	0	1.5200000	0.0000000	-0.4800000	4.6568494	3.5200000	2.0800000	-2.4800000	-2.5768494	9.5807536	8.8554718	5.9879887	2.0800000
.53	0	1.5300000	0.0000000	-0.4700000	4.6826509	3.5300000	2.1200000	-2.4700000	-2.5626509	9.6216998	8.8781583	6.0698442	2.1200000
.54	0	1.5400000	0.0000000	-0.4600000	4.7085534	3.5400000	2.1600000	-2.4600000	-2.5485534	9.6630291	8.9008950	6.1517904	2.1600000
.55	0	1.5500000	0.0000000	-0.4500000	4.7345563	3.5500000	2.2000000	-2.4500000	-2.5345563	9.7047439	8.9236819	6.2338260	2.2000000
.56	0	1.5600000	0.0000000	-0.4400000	4.7606592	3.5600000	2.2400000	-2.4400000	-2.5206592	9.7468468	8.9465190	6.3159496	2.2400000
.57	0	1.5700000	0.0000000	-0.4300000	4.7868616	3.5700000	2.2800000	-2.4300000	-2.5068616	9.7893430	8.9694061	6.3981600	2.2800000
.58	0	1.5800000	0.0000000	-0.4200000	4.8131630	3.5800000	2.3200000	-2.4200000	-2.4931630	9.8322258	8.9923431	6.4804559	2.3200000
.59	0	1.5900000	0.0000000	-0.4100000	4.8395628	3.5900000	2.3600000	-2.4100000	-2.4795628	9.8755063	9.0153302	6.5628360	2.3600000
.60	0	1.6000000	0.0000000	-0.4000000	4.8660605	3.6000000	2.4000000	-2.4000000	-2.4660605	9.9191835	9.0383671	6.6452990	2.4000000
.61	0	1.6100000	0.0000000	-0.3900000	4.8926557	3.6100000	2.4400000	-2.3900000	-2.4526557	9.9632595	9.0614539	6.7278439	2.4400000
.62	0	1.6200000	0.0000000	-0.3800000	4.9193477	3.6200000	2.4800000	-2.3800000	-2.4393477	10.0077359	9.0845904	6.8104694	2.4800000
.63	0	1.6300000	0.0000000	-0.3700000	4.9461361	3.6300000	2.5200000	-2.3700000	-2.4261361	10.0526145	9.1077767	6.8931743	2.5200000
.64	0	1.6400000	0.0000000	-0.3600000	4.9730204	3.6400000	2.5600000	-2.3599999	-2.4130204	10.0978968	9.1310126	6.9759574	2.5600000
.65	0	1.6500000	0.0000000	-0.3500000	5.0000000	3.6500000	2.6000000	-2.3500000	-2.4000000	10.1435844	9.1542981	7.0588178	2.6000000
.66	0	1.6600000	0.0000000	-0.3400000	5.0270743	3.6600000	2.6400000	-2.3400000	-2.3870743	10.1896786	9.1776332	7.1417541	2.6400000
.67	0	1.6700000	0.0000000	-0.3300000	5.0542428	3.6700000	2.6800000	-2.3300000	-2.3742428	10.2361806	9.2010177	7.2247654	2.6800000
.68	0	1.6800000	0.0000000	-0.3200000	5.0815049	3.6800000	2.7200000	-2.3200000	-2.3615049	10.2830915	9.2244516	7.3078506	2.7200000
.69	0	1.6900000	0.0000000	-0.3100000	5.1088604	3.6900000	2.7600000	-2.3100000	-2.3488604	10.3304125	9.2479349	7.3910085	2.7600000
.70	0	1.7000000	0.0000000	-0.3000000	5.1363083	3.7000000	2.8000000	-2.3000000	-2.3363083	10.3781438	9.2714674	7.4742383	2.8000000
.71	0	1.7100000	0.0000000	-0.2900000	5.1638482	3.7100000	2.8400000	-2.2900000	-2.3238482	10.4262868	9.2950492	7.5575387	2.8400000
.72	0	1.7200000	0.0000000	-0.2800000	5.1914797	3.7200000	2.8800000	-2.2800000	-2.3114797	10.4748419	9.3186800	7.6409090	2.8800000
.73	0	1.7300000	0.0000000	-0.2700000	5.2192020	3.7300000	2.9200000	-2.2700000	-2.2992020	10.5238094	9.3423600	7.7243480	2.9200000
.74	0	1.7400000	0.0000000	-0.2600000	5.2470147	3.7400000	2.9600000	-2.2600000	-2.2870147	10.5731898	9.3660890	7.8078548	2.9600000
.75	0	1.7500000	0.0000000	-0.2500000	5.2749172	3.7500000	3.0000000	-2.2500000	-2.2749172	10.6229833	9.3898669	7.8914284	3.0000000
.76	0	1.7600000	0.0000000	-0.2400000	5.3029089	3.7600000	3.0400000	-2.2400000	-2.2629089	10.6731893	9.4136936	7.9750679	3.0400000
.77	0	1.7700000	0.0000000	-0.2300000	5.3309893	3.7700000	3.0800000	-2.2300000	-2.2509893	10.7238094	9.4375692	8.0587724	3.0800000
.78	0	1.7800000	0.0000000	-0.2200000	5.3591578	3.7800000	3.1200000	-2.2200000	-2.2391578	10.7748424	9.4614933	8.1425409	3.1200000
.79	0	1.7900000	0.0000000	-0.2100000	5.3874138	3.7900000	3.1600000	-2.2100000	-2.2274138	10.8262868	9.4854664	8.2263727	3.1600000
.80	0	1.8000000	0.0000000	-0.2000000	5.4157568	3.8000000	3.2000000	-2.2000000	-2.2157568	10.8781438	9.5094879	8.3102667	3.2000000
.81	0	1.8100000	0.0000000	-0.1900000	5.4441861	3.8100000	3.2400000	-2.1900000	-2.2041861	10.9304123	9.5335579	8.3942222	3.2400000
.82	0	1.8200000	0.0000000	-0.1800000	5.4727013	3.8200000	3.2800000	-2.1800000	-2.1927013	10.9830915	9.5576764	8.4782382	3.2800000
.83	0	1.8300000	0.0000000	-0.1700000	5.5013018	3.8300000	3.3200000	-2.1700000	-2.1813018	11.0361806	9.5818432	8.5623140	3.3200000
.84	0	1.8400000	0.0000000	-0.1600000	5.5299870	3.8400000	3.3600000	-2.1600000	-2.1699870	11.0896786	9.6060582	8.6464488	3.3600000
.85	0	1.8500000	0.0000000	-0.1500000	5.5587562	3.8500000	3.4000000	-2.1500000	-2.1587562	11.1435844	9.6303215	8.7306416	3.4000000
.86	0	1.8600000	0.0000000	-0.1400000	5.5876090	3.8600000	3.4400000	-2.1400000	-2.1476090	11.1978966	9.6546329	8.8148918	3.4400000
.87	0	1.8700000	0.0000000	-0.1300000	5.6165448	3.8700000	3.4800000	-2.1300000	-2.1365448	11.2526145	9.6789923	8.8991986	3.4800000
.88	0	1.8800000	0.0000000	-0.1200000	5.6455630	3.8800000	3.5200000	-2.1200000	-2.1255630	11.3077359	9.7033997	8.9835611	3.5200000
.89	0	1.8900000	0.0000000	-0.1100000	5.6746630	3.8900000	3.5600000	-2.1100000	-2.1146630	11.3632595	9.7278550	9.0679787	3.5600000
.90	0	1.9000000	0.0000000	-0.1000000	5.7038442	3.9000000	3.6000000	-2.1000000	-2.1038442	11.4191835	9.7523581	9.1524505	3.6000000
.91	0	1.9100000	0.0000000	-0.0900000	5.7331061	3.9100000	3.6400000	-2.0900000	-2.0931061	11.4755063	9.7769089	9.2369759	3.6400000
.92	0	1.9200000	0.0000000	-0.0800000	5.7624482	3.9200000	3.6800000	-2.0800000	-2.0824482	11.5322258	9.8015073	9.3215542	3.6800000
.93	0	1.9300000	0.0000000	-0.0700000	5.7918697	3.9300000	3.7200000	-2.0700000	-2.0718697	11.5893400	9.8261534	9.4061846	3.7200000
.94	0	1.9400000	0.0000000	-0.0600000	5.8213703	3.9400000	3.7600000	-2.0600000	-2.0613703	11.6468468	9.8508468	9.4908664	3.7600000
.95	0	1.9500000	0.0000000	-0.0500000	5.8509492	3.9500000	3.8000000	-2.0500000	-2.0509492	11.7047439	9.8755877	9.5755990	3.8000000
.96	0	1.9600000	0.0000000	-0.0400000	5.8806060	3.9600000	3.8400000	-2.0400000	-2.0406060	11.7630291	9.9003759	9.6603816	3.8400000
.97	0	1.9700000	0.0000000	-0.0300000	5.9103400	3.9700000	3.8800000	-2.0300000	-2.0303400	11.8216998	9.9252113	9.7452137	3.8800000
.98	0	1.9800000	0.0000000	-0.0200000	5.9401507	3.9800000	3.9200000	-2.0200000	-2.0201507	11.8807536	9.9500938	9.8300945	3.9200000
.99	0	1.9900000	0.0000000	-0.0100000	5.9700375	3.9900000	3.9600000	-2.0100000	-2.0100375	11.9401879	9.9750234	9.9150235	3.9600000
1.00	0	2.0000000	0.0000000	0.0000000	6.0000000	4.0000000	4.0000000	-2.0000000	-2.0000000	12.0000000	10.0000000	10.0000000	4.0000000

J_{K_{-1},K_1} / J_τ / κ	$4_{0,4}$ 4_{-4}	$4_{1,4}$ 4_{-3}	$4_{1,3}$ 4_{-2}	$4_{2,3}$ 4_{-1}	$4_{2,2}$ 4_{0}	$4_{3,2}$ 4_{1}	$4_{3,1}$ 4_{2}	$4_{4,1}$ 4_{3}	$4_{4,0}$ 4_{4}	$3_{0,3}$ 3_{-3}	$3_{1,3}$ 3_{-2}	$3_{1,2}$ 3_{-1}
.00	-14.4222051	-14.3808315	-5.2915026	-4.3808315	0.0000000	4.3808315	5.2915026	14.3808315	14.4222051	-7.8989794	-7.7459666	-1.8989794
.01	-14.3915415	-14.3522125	-5.1918427	-4.3095726	0.1384601	4.4522125	5.3918427	14.4095726	14.4530813	-7.8735759	-7.7259925	-1.8245871
.02	-14.3610864	-14.3237147	-5.0928631	-4.2384366	0.2769117	4.5237147	5.4928631	14.4384366	14.4841746	-7.8483734	-7.7060699	-1.7504020
.03	-14.3308358	-14.2953374	-4.9945632	-4.1674245	0.4153464	4.5953374	5.5945632	14.4674245	14.5154893	-7.8233690	-7.6861990	-1.6764272
.04	-14.3007857	-14.2670798	-4.8969425	-4.0965369	0.5537557	4.6670798	5.6969425	14.4965369	14.5470299	-7.7985597	-7.6663797	-1.6026659
.05	-14.2709322	-14.2389409	-4.8000000	-4.0257749	0.6921310	4.7389409	5.8000000	14.5257749	14.5788012	-7.7739426	-7.6466121	-1.5291212
.06	-14.2412717	-14.2109200	-4.7037345	-3.9551391	0.8304638	4.8109200	5.9037345	14.5551391	14.6108070	-7.7495148	-7.6268961	-1.4557964
.07	-14.2118005	-14.1830163	-4.6081446	-3.8846304	0.9687455	4.8830163	6.0081446	14.5846304	14.6430549	-7.7252733	-7.6072317	-1.3826947
.08	-14.1825119	-14.1552289	-4.5132287	-3.8142498	1.1069289	4.9552289	6.1132287	14.6142498	14.6755451	-7.7012155	-7.5876189	-1.3098195
.09	-14.1534117	-14.1275570	-4.4189848	-3.7439980	1.2451212	5.0275570	6.2189848	14.6439980	14.7082904	-7.6773384	-7.5680578	-1.2371739
.10	-14.1244874	-14.1000000	-4.3254107	-3.6738759	1.3831759	5.1000000	6.3254107	14.6738759	14.7412896	-7.6536394	-7.5485482	-1.1647615
.11	-14.0957387	-14.0725568	-4.2325041	-3.6038844	1.5211886	5.1725568	6.4325041	14.7038844	14.7745501	-7.6301158	-7.5290902	-1.0925854
.12	-14.0671626	-14.0452267	-4.1402622	-3.5340243	1.6590846	5.2452267	6.5402621	14.7340243	14.8080780	-7.6067648	-7.5096838	-1.0206499
.13	-14.0387558	-14.0180091	-4.0486820	-3.4642964	1.7968769	5.3180091	6.6486820	14.7642964	14.8418788	-7.5835838	-7.4903290	-0.9489564
.14	-14.0105153	-13.9909029	-3.9577607	-3.3947017	1.9345566	5.3909029	6.7577607	14.7947017	14.8759587	-7.5605703	-7.4710257	-0.8775107
.15	-13.9824383	-13.9639075	-3.8674947	-3.3252410	2.0721145	5.4639075	6.8674947	14.8252410	14.9103238	-7.5377216	-7.4517739	-0.8063142
.16	-13.9545219	-13.9370221	-3.7778806	-3.2559151	2.2095414	5.5370221	6.9778806	14.8559151	14.9449805	-7.5150352	-7.4325737	-0.7353720
.17	-13.9267632	-13.9102459	-3.6889255	-3.1867250	2.3468280	5.6102459	7.0889255	14.8867250	14.9799258	-7.4925086	-7.4134250	-0.6646687
.18	-13.8991598	-13.8835780	-3.6005925	-3.1176715	2.4839650	5.6835780	7.2005925	14.9176715	15.0151942	-7.4701394	-7.3943278	-0.5942250
.19	-13.8717086	-13.8570178	-3.5129104	-3.0487554	2.6209426	5.7570178	7.3129104	14.9487554	15.0507660	-7.4479255	-7.3752820	-0.5241032
.20	-13.8444074	-13.8305645	-3.4258639	-2.9799777	2.7577514	5.8305645	7.4258639	14.9799777	15.0866559	-7.4258639	-7.3562877	-0.4542114
.21	-13.8172535	-13.8042173	-3.3394485	-2.9113391	2.8943815	5.9042173	7.5394485	15.0113391	15.1228720	-7.4039526	-7.3373449	-0.3845513
.22	-13.7902246	-13.7779755	-3.2536593	-2.8428407	3.0308229	5.9779755	7.6536593	15.0428407	15.1594216	-7.3821893	-7.3184534	-0.3152463
.23	-13.7633781	-13.7518382	-3.1684915	-2.7744831	3.1670656	6.0518382	7.7684915	15.0744831	15.1963125	-7.3605716	-7.2996133	-0.2461801
.24	-13.7366519	-13.7258047	-3.0839401	-2.7062674	3.3030993	6.1258047	7.8839401	15.1062674	15.2335526	-7.3390973	-7.2808246	-0.1773964
.25	-13.7100643	-13.6998743	-3.0000000	-2.6381944	3.4389136	6.1998743	8.0000000	15.1381944	15.2711500	-7.3177643	-7.2620873	-0.1088981
.26	-13.6836312	-13.6740463	-2.9166656	-2.5702649	3.5744981	6.2740463	8.1166656	15.1702649	15.3091048	-7.2965703	-7.2434013	-0.0406981
.27	-13.6572924	-13.6483198	-2.8339316	-2.5024798	3.7098419	6.3483198	8.2339316	15.2024798	15.3474435	-7.2755139	-7.2247665	0.0272231
.28	-13.6311050	-13.6226942	-2.7517925	-2.4348401	3.8449340	6.4226942	8.3517925	15.2348401	15.3861682	-7.2545910	-7.2061831	0.0948403
.29	-13.6050471	-13.5971687	-2.6702423	-2.3673465	3.9797636	6.4971687	8.4702423	15.2673465	15.4252740	-7.2338015	-7.1876508	0.1621567
.30	-13.5791167	-13.5717426	-2.5892754	-2.3000000	4.1143191	6.5717426	8.5892754	15.3000000	15.4664976	-7.2131427	-7.1691698	0.2291686
.31	-13.5533119	-13.5464152	-2.5088858	-2.2328013	4.2485892	6.6464152	8.7088858	15.3328013	15.5047226	-7.1926126	-7.1507399	0.2958725
.32	-13.5276307	-13.5211857	-2.4290674	-2.1657515	4.3825621	6.7211857	8.8290674	15.3657515	15.5450685	-7.1722092	-7.1323612	0.3622647
.33	-13.5020713	-13.4960534	-2.3498141	-2.0988513	4.5162260	6.7960534	8.9498141	15.3988513	15.5858452	-7.1519307	-7.1140337	0.4283415
.34	-13.4766318	-13.4710176	-2.2711198	-2.0321016	4.6495687	6.8710176	9.0711198	15.4321016	15.6270631	-7.1317750	-7.0957571	0.4940993
.35	-13.4513107	-13.4460776	-2.1929781	-1.9655034	4.7825778	6.9460776	9.1929781	15.4655034	15.6687328	-7.1117403	-7.0775317	0.5595346
.36	-13.4261060	-13.4212328	-2.1153827	-1.8990574	4.9152409	7.0212328	9.3153827	15.4990574	15.7108650	-7.0918248	-7.0593572	0.6246437
.37	-13.4010162	-13.3964823	-2.0383272	-1.8327645	5.0475452	7.0964823	9.4383272	15.5327645	15.7534710	-7.0720266	-7.0412338	0.6894231
.38	-13.3760396	-13.3718255	-1.9618052	-1.7666256	5.1794775	7.1718255	9.5618052	15.5666256	15.7965621	-7.0523440	-7.0231613	0.7538692
.39	-13.3511747	-13.3472617	-1.8858102	-1.7006415	5.3110247	7.2472617	9.6858102	15.6006415	15.8401499	-7.0327752	-7.0051396	0.8179787
.40	-13.3264197	-13.3227903	-1.8103356	-1.6348132	5.4421733	7.3227903	9.8103356	15.6348132	15.8842464	-7.0133185	-6.9871689	0.8817479
.41	-13.3017734	-13.2984105	-1.7353748	-1.5691441	5.5729096	7.3984105	9.9353748	15.6691441	15.9288638	-6.9939722	-6.9692490	0.9451735
.42	-13.2772341	-13.2741216	-1.6609214	-1.5036271	5.7032195	7.4741216	10.0609214	15.7036271	15.9740145	-6.9747345	-6.9513798	1.0082522
.43	-13.2528003	-13.2499230	-1.5869686	-1.4382710	5.8330889	7.5499230	10.1869686	15.7382710	16.0197114	-6.9556039	-6.9335614	1.0709905
.44	-13.2284708	-13.2258141	-1.5135099	-1.3730741	5.9625033	7.6258141	10.3135099	15.7730741	16.0659635	-6.9365788	-6.9157937	1.1333551
.45	-13.2042441	-13.2017941	-1.4405386	-1.3080371	6.0914480	7.7017941	10.4405386	15.8080371	16.1127960	-6.9176575	-6.8980766	1.1953720
.46	-13.1801188	-13.1778623	-1.3680482	-1.2431610	6.2199081	7.7778623	10.5680482	15.8431610	16.1602106	-6.8988386	-6.8804102	1.2570307
.47	-13.1560937	-13.1540182	-1.2960320	-1.1784465	6.3478685	7.8540182	10.6960320	15.8784465	16.2082251	-6.8801203	-6.8627943	1.3183254
.48	-13.1321674	-13.1302611	-1.2244833	-1.1138945	6.4753137	7.9302611	10.8244833	15.9138945	16.2568536	-6.8615014	-6.8452290	1.3792538
.49	-13.1083387	-13.1065903	-1.1533957	-1.0495058	6.6022282	8.0065903	10.9533957	15.9495058	16.3061104	-6.8429802	-6.8277141	1.4398122
.50	-13.0846063	-13.0830052	-1.0827625	-0.9852813	6.7285960	8.0830052	11.0827625	15.9852813	16.3560102	-6.8245553	-6.8102496	1.5000000

$J_{K_{-1}K_1}$ / J_τ — κ	$4_{0,4}$ / 4_{-4}	$4_{1,4}$ / 4_{-3}	$4_{1,3}$ / 4_{-2}	$4_{2,3}$ / 4_{-1}	$4_{2,2}$ / 4_0	$4_{3,2}$ / 4_1	$4_{3,1}$ / 4_2	$4_{4,1}$ / 4_3	$4_{4,0}$ / 4_4	$3_{0,3}$ / 3_{-3}	$3_{1,3}$ / 3_{-2}	$3_{1,2}$ / 3_{-1}
.50	-13.0846063	-13.0830052	-1.0827625	-0.9852813	6.7285960	8.0830052	11.0827625	15.9852813	16.3560102	-6.8245553	-6.8102496	1.5000000
.51	-13.0609691	-13.0595051	-1.0125771	-0.9212218	6.8540012	8.1595051	11.2125771	16.0212218	16.4066678	-6.8092292	-6.7864819	1.6192443
.52	-13.0374258	-13.0360895	-0.9418332	-0.8573281	6.9796274	8.2360895	11.3418332	16.0573281	16.4575180	-6.7940081	-6.7754013	1.6380373
.53	-13.0139754	-13.0127576	-0.8735241	-0.7936015	7.1042583	8.3127576	11.4735241	16.0936015	16.5097180	-6.7698442	-6.7581583	1.6783001
.54	-12.9906166	-12.9895088	-0.8046434	-0.7300415	7.2282772	8.3895088	11.6046434	16.1300415	16.5623394	-6.7517904	-6.7408950	1.7369708
.55	-12.9673484	-12.9663426	-0.7361847	-0.6666501	7.3516672	8.4663426	11.7361847	16.1666501	16.6165811	-6.7338260	-6.7236819	1.7952560
.56	-12.9441698	-12.9432583	-0.6681416	-0.6034278	7.4744116	8.5432583	11.8681416	16.2034278	16.6697581	-6.7159496	-6.7065190	1.8531531
.57	-12.9210796	-12.9202553	-0.6005079	-0.5403754	7.5964931	8.6202553	12.0005079	16.2403754	16.7245864	-6.6981600	-6.6894061	1.9106599
.58	-12.8980768	-12.8973330	-0.5332771	-0.4774936	7.7178946	8.6973330	12.1332771	16.2774936	16.7801821	-6.6804559	-6.6723431	1.9677741
.59	-12.8751605	-12.8744907	-0.4664432	-0.4147833	7.8385988	8.7744907	12.2664432	16.3147833	16.8365616	-6.6628360	-6.6553302	2.0244936
.60	-12.8523296	-12.8517279	-0.4000000	-0.3522452	7.9585883	8.8517279	12.4000000	16.3522452	16.8937412	-6.6452990	-6.6383671	2.0808164
.61	-12.8295832	-12.8290440	-0.3339412	-0.2898800	8.0765710	8.9290440	12.5339412	16.3898800	16.9517374	-6.6278439	-6.6214539	2.1367404
.62	-12.8069203	-12.8064384	-0.2682887	-0.2276616	8.1943536	9.0064384	12.6682887	16.4276616	17.0105666	-6.6104694	-6.6045904	2.1922649
.63	-12.7843400	-12.7839105	-0.2031543	-0.1655619	8.3140944	9.0839105	12.8031543	16.4655619	17.0702455	-6.5931743	-6.5877767	2.2473854
.64	-12.7618414	-12.7614597	-0.1380119	-0.1036304	8.4310509	9.1614597	12.9380119	16.5036304	17.1307904	-6.5759574	-6.5710126	2.3021031
.65	-12.7394236	-12.7390854	-0.0734313	-0.0421649	8.5472056	9.2390854	13.0734313	16.5421649	17.1922180	-6.5588178	-6.5542981	2.3564155
.66	-12.7170858	-12.7167871	-0.0092057	0.0193236	8.6625413	9.3167871	13.2092236	16.5806763	17.2545444	-6.5417541	-6.5376533	2.4103221
.67	-12.6948271	-12.6945642	0.0546708	0.0806348	8.7770410	9.3945642	13.3453291	16.6194490	17.3177806	-6.5247654	-6.5210177	2.4638193
.68	-12.6726466	-12.6724161	0.1182038	0.1417677	8.8906876	9.4724161	13.4817961	16.6582566	17.3819590	-6.5078500	-6.5044349	2.5169084
.69	-12.6505436	-12.6503423	0.1813989	0.2027216	9.0034644	9.5503423	13.6186010	16.6972181	17.4470791	-6.4910085	-6.4879349	2.5695876
.70	-12.6285173	-12.6283421	0.2442616	0.2634958	9.1153551	9.6283421	13.7557383	16.7365041	17.5131622	-6.4742393	-6.4714674	2.6218561
.71	-12.6065668	-12.6064151	0.3067975	0.3240824	9.2263551	9.7064151	13.8932024	16.7759102	17.5802235	-6.4575387	-6.4550492	2.6737131
.72	-12.5846905	-12.5845604	0.3690098	0.3844753	9.3364132	9.7845604	14.0309882	16.8154975	17.6482782	-6.4409090	-6.4386800	2.7251580
.73	-12.5628905	-12.5627784	0.4309836	0.4447353	9.4455494	9.8627784	14.1690901	16.8552566	17.7173411	-6.4243480	-6.4223600	2.7761905
.74	-12.5411632	-12.5410675	0.4924968	0.5047818	9.5537817	9.9410675	14.3075031	16.8952181	17.7874264	-6.4078548	-6.4060890	2.8268101
.75	-12.5195087	-12.5194276	0.5537780	0.5646472	9.6609607	10.0194276	14.4462219	16.9353527	17.8585480	-6.3914284	-6.3898669	2.8770166
.76	-12.4979265	-12.4978582	0.6147584	0.6243287	9.7672015	10.0978582	14.5852415	16.9756712	17.9307193	-6.3750579	-6.3736936	2.9268101
.77	-12.4764158	-12.4763586	0.6754490	0.6838258	9.8724626	10.1763586	14.7245569	17.0161741	18.0039531	-6.3587724	-6.3575692	2.9761905
.78	-12.4549759	-12.4549284	0.7358369	0.7431378	9.9767140	10.2549284	14.8641630	17.0568621	18.0782618	-6.3425409	-6.3414935	3.0251580
.79	-12.4336062	-12.4335671	0.7959448	0.8022641	10.0799492	10.3335671	15.0040551	17.0977358	18.1536569	-6.3263727	-6.3254664	3.0737131
.80	-12.4123060	-12.4122740	0.8557715	0.8612040	10.1821566	10.4122740	15.1442284	17.1380759	18.2301493	-6.3102667	-6.3094879	3.1218561
.81	-12.3910746	-12.3910489	0.9153218	0.9199569	10.2833252	10.4910489	15.2847110	17.1804050	18.3077494	-6.2942222	-6.2935575	3.1695876
.82	-12.3699116	-12.3698910	0.9746003	0.9785063	10.3834910	10.5698910	15.4253996	17.2214776	18.3864667	-6.2782382	-6.2776764	3.2169084
.83	-12.3488162	-12.3487991	1.0336601	1.0368994	10.4825063	10.6487999	15.5663883	17.2631105	18.4663098	-6.2623140	-6.2618432	3.2638193
.84	-12.3277878	-12.3277751	1.0923601	1.0950879	10.5805011	10.7277751	15.7076398	17.3049120	18.5472866	-6.2464488	-6.2460582	3.3103221
.85	-12.3068258	-12.3068161	1.1508503	1.1530870	10.6774215	10.8068161	15.8491496	17.3469129	18.6294043	-6.2306416	-6.2303215	3.3564155
.86	-12.2859298	-12.2859224	1.2090866	1.2108962	10.7732608	10.8859224	15.9909133	17.3891007	18.7126689	-6.2148918	-6.2146318	3.4021031
.87	-12.2650990	-12.2650936	1.2670732	1.2685149	10.8680132	10.9650936	16.1329267	17.4314850	18.7970858	-6.1991986	-6.1989907	3.4473854
.88	-12.2443329	-12.2443290	1.3248143	1.3259427	10.9616738	11.0443290	16.2751856	17.4740572	18.8826537	-6.1835594	-6.1833967	3.4922649
.89	-12.2236311	-12.2236283	1.3823141	1.3831790	11.0542386	11.1236283	16.4176858	17.5168209	18.9693924	-6.1679787	-6.1678550	3.5367404
.90	-12.2029929	-12.2029808	1.4395767	1.4402233	11.1457049	11.2029800	16.5604222	17.5597766	19.0572879	-6.1524505	-6.1523581	3.5808164
.91	-12.1824178	-12.1824166	1.4966060	1.4970750	11.2360705	11.2824166	16.7033939	17.6029249	19.1463472	-6.1369759	-6.1369089	3.6244936
.92	-12.1619053	-12.1619046	1.5534059	1.5537337	11.3253345	11.3619046	16.8465940	17.6462662	19.2365707	-6.1215542	-6.1215047	3.6677741
.93	-12.1414549	-12.1414545	1.6099804	1.6101990	11.4134970	11.4414545	16.9900195	17.6898009	19.3279578	-6.1061846	-6.1061534	3.7106599
.94	-12.1210661	-12.1210472	1.6663333	1.6664702	11.5005590	11.5210472	17.1336666	17.7335207	19.4205070	-6.0908664	-6.0908468	3.7531531
.95	-12.1007384	-12.1007383	1.7224682	1.7225471	11.5865225	11.6003783	17.2775317	17.7774528	19.5142159	-6.0755990	-6.0755877	3.7952560
.96	-12.0804714	-12.0804713	1.7783389	1.7784290	11.6713905	11.6804713	17.4216110	17.8215809	19.6090976	-6.0603759	-6.0603759	3.8369708
.97	-12.0602644	-12.0602644	1.8340989	1.8341157	11.7551668	11.7603982	17.5659982	17.8659017	19.7050992	-6.0452133	-6.0452113	3.8783001
.98	-12.0401172	-12.0401172	1.8896617	1.8896017	11.8378668	11.8409667	17.7105980	17.9103983	19.8022660	-6.0300943	-6.0300943	3.9192443
.99	-12.0200293	-12.0200293	1.9449667	1.9449635	11.9194653	11.9200292	17.8553989	17.9550983	19.9005638	-6.0150235	-6.0150234	3.9598120
1.00	-12.0000000	-12.0000000	2.0000000	2.0000000	12.0000000	12.0000000	18.0000000	18.0000000	20.0000000	-6.0000000	-6.0000000	4.0000000

$6_{6,0}$ / 6_6	$5_{0,5}$ / 5_{-5}	$5_{1,5}$ / 5_{-4}	$5_{1,4}$ / 5_{-3}	$5_{2,4}$ / 5_{-2}	$5_{2,3}$ / 5_{-1}	$5_{3,3}$ / 5_0	$5_{3,2}$ / 5_1	$5_{4,2}$ / 5_2	$5_{4,1}$ / 5_3	$5_{5,1}$ / 5_4	$5_{5,0}$ / 5_5	J_{K_{-1},K_1} / J_τ ; κ
33.567821	-22.9829232	-22.9785705	-10.7258610	-10.3230480	-2.7375677	0.0000000	2.7375677	10.3230480	10.7258610	22.9785705	22.9829232	.00
33.615419	-22.9475152	-22.9425832	-10.6106840	-10.2377652	-2.5446410	0.1272020	2.9345261	10.4377652	10.8420803	23.0147054	23.0254195	.01
33.651583	-22.9121072	-22.9065959	-10.4963070	-10.1524824	-2.3446439	0.2545399	3.1335397	10.5524824	10.9591976	23.0513443	23.0627677	.02
33.700283	-22.8766992	-22.8706086	-10.3828933	-10.0671996	-2.1490124	0.3817995	3.3325482	10.6671996	11.0772326	23.0881691	23.1003451	.03
33.744939	-22.8412912	-22.8346213	-10.2703675	-9.9819168	-1.9543449	0.5090468	3.5322123	10.7819168	11.1962954	23.1251822	23.1381552	.04
33.789953	-22.8058852	-22.7986340	-10.1587149	-9.8966340	-1.7604680	0.6362775	3.7325125	10.8966340	11.3163532	23.1623856	23.1762024	.05
33.835030	-22.7700208	-22.7632691	-10.0479209	-9.7985383	-1.5674004	0.7634876	3.9334293	10.9985383	11.4374212	23.1997815	23.2144915	.06
33.880475	-22.7343539	-22.7280450	-9.9379705	-9.7007884	-1.3751609	0.8906728	4.1349430	11.1007884	11.5595148	23.2373722	23.2530274	.07
33.926193	-22.6988811	-22.6929890	-9.8288493	-9.6033840	-1.1837679	1.0178291	4.3370342	11.2033840	11.6826490	23.2751599	23.2918151	.08
33.972188	-22.6635991	-22.6580992	-9.7205429	-9.5063249	-0.9932397	1.1449523	4.5396830	11.3063249	11.8068389	23.3131469	23.3308549	.09
34.018465	-22.6285049	-22.6233738	-9.6130369	-9.4096109	-0.8035945	1.2720382	4.7428699	11.4096109	11.9320994	23.3513355	23.3701669	.10
34.065031	-22.5935954	-22.5888109	-9.5063108	-9.3132415	-0.6148500	1.3990827	4.9465752	11.5132415	12.0584354	23.3897282	23.4097420	.11
34.111891	-22.5588675	-22.5544088	-9.4003702	-9.2172165	-0.4270237	1.5260817	5.1507793	11.6172165	12.1858913	23.4283271	23.4495908	.12
34.159051	-22.5243184	-22.5201658	-9.2951818	-9.1215355	-0.2401330	1.6530108	5.3554624	11.7215355	12.3144515	23.4671349	23.4897193	.13
34.206516	-22.4899452	-22.4860800	-9.1907387	-9.0261977	-0.0541948	1.7799261	5.5606049	11.8261977	12.4441401	23.5061538	23.5301337	.14
34.254294	-22.4557454	-22.4521497	-9.0870274	-8.9312031	0.1307744	1.9067632	5.7661870	11.9312031	12.5749710	23.5453864	23.5708404	.15
34.302389	-22.4217161	-22.4183733	-8.9840350	-8.8365511	0.3147584	2.0335580	5.9721890	12.0365511	12.7069576	23.5848353	23.6118459	.16
34.350810	-22.3878548	-22.3847492	-8.8817485	-8.7422411	0.4977415	2.1602463	6.1785912	12.1422411	12.8401132	23.6245028	23.6531573	.17
34.399562	-22.3541590	-22.3512756	-8.7801552	-8.6482725	0.6797083	2.2868838	6.3853776	12.2482725	12.9744506	23.6643917	23.6947816	.18
34.448654	-22.3206262	-22.3179510	-8.6792427	-8.5546449	0.8606439	2.4134464	6.5925165	12.3546449	13.1099823	23.7045045	23.7367261	.19
34.498092	-22.2872541	-22.2847737	-8.5789907	-8.4613574	1.0405340	2.5399298	6.8000000	12.4613574	13.2467200	23.7448439	23.7789997	.20
34.547884	-22.2540403	-22.2517422	-8.4794111	-8.3684096	1.2193647	2.6663297	7.0078040	12.5684096	13.3846755	23.7854125	23.8216071	.21
34.598039	-22.2209825	-22.2188550	-8.3804683	-8.2758006	1.3971228	2.7926419	7.2159085	12.6758006	13.5238597	23.8262213	23.8645597	.22
34.648565	-22.1880786	-22.1861105	-8.2821585	-8.1835299	1.5737955	2.9188621	7.4242934	12.7835299	13.6642831	23.8672483	23.9078650	.23
34.699470	-22.1553264	-22.1535071	-8.1844706	-8.0915966	1.7493708	3.0449860	7.6329385	12.8915966	13.8059556	23.9085210	23.9515321	.24
34.750764	-22.1227238	-22.1210435	-8.0873932	-8.0000000	1.9238372	3.1710094	7.8418233	13.0000000	13.9488866	23.9500340	23.9955699	.25
34.802456	-22.0902688	-22.0887181	-7.9909156	-7.9087392	2.0971840	3.2969279	8.0509274	13.1087392	14.0930848	23.9917902	24.0398882	.26
34.854555	-22.0579594	-22.0565295	-7.8950271	-7.8178134	2.2694010	3.4227371	8.2602302	13.2178134	14.2385583	24.0337969	24.0847969	.27
34.907073	-22.0257936	-22.0244763	-7.7997172	-7.7272218	2.4404789	3.5484328	8.4697108	13.3272218	14.3853146	24.0760434	24.1300063	.28
34.960019	-21.9937695	-21.9925570	-7.7049757	-7.6369635	2.6104090	3.6740106	8.6793443	13.4369635	14.5333664	24.1185463	24.1756204	.29
35.013406	-21.9618853	-21.9607703	-7.6107925	-7.5470374	2.7791835	3.7994661	8.8891213	13.5470374	14.6827017	24.1613042	24.2216712	.30
35.067244	-21.9301391	-21.9291149	-7.5171580	-7.4574428	2.9467952	3.9247903	9.0990085	13.6574428	14.8333439	24.2043199	24.2681494	.31
35.121545	-21.8985293	-21.8975893	-7.4240624	-7.3681786	3.1132378	4.0499927	9.3089883	13.7681786	14.9852915	24.2475965	24.3150741	.32
35.176323	-21.8670541	-21.8661922	-7.3314965	-7.2792428	3.2785508	4.1750550	9.5190385	13.8792428	15.1385482	24.2911372	24.3624580	.33
35.231591	-21.8357117	-21.8349224	-7.2394511	-7.1906373	3.4425945	4.2999773	9.7291369	13.9906373	15.2931172	24.3349451	24.4103142	.34
35.287363	-21.8045006	-21.8037785	-7.1479173	-7.1023582	3.6055000	4.4247520	9.9392610	14.1023582	15.4490005	24.3790233	24.4586563	.35
35.343653	-21.7734192	-21.7727593	-7.0568862	-7.0144053	3.7672193	4.5493842	10.1493877	14.2144053	15.6061998	24.4233751	24.5074985	.36
35.400476	-21.7424658	-21.7418635	-6.9663494	-6.9267774	3.9277502	4.6738599	10.3594937	14.3267774	15.7647155	24.4680036	24.5568557	.37
35.457849	-21.7116390	-21.7110899	-6.8762986	-6.8394736	4.0870914	4.7981177	10.5695562	14.4394736	15.9245475	24.5129121	24.6067493	.38
35.515788	-21.6809371	-21.6804372	-6.7867255	-6.7524926	4.2452422	4.9223331	10.7795481	14.5524926	16.0856949	24.5581040	24.6571773	.39
35.574312	-21.6503589	-21.6499043	-6.6976223	-6.6658333	4.4022031	5.0463326	10.9894437	14.6658333	16.2481557	24.6035826	24.7081746	.40
35.633437	-21.6199027	-21.6194899	-6.6089812	-6.5794943	4.5579752	5.1701386	11.1992287	14.7794943	16.4119275	24.6493513	24.7597524	.41
35.693165	-21.5895673	-21.5891027	-6.5207945	-6.4934746	4.7125605	5.2937795	11.4088654	14.8934746	16.5770068	24.6954134	24.8119291	.42
35.753575	-21.5593512	-21.5590122	-6.4330540	-6.4077728	4.8659618	5.4172398	11.6183316	15.0077728	16.7433894	24.7417724	24.8647234	.43
35.814628	-21.5292531	-21.5289664	-6.3457554	-6.3223877	5.0181827	5.5405147	11.8276002	15.1223877	16.9110703	24.7884318	24.9181552	.44
35.876368	-21.4992716	-21.4989949	-6.2588886	-6.2373181	5.1692276	5.6635997	12.0366436	15.2373181	17.0800439	24.8353951	24.9722450	.45
35.938817	-21.4694055	-21.4691561	-6.1724479	-6.1525624	5.3191018	5.7864901	12.2454337	15.3525624	17.2503037	24.8826659	25.0270141	.46
36.002002	-21.4396535	-21.4394291	-6.0864264	-6.0681196	5.4678110	5.9091813	12.4539413	15.4681196	17.4218425	24.9302247	25.0824809	.47
36.065948	-21.4100144	-21.4098128	-6.0008177	-5.9839881	5.6155621	6.0316685	12.6623368	15.5839881	17.5946334	24.9780868	25.1386583	.48
36.130684	-21.3804869	-21.3803061	-5.9156154	-5.9001656	5.7621960	6.1539386	12.8705551	15.7001656	17.7687245	25.0263589	25.1956258	.49
36.196236	-21.3510698	-21.3509080	-5.8308132	-5.8166538	5.9070198	6.2760123	13.0774679	15.8166538	17.9440500	25.0748956	25.2533452	.50

κ (J_{K_{-1},K_1} / J_r)	$6_{6,0}$ / 6_6	$5_{0,5}$ / 5_{-5}	$5_{1,5}$ / 5_{-4}	$5_{1,4}$ / 5_{-3}	$5_{2,4}$ / 5_{-2}	$5_{2,3}$ / 5_{-1}	$5_{3,3}$ / 5_0	$5_{3,2}$ / 5_1	$5_{4,2}$ / 5_2	$5_{4,1}$ / 5_3	$5_{5,1}$ / 5_4	$5_{5,0}$ / 5_5
.50	36.196238	-21.3510698	-21.3509080	-5.8308132	-5.8166538	5.9070198	6.2760123	13.0774679	15.8166538	17.9440500	25.0748956	25.2533452
.51	36.262641	-21.3217620	-21.3216175	-5.7464052	-5.7334482	6.0511432	6.3978594	13.2845399	15.9334482	18.1206188	25.1237580	25.3118653
.52	36.329927	-21.2925624	-21.2924335	-5.6623854	-5.6505483	6.1941419	6.5194837	13.4911720	16.0505483	18.2984204	25.1729498	25.3712134
.53	36.398130	-21.2634697	-21.2633551	-5.5787480	-5.5679528	6.3360259	6.6408803	13.6973300	16.1679528	18.4774437	25.2224748	25.4314179
.54	36.467286	-21.2344829	-21.2343682	-5.4954874	-5.4856602	6.4768057	6.7620444	13.9029787	16.2856602	18.6576772	25.2723368	25.4925086
.55	36.537434	-21.2056010	-21.2055109	-5.4125982	-5.4036691	6.6164924	6.8829713	14.1080819	16.4036691	18.8391006	25.3225395	25.5545163
.56	36.608614	-21.1768273	-21.1767372	-5.3300750	-5.3219778	6.7550975	7.0036562	14.3126221	16.5219778	19.0217255	25.3730870	25.6174729
.57	36.680870	-21.1481473	-21.1480772	-5.2479126	-5.2405849	6.8926331	7.1240941	14.5165007	16.6405849	19.2055141	25.4239830	25.6814119
.58	36.754248	-21.1195734	-21.1195105	-5.1661060	-5.1594890	7.0291118	7.2442203	14.7197382	16.7594890	19.3904616	25.4752310	25.7463618
.59	36.828795	-21.0911003	-21.0910463	-5.0846501	-5.0786884	7.1645465	7.3642099	14.9222736	16.8786884	19.5765538	25.5268363	25.8123765
.60	36.904563	-21.0627268	-21.0626797	-5.0035402	-4.9981816	7.2989505	7.4838781	15.1240649	16.9981816	19.7637763	25.5788015	25.8794753
.61	36.981607	-21.0344521	-21.0344110	-4.9227716	-4.9179671	7.4323374	7.6032798	15.3250687	17.1179671	19.9522147	25.6311311	25.9477029
.62	37.059982	-21.0062752	-21.0062395	-4.8423397	-4.8380433	7.5647212	7.7224103	15.5252404	17.2380433	20.1415539	25.6838291	26.0170057
.63	37.139751	-20.9781950	-20.9781641	-4.7622408	-4.7584085	7.6961162	7.8412647	15.7245342	17.3584085	20.3320787	25.7368994	26.0873957
.64	37.220978	-20.9502108	-20.9501842	-4.6824681	-4.6790613	7.8265371	7.9598379	15.9229029	17.4790613	20.5236736	25.7903462	26.1595652
.65	37.303729	-20.9223218	-20.9222987	-4.6030200	-4.6000000	7.9559984	8.0781252	16.1202982	17.6000000	20.7163231	25.8441735	26.2327217
.66	37.388078	-20.8945245	-20.8945069	-4.5238913	-4.5212229	8.0845152	8.1961215	16.3165703	17.7212229	20.9100113	25.8983854	26.3072209
.67	37.474101	-20.8668247	-20.8668102	-4.4450781	-4.4427285	8.2121025	8.3138219	16.5119683	17.8427285	21.1047221	25.9529860	26.3831097
.68	37.561878	-20.8392152	-20.8392011	-4.3665764	-4.3645152	8.3387756	8.4312216	16.7061490	17.9645152	21.3004396	26.0079795	26.4604363
.69	37.651494	-20.8116973	-20.8116854	-4.2883825	-4.2865812	8.4645498	8.5483155	16.8991323	18.0865812	21.4971474	26.0633695	26.5392502
.70	37.743039	-20.7842601	-20.7842601	-4.2104925	-4.2089250	8.5894407	8.6650986	17.0908906	18.2089250	21.6948293	26.1191614	26.6196019
.71	37.836608	-20.7569244	-20.7569224	-4.1329030	-4.1315448	8.7134637	8.7815662	17.2813594	18.3315448	21.8934690	26.1753582	26.7015435
.72	37.932300	-20.7296846	-20.7296777	-4.0556102	-4.0544391	8.8366344	8.8977132	17.4704826	18.4544391	22.0930501	26.2319645	26.7851276
.73	38.030221	-20.7025247	-20.7025190	-3.9786108	-3.9776061	8.9589683	9.0135346	17.6582027	18.5776061	22.2935563	26.2889843	26.8704080
.74	38.130481	-20.6754542	-20.6754523	-3.9019014	-3.9010441	9.0804810	9.1290256	17.8444050	18.7010441	22.4949712	26.3464220	26.9574393
.75	38.233196	-20.6484667	-20.6484629	-3.8254787	-3.8247516	9.2011880	9.2441812	18.0292020	18.8247516	22.6972787	26.4042817	27.0462767
.76	38.338486	-20.6215640	-20.6215577	-3.7493196	-3.7487526	9.3211047	9.3589964	18.2123637	18.9487268	22.9004623	26.4625675	27.1169758
.77	38.446478	-20.5947571	-20.5947677	-3.6734808	-3.6729679	9.4402466	9.4734465	18.3938882	19.0729679	23.1045061	26.5212837	27.2295926
.78	38.557304	-20.5680209	-20.5680209	-3.5979494	-3.5974734	9.5586289	9.5875864	18.5737164	19.1974734	23.3093939	26.5804345	27.3241830
.79	38.671100	-20.5413768	-20.5413753	-3.5225924	-3.5222414	9.6762670	9.7013513	18.7517895	19.3222414	23.5151098	26.6400240	27.4208028
.80	38.788007	-20.5148139	-20.5148127	-3.4475569	-3.4472704	9.7931758	9.8147563	18.9280493	19.4472704	23.7216380	26.7000564	27.5195076
.81	38.908170	-20.4883334	-20.4883334	-3.3727901	-3.3725585	9.9093704	9.9277965	19.1024382	19.5725585	23.9289629	26.7605359	27.6203519
.82	39.031738	-20.4619346	-20.4619339	-3.2982893	-3.2981041	10.0248656	10.0404672	19.2748998	19.6981041	24.1370689	26.8214667	27.7036504
.83	39.158863	-20.4356168	-20.4356163	-3.2240517	-3.2239055	10.1396760	10.1527634	19.4453290	19.8239055	24.3459277	26.8828393	27.8236374
.84	39.289698	-20.4093794	-20.4093794	-3.1500748	-3.1499610	10.2538162	10.2646805	19.6138224	19.9499610	24.5555631	26.9446985	27.9362523
.85	39.424399	-20.3832214	-20.3832217	-3.0763560	-3.0762504	10.3673006	10.3762136	19.7801787	20.0762504	24.7659211	27.0070078	28.0461772
.86	39.563119	-20.3571429	-20.3571437	-3.0028929	-3.0028272	10.4801432	10.4873580	19.9443989	20.2028272	24.9769998	27.0697848	28.1584940
.87	39.706012	-20.3311429	-20.3311421	-2.9296345	-2.9296345	10.5923580	10.5981089	20.1064365	20.3296345	25.1887848	27.1330337	28.2732464
.88	39.853228	-20.3052204	-20.3052211	-2.8567239	-2.8566890	10.7039588	10.7084618	20.2662484	20.4566890	25.4012615	27.1967584	28.3904755
.89	40.004911	-20.2793751	-20.2793750	-2.7840135	-2.7839890	10.8149592	10.8184100	20.4237945	20.5839890	25.6144158	27.2609630	28.5102189
.90	40.161202	-20.2536064	-20.2536063	-2.7115493	-2.7115327	10.9253726	10.9279548	20.5790387	20.7115327	25.8282337	27.3254515	28.6325106
.91	40.322230	-20.2279136	-20.2279135	-2.6393393	-2.6393185	11.0352490	11.0370985	20.7319485	20.8393185	26.0427015	27.3908278	28.7573807
.92	40.488111	-20.2022961	-20.2022935	-2.5673185	-2.5673185	11.1452726	11.1458002	20.8824959	20.9673446	26.2578057	27.4564958	28.8848553
.93	40.659539	-20.1767534	-20.1767534	-2.4955132	-2.4956112	11.2533205	11.2540939	21.0306573	21.0956009	26.4735329	27.5226595	29.0149559
.94	40.834889	-20.1512849	-20.1512849	-2.4241130	-2.4241109	11.3614148	11.3619622	21.1764135	21.2241109	26.6898700	27.5893227	29.1476994
.95	41.015949	-20.1258900	-20.1258900	-2.3528487	-2.3528477	11.4690056	11.4694009	21.3197506	21.3528477	26.9068043	27.6564891	29.2830981
.96	41.202217	-20.1005681	-20.1005681	-2.2818184	-2.2818184	11.5762649	11.5764056	21.4606592	21.4818180	27.1243232	27.7241625	29.4211592
.97	41.393738	-20.0753188	-20.0753188	-2.2110202	-2.2110201	11.6829045	11.6832045	21.5991351	21.6110201	27.3424142	27.7923467	29.5618851
.98	41.590539	-20.0501413	-20.0501413	-2.1404522	-2.1404522	11.7890959	11.7897760	21.7351791	21.7404522	27.5610627	27.8610455	29.7051135
.99	41.792630	-20.0250352	-20.0250352	-2.0701127	-2.0701127	11.8947733	11.8948733	21.8687971	21.8701127	27.7802643	27.9302643	29.8503653
1.00	42.0000000	-20.0000000	-20.0000000	-2.0000000	-2.0000000	12.0000000	12.0000000	22.0000000	22.0000000	28.0000000	28.0000000	30.0000000

κ / $J_{K_{-1}K_1}$ / J_τ	$6_{0,6}$ 6_{-6}	$6_{1,6}$ 6_{-5}	$6_{1,5}$ 6_{-4}	$6_{2,5}$ 6_{-3}	$6_{2,4}$ 6_{-2}	$6_{3,4}$ 6_{-1}	$6_{3,3}$ 6_0	$6_{4,3}$ 6_1	$6_{4,2}$ 6_2	$6_{5,2}$ 6_3	$6_{5,1}$ 6_4	$6_{6,1}$ 6_5
.00	-33.567821	-33.565539	-18.330302	-18.228650	-7.014366	-5.663111	0.000000	5.663111	7.014366	18.228650	18.330302	33.565539
.01	-33.524155	-33.522047	-18.199334	-18.103844	-6.779050	-5.481991	0.297136	5.844584	7.251477	18.354039	18.462197	33.609259
.02	-33.480727	-33.478780	-18.069272	-17.979616	-6.545531	-5.301233	0.594232	6.026405	7.490377	18.480014	18.595039	33.653211
.03	-33.437531	-33.435735	-17.940098	-17.855963	-6.313812	-5.120344	0.891250	6.208566	7.731060	18.606580	18.728847	33.697396
.04	-33.394566	-33.392909	-17.811793	-17.732881	-6.083893	-4.940828	1.188148	6.391061	7.973520	18.733738	18.863644	33.741820
.05	-33.351827	-33.350300	-17.684341	-17.610366	-5.855772	-4.761195	1.484888	6.573881	8.217745	18.861495	18.999452	33.786485
.06	-33.309310	-33.307903	-17.557722	-17.488414	-5.629442	-4.583097	1.781438	6.757030	8.463746	18.989852	19.136292	33.831394
.07	-33.267012	-33.265737	-17.431917	-17.367677	-5.404912	-4.404814	2.077710	6.940070	8.711449	19.118804	19.274188	33.876551
.08	-33.224930	-33.223739	-17.306922	-17.246184	-5.182164	-4.224646	2.373758	7.124224	8.960902	19.248385	19.413163	33.921959
.09	-33.183061	-33.181965	-17.182707	-17.125897	-4.961194	-4.046601	2.669464	7.308274	9.212068	19.378567	19.553243	33.967623
.10	-33.141401	-33.140394	-17.059263	-17.006158	-4.741995	-3.868306	2.964811	7.492612	9.464931	19.509364	19.694451	34.013546
.11	-33.099948	-33.099023	-16.936573	-16.886962	-4.524557	-3.691757	3.259758	7.677231	9.719474	19.640780	19.836815	34.059731
.12	-33.058698	-33.057849	-16.814624	-16.768306	-4.308866	-3.514969	3.554264	7.862123	9.975676	19.772819	19.980360	34.106183
.13	-33.017649	-33.016870	-16.693402	-16.650184	-4.094919	-3.338612	3.848288	8.047279	10.233517	19.905482	20.125113	34.152905
.14	-32.976797	-32.976083	-16.572892	-16.532594	-3.882694	-3.162691	4.141790	8.232693	10.492474	20.038774	20.271102	34.199901
.15	-32.936140	-32.935486	-16.453083	-16.415531	-3.672180	-2.987212	4.434727	8.418355	10.754026	20.172698	20.418355	34.247176
.16	-32.895477	-32.895077	-16.334511	-16.298991	-3.463361	-2.812180	4.727057	8.604257	11.016647	20.307258	20.566902	34.294734
.17	-32.855400	-32.854853	-16.215511	-16.182971	-3.256221	-2.637602	5.018739	8.790392	11.280812	20.442455	20.716772	34.342579
.18	-32.815312	-32.814812	-16.097290	-16.065967	-3.050744	-2.463481	5.309730	8.976750	11.546494	20.578293	20.867994	34.390715
.19	-32.775408	-32.774951	-15.980589	-15.952471	-2.846912	-2.289824	5.599987	9.163324	11.813666	20.714776	21.020601	34.439146
.20	-32.735686	-32.735269	-15.864092	-15.837783	-2.644706	-2.116635	5.889468	9.350104	12.082300	20.851905	21.174624	34.487879
.21	-32.696144	-32.695764	-15.748224	-15.723949	-2.444106	-1.943919	6.178130	9.537008	12.352365	20.989684	21.330094	34.536916
.22	-32.656779	-32.656433	-15.632973	-15.610514	-2.245092	-1.771682	6.465929	9.724251	12.623832	21.128116	21.487044	34.586262
.23	-32.617589	-32.617274	-15.518329	-15.497523	-2.047646	-1.599927	6.752821	9.911509	12.896669	21.267202	21.645509	34.635921
.24	-32.578572	-32.578286	-15.404283	-15.385024	-1.851744	-1.428660	7.038764	10.099119	13.170846	21.406947	21.805519	34.685904
.25	-32.539725	-32.539466	-15.290823	-15.273012	-1.657366	-1.257885	7.323712	10.286802	13.446327	21.547351	21.967110	34.736210
.26	-32.501047	-32.500812	-15.177941	-15.161483	-1.464490	-1.087606	7.607623	10.474638	13.723082	21.688419	22.130317	34.786844
.27	-32.462536	-32.462322	-15.065627	-15.050433	-1.273095	-0.917228	7.890452	10.662618	14.001075	21.830151	22.295175	34.837814
.28	-32.424188	-32.423995	-14.953872	-14.939858	-1.083156	-0.748555	8.172155	10.850733	14.280272	21.972550	22.461717	34.889124
.29	-32.386003	-32.385828	-14.842669	-14.829754	-0.894653	-0.579790	8.452688	11.038973	14.560637	22.115619	22.629980	34.940780
.30	-32.347979	-32.347821	-14.732007	-14.720117	-0.707562	-0.411538	8.732007	11.227330	14.842135	22.259359	22.800000	34.992787
.31	-32.310113	-32.309970	-14.621879	-14.610944	-0.521860	-0.243802	9.010068	11.415793	15.124729	22.403773	22.971810	35.045151
.32	-32.272403	-32.272275	-14.512277	-14.502231	-0.337524	-0.076587	9.286828	11.604352	15.408382	22.548862	23.145448	35.097878
.33	-32.234848	-32.234733	-14.403193	-14.393973	-0.154532	0.090104	9.562244	11.792999	15.693058	22.694628	23.320958	35.150974
.34	-32.197447	-32.197343	-14.294620	-14.286167	0.027138	0.256270	9.836272	11.981722	15.978718	22.841073	23.498348	35.204445
.35	-32.160196	-32.160103	-14.186550	-14.178809	0.207512	0.421905	10.108870	12.170512	16.265321	22.988198	23.677680	35.258296
.36	-32.123095	-32.123011	-14.078975	-14.071895	0.386610	0.587006	10.379996	12.359359	16.552831	23.136005	23.858979	35.312536
.37	-32.086142	-32.086067	-13.971890	-13.965421	0.564457	0.751572	10.649609	12.548252	16.841208	23.284495	24.042281	35.367169
.38	-32.049335	-32.049268	-13.865287	-13.859384	0.741074	0.915598	10.917668	12.737181	17.130411	23.433669	24.227618	35.422202
.39	-32.012672	-32.012613	-13.759159	-13.753780	0.916483	1.079082	11.184134	12.926136	17.420400	23.583530	24.415025	35.477643
.40	-31.976153	-31.976099	-13.653501	-13.648605	1.090706	1.242022	11.448969	13.115106	17.711134	23.734077	24.604532	35.533498
.41	-31.939775	-31.939727	-13.548305	-13.543856	1.263766	1.404415	11.712134	13.304081	18.002557	23.885311	24.794171	35.589775
.42	-31.903536	-31.903494	-13.443566	-13.439528	1.435683	1.566259	11.973592	13.493098	18.294768	24.037248	24.989773	35.646479
.43	-31.867434	-31.867399	-13.339278	-13.335617	1.606470	1.727651	12.233275	13.681949	18.587582	24.189847	25.185966	35.703619
.44	-31.831473	-31.831440	-13.235435	-13.232123	1.776175	1.888290	12.491257	13.870920	18.880669	24.343150	25.384178	35.761203
.45	-31.795645	-31.795616	-13.132032	-13.129040	1.944791	2.048473	12.747395	14.059802	19.174485	24.497143	25.584636	35.819237
.46	-31.759952	-31.759926	-13.029062	-13.026363	2.112348	2.208098	13.001697	14.248633	19.468785	24.651827	25.787364	35.877729
.47	-31.724390	-31.724368	-12.926521	-12.924090	2.278866	2.367165	13.254134	14.437401	19.763269	24.807202	25.992386	35.936688
.48	-31.688960	-31.688940	-12.824403	-12.822218	2.444363	2.525671	13.504680	14.626096	20.058648	24.963269	26.199723	35.996122
.49	-31.653660	-31.653643	-12.722703	-12.720743	2.608860	2.683616	13.753309	14.814704	20.354695	25.120027	26.409394	36.056038
.50	-31.618489	-31.618473	-12.621416	-12.619661	2.772374	2.840997	14.000000	15.003214	20.649876	25.277476	26.621416	36.116446

$J_{K_{-1}K_1}$ / J_τ / κ	$6_{0,6}$ / 6_{-6}	$6_{1,6}$ / 6_{-5}	$6_{1,5}$ / 6_{-4}	$6_{2,5}$ / 6_{-3}	$6_{2,4}$ / 6_{-2}	$6_{3,4}$ / 6_{-1}	$6_{3,3}$ / 6_0	$6_{4,3}$ / 6_1	$6_{4,2}$ / 6_2	$6_{5,2}$ / 6_3	$6_{5,1}$ / 6_4	$6_{6,1}$ / 6_5
.50	-31.618489	-31.618473	-12.621416	-12.619661	2.772314	2.840997	14.000000	15.003214	20.649876	25.277476	26.621416	36.116446
.51	-31.583485	-31.583431	-12.520538	-12.519963	2.934924	2.997813	14.116931	15.241908	20.743208	25.435617	26.835806	36.177353
.52	-31.548515	-31.548515	-12.420526	-12.418664	3.096529	3.154065	14.487486	15.379894	21.140573	25.594703	27.051740	36.238769
.53	-31.513733	-31.513723	-12.319988	-12.318742	3.257206	3.309750	14.728247	15.568038	21.538396	25.753972	27.273592	36.300703
.54	-31.479063	-31.479054	-12.220306	-12.219200	3.416971	3.464868	14.967003	15.756036	21.834805	25.914185	27.493302	36.363163
.55	-31.444515	-31.444507	-12.121015	-12.120035	3.575842	3.619419	15.203742	15.943875	22.131238	26.075088	27.717272	36.426159
.56	-31.410088	-31.410082	-12.022109	-12.021243	3.733835	3.773401	15.438456	16.131542	22.427638	26.236680	27.943652	36.489700
.57	-31.375781	-31.375776	-11.923585	-11.922821	3.890967	3.926816	15.671139	16.319024	22.723943	26.398960	28.172445	36.553796
.58	-31.341593	-31.341588	-11.825438	-11.824766	4.047251	4.079661	15.901788	16.506309	23.020093	26.561927	28.403649	36.618457
.59	-31.307523	-31.307519	-11.727664	-11.727075	4.202705	4.231938	16.130402	16.693382	23.316021	26.725580	28.637262	36.683693
.60	-31.273569	-31.273565	-11.630260	-11.629745	4.357143	4.383646	16.356982	16.880231	23.611662	26.889918	28.873277	36.749514
.61	-31.239730	-31.239727	-11.533273	-11.532773	4.511177	4.534786	16.581534	17.066842	23.906946	27.054934	29.111584	36.815731
.62	-31.206006	-31.206003	-11.436544	-11.436154	4.664225	4.685358	16.804064	17.253202	24.201798	27.220645	29.352479	36.882952
.63	-31.172395	-31.172393	-11.340225	-11.339887	4.816497	4.835362	17.024582	17.439297	24.495445	27.387031	29.595643	36.950590
.64	-31.138896	-31.138894	-11.244260	-11.243969	4.968009	4.984798	17.243098	17.625113	24.789009	27.554096	29.841162	37.018856
.65	-31.105508	-31.105507	-11.148646	-11.148396	5.118772	5.133668	17.459626	17.810635	25.083006	27.721838	30.089019	37.087760
.66	-31.072231	-31.072230	-11.053166	-11.053166	5.267724	5.281972	17.674103	17.995850	25.375355	27.890227	30.339196	37.157315
.67	-31.039063	-31.039062	-10.958457	-10.958274	5.418102	5.429711	17.886787	18.180744	25.666858	28.059350	30.591669	37.227530
.68	-31.006002	-31.006002	-10.863875	-10.863720	5.566694	5.576886	18.097458	18.365307	25.957420	28.229115	30.846416	37.298419
.69	-30.973050	-30.973049	-10.769630	-10.769499	5.714585	5.723498	18.306218	18.549507	26.246970	28.399550	31.103411	37.369992
.70	-30.940203	-30.940202	-10.675719	-10.675491	5.861786	5.869549	18.513091	18.733347	26.535377	28.570655	31.362628	37.442262
.71	-30.907462	-30.907461	-10.582140	-10.582100	6.008309	6.015039	18.718102	18.916806	26.823544	28.742421	31.624041	37.515231
.72	-30.874825	-30.874824	-10.488888	-10.488812	6.154144	6.159971	18.921279	19.099870	27.108360	28.914853	31.887608	37.588941
.73	-30.842291	-30.842291	-10.395961	-10.395930	6.299360	6.304345	19.122650	19.282522	27.392709	29.087945	32.153311	37.663376
.74	-30.809860	-30.809860	-10.303356	-10.303305	6.443908	6.448164	19.322244	19.464748	27.675471	29.261696	32.421112	37.738556
.75	-30.777531	-30.777531	-10.211070	-10.211028	6.587816	6.591428	19.520093	19.646532	27.956518	29.436102	32.690977	37.814496
.76	-30.745303	-30.745303	-10.119100	-10.119067	6.731095	6.734141	19.716229	19.827857	28.235720	29.611161	32.962871	37.891209
.77	-30.713174	-30.713174	-10.027444	-10.027417	6.873753	6.876304	19.910684	20.008710	28.512942	29.786870	33.236760	37.968707
.78	-30.681145	-30.681145	-9.936098	-9.936076	7.015799	7.017919	20.103492	20.189072	28.788041	29.963226	33.512617	38.047004
.79	-30.649214	-30.649214	-9.845059	-9.845043	7.157240	7.158987	20.294688	20.368929	29.069873	30.140226	33.790371	38.126113
.80	-30.617380	-30.617380	-9.754326	-9.754313	7.298086	7.299513	20.484306	20.548265	29.331287	30.317867	34.070020	38.206048
.81	-30.585643	-30.585643	-9.663896	-9.663886	7.438343	7.439496	20.672382	20.727062	29.599130	30.496146	34.351513	38.286823
.82	-30.554002	-30.554002	-9.573765	-9.573758	7.578019	7.578941	20.858952	20.905306	29.864244	30.675060	34.634813	38.368452
.83	-30.522456	-30.522456	-9.483932	-9.483927	7.717121	7.717850	21.044051	21.082978	30.126471	30.854605	34.919881	38.450948
.84	-30.491004	-30.491004	-9.394394	-9.394390	7.855657	7.856225	21.227715	21.260364	30.385648	31.034779	35.206678	38.534326
.85	-30.459646	-30.459646	-9.305148	-9.305145	7.993633	7.994081	21.409981	21.436546	30.641613	31.215577	35.495167	38.618599
.86	-30.428381	-30.428381	-9.216191	-9.216191	8.131051	8.131384	21.590885	21.612407	30.894205	31.396993	35.785307	38.785307
.87	-30.397207	-30.397207	-9.127525	-9.127523	8.267921	8.268162	21.770462	21.787631	31.143263	31.577033	36.077062	38.789891
.88	-30.366125	-30.366125	-9.039142	-9.039141	8.404266	8.404440	21.948762	21.962322	31.388432	31.761684	36.370384	38.876348
.89	-30.335133	-30.335133	-8.951042	-8.951042	8.540065	8.540187	22.125780	22.136102	31.630156	31.944946	36.665262	38.964939
.90	-30.304231	-30.304231	-8.863223	-8.863223	8.675334	8.675417	22.301591	22.309316	31.867694	32.128813	36.961632	39.053907
.91	-30.273418	-30.273418	-8.775683	-8.775683	8.810080	8.810133	22.476216	22.481825	32.101108	32.313284	37.259466	39.143857
.92	-30.242694	-30.242694	-8.688419	-8.688419	8.944306	8.944359	22.649694	22.653614	32.332007	32.498354	37.558727	39.234805
.93	-30.212057	-30.212057	-8.601429	-8.601429	9.078018	9.078038	22.822049	22.824665	32.555066	32.684019	37.859380	39.326763
.94	-30.181507	-30.181507	-8.514711	-8.514711	9.211222	9.211231	22.993322	22.994963	32.775596	32.870275	38.161388	39.419748
.95	-30.151044	-30.151044	-8.428263	-8.428263	9.343921	9.343926	23.163545	23.164490	32.991173	33.057118	38.464717	39.513772
.96	-30.120666	-30.120666	-8.342083	-8.342168	9.476119	9.476119	23.332749	23.333231	33.208322	33.244544	38.769333	39.608852
.97	-30.090374	-30.090374	-8.256168	-8.256168	9.607823	9.607823	23.500966	23.501168	33.410392	33.432550	39.075003	39.705008
.98	-30.060165	-30.060165	-8.170518	-8.170518	9.739035	9.739035	23.668286	23.668286	33.611083	33.621135	39.382291	39.802231
.99	-30.030041	-30.030041	-8.085129	-8.085129	9.869759	9.869759	23.834561	23.834569	33.807651	33.810281	39.690567	39.900560
1.00	-30.000000	-30.000000	-8.000000	-8.000000	10.000000	10.000000	24.000000	24.000000	34.000000	34.000000	40.000000	40.000000

$K_{-1}K_1$ / J_τ / κ	$\tau_{2,6}$ / τ_{-4}	$\tau_{2,5}$ / τ_{-3}	$\tau_{3,5}$ / τ_{-2}	$\tau_{3,4}$ / τ_{-1}	$\tau_{4,4}$ / τ_0	$\tau_{4,3}$ / τ_1	$\tau_{5,3}$ / τ_2	$\tau_{5,2}$ / τ_3	$\tau_{6,2}$ / τ_4	$\tau_{6,1}$ / τ_5	$\tau_{7,1}$ / τ_6	$\tau_{7,0}$ / τ_7
.00	-28.000000	-13.416079	-12.888603	-3.539564	0.000000	3.539564	12.888603	13.416079	28.000000	28.027676	46.150665	46.151161
.01	-27.852629	-13.157583	-12.655793	-3.160624	0.264486	3.920105	13.118324	13.676825	28.148142	28.178025	46.201501	46.202045
.02	-27.706020	-12.901296	-12.431708	-2.783346	0.528996	4.306784	13.350128	13.938790	28.297163	28.330263	46.253458	46.253546
.03	-27.560166	-12.647180	-12.204190	-2.407984	0.793136	4.686394	13.580292	14.205225	28.446774	28.481557	46.303991	46.304046
.04	-27.415057	-12.395193	-11.978233	-2.034021	1.057776	5.070705	13.812531	14.472956	28.597281	28.634779	46.355653	46.356370
.05	-27.270686	-12.145295	-11.752785	-1.662091	1.322091	5.457018	14.045600	14.743092	28.748999	28.788999	46.408383	46.408383
.06	-27.127045	-11.897446	-11.528201	-1.292060	1.586321	5.844613	14.279493	15.015671	28.900723	28.944237	46.459831	46.460691
.07	-26.984124	-11.651607	-11.304484	-0.923983	1.850448	6.233386	14.514205	15.290728	29.053675	29.100515	46.512356	46.513207
.08	-26.841917	-11.407738	-11.081634	-0.557916	2.114456	6.623386	14.749730	15.568298	29.207460	29.257857	46.565178	46.566207
.09	-26.700414	-11.165798	-10.859654	-0.193911	2.378326	7.014432	14.986063	15.848414	29.362088	29.416285	46.618301	46.619426
.10	-26.559609	-10.925750	-10.638546	0.167980	2.642043	7.406495	15.223196	16.131110	29.517566	29.575824	46.671731	46.672959
.11	-26.419494	-10.687555	-10.418311	0.527710	2.905588	7.799509	15.461337	16.416455	29.673195	29.736328	46.725528	46.726766
.12	-26.280061	-10.451172	-10.199949	0.885630	3.168946	8.193530	15.700335	16.703758	29.829513	29.898723	46.779793	46.780993
.13	-26.141303	-10.216560	-9.982530	1.241946	3.431940	8.588119	15.939330	16.994965	29.988445	30.061367	46.833906	46.835504
.14	-26.003209	-9.983710	-9.762851	1.593462	3.695031	8.983577	16.179595	17.288263	30.148177	30.225614	46.888610	46.890354
.15	-25.865776	-9.752553	-9.546115	1.944091	3.957724	9.379714	16.420624	17.584272	30.308051	30.391108	46.943646	46.945548
.16	-25.728996	-9.523066	-9.330254	2.292345	4.220162	9.776461	16.662408	17.883014	30.468833	30.557881	46.999019	47.001002
.17	-25.592860	-9.295213	-9.115269	2.638186	4.482329	10.173748	16.904940	18.184506	30.630531	30.725963	47.054736	47.056995
.18	-25.457362	-9.068960	-8.901159	2.981584	4.744206	10.571507	17.148209	18.488763	30.793155	30.895388	47.110801	47.113262
.19	-25.322494	-8.844275	-8.687923	3.322507	5.005779	10.969665	17.392207	18.795797	30.956715	31.066190	47.167221	47.169901
.20	-25.188251	-8.621125	-8.475561	3.660920	5.267030	11.368154	17.636924	19.105618	31.121220	31.238404	47.224020	47.226920
.21	-25.054624	-8.399430	-8.264071	3.995944	5.527913	11.767096	17.882457	19.418223	31.286681	31.412066	47.281151	47.284326
.22	-24.921608	-8.179103	-8.053452	4.328181	5.788502	12.166075	18.128150	19.733640	31.453105	31.587216	47.338673	47.342128
.23	-24.789195	-7.960571	-7.843703	4.660971	6.048690	12.564887	18.375288	20.051846	31.620504	31.763893	47.396576	47.400333
.24	-24.657379	-7.743250	-7.634822	4.989182	6.308447	12.963039	18.622481	20.372845	31.788887	31.942137	47.454865	47.458951
.25	-24.526153	-7.527314	-7.426807	5.314805	6.567890	13.363039	18.870937	20.696630	31.958263	32.121991	47.513549	47.517991
.26	-24.395512	-7.312733	-7.219656	5.637829	6.826870	13.761994	19.119749	21.023193	32.128641	32.303500	47.572634	47.577462
.27	-24.265448	-7.099482	-7.013367	5.958251	7.085416	14.160767	19.369204	21.352520	32.300032	32.486709	47.632127	47.637374
.28	-24.135955	-6.887532	-6.807938	6.276069	7.343511	14.559283	19.619236	21.684594	32.472444	32.671665	47.692037	47.697736
.29	-24.007028	-6.676860	-6.603365	6.591283	7.601140	14.957465	19.866993	22.019997	32.645888	32.858417	47.752371	47.758561
.30	-23.878660	-6.467440	-6.399647	6.903897	7.858288	15.355235	20.121302	22.356904	32.820434	33.047017	47.813136	47.819858
.31	-23.750846	-6.259247	-6.196780	7.213919	8.114093	15.752535	20.373204	22.697094	32.997060	33.237517	47.874342	47.881640
.32	-23.623578	-6.052259	-5.994761	7.521359	8.371078	16.149223	20.625683	23.039932	33.174822	33.429971	47.935996	47.943918
.33	-23.496853	-5.846452	-5.793587	7.826228	8.626691	16.545691	20.878726	23.385388	33.350854	33.624436	47.998108	48.006705
.34	-23.370663	-5.641805	-5.593254	8.128544	8.881762	16.940603	21.132323	23.733427	33.529091	33.820968	48.060685	48.070014
.35	-23.245003	-5.438295	-5.393760	8.428323	9.136276	17.335110	21.386453	24.084008	33.708726	34.019628	48.123738	48.133859
.36	-23.119868	-5.235903	-5.195100	8.725586	9.390223	17.728716	21.641105	24.437093	33.889648	34.220477	48.187276	48.198254
.37	-22.995252	-5.034608	-4.997271	9.020356	9.643578	18.121336	21.896262	24.792635	34.071674	34.423579	48.251308	48.263215
.38	-22.871149	-4.834390	-4.800269	9.312658	9.896636	18.512683	22.151908	25.150589	34.254813	34.628998	48.315844	48.328757
.39	-22.747555	-4.635231	-4.604089	9.602518	10.148481	18.903271	22.408029	25.510904	34.439073	34.836801	48.380894	48.394897
.40	-22.624465	-4.437113	-4.408729	9.889964	10.400000	19.292210	22.664607	25.873529	34.624470	35.047057	48.446470	48.461652
.41	-22.501872	-4.240016	-4.214182	10.175028	10.650877	19.680212	22.921626	26.238409	34.810994	35.259834	48.512582	48.529041
.42	-22.379772	-4.043925	-4.020446	10.457741	10.901000	20.066587	23.179068	26.605486	34.998671	35.475205	48.579240	48.597083
.43	-22.258160	-3.848823	-3.827516	10.738137	11.150657	20.451442	23.436916	26.974703	35.187502	35.693242	48.646458	48.665798
.44	-22.137030	-3.654692	-3.635388	11.016249	11.399353	20.834687	23.695998	27.345998	35.377497	35.914018	48.714245	48.735207
.45	-22.016379	-3.461519	-3.444056	11.292113	11.647717	21.216230	23.953757	27.719307	35.568661	36.137609	48.782615	48.805334
.46	-21.896201	-3.269287	-3.253516	11.565766	11.895196	21.595977	24.212714	28.094564	35.761004	36.364090	48.851580	48.876201
.47	-21.776491	-3.077982	-3.063764	11.837244	12.141708	21.973837	24.472002	28.471703	35.954552	36.593537	48.921153	48.947843
.48	-21.657246	-2.887589	-2.874794	12.106586	12.387992	22.349721	24.731601	28.850654	36.149173	36.826150	48.991347	49.020263
.49	-21.538459	-2.698099	-2.686602	12.373913	12.633024	22.722444	24.991500	29.231623	36.345150	37.061640	49.062174	49.093508
.50	-21.420127	-2.509487	-2.499183	12.639013	12.877827	23.095167	25.251658	29.613703	36.542300	37.300448	49.133651	49.167606

Energy levels for $J = 7$. Entries are the reduced energy $E(\kappa)$. The two header labels for each column are the asymmetric-top notation $J_{K_{-1}K_{1}}$ (top) and the symmetric-top notation J_{τ} (bottom). The identifying column on the right gives $J_{K_{-1}K_{1}}$ / J_{r} and the asymmetry parameter κ.

κ	$7_{2,6}$ / 7_{-4}	$7_{2,5}$ / 7_{-3}	$7_{3,5}$ / 7_{-2}	$7_{3,4}$ / 7_{-1}	$7_{4,4}$ / 7_{0}	$7_{4,3}$ / 7_{1}	$7_{5,3}$ / 7_{2}	$7_{5,2}$ / 7_{3}	$7_{6,2}$ / 7_{4}	$7_{6,1}$ / 7_{5}	$7_{7,1}$ / 7_{6}	$7_{7,0}$ / 7_{7}
.50	-21.420127	-2.509487	-2.499183	12.639013	12.877827	23.095167	25.251658	29.613703	36.542300	37.300448	49.133651	49.167606
.51	-21.302245	-2.321751	-2.312531	12.902176	13.121606	23.464554	25.512073	29.997654	36.740639	37.542530	49.205791	49.242586
.52	-21.184809	-2.134875	-2.126643	13.164612	13.364612	23.831597	25.772717	30.383118	36.940196	37.787962	49.278609	49.318483
.53	-21.067815	-1.948847	-1.941513	13.425595	13.606835	24.196207	26.033569	30.770019	37.140979	38.036817	49.352120	49.395330
.54	-20.951258	-1.763654	-1.757135	13.685135	13.848265	24.558298	26.294607	31.158273	37.342993	38.289168	49.426340	49.473168
.55	-20.853133	-1.579286	-1.573504	13.935397	14.088891	24.917786	26.555807	31.547798	37.546242	38.545088	49.501285	49.552035
.56	-20.719438	-1.395573	-1.390616	14.189038	14.328707	25.274590	26.817146	31.938507	37.750733	38.804644	49.576973	49.631976
.57	-20.604618	-1.212978	-1.208466	14.440889	14.567697	25.628631	27.078602	32.330313	37.954770	39.067903	49.653419	49.713036
.58	-20.489318	-1.031018	-1.027047	14.690989	14.805859	25.979835	27.340149	32.723125	38.165458	39.334921	49.730642	49.795184
.59	-20.374885	-0.849839	-0.846355	14.933374	15.043183	26.328130	27.601762	33.116849	38.371101	39.605777	49.808660	49.878713
.60	-20.260864	-0.669432	-0.666384	15.186081	15.279660	26.673451	27.863418	33.511388	38.581204	39.880507	49.887492	49.963439
.61	-20.147253	-0.489788	-0.487130	15.431145	15.515283	27.015736	28.125089	33.906510	38.791970	40.159168	49.967156	50.049502
.62	-20.034047	-0.310897	-0.308587	15.674602	15.750045	27.354928	28.386749	34.302510	39.004002	40.441806	50.047674	50.136967
.63	-19.921243	-0.132751	-0.130749	15.916486	15.983939	27.690976	28.648373	34.698880	39.217304	40.728463	50.129064	50.225902
.64	-19.808836	0.044660	0.046387	16.156831	16.216958	28.023835	28.909931	35.095644	39.431878	41.019172	50.211349	50.316383
.65	-19.696824	0.221345	0.222830	16.395570	16.449096	28.353466	29.171396	35.492683	39.647728	41.313965	50.294550	50.403488
.66	-19.585202	0.397311	0.398582	16.633334	16.680347	28.679836	29.432739	35.889708	39.864934	41.612882	50.378688	50.502305
.67	-19.473967	0.572566	0.573650	16.868966	16.910707	29.002902	29.693932	36.287098	40.083504	41.916380	50.463797	50.595448
.68	-19.363116	0.747119	0.748039	17.103465	17.140170	29.322697	29.954944	36.684213	40.302946	42.223038	50.549871	50.695448
.69	-19.252645	0.920977	0.921754	17.336591	17.368731	29.639157	30.215745	37.081081	40.523913	42.534328	50.636962	50.794980
.70	-19.142551	1.094147	1.094800	17.568364	17.596387	29.952293	30.476303	37.477554	40.746163	42.849751	50.725087	50.896635
.71	-19.032830	1.266637	1.267183	17.798810	17.823134	30.262107	30.736588	37.873476	40.969696	43.169297	50.814271	51.000549
.72	-18.923480	1.438454	1.438907	18.028158	18.048968	30.568607	30.996665	38.268683	41.194512	43.492950	50.904539	51.106844
.73	-18.814496	1.609604	1.609978	18.255830	18.273886	30.871810	31.256204	38.663003	41.420610	43.820686	50.995918	51.215670
.74	-18.705877	1.780095	1.780401	18.482455	18.497885	31.171737	31.515469	39.056241	41.647991	44.152476	51.088437	51.327182
.75	-18.597618	1.949932	1.950182	18.707857	18.720965	31.468417	31.774327	39.448211	41.876652	44.488283	51.182123	51.441551
.76	-18.489716	2.119125	2.119259	18.933516	18.943152	31.761183	32.032742	39.838700	42.106594	44.828065	51.277005	51.558958
.77	-18.382169	2.287673	2.287634	19.155063	19.164352	32.050617	32.290509	40.227438	42.337014	45.171716	51.373102	51.679594
.78	-18.274974	2.455689	2.455714	19.376952	19.384663	32.339343	32.547560	40.616058	42.570814	45.519766	51.470416	51.803416
.79	-18.168127	2.622876	2.622976	19.597685	19.604046	32.623434	32.804973	40.999000	42.804080	45.870766	51.569127	51.931141
.80	-18.061625	2.789540	2.789618	19.817304	19.822503	32.904503	33.061255	41.381210	43.039122	46.225926	51.669096	52.063111
.81	-17.955466	2.955587	2.955647	20.035828	20.040034	33.182612	33.316909	41.760685	43.275432	46.584774	51.770417	52.198993
.82	-17.849647	3.121023	3.121068	20.253275	20.256641	33.457823	33.571897	42.137128	43.513006	46.947242	51.873123	52.339244
.83	-17.744166	3.285852	3.285886	20.469663	20.472323	33.730203	33.826178	42.510225	43.751936	47.313256	51.977247	52.484277
.84	-17.639018	3.450081	3.450106	20.685009	20.687082	33.999821	34.079714	42.879647	43.991576	47.682741	52.082824	52.634361
.85	-17.534202	3.613714	3.613714	20.899329	20.900919	34.266751	34.332372	43.245205	44.232703	48.056617	52.189889	52.789822
.86	-17.429715	3.776757	3.776769	21.112639	21.113838	34.531065	34.583381	43.606075	44.474604	48.431806	52.298478	52.950100
.87	-17.325554	3.939214	3.939223	21.324953	21.325838	34.792800	34.835430	43.962372	44.717234	48.811225	52.408627	53.118247
.88	-17.221717	4.101091	4.101097	21.536286	21.536924	35.052153	35.085566	44.313728	44.961552	49.193793	52.520373	53.291925
.89	-17.118200	4.262392	4.262396	21.746652	21.747099	35.309080	35.335344	44.659147	45.207511	49.579424	52.633715	53.472401
.90	-17.015002	4.423122	4.423124	21.956062	21.956365	35.563701	35.582927	44.998896	45.458637	49.968037	52.748809	53.660044
.91	-16.912120	4.583286	4.583287	22.164529	22.164726	35.816094	35.830065	45.332372	45.707393	50.359547	52.865575	53.855219
.92	-16.809550	4.742888	4.742888	22.372088	22.372187	36.066336	36.076115	45.655206	45.957365	50.753870	52.984011	54.058280
.93	-16.707294	4.901986	4.901988	22.578784	22.578752	36.321033	36.325697	45.979049	46.208543	51.150924	53.104397	54.269565
.94	-16.605345	5.060424	5.060424	22.784383	22.784421	36.564775	36.568321	46.291576	46.460920	51.550627	53.226502	54.489385
.95	-16.503702	5.218367	5.218367	22.989187	22.989205	36.804935	36.807296	46.596495	46.714496	51.952896	53.350536	54.718019
.96	-16.402363	5.375765	5.375765	23.193100	23.193100	37.047346	37.048152	46.893557	46.969256	52.357653	53.476450	54.955236
.97	-16.301326	5.532624	5.532624	23.396129	23.396131	37.287986	37.288316	47.182558	47.225194	52.764818	53.604313	55.202638
.98	-16.200587	5.688946	5.688946	23.598284	23.598284	37.527077	37.527154	47.463349	47.482302	53.174314	53.734165	55.458953
.99	-16.100146	5.844737	5.844737	23.799572	23.799572	37.764243	37.764262	47.735840	47.740574	53.586066	53.866048	55.724734
1.00	-16.000000	6.000000	6.000000	24.000000	24.000000	38.000000	38.000000	48.000000	48.000000	54.000000	54.000000	56.000000

$J_{K_{-1}K_1}$ / J_τ — κ	$8_{4,4}$ / 8_0	$8_{5,4}$ / 8_1	$8_{5,3}$ / 8_2	$8_{6,3}$ / 8_3	$8_{6,2}$ / 8_4	$8_{7,2}$ / 8_5	$8_{7,1}$ / 8_6	$8_{8,1}$ / 8_7	$8_{8,0}$ / 8_8	$7_{0,7}$ / 7_{-7}	$7_{1,7}$ / 7_{-6}	$7_{1,6}$ / 7_{-5}
.00	0.000000	6.881208	8.668876	21.898573	22.072159	39.752915	39.759911	60.735550	60.735655	-46.151161	-46.150665	-28.027676
.01	0.516904	7.223494	9.100720	22.170240	22.357003	39.924465	39.930332	60.793581	60.793581	-46.100547	-46.100095	-27.878246
.02	1.033698	7.566464	9.536107	22.443229	22.644039	40.093348	40.101738	60.851687	60.851816	-46.050200	-46.049789	-27.730162
.03	1.550269	7.910094	9.975013	22.717547	22.933211	40.264966	40.274146	60.910222	60.910510	-46.000216	-45.999733	-27.582078
.04	2.066507	8.254360	10.417407	22.993197	23.224904	40.437536	40.447575	60.969074	60.969233	-45.950291	-45.949952	-27.435305
.05	2.582300	8.599239	10.863256	23.270185	23.518844	40.611071	40.622043	61.028248	61.028426	-45.900721	-45.900413	-27.289385
.06	3.097538	8.944736	11.312527	23.548315	23.815199	40.785584	40.797570	61.087750	61.087947	-45.851403	-45.851123	-27.144302
.07	3.611803	9.290736	11.765152	23.828190	24.114027	40.961090	40.974175	61.147584	61.147803	-45.802078	-45.802013	-27.000041
.08	4.125899	9.637304	12.221104	24.109215	24.415392	41.137601	41.151880	61.207757	61.207998	-45.753505	-45.753274	-26.856588
.09	4.638798	9.984384	12.680322	24.391959	24.719354	41.315133	41.330700	61.268272	61.268540	-45.704711	-45.704719	-26.713929
.10	5.150695	10.331949	13.142746	24.675324	25.025978	41.493699	41.510675	61.329136	61.329433	-45.656570	-45.656380	-26.572050
.11	5.661477	10.679975	13.608314	24.960413	25.335330	41.673314	41.691811	61.390356	61.390684	-45.608570	-45.608225	-26.430938
.12	6.171031	11.028433	14.076957	25.246861	25.647475	41.853935	41.874118	61.451882	61.452060	-45.560570	-45.560415	-26.290581
.13	6.679246	11.377293	14.548604	25.534869	25.962482	42.035543	42.057430	61.513882	61.514284	-45.512911	-45.512911	-26.150977
.14	7.186010	11.726541	15.023180	25.823818	26.280419	42.218604	42.242464	61.576201	61.576646	-45.465480	-45.465354	-26.012080
.15	7.691209	12.076135	15.500605	26.114368	26.601356	42.402567	42.428516	61.638900	61.639392	-45.418269	-45.418155	-25.873912
.16	8.194733	12.426052	15.980799	26.406259	26.925363	42.587656	42.615866	61.701984	61.702528	-45.371276	-45.371173	-25.736452
.17	8.659470	12.776265	16.463674	26.699510	27.252511	42.773887	42.804541	61.765462	61.766062	-45.324499	-45.324406	-25.599688
.18	9.196308	13.126743	16.949142	26.994120	27.582872	42.961276	42.994573	61.829339	61.830001	-45.277851	-45.277851	-25.463610
.19	9.694138	13.477460	17.437112	27.290087	27.916517	43.149841	43.185992	61.893623	61.894354	-45.231580	-45.231505	-25.328207
.20	10.189850	13.828385	17.927491	27.587407	28.253518	43.339597	43.378823	61.959321	61.959192	-45.185366	-45.185155	-25.193469
.21	10.683334	14.179489	18.420181	27.886079	28.593946	43.530562	43.573523	62.023440	62.024330	-45.139469	-45.139430	-25.059387
.22	11.174485	14.530744	18.915084	28.186099	28.937849	43.722679	43.768916	62.089990	62.090970	-45.093750	-45.093590	-24.925950
.23	11.663196	14.882118	19.412078	28.487461	29.285364	43.916189	43.966232	62.154977	62.156058	-45.048209	-45.048161	-24.793151
.24	12.149362	15.233583	19.911122	28.790162	29.636499	44.110886	44.165116	62.221410	62.222601	-45.002866	-45.002823	-24.660977
.25	12.632883	15.585107	20.412050	29.094195	29.991324	44.306863	44.365609	62.288298	62.289609	-44.957717	-44.957678	-24.529427
.26	13.113659	15.936662	20.914775	29.399554	30.349923	44.504138	44.567752	62.355648	62.357093	-44.912761	-44.912726	-24.398465
.27	13.591593	16.288215	21.419190	29.706232	30.712352	44.702729	44.771173	62.423472	62.425062	-44.867994	-44.867944	-24.268146
.28	14.066591	16.639736	21.925183	30.014222	31.078672	44.902656	44.977173	62.491776	62.493527	-44.823416	-44.823388	-24.138467
.29	14.538562	16.991195	22.432643	30.323515	31.448939	45.103937	45.184546	62.560572	62.562499	-44.779023	-44.778949	-24.009242
.30	15.007420	17.342560	22.941456	30.634102	31.823207	45.306591	45.393761	62.629870	62.631989	-44.734813	-44.734792	-23.880662
.31	15.473083	17.693800	23.451508	30.945974	32.201526	45.510639	45.605750	62.699678	62.702010	-44.690785	-44.690785	-23.752654
.32	15.935471	18.044884	23.962682	31.259119	32.583492	45.716814	45.817932	62.770009	62.772573	-44.646936	-44.646919	-23.625210
.33	16.394513	18.395780	24.474860	31.573499	32.970492	45.922988	46.033002	62.840873	62.843692	-44.603264	-44.603264	-23.498322
.34	16.850138	18.746457	24.987920	31.889184	33.361216	46.131330	46.250143	62.912280	62.915380	-44.559767	-44.559754	-23.371985
.35	17.302285	19.096883	25.501742	32.206079	33.756142	46.341145	46.469418	62.984244	62.987651	-44.516443	-44.516431	-23.246192
.36	17.750895	19.447026	26.016201	32.524197	34.155296	46.552445	46.690895	63.056775	63.060520	-44.473290	-44.473280	-23.120938
.37	18.195917	19.796855	26.531173	32.843525	34.558697	46.765270	46.914644	63.129887	63.134003	-44.430307	-44.430307	-22.996209
.38	18.637305	20.146337	27.046528	33.164047	34.966355	46.979527	47.140737	63.203592	63.208114	-44.387484	-44.387484	-22.872005
.39	19.075021	20.495442	27.562137	33.485747	35.378278	47.195527	47.369253	63.277903	63.282872	-44.344841	-44.344834	-22.748320
.40	19.509032	20.844136	28.077868	33.808607	35.794464	47.413070	47.600271	63.352834	63.358293	-44.302315	-44.302349	-22.625147
.41	19.939311	21.192389	28.595587	34.132589	36.214905	47.632081	47.833875	63.428400	63.434397	-44.260030	-44.260030	-22.502479
.42	20.365840	21.540169	29.114986	34.457700	36.639586	47.852097	48.070154	63.504614	63.511202	-44.217866	-44.217862	-22.380312
.43	20.788604	21.887449	29.624434	34.763971	37.068484	48.075097	48.309200	63.581493	63.588728	-44.175861	-44.175857	-22.258639
.44	21.207598	22.234181	30.139280	35.111288	37.501571	48.299000	48.551109	63.659051	63.666999	-44.134012	-44.134009	-22.137455
.45	21.622821	22.580350	30.653549	35.439668	37.938809	48.524742	48.795980	63.737306	63.746035	-44.092319	-44.092316	-22.016755
.46	22.034281	22.925919	31.167091	35.769088	38.380154	48.752103	49.043919	63.816273	63.825860	-44.050780	-44.050780	-21.896523
.47	22.441989	23.270868	31.679240	36.099456	38.825555	48.981168	49.295025	63.896011	63.906000	-44.009393	-44.009390	-21.776783
.48	22.845965	23.615134	32.191388	36.430956	39.274954	49.211436	49.548282	63.976418	63.987983	-43.968156	-43.968156	-21.657502
.49	23.246232	23.958718	32.701155	36.763355	39.728236	49.442005	49.807254	64.057632	64.070035	-43.927068	-43.927066	-21.538684
.50	23.642820	24.301577	33.210921	37.096694	40.185479	49.678953	50.068598	64.139632	64.153586	-43.886128	-43.886123	-21.420324

$J_{K_{-1}K_1}$, J_τ, κ	$8_{4,4}$ / 8_0	$8_{5,4}$ / 8_1	$8_{5,3}$ / 8_2	$8_{6,3}$ / 8_3	$8_{6,2}$ / 8_4	$8_{7,2}$ / 8_5	$8_{7,1}$ / 8_6	$8_{8,1}$ / 8_7	$8_{8,0}$ / 8_8	$7_{0,7}$ / 7_{-7}	$7_{1,7}$ / 7_{-6}	$7_{1,6}$ / 7_{-5}
.50	23.642820	24.301577	33.210921	37.096694	40.185479	49.678953	50.068598	64.139632	64.153586	-43.886128	-43.886126	-21.420324
.51	24.035165	24.641663	33.718496	37.430948	40.646455	49.915120	50.335599	64.243341	64.231767	-43.845333	-43.845332	-21.302417
.52	24.424893	24.982514	34.172808	37.766085	41.111102	50.153109	50.605109	64.306063	64.302019	-43.804684	-43.804681	-21.184959
.53	24.810885	25.325511	34.728427	38.101962	41.578442	50.342742	50.875102	64.389760	64.388260	-43.762794	-43.763795	-21.068069
.54	25.193153	25.665175	35.230439	38.438908	42.051193	50.634637	51.151877	64.475917	64.496243	-43.723813	-43.723812	-20.951371
.55	25.571961	26.003965	35.730250	38.776525	42.526386	50.878214	51.432859	64.584505	64.584505	-43.683588	-43.683588	-20.835231
.56	25.947363	26.341855	36.227679	39.114904	43.004875	51.123692	51.718192	64.673889	64.678889	-43.643503	-43.643503	-20.719522
.57	26.319417	26.678814	36.722549	39.454012	43.486546	51.371091	52.008026	64.737341	64.764440	-43.603555	-43.603555	-20.604240
.58	26.688183	27.014816	37.214678	39.793812	43.971280	51.620429	52.302512	64.826546	64.856208	-43.563744	-43.563744	-20.489379
.59	27.053725	27.349832	37.703883	40.134269	44.458949	51.871244	52.601805	64.916631	64.949247	-43.524068	-43.524068	-20.374937
.60	27.416105	27.683837	38.189983	40.475346	44.949424	52.124995	52.906057	65.043612	65.043612	-43.484525	-43.484525	-20.260909
.61	27.775390	28.016833	38.672887	40.817015	45.442540	52.380512	53.215274	65.099910	65.138260	-43.445113	-43.445113	-20.147291
.62	28.131645	28.348706	39.152143	41.159200	45.938240	52.637531	53.530059	65.193163	65.236574	-43.405837	-43.405837	-20.034079
.63	28.484937	28.679519	39.627846	41.501897	46.436292	52.896831	53.850116	65.287533	65.335307	-43.366688	-43.366688	-19.921270
.64	28.835334	29.009219	40.099732	41.845049	46.936570	53.158172	54.175742	65.383054	65.435643	-43.327668	-43.327668	-19.808859
.65	29.182904	29.337781	40.567632	42.183613	47.438914	53.421571	54.506884	65.479759	65.537664	-43.288776	-43.288776	-19.696843
.66	29.527713	29.665183	41.031182	42.532544	47.943159	53.687043	54.844282	65.577684	65.641190	-43.250011	-43.250011	-19.585218
.67	29.869828	29.991402	41.490828	42.876793	48.449131	53.954601	55.187469	65.676665	65.747133	-43.211371	-43.211371	-19.473980
.68	30.209314	30.316416	41.945823	43.221312	48.956649	54.224259	55.536771	65.776879	65.854784	-43.172855	-43.172855	-19.363127
.69	30.546238	30.640206	42.396229	43.566051	49.465523	54.496030	55.892305	65.879152	65.964533	-43.134462	-43.134462	-19.252654
.70	30.880663	30.962750	42.841920	43.910958	49.975554	54.769557	56.254177	65.982475	66.076506	-43.096192	-43.096192	-19.142558
.71	31.212652	31.284030	43.282784	44.255978	50.486553	55.045957	56.622480	66.086343	66.190841	-43.058042	-43.058042	-19.032834
.72	31.542266	31.604027	43.718721	44.601057	50.998238	55.324136	56.997297	66.193014	66.307148	-43.020012	-43.020012	-18.923484
.73	31.869565	31.922724	44.149645	44.946137	51.510435	55.604472	57.378692	66.300593	66.427221	-42.982102	-42.982102	-18.814500
.74	32.194607	32.240105	44.575488	45.291160	52.022817	55.886973	57.766720	66.409730	66.549614	-42.944309	-42.944309	-18.705879
.75	32.517449	32.556154	44.996198	45.636063	52.535298	56.171640	58.161414	66.520475	66.675071	-42.906633	-42.906633	-18.597620
.76	32.838145	32.870857	45.411737	45.980786	53.047415	56.458505	58.562795	66.632380	66.803813	-42.869072	-42.869072	-18.489718
.77	33.156749	33.184200	45.822088	46.325261	53.559404	56.747549	58.970865	66.746990	66.936003	-42.831627	-42.831627	-18.382170
.78	33.473311	33.496170	46.227250	46.669424	54.069504	57.038705	59.385610	66.862886	67.072148	-42.794295	-42.794295	-18.274975
.79	33.787880	33.806757	46.627238	47.013204	54.578798	57.332219	59.806909	66.980601	67.212303	-42.757076	-42.757076	-18.168127
.80	34.100504	34.115949	47.022087	47.356532	55.086428	57.627855	60.234983	67.102003	67.356875	-42.719970	-42.719970	-18.061626
.81	34.411228	34.423737	47.411845	47.699537	55.590025	57.925399	60.669498	67.221756	67.506220	-42.682974	-42.682974	-17.955467
.82	34.720094	34.730113	47.796579	48.041533	56.095025	58.225434	61.110466	67.345319	67.660735	-42.646088	-42.646088	-17.849648
.83	35.027144	35.035069	48.176369	48.383054	56.599162	58.527900	61.557793	67.471797	67.820803	-42.609312	-42.609312	-17.744166
.84	35.332418	35.338600	48.551308	48.723817	57.091586	58.832447	62.011372	67.598757	67.987043	-42.572644	-42.572644	-17.639018
.85	35.635952	35.640699	48.921503	49.063741	57.584024	59.139110	62.471084	67.728770	68.159832	-42.536083	-42.536083	-17.534202
.86	35.937783	35.941362	49.287071	49.402740	58.071772	59.447976	62.936798	67.936798	68.338737	-42.499629	-42.499629	-17.429715
.87	36.237943	36.240587	49.648139	49.740730	58.554179	59.759041	63.408376	67.995750	68.527506	-42.463281	-42.463281	-17.325554
.88	36.536463	36.538371	50.004843	50.077621	59.030550	60.072303	63.885671	68.132869	68.723660	-42.427037	-42.427037	-17.221717
.89	36.833375	36.834713	50.357326	50.413324	59.500147	60.387755	64.368527	68.232514	68.926946	-42.390898	-42.390898	-17.118200
.90	37.128706	37.129614	50.705736	50.747747	59.962198	60.705392	64.856787	68.414766	69.144120	-42.354861	-42.354861	-17.015002
.91	37.422481	37.423073	51.050225	51.080794	60.415904	61.025204	65.350285	68.559710	69.369895	-42.318927	-42.318927	-16.912120
.92	37.714726	37.715093	51.390048	51.412370	60.860449	61.347194	65.848656	68.707430	69.600112	-42.283095	-42.283095	-16.809552
.93	38.005464	38.005678	51.728064	51.742378	61.295019	61.671342	66.352334	68.858016	69.856536	-42.247363	-42.247363	-16.707294
.94	38.294715	38.294830	52.061729	52.070717	61.718822	61.997642	66.860544	69.011556	70.118935	-42.211732	-42.211732	-16.605345
.95	38.582500	38.582554	52.392103	52.397287	62.131106	62.326084	67.373325	69.168140	70.395032	-42.176199	-42.176199	-16.503702
.96	38.868835	38.868857	52.719341	52.721986	62.531197	62.656657	67.890507	69.327862	70.685481	-42.140766	-42.140766	-16.402363
.97	39.153729	39.153746	53.043599	53.044710	62.919948	62.989348	68.411926	69.497065	70.980843	-42.105430	-42.105430	-16.301328
.98	39.437235	39.437226	53.364825	53.365832	63.293600	63.324158	68.937948	69.675555	71.300840	-42.070190	-42.070190	-16.200730
.99	39.719308	39.719308	53.683821	53.685781	63.653831	63.661073	69.466829	69.826788	71.647887	-42.035047	-42.035047	-16.100146
1.00	40.000000	40.000000	54.000000	54.000000	64.000000	64.000000	70.000000	70.000000	72.000000	-42.000000	-42.000000	-16.000000

$J_{K_{-1}K_1}$ J_τ κ	$9_{8,2}$ 9_6	$9_{8,1}$ 9_7	$9_{9,1}$ 9_8	$9_{9,0}$ 9_9	$8_{0,8}$ 8_{-8}	$8_{1,8}$ 8_{-7}	$8_{1,7}$ 8_{-6}	$8_{2,7}$ 8_{-5}	$8_{2,6}$ 8_{-4}	$8_{3,6}$ 8_{-3}	$8_{3,5}$ 8_{-2}	$8_{4,5}$ 8_{-1}
.00	53.502655	53.504332	77.320562	77.320583	-60.735655	-60.735550	-39.759911	-39.752915	-22.072159	-21.898573	-8.668876	-6.881208
.01	53.693628	53.695492	77.385548	77.385572	-60.678033	-60.677939	-39.590458	-39.584076	-21.789455	-21.628824	-8.240594	-6.539629
.02	53.885648	53.887718	77.450878	77.450905	-60.620511	-60.620431	-39.421956	-39.416137	-21.508843	-21.359185	-7.815889	-6.198779
.03	54.078729	54.081026	77.516557	77.516588	-60.563684	-60.563609	-39.254390	-39.249088	-21.230272	-21.091451	-7.394769	-5.858681
.04	54.272889	54.275436	77.582625	77.582625	-60.506948	-60.506581	-39.087744	-39.083744	-20.952949	-20.825016	-6.977239	-5.519355
.05	54.468142	54.470966	77.648984	77.649023	-60.450499	-60.450439	-38.922003	-38.917610	-20.679071	-20.559872	-6.563296	-5.180823
.06	54.664505	54.667634	77.715743	77.715787	-60.394335	-60.394397	-38.757159	-38.751549	-20.407618	-20.294869	-6.146348	-4.842225
.07	54.862461	54.865461	77.782877	77.782874	-60.338423	-60.338423	-38.593175	-38.585549	-20.136450	-20.033329	-5.742146	-4.504225
.08	55.060934	55.064466	77.850437	77.850582	-60.282832	-60.282789	-38.430068	-38.420816	-19.866457	-19.772116	-5.342916	-4.170199
.09	55.260434	55.264672	77.918274	77.918336	-60.227490	-60.227451	-38.267807	-38.264816	-19.599203	-19.512065	-4.943221	-3.835047
.10	55.461415	55.466100	77.986556	77.986626	-60.172415	-60.172380	-38.106383	-38.103671	-19.333693	-19.253268	-4.547038	-3.500789
.11	55.663596	55.668772	78.055234	78.055313	-60.117603	-60.117572	-37.945995	-37.943326	-19.069889	-18.995716	-4.154342	-3.167442
.12	55.866997	55.872712	78.124316	78.124404	-60.063051	-60.063023	-37.785995	-37.783771	-18.807756	-18.739403	-3.765099	-2.835025
.13	56.071636	56.077943	78.193807	78.193807	-60.008755	-60.008755	-37.627007	-37.624996	-18.547258	-18.484318	-3.379276	-2.503555
.14	56.277533	56.284491	78.263716	78.263827	-59.954712	-59.954691	-37.468808	-37.466991	-18.288363	-18.230454	-2.996835	-2.173049
.15	56.484710	56.492381	78.334050	78.334050	-59.900919	-59.900682	-37.311174	-37.309745	-18.031034	-17.977802	-2.617735	-1.843542
.16	56.693187	56.701241	78.404815	78.404954	-59.847372	-59.847552	-37.154732	-37.153251	-17.775253	-17.726354	-2.241932	-1.514592
.17	56.902385	56.912298	78.476020	78.475847	-59.794067	-59.794055	-36.998940	-36.996833	-17.520976	-17.476009	-1.869382	-1.187474
.18	57.114127	57.124381	78.547673	78.547847	-59.741003	-59.740990	-36.843679	-36.842477	-17.268179	-17.227030	-1.500035	-0.860981
.19	57.326635	57.337921	78.619781	78.619781	-59.687976	-59.687976	-36.689261	-36.688164	-17.016813	-16.979116	-1.133844	-0.535529
.20	57.540531	57.552947	78.692354	78.692571	-59.635582	-59.635572	-36.535568	-36.534597	-16.766913	-16.732410	-0.770755	-0.211131
.21	57.755839	57.769494	78.765399	78.765642	-59.583220	-59.583211	-36.382592	-36.381719	-16.518391	-16.486841	-0.410718	0.112199
.22	57.972583	57.987594	78.838926	78.839198	-59.531085	-59.531078	-36.230322	-36.229357	-16.271242	-16.242420	-0.053678	0.434449
.23	58.190788	58.207282	78.912944	78.913247	-59.479176	-59.479169	-36.078749	-36.078048	-16.025359	-15.999138	0.300417	0.755608
.24	58.410478	58.428596	78.987463	78.987861	-59.427481	-59.427481	-35.927860	-35.923864	-15.780968	-15.756986	0.651625	1.075664
.25	58.631680	58.651574	79.062492	79.062869	-59.376022	-59.376017	-35.777659	-35.777098	-15.537795	-15.515954	1.000000	1.394605
.26	58.854418	58.876256	79.138043	79.138463	-59.324677	-59.324773	-35.628197	-35.627640	-15.295903	-15.276032	1.345597	1.712421
.27	59.078721	59.102683	79.214121	79.214592	-59.273737	-59.273733	-35.479254	-35.478807	-15.055271	-15.037211	1.688472	2.029103
.28	59.304615	59.330899	79.290743	79.291268	-59.222914	-59.222912	-35.330835	-35.330640	-14.815876	-14.799482	2.028681	2.344640
.29	59.532129	59.560949	79.367918	79.368503	-59.172300	-59.172297	-35.183467	-35.183113	-14.577701	-14.562835	2.366278	2.659025
.30	59.761290	59.792883	79.445657	79.446309	-59.121893	-59.121890	-35.036536	-35.036222	-14.340724	-14.327261	2.701318	2.972249
.31	59.992128	60.026748	79.523973	79.524699	-59.071691	-59.071689	-34.890236	-34.889957	-14.104928	-14.092750	3.033856	3.284304
.32	60.224673	60.262599	79.602877	79.603686	-59.021691	-59.021689	-34.744559	-34.744312	-13.870294	-13.859292	3.363944	3.595183
.33	60.458954	60.500490	79.682382	79.683284	-58.971889	-58.971889	-34.599404	-34.599854	-13.636808	-13.626870	3.691636	3.904869
.34	60.695004	60.740478	79.762502	79.763507	-58.922289	-58.922288	-34.455047	-34.454854	-13.404445	-13.395500	4.016983	4.213389
.35	60.932852	60.982626	79.843251	79.844369	-58.872881	-58.872881	-34.311198	-34.311028	-13.173196	-13.165146	4.340037	4.520704
.36	61.172553	61.226995	79.924641	79.925887	-58.823670	-58.823668	-34.167944	-34.167794	-12.943042	-12.935809	4.660847	4.826820
.37	61.414077	61.473655	80.006689	80.008076	-58.774648	-58.774648	-34.025878	-34.025146	-12.713969	-12.707419	4.979461	5.131733
.38	61.657519	61.722674	80.089409	80.090954	-58.725815	-58.725814	-33.883194	-33.883074	-12.485961	-12.480146	5.295928	5.435439
.39	61.902893	61.974129	80.172818	80.174427	-58.677169	-58.677169	-33.741686	-33.741585	-12.259003	-12.253801	5.610295	5.737934
.40	62.150233	62.228097	80.256930	80.258844	-58.628708	-58.628708	-33.600746	-33.600658	-12.033081	-12.028435	5.922606	6.039215
.41	62.399575	62.484661	80.341765	80.343894	-58.580430	-58.580430	-33.460370	-33.460292	-11.808183	-11.804040	6.232397	6.339228
.42	62.650953	62.743909	80.427338	80.429708	-58.532434	-58.532434	-33.320580	-33.320367	-11.584300	-11.580611	6.542239	6.637903
.43	62.904406	63.005922	80.513669	80.513369	-58.484443	-58.484443	-33.181420	-33.181282	-11.361400	-11.358123	6.847645	6.935756
.44	63.159969	63.270829	80.600776	80.603712	-58.436677	-58.436677	-33.042555	-33.042504	-11.139490	-11.136584	7.152166	7.232164
.45	63.417679	63.538703	80.688679	80.691948	-58.389113	-58.389113	-32.904369	-32.904325	-10.918552	-10.915980	7.454840	7.527351
.46	63.677576	63.809661	80.777399	80.781307	-58.341722	-58.341722	-32.766677	-32.766679	-10.698574	-10.696301	7.755706	7.821361
.47	63.939698	64.083819	80.866958	80.871007	-58.294503	-58.294503	-32.629592	-32.629559	-10.479543	-10.477539	8.054801	8.114061
.48	64.204083	64.361300	80.957377	80.961885	-58.247454	-58.247454	-32.492794	-32.492962	-10.261149	-10.259685	8.352160	8.405586
.49	64.470771	64.642230	81.048680	81.053698	-58.200574	-58.200574	-32.356904	-32.356880	-10.044280	-10.042732	8.647819	8.695893
.50	64.739802	64.926747	81.140891	81.146447	-58.153860	-58.153860	-32.221330	-32.221309	-9.828026	-9.826670	8.941810	8.984982

$J_{K_{-1}K_1}$ / J_τ / κ	$9_{8,2}$ / 9_6	$9_{8,1}$ / 9_7	$9_{9,1}$ / 9_8	$9_{9,0}$ / 9_9	$8_{0,8}$ / 8_{-8}	$8_{1,8}$ / 8_{-7}	$8_{1,7}$ / 8_{-6}	$8_{2,7}$ / 8_{-5}	$8_{2,6}$ / 8_{-4}	$8_{3,6}$ / 8_{-3}	$8_{3,5}$ / 8_{-2}	$8_{4,5}$ / 8_{-1}
.50	64.739802	64.926747	81.140891	81.146477	-58.153860	-58.153860	-32.221330	-32.221309	-9.828026	-9.826670	8.941810	8.984982
.51	65.011217	65.214995	81.230129	81.235381	-58.108031	-58.108031	-32.106633	-32.106633	-9.636276	-9.611491	9.234106	9.272686
.52	65.285055	65.505124	81.324128	81.330065	-58.060321	-58.060321	-31.986683	-31.986683	-9.451849	-9.395218	9.529212	9.555124
.53	65.561359	65.803296	81.423228	81.430943	-58.014702	-58.014702	-31.817626	-31.817631	-9.184849	-9.183751	9.814096	9.844968
.54	65.840169	66.103680	81.519334	81.527927	-57.968639	-57.968639	-31.684048	-31.684037	-8.971951	-8.971172	10.101730	10.129211
.55	66.121526	66.408454	81.616485	81.626659	-57.922734	-57.922734	-31.550956	-31.550947	-8.760118	-8.759445	10.387847	10.412251
.56	66.405473	66.717808	81.714713	81.725381	-57.876987	-57.876987	-31.418346	-31.418338	-8.549140	-8.548560	10.672474	10.694092
.57	66.692052	67.031940	81.814050	81.825941	-57.831395	-57.831395	-31.286213	-31.286206	-8.339009	-8.338510	10.955617	10.974736
.58	66.981304	67.351057	81.914530	81.927787	-57.785957	-57.785957	-31.154552	-31.154547	-8.129715	-8.129288	11.237361	11.254188
.59	67.273271	67.675378	82.016189	82.030972	-57.740671	-57.740671	-31.023360	-31.023356	-7.921250	-7.920885	11.517671	11.532454
.60	67.567995	68.005133	82.119063	82.155544	-57.695537	-57.695537	-30.892631	-30.892627	-7.713405	-7.713295	11.796590	11.809538
.61	67.865518	68.340540	82.223193	82.241554	-57.650553	-57.650553	-30.762361	-30.762344	-7.506772	-7.505005	12.076140	12.085445
.62	68.165881	68.681909	82.328619	82.349518	-57.605717	-57.605717	-30.632547	-30.632544	-7.300743	-7.300520	12.350343	12.360180
.63	68.469126	69.029436	82.435384	82.435318	-57.561028	-57.561028	-30.503180	-30.503180	-7.095509	-7.095322	12.625219	12.633749
.64	68.775294	69.383410	82.543532	82.569150	-57.516485	-57.516485	-30.374265	-30.374263	-6.891063	-6.890907	12.898789	12.906159
.65	69.084424	69.744104	82.653110	82.681738	-57.472086	-57.472086	-30.245789	-30.245788	-6.687397	-6.687267	13.171072	13.177415
.66	69.396558	70.111801	82.764169	82.796173	-57.427830	-57.427830	-30.117752	-30.117752	-6.484504	-6.484344	13.442523	13.447523
.67	69.711734	70.486787	82.876758	82.912554	-57.383715	-57.383715	-29.990149	-29.990148	-6.282377	-6.282287	13.711851	13.716490
.68	70.029991	70.869351	82.990933	83.030990	-57.339741	-57.339741	-29.862977	-29.864230	-6.081007	-6.080934	13.980382	13.984322
.69	70.351367	71.259785	83.106750	83.151599	-57.295907	-57.295907	-29.736231	-29.736230	-5.880388	-5.880329	14.247696	14.251026
.70	70.675898	71.658378	83.224269	83.274510	-57.252210	-57.252210	-29.609908	-29.609907	-5.680513	-5.680465	14.513810	14.516763
.71	71.003624	72.069174	83.343546	83.398510	-57.208650	-57.208650	-29.484504	-29.484504	-5.482949	-5.482938	14.778739	14.781079
.72	71.334571	72.481172	83.444663	83.527825	-57.165226	-57.165226	-29.358583	-29.358583	-5.285825	-5.285818	15.042447	15.044442
.73	71.668781	72.910919	83.587680	83.658561	-57.121936	-57.121936	-29.233438	-29.233437	-5.085261	-5.085261	15.305099	15.306706
.74	72.006284	73.339906	83.712666	83.792266	-57.078779	-57.078779	-29.108769	-29.108768	-4.888319	-4.888300	15.566559	15.567897
.75	72.347109	73.783368	83.839699	83.929155	-57.035754	-57.035754	-28.984504	-28.984504	-4.692064	-4.692049	15.826891	15.827964
.76	72.691288	74.236516	83.968860	84.069467	-56.992860	-56.992860	-28.860640	-28.860640	-4.496513	-4.496502	16.086107	16.086974
.77	73.038847	74.699535	84.100233	84.213467	-56.950096	-56.950096	-28.737175	-28.737175	-4.301661	-4.301652	16.344220	16.344914
.78	73.389814	75.172583	84.233907	84.361454	-56.907461	-56.907461	-28.614103	-28.614103	-4.107502	-4.107495	16.601242	16.601792
.79	73.744211	75.655783	84.369778	84.513762	-56.864953	-56.864953	-28.491422	-28.491422	-3.914028	-3.914024	16.857116	16.857616
.80	74.102063	76.149223	84.508530	84.670766	-56.822572	-56.822572	-28.369129	-28.369129	-3.721236	-3.721232	17.112056	17.112394
.81	74.463188	76.652955	84.649679	84.832890	-56.780316	-56.780316	-28.247221	-28.247221	-3.529118	-3.529113	17.365876	17.366135
.82	74.828207	77.166993	84.793528	85.000611	-56.738185	-56.738185	-28.125693	-28.125693	-3.337669	-3.337667	17.618647	17.618840
.83	75.196535	77.691310	84.940198	85.174466	-56.696177	-56.696177	-28.004544	-28.004544	-3.146898	-3.146882	17.870381	17.870525
.84	75.568387	78.225840	85.089778	85.355063	-56.654291	-56.654291	-27.883770	-27.883770	-2.956757	-2.956756	18.121089	18.121194
.85	75.943775	78.770482	85.242420	85.543088	-56.612528	-56.612528	-27.763367	-27.763367	-2.767282	-2.767281	18.370780	18.370855
.86	76.322709	79.325095	85.398244	85.739313	-56.570884	-56.570884	-27.643334	-27.643334	-2.578454	-2.578454	18.619517	18.619517
.87	76.705196	79.889508	85.557385	85.944608	-56.529360	-56.529360	-27.523666	-27.523666	-2.390268	-2.390263	18.867150	18.867186
.88	77.091242	80.463516	85.719982	86.159947	-56.487954	-56.487954	-27.404361	-27.404361	-2.202719	-2.202719	19.113846	19.113670
.89	77.480849	81.046889	85.886183	86.386414	-56.446666	-56.446666	-27.285417	-27.285417	-2.015802	-2.015802	19.359582	19.359578
.90	77.874017	81.639373	86.056141	86.625206	-56.405495	-56.405495	-27.166826	-27.166826	-1.829511	-1.829511	19.604306	19.604315
.91	78.270746	82.240693	86.230014	86.877636	-56.364439	-56.364439	-27.048596	-27.048596	-1.643842	-1.643842	19.848086	19.848091
.92	78.671030	82.850560	86.407967	87.145111	-56.323498	-56.323498	-26.930715	-26.930715	-1.458789	-1.458789	20.090910	20.090910
.93	79.074863	83.468672	86.590171	87.429125	-56.282671	-56.282671	-26.813182	-26.813182	-1.274348	-1.274348	20.332786	20.332788
.94	79.482235	84.094718	86.776801	87.731213	-56.241958	-56.241958	-26.695996	-26.695996	-1.090514	-1.090514	20.573722	20.573723
.95	79.893136	84.728384	86.968039	88.052907	-56.201356	-56.201356	-26.579153	-26.579153	-0.907283	-0.907283	20.813725	20.813725
.96	80.307551	85.369353	87.164071	88.395673	-56.160865	-56.160865	-26.462651	-26.462651	-0.724649	-0.724649	21.052802	21.052802
.97	80.725464	86.017308	87.365087	88.760831	-56.120485	-56.120485	-26.346487	-26.346487	-0.542608	-0.542608	21.290961	21.290961
.98	81.146857	86.671938	87.571283	89.149473	-56.080215	-56.080215	-26.230659	-26.230659	-0.361288	-0.361288	21.528208	21.528208
.99	81.571929	87.332704	87.782284	89.562268	-56.040064	-56.040064	-26.115164	-26.115164	-0.180614	-0.180614	21.764553	21.764553
1.00	82.000000	88.000000	88.000000	90.000000	-56.000000	-56.000000	-26.000000	-26.000000	0.000000	0.000000	22.000000	22.000000

J_{K_{-1},K_1} / J_{τ} / κ	$9_{2,8}$ / 9_{-6}	$9_{2,7}$ / 9_{-5}	$9_{3,7}$ / 9_{-4}	$9_{3,6}$ / 9_{-3}	$9_{4,6}$ / 9_{-2}	$9_{4,5}$ / 9_{-1}	$9_{5,5}$ / 9_0	$9_{5,4}$ / 9_1	$9_{6,4}$ / 9_2	$9_{6,3}$ / 9_3	$9_{7,3}$ / 9_4	$9_{7,2}$ / 9_5
.00	-53.502655	-32.876683	-32.825762	-16.009233	-15.279589	-4.316098	0.000000	4.316098	15.279509	16.009233	32.825762	32.876883
.01	-53.312712	-32.564788	-32.518021	-15.552621	-14.870834	-3.694871	0.453221	4.940251	15.689918	16.470100	33.135175	33.193917
.02	-53.123786	-32.254688	-32.211935	-15.100184	-14.463667	-3.076217	0.906378	5.567182	16.101806	16.935304	33.446278	33.507237
.03	-52.935861	-31.946544	-31.907488	-14.651838	-14.058102	-2.461775	1.359457	6.196740	16.515233	17.404922	33.759087	33.825587
.04	-52.748923	-31.640315	-31.604662	-14.207498	-13.654042	-1.850186	1.812404	6.828770	16.930185	17.879031	34.073620	34.146117
.05	-52.562958	-31.335962	-31.303441	-13.767080	-13.251825	-1.242082	2.265174	7.463118	17.346642	18.357701	34.389892	34.469870
.06	-52.377953	-31.033450	-31.003806	-13.330499	-12.851137	-0.637593	2.717723	8.099625	17.764584	18.841002	34.707922	34.793917
.07	-52.193894	-30.732742	-30.705743	-12.897671	-12.452095	-0.036643	3.170008	8.738131	18.183991	19.328998	35.027765	35.121295
.08	-52.010768	-30.433804	-30.409233	-12.468510	-12.054708	0.560049	3.621985	9.378475	18.604842	19.821749	35.349118	35.451067
.09	-51.828562	-30.136604	-30.114260	-12.042934	-11.658944	1.152971	4.073611	10.020495	19.027161	20.319309	35.672171	35.783293
.10	-51.647264	-29.841109	-29.820807	-11.620860	-11.264929	1.741817	4.524843	10.664026	19.450779	20.821727	35.997938	36.118033
.11	-51.466862	-29.547289	-29.528858	-11.202207	-10.872553	2.326889	4.975636	11.309074	19.875317	21.329045	36.326804	36.455753
.12	-51.287343	-29.255114	-29.238396	-10.786894	-10.481854	2.906889	5.425950	11.954960	20.302200	21.841299	36.653910	36.795318
.13	-51.108696	-28.964556	-28.949405	-10.374842	-10.092842	3.482937	5.875741	12.602029	20.730903	22.358517	36.984692	37.137999
.14	-50.930909	-28.675587	-28.661869	-9.965974	-9.705520	4.054554	6.324967	13.249941	21.158896	22.880720	37.317358	37.483468
.15	-50.753972	-28.388181	-28.375771	-9.560216	-9.319889	4.621672	6.773587	13.898527	21.589150	23.407920	37.651923	37.831801
.16	-50.577872	-28.102312	-28.091096	-9.157493	-8.935951	5.184230	7.221560	14.547616	22.020636	23.940312	37.988400	38.183074
.17	-50.402600	-27.817955	-27.807828	-8.757734	-8.553708	5.742177	7.668843	15.197035	22.453323	24.477322	38.326804	38.537370
.18	-50.228144	-27.535086	-27.525951	-8.360869	-8.173161	6.295468	8.115398	15.846412	22.887178	25.019507	38.667148	38.893974
.19	-50.054495	-27.253682	-27.245449	-7.966830	-7.794308	6.844069	8.561184	16.496172	23.322168	25.566654	39.009445	39.255371
.20	-49.881642	-26.973719	-26.966309	-7.575551	-7.417149	7.387305	9.006161	17.145540	23.758260	26.118732	39.353207	39.612503
.21	-49.709575	-26.695177	-26.688513	-7.186968	-7.041682	7.927109	9.450291	17.794538	24.195418	26.675700	39.699947	39.986514
.22	-49.538284	-26.418034	-26.412048	-6.801020	-6.667904	8.461522	9.893536	18.442989	24.633503	27.237508	40.048175	40.354039
.23	-49.367761	-26.142270	-26.136899	-6.417647	-6.295813	8.991195	10.335857	19.090714	25.072286	27.804098	40.398402	40.721564
.24	-49.197995	-25.867864	-25.863050	-6.036790	-5.925405	9.516136	10.777217	19.737531	25.512022	28.375400	40.750636	41.109556
.25	-49.028977	-25.594797	-25.590488	-5.658394	-5.556676	10.036361	11.217580	20.383259	25.953974	28.951339	41.104893	41.491334
.26	-48.860699	-25.323051	-25.319198	-5.282604	-5.189621	10.551895	11.656910	21.027714	26.395902	29.531829	41.461174	41.877006
.27	-48.693151	-25.052607	-25.049166	-4.909166	-4.824234	11.062769	12.095112	21.670712	26.838665	30.116774	41.819491	42.264668
.28	-48.526326	-24.783448	-24.780339	-4.538039	-4.460511	11.569020	12.532332	22.312068	27.282221	30.706174	42.179318	42.658127
.29	-48.360214	-24.515555	-24.512822	-4.169362	-4.098444	12.070694	12.968356	22.951556	27.726529	31.299617	42.542253	43.050530
.30	-48.194807	-24.249911	-24.246482	-3.801495	-3.738027	12.567840	13.403211	23.589108	28.171543	31.897287	42.906710	43.460931
.31	-48.030096	-23.985505	-23.981346	-3.436791	-3.379252	13.060515	13.836865	24.224417	28.617220	32.498958	43.273223	43.867812
.32	-47.866075	-23.719316	-23.717400	-3.074208	-3.022113	13.548778	14.269287	24.857336	29.063515	33.104499	43.641797	44.279296
.33	-47.702735	-23.456328	-23.454632	-2.713703	-2.666600	14.032936	14.700446	25.487677	29.510187	33.713771	44.012431	44.695757
.34	-47.540069	-23.194528	-23.193028	-2.355236	-2.312706	14.512336	15.130314	26.115254	29.957771	34.326631	44.385129	45.116570
.35	-47.378068	-22.933001	-22.932576	-1.998768	-1.960422	14.987771	15.558861	26.739881	30.405638	34.942929	44.759888	45.542609
.36	-47.216725	-22.674433	-22.672265	-1.644262	-1.609739	15.459075	15.986061	27.361372	30.853932	35.562509	45.136709	45.973147
.37	-47.056034	-22.416108	-22.415080	-1.291681	-1.260647	15.926326	16.411885	27.979548	31.302604	36.185209	45.515588	46.401182
.38	-46.895984	-22.158011	-22.159011	-0.940992	-0.913137	16.389216	16.836310	28.594227	31.751622	36.810666	45.896461	46.837167
.39	-46.736576	-21.902838	-21.902046	-0.592160	-0.567199	16.848986	17.259300	29.205235	32.200882	37.439306	46.279503	47.205937
.40	-46.577796	-21.647866	-21.647172	-0.245154	-0.222823	17.304557	17.680860	29.812400	32.650386	38.070357	46.664528	47.751716
.41	-46.419638	-21.393985	-21.393379	0.100057	0.120001	17.756398	18.100939	30.415555	33.100062	38.703837	47.051587	48.212140
.42	-46.262098	-21.141183	-21.140654	0.443503	0.461285	18.204940	18.519525	31.014551	33.549869	39.339563	47.440671	48.674350
.43	-46.105167	-20.889398	-20.888988	0.785214	0.801038	18.649216	18.936598	31.609201	33.999722	39.977345	47.831768	49.144450
.44	-45.948840	-20.638767	-20.638367	1.125217	1.139273	19.090358	19.352137	32.199190	34.449589	40.616947	48.224868	49.620524
.45	-45.793110	-20.389129	-20.388783	1.463539	1.475999	19.528096	19.766124	32.784971	34.899431	41.258304	48.619954	50.102656
.46	-45.637942	-20.140523	-20.140224	1.800204	1.811373	19.962580	20.178542	33.365815	35.349165	41.901784	49.016023	50.585479
.47	-45.483348	-19.892939	-19.892680	2.135269	2.145391	20.393580	20.589440	33.941898	35.798812	42.546189	49.414673	51.085372
.48	-45.329443	-19.646362	-19.646140	2.468666	2.477242	20.821681	20.998605	34.512824	36.248117	43.190193	49.816071	51.586070
.49	-45.176042	-19.400784	-19.400594	2.805510	2.808049	21.246591	21.406221	35.078786	36.697221	43.836100	50.219829	52.093048
.50	-45.023208	-19.156195	-19.156032	3.130793	3.137404	21.668482	21.812208	35.639603	37.146000	44.482612	50.624580	52.606335

$J_{K_{-1}K_1}$ / J_τ / κ	$9_{2,8}$ / 9_{-6}	$9_{2,7}$ / 9_{-5}	$9_{3,7}$ / 9_{-4}	$9_{3,6}$ / 9_{-3}	$9_{4,6}$ / 9_{-2}	$9_{4,5}$ / 9_{-1}	$9_{5,5}$ / 9_{0}	$9_{5,4}$ / 9_{1}	$9_{6,4}$ / 9_{2}	$9_{6,3}$ / 9_{3}	$9_{7,3}$ / 9_{4}	$9_{7,2}$ / 9_{5}
.50	-45.023208	-19.156195	-19.156032	3.130793	3.137404	21.668482	21.812208	35.639603	37.146000	44.482612	50.624580	52.606335
.51	-44.870935	-18.912503	-18.913583	3.459536	3.465320	22.087428	22.216055	36.195203	37.594397	45.129501	51.031196	53.125941
.52	-44.719518	-18.664949	-18.664949	3.789108	3.794808	22.503499	22.615055	36.745290	38.047554	45.424459	51.439652	53.651865
.53	-44.568052	-18.428253	-18.428351	4.118608	4.118808	22.919643	23.020282	37.290581	38.489813	45.718459	51.849832	54.183423
.54	-44.417431	-18.187515	-18.187430	4.436733	4.440546	23.327286	23.419643	37.830182	38.936715	46.070060	52.261963	54.722589
.55	-44.267350	-17.947715	-17.947644	4.759519	4.762820	23.735133	23.817326	38.364461	39.383003	47.716051	52.675756	55.267310
.56	-44.117803	-17.708844	-17.708784	5.080664	5.083712	24.140366	24.213322	38.893364	39.828616	48.361181	53.091260	55.818193
.57	-43.968765	-17.470893	-17.470843	5.400784	5.403235	24.543044	24.607627	39.416898	40.273498	49.005182	53.508439	56.375162
.58	-43.820292	-17.233853	-17.238811	5.719298	5.721399	24.943226	25.000234	39.935083	40.717588	49.647776	53.927253	56.938123
.59	-43.672318	-16.997714	-16.997679	6.036427	6.038217	25.340967	25.391141	40.447954	41.160829	50.288679	54.347660	57.506969
.60	-43.524859	-16.762468	-16.762439	6.352172	6.353701	25.736320	25.780343	40.955553	41.603162	50.927597	54.769615	58.081579
.61	-43.377909	-16.528106	-16.528042	6.668961	6.669584	26.129338	26.167838	41.457436	42.044529	51.564230	55.193073	58.661813
.62	-43.231463	-16.294622	-16.294603	6.990564	6.992499	26.519805	26.553476	41.953790	42.484904	52.199040	55.618272	59.242622
.63	-43.085518	-16.061988	-16.061988	7.291134	7.292256	26.908562	26.937705	42.447135	42.924135	52.829402	56.044295	59.835242
.64	-42.940068	-15.830246	-15.830233	7.601741	7.602514	27.294861	27.320077	42.934488	43.362260	53.457303	56.471952	60.434688
.65	-42.795109	-15.599340	-15.599329	7.910850	7.911494	27.679010	27.700741	43.416746	43.799190	54.081645	56.900897	61.015774
.66	-42.650637	-15.369268	-15.369268	8.218673	8.217207	28.061051	28.079701	43.894197	44.234871	54.702096	57.331070	61.641992
.67	-42.506647	-15.140050	-15.140043	8.525223	8.525665	28.441023	28.456958	44.366941	44.669247	55.318321	57.762408	62.251925
.68	-42.363134	-14.911651	-14.911645	8.830515	8.830877	28.818965	28.832517	44.835008	45.102265	55.929982	58.194843	62.866540
.69	-42.220095	-14.684072	-14.684068	9.134560	9.134855	29.194913	29.206382	45.298742	45.534872	56.536744	58.628306	63.485193
.70	-42.077525	-14.457307	-14.457304	9.437610	9.437610	29.569902	29.578556	45.758019	45.964015	57.138273	59.062725	64.107624
.71	-41.935420	-14.231348	-14.231346	9.739059	9.739153	29.940946	29.949046	46.213030	46.392644	57.734243	59.498021	64.731562
.72	-41.793775	-14.006188	-14.006188	10.039338	10.039493	30.311135	30.317850	46.663892	46.819644	58.324334	59.934115	65.362710
.73	-41.652588	-13.781819	-13.781818	10.338514	10.338641	30.679441	30.684998	47.110719	47.245165	58.908243	60.370923	65.994789
.74	-41.511854	-13.558234	-13.558234	10.636512	10.636608	31.045911	31.050474	47.553624	47.669861	59.485680	60.808357	66.629651
.75	-41.371569	-13.335430	-13.335429	10.933228	10.933404	31.410575	31.414292	47.992723	48.091054	60.056375	61.246326	67.266361
.76	-41.231729	-13.113395	-13.113395	11.228981	11.229039	31.773457	31.776462	48.428125	48.511401	60.620081	61.684732	67.905155
.77	-41.092331	-12.892125	-12.892125	11.523478	11.523523	32.134584	32.136991	48.859941	48.929690	61.176582	62.123446	68.545443
.78	-40.953370	-12.671613	-12.671613	11.816831	11.816865	32.493978	32.495889	49.288277	49.346690	61.725687	62.562453	69.186806
.79	-40.814843	-12.451853	-12.451853	12.109050	12.109050	32.851664	32.853164	49.713237	49.761555	62.267244	63.001552	69.828792
.80	-40.676747	-12.232837	-12.232837	12.400164	12.400164	33.207662	33.208828	50.134922	50.174519	62.801233	63.440661	70.470911
.81	-40.539077	-12.014561	-12.014561	12.690127	12.690127	33.563356	33.563356	50.553160	50.585240	63.326136	63.879838	71.113350
.82	-40.401830	-11.797018	-11.797018	12.979013	12.979013	33.914681	33.916785	50.968849	50.994504	63.845633	64.318620	71.753346
.83	-40.265004	-11.580201	-11.580201	13.266785	13.266785	34.265739	34.266792	51.381278	51.401675	64.356199	64.756817	72.392278
.84	-40.128593	-11.364105	-11.364105	13.553486	13.553486	34.615189	34.615557	51.790786	51.806719	64.859019	65.194713	73.029261
.85	-39.992595	-11.148723	-11.148723	13.839104	13.839104	34.963048	34.963312	52.197468	52.209715	65.354170	65.631967	73.662937
.86	-39.857007	-10.934051	-10.934051	14.123653	14.123653	35.309331	35.309554	52.601197	52.610467	65.841767	66.068432	74.292642
.87	-39.721825	-10.720082	-10.720082	14.407147	14.407147	35.654057	35.654183	53.002643	53.009481	66.321959	66.503955	74.917426
.88	-39.587046	-10.506810	-10.506810	14.689589	14.689589	35.997233	35.997323	53.401277	53.406231	66.794929	66.938579	75.536232
.89	-39.452667	-10.294231	-10.294231	14.970991	14.970991	36.338893	36.338947	53.797360	53.800825	67.260885	67.371537	76.147901
.90	-39.318684	-10.082337	-10.082337	15.251361	15.251361	36.679033	36.679046	54.190952	54.193305	67.720061	67.803260	76.751164
.91	-39.185095	-9.871125	-9.871125	15.530705	15.530705	37.017673	37.017693	54.582108	54.583643	68.172711	68.233371	77.344645
.92	-39.051896	-9.660589	-9.660589	15.809032	15.809032	37.354807	37.354807	54.971832	54.971832	68.619104	68.661687	77.926870
.93	-38.919084	-9.450723	-9.450723	16.086351	16.086351	37.690511	37.690511	55.357314	55.357269	69.059524	69.088020	78.496292
.94	-38.786657	-9.241522	-9.241522	16.362670	16.362670	38.024727	38.022727	55.741453	55.741453	69.494262	69.512178	79.051320
.95	-38.654611	-9.032980	-9.032980	16.637996	16.637996	38.357493	38.357494	56.123335	56.123478	69.923615	69.933960	79.590371
.96	-38.522943	-8.825094	-8.825094	16.912337	16.912337	38.688824	38.688825	56.502997	56.503054	70.347082	70.353164	80.111935
.97	-38.391651	-8.617858	-8.617858	17.185700	17.185700	39.019730	39.018825	56.880468	56.880486	70.767361	70.769582	80.614651
.98	-38.260732	-8.411587	-8.411587	17.458724	17.458924	39.347130	39.374326	57.255947	57.255980	71.182348	71.183004	81.097328
.99	-38.130182	-8.205316	-8.205316	17.729524	17.729524	39.673857	39.674326	57.629947	57.629947	71.593348	71.591418	81.552324
1.00	-38.000000	-8.000000	-8.000000	18.000000	18.000000	40.000000	40.000000	58.000000	58.000000	72.000000	72.000000	82.000000

Column headings are given as J_{K_{-1},K_1} (top) and J_τ (bottom). The first column is κ.

κ	$10_{6,4}$ / 10_2	$10_{7,4}$ / 10_3	$10_{7,3}$ / 10_4	$10_{8,3}$ / 10_5	$10_{8,2}$ / 10_6	$10_{9,2}$ / 10_7	$10_{9,1}$ / 10_8	$10_{10,1}$ / 10_9	$10_{10,0}$ / 10_{10}	$9_{0,9}$ / 9_{-9}	$9_{1,9}$ / 9_{-8}	$9_{1,8}$ / 9_{-7}
.00	10.273643	25.435926	25.689636	45.730373	45.744324	69.253079	69.253465	95.905725	95.905729	−77.320583	−77.320562	−53.504332
.01	10.959151	25.904954	26.180856	46.025548	46.091014	69.465810	69.465618	95.955730	95.970570	−77.255913	−77.255914	−53.314221
.02	11.650686	26.376336	26.676121	46.322746	46.437748	69.681618	69.681950	96.050230	96.050230	−77.191614	−77.191610	−53.125141
.03	12.348484	26.849684	27.175539	46.772762	46.790748	69.892868	69.893420	96.123048	96.123055	−77.127629	−77.127614	−52.937078
.04	13.051969	27.326172	27.679224	47.122893	47.143873	70.108490	70.109111	96.196264	96.196264	−77.063993	−77.063953	−52.750015
.05	13.760750	27.804630	28.187291	47.499229	47.499229	70.325322	70.325384	96.269885	96.269885	−77.000622	−77.000610	−52.563937
.06	14.475625	28.285447	28.699858	47.831238	47.856860	70.543384	70.544168	96.343903	96.343903	−76.937582	−76.937582	−52.378830
.07	15.196079	28.768622	29.217048	48.188526	48.216358	70.763574	70.765354	96.418320	96.418176	−76.874874	−76.874874	−52.194679
.08	15.921984	29.254151	29.738984	48.547928	48.579137	70.983272	70.984260	96.493164	96.493176	−76.812461	−76.812453	−52.011470
.09	16.653203	29.742030	30.265791	48.943883	48.943883	71.205140	71.206248	96.568431	96.568445	−76.750350	−76.750350	−51.829190
.10	17.389586	30.232251	30.797600	49.273186	49.311103	71.428318	71.429560	96.644128	96.644145	−76.688530	−76.688536	−51.647825
.11	18.130979	30.724809	31.334538	49.630899	49.680894	71.652829	71.654219	96.720264	96.720264	−76.627011	−76.627011	−51.467362
.12	18.877194	31.219693	31.876738	50.007240	50.053194	71.878693	71.880250	96.796845	96.796866	−76.565785	−76.565781	−51.287789
.13	19.628070	31.716892	32.424329	50.380334	50.428294	72.105935	72.107676	96.873878	96.873902	−76.504840	−76.504840	−51.109094
.14	20.383414	32.216394	32.977444	50.750319	50.805881	72.334537	72.336524	96.951373	96.951400	−76.444177	−76.444177	−50.931263
.15	21.143031	32.718185	33.536213	51.125317	51.186358	72.564644	72.566820	97.029336	97.029366	−76.383792	−76.383792	−50.754286
.16	21.906719	33.222249	34.100764	51.502658	51.569682	72.796155	72.798590	97.107776	97.107811	−76.323682	−76.323682	−50.578151
.17	22.674267	33.728568	34.671225	51.882374	51.955923	73.029149	73.031864	97.186702	97.186741	−76.263843	−76.263843	−50.402848
.18	23.445462	34.237123	35.247719	52.264493	52.345159	73.263640	73.266669	97.266156	97.266156	−76.204271	−76.204211	−50.228364
.19	24.220081	34.747891	35.830366	52.649045	52.737467	73.499658	73.503037	97.346046	97.346096	−76.144963	−76.144961	−50.054689
.20	24.997099	35.260851	36.419280	53.036059	53.132951	73.737230	73.740999	97.426482	97.426516	−76.085914	−76.085914	−49.881814
.21	25.778634	35.775976	37.014571	53.425565	53.531636	73.976385	73.980586	97.507441	97.507507	−76.027124	−76.027124	−49.709727
.22	26.562201	36.293240	37.616338	53.817593	53.913674	74.217153	74.221832	97.588933	97.588933	−75.968590	−75.968589	−49.538412
.23	27.348210	36.812614	38.224677	54.212172	54.339139	74.459561	74.464773	97.670968	97.671051	−75.910306	−75.910304	−49.367879
.24	28.136468	37.334066	38.839671	54.609172	54.748132	74.703642	74.708404	97.753650	97.753650	−75.852269	−75.852269	−49.198099
.25	28.926727	37.857564	39.461395	55.009098	55.160755	74.949426	74.955883	97.836707	97.836814	−75.794478	−75.794477	−49.029069
.26	29.718737	38.383072	40.089911	55.411503	55.577119	75.196947	75.204130	97.920435	97.920556	−75.736929	−75.736928	−48.860727
.27	30.512242	38.910555	40.725271	55.816573	55.997338	75.445236	75.454224	98.004750	98.004886	−75.679619	−75.679619	−48.693072
.28	31.306985	39.439971	41.367511	56.224336	56.421530	75.697329	75.706129	98.089664	98.094886	−75.622545	−75.622618	−48.526387
.29	32.102704	39.971281	42.016656	56.634820	56.849823	75.950260	75.960129	98.175190	98.175364	−75.565705	−75.565705	−48.360267
.30	32.899135	40.504442	42.672715	57.048051	57.282346	76.205066	76.216030	98.261341	98.261537	−75.509097	−75.509096	−48.194853
.31	33.696010	41.039407	43.335683	57.464005	57.719236	76.461784	76.473970	98.348130	98.348352	−75.452716	−75.452561	−48.030117
.32	34.493056	41.576129	44.005532	57.882857	58.160636	76.720451	76.733970	98.435571	98.435821	−75.396561	−75.396561	−47.866111
.33	35.289998	42.114560	44.682230	58.304482	58.606936	76.981107	76.996110	98.523678	98.523961	−75.340629	−75.340629	−47.702766
.34	36.086557	42.654647	45.365719	58.728953	59.057571	77.243792	77.260443	98.612467	98.612786	−75.284918	−75.284917	−47.540095
.35	36.882451	43.196336	46.055928	59.156293	59.513422	77.508548	77.527020	98.701952	98.702312	−75.229424	−75.229424	−47.378091
.36	37.677393	43.739573	46.752769	59.586524	59.974416	77.774413	77.795904	98.792160	98.792556	−75.174147	−75.174147	−47.216745
.37	38.471093	44.284299	47.456137	60.019666	60.439966	78.041413	78.065904	98.883555	98.883555	−75.119082	−75.119082	−47.056051
.38	39.263359	44.830454	48.165961	60.455662	60.912541	78.315671	78.340853	98.974750	98.975267	−75.064229	−75.064229	−46.896001
.39	40.053593	45.377976	48.881952	60.894758	61.390036	78.589147	78.617055	99.067168	99.067711	−75.009584	−75.009584	−46.736589
.40	40.841795	45.926801	49.604109	61.336407	61.873407	78.864919	78.895842	99.160409	99.161067	−74.955146	−74.955146	−46.577807
.41	41.627564	46.476863	50.332214	61.781706	62.362850	79.143035	79.172291	99.254432	99.255174	−74.900912	−74.900912	−46.419648
.42	42.410594	47.028094	51.066084	62.229661	62.858655	79.423545	79.461485	99.349278	99.350115	−74.846880	−74.846880	−46.262106
.43	43.190578	47.580424	51.805523	62.680620	63.360757	79.706500	79.748513	99.444968	99.445911	−74.793047	−74.793047	−46.105174
.44	43.967209	48.133779	52.550322	63.134668	63.869633	79.991954	80.033468	99.541522	99.542568	−74.739413	−74.739413	−45.948846
.45	44.740180	48.688087	53.300257	63.591584	64.385403	80.289954	80.331447	99.638965	99.640165	−74.685974	−74.685974	−45.793115
.46	45.509183	49.243271	54.055095	64.051602	64.908261	80.593574	80.637574	99.736473	99.738637	−74.632729	−74.632729	−45.637977
.47	46.273491	49.799353	54.814850	64.514815	65.438261	80.901574	80.948636	99.836611	99.838137	−74.579676	−74.579676	−45.483422
.48	47.031407	50.356255	55.579425	64.980861	65.980725	81.199352	81.226609	99.936865	99.938685	−74.526813	−74.526813	−45.329446
.49	47.789354	50.913288	56.346510	65.449829	66.521591	81.489636	81.535798	100.038108	100.040051	−74.474137	−74.474137	−45.176044
.50	48.539478	51.471176	57.118390	65.921956	67.074940	81.760262	81.846601	100.140369	100.142561	−74.421647	−74.421647	−45.023210

J_{K_{-1},K_1} / J_τ / κ	$10_{6,4}$ / 10_2	$10_{7,4}$ / 10_3	$10_{7,3}$ / 10_4	$10_{8,3}$ / 10_5	$10_{8,2}$ / 10_6	$10_{9,2}$ / 10_7	$10_{9,1}$ / 10_8	$10_{10,1}$ / 10_9	$10_{10,0}$ / 10_{10}	$9_{0,9}$ / 9_{-9}	$9_{1,9}$ / 9_{-8}	$9_{1,8}$ / 9_{-7}
.50	48.539478	51.471176	57.118390	65.921956	67.074040	81.760262	81.845601	100.140369	100.142561	−74.421647	−74.421647	−45.029210
.51	49.284159	52.029530	57.893834	66.397099	67.636397	82.164744	82.159162	100.243367	100.246151	−74.369341	−74.369341	−44.791641
.52	50.023126	52.588267	58.642314	66.876349	68.204144	82.476532	82.476132	100.348063	100.350856	−74.317218	−74.317218	−44.719220
.53	50.756127	53.147387	59.446016	67.356392	68.784314	82.682831	82.798170	100.453560	100.456712	−74.265275	−74.265275	−44.568053
.54	51.482889	53.706511	60.238522	67.840512	69.371155	82.996468	83.123951	100.562201	100.563761	−74.213511	−74.213511	−44.417432
.55	52.203211	54.265841	61.025140	68.327590	69.966713	83.313273	83.454156	100.668022	100.672042	−74.161924	−74.161924	−44.267350
.56	52.916872	54.825186	61.813729	68.817603	70.571134	83.633317	83.788984	100.777060	100.781602	−74.110512	−74.110512	−44.117803
.57	53.623682	55.384449	62.603936	69.310525	71.184516	83.956669	84.128643	100.887544	100.892487	−74.059273	−74.059273	−43.968786
.58	54.323470	55.943534	63.395401	69.806326	71.806930	84.283401	84.473358	100.998945	101.004749	−74.008207	−74.008207	−43.820293
.59	55.016091	56.502343	64.187746	70.304972	72.438426	84.613586	84.823368	101.111876	101.118440	−73.957310	−73.957310	−43.672319
.60	55.701423	57.060777	64.980584	70.806425	73.079023	84.947298	85.178932	101.226194	101.233620	−73.906583	−73.906583	−43.524859
.61	56.379369	57.618736	65.773513	71.310644	73.728716	85.284519	85.540321	101.341945	101.350353	−73.856022	−73.856022	−43.377900
.62	57.049860	58.176120	66.566115	71.817582	74.387466	85.625599	85.907830	101.459180	101.468697	−73.805627	−73.805627	−43.231463
.63	57.712853	58.732825	67.357959	72.327188	75.055208	85.970342	86.281770	101.577952	101.588733	−73.755396	−73.755396	−43.085518
.64	58.368329	59.288751	68.148593	72.839409	75.731842	86.318915	86.662473	101.698317	101.710536	−73.705327	−73.705327	−42.940068
.65	59.016299	59.843793	68.937554	73.354182	76.417240	86.671397	87.050294	101.820133	101.834189	−73.655420	−73.655420	−42.795109
.66	59.656795	60.397850	69.724538	73.871444	77.111241	87.027864	87.445606	101.944063	101.959786	−73.605672	−73.605672	−42.650634
.67	60.289876	60.950816	70.508503	74.391123	77.813655	87.388397	87.848809	102.065574	102.087424	−73.556082	−73.556082	−42.506647
.68	60.915621	61.502589	71.289471	74.913145	78.524259	87.753073	88.260319	102.196793	102.217212	−73.506649	−73.506649	−42.363134
.69	61.534131	62.053065	72.066726	75.437428	79.242802	88.121970	88.680579	102.326216	102.349270	−73.457371	−73.457371	−42.220095
.70	62.145127	62.602141	72.839716	75.963384	79.969002	88.495167	89.110046	102.457499	102.483727	−73.408247	−73.408247	−42.077525
.71	62.749943	63.149715	73.607875	76.492420	80.702551	88.872741	89.549201	102.590866	102.620728	−73.359275	−73.359275	−41.935420
.72	63.347531	63.695687	74.370625	77.022937	81.443109	89.254731	89.998537	102.726403	102.760430	−73.310455	−73.310455	−41.793775
.73	63.938453	64.239954	75.127377	77.555328	82.190306	89.641329	90.458562	102.864204	102.903010	−73.261784	−73.261784	−41.652588
.74	64.522882	64.782419	75.877542	78.089480	82.943751	90.032490	90.929790	103.004662	103.048662	−73.213263	−73.213263	−41.511854
.75	65.100999	65.322985	76.620529	78.625271	83.703015	90.428336	91.412739	103.146994	103.197606	−73.164888	−73.164888	−41.371569
.76	65.672988	65.861555	77.355755	79.162574	84.467642	90.828929	91.907920	103.292193	103.350084	−73.116660	−73.116660	−41.231729
.77	66.239039	66.398036	78.082652	79.701251	85.237143	91.234342	92.415837	103.440096	103.506373	−73.068576	−73.068576	−41.092331
.78	66.799342	66.932338	78.800676	80.241158	86.010991	91.644644	92.936969	103.590814	103.666782	−73.020636	−73.020636	−40.953370
.79	67.354089	67.464373	79.509314	80.782141	86.788619	92.059314	93.471770	103.744483	103.831662	−72.972838	−72.972838	−40.814843
.80	67.903469	67.994055	80.208097	81.324036	87.569415	92.480168	94.020650	103.901244	104.001413	−72.925182	−72.925182	−40.676747
.81	68.447465	68.521303	80.896607	81.866671	88.352714	92.905133	94.583973	104.061254	104.176490	−72.877665	−72.877665	−40.539077
.82	68.986861	69.046039	81.574487	82.409863	89.137789	93.335989	95.162044	104.224655	104.357415	−72.830287	−72.830287	−40.401830
.83	69.521232	69.568188	82.241448	82.953416	89.923840	93.771648	95.755101	104.391635	104.544789	−72.783047	−72.783047	−40.265004
.84	70.050948	70.087680	82.897280	83.497127	90.709985	94.212538	96.363307	104.562162	104.739030	−72.735944	−72.735944	−40.128593
.85	70.576170	70.604449	83.541855	84.040778	91.495239	94.658706	96.986746	104.737034	104.941757	−72.688976	−72.688976	−39.992595
.86	71.097055	71.118434	84.175127	84.584319	92.278500	95.110190	97.625418	104.915852	105.153037	−72.642142	−72.642142	−39.856979
.87	71.613747	71.629578	84.797138	85.126969	93.058526	95.567027	98.279241	105.099031	105.374336	−72.595442	−72.595442	−39.721825
.88	72.126187	72.137831	85.408013	85.669012	93.833913	96.029241	98.948604	105.286709	105.606301	−72.548874	−72.548874	−39.587131
.89	72.655102	72.643145	86.007957	86.209999	94.603071	96.496877	99.631600	105.479397	105.851850	−72.502437	−72.502437	−39.452647
.90	73.140013	73.145480	86.597249	86.749646	95.364197	96.969936	100.329578	105.677082	106.112206	−72.456130	−72.456130	−39.318684
.91	73.641232	73.644801	87.176233	87.287656	96.115260	97.448442	101.041605	105.880124	106.386750	−72.409953	−72.409953	−39.185095
.92	74.138861	74.141077	87.745308	87.823717	96.853983	97.932438	101.767246	106.088005	106.674500	−72.363903	−72.363903	−39.051896
.93	74.632994	74.634285	88.304921	88.357502	97.577851	98.421829	102.506023	106.303425	106.995269	−72.317980	−72.317980	−38.919084
.94	75.123713	75.124406	88.855554	88.888671	98.284137	98.916714	103.257427	106.524197	107.333269	−72.272184	−72.272184	−38.786657
.95	75.611096	75.611428	89.397717	89.416868	98.969973	99.417054	104.020900	106.751747	107.697413	−72.226513	−72.226513	−38.654611
.96	76.095208	76.095343	89.931936	89.941725	99.632460	99.922838	104.795879	106.984631	108.090513	−72.180965	−72.180965	−38.522945
.97	76.575847	76.576151	90.458451	90.463172	100.268335	100.433822	105.581879	107.223809	108.520531	−72.135541	−72.135541	−38.391651
.98	77.053847	77.053856	90.978672	90.978883	100.876665	100.950066	106.378253	107.477035	108.974009	−72.090240	−72.090240	−38.260732
.99	77.528466	77.528467	91.492252	91.492403	101.454090	101.472660	107.184474	107.734323	109.468570	−72.045059	−72.045059	−38.130182
1.00	78.000000	78.000000	92.000000	92.000000	102.000000	102.000000	108.000000	108.000000	110.000000	−72.000000	−72.000000	−38.000000

$J_{K_{-1}K_1}$ J_{τ} κ	$10_{6,5}$ 10_1	$10_{5,5}$ 10_0	$10_{5,6}$ 10_{-1}	$10_{4,6}$ 10_{-2}	$10_{4,7}$ 10_{-3}	$10_{3,7}$ 10_{-4}	$10_{3,8}$ 10_{-5}	$10_{2,8}$ 10_{-6}	$10_{2,9}$ 10_{-7}	$10_{1,9}$ 10_{-8}	$10_{1,10}$ 10_{-9}	$10_{0,10}$ 10_{-10}
.00	8.052907	0.000000	-8.492907	-10.273643	-25.435926	-25.689636	-45.730373	-45.744324	-69.253079	-69.253465	-95.905725	-95.905729
.01	8.604039		-7.448271	-9.592624	-24.952436	-25.128911	-45.185777	-45.413458	-69.032216	-69.032403	-95.833409	-95.833416
.02		1.598439		-8.920896	-24.504915		-44.706307	-45.075777	-68.832578	-68.832578	-95.762721	-95.762724
.03	9.721626	2.394050	-6.394870	-8.253717	-24.042919	-24.229204	-44.368688	-44.716501	-68.623806	-68.623806	-95.691769	-95.691769
.04	10.279986	3.190930	-5.844806	-7.592688	-23.583253	-23.763119		-44.377855	-68.415890	-68.416127	-95.621171	-95.621171
.05	10.839312	3.986836	-5.296133	-6.937806	-23.125910	-23.290623	-44.032896	-44.041134	-68.209315	-68.209525	-95.550923	-95.550923
.06	11.399546	4.781524	-4.748898	-6.289058	-22.670879	-22.821567	-43.698907	-43.706304	-68.003799	-68.003984	-95.481025	-95.481027
.07	11.960608	5.574752	-4.203147	-5.644618	-22.218151	-22.355886	-43.366699	-43.373134	-67.799325	-67.799489	-95.411469	-95.411471
.08	12.522503	6.366277	-3.658925	-5.009851	-21.767715	-21.893497	-43.036247	-43.042194	-67.595880	-67.596024	-95.342251	-95.342252
.09	13.085104	7.155857	-3.116275	-4.379309	-21.319559	-21.434321	-42.712855	-42.712855	-67.393448	-67.393575	-95.273366	-95.273367
.10	13.648373	7.943250	-2.575238	-3.754737	-20.873671	-20.978282	-42.380522	-42.385287	-67.192017	-67.192128	-95.204809	-95.204810
.11	14.212246	8.728218	-2.035857	-3.130069	-20.430137	-20.525306	-42.055205	-42.059464	-66.991572	-66.991572	-95.136575	-95.136576
.12	14.776660	9.510522	-1.498169	-2.523232	-19.988645	-20.075324	-41.731559	-41.735359	-66.792100	-66.792186	-95.068658	-95.068662
.13	15.341552	10.289924	-0.962212	-1.916144	-19.549480	-19.628267	-41.409552	-41.412946	-66.593587	-66.593663	-95.001063	-95.001063
.14	15.906858	11.066190	-0.428022	-1.314721	-19.112528	-19.184070	-41.089173	-41.092199	-66.396022	-66.396088	-94.933775	-94.933776
.15	16.472511	11.839089	0.104365	-0.718867	-18.677773	-18.742672	-40.770400	-40.773094	-66.199392	-66.199450	-94.866795	-94.866795
.16	17.038448	12.608392	0.634918	-0.128487	-18.245199	-18.304013	-40.453211	-40.455607	-66.003683	-66.003734	-94.800117	-94.800117
.17	17.604600	13.373875	1.163606	0.456521	-17.814791	-17.868034	-40.137586	-40.139715	-65.808886	-65.808930	-94.733738	-94.733738
.18	18.170904	14.135317	1.690397	1.036262	-17.386532	-17.434682	-39.823507	-39.825396	-65.614986	-65.615025	-94.667654	-94.667654
.19	18.737290	14.892506	2.215265	1.610845	-16.960406	-17.003902	-39.510953	-39.512628	-65.421975	-65.422008	-94.601862	-94.601862
.20	19.303692	15.645232	2.738181	2.180377	-16.536396	-16.575644	-39.199906	-39.201389	-65.229839	-65.229868	-94.536357	-94.536357
.21	19.870044	16.393296	3.259121	2.744969	-16.114483	-16.149860	-38.890347	-38.891659	-65.038568	-65.038593	-94.471136	-94.471137
.22	20.436276	17.136503	3.778062	3.304732	-15.694652	-15.726501	-38.582259	-38.583418	-64.848151	-64.848173	-94.406196	-94.406196
.23	21.002322	17.874670	4.294981	3.859777	-15.276884	-15.305524	-38.275622	-38.276645	-64.658578	-64.658597	-94.341534	-94.341534
.24	21.568113	18.607623	4.809857	4.410216	-14.861611	-14.886885	-37.971322	-37.971322	-64.469854	-64.469854	-94.277144	-94.277144
.25	22.133581	19.335198	5.322672	4.956158	-14.447466	-14.470541	-37.666637	-37.667430	-64.281921	-64.281935	-94.213026	-94.213026
.26	22.698660	20.057242	5.833408	5.497712	-14.035780	-14.056453	-37.364253	-37.364950	-64.094818	-64.094830	-94.149174	-94.149175
.27	23.263286	20.773616	6.342049	6.034985	-13.626086	-13.644583	-37.063865	-37.063865	-63.908528	-63.908528	-94.085587	-94.085587
.28	23.827375	21.484192	6.848581	6.568083	-13.218364	-13.234892	-36.763621	-36.764157	-63.723021	-63.723021	-94.022261	-94.022261
.29	24.390816	22.188858	7.352960	7.097109	-12.812596	-12.827345	-36.465340	-36.465809	-63.538291	-63.538291	-93.959192	-93.959192
.30	24.953718	22.887515	7.855265	7.622164	-12.408765	-12.421908	-36.168395	-36.168805	-63.354345	-63.354352	-93.896378	-93.896378
.31	25.515813	23.580077	8.355596	8.143346	-12.006803	-12.018547	-35.873728	-35.873756	-63.171171	-63.171175	-93.833816	-93.833816
.32	26.077155	24.266477	8.853374	8.660751	-11.606839	-11.617230	-35.578631	-35.578762	-62.988744	-62.988749	-93.771504	-93.771504
.33	26.637619	24.946659	9.349191	9.174473	-11.208708	-11.217926	-35.285292	-35.285262	-62.807076	-62.807072	-93.709437	-93.709437
.34	27.197159	25.620584	9.842842	9.684602	-10.812439	-10.820604	-34.993667	-34.993902	-62.626140	-62.626143	-93.647614	-93.647614
.35	27.755710	26.288230	10.334321	10.191224	-10.418016	-10.425236	-34.703174	-34.703378	-62.445940	-62.445943	-93.586031	-93.586031
.36	28.313594	26.949587	10.823625	10.694426	-10.025420	-10.031795	-34.413929	-34.414105	-62.266467	-62.266467	-93.524687	-93.524687
.37	28.869501	27.604661	11.310755	11.194087	-9.634633	-9.640250	-34.125916	-34.126068	-62.087707	-62.087707	-93.463578	-93.463578
.38	29.424801	28.253471	11.795699	11.690887	-9.245637	-9.250750	-33.839254	-33.839264	-61.909655	-61.909655	-93.402702	-93.402702
.39	29.978770	28.896052	12.278468	12.184303	-8.858416	-8.862757	-33.553535	-33.553648	-61.732302	-61.732303	-93.342056	-93.342056
.40	30.531441	29.532448	12.759059	12.674605	-8.472950	-8.476755	-33.269141	-33.269238	-61.555645	-61.555645	-93.281638	-93.281638
.41	31.082754	30.162718	13.237474	13.161866	-8.089223	-8.092251	-32.985927	-32.986010	-61.379671	-61.379672	-93.221445	-93.221445
.42	31.632652	30.786930	13.713716	13.646152	-7.707217	-7.710123	-32.703880	-32.703951	-61.204377	-61.204377	-93.161475	-93.161475
.43	32.181078	31.405163	14.187790	14.127528	-7.326915	-7.329447	-32.422988	-32.423048	-61.029752	-61.029752	-93.101726	-93.101726
.44	32.727977	32.017504	14.659700	14.606056	-6.948300	-6.950502	-32.143290	-32.143290	-60.855792	-60.855792	-93.042195	-93.042195
.45	33.273295	32.624050	15.129452	15.081795	-6.571356	-6.573266	-31.864620	-31.864663	-60.682488	-60.682489	-92.982881	-92.982881
.46	33.816979	33.224054	15.597054	15.554803	-6.196066	-6.197840	-31.587120	-31.587156	-60.509835	-60.509836	-92.923780	-92.923780
.47	34.358978	33.820171	16.062551	16.025144	-5.822440	-5.823840	-31.310805	-31.311165	-60.337827	-60.337827	-92.864890	-92.864890
.48	34.899243	34.409968	16.525833	16.492842	-5.450481	-5.451610	-31.035455	-31.035456	-60.166452	-60.164452	-92.806210	-92.806210
.49	35.437726	34.994411	16.987029	16.957976	-5.079953	-5.081009	-30.761238	-30.761238	-59.995710	-59.995710	-92.747738	-92.747738
.50	35.974381	35.573619	17.446108	17.420584	-4.711115	-4.712020	-30.488076	-30.488094	-59.825591	-59.825591	-92.689470	-92.689470

κ ($J_{K_{-1}K_1}$ / J_τ)	$10_{0,10}$ / 10_{-10}	$10_{1,10}$ / 10_{-9}	$10_{1,9}$ / 10_{-8}	$10_{2,9}$ / 10_{-7}	$10_{2,8}$ / 10_{-6}	$10_{3,8}$ / 10_{-5}	$10_{3,7}$ / 10_{-4}	$10_{4,7}$ / 10_{-3}	$10_{4,6}$ / 10_{-2}	$10_{5,6}$ / 10_{-1}	$10_{5,5}$ / 10_0	$10_{6,5}$ / 10_1
.50	-92.689470	-92.689470	-59.825591	-59.825591	-30.488094	-30.488076	-4.712020	-4.711115	17.420584	17.446108	35.573619	35.974381
.51	-92.639406	-92.639406	-59.745690	-59.745690	-30.216970	-30.216970	-4.344623	-4.343856	17.880712	17.903080	36.147816	36.509363
.52	-92.573543	-92.573543	-59.487200	-59.487200	-29.944985	-29.944985	-3.978804	-3.978144	18.338405	18.357957	36.716825	37.042716
.53	-92.515880	-92.515880	-59.318916	-59.318916	-29.674999	-29.674999	-3.614538	-3.613979	18.793704	18.810749	37.281070	37.572943
.54	-92.458413	-92.458413	-59.151231	-59.151231	-29.406035	-29.406035	-3.251816	-3.251342	19.246651	19.261469	37.840574	38.101860
.55	-92.401142	-92.401142	-58.984140	-58.984140	-29.138109	-29.138109	-2.890617	-2.890218	19.697284	19.710128	38.395840	38.628746
.56	-92.344064	-92.344064	-58.817637	-58.817637	-28.871186	-28.871186	-2.530927	-2.530591	20.145640	20.156740	38.945851	39.153565
.57	-92.287178	-92.287178	-58.651716	-58.651716	-28.605264	-28.605264	-2.172729	-2.172447	20.591755	20.601318	39.491866	39.676285
.58	-92.230481	-92.230481	-58.486372	-58.486372	-28.340333	-28.340333	-1.816008	-1.815773	21.035664	21.043875	40.033623	40.196874
.59	-92.173973	-92.173973	-58.321599	-58.321599	-28.076384	-28.076384	-1.460749	-1.460553	21.477399	21.484424	40.571233	40.715304
.60	-92.117650	-92.117650	-58.157391	-58.157391	-27.813408	-27.813408	-1.106937	-1.106775	21.916991	21.922981	41.104812	41.231547
.61	-92.061512	-92.061512	-57.993745	-57.993745	-27.551395	-27.551395	-0.754657	-0.754423	22.354481	22.359558	41.634467	41.745579
.62	-92.005557	-92.005557	-57.830653	-57.830653	-27.290336	-27.290336	-0.403595	-0.403486	22.789868	22.794172	42.160302	42.257376
.63	-91.949783	-91.949783	-57.668111	-57.668111	-27.030222	-27.030222	-0.054038	-0.053948	23.223211	23.226836	42.682421	42.766919
.64	-91.894188	-91.894188	-57.506115	-57.506115	-26.771045	-26.771045	0.294128	0.294201	23.654525	23.657565	43.200919	43.274188
.65	-91.838771	-91.838771	-57.344658	-57.344658	-26.512795	-26.512795	0.640918	0.640976	24.083837	24.086375	43.715891	43.779165
.66	-91.783530	-91.783530	-57.183737	-57.183737	-26.255465	-26.255465	0.986345	0.986390	24.511172	24.513200	44.227428	44.281838
.67	-91.728463	-91.728463	-57.023346	-57.023346	-25.999046	-25.999046	1.330418	1.330456	24.936554	24.938296	44.735614	44.782192
.68	-91.673570	-91.673570	-56.863480	-56.863480	-25.743529	-25.743529	1.673334	1.673372	25.359880	25.361787	45.240734	45.280009
.69	-91.618848	-91.618848	-56.704135	-56.704135	-25.488907	-25.488907	2.014564	2.014588	25.781550	25.782719	45.742264	45.775902
.70	-91.564295	-91.564295	-56.545306	-56.545306	-25.235171	-25.235171	2.354661	2.354670	26.201209	26.201978	46.240881	46.269222
.71	-91.509911	-91.509911	-56.386989	-56.386989	-24.982314	-24.982314	2.693456	2.693470	26.619003	26.619763	46.736455	46.760230
.72	-91.455694	-91.455694	-56.229179	-56.229179	-24.730327	-24.730327	3.030961	3.030972	27.034952	27.035565	47.229050	47.248864
.73	-91.401643	-91.401643	-56.071871	-56.071871	-24.479204	-24.479204	3.367188	3.367197	27.449077	27.449563	47.718742	47.735140
.74	-91.347755	-91.347755	-55.915062	-55.915062	-24.228937	-24.228937	3.702149	3.702155	27.861396	27.861816	48.205579	48.219059
.75	-91.294030	-91.294030	-55.758747	-55.758747	-23.979517	-23.979517	4.035853	4.035858	28.271927	28.272227	48.689624	48.700622
.76	-91.240467	-91.240467	-55.602921	-55.602921	-23.730938	-23.730938	4.368314	4.368317	28.680690	28.680921	49.170931	49.179831
.77	-91.187063	-91.187063	-55.447582	-55.447582	-23.483194	-23.483194	4.699541	4.699543	29.087877	29.087877	49.649551	49.656689
.78	-91.133817	-91.133817	-55.292723	-55.292723	-23.236275	-23.236275	5.029544	5.029546	29.493110	29.493110	50.125555	50.131204
.79	-91.080729	-91.080729	-55.138342	-55.138342	-22.990176	-22.990176	5.358336	5.358337	29.896533	29.896634	50.598921	50.603300
.80	-91.027797	-91.027797	-54.984435	-54.984435	-22.744890	-22.744890	5.685926	5.685927	30.298163	30.298463	51.069760	51.073226
.81	-90.975019	-90.975019	-54.830997	-54.830997	-22.500410	-22.500410	6.012324	6.012324	30.698558	30.698612	51.538091	51.540701
.82	-90.922394	-90.922394	-54.678024	-54.678024	-22.256728	-22.256728	6.337540	6.337541	31.097056	31.097095	52.003952	52.005964
.83	-90.869922	-90.869922	-54.525514	-54.525514	-22.013840	-22.013840	6.661585	6.661585	31.493773	31.493899	52.467378	52.469878
.84	-90.817600	-90.817600	-54.373461	-54.373461	-21.771737	-21.771737	6.984468	6.984468	31.889099	31.889118	52.928404	52.929502
.85	-90.765427	-90.765427	-54.221862	-54.221862	-21.530414	-21.530414	7.306198	7.306198	32.282673	32.282686	53.387062	53.387936
.86	-90.713403	-90.713403	-54.070714	-54.070714	-21.289864	-21.289864	7.626786	7.626786	32.674636	32.674657	53.843932	53.843936
.87	-90.661527	-90.661527	-53.920013	-53.920013	-21.050081	-21.050081	7.946459	7.946459	33.064539	33.064539	54.298545	54.298939
.88	-90.609796	-90.609796	-53.769756	-53.769756	-20.811059	-20.811059	8.264569	8.264569	33.453773	33.453776	54.749123	54.749375
.89	-90.558210	-90.558210	-53.619938	-53.619938	-20.572791	-20.572791	8.581783	8.581783	33.840977	33.840977	55.198596	55.198763
.90	-90.506767	-90.506767	-53.470556	-53.470556	-20.335273	-20.335273	8.897891	8.897891	34.226622	34.226623	55.645866	55.645935
.91	-90.455467	-90.455467	-53.321607	-53.321607	-20.098497	-20.098497	9.212901	9.212902	34.610720	34.610720	56.090866	56.090925
.92	-90.404308	-90.404308	-53.173088	-53.173088	-19.862458	-19.862458	9.526823	9.526823	34.993283	34.993284	56.533707	56.533936
.93	-90.353290	-90.353290	-53.024994	-53.024994	-19.627150	-19.627150	9.839664	9.839664	35.374325	35.374326	56.974385	56.974402
.94	-90.302411	-90.302411	-52.877324	-52.877324	-19.392567	-19.392567	10.151434	10.151434	35.753858	35.753858	57.412914	57.412917
.95	-90.251670	-90.251670	-52.730073	-52.730073	-19.158705	-19.158705	10.462139	10.462139	36.131892	36.131892	57.849314	57.849317
.96	-90.201066	-90.201066	-52.583238	-52.583238	-18.925557	-18.925557	10.771790	10.771790	36.508440	36.508440	58.283604	58.283603
.97	-90.150598	-90.150598	-52.436817	-52.436817	-18.693118	-18.693118	11.080393	11.080393	36.883514	36.883514	58.715801	58.715801
.98	-90.100265	-90.100265	-52.290805	-52.290805	-18.461382	-18.461382	11.387968	11.387968	37.257124	37.257283	59.145922	59.145922
.99	-90.050066	-90.050066	-52.145200	-52.145200	-18.230344	-18.230344	11.694490	11.694490	37.629283	37.629283	59.573983	59.573983
1.00	-90.000000	-90.000000	-52.000000	-52.000000	-18.000000	-18.000000	12.000000	12.000000	38.000000	38.000000	60.000000	60.000000

J_{K_{-1},K_1} / J_τ → κ	$11_{6,6}$ / 11_0	$11_{6,5}$ / 11_1	$11_{7,5}$ / 11_2	$11_{7,4}$ / 11_3	$11_{8,4}$ / 11_4	$11_{8,3}$ / 11_5	$11_{9,3}$ / 11_6	$11_{9,2}$ / 11_7	$11_{10,2}$ / 11_8	$11_{10,1}$ / 11_9	$11_{11,1}$ / 11_{10}	$11_{11,0}$ / 11_{11}
.00	0.000000	5.073416	17.590939	18.528459	37.498959	37.578159	60.633351	60.636950	87.004758	87.004844	116.491010	116.491011
.01	0.694530	6.002491	18.234087	19.246058	38.028053	38.108034	61.013604	61.017653	87.237653	87.238086	116.570146	116.570146
.02	1.389969	6.936016	18.879736	19.970846	38.545066	38.641635	61.396015	61.400564	87.472479	87.472591	116.649696	116.649698
.03	2.083227	7.873687	19.527845	20.702957	39.072555	39.179057	61.780614	61.785224	87.708580	87.708380	116.729669	116.729669
.04	2.777213	8.815195	20.178371	21.442519	39.605034	39.720394	62.167435	62.173169	87.945325	87.945471	116.810070	116.810070
.05	3.470836	9.760223	20.831268	22.189651	40.136530	40.265749	62.556941	62.562941	88.183720	88.183887	116.890905	116.890904
.06	4.164007	10.708453	21.486486	22.944459	40.673064	40.815229	62.947876	62.955082	88.423458	88.423649	116.972180	116.972182
.07	4.856636	11.659557	22.143976	23.707040	41.212661	41.369360	63.341565	63.349635	88.664561	88.664561	117.053905	117.053909
.08	5.548635	12.613209	22.803682	24.477475	41.755343	41.927007	63.737614	63.746645	88.907049	88.907296	117.136086	117.136089
.09	6.239915	13.569075	23.465159	25.255832	42.301130	42.489543	64.136139	64.146139	89.150932	89.151228	117.218731	117.218734
.10	6.930390	14.526822	24.129518	26.042164	42.850043	43.056677	64.536936	64.548224	89.396279	89.396597	117.301846	117.301850
.11	7.619972	15.486109	24.795529	26.836503	43.402100	43.628540	64.940284	64.952891	89.643066	89.643428	117.385441	117.385445
.12	8.308576	16.446498	25.463517	27.638868	43.957310	44.205272	65.345146	65.360213	89.891335	89.891335	117.469528	117.469532
.13	8.996518	17.407746	26.133516	28.449840	44.515447	44.776307	65.751416	65.764746	90.141118	90.141118	117.554115	117.554120
.14	9.682515	18.369808	26.805159	29.267642	45.077309	45.373915	66.165538	66.183037	90.392419	90.392947	117.639181	117.639187
.15	10.367684	19.331836	27.478674	30.093987	45.642106	45.966132	66.579159	66.593655	90.645287	90.645885	117.724775	117.724782
.16	11.051545	20.293682	28.153888	30.928224	46.210121	46.563324	66.995449	67.017158	90.899742	90.900419	117.810891	117.810899
.17	11.734020	21.254996	28.830727	31.770271	46.781364	47.167158	67.414450	67.438610	91.155813	91.156579	117.897538	117.897548
.18	12.415029	22.215424	29.509113	32.620120	47.355842	47.776158	67.836206	67.863206	91.413528	91.414395	117.984727	117.984738
.19	13.094497	23.174614	30.188966	33.477347	47.933561	48.391148	68.260759	68.290630	91.672919	91.673898	118.072466	118.072479
.20	13.772351	24.133211	30.870204	34.342103	48.514525	49.012763	68.688154	68.721341	91.934015	91.935122	118.160766	118.160781
.21	14.448516	25.087861	31.552745	35.214352	49.098194	49.640042	69.118435	69.154707	92.196814	92.198099	118.249638	118.249638
.22	15.122924	26.037493	32.236502	36.093182	49.685118	50.272098	69.551449	69.592547	92.461128	92.462598	118.339093	118.339093
.23	15.795503	26.991875	32.921387	36.979182	50.276894	50.915652	69.987838	70.033206	92.727861	92.729455	118.429142	118.429165
.24	16.466189	27.939533	33.607313	37.871795	50.870831	51.563581	70.427852	70.477352	92.996108	92.997907	118.519827	118.519822
.25	17.134917	28.883816	34.294187	38.770817	51.467997	52.218658	70.869338	70.925076	93.266259	93.268259	118.611069	118.611069
.26	17.801624	29.824370	34.981916	39.675990	52.068382	52.881085	71.314742	71.376476	93.538266	93.540552	118.702971	118.702971
.27	18.466247	30.760845	35.670408	40.587046	52.671809	53.551064	71.763312	71.831653	93.812251	93.814827	118.795515	118.795515
.28	19.128731	31.692895	36.359565	41.503698	53.278748	54.228794	72.215098	72.290716	94.088226	94.091128	118.888716	118.888716
.29	19.789020	32.620176	37.049292	42.425649	53.888603	54.914472	72.670649	72.753777	94.366231	94.369500	118.982587	118.982638
.30	20.447060	33.542354	37.739488	43.352590	54.501783	55.608293	73.128511	73.220954	94.646309	94.649989	119.077142	119.077200
.31	21.102800	34.454912	38.430866	44.284200	55.117993	56.309419	73.590236	73.692374	94.928532	94.932643	119.172395	119.172460
.32	21.756190	35.370100	39.120894	45.220147	55.737292	57.021098	74.055373	74.168168	95.212855	95.217515	119.268363	119.268439
.33	22.407186	36.275043	39.811900	46.160089	56.359649	57.740330	74.523972	74.648476	95.499415	95.504567	119.365060	119.365160
.34	23.055743	37.173636	40.502972	47.103677	56.985027	58.468594	74.996081	75.133445	95.788230	95.794124	119.462504	119.462604
.35	23.701820	38.065599	41.194007	48.050551	57.613387	59.205733	75.471752	75.623229	96.079347	96.085974	119.560711	119.560625
.36	24.345378	38.950671	41.884900	49.000343	58.244685	59.951973	75.951072	76.117992	96.372819	96.380267	119.659699	119.659699
.37	24.986381	39.828607	42.575547	49.952678	58.878875	60.707422	76.433971	76.617907	96.668697	96.677067	119.759486	119.759635
.38	25.624795	40.699182	43.265844	50.907172	59.515905	61.472168	76.920617	77.123156	96.967037	96.976440	119.860091	119.860261
.39	26.260589	41.562194	43.955685	51.863434	60.155721	62.246275	77.410019	77.653930	97.267893	97.278455	119.961534	119.961534
.40	26.893734	42.417463	44.644965	52.821064	60.798265	63.029785	77.905225	78.150433	97.571325	97.583186	120.063836	120.064059
.41	27.524204	43.264832	45.333578	53.779655	61.443474	63.822712	78.402376	78.672877	97.877391	97.890709	120.167019	120.167273
.42	28.151975	44.104172	46.021420	54.738792	62.091028	64.625046	78.905223	79.200487	98.186155	98.201106	120.271104	120.271104
.43	28.777025	44.935378	46.708386	55.698053	62.741621	65.436747	79.411109	79.736497	98.497679	98.514461	120.376114	120.376446
.44	29.399335	45.758372	47.394371	56.657005	63.394414	66.257745	79.920975	80.281577	98.812030	98.830864	120.482075	120.482075
.45	30.018889	46.573101	48.079272	57.615210	64.049583	67.087943	80.434866	80.826724	99.129276	99.150411	120.589012	120.589445
.46	30.635672	47.379541	48.762986	58.572221	64.707045	67.927594	80.952818	81.382470	99.449489	99.473201	120.696951	120.696951
.47	31.249671	48.177689	49.445411	59.527581	65.366715	68.775818	81.474870	81.945617	99.772740	99.799342	120.805919	120.806485
.48	31.860876	48.967570	50.126447	60.480829	66.028501	69.631818	82.001414	82.516641	100.099606	100.128606	120.915948	120.915948
.49	32.469280	49.749800	50.805592	61.430817	66.692501	70.497892	82.531414	83.096157	100.428644	100.462734	121.027063	121.027803
.50	33.074876	50.522746	51.483950	62.379095	67.358039	71.371402	83.065969	83.683066	100.761495	100.799033	121.139299	121.140146

$J_{K_{-1}K_1}$	$11_{6,6}$	$11_{6,5}$	$11_{7,5}$	$11_{7,4}$	$11_{8,4}$	$11_{8,3}$	$11_{9,3}$	$11_{9,2}$	$11_{10,2}$	$11_{10,1}$	$11_{11,1}$	$11_{11,0}$
J_τ	11_0	11_1	11_2	11_3	11_4	11_5	11_6	11_7	11_8	11_9	11_{10}	11_{11}
κ												
.50	33.074876	50.522746	51.483950	62.379095	67.358039	71.371402	83.065969	83.683066	100.761495	100.799033	121.139299	121.140146
.51	33.677661	51.288199	52.160224	63.320160	68.026756	72.253078	83.604751	84.279166	101.097682	101.139778	121.252690	121.253658
.52	34.277631	52.045700	52.834720	64.263180	68.694847	73.142248	84.147784	84.884299	101.437311	101.484516	121.367190	121.368279
.53	34.874787	52.795376	53.507343	65.198684	69.365706	74.039190	84.693190	85.499190	101.730590	101.833110	121.481075	121.484343
.54	35.469130	53.537365	54.178005	66.129173	70.038049	74.942956	85.246686	86.125013	102.127251	102.186599	121.600144	121.601595
.55	36.060663	54.271820	54.846616	67.054156	70.711756	75.853373	85.802587	86.757278	102.477748	102.544287	121.718517	121.720179
.56	36.649391	54.998904	55.513091	67.973145	71.386703	76.770050	86.362600	87.401935	102.832058	102.906656	121.838237	121.840140
.57	37.235319	55.718787	56.177346	68.885660	72.062761	77.692570	86.927332	88.057320	103.190281	103.273912	121.959349	121.961529
.58	37.818456	56.431649	56.839301	69.791231	72.739798	78.620499	87.496182	88.723756	103.552520	103.646273	122.081899	122.084399
.59	38.398810	57.137670	57.498878	70.689401	73.417678	79.553377	88.069346	89.401555	103.918882	104.023979	122.205937	122.208804
.60	38.976392	57.837037	58.156003	71.579733	74.096261	80.490729	88.646813	90.093100	104.289475	104.407283	122.331516	122.334806
.61	39.551211	58.529936	58.810605	72.461812	74.775404	81.432056	89.227748	90.797932	104.664903	104.796464	122.458569	122.462460
.62	40.123282	59.216554	59.462614	73.335064	75.454775	82.377404	89.811568	91.505932	105.044807	105.191818	122.587522	122.591859
.63	40.692577	59.897077	60.111987	74.199692	76.134127	83.324532	90.404844	92.231847	105.427772	105.593669	122.718069	122.723054
.64	41.259231	60.571687	60.758802	75.054818	76.814698	84.274575	90.999303	92.970301	105.816436	106.002365	122.850399	122.856131
.65	41.823139	61.240564	61.402643	75.900349	77.494570	85.226376	91.597920	93.721418	106.209919	106.418286	122.984582	122.991177
.66	42.384356	61.903882	62.043494	76.736048	78.174230	86.179319	92.200646	94.485275	106.579211	106.841838	123.120690	123.128285
.67	42.942900	62.561811	62.681645	77.561726	78.853515	87.132763	92.807423	95.261998	107.011847	107.273466	123.258804	123.267556
.68	43.498787	63.214516	63.316871	78.377241	79.532259	88.086032	93.418185	96.051254	107.420550	107.713646	123.409090	123.409049
.69	44.052035	63.862155	63.949128	79.182504	80.210293	89.038423	94.032856	96.853256	107.834589	108.162897	123.541385	123.553036
.70	44.602663	64.504880	64.578377	79.977471	80.887447	89.989197	94.651354	97.667757	108.254097	108.621774	123.686037	123.699497
.71	45.150690	65.142836	65.204584	80.762151	81.563550	90.937651	95.273440	98.494972	108.679211	109.090877	123.833062	123.848627
.72	45.696134	65.776161	65.827717	81.536600	82.239426	91.883904	95.899441	99.333345	109.110017	109.570129	123.982569	124.000586
.73	46.239016	66.404988	66.447749	82.299426	82.913805	92.823904	96.528811	100.183876	109.546804	110.060378	124.134671	124.155550
.74	46.779355	67.029440	67.064655	83.052244	83.583805	93.760115	97.161568	101.045695	109.989560	110.566191	124.289494	124.313715
.75	47.317167	67.649635	67.678415	83.797761	84.253957	94.690483	97.797574	101.918375	110.438473	111.083062	124.447169	124.475302
.76	47.852478	68.265685	68.289013	84.534681	84.922185	95.614061	98.436679	102.801411	110.893681	111.613798	124.607837	124.640555
.77	48.385305	68.877694	68.896435	85.260243	85.588316	96.529278	99.078718	103.694240	111.355369	112.159244	124.771650	124.809752
.78	48.915669	69.485760	69.500671	85.976712	86.252177	97.436947	99.723513	104.596243	111.823530	112.720255	124.938770	124.983205
.79	49.443590	70.089974	70.101714	86.684369	86.913600	98.334276	100.370872	105.506737	112.298441	113.298441	125.109374	125.161273
.80	49.969088	70.690426	70.699562	87.383511	87.572419	99.220880	101.020586	106.424979	112.780184	113.892493	125.283647	125.344363
.81	50.492183	71.287192	71.294212	88.074439	88.228469	100.095801	101.672429	107.342321	113.268475	114.502545	125.461794	125.532942
.82	51.012846	71.880565	71.885562	88.755967	88.881553	100.957228	102.326386	108.283361	113.764675	115.137381	125.644010	125.727553
.83	51.531095	72.469969	72.473597	89.432887	89.531834	101.807015	102.981513	109.217677	114.267665	115.789057	125.830590	125.928826
.84	52.047252	73.056106	73.059014	90.101019	90.178450	102.641708	103.638209	110.157969	114.777972	116.461136	126.021726	126.137496
.85	52.560935	73.638833	73.640923	90.762154	90.821895	103.461567	104.295944	111.101059	115.295704	117.154177	126.217710	126.354427
.86	53.072314	74.218200	74.219969	91.416580	91.461836	104.266085	104.954390	112.046655	115.820690	117.868608	126.418834	126.580639
.87	53.581408	74.794260	74.795266	92.064575	92.098147	105.054908	105.613199	112.990084	116.353836	118.604709	126.625413	126.817340
.88	54.088236	75.367062	75.367700	92.706339	92.730708	105.827763	106.270985	113.932763	116.894415	119.362600	126.836230	127.064762
.89	54.592818	75.936648	75.937077	93.342301	93.359411	106.584584	106.930372	114.871643	117.442775	120.142232	127.056315	127.328247
.90	55.095170	76.503063	76.503326	93.972513	93.984158	107.326172	107.545997	115.804019	117.999982	120.943390	127.281395	127.606222
.91	55.595312	77.066341	77.066495	94.597249	94.604858	108.052027	108.244126	116.728369	118.563096	121.765699	127.513445	127.902338
.92	56.093262	77.626525	77.626606	95.216705	95.221435	108.762916	108.898551	117.640397	119.135162	122.608635	127.752914	128.219494
.93	56.589037	78.183635	78.183679	95.831061	95.833820	109.459436	109.550656	118.536880	119.715217	123.471547	128.000284	128.561096
.94	57.082655	78.737735	78.737735	96.440478	96.441960	110.142296	110.199580	119.199686	120.199686	124.353677	128.256067	128.931069
.95	57.574133	79.288791	79.288799	97.045098	97.045808	110.812286	110.845656	120.262666	120.899385	125.254184	128.520807	129.333829
.96	58.063487	79.836888	79.836891	97.645046	97.645335	111.470262	111.487347	121.089661	121.503514	126.126167	128.795007	129.774142
.97	58.550735	80.382034	80.382034	98.240428	98.240428	112.115115	112.124314	121.529894	121.702869	127.079490	129.079490	130.211636
.98	59.035892	80.924252	80.924252	98.831336	98.831154	112.747334	112.757699	122.324314	122.715862	128.056800	129.374672	130.786521
.99	59.518970	81.463566	81.463566	99.417841	99.417841	113.373952	113.381348	123.337148	123.363946	129.021547	129.681282	131.366799
1.00	60.000000	82.000000	82.000000	100.000000	100.000000	114.000000	114.000000	124.000000	124.000000	130.000000	130.000000	132.000000

J_{K_{-1},K_1} / J_τ / κ	$12_{12,0}$ 12_{12}	$11_{0,11}$ 11_{-11}	$11_{1,11}$ 11_{-10}	$11_{1,10}$ 11_{-9}	$11_{2,10}$ 11_{-8}	$11_{2,9}$ 11_{-7}	$11_{3,9}$ 11_{-6}	$11_{3,8}$ 11_{-5}	$11_{4,8}$ 11_{-4}	$11_{4,7}$ 11_{-3}	$11_{5,7}$ 11_{-2}	$11_{5,6}$ 11_{-1}
.00	139.076386	-116.491011	-116.491011	-87.004844	-87.004758	-60.636950	-60.633351	-37.578159	-37.498959	-18.528549	-17.590939	-5.073416
.01	139.162600	-116.412284	-116.412283	-86.772847	-86.772772	-60.258421	-60.255223	-37.051923	-36.980289	-17.817907	-16.950331	-4.149087
.02	139.249265	-116.333960	-116.333960	-86.542075	-86.542009	-59.882028	-59.879190	-36.529240	-36.464505	-17.114255	-16.312298	-3.229792
.03	139.336387	-116.256032	-116.256032	-86.312511	-86.312453	-59.507739	-59.505222	-36.010030	-35.951580	-16.417352	-15.676874	-2.315810
.04	139.423975	-116.178496	-116.178495	-86.084136	-86.084086	-59.135519	-59.133288	-35.494215	-35.441487	-15.727046	-15.044091	-1.407408
.05	139.512034	-116.101345	-116.101344	-85.856933	-85.856890	-58.765336	-58.763361	-34.981721	-34.934197	-15.043180	-14.413977	-0.504843
.06	139.600573	-116.024574	-116.024573	-85.630886	-85.630848	-58.397158	-58.395412	-34.472477	-34.429682	-14.365598	-13.786551	0.391631
.07	139.689599	-115.948177	-115.948177	-85.405977	-85.405944	-58.030956	-58.029413	-33.966415	-33.927914	-13.694144	-13.161856	1.281809
.08	139.779120	-115.872150	-115.872150	-85.182192	-85.182163	-57.666699	-57.665336	-33.463361	-33.428865	-13.028663	-12.539896	2.165451
.09	139.869144	-115.796487	-115.796487	-84.959513	-84.959488	-57.304359	-57.303157	-32.963579	-32.932506	-12.369001	-11.920695	3.042366
.10	139.959680	-115.721183	-115.721183	-84.737926	-84.737904	-56.943907	-56.942848	-32.466682	-32.438808	-11.715006	-11.304270	3.912570
.11	140.050736	-115.646234	-115.646234	-84.517415	-84.517395	-56.585316	-56.584383	-31.972723	-31.947742	-11.066528	-10.690637	4.775297
.12	140.142322	-115.571634	-115.571633	-84.297967	-84.297950	-56.228559	-56.227739	-31.481645	-31.459281	-10.423423	-10.079808	5.631003
.13	140.234445	-115.497378	-115.497378	-84.079562	-84.079552	-55.873611	-55.872996	-30.993397	-30.973395	-9.785547	-9.471793	6.479350
.14	140.327115	-115.423463	-115.423463	-83.862198	-83.862186	-55.520046	-55.519814	-30.507926	-30.490057	-9.152760	-8.866600	7.320257
.15	140.420343	-115.349884	-115.349884	-83.645851	-83.645841	-55.169040	-55.168486	-30.025184	-30.009237	-8.524929	-8.264238	8.153610
.16	140.514137	-115.276636	-115.276636	-83.430510	-83.430501	-54.819368	-54.818883	-29.545124	-29.530908	-7.901921	-7.664709	8.979352
.17	140.608509	-115.203715	-115.203715	-83.216163	-83.216155	-54.471407	-54.470983	-29.067700	-29.055042	-7.283611	-7.068018	9.797433
.18	140.703468	-115.131117	-115.131117	-83.002796	-83.002789	-54.125135	-54.124764	-28.592869	-28.581610	-6.669875	-6.474164	10.607828
.19	140.799025	-115.058837	-115.058837	-82.790397	-82.790392	-53.780528	-53.780205	-28.120588	-28.110586	-6.060595	-5.883149	11.410529
.20	140.895192	-114.986873	-114.986873	-82.578954	-82.578950	-53.437565	-53.437284	-27.650816	-27.641942	-5.455657	-5.294968	12.205545
.21	140.991980	-114.915219	-114.915219	-82.368455	-82.368451	-53.096225	-53.095980	-27.183515	-27.175650	-4.854952	-4.709619	12.992906
.22	141.089401	-114.843829	-114.843829	-82.158888	-82.158885	-52.756247	-52.756274	-26.718646	-26.711685	-4.258324	-4.127096	13.772407
.23	141.187466	-114.772829	-114.772829	-81.950242	-81.950242	-52.417829	-52.417580	-26.256173	-26.250199	-3.665639	-3.547503	14.544865
.24	141.286190	-114.702085	-114.702085	-81.742505	-81.742502	-52.081735	-52.081575	-25.796059	-25.790627	-3.077200	-2.970502	15.309593
.25	141.385585	-114.631637	-114.631637	-81.535666	-81.535663	-51.746682	-51.746543	-25.338270	-25.333482	-2.492414	-2.396412	16.066944
.26	141.485663	-114.561482	-114.561482	-81.329714	-81.329712	-51.413031	-51.413032	-24.882774	-24.878554	-1.911374	-1.825115	16.817016
.27	141.586440	-114.491615	-114.491615	-81.124639	-81.124637	-51.080126	-51.081022	-24.429537	-24.425837	-1.333996	-1.256598	17.559922
.28	141.687930	-114.422033	-114.422033	-80.920430	-80.920429	-50.750586	-50.750497	-23.978529	-23.975429	-0.760197	-0.690849	18.295783
.29	141.790148	-114.352734	-114.352734	-80.717078	-80.717077	-50.421515	-50.421438	-23.529718	-23.526868	-0.189900	-0.127854	19.024731
.30	141.893109	-114.283714	-114.283714	-80.514572	-80.514571	-50.093894	-50.093828	-23.083076	-23.080582	0.376971	0.432400	19.746902
.31	141.996829	-114.214969	-114.214969	-80.312902	-80.312901	-49.767677	-49.767651	-22.638573	-22.636394	0.940488	0.989751	20.462437
.32	142.101324	-114.146497	-114.146497	-80.112058	-80.112058	-49.449567	-49.449526	-22.196372	-22.194481	1.500687	1.544291	21.171308
.33	142.206610	-114.078294	-114.078294	-79.912035	-79.912035	-49.129526	-49.119546	-21.756372	-21.754219	2.057730	2.096886	21.874193
.34	142.312716	-114.010357	-114.010357	-79.712815	-79.712814	-48.797581	-48.797546	-21.317625	-21.316186	2.611583	2.646344	22.570713
.35	142.419646	-113.942684	-113.942684	-79.514396	-79.514395	-48.476964	-48.476954	-20.881407	-20.880159	3.162340	3.193147	23.261198
.36	142.527426	-113.875271	-113.875271	-79.316767	-79.316767	-48.157699	-48.157674	-20.447197	-20.446116	3.710057	3.737313	23.945799
.37	142.636074	-113.808116	-113.808116	-79.119919	-79.119919	-47.839771	-47.839751	-20.014968	-20.014034	4.254791	4.278862	24.624667
.38	142.745613	-113.741215	-113.741215	-78.923845	-78.923845	-47.523169	-47.523151	-19.584698	-19.583892	4.796593	4.817812	25.297953
.39	142.856064	-113.674567	-113.674567	-78.728534	-78.728534	-47.207858	-47.207858	-19.156363	-19.155363	5.335514	5.354185	25.965804
.40	142.967450	-113.608167	-113.608167	-78.533980	-78.533980	-46.893871	-46.893858	-18.729941	-18.729345	5.871604	5.888000	26.628364
.41	143.079794	-113.542015	-113.542015	-78.340174	-78.340174	-46.581144	-46.581144	-18.305468	-18.304897	6.404909	6.419279	27.285775
.42	143.193120	-113.476106	-113.476106	-78.147107	-78.147107	-46.269692	-46.269683	-17.882743	-17.882304	6.935473	6.948041	27.938176
.43	143.307456	-113.410438	-113.410438	-77.954773	-77.954773	-45.959488	-45.959484	-17.461926	-17.461553	7.463340	7.474309	28.585701
.44	143.422827	-113.345010	-113.345010	-77.763163	-77.763163	-45.650523	-45.650517	-17.042934	-17.042611	7.988551	7.998103	29.228479
.45	143.539262	-113.279818	-113.279818	-77.572270	-77.572269	-45.342784	-45.342779	-16.625748	-16.625479	8.511146	8.519445	29.866638
.46	143.656790	-113.214860	-113.214860	-77.382086	-77.382086	-45.036258	-45.036255	-16.210749	-16.210241	9.031163	9.038357	30.500497
.47	143.775443	-113.150133	-113.150133	-77.192604	-77.192604	-44.730932	-44.730928	-15.796715	-15.796523	9.548639	9.554860	31.129594
.48	143.895252	-113.085636	-113.085636	-77.003817	-77.003817	-44.426747	-44.426747	-15.384861	-15.384667	10.063610	10.068670	31.754360
.49	144.016252	-113.021139	-113.021139	-76.815717	-76.815717	-44.123830	-44.123830	-14.974826	-14.974615	10.576130	10.580275	32.375429
.50	144.138478	-112.957319	-112.957319	-76.628301	-76.628301	-43.822034	-43.822034	-14.566224	-14.566110	11.086171	11.090131	32.992217

J_{τ_{-1},τ_1} / J_{τ_1}; κ	$12_{12,0}$ / 12_{12}	$11_{0,11}$ / 11_{-11}	$11_{1,11}$ / 11_{-10}	$11_{1,10}$ / 11_{-9}	$11_{2,10}$ / 11_{-8}	$11_{2,9}$ / 11_{-7}	$11_{3,9}$ / 11_{-6}	$11_{3,8}$ / 11_{-5}	$11_{4,8}$ / 11_{-4}	$11_{4,7}$ / 11_{-3}	$11_{5,7}$ / 11_{-2}	$11_{5,6}$ / 11_{-1}
.50	144.138478	-112.957319	-112.957319	-76.628301	-76.628301	-43.822034	-43.822032	-14.566224	-14.561110	11.086171	11.090131	32.992217
.51	144.261968	-112.893495	-112.893495	-76.441557	-76.441557	-43.521388	-43.521386	-14.159469	-14.159373	11.593827	11.597215	33.605046
.52	144.386762	-112.829892	-112.829892	-76.255481	-76.255481	-43.221881	-43.221804	-13.754388	-13.754309	12.099108	12.101997	34.214011
.53	144.512901	-112.766506	-112.766506	-76.070065	-76.070065	-42.923505	-42.923505	-13.351001	-13.350388	12.602062	12.604500	34.819201
.54	144.640429	-112.703336	-112.703336	-75.885304	-75.885304	-42.626246	-42.626242	-12.949181	-12.949127	13.102662	13.104705	35.420703
.55	144.769394	-112.640380	-112.640380	-75.701190	-75.701190	-42.330095	-42.330094	-12.549022	-12.548977	13.600993	13.602753	36.018599
.56	144.899844	-112.577635	-112.577635	-75.517718	-75.517718	-42.035039	-42.035038	-12.150470	-12.150433	14.097063	14.098544	36.612966
.57	145.031831	-112.515101	-112.515101	-75.334882	-75.334882	-41.741069	-41.741068	-11.753509	-11.753479	14.590899	14.592141	37.203881
.58	145.165412	-112.452773	-112.452773	-75.152675	-75.152675	-41.448173	-41.448173	-11.358124	-11.358124	15.082525	15.083563	37.791413
.59	145.300645	-112.390652	-112.390652	-74.971092	-74.971092	-41.156343	-41.156342	-10.964299	-10.964279	15.571967	15.572831	38.375629
.60	145.437594	-112.328734	-112.328734	-74.790126	-74.790126	-40.865566	-40.865566	-10.572018	-10.572000	16.059250	16.059966	38.956595
.61	145.576325	-112.267018	-112.267018	-74.609771	-74.609771	-40.575835	-40.575835	-10.181267	-10.181305	16.544396	16.544984	39.534319
.62	145.716910	-112.205501	-112.205501	-74.430023	-74.430023	-40.287138	-40.287138	-9.792024	-9.792024	17.027584	17.027901	40.109103
.63	145.859418	-112.144183	-112.144183	-74.250875	-74.250875	-39.999461	-39.999461	-9.404297	-9.404290	17.508371	17.508768	40.680579
.64	146.003959	-112.083061	-112.083061	-74.072322	-74.072322	-39.712811	-39.712811	-9.018043	-9.018043	17.987244	17.987567	41.249121
.65	146.150594	-112.022134	-112.022134	-73.894359	-73.894359	-39.427161	-39.427161	-8.633274	-8.632674	18.464069	18.464387	41.814687
.66	146.299429	-111.961399	-111.961399	-73.716979	-73.716979	-39.142508	-39.142508	-8.249957	-8.249954	18.938867	18.939077	42.377327
.67	146.450568	-111.900855	-111.900855	-73.540178	-73.540178	-38.858844	-38.858844	-7.868086	-7.868086	19.411659	19.411826	42.937078
.68	146.604124	-111.840501	-111.840501	-73.363951	-73.363951	-38.576158	-38.576158	-7.487647	-7.487645	19.882463	19.882597	43.494002
.69	146.760220	-111.780334	-111.780334	-73.188292	-73.188292	-38.294443	-38.294443	-7.108627	-7.108355	20.351301	20.351461	44.048122
.70	146.918991	-111.720352	-111.720352	-73.013197	-73.013197	-38.013689	-38.013688	-6.731013	-6.731012	20.818190	20.818272	44.599484
.71	147.080584	-111.660556	-111.660556	-72.838660	-72.838660	-37.733888	-37.733888	-6.354792	-6.354792	21.283213	21.283229	45.148133
.72	147.245160	-111.600941	-111.600941	-72.664676	-72.664676	-37.455033	-37.455033	-5.979952	-5.979952	21.746196	21.746246	45.694077
.73	147.412900	-111.541508	-111.541508	-72.491242	-72.491242	-37.177112	-37.177112	-5.606482	-5.606482	22.207388	22.207388	46.237380
.74	147.584002	-111.482254	-111.482254	-72.318351	-72.318351	-36.900121	-36.900121	-5.234368	-5.234368	22.666628	22.666657	46.778063
.75	147.758687	-111.423178	-111.423178	-72.145999	-72.145999	-36.624050	-36.624050	-4.863599	-4.863599	23.124047	23.124069	47.316159
.76	147.937202	-111.364279	-111.364279	-71.974182	-71.974182	-36.348891	-36.348891	-4.494163	-4.494163	23.579624	23.579624	47.851698
.77	148.119826	-111.305554	-111.305554	-71.802895	-71.802895	-36.074636	-36.074636	-4.126043	-4.126043	24.033376	24.033387	48.384708
.78	148.306875	-111.247003	-111.247003	-71.632134	-71.632134	-35.801278	-35.801278	-3.759244	-3.759244	24.485318	24.485318	48.915217
.79	148.498707	-111.188624	-111.188624	-71.461894	-71.461894	-35.528810	-35.528810	-3.393738	-3.393738	24.935467	24.935473	49.443252
.80	148.695733	-111.130415	-111.130415	-71.292170	-71.292170	-35.257225	-35.257225	-3.029521	-3.029521	25.383838	25.383838	49.968833
.81	148.898425	-111.072376	-111.072376	-71.122959	-71.122959	-34.986516	-34.986516	-2.666581	-2.666581	25.830447	25.830450	50.492002
.82	149.107333	-111.014504	-111.014504	-70.954256	-70.954256	-34.716665	-34.716665	-2.304907	-2.304907	26.275312	26.275312	51.012766
.83	149.323096	-110.956798	-110.956798	-70.786058	-70.786058	-34.447679	-34.447679	-1.944489	-1.944489	26.718440	26.718441	51.531154
.84	149.546470	-110.899258	-110.899258	-70.618359	-70.618359	-34.179546	-34.179546	-1.585316	-1.585316	27.159853	27.159853	52.047189
.85	149.778347	-110.841881	-110.841881	-70.451156	-70.451156	-33.912259	-33.912259	-1.227379	-1.227379	27.599562	27.599563	52.560893
.86	150.019796	-110.784666	-110.784666	-70.284445	-70.284445	-33.645811	-33.645811	-0.870667	-0.870667	28.037583	28.037584	53.072287
.87	150.272102	-110.727613	-110.727613	-70.118221	-70.118221	-33.380196	-33.380196	-0.515170	-0.515170	28.473929	28.473929	53.581391
.88	150.536824	-110.670719	-110.670719	-69.952482	-69.952482	-33.115406	-33.115406	-0.160878	-0.160878	28.908614	28.908614	54.088226
.89	150.815862	-110.613983	-110.613983	-69.787222	-69.787222	-32.851434	-32.851434	0.192217	0.192217	29.341652	29.341652	54.592811
.90	151.111552	-110.557405	-110.557405	-69.622439	-69.622439	-32.588276	-32.588276	0.544127	0.544127	29.773055	29.773055	55.095167
.91	151.426749	-110.500982	-110.500982	-69.458129	-69.458129	-32.325923	-32.325923	0.894861	0.894861	30.202837	30.202837	55.595311
.92	151.765056	-110.444714	-110.444714	-69.294287	-69.294287	-32.064370	-32.064370	1.244426	1.244426	30.631011	30.631011	56.093261
.93	152.130768	-110.388600	-110.388600	-69.130910	-69.130910	-31.803610	-31.803610	1.592834	1.592834	31.057590	31.057590	56.589037
.94	152.529198	-110.332638	-110.332638	-68.967995	-68.967995	-31.543637	-31.543637	1.940092	1.940092	31.482586	31.482586	57.082655
.95	152.966658	-110.276827	-110.276827	-68.805538	-68.805538	-31.284446	-31.284446	2.286210	2.286210	31.906011	31.906011	57.574133
.96	153.450411	-110.221166	-110.221166	-68.643535	-68.643535	-31.026030	-31.026030	2.631197	2.631197	32.327818	32.327818	58.063765
.97	153.988376	-110.165654	-110.165654	-68.481983	-68.481983	-30.768383	-30.768383	2.975075	2.975075	32.748088	32.748088	58.550735
.98	154.589450	-110.110290	-110.110290	-68.320879	-68.320879	-30.511500	-30.511500	3.317810	3.317810	33.166984	33.166984	59.035892
.99	155.255374	-110.055072	-110.055072	-68.160219	-68.160219	-30.255374	-30.255374	3.659453	3.659453	33.584248	33.584248	59.518975
1.00	156.000000	-110.000000	-110.000000	-68.000000	-68.000000	-30.000000	-30.000000	4.000000	4.000000	34.000000	34.000000	60.000000

J_{-1,K_1} / J_r κ	$12_{6,6}$ 12_0	$12_{7,6}$ 12_1	$12_{7,5}$ 12_2	$12_{8,5}$ 12_3	$12_{8,4}$ 12_4	$12_{9,4}$ 12_5	$12_{9,3}$ 12_6	$12_{10,3}$ 12_7	$12_{10,2}$ 12_8	$12_{11,2}$ 12_9	$12_{11,1}$ 12_{10}	$12_{12,1}$ 12_{11}
.00	0.000000	9.188833	11.839818	28.869165	29.209203	51.539141	51.562056	77.539946	77.540834	106.757524	106.757543	139.076386
.01	1.141872	10.009771	12.838631	29.587693	29.961313	52.110339	52.136039	77.955139	77.956153	107.011901	107.011923	139.162600
.02	2.283284	10.832409	13.846763	30.309919	30.720012	52.846763	52.713673	78.372669	78.373825	107.267645	107.267670	139.249265
.03	3.423375	11.656640	14.864079	31.035839	31.485495	53.262796	53.295035	78.792572	78.793890	107.524778	107.524807	139.336387
.04	4.562284	12.482356	15.890416	31.765445	32.257765	53.844141	53.880205	79.214885	79.216385	107.783321	107.783355	139.423974
.05	5.700153	13.309443	16.925584	32.498725	33.037630	54.428966	54.469264	79.639647	79.641352	108.043296	108.043335	139.512033
.06	6.835124	14.137280	17.969365	33.235664	33.824595	55.017284	55.062299	80.066894	80.068333	108.304726	108.304772	139.600572
.07	7.961356	14.967538	19.021365	33.976464	34.619169	55.609169	55.659401	80.496668	80.498370	108.567688	108.567735	139.689598
.08	9.096355	15.797795	20.081765	34.720444	35.421931	56.204656	56.260664	80.929009	80.931508	108.832107	108.832160	139.779119
.09	10.221714	16.629217	21.149824	35.468239	36.232530	56.803789	56.866189	81.363959	81.366793	109.097986	109.098056	139.869144
.10	11.342973	17.461425	22.225377	36.219601	37.051857	57.406610	57.476078	81.801560	81.804772	109.365479	109.365558	139.959679
.11	12.459694	18.294297	23.308090	36.974498	37.878174	58.013166	58.090442	82.245495	82.245495	109.634641	109.634641	140.050735
.12	13.571443	19.127710	24.397609	37.732893	38.714920	58.623497	58.709396	82.684893	82.689010	109.905237	109.905332	140.142320
.13	14.677798	19.961539	25.493564	38.494746	39.559998	59.237564	59.333059	83.130715	83.135373	110.177537	110.177650	140.234444
.14	15.778342	20.795661	26.595569	39.260014	40.414186	59.855663	59.961559	83.579371	83.584635	110.451509	110.451649	140.327114
.15	16.872673	21.629948	27.703222	40.028649	41.277186	60.477580	60.595097	84.030908	84.036855	110.727173	110.727173	140.420354
.16	17.960149	22.464515	28.810708	40.800600	42.150050	61.103447	61.233608	84.485375	84.492089	111.004558	111.004741	140.514135
.17	19.040149	23.299028	29.913808	41.575809	43.033000	61.733299	61.877445	84.942824	84.950664	111.283695	111.283905	140.608506
.18	20.114560	24.132541	31.055881	42.354217	43.927000	62.367177	62.526693	85.403307	85.411850	111.564616	111.564957	140.703465
.19	21.180296	24.966226	32.181884	43.135760	44.832000	63.005120	63.181515	85.866875	85.866504	111.847354	111.847631	140.799022
.20	22.238038	25.799443	33.311363	43.920369	45.742054	63.647164	63.842083	86.333585	86.344430	112.131944	112.132260	140.895188
.21	23.287491	26.632067	34.443858	44.707971	46.665234	64.293347	64.508575	86.803491	86.815701	112.418419	112.418782	140.991975
.22	24.328387	27.463970	35.578899	45.498489	47.598764	64.943702	65.181180	87.276651	87.290389	112.706816	112.706824	141.087406
.23	25.360481	28.295028	36.716011	46.291842	48.542699	65.598261	65.860406	87.752977	87.768459	112.997637	112.997637	141.187420
.24	26.383559	29.125117	37.854715	47.087944	49.497064	66.257056	66.545530	88.232965	88.250332	113.289526	113.290069	141.286183
.25	27.397436	29.954113	38.994522	47.886706	50.461857	66.920115	67.237699	88.716240	88.735752	113.583917	113.584538	141.385576
.26	28.401766	30.781894	40.134940	48.688034	51.437043	67.587415	67.936829	89.203011	89.224921	113.880387	113.881095	141.485654
.27	29.396996	31.608340	41.275470	49.491829	52.422553	68.259129	68.643158	89.693339	89.717931	114.179786	114.179786	141.586429
.28	30.382463	32.433331	42.415610	50.297990	53.418287	68.935129	69.356932	90.187291	90.214880	114.479730	114.480654	141.687918
.29	31.358295	33.256750	43.554853	51.106410	54.424110	69.615481	70.078407	90.684934	90.715869	114.782692	114.783746	141.790133
.30	32.324462	34.078482	44.692684	51.916979	55.439854	70.300203	70.807849	91.186333	91.221007	115.089113	115.089113	141.893092
.31	33.280961	34.898415	45.828590	52.729584	56.465518	70.989306	71.545532	91.691560	91.730560	115.395433	115.396807	141.996809
.32	34.227822	35.716436	46.962048	53.544105	57.500266	71.682308	72.291483	92.200786	92.244183	115.705070	115.706807	142.101303
.33	35.165097	36.532439	48.092539	54.360411	58.544266	72.379497	73.046763	92.713775	92.762466	116.017591	116.019373	142.206588
.34	36.092868	37.346317	49.219536	55.178411	59.597521	73.082965	73.810896	93.230909	93.285385	116.332332	116.334363	142.312685
.35	37.011236	38.157967	50.342516	55.997941	60.659203	73.789637	74.584442	93.752157	93.813082	116.649587	116.651901	142.419610
.36	37.920328	38.967290	51.460956	56.818840	61.729126	74.500692	75.367703	94.277596	94.345702	116.969414	116.972050	142.527384
.37	38.820284	39.774189	52.574333	57.641094	62.806907	75.216119	76.160985	94.807302	94.883403	117.291873	117.294874	142.636027
.38	39.711266	40.578571	53.682132	58.464444	63.892143	75.935899	76.964593	95.341351	95.426351	117.617025	117.620442	142.745558
.39	40.593446	41.380345	54.783843	59.288788	64.984403	76.660010	77.778824	95.879821	95.974720	117.944934	117.948824	142.856000
.40	41.467010	42.179426	55.878965	60.113983	66.083237	77.388425	78.603972	96.472791	96.528697	118.275666	118.280093	142.967375
.41	42.332153	42.975729	56.967011	60.939881	67.188174	78.121111	79.440133	96.920908	97.087698	118.609291	118.614328	143.079708
.42	43.189076	43.769177	58.047509	61.766334	68.299374	78.858119	80.288133	97.522548	97.654278	118.945879	118.951611	143.193021
.43	44.037987	44.559693	59.120006	62.593006	69.414779	79.599032	81.147662	98.070495	98.226315	119.285506	119.292026	143.307343
.44	44.879095	45.347206	60.184069	63.420293	70.534613	80.344373	82.019135	98.641261	98.804829	119.628247	119.635663	143.422694
.45	45.712612	46.131649	61.239296	64.247490	71.658886	81.093692	82.902754	99.207926	99.390072	119.974183	119.982618	143.539107
.46	46.538751	46.912958	62.285309	65.074625	72.786638	81.847007	83.798690	99.779232	99.982312	120.323937	120.332917	143.656611
.47	47.357719	47.691074	63.321766	65.901537	73.912297	82.604308	84.707079	100.356273	100.581835	120.675975	120.686885	143.775236
.48	48.169726	48.465940	64.351361	66.728068	75.050273	83.365457	85.628020	100.938113	101.188944	121.030958	121.043087	143.895014
.49	48.974974	49.237507	65.364827	67.554057	76.184962	84.130590	86.561571	101.525072	101.804155	121.388802	121.405494	144.015969
.50	49.773662	50.005726	66.379344	68.379344	77.320742	84.899017	87.507743	102.117514	102.427229	121.754812	121.770854	144.138156

$J_{K_{-1}K_1}$ / J_τ / κ	$12_{6,6}$ / 12_0	$12_{7,6}$ / 12_1	$12_{7,5}$ / 12_2	$12_{8,5}$ / 12_3	$12_{8,4}$ / 12_4	$12_{9,4}$ / 12_5	$12_{9,3}$ / 12_6	$12_{10,3}$ / 12_7	$12_{10,2}$ / 12_8	$12_{11,2}$ / 12_9	$12_{11,1}$ / 12_{10}	$12_{12,1}$ / 12_{11}
.50	49.773662	50.005726	66.370936	68.379344	77.320742	84.899017	87.507743	102.117514	102.427229	121.754812	121.770854	144.138156
.51	50.565983	50.770555	67.366506	69.203767	78.456978	85.672138	88.465601	102.715225	103.059110	122.121781	122.104903	144.361528
.52	51.565725	51.531955	68.351397	70.027165	79.593016	86.446949	89.437759	103.318376	103.699987	122.472601	122.497990	144.481528
.53	52.132269	52.288810	69.325515	70.849377	80.728188	87.228113	90.420972	103.927035	104.350266	122.867601	122.890966	144.512398
.54	52.906588	53.044323	70.288810	71.670244	81.861809	88.008378	91.417184	104.541270	105.010372	123.246231	123.287052	144.639846
.55	53.675249	53.795236	71.241276	72.469605	82.993176	88.793847	92.405923	105.161146	105.680754	123.629272	123.659773	144.768716
.56	54.438413	54.542601	72.182949	73.307304	84.121575	89.593176	93.444304	105.786723	106.361880	124.016626	124.051314	144.899057
.57	55.196232	55.286398	73.113904	74.123184	85.246273	90.379611	94.474988	106.418058	107.054239	124.408421	124.439660	145.030947
.58	55.948852	56.026610	74.034255	74.937012	86.365527	91.167609	95.516582	107.055202	107.758335	124.804788	124.826907	145.164350
.59	56.696411	56.763225	74.944146	75.748875	87.481582	91.964141	96.568646	107.698203	108.474689	125.205864	125.256907	145.299410
.60	57.439039	57.496386	75.843735	76.558386	88.590676	92.763037	97.630691	108.347101	109.203833	125.611794	125.669861	145.436156
.61	58.176863	58.225627	76.733284	77.365478	89.693043	93.564123	98.702180	109.001931	109.946306	126.022725	126.089390	145.714960
.62	58.910000	58.951403	77.612957	78.110010	90.787918	94.367211	99.782561	109.662720	110.702648	126.437884	126.506758	145.857154
.63	59.638559	59.673561	78.483017	78.971944	91.874543	95.172114	100.871195	110.329490	111.472394	126.857298	126.945758	146.001307
.64	60.362648	60.392174	79.343722	79.770848	92.952174	95.978628	101.967434	111.002021	112.259067	127.287087	127.384454	146.147497
.65	61.082366	61.107033	80.195338	80.566887	94.020089	96.786544	103.070582	111.681006	113.060168	127.716916	127.830447	146.295810
.66	61.797807	61.818360	81.038141	81.359843	95.124695	97.594695	104.179942	112.365750	113.877167	128.157972	128.284144	146.464335
.67	62.509058	62.526091	81.872408	82.149709	96.177695	98.405714	105.294624	113.056464	114.710492	128.602338	128.745991	146.599169
.68	63.216203	63.230240	82.698418	82.936939	97.158817	99.215611	106.413925	113.753215	115.560021	129.052902	129.216476	146.754413
.69	63.919322	63.930819	83.516447	83.719039	98.181383	100.027805	107.536945	114.455680	116.427565	129.509859	129.696134	146.912179
.70	64.618486	64.627843	84.326769	84.498522	99.191260	100.839345	108.662779	115.164089	117.311866	129.973408	130.185551	147.072583
.71	65.313767	65.321331	85.129649	85.274363	100.188044	101.650881	109.790468	115.878281	118.212557	130.444991	130.684298	147.225753
.72	66.005230	66.011300	85.925346	86.046535	101.171416	102.462156	110.919004	116.598175	119.123911	130.921911	131.196298	147.401825
.73	66.692938	66.697770	86.714109	86.814894	102.141141	103.272193	112.047320	117.323418	120.047389	131.405691	131.719105	147.570944
.74	67.376947	67.380763	87.496176	87.579385	103.097078	104.082856	113.174286	118.054663	121.023308	131.897715	132.254638	147.743270
.75	68.057315	68.060301	88.271775	88.339941	104.039174	104.891738	114.298709	118.791014	121.994270	132.397406	132.803821	147.918973
.76	68.734093	68.736407	89.041122	89.096498	104.967466	105.699274	115.419325	119.532575	122.981911	132.904991	133.367658	148.098236
.77	69.407330	69.409104	89.804423	89.848648	105.882095	106.506182	116.534800	120.279177	123.985755	133.420701	133.947237	148.281260
.78	70.077074	70.078419	90.561419	90.597407	106.783202	107.309178	117.643731	121.030629	125.005212	133.944766	134.543730	148.468361
.79	70.743369	70.744375	91.313638	91.341669	107.671117	108.109380	118.744647	121.786717	126.039582	134.477460	135.158335	148.659473
.80	71.406256	71.407060	92.059902	92.081754	108.546152	108.910304	119.836018	122.547202	127.088055	135.018894	135.792525	148.855152
.81	72.065777	72.066318	92.800816	92.817634	109.408690	109.706865	120.916270	123.311818	128.149710	135.569227	136.447525	149.055575
.82	72.721970	72.722356	93.536524	93.553269	110.259154	110.501840	121.983804	124.080243	129.223048	136.129227	137.124814	149.261043
.83	73.374870	73.375142	94.267162	94.276695	111.097996	111.290586	123.012655	124.852580	130.308304	136.703522	137.825803	149.471887
.84	74.024514	74.024701	94.992854	94.999850	111.925690	112.077096	124.037075	125.627382	131.402802	137.277577	138.551893	149.688466
.85	74.670935	74.671060	95.713712	95.713745	112.742715	112.859958	125.094377	126.405287	132.505574	137.866553	139.304411	149.911171
.86	75.314164	75.314246	96.433782	96.433782	113.549286	113.638610	126.095688	127.185536	133.615013	138.465673	140.093380	150.140433
.87	75.954234	75.954286	97.143764	97.143764	114.346685	114.412910	127.077142	127.967660	134.729304	139.075130	140.893380	150.376318
.88	76.591286	76.591440	97.848286	97.849901	115.144566	115.182625	128.037808	128.751148	135.846375	139.695109	141.731673	150.620538
.89	77.225011	77.225029	98.550768	98.551806	115.913642	115.947535	128.977031	129.535442	136.963828	140.325779	142.599987	150.872452
.90	77.855775	77.855786	99.248856	99.249495	116.684331	116.707438	129.894462	130.319932	138.078862	140.967297	143.498562	151.133070
.91	78.483494	78.483499	99.942615	99.942690	117.447048	117.462064	130.790078	131.104996	139.203171	141.619800	144.427323	151.403056
.92	79.108192	79.108195	100.632206	100.632132	118.202081	118.211494	131.664181	131.891181	140.287171	142.283411	145.385877	151.683132
.93	79.729896	79.729897	101.317381	101.317396	118.949833	118.949833	132.528667	132.673671	141.372963	142.958232	146.373530	151.974082
.94	80.348630	80.348631	101.998492	101.998540	119.690573	119.693528	133.350546	133.445739	142.438101	143.644344	147.389323	152.276754
.95	80.964419	80.964420	102.675484	102.675503	120.424584	120.425648	134.164792	134.220095	143.476570	144.341809	148.432076	152.592058
.96	81.577287	81.577287	103.348043	103.348043	121.152031	121.152609	134.963391	134.989766	144.480832	145.050666	149.500439	152.920591
.97	82.187257	82.187257	104.017271	104.017272	121.873349	121.873349	135.733715	135.753715	145.442736	145.770931	150.592948	153.261831
.98	82.794350	82.794350	104.682141	104.682141	122.588130	122.588165	136.507303	136.510847	146.354182	146.502598	151.608074	153.623826
.99	83.398591	83.398591	105.343040	105.343040	123.297046	123.297046	137.259565	137.260007	147.208164	147.245638	152.643831	154.000000
1.00	84.000000	84.000000	106.000000	106.000000	124.000000	124.000000	138.000000	138.000000	148.000000	148.000000	154.000000	154.000000

$J_{K_{-1}K_1}$ / J_τ κ	$12_{0,12}$ / 12_{-12}	$12_{1,12}$ / 12_{-11}	$12_{1,11}$ / 12_{-10}	$12_{2,11}$ / 12_{-9}	$12_{2,10}$ / 12_{-8}	$12_{3,10}$ / 12_{-7}	$12_{3,9}$ / 12_{-6}	$12_{4,9}$ / 12_{-5}	$12_{4,8}$ / 12_{-4}	$12_{5,8}$ / 12_{-3}	$12_{5,7}$ / 12_{-2}	$12_{6,7}$ / 12_{-1}
.00	-139.076386	-139.076386	-106.757543	-106.757524	-77.540834	-77.539946	-51.562056	-51.539141	-29.209203	-28.869165	-11.839818	-9.188833
.01	-138.990616	-138.990616	-106.504510	-106.504493	-77.127833	-77.127054	-50.991654	-50.971238	-28.463490	-28.154340	-10.850429	-8.369700
.02	-138.908784	-138.908784	-106.252608	-106.252605	-76.716431	-76.714893	-50.415459	-50.406854	-27.723992	-27.443216	-9.870502	-7.552474
.03	-138.820382	-138.820382	-106.002605	-106.002605	-76.308196	-76.301111	-49.840699	-49.846384	-26.990705	-26.735984	-8.900205	-6.737254
.04	-138.735905	-138.735905	-105.753294	-105.753284	-75.902371	-75.901862	-49.361241	-49.286884	-26.262933	-26.032061	-7.933940	-5.924137
.05	-138.651846	-138.651846	-105.505453	-105.505445	-75.498285	-75.497832	-48.744488	-48.731744	-25.541037	-25.332017	-6.988241	-5.113216
.06	-138.568201	-138.568201	-105.258864	-105.258858	-75.096345	-75.095950	-48.190993	-48.179692	-24.824685	-24.635650	-6.046579	-4.304582
.07	-138.484961	-138.484961	-105.013507	-105.013501	-74.696520	-74.696176	-47.640699	-47.630687	-24.113726	-23.942947	-5.114397	-3.498322
.08	-138.402123	-138.402123	-104.769362	-104.769358	-74.298778	-74.298170	-47.093520	-47.084690	-23.408014	-23.253947	-4.191619	-2.694521
.09	-138.319681	-138.319681	-104.526426	-104.526421	-73.903090	-73.902830	-46.549495	-46.541662	-22.707411	-22.568481	-3.278148	-1.893259
.10	-138.237628	-138.237628	-104.284667	-104.284663	-73.509426	-73.509201	-46.008481	-46.001563	-22.011785	-21.886685	-2.373846	-1.094615
.11	-138.155960	-138.155960	-104.044075	-104.044071	-73.117305	-73.117057	-45.470459	-45.464356	-21.320810	-21.208488	-1.478640	-0.298663
.12	-138.074672	-138.074672	-103.804633	-103.804631	-72.727851	-72.727851	-44.932610	-44.930067	-20.634964	-20.533870	-0.592323	0.494526
.13	-137.993758	-137.993758	-103.566326	-103.566323	-72.340301	-72.340857	-44.403220	-44.398007	-19.953603	-19.862381	0.285244	1.284885
.14	-137.913214	-137.913214	-103.329139	-103.329137	-71.954634	-71.954334	-43.873875	-43.869712	-19.276605	-19.195286	1.154236	2.072349
.15	-137.833034	-137.833034	-103.093057	-103.093057	-71.570507	-71.570398	-43.347358	-43.343055	-18.604076	-18.531271	2.014831	2.856858
.16	-137.753214	-137.753214	-102.858065	-102.858064	-71.188417	-71.188323	-42.823610	-42.820403	-17.935846	-17.870743	2.867214	3.638355
.17	-137.673750	-137.673750	-102.624150	-102.624149	-70.808106	-70.808086	-42.302588	-42.299773	-17.271819	-17.213673	3.711579	4.416787
.18	-137.594636	-137.594636	-102.391298	-102.391297	-70.429731	-70.429061	-41.784254	-41.781796	-16.611904	-16.560037	4.548119	5.192103
.19	-137.515868	-137.515868	-102.159494	-102.159493	-70.053087	-70.053027	-41.268569	-41.266641	-15.956012	-15.909805	5.377032	5.964256
.20	-137.437442	-137.437442	-101.928726	-101.928726	-69.678211	-69.678159	-40.755496	-40.753621	-15.304062	-15.262950	6.198515	6.733204
.21	-137.359354	-137.359354	-101.698981	-101.698981	-69.305030	-69.305037	-40.244049	-40.243365	-14.655973	-14.619442	7.012765	7.498907
.22	-137.281599	-137.281599	-101.470246	-101.470246	-68.933938	-68.935960	-39.731590	-39.735180	-14.011676	-13.979342	7.819733	8.261439
.23	-137.204173	-137.204173	-101.242509	-101.242509	-68.563970	-68.563938	-39.233590	-39.231590	-13.371676	-13.342552	8.620343	9.020343
.24	-137.127073	-137.127073	-101.015757	-101.015757	-68.195946	-68.195495	-38.728611	-38.727538	-12.734127	-12.708708	9.414053	9.776201
.25	-137.050294	-137.050294	-100.789978	-100.789978	-67.829583	-67.829560	-38.228072	-38.227142	-12.100753	-12.078292	10.201291	10.528596
.26	-136.973833	-136.973833	-100.565161	-100.565161	-67.464860	-67.464860	-37.729940	-37.729163	-11.470890	-11.451072	10.982237	11.277598
.27	-136.897685	-136.897685	-100.341295	-100.341294	-67.101758	-67.101758	-37.234186	-37.233492	-10.844478	-10.827018	11.757066	12.023187
.28	-136.821848	-136.821848	-100.118367	-100.118367	-66.740256	-66.740256	-36.740779	-36.740779	-10.221457	-10.206099	12.525949	12.765347
.29	-136.746316	-136.746316	-99.896367	-99.896367	-66.380335	-66.380323	-36.249689	-36.249173	-9.601772	-9.588283	13.289049	13.504062
.30	-136.671088	-136.671088	-99.675284	-99.675284	-66.021968	-66.021968	-35.760887	-35.760884	-8.973540	-8.361530	14.046524	14.239322
.31	-136.596159	-136.596159	-99.455108	-99.455108	-65.665165	-65.665165	-35.274346	-35.271966	-8.372190	-7.753144	14.798524	14.971119
.32	-136.521526	-136.521526	-99.235828	-99.235828	-65.309837	-65.309837	-34.790037	-34.789712	-7.762191	-7.147431	15.545203	15.699447
.33	-136.447185	-136.447185	-99.017433	-99.017433	-64.956103	-64.956103	-34.307035	-34.307656	-7.155323	-6.544665	16.286692	16.424302
.34	-136.373134	-136.373134	-98.799915	-98.799915	-64.603814	-64.603814	-33.828012	-33.827774	-6.551537	-5.944817	17.023110	17.145684
.35	-136.299369	-136.299369	-98.583262	-98.583262	-64.253009	-64.253009	-33.350243	-33.350040	-5.950790	-5.347856	17.754645	17.863593
.36	-136.225887	-136.225887	-98.367466	-98.367466	-63.903658	-63.903658	-32.874602	-32.874430	-5.353037	-4.753751	18.481359	18.578035
.37	-136.152684	-136.152684	-98.152517	-98.152517	-63.555746	-63.555746	-32.401066	-32.400400	-4.758236	-4.162473	19.203389	19.289014
.38	-136.079758	-136.079758	-97.938406	-97.938406	-63.209264	-63.209264	-31.929409	-31.929486	-4.166348	-3.573991	19.920847	19.996538
.39	-136.007106	-136.007106	-97.725123	-97.725123	-62.864193	-62.864193	-31.460109	-31.460105	-3.577333	-3.573991	20.633840	20.700618
.40	-135.934724	-135.934724	-97.512659	-97.512659	-62.520518	-62.520516	-30.992842	-30.990254	-2.991152	-2.988277	21.342469	21.401263
.41	-135.862611	-135.862611	-97.301006	-97.301006	-62.178221	-62.178221	-30.527485	-30.527413	-2.407709	-2.405301	22.046831	22.098468
.42	-135.790762	-135.790762	-97.090156	-97.090156	-61.837289	-61.837289	-30.064116	-30.064055	-1.827148	-1.825035	22.747019	22.792308
.43	-135.719175	-135.719175	-96.880099	-96.880099	-61.497707	-61.497707	-29.602317	-29.602214	-1.249254	-1.247449	23.443119	23.482777
.44	-135.647848	-135.647848	-96.670827	-96.670827	-61.159463	-61.159463	-29.143257	-29.143214	-0.674054	-0.672515	24.135215	24.169793
.45	-135.576778	-135.576778	-96.462333	-96.462333	-60.822540	-60.822540	-28.685588	-28.685248	-0.101515	-0.100206	24.823387	24.853496
.46	-135.505962	-135.505962	-96.254607	-96.254607	-60.486926	-60.486926	-28.230094	-28.230000	0.468395	0.469504	25.507711	25.533864
.47	-135.435397	-135.435397	-96.047643	-96.047643	-60.152607	-60.152607	-27.776347	-27.776323	1.035708	1.036646	26.188529	26.210919
.48	-135.365081	-135.365081	-95.841432	-95.841432	-59.819570	-59.819570	-27.324446	-27.324446	1.600454	1.601245	26.865099	26.884682
.49	-135.295011	-135.295011	-95.635966	-95.635966	-59.487800	-59.487800	-26.874424	-26.874408	2.162662	2.163326	27.538297	27.555125
.50	-135.225186	-135.225186	-95.431239	-95.431239	-59.157287	-59.157287	-26.426210	-26.426196	2.722360	2.722917	28.207915	28.222421

κ	$12_{0,12}$ / 12_{-12}	$12_{1,12}$ / 12_{-11}	$12_{1,11}$ / 12_{-10}	$12_{2,11}$ / 12_{-9}	$12_{2,10}$ / 12_{-8}	$12_{3,10}$ / 12_{-7}	$12_{3,9}$ / 12_{-6}	$12_{4,9}$ / 12_{-5}	$12_{4,8}$ / 12_{-4}	$12_{5,8}$ / 12_{-3}	$12_{5,7}$ / 12_{-2}	$12_{6,7}$ / 12_{-1}
.50	-135.225186	-135.225186	-95.431239	-95.431239	-59.157287	-59.157287	-26.426210	-26.426196	2.722360	2.722917	28.207915	28.222421
.51	-135.155602	-135.155602	-95.252243	-95.252243	-58.826016	-58.826016	-25.979801	-25.979800	3.279578	3.280043	28.874014	28.886446
.52	-135.086257	-135.086257	-95.073970	-95.073970	-58.499976	-58.499976	-25.535180	-25.535177	3.834341	3.834729	29.536650	29.547272
.53	-135.017148	-135.017148	-94.821414	-94.821414	-58.173154	-58.173154	-25.092322	-25.092322	4.386678	4.387000	30.195878	30.204925
.54	-134.948275	-134.948275	-94.619567	-94.619567	-57.847539	-57.847539	-24.651229	-24.651224	4.936814	4.936880	30.851749	30.859430
.55	-134.879633	-134.879633	-94.418422	-94.418422	-57.523118	-57.523118	-24.211864	-24.211859	5.484175	5.484394	31.504315	31.510813
.56	-134.811221	-134.811221	-94.217973	-94.217973	-57.199879	-57.199879	-23.774216	-23.774212	6.029386	6.029566	32.153621	32.159100
.57	-134.743037	-134.743037	-94.018213	-94.018213	-56.877872	-56.877872	-23.338267	-23.338265	6.572272	6.572418	32.799715	32.804318
.58	-134.675078	-134.675078	-93.819135	-93.819135	-56.556905	-56.556905	-22.904003	-22.904001	7.112855	7.112975	33.442640	33.446649
.59	-134.607343	-134.607343	-93.620733	-93.620733	-56.237147	-56.237147	-22.471406	-22.471404	7.651161	7.651258	34.082439	34.085649
.60	-134.539828	-134.539828	-93.423001	-93.423001	-55.918527	-55.918527	-22.040459	-22.040458	8.187212	8.187290	34.719152	34.721816
.61	-134.472533	-134.472533	-93.225932	-93.225932	-55.601034	-55.601034	-21.611149	-21.611148	8.721200	8.721092	35.352818	35.355019
.62	-134.405456	-134.405456	-93.029520	-93.029520	-55.284658	-55.284658	-21.183458	-21.183457	9.252638	9.252638	35.983475	35.985225
.63	-134.338593	-134.338593	-92.833759	-92.833759	-54.969389	-54.969389	-20.757372	-20.757371	9.782057	9.782096	36.611160	36.612641
.64	-134.271944	-134.271944	-92.638642	-92.638642	-54.655215	-54.655215	-20.332875	-20.332875	10.309300	10.309304	37.235907	37.237112
.65	-134.205505	-134.205505	-92.444165	-92.444165	-54.342128	-54.342128	-19.909953	-19.909953	10.834344	10.834439	37.857750	37.858726
.66	-134.139276	-134.139276	-92.250321	-92.250321	-54.030117	-54.030117	-19.488592	-19.488592	11.357394	11.357413	38.476623	38.477309
.67	-134.073255	-134.073255	-92.057105	-92.057105	-53.719172	-53.719172	-19.068776	-19.068776	11.878268	11.878268	39.092680	39.093405
.68	-134.007439	-134.007439	-91.864510	-91.864510	-53.409284	-53.409284	-18.650493	-18.650493	12.397064	12.397064	39.706683	39.706683
.69	-133.941826	-133.941826	-91.672532	-91.672532	-53.100443	-53.100443	-18.233727	-18.233727	12.913777	12.913786	40.316730	40.317125
.70	-133.876416	-133.876416	-91.481164	-91.481164	-52.792640	-52.792640	-17.818465	-17.818465	13.428452	13.428458	40.924529	40.924839
.71	-133.811205	-133.811205	-91.290402	-91.290402	-52.485866	-52.485866	-17.404695	-17.404695	13.941098	13.941103	41.529848	41.529848
.72	-133.746194	-133.746194	-91.100241	-91.100241	-52.180112	-52.180112	-16.992402	-16.992402	14.451734	14.451737	42.131993	42.132179
.73	-133.681378	-133.681378	-90.910674	-90.910674	-51.875368	-51.875368	-16.581573	-16.581573	14.960378	14.960380	42.731713	42.731854
.74	-133.616758	-133.616758	-90.721696	-90.721696	-51.571626	-51.571626	-16.172196	-16.172196	15.467048	15.467050	43.328792	43.328899
.75	-133.552331	-133.552331	-90.533304	-90.533304	-51.268878	-51.268878	-15.764259	-15.764259	15.971761	15.971763	43.923257	43.923337
.76	-133.488095	-133.488095	-90.345491	-90.345491	-50.967115	-50.967115	-15.357748	-15.357748	16.474536	16.474536	44.515312	44.515192
.77	-133.424050	-133.424050	-90.158253	-90.158253	-50.666327	-50.666327	-14.952651	-14.952651	16.975389	16.975389	45.104488	45.104486
.78	-133.360192	-133.360192	-89.971585	-89.971585	-50.366508	-50.366508	-14.548956	-14.548956	17.474336	17.474336	45.691211	45.691245
.79	-133.296522	-133.296522	-89.785481	-89.785481	-50.067649	-50.067649	-14.146652	-14.146652	17.971394	17.971395	46.275485	46.275485
.80	-133.233037	-133.233037	-89.599939	-89.599939	-49.769742	-49.769742	-13.745726	-13.745726	18.466580	18.466580	46.857234	46.857234
.81	-133.169735	-133.169735	-89.414951	-89.414951	-49.472778	-49.472778	-13.346167	-13.346167	18.959909	18.959910	47.436502	47.436513
.82	-133.106616	-133.106616	-89.230515	-89.230515	-49.176751	-49.176751	-12.947964	-12.947964	19.451398	19.451398	48.013335	48.013335
.83	-133.043677	-133.043677	-89.046625	-89.046625	-48.881652	-48.881652	-12.551105	-12.551105	19.941061	19.941061	48.587738	48.587744
.84	-132.980918	-132.980918	-88.863278	-88.863278	-48.587473	-48.587473	-12.155579	-12.155579	20.428914	20.428914	49.159737	49.159737
.85	-132.918336	-132.918336	-88.680467	-88.680467	-48.294208	-48.294208	-11.761375	-11.761375	20.914971	20.914971	49.729341	49.729343
.86	-132.855930	-132.855930	-88.498191	-88.498191	-48.001849	-48.001849	-11.368483	-11.368483	21.393761	21.393761	50.296583	50.296583
.87	-132.793700	-132.793700	-88.316443	-88.316443	-47.710388	-47.710388	-10.976892	-10.976892	21.881765	21.881761	50.861476	50.861477
.88	-132.731642	-132.731642	-88.135219	-88.135219	-47.419818	-47.419818	-10.586591	-10.586591	22.362521	22.362521	51.424042	51.424042
.89	-132.669757	-132.669757	-87.954517	-87.954517	-47.130132	-47.130132	-10.197570	-10.197570	22.841544	22.841544	51.984300	51.984300
.90	-132.608043	-132.608043	-87.774330	-87.774330	-46.841324	-46.841324	-9.809819	-9.809819	23.318845	23.318845	52.542269	52.542269
.91	-132.546498	-132.546498	-87.594657	-87.594657	-46.553385	-46.553385	-9.423327	-9.423327	23.794435	23.794435	53.097967	53.097967
.92	-132.485121	-132.485121	-87.415491	-87.415491	-46.266310	-46.266310	-9.038086	-9.038086	24.268330	24.268330	53.651413	53.651413
.93	-132.423910	-132.423910	-87.236830	-87.236830	-45.980092	-45.980092	-8.654084	-8.654084	24.740542	24.740542	54.202624	54.202624
.94	-132.362865	-132.362865	-87.058669	-87.058669	-45.694724	-45.694724	-8.271313	-8.271313	25.211085	25.211085	54.751619	54.751619
.95	-132.301985	-132.301985	-86.881004	-86.881004	-45.410198	-45.410198	-7.889763	-7.889763	25.679971	25.679971	55.298415	55.298415
.96	-132.241267	-132.241267	-86.703833	-86.703833	-45.126530	-45.126530	-7.509424	-7.509424	26.147214	26.147214	55.843029	55.843029
.97	-132.180711	-132.180711	-86.527155	-86.527155	-44.843653	-44.843653	-7.130287	-7.130287	26.612826	26.612826	56.385478	56.385478
.98	-132.120315	-132.120315	-86.350953	-86.350953	-44.561619	-44.561619	-6.752344	-6.752344	27.076819	27.076819	56.925778	56.925778
.99	-132.060078	-132.060078	-86.175237	-86.175237	-44.280403	-44.280403	-6.375584	-6.375584	27.539206	27.539206	57.463947	57.463947
1.00	-132.000000	-132.000000	-86.000000	-86.000000	-44.000000	-44.000000	-6.000000	-6.000000	28.000000	28.000000	58.000000	58.000000

Transition Strengths for Rotational Transitions

Intensity of a transition between rotational levels J_{kl} and J'_{mn} is proportional to

$$(\mu_x)^2 \; {}^xS_{J_{kl}J'_{mn}}(\kappa) \; = \; (2J + 1)|(\mu_x)_{J_{kl}J'_{mn}}|^2$$

Here μ_x is the dipole moment along one of the principal axes of inertia ($x = a$, b, or c), and S is the quantity tabulated here as a function of initial and final state and of the asymmetry parameter κ. However, each value has been multiplied by 10^4 to eliminate decimal points. The upper sign for values of K applies to transition subbranches listed in the two left-hand columns, and the lower sign to those in the right-hand columns. The axis along which a dipole moment is required to produce a given transition is indicated by a superscript to the left of the subbranch designation. Thus ${}^cQ_{10}$ indicates a Q branch ($\Delta J = 0$) with a change in K_{-1} of 1, a change in K_1 of 0, and that a dipole moment μ_c along the c axis is required for the transition. For further discussion see Chap. 4. (Tables in this appendix are taken from Cross, Hainer, and King [122].)

STRENGTHS FOR ROTATIONAL TRANSITIONS

Symmetric-rotor Subbranches—a and c Prolate-and-oblate Subbranches

Subbranch				κ			Subbranch	
$^cQ_{1,0}$ $J + K_{-1} + K_1$ even	$^cQ_{-1,0}$ $J + K_{-1} + K_1$ odd	∓ 1	∓ 0.5	0	± 0.5	± 1	$^aQ_{0,1}$ $J + K_{-1} + K_1$ even	$^aQ_{0,-1}$ $J + K_{-1} + K_1$ odd
$1_{0,1}$	$1_{1,1}$	15000	15000	15000	15000	15000	$1_{1,0}$	$1_{1,1}$
$2_{0,2}$	$2_{1,2}$	25000	28223	31100	32845	33333	$2_{2,0}$	$2_{2,1}$
$3_{0,3}$	$3_{1,3}$	35000	45104	50431	52155	52500	$3_{3,0}$	$3_{3,1}$
$4_{0,4}$	$4_{1,4}$	45000	64494	70244	71708	72000	$4_{4,0}$	$4_{4,1}$
$5_{0,5}$	$5_{1,5}$	55000	84696	90073	91399	91667	$5_{5,0}$	$5_{5,1}$
$6_{0,6}$	$6_{1,6}$	65000	104928	109923	111174	111429	$6_{6,0}$	$6_{6,1}$
$7_{0,7}$	$7_{1,7}$	75000	125065	129799	131004	131250	$7_{7,0}$	$7_{7,1}$
$8_{0,8}$	$8_{1,8}$	85000	145135	149698	150871	151111	$8_{8,0}$	$8_{8,1}$
$9_{0,9}$	$9_{1,9}$	95000	165170	169614	170764	171000	$9_{9,0}$	$9_{9,1}$
$10_{0,10}$	$10_{1,10}$	105000	185187	189544	190677	190909	$10_{10,0}$	$10_{10,1}$
$11_{0,11}$	$11_{1,11}$	115000	205194	209484	210603	210834	$11_{11,0}$	$11_{11,1}$
$12_{0,12}$	$12_{1,12}$	125000	225195	229434	230542	230769	$12_{12,0}$	$12_{12,1}$
$2_{1,1}$	$2_{2,1}$	8333	8333	8333	8333	8333	$2_{1,1}$	$2_{1,2}$
$3_{1,2}$	$3_{2,2}$	14583	16278	18811	21875	23333	$3_{2,1}$	$3_{2,2}$
$4_{1,3}$	$4_{2,3}$	20250	26168	34242	39363	40500	$4_{3,1}$	$4_{3,2}$
$5_{1,4}$	$5_{2,4}$	25667	39338	52949	57742	58667	$5_{4,1}$	$5_{4,2}$
$6_{1,5}$	$6_{2,5}$	30952	56179	72319	76548	77381	$6_{5,1}$	$6_{5,2}$
$7_{1,6}$	$7_{2,6}$	36161	75597	91744	95646	96429	$7_{6,1}$	$7_{6,2}$
$8_{1,7}$	$8_{2,7}$	41319	95950	111231	114943	115694	$8_{7,1}$	$8_{7,2}$
$9_{1,8}$	$9_{2,8}$	46444	116333	130792	134381	135111	$9_{8,1}$	$9_{8,2}$
$10_{1,9}$	$10_{2,9}$	51545	136551	150418	153921	154636	$10_{9,1}$	$10_{9,2}$
$11_{1,10}$	$11_{2,10}$	56629	156642	170100	173540	174242	$11_{10,1}$	$11_{10,2}$
$12_{1,11}$	$12_{2,11}$	61699	176660	189825	193216	193910	$12_{11,1}$	$12_{11,2}$

Block 1 (10 rows)

Left	I	II	III	IV	V	Mid (a)	Mid (b)	Right
$3_{2,1}$	8750	7403	6406	5944	5833	$3_{3,1}$	$3_{1,2}$	$3_{1,3}$
$4_{2,2}$	15750	13221	13196	15598	18000	$4_{3,2}$	$4_{2,2}$	$4_{2,3}$
$5_{2,3}$	22000	19105	23397	30662	33000	$5_{3,3}$	$5_{3,2}$	$5_{3,3}$
$6_{2,4}$	27857	26374	38620	47709	49524	$6_{3,4}$	$6_{4,2}$	$6_{4,3}$
$7_{2,5}$	33482	36237	57062	65399	66964	$7_{3,5}$	$7_{5,2}$	$7_{5,3}$
$8_{2,6}$	38958	49682	76155	83565	85000	$8_{3,6}$	$8_{6,2}$	$8_{6,3}$
$9_{2,7}$	44333	66864	95251	102089	103444	$9_{3,7}$	$9_{7,2}$	$9_{7,3}$
$10_{2,8}$	49636	86630	114393	120880	122183	$10_{3,8}$	$10_{8,2}$	$10_{8,3}$
$11_{2,9}$	54886	107332	133621	139873	141136	$11_{3,9}$	$11_{9,2}$	$11_{9,3}$
$12_{2,10}$	60096	128002	152940	159022	160256	$12_{3,10}$	$12_{10,2}$	$12_{10,3}$

Block 2 (9 rows)

Left	I	II	III	IV	V	Mid (a)	Mid (b)	Right
$4_{3,1}$	9000	7587	6026	4847	4500	$4_{4,1}$	$4_{1,3}$	$4_{1,4}$
$5_{3,2}$	16500	13464	11058	11750	14667	$5_{4,2}$	$5_{2,3}$	$5_{2,4}$
$6_{3,3}$	23214	18339	17488	23981	27857	$6_{4,3}$	$6_{3,3}$	$6_{3,4}$
$7_{3,4}$	29464	22914	27745	39794	42857	$7_{4,4}$	$7_{4,3}$	$7_{4,4}$
$8_{3,5}$	35417	28185	43063	56506	59028	$8_{4,5}$	$8_{5,3}$	$8_{5,4}$
$9_{3,6}$	41167	35293	61523	73754	76000	$9_{4,6}$	$9_{6,3}$	$9_{6,4}$
$10_{3,7}$	46773	45350	80547	91464	93546	$10_{4,7}$	$10_{7,3}$	$10_{7,4}$
$11_{3,8}$	52273	59213	99473	109542	111515	$11_{4,8}$	$11_{8,3}$	$11_{8,4}$
$12_{3,9}$	57692	76888	118383	127913	129808	$12_{4,9}$	$12_{9,3}$	$12_{9,4}$

Block 3 (8 rows)

Left	I	II	III	IV	V	Mid (a)	Mid (b)	Right
$5_{4,1}$	9167	7777	6127	4374	3667	$5_{5,1}$	$5_{1,4}$	$5_{1,5}$
$6_{4,2}$	17024	14084	10758	9464	12381	$6_{5,2}$	$6_{2,4}$	$6_{2,5}$
$7_{4,3}$	24107	19340	15156	18769	24107	$7_{5,3}$	$7_{3,4}$	$7_{3,5}$
$8_{4,4}$	30694	23768	21441	33034	37778	$8_{5,4}$	$8_{4,4}$	$8_{4,5}$
$9_{4,5}$	36944	27638	31860	49002	52778	$9_{5,5}$	$9_{5,4}$	$9_{5,5}$
$10_{4,6}$	42955	31542	47402	65474	68727	$10_{5,6}$	$10_{6,4}$	$10_{6,5}$
$11_{4,7}$	48788	36457	66028	82425	85379	$11_{5,7}$	$11_{7,4}$	$11_{7,5}$
$12_{4,8}$	54487	43527	85120	99805	102564	$12_{5,8}$	$12_{8,4}$	$12_{8,5}$

STRENGTHS FOR ROTATIONAL TRANSITIONS

SYMMETRIC-ROTOR SUBBRANCHES—a AND c PROLATE-AND-OBLATE SUBBRANCHES

Subbranch				κ			Subbranch	
$^cQ_{1,0}$ $J + K_{-1} + K_1$ even	$^cQ_{-1,0}$ $J + K_{-1} + K_1$ odd	∓ 1	∓ 0.5	0	± 0.5	± 1	$^aQ_{0,1}$ $J + K_{-1} + K_1$ even	$^aQ_{0,-1}$ $J + K_{-1} + K_1$ odd
$6_{6,1}$	$6_{6,1}$	9286	7913	6271	4244	3095	$6_{1,5}$	$6_{1,6}$
$7_{6,2}$	$7_{6,2}$	17411	14552	11116	8220	10714	$7_{2,5}$	$7_{2,6}$
$8_{6,3}$	$8_{6,3}$	24792	20246	14956	14996	21250	$8_{3,5}$	$8_{3,6}$
$9_{6,4}$	$9_{6,4}$	31667	25157	18940	27035	33778	$9_{4,5}$	$9_{4,6}$
$10_{6,5}$	$10_{6,5}$	38182	29364	25162	42308	47727	$10_{5,5}$	$10_{5,6}$
$11_{6,6}$	$11_{6,6}$	44432	32945	35783	58220	62727	$11_{6,5}$	$11_{6,6}$
$12_{6,7}$	$12_{6,7}$	50481	36135	51610	74526	78526	$12_{7,5}$	$12_{7,6}$
$7_{6,1}$	$7_{7,1}$	9375	8011	6383	4273	2679	$7_{1,6}$	$7_{1,7}$
$8_{6,2}$	$8_{7,2}$	17708	14899	11514	7682	9444	$8_{2,6}$	$8_{2,7}$
$9_{6,3}$	$9_{7,3}$	25333	20931	15562	12549	19000	$9_{3,6}$	$9_{3,7}$
$10_{6,4}$	$10_{7,4}$	32455	26254	18837	21925	30545	$10_{4,6}$	$10_{4,7}$
$11_{6,5}$	$11_{7,5}$	39205	30943	22510	36052	43561	$11_{5,6}$	$11_{5,7}$
$12_{6,6}$	$12_{7,6}$	45673	35027	28709	51607	57692	$12_{6,6}$	$12_{6,7}$
$8_{7,1}$	$8_{8,1}$	9444	8087	6468	4346	2361	$8_{1,7}$	$8_{1,8}$
$9_{7,2}$	$9_{8,2}$	17944	15167	11832	7594	8444	$9_{2,7}$	$9_{2,8}$
$10_{7,3}$	$10_{8,3}$	25773	21462	16215	11172	17182	$10_{3,7}$	$10_{3,8}$
$11_{7,4}$	$11_{8,4}$	33106	27105	19675	18011	27879	$11_{4,7}$	$11_{4,8}$
$12_{7,5}$	$12_{8,5}$	40064	32174	22501	30163	40064	$12_{5,7}$	$12_{6,8}$
$9_{8,1}$	$9_{9,1}$	9500	8146	6535	4418	2111	$9_{1,8}$	$9_{1,9}$
$10_{8,2}$	$10_{9,2}$	18136	15380	12081	7731	7636	$10_{2,8}$	$10_{2,9}$
$11_{8,3}$	$11_{9,3}$	26136	21887	16748	10598	15682	$11_{3,8}$	$11_{3,9}$
$12_{8,4}$	$12_{9,4}$	33654	27788	20570	15383	25641	$12_{4,8}$	$12_{4,9}$

$J+K_{-1}+K_1$ even	$^{c}P_{-1,0}$ $J+K_{-1}+K_1$ even	∓1	∓0.5	0	±0.5	±1	$J+K_{-1}+K_1$ even	$^{a}R_{0,1}$ $J+K_{-1}+K_1$ even	$^{a}P_{0,-1}$ $J+K_{-1}+K_1$ even
$10_{10,1}$		9545	8194	6588	4479	1909	$10_{1,9}$		$10_{1,10}$
$11_{10,2}$		18295	15554	12280	7933	6970	$11_{2,9}$		$11_{2,10}$
$12_{10,3}$		26442	22237	17176	10563	14423	$12_{3,9}$		$12_{3,10}$
$11_{11,1}$		9583	8234	6631	4530	1742	$11_{1,10}$		$11_{1,11}$
$12_{11,2}$		18429	15699	12443	8126	6410	$12_{2,10}$		$12_{2,11}$
$12_{12,1}$		9615	8268	6667	4571	1603	$12_{1,11}$		$12_{1,12}$

$^{c}R_{1,0}$ $J+K_{-1}+K_1$ even	$^{c}P_{-1,0}$ $J+K_{-1}+K_1$ even	∓1	∓0.5	0	±0.5	±1	$^{a}R_{0,1}$ $J+K_{-1}+K_1$ even	$^{a}P_{0,-1}$ $J+K_{-1}+K_1$ even
$0_{0,0}$	$1_{1,0}$	10000	10000	10000	10000	10000	$0_{0,0}$	$1_{0,1}$
$1_{1,0}$	$2_{2,0}$	15000	16934	18660	19707	20000	$1_{0,1}$	$2_{0,2}$
$2_{2,0}$	$3_{3,0}$	25000	25893	27201	29029	30000	$2_{0,2}$	$3_{0,3}$
$3_{3,0}$	$4_{4,0}$	35000	35773	36728	38312	40000	$3_{0,3}$	$4_{0,4}$
$4_{4,0}$	$5_{5,0}$	45000	45745	46619	47897	50000	$4_{0,4}$	$5_{0,5}$
$5_{5,0}$	$6_{6,0}$	55000	55730	56582	57727	60000	$5_{0,5}$	$6_{0,6}$
$6_{6,0}$	$7_{7,0}$	65000	65721	66562	67660	70000	$6_{0,6}$	$7_{0,7}$
$7_{7,0}$	$8_{8,0}$	75000	75714	76549	77628	80000	$7_{0,7}$	$8_{0,8}$
$8_{8,0}$	$9_{9,0}$	85000	85708	86539	87610	90000	$8_{0,8}$	$9_{0,9}$
$9_{9,0}$	$10_{10,0}$	95000	95704	96531	97597	100000	$9_{0,9}$	$10_{0,10}$
$10_{10,0}$	$11_{11,0}$	105000	105701	106525	107588	110000	$10_{0,10}$	$11_{0,11}$
$11_{11,0}$	$12_{12,0}$	115000	115698	116519	117580	120000	$11_{0,11}$	$12_{0,12}$

STRENGTHS FOR ROTATIONAL TRANSITIONS

SYMMETRIC-ROTOR SUBBRANCHES—a AND c PROLATE-AND-OBLATE SUBBRANCHES

Subbranch		κ					Subbranch	
$^cR_{1,0}$ $J+K_{-1}+K_1$ even	$^cP_{-1,0}$ $J+K_{-1}+K_1$ even	∓ 1	∓ 0.5	0	± 0.5	± 1	$^aR_{0,1}$ $J+K_{-1}+K_1$ even	$^aP_{0,-1}$ $J+K_{-1}+K_1$ even
$1_{0,1}$	$2_{1,1}$	15000	15000	15000	15000	15000	$1_{1,0}$	$2_{1,1}$
$2_{1,1}$	$3_{2,1}$	16667	22500	25581	26509	26667	$2_{1,1}$	$3_{1,2}$
$3_{2,1}$	$4_{3,1}$	26250	29261	33801	36902	37500	$3_{1,2}$	$4_{1,3}$
$4_{3,1}$	$5_{4,1}$	36000	38400	41758	46530	48000	$4_{1,3}$	$5_{1,4}$
$5_{4,1}$	$6_{5,1}$	45833	48106	50867	55604	58333	$5_{1,4}$	$6_{1,5}$
$6_{5,1}$	$7_{6,1}$	55714	57930	60533	64605	68571	$6_{1,5}$	$7_{1,6}$
$7_{6,1}$	$8_{7,1}$	65625	67805	70356	73938	78750	$7_{1,6}$	$8_{1,7}$
$8_{7,1}$	$9_{8,1}$	75556	77710	80235	83593	88889	$8_{1,7}$	$9_{1,8}$
$9_{8,1}$	$10_{9,1}$	85500	87636	90142	93412	99000	$9_{1,8}$	$10_{1,9}$
$10_{9,1}$	$11_{10,1}$	95455	97576	100068	103301	109091	$10_{1,9}$	$11_{1,10}$
$11_{10,1}$	$12_{11,1}$	105416	107526	110008	113219	119166	$11_{1,10}$	$12_{1,11}$
$2_{0,2}$	$3_{1,2}$	20000	18636	17345	16724	16667	$2_{2,0}$	$3_{2,1}$
$3_{1,2}$	$4_{2,2}$	18750	29055	30992	30230	30000	$3_{2,1}$	$4_{2,2}$
$4_{2,2}$	$5_{3,2}$	28000	34387	41441	42462	42000	$4_{2,2}$	$5_{2,3}$
$5_{3,2}$	$6_{4,2}$	37500	41961	49227	53738	53333	$5_{2,3}$	$6_{2,4}$
$6_{4,2}$	$7_{5,2}$	47143	51182	56697	64087	64286	$6_{2,4}$	$7_{2,5}$
$7_{5,2}$	$8_{6,2}$	56875	60756	65450	73564	75000	$7_{2,5}$	$8_{2,6}$
$8_{6,2}$	$9_{7,2}$	66667	70451	74899	82413	85556	$8_{2,6}$	$9_{2,7}$
$9_{7,2}$	$10_{8,2}$	76500	80218	84567	91174	96000	$9_{2,7}$	$10_{2,8}$
$10_{8,2}$	$11_{9,2}$	86364	90031	94328	100297	106364	$10_{2,8}$	$11_{2,9}$
$11_{9,2}$	$12_{10,2}$	96250	99880	104134	109796	116666	$11_{2,9}$	$12_{2,10}$
$3_{0,3}$	$4_{1,3}$	25000	20331	18001	17567	17500	$3_{3,0}$	$4_{3,1}$
$4_{1,3}$	$5_{2,3}$	21000	34848	33475	32109	32000	$4_{3,1}$	$5_{3,2}$

Block 1

$5_{2,3}$	$6_{3,3}$	$5_{3,2}$	45000	45219	47032	41218	30000	$5_{3,2}$	$6_{3,3}$	$6_{3,3}$
$6_{3,3}$	$7_{3,4}$	$6_{3,3}$	57143	57683	57381	46575	39286	$6_{3,3}$	$7_{3,4}$	$7_{4,3}$
$7_{4,3}$	$8_{3,5}$	$7_{3,4}$	68750	69691	64788	54876	48750	$7_{3,4}$	$8_{3,5}$	$8_{5,3}$
$8_{5,3}$	$9_{3,6}$	$8_{3,5}$	80000	80981	71834	64092	58333	$8_{3,5}$	$9_{3,6}$	$9_{6,3}$
$9_{6,3}$	$10_{3,7}$	$9_{3,6}$	91000	91312	80274	73557	68000	$9_{3,6}$	$10_{3,7}$	$10_{7,3}$
$10_{7,3}$	$11_{3,8}$	$10_{3,7}$	101818	100665	89526	83147	77727	$10_{3,7}$	$11_{3,8}$	$11_{8,3}$
$11_{8,3}$	$12_{3,9}$	$11_{3,8}$	112500	109320	90948	92819	87500	$11_{3,8}$	$12_{3,9}$	$12_{9,3}$

Block 2

$4_{0,4}$	$5_{4,1}$	$4_{4,0}$	18000	18082	18478	20650	30000	$4_{4,0}$	$5_{4,1}$	$5_{1,4}$
$5_{1,4}$	$6_{4,2}$	$5_{4,1}$	33333	33475	34370	38686	23333	$5_{4,1}$	$6_{4,2}$	$6_{2,4}$
$6_{2,4}$	$7_{4,3}$	$6_{4,2}$	47143	47326	49439	48639	32143	$6_{4,2}$	$7_{4,3}$	$7_{3,4}$
$7_{3,4}$	$8_{4,4}$	$7_{4,3}$	60000	60238	63082	52676	41250	$7_{4,3}$	$8_{4,4}$	$8_{4,4}$
$8_{4,4}$	$9_{4,5}$	$8_{4,4}$	72222	72645	73357	59229	50556	$8_{4,4}$	$9_{4,5}$	$9_{5,4}$
$9_{5,4}$	$10_{4,6}$	$9_{4,5}$	84000	84877	80428	67888	60000	$9_{4,5}$	$10_{4,6}$	$10_{6,4}$
$10_{6,4}$	$11_{4,7}$	$10_{4,6}$	95455	96881	87093	77064	69546	$10_{4,6}$	$11_{4,7}$	$11_{7,4}$
$11_{7,4}$	$12_{4,8}$	$11_{4,7}$	106666	108226	95254	86444	79167	$11_{4,7}$	$12_{4,8}$	$12_{8,4}$

Block 3

$5_{0,5}$	$6_{5,1}$	$5_{5,0}$	18333	18422	18847	20660	35000	$5_{5,0}$	$6_{5,1}$	$6_{1,5}$
$6_{1,5}$	$7_{5,2}$	$6_{5,1}$	34286	34447	35224	40254	25714	$6_{5,1}$	$7_{5,2}$	$7_{2,5}$
$7_{2,5}$	$8_{5,3}$	$7_{5,2}$	48750	48971	50352	54914	34375	$7_{5,2}$	$8_{5,3}$	$8_{3,5}$
$8_{3,5}$	$9_{5,4}$	$8_{5,3}$	62222	62490	65354	60334	43333	$8_{5,3}$	$9_{5,4}$	$9_{4,5}$
$9_{4,5}$	$10_{5,5}$	$9_{5,4}$	75000	75309	79136	64543	52500	$9_{5,4}$	$10_{5,5}$	$10_{5,5}$
$10_{5,5}$	$11_{5,6}$	$10_{5,5}$	87273	87664	89354	72156	61818	$10_{5,5}$	$11_{5,6}$	$11_{6,5}$
$11_{6,5}$	$12_{5,7}$	$11_{5,6}$	99167	99820	96120	80944	71250	$11_{5,6}$	$12_{5,7}$	$12_{7,5}$

Block 4

$6_{0,6}$	$7_{6,1}$	$6_{6,0}$	18571	18664	19108	20793	40000	$6_{6,0}$	$7_{6,1}$	$7_{1,6}$
$7_{1,6}$	$8_{6,2}$	$7_{6,1}$	35000	35171	35988	40367	28125	$7_{6,1}$	$8_{6,2}$	$8_{2,6}$
$8_{2,6}$	$9_{6,3}$	$8_{6,2}$	50000	50241	51410	58807	36667	$8_{6,2}$	$9_{6,3}$	$9_{3,6}$
$9_{3,6}$	$10_{6,4}$	$9_{6,3}$	64000	64301	66193	68406	45500	$9_{6,3}$	$10_{6,4}$	$10_{4,6}$
$10_{4,6}$	$11_{6,5}$	$10_{6,4}$	77273	77627	81252	71334	54546	$10_{6,4}$	$11_{6,5}$	$11_{5,6}$
$11_{5,6}$	$12_{6,6}$	$11_{6,5}$	90000	90399	95192	77023	63750	$11_{6,5}$	$12_{6,6}$	$12_{6,6}$

STRENGTHS FOR ROTATIONAL TRANSITIONS

SYMMETRIC-ROTOR SUBBRANCHES—a AND c PROLATE-AND-OBLATE SUBBRANCHES

Subbranch $^cR_{1,0}$ $J + K_{-1} + K_1$ even	$^cP_{-1,0}$ $J + K_{-1} + K_1$ even	∓ 1	∓ 0.5	0	± 0.5	± 1	Subbranch $^aR_{0,1}$ $J + K_{-1} + K_1$ even	$^aP_{0,-1}$ $J + K_{-1} + K_1$ even
$7_{0,7}$	$8_{1,7}$	45000	20990	19300	18844	18750	$7_{7,0}$	$8_{7,1}$
$8_{1,7}$	$9_{2,7}$	30556	40255	36587	35733	35556	$8_{7,1}$	$9_{7,2}$
$9_{2,7}$	$10_{3,7}$	39000	60147	52457	51252	51000	$9_{7,2}$	$10_{7,3}$
$10_{3,7}$	$11_{4,7}$	47727	75043	67350	65775	65455	$10_{7,3}$	$11_{7,4}$
$11_{4,7}$	$12_{5,7}$	56667	79651	81971	79549	79167	$11_{7,4}$	$12_{7,5}$
$8_{0,8}$	$9_{1,8}$	50000	21170	19449	18985	18889	$8_{8,0}$	$9_{8,1}$
$9_{1,8}$	$10_{2,8}$	33000	40430	37059	36182	36000	$9_{8,1}$	$10_{8,2}$
$10_{2,8}$	$11_{3,8}$	41364	59963	53330	52078	51818	$10_{8,2}$	$11_{8,3}$
$11_{3,8}$	$12_{4,8}$	50000	78946	68596	66999	66667	$11_{8,3}$	$12_{8,4}$
$9_{0,9}$	$10_{1,9}$	55000	21315	19566	19097	19000	$9_{9,0}$	$10_{9,1}$
$10_{1,9}$	$11_{2,9}$	35454	40779	37443	36548	36364	$10_{9,1}$	$11_{9,2}$
$11_{2,9}$	$12_{3,9}$	43750	59640	54051	52766	52500	$11_{9,2}$	$12_{9,3}$
$10_{0,10}$	$11_{1,10}$	60000	21432	19662	19189	19091	$10_{10,0}$	$11_{10,1}$
$11_{1,10}$	$12_{2,10}$	37917	41138	37761	36854	36667	$11_{10,1}$	$12_{10,2}$
$11_{0,11}$	$12_{1,11}$	65000	21527	19742	19265	19167	$11_{11,0}$	$12_{11,1}$

Subbranch $^cR_{1,0}$ $J + K_{-1} + K_1$ odd	$^cP_{-1,0}$ $J + K_{-1} + K_1$ odd	∓ 1	∓ 0.5	0	± 0.5	± 1	Subbranch $^aR_{0,1}$ $J + K_{-1} + K_1$ odd	$^aP_{0,-1}$ $J + K_{-1} + K_1$ odd
$1_{1,1}$	$2_{2,1}$	15000	15000	15000	15000	15000	$1_{1,1}$	$2_{1,2}$
$2_{2,1}$	$3_{3,1}$	25000	25710	26243	26564	26667	$2_{1,2}$	$3_{1,3}$

Block 1

Upper	Upper′						Lower′	Lower
$4_{1,4}$	$3_{1,3}$	35000	35758	36540	37210	37500	$4_{4,1}$	$3_{3,1}$
$5_{1,5}$	$4_{1,4}$	45000	45743	46583	47478	48000	$5_{5,1}$	$4_{4,1}$
$6_{1,6}$	$5_{1,5}$	55000	55730	56576	57578	58333	$6_{6,1}$	$5_{5,1}$
$7_{1,7}$	$6_{1,6}$	65000	65721	66561	67607	68571	$7_{7,1}$	$6_{6,1}$
$8_{1,8}$	$7_{1,7}$	75000	75714	76550	77609	78750	$8_{8,1}$	$7_{7,1}$
$9_{1,9}$	$8_{1,8}$	85000	85708	86539	87603	88889	$9_{9,1}$	$8_{8,1}$
$10_{1,10}$	$9_{1,9}$	95000	95704	96531	97595	99000	$10_{10,1}$	$9_{9,1}$
$11_{1,11}$	$10_{1,10}$	105000	105701	106525	107587	109091	$11_{11,1}$	$10_{10,1}$
$12_{1,12}$	$11_{1,11}$	115000	115698	116519	117580	119166	$12_{12,1}$	$11_{11,1}$

Block 2

Upper	Upper′						Lower′	Lower
$3_{2,2}$	$2_{2,1}$	16667	16667	16667	16667	16667	$2_{1,2}$	$3_{2,2}$
$4_{2,3}$	$3_{2,2}$	26250	28258	29391	29882	30000	$3_{2,2}$	$4_{3,2}$
$5_{2,4}$	$4_{2,3}$	36000	38290	40354	41637	42000	$4_{3,2}$	$5_{4,2}$
$6_{2,5}$	$5_{2,4}$	45833	48094	50537	52600	53333	$5_{4,2}$	$6_{5,2}$
$7_{2,6}$	$6_{2,5}$	55714	57929	60461	63088	64286	$6_{5,2}$	$7_{6,2}$
$8_{2,7}$	$7_{2,6}$	65625	67805	70340	73291	75000	$7_{6,2}$	$8_{7,2}$
$9_{2,8}$	$8_{2,7}$	75556	77710	80231	83338	85556	$8_{7,2}$	$9_{8,2}$
$10_{2,9}$	$9_{2,8}$	85500	87636	90142	93314	96000	$9_{8,2}$	$10_{9,2}$
$11_{2,10}$	$10_{2,9}$	95455	97576	100068	103262	106364	$10_{9,2}$	$11_{10,2}$
$12_{2,11}$	$11_{2,10}$	105416	107526	110008	113205	116666	$11_{10,2}$	$12_{11,2}$

Block 3

Upper	Upper′						Lower′	Lower
$4_{3,2}$	$3_{3,1}$	18750	18207	17796	17564	17500	$4_{2,3}$	$3_{1,3}$
$5_{3,3}$	$4_{3,2}$	28000	31148	32063	32074	32000	$5_{3,3}$	$4_{2,3}$
$6_{3,4}$	$5_{3,3}$	37500	41486	44187	45001	45000	$6_{4,3}$	$5_{3,3}$
$7_{3,5}$	$6_{3,4}$	47143	51127	54949	56948	57143	$7_{5,3}$	$6_{4,3}$
$8_{3,6}$	$7_{3,5}$	56875	60749	64999	68208	68750	$8_{6,3}$	$7_{5,3}$
$9_{3,7}$	$8_{3,6}$	66667	70450	74791	78959	80000	$9_{7,3}$	$8_{6,3}$
$10_{3,8}$	$9_{3,7}$	76500	80217	84543	89339	91000	$10_{8,3}$	$9_{7,3}$
$11_{3,9}$	$10_{3,8}$	86364	90031	94320	99469	101818	$11_{9,3}$	$10_{8,3}$
$12_{3,10}$	$11_{3,9}$	96250	99880	104133	109453	112500	$12_{10,3}$	$11_{9,3}$

STRENGTHS FOR ROTATIONAL TRANSITIONS

SYMMETRIC-ROTOR SUBBRANCHES—a AND c PROLATE-AND-OBLATE SUBBRANCHES

Subbranch		κ					Subbranch	
$^cR_{1,0}$ $J + K_{-1} + K_1$ odd	$^cP_{-1,0}$ $J + K_{-1} + K_1$ odd	∓ 1	∓ 0.5	0	± 0.5	± 1	$^aR_{0,1}$ $J + K_{-1} + K_1$ odd	$^aP_{0,-1}$ $J + K_{-1} + K_1$ odd
$4_{1,4}$	$5_{2,4}$	21000	19363	18449	18082	18000	$4_{4,1}$	$5_{4,2}$
$5_{2,4}$	$6_{3,4}$	30000	33887	33934	33473	33333	$5_{4,2}$	$6_{4,3}$
$6_{3,4}$	$7_{4,4}$	39286	45000	47370	47311	47143	$6_{4,3}$	$7_{4,4}$
$7_{4,4}$	$8_{5,4}$	48750	54655	59178	60145	60000	$7_{4,4}$	$8_{4,5}$
$8_{5,4}$	$9_{6,4}$	58333	64063	69788	72255	72222	$8_{4,5}$	$9_{4,6}$
$9_{6,4}$	$10_{7,4}$	68000	73554	79716	83787	84000	$9_{4,6}$	$10_{4,7}$
$10_{7,4}$	$11_{8,4}$	77727	83147	89384	94830	95455	$10_{4,7}$	$11_{4,8}$
$11_{8,4}$	$12_{9,4}$	87500	92819	99012	105459	106666	$11_{4,8}$	$12_{4,9}$
$5_{1,5}$	$6_{2,5}$	23333	20137	18843	18422	18333	$5_{5,1}$	$6_{5,2}$
$6_{2,5}$	$7_{3,5}$	32143	36189	35151	34447	34286	$6_{5,2}$	$7_{5,3}$
$7_{3,5}$	$8_{4,5}$	41250	48511	49684	48970	48750	$7_{5,3}$	$8_{5,4}$
$8_{4,5}$	$9_{5,5}$	50556	58512	62686	62483	62222	$8_{5,4}$	$9_{5,5}$
$9_{5,5}$	$10_{6,5}$	60000	67785	74286	75273	75000	$9_{5,5}$	$10_{5,6}$
$10_{6,5}$	$11_{7,5}$	69546	77050	84474	87504	87273	$10_{5,6}$	$11_{5,7}$
$11_{7,5}$	$12_{8,5}$	79167	86442	94594	99259	99167	$11_{5,7}$	$12_{5,8}$
$6_{1,6}$	$7_{2,6}$	25714	20629	19107	18664	18571	$6_{6,1}$	$7_{6,2}$
$7_{2,6}$	$8_{3,6}$	34375	37948	35978	35171	35000	$7_{6,2}$	$8_{6,3}$
$8_{3,6}$	$9_{4,6}$	43333	51721	51284	50241	50000	$8_{6,3}$	$9_{6,4}$
$9_{4,6}$	$10_{5,6}$	52500	62496	65297	64301	64000	$9_{6,4}$	$10_{6,5}$
$10_{5,6}$	$11_{6,6}$	61818	71831	78026	77624	77273	$10_{6,5}$	$11_{6,6}$
$11_{6,6}$	$12_{7,6}$	71250	80896	89476	90385	90000	$11_{6,6}$	$12_{6,7}$

$8_{7,2}$	$7_{7,1}$	18750	18844	19300	20944	28125	$8_{2,7}$	$7_{1,7}$
$9_{7,3}$	$8_{7,2}$	35556	35733	36585	39200	36667	$9_{3,7}$	$8_{2,7}$
$10_{7,4}$	$9_{7,3}$	51000	51252	52436	54420	45500	$10_{4,7}$	$9_{3,7}$
$11_{7,5}$	$10_{7,4}$	65455	65775	67167	66365	54546	$11_{5,7}$	$10_{4,7}$
$12_{7,6}$	$11_{7,5}$	79167	79549	80851	76087	63750	$12_{6,7}$	$11_{5,7}$
$9_{8,2}$	$8_{8,1}$	18889	18985	19449	21157	30556	$9_{2,8}$	$8_{1,8}$
$10_{8,3}$	$9_{8,2}$	36000	36182	37061	40063	39000	$10_{3,8}$	$9_{2,8}$
$11_{8,4}$	$10_{8,3}$	51818	52078	53327	56523	47727	$11_{4,8}$	$10_{3,8}$
$12_{8,5}$	$11_{8,4}$	66667	66999	68563	69870	56667	$12_{5,8}$	$11_{4,8}$
$10_{9,2}$	$9_{9,1}$	19000	19097	19566	21311	33000	$10_{2,9}$	$9_{1,9}$
$11_{9,3}$	$10_{9,2}$	36364	36548	37443	40664	41364	$11_{3,9}$	$10_{2,9}$
$12_{9,4}$	$11_{9,3}$	52500	52766	54050	58070	50000	$12_{4,9}$	$11_{3,9}$
$11_{10,2}$	$10_{10,1}$	19091	19189	19662	21431	35454	$11_{2,10}$	$10_{1,10}$
$12_{10,3}$	$11_{10,2}$	36667	36854	37761	41102	43750	$12_{3,10}$	$11_{2,10}$
$12_{11,2}$	$11_{11,1}$	19167	19265	19742	21526	37917	$12_{2,11}$	$11_{1,11}$

STRENGTHS FOR ROTATIONAL TRANSITIONS

Symmetric-rotor Subbranches—a and c Prolate-or-oblate Subbranches

Subbranch		κ					Subbranch	
$^cQ_{-1,2}$ $J+K_{-1}+K_1$ even	$^cQ_{1,-2}$ $J+K_{-1}+K_1$ odd	∓ 1	∓ 0.5	0	± 0.5	± 1	$^aQ_{2,-1}$ $J+K_{-1}+K_1$ even	$^aQ_{-2,1}$ $J+K_{-1}+K_1$ odd
$2_{2,0}$	$2_{1,2}$	8333	5110	2233	488		$2_{0,2}$	$2_{2,1}$
$3_{2,1}$	$3_{1,3}$	14583	5722	1328	165		$3_{1,2}$	$3_{3,1}$
$4_{2,2}$	$4_{1,4}$	20250	4363	650	78		$4_{2,2}$	$4_{4,1}$
$5_{2,3}$	$5_{1,5}$	25667	2859	374	54		$5_{3,2}$	$5_{5,1}$
$6_{2,4}$	$6_{1,6}$	30952	1843	266	43		$6_{4,2}$	$6_{6,1}$
$7_{2,5}$	$7_{1,7}$	36161	1262	218	35		$7_{5,2}$	$7_{7,1}$
$8_{2,6}$	$8_{1,8}$	41319	945	183	30		$8_{6,2}$	$8_{8,1}$
$9_{2,7}$	$9_{1,9}$	46444	770	160	26		$9_{7,2}$	$9_{9,1}$
$10_{2,8}$	$10_{1,10}$	51545	664	141	23		$10_{8,2}$	$10_{10,1}$
$11_{2,9}$	$11_{1,11}$	56629	590	125	21		$11_{9,2}$	$11_{11,1}$
$12_{2,10}$	$12_{1,12}$	61699	533	115	19		$12_{10,2}$	$12_{12,1}$
$3_{3,0}$	$3_{2,2}$	8750	7055	4522	1458		$3_{0,3}$	$3_{2,2}$
$4_{3,1}$	$4_{2,3}$	15750	11214	4568	638		$4_{1,3}$	$4_{3,2}$
$5_{3,2}$	$5_{2,4}$	22000	12576	2754	274		$5_{2,3}$	$5_{4,2}$
$6_{3,3}$	$6_{2,5}$	27587	11283	1492	171		$6_{3,3}$	$6_{5,2}$
$7_{3,4}$	$7_{2,6}$	33482	8559	925	132		$7_{4,3}$	$7_{6,2}$
$8_{3,5}$	$8_{2,7}$	38958	5932	685	108		$8_{5,3}$	$8_{7,2}$
$9_{3,6}$	$9_{2,8}$	44333	4077	567	92		$9_{6,3}$	$9_{8,2}$
$10_{3,7}$	$10_{2,9}$	49636	2945	490	80		$10_{7,3}$	$10_{9,2}$
$11_{3,8}$	$11_{2,10}$	54886	2294	433	71		$11_{8,5}$	$11_{10,2}$
$12_{3,9}$	$12_{2,11}$	60096	1917	387	64		$12_{9,3}$	$12_{11,2}$
$4_{4,0}$	$4_{3,2}$	9000	7558	5617	2547		$4_{0,4}$	$4_{2,3}$
$5_{4,1}$	$5_{3,3}$	16500	13242	7983	1599		$5_{1,4}$	$5_{3,3}$

$6_{4,2}$	$6_{3,4}$	23214	17320	6820	681		$6_{2,4}$	$6_{4,3}$
$7_{4,3}$	$7_{3,5}$	29464	19464	4223	374		$7_{3,4}$	$7_{5,3}$
$8_{4,4}$	$8_{3,6}$	35417	19178	2433	273		$8_{4,4}$	$8_{6,3}$
$9_{4,5}$	$9_{3,7}$	41167	16526	1579	222		$9_{5,4}$	$9_{7,3}$
$10_{4,6}$	$10_{3,8}$	46773	12665	1205	188		$10_{6,4}$	$10_{8,3}$
$11_{4,7}$	$11_{3,9}$	52273	9080	1014	163		$11_{7,4}$	$11_{9,3}$
$12_{4,8}$	$12_{3,10}$	57692	6485	888	144		$12_{8,4}$	$12_{10,3}$
$5_{5,0}$	$5_{4,2}$	9167	7775	6052	3368		$5_{0,5}$	$5_{2,4}$
$6_{5,1}$	$6_{4,3}$	17024	14062	9982	3054		$6_{1,5}$	$6_{3,4}$
$7_{5,2}$	$7_{4,4}$	24107	19225	11103	1459		$7_{2,5}$	$7_{4,4}$
$8_{5,3}$	$8_{4,5}$	30694	23287	9000	720		$8_{3,5}$	$8_{5,4}$
$9_{5,4}$	$9_{4,6}$	36944	26001	5708	481		$9_{4,5}$	$9_{6,4}$
$10_{5,5}$	$10_{4,7}$	42955	26852	3433	382		$10_{5,5}$	$10_{7,4}$
$11_{5,6}$	$11_{4,8}$	48788	25327	2306	321		$11_{6,5}$	$11_{8,4}$
$12_{5,7}$	$12_{4,9}$	54487	21546	1796	277		$12_{7,5}$	$12_{9,4}$
$6_{6,0}$	$6_{5,2}$	9286	7912	6257	3863		$6_{0,6}$	$6_{2,5}$
$7_{6,1}$	$7_{5,3}$	17411	14550	10952	4657		$7_{1,6}$	$7_{3,5}$
$8_{6,2}$	$8_{5,4}$	24792	20233	13841	2772		$8_{2,6}$	$8_{4,5}$
$9_{6,3}$	$9_{5,5}$	31667	25098	14023	1322		$9_{3,6}$	$9_{5,5}$
$10_{6,4}$	$10_{5,6}$	38182	29140	11121	785		$10_{4,6}$	$10_{6,5}$
$11_{6,5}$	$11_{5,7}$	44432	32202	7197	595		$11_{5,6}$	$11_{7,5}$
$12_{6,6}$	$12_{5,8}$	50481	33921	4472	494		$12_{6,6}$	$12_{8,5}$
$7_{7,0}$	$7_{6,2}$	9375	8011	6381	4141		$7_{0,7}$	$7_{2,6}$
$8_{7,1}$	$8_{6,3}$	17708	14899	11480	5982		$8_{1,7}$	$8_{3,6}$
$9_{7,2}$	$9_{6,4}$	25333	20930	15306	4603		$9_{2,7}$	$9_{4,6}$
$10_{7,3}$	$10_{6,5}$	32455	26247	17406	2348		$10_{3,7}$	$10_{5,6}$
$11_{7,4}$	$11_{6,6}$	39205	30913	16805	1258		$11_{4,7}$	$11_{6,6}$
$12_{7,5}$	$12_{6,7}$	45673	34922	13192	878		$12_{5,7}$	$12_{7,6}$

STRENGTHS FOR ROTATIONAL TRANSITIONS

SYMMETRIC-ROTOR SUBBRANCHES—a AND c PROLATE-OR-OBLATE SUBBRANCHES

Subbranch		κ					Subbranch	
$^cQ_{-1,2} + K_1$ even — $J + K_{-1} + K_1$ even	$^cQ_{1,-2} + K_1$ odd — $J + K_{-1} + K_1$ odd	∓ 1	∓ 0.5	0	± 0.5	± 1	$^aQ_{2,-1} + K_1$ even — $J + K_{-1} + K_1$ even	$^aQ_{-2,1} + K_1$ odd — $J + K_{-1} + K_1$ odd
$8_{8,0}$	$8_{7,2}$	9444	8087	6468	4302		$8_{0,8}$	$8_{2,7}$
$9_{8,1}$	$9_{7,3}$	17944	15167	11825	6888		$9_{1,8}$	$9_{3,7}$
$10_{8,2}$	$10_{7,4}$	25773	21462	16158	6602		$10_{2,8}$	$10_{4,7}$
$11_{8,3}$	$11_{7,5}$	33106	27105	19325	3955		$11_{3,8}$	$11_{5,7}$
$12_{8,4}$	$12_{7,6}$	40064	32170	20771	2027		$12_{4,8}$	$12_{6,7}$
$9_{9,0}$	$9_{8,2}$	9500	8146	6535	4403		$9_{0,9}$	$9_{2,8}$
$10_{9,1}$	$10_{8,3}$	18136	15380	12079	7464		$10_{1,9}$	$10_{3,8}$
$11_{9,2}$	$11_{8,4}$	26136	21887	16736	8319		$11_{2,9}$	$11_{4,8}$
$12_{9,3}$	$12_{8,5}$	33654	27788	20487	6108		$12_{3,9}$	$12_{5,8}$
$10_{10,0}$	$10_{9,2}$	9545	8194	6588	4474		$10_{0,10}$	$10_{2,9}$
$11_{10,1}$	$11_{9,3}$	18295	15554	12279	7835		$11_{1,10}$	$11_{3,9}$
$12_{10,2}$	$12_{9,4}$	26442	22237	17173	9567		$12_{2,10}$	$12_{4,9}$
$11_{11,0}$	$11_{10,2}$	9583	8234	6631	4528		$11_{0,11}$	$11_{2,10}$
$12_{11,1}$	$12_{10,3}$	18429	15699	12443	8090		$12_{1,11}$	$12_{3,10}$
$12_{12,0}$	$12_{11,2}$	9615	8268	6667	4571		$12_{0,12}$	$12_{2,11}$

$^cR_{-1,2} + K_1$ even — $J + K_{-1} + K_1$ even	$^cP_{1,-2} + K_1$ even — $J + K_{-1} + K_1$ even	∓ 1	∓ 0.5	0	± 0.5	± 1	$^aR_{2,-1} + K_1$ even — $J + K_{-1} + K_1$ even	$^aP_{-2,1} + K_1$ even — $J + K_{-1} + K_1$ even
$1_{1,0}$	$2_{0,2}$	5000	3066	1340	293		$1_{0,1}$	$2_{2,0}$
$2_{1,1}$	$3_{0,3}$	10000	4167	1086	157		$2_{1,1}$	$3_{3,0}$

Block 1

Left outer	Left inner	D	C	B	A	Right inner	Right outer
$3_{1,2}$	$4_{0,4}$	15000	3944	800	123	$3_{2,1}$	$4_{4,0}$
$4_{1,3}$	$5_{0,5}$	20000	3386	696	117	$4_{3,1}$	$5_{5,0}$
$5_{1,4}$	$6_{0,6}$	25000	2976	667	114	$5_{4,1}$	$6_{6,0}$
$6_{1,5}$	$7_{0,7}$	30000	2770	656	112	$6_{5,1}$	$7_{7,0}$
$7_{1,6}$	$8_{0,8}$	35000	2686	649	111	$7_{6,1}$	$8_{8,0}$
$8_{1,7}$	$9_{0,9}$	40000	2652	644	110	$8_{7,1}$	$9_{9,0}$
$9_{1,8}$	$10_{0,10}$	45000	2634	640	109	$9_{8,1}$	$10_{10,0}$
$10_{1,9}$	$11_{0,11}$	50000	2621	637	109	$10_{9,1}$	$11_{11,0}$
$11_{1,10}$	$12_{0,12}$	55000	2610	634	108	$11_{10,1}$	$12_{12,0}$

Block 2

Left outer	Left inner	D	C	B	A	Right inner	Right outer
$2_{2,0}$	$3_{1,2}$	1667	2062	1905	776	$2_{0,2}$	$3_{2,1}$
$3_{2,1}$	$4_{1,3}$	3750	5114	2884	480	$3_{1,2}$	$4_{3,1}$
$4_{2,2}$	$5_{1,4}$	6000	7788	2336	310	$4_{2,2}$	$5_{4,1}$
$5_{2,3}$	$6_{1,5}$	8333	8748	1768	268	$5_{3,2}$	$6_{5,1}$
$6_{2,4}$	$7_{1,6}$	10714	8172	1529	254	$6_{4,2}$	$7_{6,1}$
$7_{2,5}$	$8_{1,7}$	13125	7135	1445	246	$7_{5,2}$	$8_{7,1}$
$8_{2,6}$	$9_{1,8}$	15556	6332	1406	240	$8_{6,2}$	$9_{8,1}$
$9_{2,7}$	$10_{1,9}$	18000	5885	1380	235	$9_{7,2}$	$10_{9,1}$
$10_{2,8}$	$11_{1,10}$	20455	5673	1360	232	$10_{8,2}$	$11_{10,1}$
$11_{2,9}$	$12_{1,11}$	22917	5570	1344	229	$11_{9,2}$	$12_{11,1}$

Block 3

Left outer	Left inner	D	C	B	A	Right inner	Right outer
$3_{3,0}$	$4_{2,2}$	1250	1176	1316	1061	$3_{0,3}$	$4_{2,2}$
$4_{3,1}$	$5_{2,3}$	3000	3166	3516	1032	$4_{1,3}$	$5_{3,2}$
$5_{3,2}$	$6_{2,4}$	5000	6089	4448	613	$5_{2,3}$	$6_{4,2}$
$6_{3,3}$	$7_{2,5}$	7143	9630	3653	469	$6_{3,3}$	$7_{5,2}$
$7_{3,4}$	$8_{2,6}$	9375	12493	2828	426	$7_{4,3}$	$8_{6,2}$
$8_{3,5}$	$9_{2,7}$	11667	13383	2447	404	$8_{5,3}$	$9_{7,2}$
$9_{3,6}$	$10_{2,8}$	14000	12500	2301	389	$9_{6,3}$	$10_{8,2}$
$10_{3,7}$	$11_{2,9}$	16364	11048	2226	378	$10_{7,2}$	$11_{9,2}$
$11_{3,8}$	$12_{2,10}$	18750	9884	2174	370	$11_{8,3}$	$12_{10,2}$

STRENGTHS FOR ROTATIONAL TRANSITIONS

SYMMETRIC-ROTOR SUBBRANCHES—a AND c PROLATE-OR-OBLATE SUBBRANCHES

Subbranch		κ					Subbranch	
$^cR_{-1,2}$ $J + K_{-1} + K_1$ even	$^cP_{1,-2}$ $J + K_{-1} + K_1$ even	∓ 1	∓ 0.5	0	± 0.5	± 1	$^aR_{2,-1}$ $J + K_{-1} + K_1$ even	$^aP_{-2,1}$ $J + K_{-1} + K_1$ even
$4_{4,0}$	$5_{3,2}$	1000	882	849	963		$4_{0,4}$	$5_{2,3}$
$5_{4,1}$	$6_{3,3}$	2500	2272	2651	1677		$5_{1,4}$	$6_{3,3}$
$6_{4,2}$	$7_{3,4}$	4286	4145	5108	1110		$6_{2,4}$	$7_{4,3}$
$7_{4,3}$	$8_{3,5}$	6250	6720	6025	747		$7_{3,4}$	$8_{5,3}$
$8_{4,4}$	$9_{3,6}$	8333	10196	5011	636		$8_{4,4}$	$9_{6,3}$
$9_{4,5}$	$10_{3,7}$	10500	14160	3947	590		$9_{5,4}$	$10_{7,3}$
$10_{4,6}$	$11_{3,8}$	12727	17215	3427	561		$10_{6,4}$	$11_{8,3}$
$11_{4,7}$	$12_{3,9}$	15000	18047	3214	540		$11_{7,4}$	$12_{9,3}$
$5_{5,0}$	$6_{4,2}$	833	730	638	723		$5_{0,5}$	$6_{2,4}$
$6_{5,1}$	$7_{4,3}$	2143	1898	1863	2013		$6_{1,5}$	$7_{3,4}$
$7_{5,2}$	$8_{4,4}$	3750	3384	4016	1833		$7_{2,5}$	$8_{4,4}$
$8_{5,3}$	$9_{4,5}$	5556	5182	6701	1165		$8_{3,5}$	$9_{5,4}$
$9_{5,4}$	$10_{4,6}$	7500	7437	7610	901		$9_{4,5}$	$10_{6,4}$
$10_{5,5}$	$11_{4,7}$	9545	10443	6398	809		$10_{5,5}$	$11_{7,4}$
$11_{5,6}$	$12_{4,8}$	11667	14380	5109	758		$11_{6,5}$	$12_{8,4}$
$6_{6,0}$	$7_{5,2}$	714	625	533	530		$6_{0,6}$	$7_{2,5}$
$7_{6,1}$	$8_{5,3}$	1875	1652	1470	1860		$7_{1,6}$	$8_{3,5}$
$8_{6,2}$	$9_{5,4}$	3333	2968	2958	2593		$8_{2,6}$	$9_{4,5}$
$9_{6,3}$	$10_{5,5}$	5000	4516	5406	1801		$9_{3,6}$	$10_{5,5}$
$10_{6,4}$	$11_{5,6}$	6818	6288	8298	1256		$10_{4,6}$	$11_{6,5}$
$11_{6,5}$	$12_{5,7}$	8750	8348	9203	1069		$11_{5,6}$	$12_{7,5}$

$7_{7,0}$	$8_{6,2}$	625	546	463	411	$7_{0,7}$	$8_{2,6}$
$8_{7,1}$	$9_{6,3}$	1667	1465	1267	1486	$8_{1,7}$	$9_{3,6}$
$9_{7,2}$	$10_{6,4}$	3000	2656	2402	2962	$9_{2,7}$	$10_{4,6}$
$10_{7,3}$	$11_{6,5}$	4545	4065	4106	2664	$10_{3,7}$	$11_{5,6}$
$11_{7,4}$	$12_{6,6}$	6250	5659	6816	1772	$11_{4,7}$	$12_{6,6}$
$8_{8,0}$	$9_{7,2}$	556	485	411	341	$8_{0,8}$	$9_{2,7}$
$9_{8,1}$	$10_{7,3}$	1500	1316	1129	1154	$9_{1,8}$	$10_{3,7}$
$10_{8,2}$	$11_{7,4}$	2727	2407	2110	2759	$10_{2,8}$	$11_{4,7}$
$11_{8,3}$	$12_{7,5}$	4167	3705	3398	3519	$11_{3,8}$	$12_{6,7}$
$9_{9,0}$	$10_{8,2}$	500	436	369	298	$9_{0,9}$	$10_{2,8}$
$10_{9,1}$	$11_{8,3}$	1364	1195	1021	932	$10_{1,9}$	$11_{3,8}$
$11_{9,2}$	$12_{8,4}$	2500	2201	1910	2276	$11_{2,9}$	$12_{4,8}$
$10_{10,0}$	$11_{9,2}$	455	397	335	268	$10_{0,10}$	$11_{2,9}$
$11_{10,1}$	$12_{9,3}$	1250	1094	933	797	$11_{1,10}$	$12_{2,9}$
$11_{11,0}$	$12_{10,2}$	417	363	307	244	$11_{0,11}$	$12_{2,10}$

STRENGTHS FOR ROTATIONAL TRANSITIONS

SYMMETRIC-ROTOR SUBBRANCHES—a AND c PROLATE-OR-OBLATE SUBBRANCHES

Subbranch				κ			Subbranch	
$^cR_{-1,2}$ $J + K_{-1} + K_1$ odd	$^cP_{1,-2}$ $J + K_{-1} - K_1$ odd	∓ 1	∓ 0.5	0	± 0.5	± 1	$^aR_{2,-1}$ $J + K_{-1} + K_1$ odd	$^aP_{-2,1}$ $J + K_{-1} + K_1$ odd
$2_{2,1}$	$3_{1,3}$	1667	956	423	103		$2_{1,2}$	$3_{3,1}$
$3_{2,2}$	$4_{1,4}$	3750	1742	609	118		$3_{2,2}$	$4_{4,1}$
$4_{2,3}$	$5_{1,5}$	6000	2228	657	116		$4_{3,2}$	$5_{5,1}$
$5_{2,4}$	$6_{1,6}$	8333	2480	661	114		$5_{4,2}$	$6_{6,1}$
$6_{2,5}$	$7_{1,7}$	10714	2590	655	112		$6_{5,2}$	$7_{7,1}$
$7_{2,6}$	$8_{1,8}$	13125	2627	649	111		$7_{6,2}$	$8_{8,1}$
$8_{2,7}$	$9_{1,9}$	15556	2633	644	110		$8_{7,2}$	$9_{9,1}$
$9_{2,8}$	$10_{1,10}$	18000	2627	640	109		$9_{8,2}$	$10_{10,1}$
$10_{2,9}$	$11_{1,11}$	20455	2619	637	109		$10_{9,2}$	$11_{11,1}$
$11_{2,10}$	$12_{1,12}$	22917	2610	634	108		$11_{10,2}$	$12_{12,1}$
$3_{3,1}$	$4_{2,3}$	1250	1025	643	213		$3_{1,3}$	$4_{3,2}$
$4_{3,2}$	$5_{2,4}$	3000	2317	1159	269		$4_{2,3}$	$5_{4,2}$
$5_{3,3}$	$6_{2,5}$	5000	3522	1389	265		$5_{3,3}$	$6_{5,2}$
$6_{3,4}$	$7_{2,6}$	7143	4450	1442	254		$6_{4,3}$	$7_{6,2}$
$7_{3,5}$	$8_{2,7}$	9375	5049	1429	246		$7_{5,3}$	$8_{7,2}$
$8_{3,6}$	$9_{2,8}$	11667	5372	1403	240		$8_{6,3}$	$9_{8,2}$
$9_{3,7}$	$10_{2,9}$	14000	5507	1379	235		$9_{7,3}$	$10_{9,2}$
$10_{3,8}$	$11_{2,10}$	16364	5539	1360	232		$10_{8,3}$	$11_{10,2}$
$11_{3,9}$	$12_{2,11}$	18750	5523	1344	229		$11_{9,3}$	$12_{11,2}$
$4_{4,1}$	$5_{3,3}$	1000	869	664	300		$4_{1,4}$	$5_{3,3}$
$5_{4,2}$	$6_{3,4}$	2500	2168	1455	440		$5_{2,4}$	$6_{4,3}$
$6_{4,3}$	$7_{3,5}$	4286	3662	2018	447		$6_{3,4}$	$7_{5,3}$
$7_{4,4}$	$8_{3,6}$	6250	5157	2266	424		$7_{4,4}$	$8_{6,3}$

$9_{7,3}$	$8_{5,4}$	404	2309	6471	8333	$9_{3,7}$	$8_{4,5}$
$10_{8,3}$	$9_{6,4}$	389	2272	7475	10500	$10_{3,8}$	$9_{4,6}$
$11_{9,3}$	$10_{7,4}$	378	2220	8130	12727	$11_{3,9}$	$10_{4,7}$
$12_{10,3}$	$11_{8,4}$	370	2173	8481	15000	$12_{3,10}$	$11_{4,8}$
$6_{3,4}$	$5_{1,5}$	346	601	729	833	$6_{4,3}$	$5_{5,1}$
$7_{4,4}$	$6_{2,5}$	603	1489	1889	2143	$7_{4,4}$	$6_{5,2}$
$8_{5,4}$	$7_{3,5}$	656	2360	3329	3750	$8_{4,5}$	$7_{6,3}$
$9_{6,4}$	$8_{4,5}$	626	2953	4947	5556	$9_{4,6}$	$8_{6,4}$
$10_{7,4}$	$9_{5,5}$	589	3209	6629	7500	$10_{4,7}$	$9_{6,5}$
$11_{8,4}$	$10_{6,5}$	561	3235	8239	9545	$11_{4,8}$	$10_{6,6}$
$12_{9,4}$	$11_{7,5}$	540	3171	9627	11667	$12_{4,9}$	$11_{6,7}$
$7_{3,5}$	$6_{1,6}$	357	527	625	714	$7_{5,3}$	$6_{6,1}$
$8_{4,5}$	$7_{2,6}$	726	1386	1651	1875	$8_{5,4}$	$7_{6,2}$
$9_{5,5}$	$8_{3,6}$	873	2400	2962	3333	$9_{5,5}$	$8_{6,3}$
$10_{6,5}$	$9_{4,6}$	858	3326	4489	5000	$10_{5,6}$	$9_{6,4}$
$11_{7,5}$	$10_{5,6}$	804	3940	6181	6818	$11_{5,7}$	$10_{6,5}$
$12_{8,5}$	$11_{6,6}$	758	4195	7975	8750	$12_{5,8}$	$11_{6,6}$
$8_{3,6}$	$7_{1,7}$	341	462	546	625	$8_{6,3}$	$7_{7,1}$
$9_{4,6}$	$8_{2,7}$	793	1251	1465	1667	$9_{6,4}$	$8_{7,2}$
$10_{5,6}$	$9_{3,7}$	1069	2267	2656	3000	$10_{6,5}$	$9_{7,3}$
$11_{6,6}$	$10_{4,7}$	1110	3368	4062	4545	$11_{6,6}$	$10_{7,4}$
$12_{7,6}$	$11_{5,7}$	1050	4336	5646	6250	$12_{6,7}$	$11_{7,5}$
$9_{3,7}$	$8_{1,8}$	317	411	485	556	$9_{7,3}$	$8_{8,1}$
$10_{4,7}$	$9_{2,8}$	808	1126	1316	1500	$10_{7,4}$	$9_{8,2}$
$11_{5,7}$	$10_{3,8}$	1216	2082	2407	2727	$11_{7,5}$	$10_{8,3}$
$12_{6,7}$	$11_{4,8}$	1361	3209	3704	4167	$12_{7,6}$	$11_{8,4}$

STRENGTHS FOR ROTATIONAL TRANSITIONS

SYMMETRIC-ROTOR SUBBRANCHES—a AND c PROLATE-OR-OBLATE SUBBRANCHES

Subbranch	Subbranch	κ					Subbranch	Subbranch
${}^cR_{-1,2}$ $J + K_{-1} + K_1$ odd	${}^cP_{1,-2}$ $J + K_{-1} + K_1$ odd	∓ 1	∓ 0.5	0	± 0.5	± 1	${}^aR_{2,-1}$ $J + K_{-1} + K_1$ odd	${}^aP_{-2,1}$ $J + K_{-1} + K_1$ odd
$9_{9,1}$	$10_{8,3}$	500	436	369	290		$9_{1,9}$	$10_{3,8}$
$10_{9,2}$	$11_{8,4}$	1364	1195	1021	786		$10_{1,9}$	$11_{4,8}$
$11_{9,3}$	$12_{8,5}$	2500	2201	1904	1296		$11_{1,9}$	$12_{5,8}$
$10_{10,1}$	$11_{9,3}$	455	397	335	265		$10_{1,10}$	$11_{3,9}$
$11_{10,2}$	$12_{9,4}$	1250	1094	931	741		$11_{2,10}$	$12_{4,9}$
$11_{11,1}$	$12_{10,3}$	417	363	307	243		$11_{1,11}$	$12_{3,10}$

SYMMETRIC-ROTOR SUBBRANCHES—b PROLATE-AND-OBLATE SUBBRANCHES

Subbranch	Subbranch	κ					Subbranch	Subbranch
${}^bQ_{-1,1}$ $J + K_{-1} + K_1$ even	${}^bQ_{1,-1}$ $J + K_{-1} + K_1$ even	∓ 1	∓ 0.5	0	± 0.5	± 1	${}^bQ_{1,-1}$ $J + K_{-1} + K_1$ even	${}^bQ_{-1,1}$ $J + K_{-1} + K_1$ even
$1_{1,0}$	$1_{0,1}$	15000	15000	15000	15000	15000	$1_{0,1}$	$1_{1,0}$
$2_{1,1}$	$2_{0,2}$	25000	21289	16667	12044	8333	$2_{1,1}$	$2_{2,0}$
$3_{1,2}$	$3_{0,3}$	35000	23196	14583	10583	8750	$3_{2,1}$	$3_{3,0}$
$4_{1,3}$	$4_{0,4}$	45000	22157	13527	10617	9000	$4_{3,1}$	$4_{4,0}$
$5_{1,4}$	$5_{0,5}$	55000	20634	13413	10753	9167	$5_{4,1}$	$5_{5,0}$
$6_{1,5}$	$6_{0,6}$	65000	19779	13484	10861	9286	$6_{5,1}$	$6_{6,0}$
$7_{1,6}$	$7_{0,7}$	75000	19511	13559	10943	9375	$7_{6,1}$	$7_{7,0}$

$8_{8,0}$	$8_{7,1}$	9444	11008	13620	19487	85000	$8_{1,7}$	$8_{8,8}$
$9_{9,0}$	$9_{8,1}$	9500	11060	13669	19524	95000	$9_{1,8}$	$9_{9,9}$
$10_{10,0}$	$10_{9,1}$	9545	11103	13710	19565	105000	$10_{1,9}$	$10_{0,10}$
$11_{11,0}$	$11_{10,1}$	9583	11139	13744	19604	115000	$11_{1,10}$	$11_{0,11}$
$12_{12,0}$	$12_{11,1}$	9615	11170	13774	19633	125000	$12_{1,11}$	$12_{0,12}$
$2_{1,1}$	$2_{0,2}$	25000	21289	16667	12044	8333	$2_{2,0}$	$2_{1,1}$
$3_{2,1}$	$3_{1,2}$	14583	24417	28872	24417	14583	$3_{2,1}$	$3_{1,2}$
$4_{3,1}$	$4_{2,2}$	15750	20622	31154	36119	20250	$4_{2,2}$	$4_{1,3}$
$5_{4,1}$	$5_{3,2}$	16500	20038	28164	43650	25667	$5_{2,3}$	$5_{1,4}$
$6_{5,1}$	$6_{4,2}$	17024	20356	26402	45529	30952	$6_{2,4}$	$6_{1,5}$
$7_{6,1}$	$7_{5,2}$	17411	20670	26163	43602	36161	$7_{2,5}$	$7_{1,6}$
$8_{7,1}$	$8_{6,2}$	17708	20926	26300	41002	41319	$8_{2,6}$	$8_{1,7}$
$9_{8,1}$	$9_{7,2}$	17944	21134	26465	39408	46444	$9_{2,7}$	$9_{1,8}$
$10_{9,1}$	$10_{8,2}$	18136	21307	26611	38815	51545	$10_{2,8}$	$10_{1,9}$
$11_{10,1}$	$11_{9,2}$	18295	21452	26737	38701	56629	$11_{2,9}$	$11_{1,10}$
$12_{11,1}$	$12_{10,2}$	18429	21576	26846	38736	61699	$12_{2,10}$	$12_{1,11}$
$3_{1,2}$	$3_{0,3}$	35000	23196	14583	10583	8750	$3_{3,0}$	$3_{2,1}$
$4_{2,2}$	$4_{1,3}$	20250	36119	31154	20622	15750	$4_{3,1}$	$4_{2,2}$
$5_{3,2}$	$5_{2,3}$	22000	32340	44017	32340	22000	$5_{3,2}$	$5_{2,3}$
$6_{4,2}$	$6_{3,3}$	23214	29422	45920	45986	27857	$6_{3,3}$	$6_{2,4}$
$7_{5,2}$	$7_{4,3}$	24107	29481	41862	58783	33482	$7_{3,4}$	$7_{2,5}$
$8_{6,2}$	$8_{5,3}$	24792	29932	39333	66715	38958	$8_{3,5}$	$8_{2,6}$
$9_{7,2}$	$9_{6,3}$	25333	30348	38859	68174	44333	$9_{3,6}$	$9_{2,7}$
$10_{8,2}$	$10_{7,3}$	25773	30705	38980	65282	49636	$10_{3,7}$	$10_{2,8}$
$11_{9,2}$	$11_{8,3}$	26136	31011	39182	61636	54886	$11_{3,8}$	$11_{2,9}$
$12_{10,2}$	$12_{9,3}$	26442	31275	39377	59285	60096	$12_{3,9}$	$12_{2,10}$

STRENGTHS FOR ROTATIONAL TRANSITIONS

SYMMETRIC-ROTOR SUBBRANCHES—b PROLATE-AND-OBLATE SUBBRANCHES

Subbranch		κ					Subbranch	
$J+K_{-1}+K_1$ even $^bQ_{-1,1}$	$J+K_{-1}+K_1$ even $^bQ_{1,-1}$	∓ 1	∓ 0.5	0	± 0.5	± 1	$J+K_{-1}+K_1$ even $^bQ_{1,-1}$	$J+K_{-1}+K_1$ even $^bQ_{-1,1}$
$4_{4,0}$	$4_{3,1}$	9000	10617	13527	22157	45000	$4_{0,4}$	$4_{1,3}$
$5_{4,1}$	$5_{3,2}$	16500	20038	28164	43650	25667	$5_{1,4}$	$5_{2,3}$
$6_{4,2}$	$6_{3,3}$	23214	29422	45920	45986	27857	$6_{2,4}$	$6_{3,3}$
$7_{4,3}$	$7_{3,4}$	29464	39987	59402	39987	29464	$7_{3,4}$	$7_{4,3}$
$8_{4,4}$	$8_{3,5}$	35417	52950	60829	38601	30694	$8_{4,4}$	$8_{5,3}$
$9_{4,5}$	$9_{3,6}$	41167	67954	55712	38960	31667	$9_{5,4}$	$9_{6,3}$
$10_{4,6}$	$10_{3,7}$	46773	81732	52398	39466	32455	$10_{6,4}$	$10_{7,3}$
$11_{4,7}$	$11_{3,8}$	52273	89952	51626	39938	33106	$11_{7,4}$	$11_{8,3}$
$12_{4,8}$	$12_{3,9}$	57692	90961	51673	40360	33654	$12_{8,4}$	$12_{9,3}$
$5_{5,0}$	$5_{4,1}$	9167	10753	13413	20634	55000	$5_{0,5}$	$5_{1,4}$
$6_{5,1}$	$6_{4,2}$	17024	20356	26402	45529	30952	$6_{1,5}$	$6_{2,4}$
$7_{5,2}$	$7_{4,3}$	24107	29481	41862	58783	33482	$7_{2,5}$	$7_{3,4}$
$8_{5,3}$	$8_{4,4}$	30694	38601	60829	52950	35417	$8_{3,5}$	$8_{4,4}$
$9_{5,4}$	$9_{4,5}$	36944	48332	74882	48332	36944	$9_{4,5}$	$9_{5,4}$
$10_{5,5}$	$10_{4,6}$	42955	59745	75829	47998	38182	$10_{5,5}$	$10_{6,4}$
$11_{5,6}$	$11_{4,7}$	48788	73909	69690	48463	39205	$11_{6,5}$	$11_{7,4}$
$12_{5,7}$	$12_{4,8}$	54487	90148	65598	48989	40064	$12_{7,5}$	$12_{8,4}$
$6_{6,0}$	$6_{5,1}$	9286	10861	13484	19779	65000	$6_{0,6}$	$6_{1,5}$
$7_{6,1}$	$7_{5,2}$	17411	20670	26163	43602	36161	$7_{1,6}$	$7_{2,5}$
$8_{6,2}$	$8_{5,3}$	24792	29932	39333	66715	38958	$8_{2,6}$	$8_{3,5}$
$9_{6,3}$	$9_{5,4}$	31667	38960	55712	67954	41167	$9_{3,6}$	$9_{4,5}$
$10_{6,4}$	$10_{5,5}$	38182	47998	75829	59745	42955	$10_{4,6}$	$10_{5,5}$

$11_{6,5}$	$11_{5,6}$	44432	57343	90410	57343	44432	$11_{6,6}$	$11_{6,5}$
$12_{6,6}$	$12_{5,7}$	50481	67590	90893	57486	45673	$12_{6,6}$	$12_{6,5}$
$7_{7,0}$	$7_{6,1}$	9375	10943	13559	19511	75000	$7_{0,7}$	$7_{1,6}$
$8_{7,1}$	$8_{6,2}$	17708	20926	26300	41002	41319	$8_{1,7}$	$8_{2,6}$
$9_{7,2}$	$9_{6,3}$	25333	30348	38859	68174	44333	$9_{2,7}$	$9_{3,6}$
$10_{7,3}$	$10_{6,4}$	32455	39466	52398	81732	46773	$10_{3,7}$	$10_{4,6}$
$11_{7,4}$	$11_{6,5}$	39205	48463	69690	73909	48788	$11_{4,7}$	$11_{5,6}$
$12_{7,5}$	$12_{6,6}$	45673	57486	90893	67590	50481	$12_{5,7}$	$12_{6,6}$
$8_{8,0}$	$8_{7,1}$	9444	11008	13620	19487	85000	$8_{0,8}$	$8_{1,7}$
$9_{8,1}$	$9_{7,2}$	17944	21134	26465	39408	46444	$9_{1,8}$	$9_{2,7}$
$10_{8,2}$	$10_{7,3}$	25773	30705	38980	65282	49636	$10_{2,8}$	$10_{3,7}$
$11_{8,3}$	$11_{7,4}$	33106	39938	51626	89952	52273	$11_{3,8}$	$11_{4,7}$
$12_{8,4}$	$12_{7,5}$	40064	48989	65598	90148	54487	$12_{4,8}$	$12_{5,7}$
$9_{9,0}$	$9_{8,1}$	9500	11060	13669	19524	95000	$9_{0,9}$	$9_{1,8}$
$10_{9,1}$	$10_{8,2}$	18136	21307	26611	38815	51545	$10_{1,9}$	$10_{2,8}$
$11_{9,2}$	$11_{8,3}$	26136	31011	39182	61636	54886	$11_{2,9}$	$11_{3,8}$
$12_{9,3}$	$12_{8,4}$	33654	40360	51673	90961	57692	$12_{3,9}$	$12_{4,8}$
$10_{10,0}$	$10_{9,1}$	9545	11103	13710	19565	105000	$10_{0,10}$	$10_{1,9}$
$11_{10,1}$	$11_{9,2}$	18295	21452	26737	38701	56629	$11_{1,10}$	$11_{2,9}$
$12_{10,2}$	$12_{9,3}$	26442	31275	39377	59285	60096	$12_{2,10}$	$12_{3,9}$
$11_{11,0}$	$11_{10,1}$	9583	11139	13744	19604	115000	$11_{0,11}$	$11_{1,10}$
$12_{11,1}$	$12_{10,2}$	18429	21576	26846	38736	61699	$12_{1,11}$	$12_{2,10}$
$12_{12,0}$	$12_{11,1}$	9615	11170	13774	19633	125000	$12_{0,12}$	$12_{1,11}$

STRENGTHS FOR ROTATIONAL TRANSITIONS

SYMMETRIC-ROTOR SUBBRANCHES—b PROLATE-AND-OBLATE SUBBRANCHES

Subbranch		κ					Subbranch	
${}^bQ_{-1,1}$ $J+K_{-1}+K_1$ odd	${}^bQ_{1,-1}$ $J+K_{-1}+K_1$ odd	∓ 1	∓ 0.5	0	± 0.5	± 1	${}^bQ_{1,-1}$ $J+K_{-1}+K_1$ odd	${}^bQ_{-1,1}$ $J+K_{-1}+K_1$ odd
$2_{2,1}$	$2_{1,2}$	8333	8333	8333	8333	8333	$2_{1,2}$	$2_{2,1}$
$3_{2,2}$	$3_{1,3}$	14583	13160	11667	10173	8750	$3_{2,2}$	$3_{3,1}$
$4_{2,3}$	$4_{1,4}$	20250	16126	12886	10584	9000	$4_{3,2}$	$4_{4,1}$
$5_{2,4}$	$5_{1,5}$	25667	17823	13300	10751	9167	$5_{4,2}$	$5_{5,1}$
$6_{2,5}$	$6_{1,6}$	30952	18716	13464	10860	9286	$6_{5,2}$	$6_{6,1}$
$7_{2,6}$	$7_{1,7}$	36161	19158	13555	10943	9375	$7_{6,2}$	$7_{7,1}$
$8_{2,7}$	$8_{1,8}$	41319	19374	13619	11008	9444	$8_{7,2}$	$8_{8,1}$
$9_{2,8}$	$9_{1,9}$	46444	19487	13669	11060	9500	$9_{8,2}$	$9_{9,1}$
$10_{2,9}$	$10_{1,10}$	51545	19553	13710	11103	9545	$10_{9,2}$	$10_{10,1}$
$11_{2,10}$	$11_{1,11}$	56629	19598	13744	11139	9583	$11_{10,2}$	$11_{11,1}$
$12_{2,11}$	$12_{1,12}$	61699	19632	13774	11170	9615	$12_{11,2}$	$12_{12,1}$
$3_{3,1}$	$3_{2,2}$	8750	10173	11667	13160	14583	$3_{1,3}$	$3_{2,2}$
$4_{3,2}$	$4_{2,3}$	15750	18280	19208	18280	15750	$4_{2,3}$	$4_{3,2}$
$5_{3,3}$	$5_{2,4}$	22000	24936	23333	19781	16500	$5_{3,3}$	$5_{4,2}$
$6_{3,4}$	$6_{2,5}$	27857	30089	25173	20331	17024	$6_{4,3}$	$6_{5,2}$
$7_{3,5}$	$7_{2,6}$	33482	33722	25914	20668	17411	$7_{5,3}$	$7_{6,2}$
$8_{3,6}$	$8_{2,7}$	38958	36030	26251	20926	17708	$8_{6,3}$	$8_{7,2}$
$9_{3,7}$	$9_{2,8}$	44333	37360	26455	21134	17944	$9_{7,3}$	$9_{8,2}$
$10_{3,8}$	$10_{2,9}$	49636	38072	26609	21307	18136	$10_{8,3}$	$10_{9,2}$
$11_{3,9}$	$11_{2,10}$	54886	38443	26737	21452	18295	$11_{9,3}$	$11_{10,2}$
$12_{3,10}$	$12_{2,11}$	60096	38646	26846	21576	18429	$12_{10,3}$	$12_{11,2}$
$4_{4,1}$	$4_{3,2}$	9000	10584	12886	16126	20250	$4_{1,4}$	$4_{2,3}$
$5_{4,2}$	$5_{3,3}$	16500	19781	23333	24936	22000	$5_{2,4}$	$5_{3,3}$

Group 1

$6_{4,3}$	$6_{3,4}$	23214	28237	30910	28237	23214	$6_{3,4}$	$6_{4,3}$
$7_{4,4}$	$7_{3,5}$	29464	35974	35396	29347	24107	$7_{4,4}$	$7_{5,3}$
$8_{4,5}$	$8_{3,6}$	35417	42717	37550	29917	24792	$8_{5,4}$	$8_{6,3}$
$9_{4,6}$	$9_{3,7}$	41167	48149	38467	30347	25333	$9_{6,4}$	$9_{7,3}$
$10_{4,7}$	$10_{3,8}$	46773	52121	38896	30705	25773	$10_{7,4}$	$10_{8,3}$
$11_{4,8}$	$11_{3,9}$	52273	54738	39163	31011	26136	$11_{8,4}$	$11_{9,3}$
$12_{4,9}$	$12_{3,10}$	57692	56299	39373	31275	26442	$12_{9,4}$	$12_{10,3}$

Group 2

$5_{5,1}$	$5_{4,2}$	9167	10751	13300	17823	25667	$5_{1,5}$	$5_{2,4}$
$6_{5,2}$	$6_{4,3}$	17024	20331	25173	30089	27857	$6_{2,5}$	$6_{3,4}$
$7_{5,3}$	$7_{4,4}$	24107	29347	35396	35974	29464	$7_{3,5}$	$7_{4,4}$
$8_{5,4}$	$8_{4,5}$	30694	38050	43064	38050	30694	$8_{4,5}$	$8_{5,4}$
$9_{5,5}$	$9_{4,6}$	36944	46432	47757	38893	31667	$9_{5,5}$	$9_{6,4}$
$10_{5,6}$	$10_{4,7}$	42955	54266	50083	39458	32455	$10_{6,5}$	$10_{7,4}$
$11_{5,7}$	$11_{4,8}$	48788	61181	51087	39937	33106	$11_{7,5}$	$11_{8,4}$
$12_{5,8}$	$12_{4,9}$	54487	66823	51551	40360	33654	$12_{8,5}$	$12_{9,4}$

Group 3

$6_{6,1}$	$6_{5,2}$	9286	10860	13464	18716	30952	$6_{1,6}$	$6_{2,5}$
$7_{6,2}$	$7_{5,3}$	17411	20668	25914	33722	33482	$7_{2,6}$	$7_{3,5}$
$8_{6,3}$	$8_{5,4}$	24792	29917	37550	42717	35417	$8_{3,6}$	$8_{4,5}$
$9_{6,4}$	$9_{5,5}$	31667	38893	47757	46432	36944	$9_{4,6}$	$9_{5,5}$
$10_{6,5}$	$10_{5,6}$	38182	47745	55515	47745	38182	$10_{5,6}$	$10_{6,5}$
$11_{6,6}$	$11_{5,7}$	44432	56495	60341	48430	39205	$11_{6,6}$	$11_{7,5}$
$12_{6,7}$	$12_{5,8}$	50481	65013	62764	48985	40064	$12_{7,6}$	$12_{8,5}$

Group 4

$7_{7,1}$	$7_{6,2}$	9375	10943	13555	19158	36161	$7_{1,7}$	$7_{2,6}$
$8_{7,2}$	$8_{6,3}$	17708	20926	26251	36030	38958	$8_{2,7}$	$8_{3,6}$
$9_{7,3}$	$9_{6,4}$	25333	30347	38467	48149	41167	$9_{3,7}$	$9_{4,6}$
$10_{7,4}$	$10_{6,5}$	32455	39458	50083	54266	42955	$10_{4,7}$	$10_{5,6}$
$11_{7,5}$	$11_{6,6}$	39205	48430	60341	56495	44432	$11_{5,7}$	$11_{6,6}$
$12_{7,6}$	$12_{6,7}$	45673	57370	68182	57370	45673	$12_{6,7}$	$12_{7,6}$

STRENGTHS FOR ROTATIONAL TRANSITIONS

SYMMETRIC-ROTOR SUBBRANCHES—b PROLATE-AND-OBLATE SUBBRANCHES

Subbranch			κ				Subbranch	
$^bQ_{-1,1}$ $J+K_{-1}+K_1$ even	$^bQ_{-1,1}$ $J+K_{-1}+K_1$ odd	∓ 1	∓ 0.5	0	± 0.5	± 1	$^bQ_{-1,1}$ $J+K_{-1}+K_1$ odd	$^bQ_{-1,1}$ $J+K_{-1}+K_1$ odd
$8_{8,1}$	$8_{7,2}$	9444	11008	13619	19374	41319	$8_{1,8}$	$8_{2,7}$
$9_{8,2}$	$9_{7,3}$	17944	21134	26455	37360	44333	$9_{2,8}$	$9_{3,7}$
$10_{8,3}$	$10_{7,4}$	25773	30705	38896	52121	46773	$10_{3,8}$	$10_{4,7}$
$11_{8,4}$	$11_{7,5}$	33106	39937	51087	61181	48788	$11_{4,8}$	$11_{5,7}$
$12_{8,5}$	$12_{7,6}$	40064	48985	62764	65013	50481	$12_{5,8}$	$12_{6,7}$
$9_{9,1}$	$9_{8,2}$	9500	11060	13669	19487	46444	$9_{1,9}$	$9_{2,8}$
$10_{9,2}$	$10_{8,3}$	18136	21307	26609	38072	49636	$10_{2,9}$	$10_{3,8}$
$11_{9,3}$	$11_{8,4}$	26136	31011	39163	54738	52273	$11_{3,9}$	$11_{4,8}$
$12_{9,4}$	$12_{8,5}$	33654	40360	51551	66823	54487	$12_{4,9}$	$12_{5,8}$
$10_{10,1}$	$10_{9,2}$	9545	11103	13710	19553	51545	$10_{1,10}$	$10_{2,9}$
$11_{10,2}$	$11_{9,3}$	18295	21452	26737	38443	54886	$11_{2,10}$	$11_{3,9}$
$12_{10,3}$	$12_{9,4}$	26442	31275	39373	56299	57692	$12_{3,10}$	$12_{4,9}$
$11_{11,1}$	$11_{10,2}$	9583	11139	13744	19598	56629	$11_{1,11}$	$11_{2,10}$
$12_{11,2}$	$12_{10,3}$	18429	21576	26846	38646	60096	$12_{2,11}$	$12_{3,10}$
$12_{12,1}$	$12_{11,2}$	9615	11170	13774	19632	61699	$12_{1,12}$	$12_{2,11}$

Subbranch			κ				Subbranch	
$^bR_{1,1}$ $J+K_{-1}+K_1$ even	$^bP_{-1,-1}$ $J+K_{-1}+K_1$ odd	∓ 1	∓ 0.5	0	± 0.5	± 1	$^bR_{1,1}$ $J+K_{-1}+K_1$ even	$^bP_{-1,-1}$ $J+K_{-1}+K_1$ odd
$0_{0,0}$	$1_{1,1}$	10000	10000	10000	10000	10000	$0_{0,0}$	$1_{1,1}$
$1_{1,0}$	$2_{2,1}$	15000	15000	15000	15000	15000	$1_{0,1}$	$2_{1,2}$

Table of coefficients (page displayed with rotated row/column index labels). The five central columns contain the numerical values; the columns labelled *L* (left indices) and *R* (right indices) give the index labels printed to the left and right of the data.

Group 1

L						R
$2_{0,2}$	20000	21383	22847	24086	25000	$3_{1,3}$
$3_{0,3}$	25000	29584	32533	34083	35000	$4_{1,4}$
$4_{0,4}$	30000	39100	42585	44117	45000	$5_{1,5}$
$5_{0,5}$	35000	49126	52653	54140	55000	$6_{1,6}$
$6_{0,6}$	40000	59250	62702	64155	65000	$7_{1,7}$
$7_{0,7}$	45000	69364	72737	74165	75000	$8_{1,8}$
$8_{0,8}$	50000	79453	82763	84173	85000	$9_{1,9}$
$9_{0,9}$	55000	89522	92782	94179	95000	$10_{1,10}$
$10_{0,10}$	60000	99576	102798	104184	105000	$11_{1,11}$
$11_{0,11}$	65000	109620	112810	114188	115000	$12_{1,12}$

Outer left labels (group 1): $2_{2,0}$, $3_{3,0}$, $4_{4,0}$, $5_{5,0}$, $6_{6,0}$, $7_{7,0}$, $8_{8,0}$, $9_{9,0}$, $10_{10,0}$, $11_{11,0}$
Outer right labels (group 1): $3_{3,1}$, $4_{4,1}$, $5_{5,1}$, $6_{6,1}$, $7_{7,1}$, $8_{8,1}$, $9_{9,1}$, $10_{10,1}$, $11_{11,1}$, $12_{12,1}$

Group 2

L						R
$1_{1,0}$	15000	15000	15000	15000	15000	$2_{2,1}$
$2_{1,1}$	16667	16667	16667	16667	16667	$3_{2,2}$
$3_{1,2}$	18750	19563	21079	23549	26250	$4_{2,3}$
$4_{1,3}$	21000	23919	28748	33165	36000	$5_{2,4}$
$5_{1,4}$	23333	30161	38409	43122	45833	$6_{2,5}$
$6_{1,5}$	25714	38383	48508	53091	55714	$7_{2,6}$
$7_{1,6}$	28125	48001	58609	63059	65625	$8_{2,7}$
$8_{1,7}$	30556	58192	68678	73030	75556	$9_{2,8}$
$9_{1,8}$	33000	68479	78720	83002	85500	$10_{2,9}$
$10_{1,9}$	35454	78723	88749	92979	95455	$11_{2,10}$
$11_{1,10}$	37917	88911	98767	102958	105416	$12_{2,11}$

Outer left labels (group 2): $2_{1,2}$, $3_{2,2}$, $4_{3,2}$, $5_{4,2}$, $6_{5,2}$, $7_{6,2}$, $8_{7,2}$, $9_{8,2}$, $10_{9,2}$, $11_{10,2}$, $12_{11,2}$
Outer right labels (group 2): $2_{1,2}$, $3_{2,2}$, $4_{3,2}$, $5_{4,2}$, $6_{5,2}$, $7_{6,2}$, $8_{7,2}$, $9_{8,2}$, $10_{9,2}$, $11_{10,2}$, $12_{11,2}$

Group 3

L						R
$2_{2,0}$	25000	24086	22847	21383	20000	$3_{3,1}$
$3_{2,1}$	26250	23549	21079	19563	18750	$4_{3,2}$
$4_{2,2}$	28000	23609	22028	23609	28000	$5_{3,3}$
$5_{2,3}$	30000	24633	26305	32338	37500	$6_{3,4}$
$6_{2,4}$	32143	27060	34093	42259	47143	$7_{3,5}$
$7_{2,5}$	34375	31293	43935	52226	56875	$8_{3,6}$
$8_{2,6}$	36667	37664	54199	62172	66667	$9_{3,7}$
$9_{2,7}$	39000	46127	64411	72170	76500	$10_{3,8}$
$10_{2,8}$	41364	56030	74550	82050	86364	$11_{3,9}$
$11_{2,9}$	43750	66512	84638	91993	96250	$12_{3,10}$

Outer left labels (group 3): $3_{1,3}$, $4_{2,3}$, $5_{3,3}$, $6_{4,3}$, $7_{5,3}$, $8_{6,3}$, $9_{7,3}$, $10_{8,3}$, $11_{9,3}$, $12_{10,3}$
Outer right labels (group 3): $3_{3,1}$, $4_{3,2}$, $5_{3,3}$, $6_{3,4}$, $7_{3,5}$, $8_{3,6}$, $9_{3,7}$, $10_{3,8}$, $11_{3,9}$, $12_{3,10}$

STRENGTHS FOR ROTATIONAL TRANSITIONS

Symmetric-rotor Subbranches—b Prolate-and-oblate Subbranches

Subbranch		κ					Subbranch	
${}^bR_{1,1}$ $J+K_{-1}+K_1$ even	${}^bP_{-1,-1}$ $J+K_{-1}+K_1$ odd	∓ 1	∓ 0.5	0	± 0.5	± 1	${}^bR_{1,1}$ $J+K_{-1}+K_1$ even	${}^bP_{-1,-1}$ $J+K_{-1}+K_1$ odd
$3_{0,3}$	$4_{1,4}$	25000	29584	32533	34083	35000	$3_{3,0}$	$4_{4,1}$
$4_{1,3}$	$5_{2,4}$	21000	23919	28748	33165	36000	$4_{3,1}$	$5_{4,2}$
$5_{2,3}$	$6_{3,4}$	30000	24633	26305	32338	37500	$5_{3,2}$	$6_{4,3}$
$6_{3,3}$	$7_{4,4}$	39286	31500	26801	31500	39286	$6_{3,3}$	$7_{4,4}$
$7_{4,3}$	$8_{5,4}$	48750	41277	31054	31018	41250	$7_{3,4}$	$8_{4,5}$
$8_{5,3}$	$9_{6,4}$	58333	51336	39046	31553	43333	$8_{3,5}$	$9_{4,6}$
$9_{6,3}$	$10_{7,4}$	68000	61325	49147	33704	45500	$9_{3,6}$	$10_{4,7}$
$10_{7,3}$	$11_{8,4}$	77727	71271	59638	37919	47727	$10_{3,7}$	$11_{4,8}$
$11_{8,3}$	$12_{9,4}$	87500	81200	70013	44472	50000	$11_{3,8}$	$12_{4,9}$
$4_{0,4}$	$5_{1,5}$	30000	39100	42585	44117	45000	$4_{4,0}$	$5_{5,1}$
$5_{1,4}$	$6_{2,5}$	23333	30161	38409	43122	45833	$5_{4,1}$	$6_{5,2}$
$6_{2,4}$	$7_{3,5}$	32143	27060	34093	42259	47143	$6_{4,2}$	$7_{5,3}$
$7_{3,4}$	$8_{4,5}$	41250	31018	31054	41277	48750	$7_{4,3}$	$8_{5,4}$
$8_{4,4}$	$9_{5,5}$	50556	40057	31211	40057	50556	$8_{4,4}$	$9_{5,5}$
$9_{5,4}$	$10_{6,5}$	60000	50269	35484	38709	52500	$9_{4,5}$	$10_{5,6}$
$10_{6,4}$	$11_{7,5}$	69546	60393	43709	37729	54546	$10_{4,6}$	$11_{5,7}$
$11_{7,4}$	$12_{8,5}$	79167	70407	54102	37878	56667	$11_{4,7}$	$12_{5,8}$
$5_{0,5}$	$6_{1,6}$	35000	49126	52653	54140	55000	$5_{5,0}$	$6_{6,1}$
$6_{1,5}$	$7_{2,6}$	25714	38383	48508	53091	55714	$6_{5,1}$	$7_{6,2}$
$7_{2,5}$	$8_{3,6}$	34375	31293	43935	52226	56875	$7_{5,2}$	$8_{6,3}$
$8_{3,5}$	$9_{4,6}$	43333	31553	39046	51336	58333	$8_{5,3}$	$9_{6,4}$
$9_{4,5}$	$10_{5,6}$	52500	38709	35484	50269	60000	$9_{5,4}$	$10_{6,5}$

Top-left index labels

$11_{6,6}$ $12_{6,7}$
$7_{7,1}$ $8_{7,2}$ $9_{7,3}$ $10_{7,4}$ $11_{7,5}$ $12_{7,6}$
$8_{8,1}$ $9_{8,2}$ $10_{8,3}$ $11_{8,4}$ $12_{8,5}$
$9_{9,1}$ $10_{9,2}$ $11_{9,3}$ $12_{9,4}$
$10_{10,1}$ $11_{10,2}$ $12_{10,3}$
$11_{11,1}$ $12_{11,2}$
$12_{12,1}$

Top-right index labels

$10_{6,5}$ $11_{6,6}$
$6_{6,0}$ $7_{6,1}$ $8_{6,2}$ $9_{6,3}$ $10_{6,4}$ $11_{6,5}$
$7_{7,0}$ $8_{7,1}$ $9_{7,2}$ $10_{7,3}$ $11_{7,4}$
$8_{8,0}$ $9_{8,1}$ $10_{8,2}$ $11_{8,3}$
$9_{9,0}$ $10_{9,1}$ $11_{9,2}$
$10_{10,0}$ $11_{10,1}$
$11_{11,0}$

Numeric table

Label					
$11_{6,6}$	61818	48913	35365	48913	61818
$12_{6,7}$	71250	59273	39679	47228	63750
$7_{7,1}$	40000	59250	62702	64155	65000
$8_{7,2}$	28125	48001	58609	63059	65625
$9_{7,3}$	36667	37664	54199	62172	66667
$10_{7,4}$	45500	33704	49147	61325	68000
$11_{7,5}$	54546	37729	43709	60393	69546
$12_{7,6}$	63750	47228	39679	59273	71250
$8_{8,1}$	45000	69364	72737	74165	75000
$9_{8,2}$	30556	58192	68678	73030	75556
$10_{8,3}$	39000	46127	64411	72110	76500
$11_{8,4}$	47727	37919	59638	71271	77727
$12_{8,5}$	56667	37878	54102	70407	79167
$9_{9,1}$	50000	79453	82763	84173	85000
$10_{9,2}$	33000	68479	78720	83002	85500
$11_{9,3}$	41364	56030	74550	82050	86364
$12_{9,4}$	50000	44472	70013	81200	87500
$10_{10,1}$	55000	89522	92782	94179	95000
$11_{10,2}$	35454	78723	88749	92979	95455
$12_{10,3}$	43750	66512	84638	91993	96250
$11_{11,1}$	60000	99576	102798	104184	105000
$12_{11,2}$	37917	88911	98767	102958	105416
$12_{12,1}$	65000	109620	112810	114188	115000

Bottom-right index labels

$11_{6,6}$ $12_{7,6}$
$7_{1,7}$ $8_{2,7}$ $9_{3,7}$ $10_{4,7}$ $11_{6,7}$ $12_{6,7}$
$8_{1,8}$ $9_{2,8}$ $10_{3,8}$ $11_{4,8}$ $12_{5,8}$
$9_{1,9}$ $10_{2,9}$ $11_{3,9}$ $12_{4,9}$
$10_{1,10}$ $11_{2,10}$ $12_{3,10}$
$11_{1,11}$ $12_{2,11}$
$12_{1,12}$

Bottom-left index labels

$10_{6,5}$ $11_{6,6}$
$6_{0,6}$ $7_{1,6}$ $8_{2,6}$ $9_{3,6}$ $10_{4,6}$ $11_{5,6}$
$7_{0,7}$ $8_{1,7}$ $9_{2,7}$ $10_{3,7}$ $11_{4,7}$
$8_{0,8}$ $9_{1,8}$ $10_{2,8}$ $11_{3,8}$
$9_{0,9}$ $10_{1,9}$ $11_{2,9}$
$10_{0,10}$ $11_{1,10}$
$11_{0,11}$

STRENGTHS FOR ROTATIONAL TRANSITIONS

SYMMETRIC-ROTOR SUBBRANCHES—b PROLATE-AND-OBLATE SUBBRANCHES

Subbranch		κ					Subbranch	
$^bR_{-1,1}$ $J + K_{-1} + K_1$ odd	$^bP_{1,-1}$ $J + K_{-1} + K_1$ even	∓1	∓0.5	0	±0.5	±1	$^bR_{1,-1}$ $J + K_{-1} + K_1$ odd	$^bP_{-1,1}$ $J + K_{-1} + K_1$ even
$1_{1,1}$	$2_{0,2}$	5000	7226	10000	12774	15000	$1_{1,1}$	$2_{2,0}$
$2_{1,2}$	$3_{0,3}$	10000	16667	21498	23874	25000	$2_{2,1}$	$3_{3,0}$
$3_{1,3}$	$4_{0,4}$	15000	27406	32266	34065	35000	$3_{3,1}$	$4_{4,0}$
$4_{1,4}$	$5_{0,5}$	20000	38266	42535	44115	45000	$4_{4,1}$	$5_{5,0}$
$5_{1,5}$	$6_{0,6}$	25000	48829	52643	54140	55000	$5_{5,1}$	$6_{6,0}$
$6_{1,6}$	$7_{0,7}$	30000	59146	62700	64155	65000	$6_{6,1}$	$7_{7,0}$
$7_{1,7}$	$8_{0,8}$	35000	69327	72736	74165	75000	$7_{7,1}$	$8_{8,0}$
$8_{1,8}$	$9_{0,9}$	40000	79440	82762	84173	85000	$8_{8,1}$	$9_{9,0}$
$9_{1,9}$	$10_{0,10}$	45000	89517	92782	94179	95000	$9_{9,1}$	$10_{10,0}$
$10_{1,10}$	$11_{0,11}$	50000	99574	102798	104184	105000	$10_{10,1}$	$11_{11,0}$
$11_{1,11}$	$12_{0,12}$	55000	109619	112810	114188	115000	$11_{11,1}$	$12_{12,0}$
$2_{2,1}$	$3_{1,2}$	1667	2792	5168	10000	16667	$2_{1,2}$	$3_{2,1}$
$3_{2,2}$	$4_{1,3}$	3750	7602	15000	22398	26250	$3_{2,2}$	$4_{3,1}$
$4_{2,3}$	$5_{1,4}$	6000	14796	26797	33039	36000	$4_{3,2}$	$5_{4,1}$
$5_{2,4}$	$6_{1,5}$	8333	24389	37946	43109	45833	$5_{4,2}$	$6_{5,1}$
$6_{2,5}$	$7_{1,6}$	10714	35443	48405	53090	55714	$6_{5,2}$	$7_{6,1}$
$7_{2,6}$	$8_{1,7}$	13125	46736	58587	63059	65625	$7_{6,2}$	$8_{7,1}$
$8_{2,7}$	$9_{1,8}$	15556	57689	68672	73029	75556	$8_{7,2}$	$9_{8,1}$
$9_{2,8}$	$10_{1,9}$	18000	68283	78719	83002	85500	$9_{8,2}$	$10_{9,1}$
$10_{2,9}$	$11_{1,10}$	20455	78648	88749	92979	95455	$10_{9,2}$	$11_{10,1}$
$11_{2,10}$	$12_{1,11}$	22917	88882	98767	102958	105416	$11_{10,2}$	$12_{11,1}$
$3_{3,1}$	$4_{2,2}$	1250	1537	2692	6941	18750	$3_{1,3}$	$4_{2,2}$
$4_{3,2}$	$5_{2,3}$	3000	4022	8877	19900	28000	$4_{2,3}$	$5_{3,2}$

$6_{4,2}$	$5_{3,3}$	5000	7698	19335	31792	37500	$6_{2,4}$	$5_{3,3}$
$7_{5,2}$	$6_{4,3}$	7143	13138	31685	42193	47143	$7_{2,5}$	$6_{3,4}$
$8_{6,2}$	$7_{5,3}$	9375	20912	43306	52219	56875	$8_{2,6}$	$7_{3,5}$
$9_{7,2}$	$8_{6,3}$	11667	31041	54046	62172	66667	$9_{2,7}$	$8_{3,6}$
$10_{8,2}$	$9_{7,3}$	14000	42620	64375	72111	76500	$10_{2,8}$	$9_{3,7}$
$11_{9,2}$	$10_{8,3}$	16364	54434	74542	82050	86364	$11_{2,9}$	$10_{3,8}$
$12_{10,2}$	$11_{9,3}$	18750	65840	84638	91993	96250	$12_{2,10}$	$11_{3,9}$

$5_{2,3}$	$4_{1,4}$	1000	1162	1666	4522	21000	$5_{3,2}$	$4_{4,1}$
$6_{3,3}$	$5_{2,4}$	2500	2920	5238	16127	30000	$6_{3,3}$	$5_{4,2}$
$7_{4,3}$	$6_{3,4}$	4286	5148	12183	29700	39286	$7_{3,4}$	$6_{4,3}$
$8_{5,3}$	$7_{4,4}$	6250	8062	23299	41022	48750	$8_{3,5}$	$7_{4,4}$
$9_{6,3}$	$8_{5,4}$	8333	12161	36249	51302	58333	$9_{3,6}$	$8_{4,5}$
$10_{7,3}$	$9_{6,4}$	10500	18094	48371	61321	68000	$10_{3,7}$	$9_{4,6}$
$11_{8,3}$	$10_{7,4}$	12727	26418	59438	71271	77727	$11_{3,8}$	$10_{4,7}$
$12_{9,3}$	$11_{8,4}$	15000	37104	69962	81200	87500	$12_{3,9}$	$11_{4,8}$

$6_{2,4}$	$5_{1,5}$	833	966	1253	2984	23333	$6_{4,2}$	$5_{5,1}$
$7_{3,4}$	$6_{2,5}$	2143	2475	3549	11918	32143	$7_{4,3}$	$6_{5,2}$
$8_{4,4}$	$7_{3,5}$	3750	4309	7685	26263	41250	$8_{4,4}$	$7_{5,3}$
$9_{5,4}$	$8_{4,5}$	5556	6389	15266	39231	50556	$9_{4,5}$	$8_{5,4}$
$10_{6,4}$	$9_{5,5}$	7500	8819	27015	50150	60000	$10_{4,6}$	$9_{5,5}$
$11_{7,4}$	$10_{6,5}$	9545	11954	40562	60377	69546	$11_{4,7}$	$10_{5,6}$
$12_{8,4}$	$11_{7,5}$	11667	16365	53190	70405	79167	$12_{4,8}$	$11_{5,7}$

$7_{2,5}$	$6_{1,6}$	714	829	1052	2102	25714	$7_{5,2}$	$6_{6,1}$
$8_{3,5}$	$7_{2,6}$	1875	2170	2825	8386	34375	$8_{5,3}$	$7_{6,2}$
$9_{4,5}$	$8_{3,6}$	3333	3839	5485	21522	43333	$9_{5,4}$	$8_{6,3}$
$10_{5,5}$	$9_{4,6}$	5000	5717	10057	36378	52500	$10_{5,5}$	$9_{6,4}$
$11_{6,5}$	$10_{5,6}$	6818	7736	18198	48537	61818	$11_{5,6}$	$10_{6,5}$
$12_{7,5}$	$11_{6,6}$	8750	9912	30549	59217	71250	$12_{5,7}$	$11_{6,6}$

STRENGTHS FOR ROTATIONAL TRANSITIONS

SYMMETRIC-ROTOR SUBBRANCHES—b PROLATE-AND-OBLATE SUBBRANCHES

Subbranch ${}^bR_{-1,1}$ $J + K_{-1} + K_1$ even	Subbranch ${}^bP_{1,-3}$ $J + K_{-1} + K_1$ odd	∓ 1	∓ 0.5	0	± 0.5	± 1	Subbranch ${}^bR_{1,-1}$ $J + K_{-1} + K_1$ odd	Subbranch ${}^bP_{-1,1}$ $J + K_{-1} + K_1$ even
$7_{7,1}$	$8_{6,2}$	625	726	918	1615	28125	$7_{1,7}$	$8_{2,6}$
$8_{7,2}$	$9_{6,3}$	1667	1932	2461	5953	36667	$8_{2,7}$	$9_{3,6}$
$9_{7,3}$	$10_{6,4}$	3000	3467	4517	16387	45500	$9_{3,7}$	$10_{4,6}$
$10_{7,4}$	$11_{6,5}$	4545	5226	7432	32088	54546	$10_{4,7}$	$11_{5,6}$
$11_{7,5}$	$12_{6,6}$	6250	7134	12369	46152	63750	$11_{5,7}$	$12_{6,6}$
$8_{8,1}$	$9_{7,2}$	556	645	816	1342	30556	$8_{1,8}$	$9_{2,7}$
$9_{8,2}$	$10_{7,3}$	1500	1741	2210	4454	39000	$9_{2,8}$	$10_{3,7}$
$10_{8,3}$	$11_{7,4}$	2727	3158	4030	12003	47727	$10_{3,8}$	$11_{4,7}$
$11_{8,4}$	$12_{7,5}$	4167	4808	6264	26503	56667	$11_{4,8}$	$12_{5,7}$
$9_{9,1}$	$10_{8,2}$	500	581	734	1176	33000	$9_{1,9}$	$10_{2,8}$
$10_{9,2}$	$11_{8,3}$	1364	1583	2009	3581	41364	$10_{2,9}$	$11_{3,8}$
$11_{9,3}$	$12_{8,4}$	2500	2898	3688	8874	50000	$11_{3,9}$	$12_{4,8}$
$10_{10,1}$	$11_{9,2}$	455	528	667	1059	35454	$10_{1,10}$	$11_{2,9}$
$11_{10,2}$	$12_{9,3}$	1250	1452	1841	3076	43750	$11_{2,10}$	$12_{3,9}$
$11_{11,1}$	$12_{10,2}$	417	484	611	967	37917	$11_{1,11}$	$12_{2,10}$

SYMMETRIC-ROTOR SUBBRANCHES—b PROLATE-OR-OBLATE SUBBRANCHES

Subbranch ${}^bR_{-1,3}$ $J + K_{-1} + K_1$ even	Subbranch ${}^bP_{1,-1}$ $J + K_{-1} + K_1$ odd	∓ 1	∓ 0.5	0	± 0.5	± 1	Subbranch ${}^bR_{3,-1}$ $J + K_{-1} + K_1$ even	Subbranch ${}^bP_{-3,1}$ $J + K_{-1} + K_1$ odd
$2_{2,0}$	$3_{1,3}$	1667	1097	486	101		$2_{0,2}$	$3_{3,1}$
$3_{2,1}$	$4_{1,4}$	3750	1452	297	32		$3_{1,2}$	$4_{4,1}$

$4_{2,2}$	$5_{1,5}$	6000	1159	140	14	$4_{2,2}$	$5_{5,1}$
$5_{2,3}$	$6_{1,6}$	8333	758	77	9	$5_{3,2}$	$6_{6,1}$
$6_{2,4}$	$7_{1,7}$	10714	481	54	7	$6_{4,2}$	$7_{7,1}$
$7_{2,5}$	$8_{1,8}$	13125	323	42	6	$7_{5,2}$	$8_{8,1}$
$8_{2,6}$	$9_{1,9}$	15556	238	35	5	$8_{6,2}$	$9_{9,1}$
$9_{2,7}$	$10_{1,10}$	18000	191	30	4	$9_{7,2}$	$10_{10,1}$
$10_{2,8}$	$11_{1,11}$	20455	163	27	4	$10_{8,2}$	$11_{11,1}$
$11_{2,9}$	$12_{1,12}$	22917	144	24	3	$11_{9,2}$	$12_{12,1}$

$3_{3,0}$	$4_{2,3}$	1250	1323	1091	416	$3_{0,3}$	$4_{3,2}$
$4_{3,1}$	$5_{2,4}$	3000	2753	1252	163	$4_{1,3}$	$5_{4,2}$
$5_{3,2}$	$6_{2,5}$	5000	3538	737	62	$5_{2,3}$	$6_{5,2}$
$6_{3,3}$	$7_{2,6}$	7143	3362	375	35	$6_{3,3}$	$7_{6,2}$
$7_{3,4}$	$8_{2,7}$	9375	2573	219	26	$7_{4,3}$	$8_{7,2}$
$8_{3,5}$	$9_{2,8}$	11667	1754	155	20	$8_{5,3}$	$9_{8,2}$
$9_{3,6}$	$10_{2,9}$	14000	1174	124	17	$9_{6,3}$	$10_{9,2}$
$10_{3,7}$	$11_{2,10}$	16364	826	104	14	$10_{7,3}$	$11_{10,2}$
$11_{3,8}$	$12_{2,11}$	18750	628	90	12	$11_{8,3}$	$12_{11,2}$

$4_{4,0}$	$5_{3,3}$	1000	1144	1259	855	$4_{0,4}$	$5_{3,3}$
$5_{4,1}$	$6_{3,4}$	2500	2771	2305	514	$5_{1,4}$	$6_{4,3}$
$6_{4,2}$	$7_{3,5}$	4286	4433	2107	189	$6_{2,4}$	$7_{5,3}$
$7_{4,3}$	$8_{3,6}$	6250	5663	1258	92	$7_{3,4}$	$8_{6,3}$
$8_{4,4}$	$9_{3,7}$	8333	6007	676	62	$8_{4,4}$	$9_{7,3}$
$9_{4,5}$	$10_{3,8}$	10500	5335	411	47	$9_{5,4}$	$10_{8,3}$
$10_{4,6}$	$11_{3,9}$	12727	4082	298	38	$10_{6,4}$	$11_{9,3}$
$11_{4,7}$	$12_{3,10}$	15000	2866	241	32	$11_{7,4}$	$12_{10,3}$

STRENGTHS FOR ROTATIONAL TRANSITIONS

SYMMETRIC-ROTOR SUBBRANCHES—b PROLATE-OR-OBLATE SUBBRANCHES

Subbranch		κ					Subbranch	
${}^bR_{-1,3}$ $J + K_{-1} + K_1$ even	${}^bP_{1,-3}$ $J + K_{-1} + K_1$ odd	∓ 1	∓ 0.5	0	± 0.5	± 1	${}^bP_{3,-1}$ $J + K_{-1} + K_1$ even	${}^bP_{-3,1}$ $J + K_{-1} + K_1$ odd
$5_{5,0}$	$6_{4,3}$	833	965	1174	1186		$5_{0,5}$	$6_{3,4}$
$6_{5,1}$	$7_{4,4}$	2143	2461	2707	1158		$6_{1,5}$	$7_{4,4}$
$7_{5,2}$	$8_{4,5}$	3750	4229	3560	488		$7_{2,5}$	$8_{5,4}$
$8_{5,3}$	$9_{4,6}$	5556	6040	3008	209		$8_{3,5}$	$9_{6,4}$
$9_{5,4}$	$10_{4,7}$	7500	7603	1831	125		$9_{4,5}$	$10_{7,4}$
$10_{5,5}$	$11_{4,8}$	9545	8531	1026	92		$10_{5,5}$	$11_{8,4}$
$11_{5,6}$	$12_{4,9}$	11667	8452	644	73		$11_{6,5}$	$12_{9,4}$
$6_{6,0}$	$7_{5,3}$	714	829	1039	1321		$6_{0,6}$	$7_{3,5}$
$7_{6,1}$	$8_{5,4}$	1875	2169	2640	1942		$7_{1,6}$	$8_{4,5}$
$8_{6,2}$	$9_{5,5}$	3333	3830	4213	1086		$8_{2,6}$	$9_{5,5}$
$9_{6,3}$	$10_{5,6}$	5000	5676	4833	447		$9_{3,6}$	$10_{6,5}$
$10_{6,4}$	$11_{5,7}$	6818	7575	3937	234		$10_{4,6}$	$11_{7,5}$
$11_{6,5}$	$12_{5,8}$	8750	9349	2441	161		$11_{5,6}$	$12_{8,5}$
$7_{7,0}$	$8_{6,3}$	625	726	916	1313		$7_{0,7}$	$8_{3,6}$
$8_{7,1}$	$9_{6,4}$	1667	1932	2426	2561		$8_{1,7}$	$9_{4,6}$
$9_{7,2}$	$10_{6,5}$	3000	3466	4214	2032		$9_{2,7}$	$10_{5,6}$
$10_{7,3}$	$11_{6,6}$	4545	5221	5738	917		$10_{3,7}$	$11_{6,6}$
$11_{7,4}$	$12_{6,7}$	6250	7114	6113	427		$11_{4,7}$	$12_{7,6}$
$8_{8,0}$	$9_{7,3}$	556	645	816	1238		$8_{0,8}$	$9_{3,7}$
$9_{8,1}$	$10_{7,4}$	1500	1741	2204	2870		$9_{1,8}$	$10_{4,7}$
$10_{8,2}$	$11_{7,5}$	2727	3158	3967	3126		$10_{2,8}$	$11_{5,7}$
$11_{8,3}$	$12_{7,6}$	4167	4807	5836	1752		$11_{3,8}$	$12_{6,7}$

FORBIDDEN SUBBRANCHES—a AND c SUBBRANCHES

Subbranch		κ					Subbranch	
$J + K_{-1} + K_1$ even $^cQ_{-3,2}$	$J + K_{-1} + K_1$ odd $^cQ_{3,-2}$	∓ 1	∓ 0.5	0	± 0.5	± 1	$J + K_{-1} + K_1$ odd $^aQ_{2,-3}$	$J + K_{-1} + K_1$ even $^aQ_{-2,3}$
$3_{3,1}$	$3_{0,3}$		104	169	69		$3_{1,3}$	$3_{3,0}$
$4_{3,2}$	$4_{0,4}$		356	283	68		$4_{2,3}$	$4_{4,0}$
$5_{3,3}$	$5_{0,5}$		621	286	53		$5_{2,3}$	$5_{5,0}$
$6_{3,4}$	$6_{0,6}$		773	250	43		$6_{4,3}$	$6_{6,0}$
$7_{3,5}$	$7_{0,7}$		809	213	35		$7_{5,3}$	$7_{7,0}$
$8_{3,6}$	$8_{0,8}$		774	183	30		$8_{6,3}$	$8_{8,0}$
$9_{3,7}$	$9_{0,9}$		712	160	26		$9_{7,3}$	$9_{9,0}$
$10_{3,8}$	$10_{0,10}$		645	141	23		$10_{8,3}$	$10_{10,0}$
$11_{3,9}$	$11_{0,11}$		584	127	21		$11_{9,3}$	$11_{11,0}$
$12_{3,10}$	$12_{0,12}$		531	115	19		$12_{10,3}$	$12_{12,0}$
$9_{9,0}$	$10_{8,3}$	500	581	734	1143		$9_{0,9}$	$10_{8,8}$
$10_{9,1}$	$11_{8,4}$	1364	1583	2008	2924		$10_{1,9}$	$11_{4,8}$
$11_{9,2}$	$12_{8,5}$	2500	2898	3676	4014		$11_{2,9}$	$12_{6,8}$
$10_{10,0}$	$11_{9,3}$	455	528	667	1049		$10_{0,10}$	$11_{3,9}$
$11_{10,1}$	$12_{9,4}$	1250	1452	1841	2831		$11_{1,10}$	$12_{4,9}$
$11_{11,0}$	$12_{10,3}$	417	484	611	964		$11_{0,11}$	$12_{3,10}$

STRENGTHS FOR ROTATIONAL TRANSITIONS

Forbidden Subbranches—a and c Subbranches

Subbranch		κ					Subbranch	
$^cQ_{-2,2}$ $J + K_{-1} + K_1$ even	$^cQ_{3,-2}$ $J + K_{-1} + K_1$ odd	∓1	∓0.5	0	±0.5	±1	$^aQ_{2,-3}$ $J + K_{-1} + K_1$ odd	$^aQ_{-2,3}$ $J + K_{-1} + K_1$ even
$4_{4,1}$	$4_{1,3}$		31	164	153		$4_{1,4}$	$4_{3,1}$
$5_{4,2}$	$5_{1,4}$		174	504	193		$5_{2,4}$	$5_{4,1}$
$6_{4,3}$	$6_{1,5}$		527	708	163		$6_{3,4}$	$6_{5,1}$
$7_{4,4}$	$7_{1,6}$		1057	716	131		$7_{4,4}$	$7_{6,1}$
$8_{4,5}$	$8_{1,7}$		1557	642	108		$8_{5,4}$	$8_{7,1}$
$9_{4,6}$	$9_{1,8}$		1855	559	92		$9_{6,4}$	$9_{8,1}$
$10_{4,7}$	$10_{1,9}$		1940	489	80		$10_{7,4}$	$10_{9,1}$
$11_{4,8}$	$11_{1,10}$		1888	432	71		$11_{8,4}$	$11_{10,1}$
$12_{4,9}$	$12_{1,11}$		1770	387	64		$12_{9,4}$	$12_{11,1}$
$5_{5,1}$	$5_{2,3}$		9	79	191		$5_{1,5}$	$5_{3,2}$
$6_{5,2}$	$6_{2,4}$		52	413	342		$6_{2,5}$	$6_{4,2}$
$7_{5,3}$	$7_{2,5}$		187	919	325		$7_{3,5}$	$7_{5,2}$
$8_{5,4}$	$8_{2,6}$		515	1216	268		$8_{4,5}$	$8_{6,2}$
$9_{5,5}$	$9_{2,7}$		1123	1236	222		$9_{5,5}$	$9_{7,2}$
$10_{5,6}$	$10_{2,8}$		1924	1128	188		$10_{6,5}$	$10_{8,2}$
$11_{5,7}$	$11_{2,9}$		2650	999	163		$11_{7,5}$	$11_{9,2}$
$12_{5,8}$	$12_{2,10}$		3098	885	144		$12_{8,5}$	$12_{10,2}$
$6_{6,1}$	$6_{3,3}$		4	33	168		$6_{1,6}$	$6_{3,3}$
$7_{6,2}$	$7_{3,4}$		22	212	457		$7_{2,6}$	$7_{4,3}$
$8_{6,3}$	$8_{3,5}$		68	714	524		$8_{3,6}$	$8_{5,3}$
$9_{6,4}$	$9_{3,6}$		184	1385	456		$9_{4,6}$	$9_{6,3}$
$10_{6,5}$	$10_{3,7}$		445	1777	379		$10_{5,6}$	$10_{7,3}$

$11_{6,6}$	$11_{3,8}$	971	1816	320	$11_{6,6}$	$11_{8,3}$
$12_{6,7}$	$12_{3,9}$	1829	1681	277	$12_{7,6}$	$12_{9,3}$
$7_{7,1}$	$7_{4,3}$	3	17	114	$7_{1,7}$	$7_{3,4}$
$8_{7,2}$	$8_{4,4}$	13	98	480	$8_{2,7}$	$8_{4,4}$
$9_{7,3}$	$9_{4,5}$	36	385	714	$9_{3,7}$	$9_{5,4}$
$10_{7,4}$	$10_{4,6}$	84	1053	686	$10_{4,7}$	$10_{6,4}$
$11_{7,5}$	$11_{4,7}$	183	1885	584	$11_{5,7}$	$11_{7,4}$
$12_{7,6}$	$12_{4,8}$	384	2375	493	$12_{6,7}$	$12_{8,4}$
$8_{8,1}$	$8_{5,3}$	2	10	69	$8_{1,8}$	$8_{3,5}$
$9_{8,2}$	$9_{5,4}$	8	52	403	$9_{2,8}$	$9_{4,5}$
$10_{8,3}$	$10_{5,5}$	23	190	830	$10_{3,8}$	$10_{5,5}$
$11_{8,4}$	$11_{5,6}$	52	588	931	$11_{4,8}$	$11_{6,5}$
$12_{8,5}$	$12_{5,7}$	102	1419	832	$12_{5,8}$	$12_{7,5}$
$9_{9,1}$	$9_{6,3}$	1	7	41	$9_{1,9}$	$9_{3,6}$
$10_{9,2}$	$10_{6,4}$	6	34	282	$10_{2,9}$	$10_{4,6}$
$11_{9,3}$	$11_{6,5}$	17	106	817	$11_{3,9}$	$11_{5,6}$
$12_{9,4}$	$12_{6,6}$	36	304	1141	$12_{4,9}$	$12_{6,6}$
$10_{10,1}$	$10_{7,3}$	1	5	25	$10_{1,10}$	$10_{3,7}$
$11_{10,2}$	$11_{7,4}$	5	24	180	$11_{2,10}$	$11_{4,7}$
$12_{10,3}$	$12_{7,5}$	12	70	677	$12_{3,10}$	$12_{5,7}$
$11_{11,1}$	$11_{8,3}$	1	4	17	$11_{1,11}$	$11_{3,8}$
$12_{11,2}$	$12_{8,4}$	4	19	112	$12_{2,11}$	$12_{4,8}$
$12_{12,1}$	$12_{9,3}$	0	3	13	$12_{1,12}$	$12_{3,9}$

STRENGTHS FOR ROTATIONAL TRANSITIONS

FORBIDDEN SUBBRANCHES—a AND c SUBBRANCHES

Subbranch							Subbranch	
$^cQ_{-3,4}$ $J + K_{-1} + K_1$ even	$^cQ_{3,-4}$ $J + K_{-1} + K_1$ odd	∓1	∓0.5	0	±0.5	±1	$^aQ_{4,-3}$ $J + K_{-1} + K_1$ even	$^aQ_{-4,3}$ $J + K_{-1} + K_1$ odd
				κ				
$4_{4,0}$	$4_{1,4}$		8	10	2		$4_{0,4}$	$4_{4,1}$
$5_{4,1}$	$5_{1,5}$		24	14	1		$5_{1,4}$	$5_{5,1}$
$6_{4,2}$	$6_{1,6}$		42	9	0		$6_{2,4}$	$6_{6,1}$
$7_{4,3}$	$7_{1,7}$		51	2	0		$7_{3,4}$	$7_{7,1}$
$8_{4,4}$	$8_{1,8}$		48	1	0		$8_{4,4}$	$8_{8,1}$
$9_{4,5}$	$9_{1,9}$		35	1	0		$9_{5,4}$	$9_{9,1}$
$10_{4,6}$	$10_{1,10}$		22	0	0		$10_{6,4}$	$10_{10,1}$
$11_{4,7}$	$11_{1,11}$		12	0	0		$11_{7,4}$	$11_{11,1}$
$12_{4,8}$	$12_{1,12}$		7	0	0		$12_{8,4}$	$12_{12,1}$
$5_{5,0}$	$5_{2,4}$		7	17	7		$5_{0,5}$	$5_{4,2}$
$6_{5,1}$	$6_{2,5}$		27	39	4		$6_{1,5}$	$6_{5,2}$
$7_{5,2}$	$7_{2,6}$		62	44	1		$7_{2,5}$	$7_{6,2}$
$8_{5,3}$	$8_{2,7}$		106	28	0		$8_{3,5}$	$8_{7,2}$
$9_{5,4}$	$9_{2,8}$		142	13	0		$9_{4,5}$	$9_{8,2}$
$10_{5,5}$	$10_{2,9}$		154	5	0		$10_{5,5}$	$10_{9,2}$
$11_{5,6}$	$11_{2,10}$		137	3	0		$11_{6,5}$	$11_{10,2}$
$12_{5,7}$	$12_{2,11}$		102	1	0		$12_{7,5}$	$12_{11,2}$
$6_{6,0}$	$6_{3,4}$		4	16	14		$6_{0,6}$	$6_{4,3}$
$7_{6,1}$	$7_{3,5}$		19	53	15		$7_{1,6}$	$7_{5,3}$
$8_{6,2}$	$8_{3,6}$		50	88	5		$8_{2,6}$	$8_{6,3}$
$9_{6,3}$	$9_{3,7}$		102	88	2		$9_{3,6}$	$9_{7,3}$

$10_{8,3}$	$10_{4,6}$	1	56	170	$10_{3,8}$	$10_{6,4}$
$11_{9,3}$	$11_{5,6}$	0	27	240	$11_{3,9}$	$11_{6,5}$
$12_{10,3}$	$12_{6,6}$	0	12	289	$12_{3,10}$	$12_{6,6}$
$7_{4,4}$	$7_{0,7}$	19	13	3	$7_{4,4}$	$7_{7,0}$
$8_{5,4}$	$8_{1,7}$	30	51	12	$8_{4,5}$	$8_{7,1}$
$9_{6,4}$	$9_{2,7}$	17	108	34	$9_{4,6}$	$9_{7,2}$
$10_{7,4}$	$10_{3,7}$	5	152	74	$10_{4,7}$	$10_{7,3}$
$11_{8,4}$	$11_{4,7}$	2	144	137	$11_{4,8}$	$11_{7,4}$
$12_{9,4}$	$12_{5,7}$	1	93	223	$12_{4,9}$	$12_{7,5}$
$8_{4,5}$	$8_{0,8}$	20	10	2	$8_{5,4}$	$8_{8,0}$
$9_{5,5}$	$9_{1,8}$	46	41	8	$9_{5,5}$	$9_{8,1}$
$10_{6,5}$	$10_{2,8}$	39	102	23	$10_{5,6}$	$10_{8,2}$
$11_{7,5}$	$11_{3,8}$	15	179	51	$11_{5,7}$	$11_{8,3}$
$12_{8,5}$	$12_{4,8}$	5	129	97	$12_{5,8}$	$12_{8,4}$
$9_{4,6}$	$9_{0,9}$	19	7	1	$9_{6,4}$	$9_{9,0}$
$10_{5,6}$	$10_{1,9}$	56	32	6	$10_{6,5}$	$10_{9,1}$
$11_{6,6}$	$11_{2,9}$	67	84	17	$11_{6,6}$	$11_{9,2}$
$12_{7,6}$	$12_{3,9}$	37	169	36	$12_{6,7}$	$12_{9,3}$
$10_{4,7}$	$10_{0,10}$	17	5	1	$10_{7,4}$	$10_{10,0}$
$11_{5,7}$	$11_{1,10}$	59	24	5	$11_{7,5}$	$11_{10,1}$
$12_{6,7}$	$12_{2,10}$	93	66	12	$12_{7,6}$	$12_{10,2}$
$11_{4,8}$	$11_{0,11}$	14	4	1	$11_{8,4}$	$11_{11,0}$
$12_{5,8}$	$12_{1,11}$	57	19	4	$12_{8,5}$	$12_{11,1}$
$12_{4,9}$	$12_{0,12}$	12	3	0	$12_{9,4}$	$12_{12,0}$

STRENGTHS FOR ROTATIONAL TRANSITIONS

FORBIDDEN SUBBRANCHES—a AND c SUBBRANCHES

Subbranch		∓1	κ			±1	Subbranch	
$^{c}R_{3,-2}$ $J + K_{-1} + K_1$ even	$^{c}P_{-3,2}$ $J + K_{-1} + K_1$ even		∓0.5	0	±0.5		$^{a}R_{-2,3}$ $J + K_{-1} + K_1$ even	$^{a}P_{2,-3}$ $J + K_{-1} + K_1$ even
$2_{0,2}$	$3_{3,0}$		75	215	138		$2_{2,0}$	$3_{0,3}$
$3_{0,3}$	$4_{3,1}$		294	313	51		$3_{3,0}$	$4_{1,3}$
$4_{0,4}$	$5_{3,2}$		528	176	17		$4_{4,0}$	$5_{2,3}$
$5_{0,5}$	$6_{3,3}$		558	79	9		$5_{5,0}$	$6_{3,3}$
$6_{0,6}$	$7_{3,4}$		418	41	5		$6_{6,0}$	$7_{4,3}$
$7_{0,7}$	$8_{3,5}$		263	26	4		$7_{7,0}$	$8_{5,3}$
$8_{0,8}$	$9_{3,6}$		159	19	3		$8_{8,0}$	$9_{6,3}$
$9_{0,9}$	$10_{3,7}$		102	14	2		$9_{9,0}$	$10_{7,3}$
$10_{0,10}$	$11_{3,8}$		70	11	2		$10_{10,0}$	$11_{8,3}$
$11_{0,11}$	$12_{3,9}$		53	9	1		$11_{11,0}$	$12_{9,3}$
$3_{1,2}$	$4_{4,0}$		24	146	272		$3_{2,1}$	$4_{0,4}$
$4_{1,3}$	$5_{4,1}$		122	538	212		$4_{3,1}$	$5_{1,4}$
$5_{1,4}$	$6_{4,2}$		377	665	76		$5_{4,1}$	$6_{2,4}$
$6_{1,5}$	$7_{4,3}$		803	409	34		$6_{5,1}$	$7_{3,4}$
$7_{1,6}$	$8_{4,4}$		1171	206	21		$7_{6,1}$	$8_{4,4}$
$8_{1,7}$	$9_{4,5}$		1200	115	15		$8_{7,1}$	$9_{5,4}$
$9_{1,8}$	$10_{4,6}$		938	76	11		$9_{8,1}$	$10_{6,4}$
$10_{1,9}$	$11_{4,7}$		631	57	9		$10_{9,1}$	$11_{7,4}$
$11_{1,10}$	$12_{4,8}$		409	45	7		$11_{10,1}$	$12_{8,4}$
$4_{2,2}$	$5_{5,0}$		11	70	262		$4_{2,2}$	$5_{0,5}$
$5_{2,3}$	$6_{5,1}$		46	359	464		$5_{3,2}$	$6_{1,5}$
$6_{2,4}$	$7_{5,2}$		142	894	220		$6_{4,2}$	$7_{2,5}$

$8_{3,5}$ $9_{4,5}$ $10_{6,5}$ $11_{6,6}$ $12_{7,5}$

$6_{0,6}$ $7_{1,6}$ $8_{2,6}$ $9_{3,6}$ $10_{4,6}$ $11_{5,6}$ $12_{6,6}$

$7_{0,7}$ $8_{1,7}$ $9_{2,7}$ $10_{3,7}$ $11_{4,7}$ $12_{5,7}$

$8_{0,8}$ $9_{1,8}$ $10_{2,8}$ $11_{3,8}$ $12_{4,8}$

$9_{0,9}$ $10_{1,9}$ $11_{2,9}$ $12_{3,9}$

$7_{5,2}$ $8_{6,2}$ $9_{7,2}$ $10_{8,2}$ $11_{9,2}$

$5_{2,3}$ $6_{3,3}$ $7_{4,3}$ $8_{5,3}$ $9_{6,3}$ $10_{7,3}$ $11_{8,3}$

$6_{2,4}$ $7_{3,4}$ $8_{4,4}$ $9_{5,4}$ $10_{6,4}$ $11_{7,4}$

$7_{2,5}$ $8_{3,5}$ $9_{4,5}$ $10_{5,5}$ $11_{6,5}$

$8_{2,6}$ $9_{3,6}$ $10_{4,6}$ $11_{5,6}$

92	52	36	27	21		
184	624	487	210	108	71	53
119	574	816	435	205	126	
81	424	995	790	381		
60	292	910	1182			

1038	675	365	215	148		
39	187	610	1272	1426	963	548
27	110	338	887	1662	1825	
21	78	206	514	1183		
18	62	149	324			

364	793	1395	1869	1887		
7	27	69	157	336	684	1276
5	19	46	94	176	321	
4	15	36	69	121		
4	13	29	55			

$8_{5,3}$ $9_{5,4}$ $10_{5,5}$ $11_{5,6}$ $12_{5,7}$

$6_{6,0}$ $7_{6,1}$ $8_{6,2}$ $9_{6,3}$ $10_{6,4}$ $11_{6,5}$ $12_{6,6}$

$7_{7,0}$ $8_{7,1}$ $9_{7,2}$ $10_{7,3}$ $11_{7,4}$ $12_{7,5}$

$8_{8,0}$ $9_{8,1}$ $10_{8,2}$ $11_{8,3}$ $12_{8,4}$

$9_{9,0}$ $10_{9,1}$ $11_{9,2}$ $12_{9,3}$

$7_{2,5}$ $8_{2,6}$ $9_{2,7}$ $10_{2,8}$ $11_{2,9}$

$5_{8,2}$ $6_{3,3}$ $7_{3,4}$ $8_{3,5}$ $9_{3,6}$ $10_{3,7}$ $11_{3,8}$

$6_{4,2}$ $7_{4,3}$ $8_{4,4}$ $9_{4,5}$ $10_{4,6}$ $11_{4,7}$

$7_{5,2}$ $8_{5,3}$ $9_{5,4}$ $10_{5,5}$ $11_{5,6}$

$8_{6,2}$ $9_{6,3}$ $10_{6,4}$ $11_{6,5}$

STRENGTHS FOR ROTATIONAL TRANSITIONS

FORBIDDEN SUBBRANCHES—a AND c SUBBRANCHES

Subbranch			κ					Subbranch	
$^cR_{3,-2}$ $J + K_{-1} + K_1$ even	$^cP_{-3,2}$ $J + K_{-1} + K_1$ even	∓ 1	∓ 0.5	0	± 0.5	± 1	$^aR_{-2,3}$ $J + K_{-1} + K_1$ even	$^aP_{2,-3}$ $J + K_{-1} + K_1$ even
$9_{7,2}$	$10_{10,0}$		3	15	48		$9_{2,7}$	$10_{0,10}$
$10_{7,3}$	$11_{10,1}$		11	52	207		$10_{3,7}$	$11_{1,10}$
$11_{7,4}$	$12_{10,2}$		24	121	694		$11_{4,7}$	$12_{2,10}$
$10_{8,2}$	$11_{11,0}$		3	13	41		$10_{2,8}$	$11_{0,11}$
$11_{8,3}$	$12_{11,1}$		9	45	158		$11_{3,8}$	$12_{1,11}$
$11_{9,2}$	$12_{12,0}$		2	12	36		$11_{2,9}$	$12_{0,12}$

Subbranch			κ					Subbranch	
$^cR_{3,-2}$ $J + K_{-1} + K_1$ odd	$^cP_{-3,2}$ $J + K_{-1} + K_1$ odd	∓ 1	∓ 0.5	0	± 0.5	± 1	$^aR_{-2,3}$ $J + K_{-1} + K_1$ odd	$^aP_{2,-3}$ $J + K_{-1} + K_1$ odd
$3_{1,3}$	$4_{4,1}$		10	21	13		$3_{3,1}$	$4_{1,4}$
$4_{1,4}$	$5_{4,2}$		31	38	12		$4_{4,1}$	$5_{2,4}$
$5_{1,5}$	$6_{4,3}$		56	39	8		$5_{5,1}$	$6_{3,4}$
$6_{1,6}$	$7_{4,4}$		73	32	5		$6_{6,1}$	$7_{4,4}$
$7_{1,7}$	$8_{4,5}$		80	24	4		$7_{7,1}$	$8_{5,4}$
$8_{1,8}$	$9_{4,6}$		77	18	3		$8_{8,1}$	$9_{6,4}$
$9_{1,9}$	$10_{4,7}$		68	14	2		$9_{9,1}$	$10_{7,4}$
$10_{1,10}$	$11_{4,8}$		58	11	2		$10_{10,1}$	$11_{8,4}$
$11_{1,11}$	$12_{4,9}$		49	9	1		$11_{11,1}$	$12_{9,4}$

$5_{1,5}$	$4_{3,2}$	32	31	9	$5_{6,1}$	$4_{2,3}$
$6_{2,5}$	$5_{4,2}$	39	79	33	$6_{5,2}$	$5_{2,4}$
$7_{3,5}$	$6_{5,2}$	30	108	73	$7_{5,3}$	$6_{2,5}$
$8_{4,5}$	$7_{6,2}$	21	108	124	$8_{5,4}$	$7_{2,6}$
$9_{5,5}$	$8_{7,2}$	15	91	172	$9_{5,5}$	$8_{2,7}$
$10_{6,5}$	$9_{8,2}$	11	72	205	$10_{5,6}$	$9_{2,8}$
$11_{7,5}$	$10_{9,2}$	9	56	216	$11_{5,7}$	$10_{2,9}$
$12_{8,5}$	$11_{10,2}$	7	45	209	$12_{5,8}$	$11_{2,10}$

$6_{1,6}$	$5_{3,3}$	46	30	7	$6_{6,1}$	$5_{3,3}$
$7_{2,6}$	$6_{4,3}$	76	92	25	$7_{6,2}$	$6_{3,4}$
$8_{3,6}$	$7_{5,3}$	67	163	60	$8_{6,3}$	$7_{3,5}$
$9_{4,6}$	$8_{6,3}$	50	202	115	$9_{6,4}$	$8_{3,6}$
$10_{5,6}$	$9_{7,3}$	36	199	187	$10_{6,5}$	$9_{3,7}$
$11_{6,6}$	$10_{8,3}$	27	171	267	$11_{6,6}$	$10_{3,8}$
$12_{7,6}$	$11_{9,3}$	21	139	337	$12_{6,7}$	$11_{3,9}$

$7_{1,7}$	$6_{3,4}$	53	25	5	$7_{7,1}$	$6_{4,3}$
$8_{2,7}$	$7_{4,4}$	110	86	19	$8_{7,2}$	$7_{4,4}$
$9_{3,7}$	$8_{5,4}$	118	177	46	$9_{7,3}$	$8_{4,5}$
$10_{4,7}$	$9_{6,4}$	94	268	89	$10_{7,4}$	$9_{4,6}$
$11_{5,7}$	$10_{7,4}$	70	316	154	$11_{7,5}$	$10_{4,7}$
$12_{6,7}$	$11_{8,4}$	53	309	241	$12_{7,6}$	$11_{4,8}$

$8_{1,8}$	$7_{3,5}$	53	21	4	$8_{8,1}$	$7_{5,3}$
$9_{2,8}$	$8_{4,5}$	134	73	15	$9_{8,2}$	$8_{5,4}$
$10_{3,5}$	$9_{5,5}$	173	163	35	$10_{8,3}$	$9_{5,5}$
$11_{4,8}$	$10_{6,5}$	155	281	69	$11_{8,4}$	$10_{5,6}$
$12_{5,8}$	$11_{7,5}$	120	390	119	$12_{8,5}$	$11_{5,7}$

STRENGTHS FOR ROTATIONAL TRANSITIONS

FORBIDDEN SUBBRANCHES—a AND c SUBBRANCHES

Subbranch			κ				Subbranch	
${}^cR_{3,-2}$ $J + K_{-1} + K_1$ odd	${}^cP_{-3,2}$ $J + K_{-1} + K_1$ odd	∓1	∓0.5	0	±0.5	±1	${}^aR_{-2,3}$ $J + K_{-1} + K_1$ odd	${}^aP_{2,-3}$ $J + K_{-1} + K_1$ odd
$8_{6,3}$	$9_{9,1}$		4	18	49		$8_{3,6}$	$9_{1,9}$
$9_{6,4}$	$10_{9,2}$		13	61	143		$9_{4,6}$	$10_{2,9}$
$10_{6,5}$	$11_{9,3}$		29	140	219		$10_{5,6}$	$11_{3,9}$
$11_{6,6}$	$12_{9,4}$		54	258	224		$11_{6,6}$	$12_{4,9}$
$9_{7,3}$	$10_{10,1}$		3	15	44		$9_{3,7}$	$10_{1,10}$
$10_{7,4}$	$11_{10,2}$		11	52	140		$10_{4,7}$	$11_{2,10}$
$11_{7,5}$	$12_{10,3}$		24	119	249		$11_{6,7}$	$12_{3,10}$
$10_{8,3}$	$11_{11,1}$		3	13	40		$10_{3,8}$	$11_{1,11}$
$11_{8,4}$	$12_{11,2}$		9	45	131		$11_{4,8}$	$12_{2,11}$
$11_{9,3}$	$12_{12,1}$		2	12	36		$11_{5,9}$	$12_{1,12}$

Subbranch			κ				Subbranch	
${}^cR_{-3,4}$ $J + K_{-1} + K_1$ even	${}^cP_{3,-4}$ $J + K_{-1} + K_1$ even	∓1	∓0.5	0	±0.5	±1	${}^aR_{4,-3}$ $J + K_{-1} + K_1$ even	${}^aP_{-4,3}$ $J + K_{-1} + K_1$ even
$3_{3,0}$	$4_{0,4}$		28	18	2		$3_{0,3}$	$4_{4,0}$
$4_{3,1}$	$5_{0,5}$		78	17	0		$4_{1,3}$	$5_{5,0}$
$5_{3,2}$	$6_{0,6}$		107	8	0		$5_{2,3}$	$6_{6,0}$
$6_{3,3}$	$7_{0,7}$		96	3	0		$6_{3,3}$	$7_{7,0}$

$8_{8,0}$	$7_{4,3}$	0	2	65	$8_{0,8}$	$7_{3,4}$
$9_{9,0}$	$8_{5,3}$	0	1	39	$9_{0,9}$	$8_{3,5}$
$10_{10,0}$	$9_{6,3}$	0	1	23	$10_{0,10}$	$9_{3,6}$
$11_{11,0}$	$10_{7,3}$	0	1	14	$11_{0,11}$	$10_{3,7}$
$12_{12,0}$	$11_{8,3}$	0	0	9	$12_{0,12}$	$11_{3,8}$
$5_{4,1}$	$4_{0,4}$	7	31	11	$5_{1,4}$	$4_{4,0}$
$6_{5,1}$	$5_{1,4}$	3	62	58	$6_{1,5}$	$5_{4,1}$
$7_{6,1}$	$6_{2,4}$	1	51	152	$7_{1,6}$	$6_{4,2}$
$8_{7,1}$	$7_{3,4}$	0	25	252	$8_{1,7}$	$7_{4,3}$
$9_{8,1}$	$8_{4,4}$	0	11	290	$9_{1,8}$	$8_{4,4}$
$10_{9,1}$	$9_{5,4}$	0	6	250	$10_{1,9}$	$9_{4,5}$
$11_{10,1}$	$10_{6,4}$	0	4	175	$11_{1,10}$	$10_{4,6}$
$12_{11,1}$	$11_{7,4}$	0	3	109	$12_{1,11}$	$11_{4,7}$
$6_{4,2}$	$5_{0,5}$	15	21	3	$6_{2,4}$	$5_{5,0}$
$7_{5,2}$	$6_{1,5}$	12	83	17	$7_{2,5}$	$6_{5,1}$
$8_{6,2}$	$7_{2,5}$	4	124	65	$8_{2,6}$	$7_{5,2}$
$9_{7,2}$	$8_{3,5}$	1	96	173	$9_{2,7}$	$8_{5,3}$
$10_{8,2}$	$9_{4,5}$	1	50	335	$10_{2,8}$	$9_{5,4}$
$11_{9,2}$	$10_{5,5}$	0	23	477	$11_{2,9}$	$10_{5,5}$
$12_{10,2}$	$11_{6,5}$	0	13	517	$12_{2,10}$	$11_{5,6}$
$7_{4,3}$	$6_{0,6}$	21	9	1	$7_{3,4}$	$6_{6,0}$
$8_{5,3}$	$7_{1,6}$	28	59	5	$8_{3,5}$	$7_{6,1}$
$9_{6,3}$	$8_{2,6}$	12	151	20	$9_{3,6}$	$8_{6,2}$
$10_{7,3}$	$9_{3,6}$	4	198	60	$10_{3,7}$	$9_{6,3}$
$11_{8,3}$	$10_{4,6}$	2	151	155	$11_{3,8}$	$10_{6,4}$
$12_{9,3}$	$11_{5,6}$	1	81	329	$12_{3,9}$	$11_{6,5}$

STRENGTHS FOR ROTATIONAL TRANSITIONS

FORBIDDEN SUBBRANCHES—a AND c SUBBRANCHES

Subbranch		κ					Subbranch	
$^cR_{-3,4}$ $J + K_{-1} + K_1$ even	$^cP_{3,-4}$ $J + K_{-1} + K_1$ even	∓1	∓0.5	0	±0.5	±1	$^aR_{4,-3}$ $J + K_{-1} + K_1$ even	$^aP_{-4,3}$ $J + K_{-1} + K_1$ even
$7_{7,0}$	$8_{4,4}$		0	4	20		$7_{0,7}$	$8_{4,4}$
$8_{7,1}$	$9_{4,5}$		2	26	46		$8_{1,7}$	$9_{5,4}$
$9_{7,2}$	$10_{4,6}$		8	107	31		$9_{2,7}$	$10_{6,4}$
$10_{7,3}$	$11_{4,7}$		22	230	11		$10_{3,7}$	$11_{7,4}$
$11_{7,4}$	$12_{4,8}$		55	282	4		$11_{4,7}$	$12_{8,4}$
$8_{8,0}$	$9_{5,4}$		0	2	15		$8_{0,8}$	$9_{4,5}$
$9_{8,1}$	$10_{5,5}$		1	12	56		$9_{1,8}$	$10_{5,5}$
$10_{8,2}$	$11_{5,6}$		4	54	59		$10_{2,8}$	$11_{6,5}$
$11_{8,3}$	$12_{5,7}$		11	165	27		$11_{3,8}$	$12_{7,5}$
$9_{9,0}$	$10_{6,4}$		0	1	10		$9_{0,9}$	$10_{4,6}$
$10_{9,1}$	$11_{6,5}$		1	6	53		$10_{1,9}$	$11_{5,6}$
$11_{9,2}$	$12_{6,6}$		3	25	87		$11_{2,9}$	$12_{6,6}$
$10_{10,0}$	$11_{7,4}$		0	1	5		$10_{0,10}$	$11_{4,7}$
$11_{10,1}$	$12_{7,5}$		1	4	41		$11_{1,10}$	$12_{6,7}$
$11_{11,0}$	$12_{8,4}$		0	0	3		$11_{0,11}$	$12_{4,8}$

$^cR_{-3,4}$ $J + K_{-1} + K_1$ odd	$^cP_{3,-4}$ $J + K_{-1} + K_1$ odd	∓1	∓0.5	0	±0.5	±1	$^aR_{4,-3}$ $J + K_{-1} + K_1$ odd	$^aP_{-4,3}$ $J + K_{-1} + K_1$ odd
$4_{4,1}$	$5_{1,5}$		2	1	0		$4_{1,4}$	$5_{5,1}$
$5_{4,2}$	$6_{1,6}$		5	2	0		$5_{2,4}$	$6_{6,1}$

$6_{4,3}$	$7_{1,7}$	8	2	0	$6_{3,4}$	$7_{7,1}$
$7_{4,4}$	$8_{1,8}$	11	1	0	$7_{4,4}$	$8_{8,1}$
$8_{4,5}$	$9_{1,9}$	12	1	0	$8_{5,4}$	$9_{9,1}$
$9_{4,6}$	$10_{1,10}$	11	1	0	$9_{6,4}$	$10_{10,1}$
$10_{4,7}$	$11_{1,11}$	9	1	0	$10_{7,4}$	$11_{11,1}$
$11_{4,8}$	$12_{1,12}$	8	0	0	$11_{8,4}$	$12_{12,1}$

$5_{5,1}$	$6_{2,5}$	1	2	0	$5_{1,5}$	$6_{6,2}$
$6_{5,2}$	$7_{2,6}$	6	5	0	$6_{2,5}$	$7_{6,2}$
$7_{5,3}$	$8_{2,7}$	14	6	0	$7_{3,5}$	$8_{7,2}$
$8_{5,4}$	$9_{2,8}$	23	6	0	$8_{4,5}$	$9_{8,2}$
$9_{5,5}$	$10_{2,9}$	32	4	0	$9_{5,5}$	$10_{9,2}$
$10_{5,6}$	$11_{2,10}$	36	3	0	$10_{6,5}$	$11_{10,2}$
$11_{5,7}$	$12_{2,11}$	37	3	0	$11_{7,5}$	$12_{11,2}$

$6_{6,1}$	$7_{3,5}$	1	2	1	$6_{1,6}$	$7_{5,3}$
$7_{6,2}$	$8_{3,6}$	4	8	1	$7_{2,6}$	$8_{6,3}$
$8_{6,3}$	$9_{3,7}$	11	13	1	$8_{3,6}$	$9_{7,3}$
$9_{6,4}$	$10_{3,8}$	23	14	1	$9_{4,6}$	$10_{8,3}$
$10_{6,5}$	$11_{3,9}$	39	13	0	$10_{5,6}$	$11_{9,3}$
$11_{6,6}$	$12_{3,10}$	56	10	0	$11_{6,6}$	$12_{10,3}$

$7_{7,1}$	$8_{4,5}$	0	2	2	$7_{1,7}$	$8_{6,4}$
$8_{7,2}$	$9_{4,6}$	2	8	3	$8_{2,7}$	$9_{6,4}$
$9_{7,3}$	$10_{4,7}$	7	17	2	$9_{3,7}$	$10_{7,4}$
$10_{7,4}$	$11_{4,8}$	16	24	1	$10_{4,7}$	$11_{8,4}$
$11_{7,5}$	$12_{4,9}$	32	25	1	$11_{5,7}$	$12_{9,4}$

$8_{8,1}$	$9_{5,5}$	0	1	2	$8_{1,8}$	$9_{5,5}$
$9_{8,2}$	$10_{5,6}$	1	6	4	$9_{2,8}$	$10_{6,5}$
$10_{8,3}$	$11_{5,7}$	4	17	4	$10_{3,8}$	$11_{7,5}$
$11_{8,4}$	$12_{5,8}$	10	29	3	$11_{4,8}$	$12_{8,5}$

STRENGTHS FOR ROTATIONAL TRANSITIONS

FORBIDDEN SUBBRANCHES—a AND c SUBBRANCHES

Subbranch		∓1	∓0.5	0	±0.5	±1	Subbranch	
$^cR_{-3,4}$ $J+K_{-1}+K_1$ even	$^cP_{3,-4}$ $J+K_{-1}+K_1$ even			κ			$^aR_{4,-3}$ $J+K_{-1}+K_1$ odd	$^aP_{-4,3}$ $J+K_{-1}+K_1$ odd
$9_{9,1}$	$10_{6,5}$		0	1	2		$9_{1,9}$	$10_{6,6}$
$10_{9,2}$	$11_{6,6}$		1	5	5		$10_{2,9}$	$11_{6,6}$
$11_{9,3}$	$12_{6,7}$		3	14	7		$11_{3,9}$	$12_{7,6}$
$10_{10,1}$	$11_{7,5}$		0	1	2		$10_{1,10}$	$11_{6,7}$
$11_{10,2}$	$12_{7,6}$		1	3	6		$11_{2,10}$	$12_{6,7}$
$11_{11,1}$	$12_{8,5}$		0	0	2		$11_{1,11}$	$12_{6,8}$

FORBIDDEN SUBBRANCHES—b SUBBRANCHES

Subbranch		∓1	∓0.5	0	±0.5	±1	Subbranch	
$^bQ_{-3,3}$ $J+K_{-1}+K_1$ even	$^bQ_{3,-3}$ $J+K_{-1}+K_1$ even			κ			$^bQ_{3,-3}$ $J+K_{-1}+K_1$ even	$^bQ_{-3,3}$ $J+K_{-1}+K_1$ even
$3_{3,0}$	$3_{0,3}$		138	297	138		$3_{0,3}$	$3_{3,0}$
$4_{3,1}$	$4_{0,4}$		445	319	41		$4_{1,3}$	$4_{4,0}$
$5_{3,2}$	$5_{0,5}$		674	158	13		$5_{2,3}$	$5_{5,0}$
$6_{3,3}$	$6_{0,6}$		640	67	6		$6_{3,3}$	$6_{6,0}$
$7_{3,4}$	$7_{0,7}$		450	33	4		$7_{4,3}$	$7_{7,0}$
$8_{3,5}$	$8_{0,8}$		273	21	3		$8_{6,3}$	$8_{8,0}$

		col1	col2	col3		
$9_{9,0}$	$9_{6,3}$	2	15	162	$9_{0,9}$	$9_{3,6}$
$10_{10,0}$	$10_{7,3}$	1	11	101	$10_{0,10}$	$10_{3,7}$
$11_{11,0}$	$11_{8,3}$	1	9	69	$11_{0,11}$	$11_{3,8}$
$12_{12,0}$	$12_{9,3}$	1	7	52	$12_{0,12}$	$12_{3,9}$
$4_{3,1}$	$4_{0,4}$	445	319	41	$4_{1,3}$	$4_{4,0}$
$5_{4,1}$	$5_{1,4}$	230	846	230	$5_{1,4}$	$5_{4,1}$
$6_{5,1}$	$6_{2,4}$	70	793	684	$6_{1,5}$	$6_{4,2}$
$7_{6,1}$	$7_{3,4}$	29	422	1287	$7_{1,6}$	$7_{4,3}$
$8_{7,1}$	$8_{4,4}$	17	197	1640	$8_{1,7}$	$8_{4,4}$
$9_{8,1}$	$9_{5,4}$	11	105	1513	$9_{1,8}$	$9_{4,5}$
$10_{9,1}$	$10_{6,4}$	8	67	1105	$10_{1,9}$	$10_{4,6}$
$11_{10,1}$	$11_{7,4}$	6	49	713	$11_{1,10}$	$11_{4,7}$
$12_{11,1}$	$12_{8,4}$	5	38	449	$12_{1,11}$	$12_{4,8}$
$5_{3,2}$	$5_{0,5}$	674	158	13	$5_{2,3}$	$5_{5,0}$
$6_{4,2}$	$6_{1,5}$	684	793	70	$6_{2,4}$	$6_{5,1}$
$7_{5,2}$	$7_{2,5}$	248	1513	248	$7_{2,5}$	$7_{5,2}$
$8_{6,2}$	$8_{3,5}$	91	1360	681	$8_{2,6}$	$8_{5,3}$
$9_{7,2}$	$9_{4,5}$	48	763	1444	$9_{2,7}$	$9_{5,4}$
$10_{8,2}$	$10_{5,5}$	31	380	2306	$10_{2,8}$	$10_{5,5}$
$11_{9,2}$	$11_{6,5}$	22	212	2749	$11_{2,9}$	$11_{5,6}$
$12_{10,2}$	$12_{7,5}$	16	141	2519	$12_{2,10}$	$12_{5,7}$
$6_{3,3}$	$6_{0,6}$	640	67	6	$6_{3,3}$	$6_{6,0}$
$7_{4,3}$	$7_{1,6}$	1287	422	29	$7_{3,4}$	$7_{6,1}$
$8_{5,3}$	$8_{2,6}$	681	1360	91	$8_{3,5}$	$8_{6,2}$
$9_{6,3}$	$9_{3,6}$	245	2252	245	$9_{3,6}$	$9_{6,3}$
$10_{7,3}$	$10_{4,6}$	112	1993	592	$10_{3,7}$	$10_{6,4}$
$11_{8,3}$	$11_{5,6}$	69	1165	1278	$11_{3,8}$	$11_{6,5}$
$12_{9,3}$	$12_{6,6}$	48	609	2337	$12_{3,9}$	$12_{6,6}$

STRENGTHS FOR ROTATIONAL TRANSITIONS

Forbidden Subbranches—*b* Subbranches

Subbranch		κ					Subbranch	
$^bQ_{-3,3}$ $J+K_{-1}+K_1$ even	$^bQ_{3,-3}$ $J+K_{-1}+K_1$ even	∓1	∓0.5	0	±0.5	±1	$^bQ_{3,-3}$ $J+K_{-1}+K_1$ even	$^bQ_{-3,3}$ $J+K_{-1}+K_1$ even
$7_{7,0}$	$7_{4,3}$		4	33	450		$7_{0,7}$	$7_{3,4}$
$8_{7,1}$	$8_{4,4}$		17	197	1640		$8_{1,7}$	$8_{4,4}$
$9_{7,2}$	$9_{4,5}$		48	763	1444		$9_{2,7}$	$9_{5,4}$
$10_{7,3}$	$10_{4,6}$		112	1993	592		$10_{3,7}$	$10_{6,4}$
$11_{7,4}$	$11_{4,7}$		244	3040	244		$11_{4,7}$	$11_{7,4}$
$12_{7,5}$	$12_{4,8}$		512	2675	136		$12_{5,7}$	$12_{8,4}$
$8_{8,0}$	$8_{5,3}$		3	21	273		$8_{0,8}$	$8_{3,5}$
$9_{8,1}$	$9_{5,4}$		11	105	1513		$9_{1,8}$	$9_{4,5}$
$10_{8,2}$	$10_{5,5}$		31	380	2306		$10_{2,8}$	$10_{5,5}$
$11_{8,3}$	$11_{5,6}$		69	1165	1278		$11_{3,8}$	$11_{6,5}$
$12_{8,4}$	$12_{5,7}$		136	2675	512		$12_{4,8}$	$12_{7,5}$
$9_{9,0}$	$9_{6,3}$		2	15	162		$9_{0,9}$	$9_{3,6}$
$10_{9,1}$	$10_{6,4}$		8	67	1105		$10_{1,9}$	$10_{4,6}$
$11_{9,2}$	$11_{6,5}$		22	212	2749		$11_{2,9}$	$11_{5,6}$
$12_{9,3}$	$12_{6,6}$		48	609	2337		$12_{3,9}$	$12_{6,6}$
$10_{10,0}$	$10_{7,3}$		1	11	101		$10_{0,10}$	$10_{3,7}$
$11_{10,1}$	$11_{7,4}$		6	49	713		$11_{1,10}$	$11_{4,7}$
$12_{10,2}$	$12_{7,5}$		16	141	2519		$12_{2,10}$	$12_{5,7}$
$11_{11,0}$	$11_{8,3}$		1	9	69		$11_{0,11}$	$11_{3,8}$
$12_{11,1}$	$12_{8,4}$			38	449		$12_{1,11}$	$12_{4,8}$
$12_{12,0}$	$12_{9,3}$		1	7	52		$12_{0,12}$	$12_{3,9}$

$^bQ_{-2,3}$, $J + K_{-1} + K_1$ odd	$^bQ_{3,-3}$, $J + K_{-1} + K_1$ odd	± 1	± 0.5	0	∓ 0.5	∓ 1	$^bQ_{3,-3}$, $J + K_{-1} + K_1$ odd	$^bQ_{-2,3}$, $J + K_{-1} + K_1$ odd
$4_{4,1}$	$4_{1,4}$		10	20	10		$4_{1,4}$	$4_{4,1}$
$5_{5,1}$	$5_{2,4}$		9	33	34		$5_{1,5}$	$5_{4,2}$
$6_{6,1}$	$6_{3,4}$		6	33	58		$6_{1,6}$	$6_{4,3}$
$7_{7,1}$	$7_{4,4}$		4	26	76		$7_{1,7}$	$7_{4,4}$
$8_{8,1}$	$8_{5,4}$		3	19	82		$8_{1,8}$	$8_{4,5}$
$9_{9,1}$	$9_{6,4}$		2	14	77		$9_{1,9}$	$9_{4,6}$
$10_{10,1}$	$10_{7,4}$		1	11	68		$10_{1,10}$	$10_{4,7}$
$11_{11,1}$	$11_{8,4}$		1	9	58		$11_{1,11}$	$11_{4,8}$
$12_{12,1}$	$12_{9,4}$		1	7	48		$12_{1,12}$	$12_{4,9}$
$5_{4,2}$	$5_{1,5}$		34	33	9		$5_{2,4}$	$5_{5,1}$
$6_{5,2}$	$6_{2,5}$		36	83	36		$6_{2,5}$	$6_{5,2}$
$7_{6,2}$	$7_{3,5}$		25	108	84		$7_{2,6}$	$7_{5,3}$
$8_{7,2}$	$8_{4,5}$		16	102	144		$8_{2,7}$	$8_{5,4}$
$9_{8,2}$	$9_{5,5}$		11	83	199		$9_{2,8}$	$9_{5,5}$
$10_{9,2}$	$10_{6,5}$		8	63	233		$10_{2,9}$	$10_{5,6}$
$11_{10,2}$	$11_{7,5}$		6	48	241		$11_{2,10}$	$11_{5,7}$
$12_{11,2}$	$12_{8,5}$		5	38	228		$12_{2,11}$	$12_{5,8}$
$6_{4,3}$	$6_{1,6}$		58	33	6		$6_{3,4}$	$6_{6,1}$
$7_{5,3}$	$7_{2,6}$		84	108	25		$7_{3,5}$	$7_{6,2}$
$8_{6,3}$	$8_{3,6}$		67	187	67		$8_{3,6}$	$8_{6,3}$
$9_{7,3}$	$9_{4,6}$		45	221	136		$9_{3,7}$	$9_{6,4}$
$10_{8,3}$	$10_{5,6}$		31	206	228		$10_{3,8}$	$10_{6,5}$
$11_{9,3}$	$11_{6,6}$		22	169	328		$11_{3,9}$	$11_{6,6}$
$12_{10,3}$	$12_{7,6}$		16	132	410		$12_{3,10}$	$12_{6,7}$

STRENGTHS FOR ROTATIONAL TRANSITIONS

FORBIDDEN SUBBRANCHES—b SUBBRANCHES

Subbranch		κ					Subbranch	
$^bQ_{-2,3}$ $J + K_{-1} + K_1$ odd	$^bQ_{3,-3}$ $J + K_{-1} + K_1$ odd	∓ 1	∓ 0.5	0	± 0.5	± 1	$^bQ_{3,-3}$ $J + K_{-1} + K_1$ odd	$^bQ_{-2,3}$ $J + K_{-1} + K_1$ odd
$7_{7,1}$	$7_{4,4}$		4	26	76		$7_{1,7}$	$7_{4,4}$
$8_{7,2}$	$8_{4,5}$		16	102	144		$8_{2,7}$	$8_{5,4}$
$9_{7,3}$	$9_{4,6}$		45	221	136		$9_{3,7}$	$9_{6,4}$
$10_{7,4}$	$10_{4,7}$		98	327	98		$10_{4,7}$	$10_{7,4}$
$11_{7,5}$	$11_{4,8}$		183	367	68		$11_{5,7}$	$11_{8,4}$
$12_{7,6}$	$12_{4,9}$		299	340	48		$12_{6,7}$	$12_{9,4}$
$8_{8,1}$	$8_{5,4}$		3	19	82		$8_{1,8}$	$8_{4,5}$
$9_{8,2}$	$9_{5,5}$		11	83	199		$9_{2,8}$	$9_{5,5}$
$10_{8,3}$	$10_{5,6}$		31	206	228		$10_{3,8}$	$10_{6,5}$
$11_{8,4}$	$11_{5,7}$		68	367	183		$11_{4,8}$	$11_{7,5}$
$12_{8,5}$	$12_{5,8}$		129	498	129		$12_{5,8}$	$12_{8,5}$
$9_{9,1}$	$9_{6,4}$		2	14	77		$9_{1,9}$	$9_{4,6}$
$10_{9,2}$	$10_{6,5}$		8	63	233		$10_{2,9}$	$10_{5,6}$
$11_{9,3}$	$11_{6,6}$		22	169	328		$11_{3,9}$	$11_{6,6}$
$12_{9,4}$	$12_{6,7}$		48	340	299		$12_{4,9}$	$12_{7,6}$
$10_{10,1}$	$10_{7,4}$		1	11	68		$10_{1,10}$	$10_{4,7}$
$11_{10,2}$	$11_{7,5}$		6	48	241		$11_{2,10}$	$11_{5,7}$
$12_{10,3}$	$12_{7,6}$		16	132	410		$12_{3,10}$	$12_{6,7}$
$11_{11,1}$	$11_{8,4}$		1	9	58		$11_{1,11}$	$11_{4,8}$
$12_{11,2}$	$12_{8,5}$		5	38	228		$12_{2,11}$	$12_{5,8}$
$12_{12,1}$	$12_{9,4}$		1	7	48		$12_{1,12}$	$12_{4,9}$

$^bP_{-3,3}$ $J+K_{-1}+K_1$ even	$^bR_{3,-3}$ $J+K_{-1}+K_1$ odd	± 1	± 0.5	0	∓ 0.5	∓ 1	$^bP_{3,-3}$ $J+K_{-1}+K_1$ even	$^bR_{-3,3}$ $J+K_{-1}+K_1$ odd
$4_{4,0}$	$3_{1,3}$		13	41	38		$4_{0,4}$	$3_{3,1}$
$5_{5,0}$	$4_{2,3}$		12	62	111		$5_{0,5}$	$4_{3,2}$
$6_{6,0}$	$5_{3,3}$		9	59	175		$6_{0,6}$	$5_{3,3}$
$7_{7,0}$	$6_{4,3}$		7	50	206		$7_{0,7}$	$6_{3,4}$
$8_{8,0}$	$7_{5,3}$		6	42	209		$8_{0,8}$	$7_{3,5}$
$9_{9,0}$	$8_{6,3}$		5	35	195		$9_{0,9}$	$8_{3,6}$
$10_{10,0}$	$9_{7,3}$		4	30	177		$10_{0,10}$	$9_{3,7}$
$11_{11,0}$	$10_{8,3}$		4	27	158		$11_{0,11}$	$10_{3,8}$
$12_{12,0}$	$11_{9,3}$		3	24	142		$12_{0,12}$	$11_{3,9}$
$5_{4,1}$	$4_{1,4}$		41	63	14		$5_{1,4}$	$4_{4,1}$
$6_{5,1}$	$5_{2,4}$		44	149	77		$6_{1,5}$	$5_{4,2}$
$7_{6,1}$	$6_{3,4}$		34	182	209		$7_{1,6}$	$6_{4,3}$
$8_{7,1}$	$7_{4,4}$		26	170	370		$8_{1,7}$	$7_{4,4}$
$9_{8,1}$	$8_{5,4}$		20	146	493		$9_{1,8}$	$8_{4,5}$
$10_{9,1}$	$9_{6,4}$		17	122	549		$10_{1,9}$	$9_{4,6}$
$11_{10,1}$	$10_{7,4}$		14	104	550		$11_{1,10}$	$10_{4,7}$
$12_{11,1}$	$11_{8,4}$		12	90	519		$12_{1,11}$	$11_{4,8}$
$6_{4,2}$	$5_{1,5}$		71	43	3		$6_{2,4}$	$5_{5,1}$
$7_{5,2}$	$6_{2,5}$		97	175	23		$7_{2,5}$	$6_{5,2}$
$8_{6,2}$	$7_{3,5}$		80	305	87		$8_{2,6}$	$7_{5,3}$
$9_{7,2}$	$8_{4,5}$		61	347	234		$9_{2,7}$	$8_{5,4}$
$10_{8,2}$	$9_{5,5}$		47	324	466		$10_{2,8}$	$9_{5,5}$
$11_{9,2}$	$10_{6,5}$		38	279	715		$11_{2,9}$	$10_{5,6}$
$12_{10,2}$	$11_{7,5}$		32	237	896		$12_{2,10}$	$11_{5,7}$

STRENGTHS FOR ROTATIONAL TRANSITIONS

FORBIDDEN SUBBRANCHES—b SUBBRANCHES

Subbranch				κ			Subbranch	
$^bR_{-3,-3}$	$^bP_{-3,-3}$	∓1	∓0.5	0	±0.5	±1	$^bR_{3,-3}$	$^bP_{-3,-3}$
$J + K_{-1} + K_1$ odd	$J + K_{-1} + K_1$ even						$J + K_{-1} + K_1$ odd	$J + K_{-1} + K_1$ even
$6_{6,1}$	$7_{3,4}$		1	19	87		$6_{1,6}$	$7_{4,3}$
$7_{6,2}$	$8_{3,5}$		7	118	162		$7_{2,6}$	$8_{5,3}$
$8_{6,3}$	$9_{3,6}$		26	321	153		$8_{3,6}$	$9_{6,3}$
$9_{6,4}$	$10_{3,7}$		80	495	119		$9_{4,6}$	$10_{7,3}$
$10_{6,5}$	$11_{3,8}$		208	547	91		$10_{5,6}$	$11_{8,3}$
$11_{6,6}$	$12_{3,9}$		446	511	73		$11_{6,6}$	$12_{9,3}$
$7_{7,1}$	$8_{4,4}$		1	7	82		$7_{1,7}$	$8_{4,4}$
$8_{7,2}$	$9_{4,5}$		3	55	218		$8_{2,7}$	$9_{5,4}$
$9_{7,3}$	$10_{4,6}$		11	218	247		$9_{3,7}$	$10_{6,4}$
$10_{7,4}$	$11_{4,7}$		29	492	205		$10_{4,7}$	$11_{7,4}$
$11_{7,5}$	$12_{4,8}$		73	711	158		$11_{5,7}$	$12_{8,4}$
$8_{8,1}$	$9_{5,4}$		0	4	62		$8_{1,8}$	$9_{4,5}$
$9_{8,2}$	$10_{5,5}$		2	24	241		$9_{2,8}$	$10_{5,5}$
$10_{8,3}$	$11_{5,6}$		6	108	346		$10_{3,8}$	$11_{6,5}$
$11_{8,4}$	$12_{5,7}$		14	336	319		$11_{4,8}$	$12_{7,5}$
$9_{9,1}$	$10_{6,4}$		0	2	38		$9_{1,9}$	$10_{4,6}$
$10_{9,2}$	$11_{6,5}$		1	12	219		$10_{2,9}$	$11_{5,6}$
$11_{9,3}$	$12_{6,6}$		4	50	420		$11_{3,9}$	$12_{6,6}$
$10_{10,1}$	$11_{7,4}$		0	1	22		$10_{1,10}$	$11_{4,7}$
$11_{10,2}$	$12_{7,5}$		1	7	165		$11_{2,10}$	$12_{5,7}$
$11_{11,1}$	$12_{8,4}$		0	1	12		$11_{1,11}$	$12_{4,8}$

STRENGTHS FOR ROTATIONAL TRANSITIONS

FORBIDDEN SUBBRANCHES—b SUBBRANCHES

| Subbranch | | | κ | | | | Subbranch | |
${}^bR_{-3,5}$ $J+K_{-1}+K_1$ even	${}^bP_{3,-5}$ $J+K_{-1}+K_1$ odd	∓1	∓0.5	0	±0.5	±1	${}^bR_{5,-3}$ $J+K_{-1}+K_1$ even	${}^bP_{-5,3}$ $J+K_{-1}+K_1$ odd
$4_{4,0}$	$5_{1,5}$		2	2	0		$4_{0,4}$	$5_{5,1}$
$5_{4,1}$	$6_{1,6}$		6	3	0		$5_{1,4}$	$6_{6,1}$
$6_{4,2}$	$7_{1,7}$		11	2	0		$6_{2,4}$	$7_{7,1}$
$7_{4,3}$	$8_{1,8}$		13	1	0		$7_{3,4}$	$8_{8,1}$
$8_{4,4}$	$9_{1,9}$		12	0	0		$8_{4,4}$	$9_{9,1}$
$9_{4,5}$	$10_{1,10}$		9	0	0		$9_{5,4}$	$10_{10,1}$
$10_{4,6}$	$11_{1,11}$		5	0	0		$10_{6,4}$	$11_{11,1}$
$11_{4,7}$	$12_{1,12}$		3	0	0		$11_{7,4}$	$12_{12,1}$
$5_{5,0}$	$6_{2,5}$		2	4	2		$5_{0,5}$	$6_{5,2}$
$6_{5,1}$	$7_{2,6}$		8	9	1		$6_{1,5}$	$7_{6,2}$
$7_{5,2}$	$7_{2,7}$		18	10	0		$7_{2,5}$	$8_{7,2}$
$8_{5,3}$	$9_{2,8}$		30	6	0		$8_{3,5}$	$9_{8,2}$
$9_{5,4}$	$10_{2,9}$		40	3	0		$9_{4,5}$	$10_{9,2}$
$10_{5,5}$	$11_{2,10}$		43	1	0		$10_{5,5}$	$11_{10,2}$
$11_{5,6}$	$12_{2,11}$		37	1	0		$11_{6,5}$	$12_{11,2}$
$6_{6,0}$	$7_{3,5}$		1	5	4		$6_{0,6}$	$7_{5,3}$
$7_{6,1}$	$8_{3,6}$		5	15	4		$7_{1,6}$	$8_{6,3}$

STRENGTHS FOR ROTATIONAL TRANSITIONS

Forbidden Subbranches—*b* Subbranches

| Subbranch | | | | κ | | | Subbranch | |
$^bR_{-3,5}$ $J + K_{-1} + K_1$ even	$^vP_{3,-5}$ $J + K_{-1} + K_1$ odd	∓1	∓0.5	0	±0.5	±1	$^bR_{5,-3}$ $J + K_{-1} + K_1$ even	$^bP_{-6,3}$ $J + K_{-1} + K_1$ odd
$8_{6,2}$	$9_{3,7}$		15	24	1		$8_{2,6}$	$9_{7,3}$
$9_{6,3}$	$10_{3,8}$		31	23	0		$9_{3,6}$	$10_{8,3}$
$10_{6,4}$	$11_{3,9}$		52	14	0		$10_{4,6}$	$11_{9,3}$
$11_{6,5}$	$12_{3,10}$		73	6	0		$11_{5,6}$	$12_{10,3}$
$7_{7,0}$	$8_{4,5}$		1	4	6		$7_{0,7}$	$8_{5,4}$
$8_{7,1}$	$9_{4,6}$		3	16	9		$8_{1,7}$	$9_{6,4}$
$9_{7,2}$	$10_{4,7}$		9	33	4		$9_{2,7}$	$10_{7,4}$
$10_{7,3}$	$11_{4,8}$		22	44	1		$10_{3,7}$	$11_{8,4}$
$11_{7,4}$	$12_{4,9}$		42	39	0		$11_{4,7}$	$12_{9,4}$
$8_{8,0}$	$9_{5,5}$		0	3	7		$8_{0,8}$	$9_{6,5}$
$9_{8,1}$	$10_{5,6}$		2	13	15		$9_{1,8}$	$10_{6,5}$
$10_{8,2}$	$11_{5,7}$		6	33	11		$10_{2,8}$	$11_{7,5}$
$11_{8,3}$	$12_{5,8}$		14	57	4		$11_{3,8}$	$12_{8,5}$
$9_{9,0}$	$10_{6,5}$		0	2	8		$9_{0,9}$	$10_{5,6}$
$10_{9,1}$	$11_{6,6}$		1	9	21		$10_{1,9}$	$11_{6,6}$
$11_{9,2}$	$12_{6,7}$		4	28	22		$11_{2,9}$	$12_{7,6}$
$10_{10,0}$	$11_{7,5}$		0	1	7		$10_{0,10}$	$11_{5,7}$
$11_{10,1}$	$12_{7,6}$		1	7	24		$11_{1,10}$	$12_{6,7}$
$11_{11,0}$	$12_{8,5}$		0	1	6		$11_{0,11}$	$12_{6,8}$

Molecular Constants Involved in Microwave Spectra

In accordance with accepted practice, molecules are listed alphabetically according to their empirical formulas by the following procedures:

1. Symbols for the elements in the empirical formula for a molecule are in alphabetical order except that within a molecular formula:

   a. C for carbon precedes all other symbols. This groups all organic compounds together.

   b. In organic compounds, H precedes all other symbols except C.

   c. D (for deuterium) is regarded for purposes of listing as H^2.

2. All molecules with formulas of the form X_nY precede those of the form $X_{n+1}YZ$, etc.

To aid in identification, in some cases the usual chemical formula or the name of a compound is given in parentheses. The more abundant isotopic species of molecules are usually listed before those which are less abundant.

The table lists molecular constants which determine or are determined by microwave spectra. These include rotational constants A, B, and C (*cf.* pages 5 and 83), centrifugal stretching constants D_e (*cf.* pages 9 and 25) or D_J and D_{JK} (page 78), rotation-vibration constants α (pages 9 and 25), and l-type doubling constants (pages 33 and 79). Vibrational frequencies ω_1, ω_2, etc., usually obtained from infrared spectra, are listed in the better known or more important cases since they affect the relative intensities of excited state lines. Quadrupole coupling constants eqQ (page 150) or $eQ\frac{\partial^2 V}{\partial a^2}$, $eQ\frac{\partial^2 V}{\partial b^2}$, and $eQ\frac{\partial^2 V}{\partial c^2}$ (pages 159 to 162), are listed for each nucleus in the molecule. The molecular dipole moment μ, or its components μ_a, μ_b, and μ_c along principal axes, is also given. In the column labeled "Remarks," a number of less frequently measured molecular constants may be found for cases where they are known. These include magnetic hyperfine constants (pages 216 and 220), molecular g factors (pages 292 to 296), line-width parameters $\Delta\nu$ (page 343), and others.

References are usually given adjacent to each molecular constant. In the case of dipole moments which are listed by Wesson [336a] and which have not been measured by microwave techniques, no reference is given. Similarly references are omitted for vibrational frequencies which are given in one of Herzberg's volumes on molecular spectra ([130] or [471]) and which have not been measured by microwave techniques.

Most of this table has been compiled by Mr. G. C. Dousmanis.

MOLECULAR CONSTANTS INVOLVED IN MICROWAVE SPECTRA

Chemical symbol	Rotational constants A, B, and C, Mc	Vibrational frequencies ω in wave numbers (cm⁻¹); d indicates a degenerate vibration	Rotation-vibration constants, Mc (α, D, or q_l)	Quadrupole coupling constant eqQ, Mc	Dipole moment μ in 10⁻¹⁸ esu (Debye)	Reference for structure	Remarks
AsCl₃³⁵	$B_0 = 2147.2$ [481]	$\omega_1 = 410$, $\omega_2 = 193$, $\omega_3 = 370d$, $\omega_4 = 159d$	$\alpha_2 = 4.2$ [481]	As⁷⁵ = −173 [597]	1.97	[481]	5 lines measured [481]
AsCl₂³⁵Cl³⁷							2 lines measured [481]
AsCl³⁵Cl₂³⁷	$B_0 = 2044.7$ [481]						
AsF₃	$B_0 = 5878.971$ [266], [824]	$\omega_1 = 707$, $\omega_2 = 341$, $\omega_3 = 644d$, $\omega_4 = 274d$	$\alpha_2 = -5$ [824], $\alpha_4 = -0.16$ [824], $D_{JK} = 0.009$ [824]	As⁷⁵ = −236.23 [266], [824]	2.82 [528]	[266], [824]	Magnetic h.f.s. interaction $= \mathbf{I}\cdot\mathbf{J}\left[-0.012 - \dfrac{0.012K^2}{J(J+1)}\right]$ Mc [824]
AsH₂D				As⁷⁵ = −164 [606]	0.22 [606]	[735],[736], [606]	
AsD₃	$B_0 = 57{,}476.15$ [943]			As⁷⁵ = −165.6 [943]			
B₂¹¹B⁷⁹H₅ (B₂H₄Br₁) bromodiborane	$B_0 = 3369.65$ [447], $C_0 = 3141.48$ [447]			Br⁷⁹ = 293 [447]		[447]	
B¹¹B⁹Br⁷⁹H₅	$B_0 = 3398.62$ [447], $C_0 = 3176.05$ [447]			Br⁷⁹ = 293 [447]			
(B¹⁰ nearest Br⁷⁹) B¹⁰B¹¹Br⁷⁹H₅	$B_0 = 3523.72$ [447], $C_0 = 3278.42$ [447]			Br⁷⁹ = 293 [447]			
(B¹⁰ nearest Br⁷⁹) B₂¹¹B⁸¹H₅	$B_0 = 3350.75$ [447], $C_0 = 3124.95$ [447]			Br⁸¹ = 244 [447]			
B¹¹B⁹B⁸¹H₅ (B¹⁰ nearest Br⁸¹)	$B_0 = 3379.95$ [447], $C_0 = 3159.85$ [447]			Br⁸¹ = 244 [447]			
B¹⁰B¹¹B⁸¹H₅ (B¹¹ nearest Br⁸¹)	$B_0 + C_0 = 6766.4$ [447]			Br⁸¹ = 244 [447]			
B₄H₁₁ (pentaborane)	$B_0 = 7002.9$ [697],[939], $C_0 = 48.9 \times 10^3$ [939]				2.13 [939]	[697],[939]	Lines for other asymmetric species measured
B¹⁰B₄¹¹H₁₁ (B¹⁰ at apex)	$B_0 = 7089.8$ [697]						
B₄¹¹D₉	$B_0 = 5211.35$ [939], $C_0 = 37 \times 10^3$ [939]				2.16 [939]		

							Rotation-vibration constant
$Br^{79}Cl^{35}$	$B_e = 4570.92$ [535]	~430	$\alpha = 23.22$ [535]	$Br^{79} = 876.8$ [535]; $Cl^{35} = -103.6$ [535]	0.57 [535]	[535]	$\gamma_e = 0.0031$ Mc [938]
$Br^{81}Cl^{35}$	$B_e = 4536.14$ [535]	~430	$\alpha = 22.95$ [535]	$Br^{81} = 732.9$ [535]; $Cl^{35} = -103.6$ [535]			
$Br^{79}Cl^{37}$	$B_e = 4499.84$ [535]	~420	$\alpha = 21.94$ [535]	$Br^{79} = 876.8$ [535]; $Cl^{37} = -81.1$ [535]			
$Br^{81}Cl^{37}$	$B_e = 4365.01$ [535]	~420	$\alpha = 21.67$ [535]	$Br^{81} = 732.9$ [535]; $Cl^{37} = -81.1$ [535]			
$Br^{79}Cs$ (CsBr)	$B_e = 1081.34$ [938]	171 [772]	$\alpha = 3.718$ [938]; $D = 0.00027$ [938]	....	....	[938]	
$Br^{81}Cs$	$B_0 = 1064.59$ [938]		$\alpha = 3.631$ [938]				
$Br^{79}F$	$B_e = 10{,}706.59$ [534]	671	$\alpha = 156.3$ [534]	$Br^{79} = 1089.0$ [534]	1.29 [534]	[534]	
$Br^{81}F$	$B_e = 10{,}655.7$ [534]		$\alpha = 155.8$ [534]	$Br^{81} = 900.2$ [534]			
$Br^{79}F_3Si^{28}$ (SiF$_3$Br)	$B_0 = 1549.9$ [641]		$D_{JK} = 0.0008$ [641]	$Br^{79} = 440$ [641]		[525],[641]	
$Br^{81}F_3Si^{28}$	$B_0 = 1534.1$ [641]		$D_{JK} = 0.008$ [641]	$Br^{81} = 370$ [641]			
$Br^{79}Ge^{70}H_3$ (GeH$_3$Br)	$B_0 = 2438.57$ [521]			$Br^{81} = 380$ [521]		[521],[727]	
$Br^{79}Ge^{70}H_3$	$B_0 = 2410.17$ [521]			$Br^{81} = 321$ [521]			
$Br^{81}Ge^{70}H_3$	$B_0 = 2406.42$ [521]			$Br^{81} = 380$ [521]			
$Br^{79}Ge^{72}H_3$	$B_0 = 2378.01$ [521]			$Br^{81} = 321$ [521]			
$Br^{79}Ge^{74}H_3$	$B_0 = 2375.88$ [521]			$Br^{79} = 380$ [521]			
$Br^{81}Ge^{74}H_3$	$B_0 = 2347.46$ [521]			$Br^{81} = 321$ [521]			
$Br^{79}Ge^{76}H_3$	$B_0 = 2346.84$ [521]			$Br^{79} = 380$ [521]			
$Br^{81}Ge^{76}H_3$	$B_0 = 2318.37$ [521]			$Br^{81} = 321$ [521]			
$Br^{79}H_3Si^{28}$ (SiH$_3$Br)	$B_0 = 4321.77$ [409]			$Br^{79} = 336$ [409]	1.31 [521]	[409],[521],[727]	
$Br^{81}H_3Si^{28}$	$B_0 = 4292.62$ [409]			$Br^{81} = 278$ [409]			
$Br^{79}H_3Si^{29}$	$B_0 = 4232.96$ [409]			$Br^{79} = 336$ [409]			
$Br^{81}H_3Si^{29}$	$B_0 = 4203.70$ [409]			$Br^{81} = 278$ [409]			
$Br^{79}H_3Si^{30}$	$B_0 = 4149.39$ [409]			$Br^{79} = 336$ [409]			
$Br^{81}H_3Si^{30}$	$B_0 = 4120.09$ [409]			$Br^{81} = 278$ [409]			
$Br^{79}K^{39}$ (KBr)	$Y_{01}(\approx B_e) = 2434.947$ [799]	231	$a_e = 12.136$ [799]	$Br^{79} = 10.244$ [799]; $K^{39} = -5.003$ [799]	10.41 [799]	[799]	$\gamma_e = 0.023$ Mc [799]; variation of eqQ with vibrational state [799]
$Br^{79}K^{39}$	$Y_{01}(\approx B_e) = 2415.075$ [799]		$a_e = 11.987$ [799]	$Br^{81} = 8.555$ [799]; $K^{39} = -5.002$ [799]			
$Br^{79}Li^{7}$ (LiBr)	$Y_{01}(\approx B_e) = 16{,}650.57$ [938]	~480 [938]	$a_e = 169.09$ [938]	$Br^{79} = 37.2$ [938]	6.2 [938]	[938]	$\gamma_e = 0.022$ Mc [938]
$Br^{81}Li^{34}$	$Y_{01}(\approx B_e) = 19{,}161.51$ [938]						

MOLECULAR CONSTANTS INVOLVED IN MICROWAVE SPECTRA

Chemical symbol	Rotational constants A, B, and C, Mc	Vibrational frequencies ω in wave numbers (cm⁻¹); d indicates a degenerate vibration	Rotation-vibration constants, Mc (α, D, or q_l)	Quadrupole coupling constant eqQ, Mc	Dipole moment μ in 10^{-18} esu (Debye)	Reference for structure	Remarks
$Br^{81}Li^7$	$Y_{01}(\approx B_e) = 16{,}650.00$ [938]	....	$\alpha_e = 168.58$ [938]	$Br^{81} = 30.7$ [938]		....	$\gamma_e = 0.65$ Mc [938]
$Br^{79}Na$ (NaBr)	$Y_{01}(\approx B_e) = 4534.51$ [938]	315	$\alpha_e = 28.25$ [938] $D = 0.007$ [938]	$Br^{79} = 58$ [938]		[938]	$\gamma_e = 0.08$ Mc [938]
$Br^{81}Na$	$Y_{01}(\approx B_e) = 4509.34$ [938]	....	$\alpha_e = 28.06$ [938]			[938]	
$Br^{79}Rb^{85}$ (RbBr)	$B_e = 1424.83$ [938]	181 [938]	$\alpha_e = 5.578$ [938] $D = 0.0004$ [938]			[938]	$\gamma_e = 0.008$
$Br^{79}Rb^{87}$	$B_e = 1409.06$ [938]	....	$\alpha_e = 5.474$ [938]				
$Br^{81}Rb^{85}$	$B_e = 1406.59$ [938]	....	$\alpha_e = 5.461$ [938]				
$Br^{79}P$	$B_0 = 996.4$ [551]	....					
$Br^{81}P$	$B_0 = 974.4$ [551]	....					
$CBr^{79}F_3$ (CF₃Br)	$B_0 = 2098.06$ [520], [743]	....	$D_{JK} = 0.0013$ [743]	$Br^{79} = 619$ [520],[743]		[520],[743]	B_0 may be incorrect [947]
$CBr^{81}F_3$	$B_0 = 2078.50$ [520], [743]	....	$D_{JK} = 0.0012$ [743]	$Br^{81} = 517$ [520],[743]			B_0 may be incorrect [947]
$C^{12}Br^{79}N^{14}$ (BrCN)	$B_0 = 4120.198$ [329], [320],[754]	$\omega_1 = 580$ $\omega_2 = 368d$ $\omega_3 = 2187$	$\alpha_1 = 15.54$ [329],[754] $\alpha_2 = -11.564$ [329],[754] $D_J = 0.0009$ [531] $q_l = 3.918$ [329],[754]	$Br^{79} = 686.1$ [329],[320], [754] $N^{14} = -3.83$ [329]	2.94	[329],[320]	eqQ for Br^{79} in vibrational state (010) = 682.8 [320],[744]; αz from vibrational state (0240) = -11.528 [744]; Fermi resonance energy $W_{12} = 61.5$ cm⁻¹ [744]; $\Delta\nu = 27.1$ Mc/mm Hg [922]; $\Delta\nu \propto T^{-1.6}$ [992]
$C^{13}Br^{79}N^{14}$	$B_0 = 4073.373$ [320], [754]						
$C^{12}Br^{81}N^{14}$	$B_0 = 4096.788$ [329], [320],[754]		$\alpha_1 = 15.48$ [329],[754] $D_J = 0.0008$ [531] $q_l = 3.874$ [329],[754]	$Br^{81} = 572.27$ [329], [320],[754]		....	eqQ for Br^{81} in vibrational state (010) = 570.4 [320],[754]; αz from vibrational state (020) = -11.462 [754]; Fermi resonance energy $W_{12} = 60.5$ cm⁻¹ [754]
$C^{13}Br^{81}N^{14}$	$B_0 = 4049.608$ [320], [754]						

	Rotational constants	ω		Quadrupole coupling			
C12Br79N15	$B_0 = 3944.846$ [754]						
C12Br81N15	$B_0 = 3921.787$ [754]						
CCl35F3 (CF3Cl)	$B_0 = 3335.56$ [358]						
CCl37F3	$B_0 = 3251.51$ [358]						
C12Cl35N14 (ClCN)	$B_0 = 5970.821$ [329],[320]	$\omega_1 = 729$, $\omega_2 = 397d$, $\omega_3 = 2201$	$\alpha_z = -16.39$ [329], $q_1 = 7.500$ [329]	Cl35 = −78.05 [358], Cl37 = −61.44 [358], Cl35 = −83.33 [329],[320],[575], N14 = −3.63 [329]	2.80 [528]	[358]	$\Delta\nu = 50$ Mc/mm Hg
C12Cl35N14	$B_0 = 5939.795$ [329],[320]					[329],[320]	
C12Cl36N14	$B_0 = 5907.31$ [421]			Cl35 = 18.1[421]			
C12Cl37N14	$B_0 = 5847.252$ [329],[320]			Cl37 = −65.3 [329],[320]			
C12Cl37N14	$B_0 = 5814.710$ [329],[320]						
CCl2^34O (COCl2, phosgene)	$A_0 = 7918.75$ [859b], $B_0 = 3474.99$ [859b], $C_0 = 2412.25$ [859b]			$e\,\frac{\partial^2 V}{\partial a^2}\,Q_{Cl35} = -37.20$ [859b], $e\,\frac{\partial^2 V}{\partial c^2}\,Q_{Cl35} = 27.07$ [859b], $e\,\frac{\partial^2 V}{\partial c^2}\,Q_{Cl37} = 24.20$ [859b]		[859b]	
CCl35Cl37O	$A_0 = 7867.76$ [859b], $B_0 = 3379.94$ [859b], $C_0 = 2361.48$ [859b]						
CF3I	$B_0 = 1523.23$ [524],[897]		$D_{JK} = 0.0006$ [524]	$I127 = -2143.8$ [524],[897]	1.0 [897]	[524]	
CHBr79	$B_0 = 1247.61$ [551],[764]					[764]	
CDBr79	$B_0 = 1239.45$ [764]						
CHBr81	$B_0 = 1217.30$ [551],[764]						
CDBr81	$B_0 = 1209.51$ [764]						18 lines measured [714]
CHClF					1.29	[545],[690]	
CHCl35	$B_0 = 3301.94$ (414),[545],[690]	$\omega_1 = 3030$, $\omega_2 = 672$, $\omega_3 = 363$, $\omega_4 = 1217d$, $\omega_5 = 760d$, $\omega_6 = 261d$			1.2		

MOLECULAR CONSTANTS INVOLVED IN MICROWAVE SPECTRA

Chemical symbol	Rotational constants A, B, and C, Mc	Vibrational frequencies ω in wave numbers (cm^{-1}); d indicates a degenerate vibration	Rotation-vibration constants, Mc (α, D, or q_l)	Quadrupole coupling constant eqQ, Mc	Dipole moment μ in 10^{-18} esu (Debye)	Reference for structure	Remarks
CDCl$_3{}^{35}$	$B_0 = 3250.17$ [414], [545], [690]						
CHCl^{37}	$B_0 = 3129.51$ [690]						
C^{12}HF$_3$	$B_0 = 10,348.74$ [367], [690]				1.64 [690] [644]	[367], [690]	$\Delta\nu = 18$ Mc/mm Hg [367]
C^{13}HF$_3$	$B_0 = 10,422.00$ [690]						
C^{12}DF$_3$	$B_0 = 9921.35$ [690]						
C^{12}HN (HCN)	$B_0 = 44,315.97$ [532], [733]	$\omega_1 = 2041.2$ $\omega_2 = 711.7d$ $\omega_3 = 3368.6$	$D_J = 0.1$ [532] $q_l = 224.471 - 0.0026614J(J+1)$ [529], [761], [911], [997]	N^{14} = -4.58 [532]	3.00 [803]		Dipole moment in vibrational state (01^{1}0) = 2.96 [529]
C^{13}HN	$B_0 = 43,170.1$ [532], [733]						
C^{12}DN	$B_0 = 36,207.5$ [532], [733]		$q_l = 188.37 - 0.0022J(J+1)$ [761]	$D = 0.15$ [998]		[532], [733]	$\Delta\nu = 25$ Mc/mm Hg [413]
C^{13}DN	$B_0 = 35,587.57$ [532]						
CHN^{14}O (HCNO)	$A_0 = 9194 \times 10^4$ [479] $B_0 = 10992$ [479] $C_0 = 10991$ [479]			N^{14} = 2.00 [646]	1.59 [644]	[479]	
CDN^{14}O	$B_0 = 10199$ [479] $C_0 = 10197$ [479]						
CHN^{15}O	$B_0 = 10663$ [479] $C_0 = 10662$ [479]						
C^{12}HNS^{32} (HNCS)	$B_0 = 5903.0$ [434] $C_0 = 5828.0$ [434]			N^{14} = 1.20 [646]	1.72 [434]	[434], [794]	B_0 and C_0 are for rotational state with $K = 1$; for $K = 0$, $\frac{1}{2}(B_0 + C_0) = 5884.5$ [434] $\frac{1}{2}(B_0 + C_0)$ is for $K = 1$
C^{13}HNS^{32}	$\frac{1}{2}[B_0 + C_0] = 5847.3$ [434]						
C^{12}DNS^{32}	$B_0 = 5529.5$ [794] $C_0 = 5418.7$ [794]					[794]	B_0 and C_0 are for $K = 1$; for $K = 0$, $\frac{1}{2}(B_0 + C_0) = 5472.9$ [794] $\frac{1}{2}(B_0 + C_0)$ is for $K = 1$
C^{13}DNS^{32}	$\frac{1}{2}[B_0 + C_0] = 5459.8$ [434]						

$C^{15}HNS^{33}$	$\frac{1}{2}(B_0+C_0) = 5793.5$ [794]		$S^{33} = -27.5$ [794]			$\frac{1}{2}(B_0+C_0)$ is for $K=0$
$C^{15}HNS^{34}$	$B_0 = 5763.6$ [794] $C_0 = 5691.7$ [794]					B_0 and C_0 are for $K=1$; for $K=0$, $\frac{1}{2}(B_0+C_0) = 5726.7$ [794]
CH_3Br_7						16 lines measured [334],[403]
$CH_2Cl_2^{35}$	$A_0 = 32,001.8$ [732] $B_0 = 3320.4$ [732] $C_0 = 3065.2$ [732]		$e\dfrac{\partial^2 V}{\partial a^2} Q_{Cl^{35}} = -41.8$ [732] $e\dfrac{\partial^2 V}{\partial b^2} Q_{Cl^{35}} = 2.6$ [732] $e\dfrac{\partial^2 V}{\partial c^2} Q_{Cl^{35}} = 39.2$ [732]	1.5 1.62 [732]	[732]	
$CDHCl_2^{35}$	$A_0 = 27,198$ [732] $B_0 = 3305$ [732] $C_0 = 3027$ [732]					
$CD_2Cl_2^{35}$	$A_0 = 23,676$ [732] $B_0 = 3284$ [732] $C_0 = 2993$ [732]					
$CH_2Cl^{35}Cl^{37}$	$A_0 = 31,878.2$ [732] $B_0 = 3231.5$ [732] $C_0 = 2988.2$ [732]					
$CDHCl^{35}Cl^{37}$	$A_0 = 27,090$ [732] $B_0 = 3217$ [732] $C_0 = 2951$ [732]					
$CD_2Cl_2^{37}$	$A_0 = 23,582$ [732] $B_0 = 3197$ [732] $C_0 = 2920$ [732]					
$CH_2Cl_2^{37}$	$A_0 = 31,754$ [732] $B_0 = 3143$ [732] $C_0 = 2912$ [732]					
CH_2F_2	$A_0 = 49,138.4$ [722] $B_0 = 10,603.89$ [722] $C_0 = 9249.20$ [722]			1.96 [712]	[722]	
CH_2O	$A_0 = 282,106$ [353],[601] $B_0 = 38,834$ [353],[601] $C_0 = 34,004$ [353],[601] $\omega_1 = 2780$ $\omega_2 = 1743.6$ $\omega_3 = 1503$ $\omega_4 = 2874$ $\omega_5 = 1280$ $\omega_6 = 1167$			2.34 [601] [644]	[601]	$\Delta\nu = 10$ Mc/mm Hg [601] centrifugal distortion effects [601]

MOLECULAR CONSTANTS INVOLVED IN MICROWAVE SPECTRA

Chemical symbol	Rotational constants A, B, and C, Mc	Vibrational frequencies ω in wave numbers (cm⁻¹); d indicates a degenerate vibration	Rotation-vibration constants, Mc (α, D, or q_i)	Quadrupole coupling constant eqQ, Mc	Dipole moment μ in 10^{-18} esu (Debye)	Reference for structure	Remarks
CH_2O_2 (HCOOH)	$A_0 \approx 80 \times 10^3$ [634], $B_0 = 12{,}055.9$ [634],[796b],[991], $C_0 = 10{,}415.3$ [634],[796b],[991]	....	....	....	1.7	[991]	The rotational constants given in [796b] are incorrect. (Private communication from G. Erlandsson.)
$CHDO_2$ (HCOOD)	$B_0 = 11{,}762.4$ [991], $C_0 = 9970.3$ [991]						
$CH_2B^{11}O$ (BH₂CO)	$B_0 = 8657.2$ [461]		$D_J = 0.177$ [417], $D_{JK} = 0.36$ [461]	$B^{11} = 1.55$ [461]	1.80 [417],[546], [461]	[461]	
$CD_2B^{11}O$	$B_0 = 7336.56$ [461]		$D_{JK} = 0.24$ [461]				
$CH_2B^{10}O$	$B_0 = 8980.1$ [417], [546],[461]		$\alpha_a = -22.6$ [417],[546], $\alpha_b = -5.7$ [417],[546], $D_{JK} = 0.39$ [461]	$B^{10} = 3.4$ [461],[546]			
$CD_2B^{10}O$	$B_0 = 7530.34$ [461]		$D_{JK} = 0.29$ [461]				
$C^{12}H_3Br^{79}$	$B_0 = 9568.19$ [280], [411],[531]	$\omega_1 = 2972$ $\omega_2 = 1305.1$ $\omega_3 = 611$ $\omega_4 = 3055.9d$ $\omega_5 = 1445.3d$ $\omega_6 = 952.0d$	$\alpha_3 = 72.77$ [950a], $D_J = 0.010$ [531],[989], $D_{JK} = 0.128$ [531],[989]	$Br^{79} = 577.15$ [280], [411],[950c]	1.797 [528]	[280],[533], [318],[729]	
$C^{12}HD_2Br^{79}$	$B_0 - C_0 = 158.85$ [729]						
$C^{12}D_3Br^{79}$	$B_0 = 7714.57$ [746]	$\omega_1 = 2151$ $\omega_2 = 987$ $\omega_3 = 577$ $\omega_4 = 2293d$ $\omega_5 = 1053d$ $\omega_6 = 717d$	$D_{JK} = 0.039$ [746]	$Br^{79} = 574.6$ [746]			
$C^{12}H_3Br^{81}$	$B_0 = 9531.84$ [280], [411],[531]		$\alpha_3 = 77.32$ [950a]	$Br^{81} = 482.16$ [280], [411],[950a]			
$C^{13}H_3Br^{81}$	$B_0 = 9082.86$ [533]						
$C^{12}HD_2Br^{81}$	$B_0 - C_0 = 157.22$ [729]						

Molecule	Rotational constants (Mc)	Vibrational frequencies	Distortion / α constants	Coupling constants (Mc)		References	Notes
C¹²D₃Br⁸¹	$B_0 = 7681.23$ [746]	....	$D_{JK} = 0.039$ [746]	Br⁸¹ = 479.8 [746]	....	[928]	$\Delta\nu = 21$ Mc/mm Hg [902]
CH₃Br⁷⁹Hg¹⁹⁸ (CH₃HgBr)	$B_0 = 1142.86$ [928]	....	$D_{JK} = 0.008$ [928]	Br⁷⁹ = 350 [928]			
CH₃Br⁷⁹Hg¹⁹⁹	$B_0 = 1142.10$ [928]	....	$D_{JK} = 0.008$ [928]				
CH₃Br⁷⁹Hg²⁰⁰	$B_0 = 1141.36$ [928]	....	$D_{JK} = 0.008$ [928]				
CH₃Br⁷⁹Hg²⁰¹	$B_0 = 1139.88$ [928]	....	$D_{JK} = 0.008$ [928]	Br⁸¹ = 290 [928]			
CH₃Br⁸Hg¹⁹⁸	$B_0 = 1125.28$ [928]	....	$D_{JK} = 0.008$ [928]				
CH₃Br⁸Hg¹⁹⁹	$B_0 = 1124.51$ [928]	....	$D_{JK} = 0.008$ [928]				
CH₃Br⁸Hg²⁰⁰	$B_0 = 1123.76$ [928]	....	$D_{JK} = 0.008$ [928]				
CH₃Br⁸Hg²⁰¹	$B_0 = 1122.27$ [928]	....	$D_{JK} = 0.008$ [928]				
C¹²H₃Cl³⁵	$A_0 \approx 150 \times 10^3$ [318], [412] $B_0 = 13{,}292.840$ [280], [531],[989],[999]	$\omega_1 = 2966.2$ $\omega_2 = 1354.9$ $\omega_3 = 732.1$ $\omega_4 = 3041.8d$ $\omega_5 = 1454.6d$ $\omega_6 = 1015.0d$	$\alpha_B = 115.21$ [950a] $\alpha_B = 49.01$ [950c] $D_J = 0.0180$ [531],[913] $D_{JK} = 0.198$ [531],[989]	Cl³⁵ = −74.740 [280], [378],[575],[746],[999]	1.869 [528]	[280],[318], [362],[412], [496],[531], [729]	
C¹²H₃Cl³⁵	$B_0 = 12{,}796.2$ [362]						
C¹²HD₂Cl³⁵	$B_0 = 11{,}681.5$ [496], [729] $C_0 = 11{,}372.6$ [496], [729]						
C¹²D₃Cl³⁵	$B_0 = 10{,}841.88$ [412], [746]	$\omega_1 = 2161$ $\omega_2 = 1029$ $\omega_3 = 695$ $\omega_4 = 2286d$ $\omega_5 = 1058d$ $\omega_6 = 775d$		Cl³⁵ = −74.41 [746]			
C¹²H₃Cl³⁶		....		Cl³⁶ = −15.8 [691],[594]			D_J probably more nearly 0.018 Mc than 0.027 [783b]
C¹²H₃Cl³⁷	$B_0 = 13{,}088.137$ [280], [531],[999]		$\alpha_B = 112.30$ [950a] $\alpha_B = 48.19$ [950c] $D_J = 0.027$ [531] $D_{JK} = 0.184$ [531]	Cl³⁷ = −58.921 [280], [378]Υ[746],[950c],[999]			
C¹²H₃Cl³⁷	$B_0 = 12{,}590.0$ [362]						
C¹²HD₂Cl³⁷	$B_0 + C_0 = 24{,}674$ [496]						
C¹²D₃Cl³⁷	$B_0 = 10{,}655.43$ [412], [746]	....		Cl³⁷ = −58.58 [746]			
CH₃Cl³⁵Hg¹⁹⁸	$B_0 = 2077.44$ [928]	....	$D_{JK} = 0.022$ [928]	Cl³⁵ = −42 [928]		[928]	
CH₃Cl³⁵Hg¹⁹⁹	$B_0 = 2077.13$ [928]						
CH₃Cl³⁵Hg²⁰⁰	$B_0 = 2076.82$ [928]						

MOLECULAR CONSTANTS INVOLVED IN MICROWAVE SPECTRA

Chemical symbol	Rotational constants A, B, and C, Mc	Vibrational frequencies ω in wave numbers (cm⁻¹); d indicates a degenerate vibration	Rotation-vibration constants, Mc (α, D, or q_i)	Quadrupole coupling constant eqQ, Mc	Dipole moment μ in 10^{-18} esu (Debye)	Reference for structure	Remarks
$CH_3Cl^{35}Hg^{202}$	$B_0 = 2076.20$ [928]						
$CH_3Cl^{35}Hg^{204}$	$B_0 = 2075.59$ [928]						
$CH_3Cl^{37}Hg^{198}$	$B_0 = 2006.14$ [928]			$Cl^{35} = -33$ [928]			$\Delta\nu = 20$ Mc/mm Hg [367]
$CH_3Cl^{37}Hg^{199}$	$B_0 = 2005.79$ [928]		$D_{JK} = 0.020$ [928]				
$CH_3Cl^{37}Hg^{200}$	$B_0 = 2005.45$ [928]						
$CH_3Cl^{37}Hg^{202}$	$B_0 = 2004.76$ [928]						
$CH_3Cl^{37}Hg^{204}$	$B_0 = 2004.09$ [928]						
$CH_3Cl^{35}Si$ (CH_3SiCl_3)	$B_0 = 1769.84$ [847]						
$CH_3Cl^{35}Si$	$B_0 = 1699.79$ [847]					[847]	
$C^{12}H_3F$	$A_0 \approx 154 \times 10^3$ [318]; $B_0 = 25{,}536.12$ [367, 457],[595],[946]	$\omega_1 = 2964.5$ $\omega_2 = 1475.3$ $\omega_3 = 1048.2$ $\omega_4 = 2982.2d$ $\omega_5 = 1471.1d$ $\omega_6 = 1195.5d$	$D_J = 0.059$ [595],[946],[989] $D_{JK} = 0.445$ [457],[595],[946],[989]		1.79 [367],[803]	[280],[318],[367],[457]	
$C^{12}H_2DF$	$B_0 = 24{,}043$ [974] $C_0 = 22{,}959$ [974]		$D_J = 0.033$ [595] $D_{JK} = 0.228$ [457][595]				
$C^{12}HD_2F$	$\tfrac{1}{2}[B_0 + C_0] = 21{,}844.96$ [974]						
$C^{12}H_3F$	$B_0 = 24{,}862.37$ [367]						
$C^{12}D_3F$	$B_0 = 20{,}449.83$ [746]						
CH_3FSi^{28} (CH_3SiF_3)	$B_0 = 3715.63$ [525], [641]	ω (torsional) $= 140$ [618],[641]				[525],[641]	Potential barrier height $= 410$ cm⁻¹ [618],[641]; rotation-vibration structure for torsional vibrations [641]
CH_3HgI	$B_0 = 788$ [462]						
$C^{12}H_3I^{127}$	$A_0 \approx 150 \times 10^3$ [280]; $B_0 = 7501.31$ [280], [531],[989]	$\omega_1 = 2969.8$ $\omega_2 = 1251.5$ $\omega_3 = 532.8$ $\omega_4 = 3060.3d$ $\omega_5 = 1440.3d$ $\omega_6 = 880.1d$	$D_J = 0.0063$ [531],[989] $D_{JK} = 0.099$ [531],[989]	$I^{127} = -1934$ [280]	1.65 [528]	[280],[318], [412],[729]	

Molecule	Rotational constants	D_{JK}	Coupling constants	Dipole moment μ	References	Remarks
$C^{13}H_3I^{127}$	$B_0 = 7119.04$ [280]		$I^{127} = -1929$ [746]			Height of hindering potential barrier = 4.20 cm⁻¹ [986]
$C^{12}HD_2I^{127}$	$B_0 - C_0 = 97.67$ [729]					
$C^{12}D_3I^{127}$	$B_0 = 6040.28$ [412], [746]	$D_{JK} = 0.047$ [746]				
$C^{12}H_3I^{129}$			$I^{129} = -1422$ [389]			
$C^{12}H_3I^{131}$			$I^{131} = -973$ [836]			
CH_3NO_2	$B_0 = 10{,}542.5$ [986]; $C_0 = 5876.7$ [986]	$D_{JK} = 0.16$ [986]		3.1	[561],[817]	Moments and products of inertia [561],[817]; potential barrier height = 374.8 cm⁻¹ [561],[817]; frequencies of lines [590]
$C^{12}H_4O^{16}$ (CH_3OH)	May be computed from structure [561],[817]			μ parallel to C—O bond = 0.885 [561],[817]; μ perpendicular to C—O bond = 1.44 [561]		
$C^{12}H_4O^{16}$				μ parallel to C—O bond = 0.886 [817]; μ perpendicular to C—O bond = 1.44 [817]		Frequencies of lines [590]
$C^{12}H_4O^{18}$				μ parallel to C—O bond = 0.890 [817]; μ perpendicular to C—O bond = 1.44 [817]		Frequencies of lines [590]
$C^{12}H_3S^{32}$ (CH_3SH)	$\frac{1}{2}(B_0+C_0) = 12{,}645.6$ [640],[870]			1.26 [640]	[640],[870]	Rotation-vibration interactions for ground and two excited states of torsional motion [870]; potential barrier height = 280 cm⁻¹ [978a],[949a]
$C^{12}H_3S^{33}$	$\frac{1}{2}(B_0+C_0) = 12{,}193.8$ [870]					
$C^{12}H_3SD$	$\frac{1}{2}(B_0+C_0) = 12{,}193.1$ [870]					
$C^{12}D_3SH$	$\frac{1}{2}(B_0+C_0) = 10{,}366.2$ [870]					
$C^{12}H_3S^{34}$	$\frac{1}{2}(B_0+C_0) = 12{,}439.5$ [640],[870]		$e\dfrac{\partial^2 V}{\partial a^2}Q_{S^{33}} = -27.57$ [903]			
CH_5N (CH_3NH_2)	$A_0 = 105{,}976$ [721]; $B_0 = 22{,}604$ [721]; $C_0 = 21{,}723$ [721]			$\mu_a = 0.30$ [721]; $\mu_c = 1.23$ [721]		Rotation-inversion spectrum [721],[865],[954],[976]; potential barrier = 660 cm⁻¹ [954],[976]

MOLECULAR CONSTANTS INVOLVED IN MICROWAVE SPECTRA

Chemical symbol	Rotational constants A, B, and C, Mc	Vibrational frequencies ω in wave numbers (cm^{-1}); d indicates a degenerate vibration	Rotation-vibration constants, Mc (α, D, or q_i)	Quadrupole coupling constant eqQ, Mc	Dipole moment μ in 10^{-18} esu (Debye)	Reference for structure	Remarks
CH$_3$Si^{28} (CH$_3$SiH$_3$)...	$B_0 = 10{,}969.0$ [604]	....	....	....	0.73 [604]		Potential barrier height = 558 cm^{-1} [604],[948a]. Analysis of rotation interaction with hindered motion [948a].
CH$_3$Si^{28}D$_3$...........	$B_0 = 9622.8$ [604]						
CH$_3$Si^{29}...........	$B_0 = 10{,}885.5$ [604]						
CH$_3$Si^{29}D$_3$...........	$B_0 = 8572$ [604]						
CH$_3$Si^{30}...........	$B_0 = 10{,}806.5$ [604]						
CH$_3$Si^{30}D$_3$...........	$B_0 = 9525$ [604]						
CH$_3$Sn^{116} (CH$_3$SnH$_3$)...	$B_0 = 6910.5$ [603]				0.68 [603]	[603]	
CH$_3$Sn^{117}...........	$B_0 = 6905.3$ [603]						
CH$_3$Sn^{118}...........	$B_0 = 6900.2$ [603]						
CH$_3$Sn^{119}...........	$B_0 = 6895.1$ [603]						
CH$_3$Sn^{120}...........	$B_0 = 6890.2$ [603]						
C^{14}N (ICN)...........	$B_0 = 3225.527$ [320], [329]	$\omega_1 = 470$ $\omega_2 = 321d$ $\omega_3 = 2158$	$\alpha_1 = 9.33$ [329] $\alpha_2 = -9.52$ [329] $D_J = 0.0009$ [531] $q_i = 2.69$ [329]	I$^{127} = -2420$ [329] N$^{14} = 3.80$ [329]	3.71	[320],[329]	$\Delta\nu = 20$ Mc/mm Hg [329]; anomalies in I^{127} h.f.s. [703]
C^{12}N...........	$B_0 = 3177.035$ [320]	2170.21	$\alpha = 524.1$ [457] $D_J = 0.189$ [457],[773a]	....			
C^{13}O^{16}...........	$B_e = 57{,}897.5$ [457]	2074.81	$\alpha = 488.3$ [457] $D_e = 0.174$ [457]	....	0.10	[457]	
	$B_e = 55{,}344.9$ [457]						
COF$_2$...........	$\frac{A_0 - C_0}{2} = 2966.2$ [649]	....	....	....			Asymmetry parameter $\kappa = 0.9796$ [649]
C^{12}O^{16}S^{32} (OCS)...	$B_0 = 6081.490$ [189], [755],[823]	$\omega_1 = 859$ $\omega_2 = 527d$ $\omega_3 = 2079$	$\alpha_1 = 20.56$ [329],[755] $\alpha_2 = -10.56$ [329],[755] $\alpha_4 = 36.36$ [968] $D_J = 0.001310$ [595],[946] $q_i = 6.344$ [329],[755]	....	0.709 [146],[530] [644]	[329]	Dipole moment in vibrational state (01^{1}0) = 0.700 [530]; $\Delta\nu =$ 6.1 Mc/mm Hg [329],[706]; Fermi resonance energy $W_{11} =$ 43.2 cm^{-1} [752]; molecular g factor = -0.025 [682]; variation of line width with J and T [706],[921]

Species	B_0	Vibrational constants	Nuclear quadrupole coupling	μ		Remarks
$C^{12}O^{16}S^{32}$	$B_0 = 6061.886$ [329], [755]	$\alpha_1 = 17.94$ [714]; $\alpha_2 = -10.10$ [329]; $q_l = 6.45$ [329]	....	0.709 [530]		Dipole moment in vibrational states (01¹0) = 0.730 [418], (100) = 0.728 [418]; molecular g factor = -0.019 [806c]. Frequency ratios for Se isotopes [456]
$C^{13}O^{16}S^{32}$	$B_0 = 6043.25$ [314]	$\alpha_2 = -9.4$ [314]; $q_l = 6.7$ [314]	....			
$C^{12}O^{17}S^{32}$	$B_0 = 5883.67$ [392]	$\alpha_1 = 16.19$ [714]; $\alpha_2 = -10.16$ [714]; $q_l = 5.62$ [714]	$O^{17} = -1.32$ [688]			
$C^{12}O^{18}S^{32}$	$B_0 = 5704.83$ [329]					
$C^{12}O^{16}S^{33}$	$B_0 = 6004.905$ [327], [755],[999]	....	$S^{33} = -29.130$ [327], [681],[999]	0.709 [530]		
$C^{12}O^{16}S^{34}$	$B_0 = 5932.816$ [189], [755]	$\alpha_1 = 17.68$ [714]; $\alpha_2 = -10.37$ [329]; $q_l = 6.07$ [329]	....			
$C^{13}O^{16}S^{35}$	$B_0 = 5865.2$ [356]	....	$S^{35} = 21.90$ [356],[908]		[418]	
$C^{14}O^{16}S^{36}$	$B_0 = 5799.67$ [392]					
$C^{13}O^{16}S^{34}$	$B_0 = 5911.730$ [327]					
$C^{12}O^{16}Se^{80}$ (OCSe)	$B_0 = 4017.68$ [418]	$\alpha_1 = 13.27$ [418]; $\alpha_2 = -6.92$ [418]; $D_J = 0.0008$ [418]; $q_l = 3.15$ [418]		0.754 [418]		
$C^{12}O^{16}Se^{80}$	$B_0 = 3980.05$ [418]	....	....			
$C^{12}O^{16}Se^{74}$	$B_0 = 4095.79$ [418]	....	$Se^{75} = 946$ Mc [892]			
$C^{12}O^{16}Se^{75}$		$\alpha_2 = -7.00$ [418]; $q_l = 3.24$ [418]				
$C^{12}O^{16}Se^{76}$	$B_0 = 4068.47$ [418]	$\alpha_1 = 13.48$ [418]; $\alpha_2 = -6.98$ [418]; $q_l = 3.21$ [418]				
$C^{12}O^{16}Se^{77}$	$B_0 = 4055.30$ [418]					
$C^{12}O^{16}Se^{78}$	$B_0 = 4042.46$ [418]	$\alpha_1 = 13.40$ [418]; $\alpha_2 = -6.96$ [418]; $D_J = 0.0008$ [418]; $q_l = 3.19$ [418]				
$C^{12}O^{16}Se^{78}$	$B_0 = 4005.11$ [418]	....	....			
$C^{13}O^{16}Se^{79}$		$\alpha_1 = 13.12$ [418]; $\alpha_2 = -6.86$ [418]; $D_J = 0.0008$ [418]; $q_l = 3.12$ [418]	$Se^{79} = 752.09$ [806c]			
$C^{12}O^{16}Se^{82}$	$B_0 = 3994.01$ [418]					

MOLECULAR CONSTANTS INVOLVED IN MICROWAVE SPECTRA

Chemical symbol	Rotational constants A, B, and C, Mc	Vibrational frequencies ω in wave numbers (cm^{-1}); d indicates a degenerate vibration	Rotation-vibration constants, Mc (α, D, or q_i)	Quadrupole coupling constant eqQ, Mc	Dipole moment μ in 10^{-18} esu (Debye)	Reference for structure	Remarks
$C^{12}S^{32}$	$B_e = 24{,}584.35$ [777]	1285.1	$\alpha_e = 177.54$ [777] $D_e = 0.040$ [777]	. . . .	2.0 [777]	[777]	
$C^{12}S^{32}$	$B_e = 23{,}205.26$ [777]						
$C^{12}S^{33}$	$B_e = 24{,}381.01$ [777]			$S^{33} = 12.84$ [777]			Magnetic h.f.s. $= 0.02I\cdot J$ [777]
$C^{12}S^{34}$	$B_e = 24{,}190.20$ [777]						
CSF_8 (CF_3SF_5)	$B_0 = 1097.6$ [713]	ω(torsional) $= 94$ [713]				[713]	Potential barrier height $= 220$ cm^{-1} [713]; rotation-vibration effects for torsional vibration [713]
$CSSe$ ($SCSe$)	$B_0 \approx 2020$ [432]						Data on several isotopes questionable [432]
$CSTe^{130}$ ($SCTe$)	$B_0 = 1559.9303$ [932]		$\alpha_2 = -3.2446$ [932] $q_i = 0.6599$ [932]		0.172 [932]	[932]	
$CSTe^{122}$	$B_0 = 1584.1224$ [932]		$\alpha_2 = -3.2870$ [932] $q_i = 0.6786$ [932]				
$CSTe^{123}$	$B_0 = 1580.9261$ [932]		$\alpha_2 = -3.2818$ [932] $q_i = 0.6776$ [932]				
$CSTe^{124}$	$B_0 = 1577.7898$ [932]		$\alpha_2 = -3.2764$ [932] $q_i = 0.6752$ [932]				
$CSTe^{125}$	$B_0 = 1574.6925$ [932]		$\alpha_2 = -3.2712$ [932] $q_i = 0.6728$ [932]				
$CSTe^{126}$	$B_0 = 1571.6524$ [932]		$\alpha_2 = -3.2657$ [932] $q_i = 0.6706$ [932]				
$CSTe^{128}$	$B_0 = 1565.7022$ [932]		$\alpha_2 = -3.2551$ [932] $q_i = 0.6649$ [932]				
C_2FN^{14} (CF_3CN)	$B_0 = 2945.54$ [743]		$D_J = 0.0004$ [743] $D_{JK} = 0.0058$ [743]	$N^{14} = -4.70$ [743]		[743]	
C_2FN^{15}	$B_0 = 2855.86$ [743]		$D_J = 0.0004$ [743] $D_{JK} = 0.0056$ [743]				
C_2HCl^{35} ($HCCCl$)	$B_0 = 5684.2$ [425]			$Cl^{35} = -79.7$ [425]	0.44 [425]	[425]	
C_2DCl^{35}	$B_0 = 5187.0$ [425]			$Cl^{35} = -79.6$ $D = 0.18$ [998]			

Molecule	Rotational constants		Dipole moment	Ref.	Notes
C₂HCl³⁷......... C₂DCl³⁷..........	$B_0 = 5572.3$ [425] $B_0 = 5084.2$ [425]	Cl³⁷ = -62.7 [425] Cl³⁷ = -63.1 [425]			
C₂H₃Cl³⁵F (CH₂CFCl)....	$A_0 = 10{,}681.6$ [440] $B_0 = 5102.2$ [440] $C_0 = 3448.4$ [440]	$e\,\dfrac{\partial^2 V}{\partial a^2}\,Q_{\mathrm{Cl}}{}^{35} = -73.3$ [440] $e\,\dfrac{\partial^2 V}{\partial b^2}\,Q_{\mathrm{Cl}}{}^{35} = 39.8$ [440]			
C₂H₃F..........	$A_0 = 10{,}681.3$ [440] $B_0 = 4955.0$ [440] $C_0 = 3380.5$ [440]		1.37 [406]	[406]	
C₂H₃F (CH₂CF₂)....	$A_0 = 11{,}001$ [406] $B_0 = 10{,}427$ [406] $C_0 = 5345.1$ [406]	$D_{JK} = 0.477$ [708] $D_J = 0.003 \pm 0.002$ [708]	1.41 [708]	[708],[769a]	Dipole moments measured in excited states [708]; rotation-vibration constants [708]
C₂H₂O (H₂C₂O, ketene)	$A_0 \approx 280 \times 10^3$ [431], [708] $B_0 = 10{,}293.28$ [431], [708] $C_0 = 9915.87$ [431], [708]	$\omega_7 = 670$ [708] $\omega_8 = 570$ [708] $\omega_9 = 490$ [708]			
C₂HDO (deuterated ketene)	$A_0 \approx 195 \times 10^3$ [708] $B_0 = 9647.05$ [708] $C_0 = 9174.63$ [708]		1.42 [708]		
C₂D₂O..........	$B_0 = 9120.80$ [431], [708] $C_0 = 8552.66$ [431], [708]	$D_{JK} = 0.35$ [708]	1.44 [708]		
C₂H₃Br⁷⁹ (vinyl bromide) C₂H₃Br⁸¹	$B_0 = 4162.2$ [447] $C_0 = 3862.9$ [447] $B_0 = 4138.0$ [447] $C_0 = 3841.9$ [447]	Br⁷⁹ = 479 [447] Br⁸¹ = 399 [447]			
C₂H₃Cl³⁵ (vinyl chloride)	$A_0 = 56121$ [370] $B_0 = 6030.5$ [370] $C_0 = 5445.2$ [370]	$e\,\dfrac{\partial^2 V}{\partial a^2}\,Q_{\mathrm{Cl}}{}^{35} = -57$ [370] $e\,\dfrac{\partial^2 V}{\partial b^2}\,Q_{\mathrm{Cl}}{}^{35} = 26$ [370]	1.44		
C₂H₃Cl³⁷.........	$A_0 = 56281$ [370] $B_0 = 5903.7$ [370] $C_0 = 5341.3$ [370]				
C₂H₃Cl³⁵ (CH₃CCl₃) C₂H₃F (CF₃CH₃, methyl fluoroform)	$B_0 = 2372.6$ [GGu] $B_0 = 5185$ [268],[528]	$\omega(\text{torsional}) = 234$ [618]	2.32 [528],[690]	[690]	Potential barrier height = 1216 cm⁻¹ [363],[618]

MOLECULAR CONSTANTS INVOLVED IN MICROWAVE SPECTRA

Chemical symbol	Rotational constants A, B, and C, Mc	Vibrational frequencies ω in wave numbers (cm⁻¹); d indicates a degenerate vibration	Rotation-vibration constants, Mc (α, D, or q_i)	Quadrupole coupling constant eqQ, Mc	Dipole moment μ in 10^{-18} esu (Debye)	Reference for structure	Remarks
C₂H₃I (vinyl iodide)	$A_0 = 52 \times 10^3$ [912],[964]; $B_0 = 3258.7$ [912],[964]; $C_0 = 3066.7$ [912],[964]	. . . .		$e\frac{\partial^2 V}{\partial a^2} Q_{127} = -1650$ [912],[964]; $e\frac{\partial^2 V}{\partial b^2} Q_{127} = 767$ [912],[964]			
C₂H₃N¹⁴ (CH₃CN, methyl cyanide)	$B_0 = 9198.83$ [402],[446],[480],[803]	$\omega_4 = 918$; $\omega_7 = 1124d$; $\omega_8 = 380d$	$\alpha_4 = 46.3$ [446]; $\alpha_7 = 5.2$ [446]; $\alpha_8 = -22.5$ [446]; $D_{JK} = 0.178$ [480]; $q_7 = 4.5$ [446],[503]; $q_8 = 17.7$ [446],[503]	N¹⁴ $= -4.40$ [229],[402],[446],[480],[543]	3.92 [446],[803]	[480],[543]	
C¹³H₃C¹²N¹⁴	$B_0 = 8933.15$ [446]						
C¹²H₃C¹³N¹⁴	$B_0 = 9194.20$ [446],[480]						
C¹²H₂DN¹⁴	$B_0 = 8759.17$ [974]; $C_0 = 8608.50$ [974]						
C¹²HD₂N¹⁴	$B_0 = 8320.05$ [974]; $C_0 = 8164.42$ [974]						
C¹²D₃C¹²N¹⁴	$B_0 = 7857.93$ [480]		$D_{JK} = 0.113$ [480]				
C¹²D₃C¹³N¹⁴	$B_0 = 7448.51$ [480]		$D_{JK} = 0.110$ [480]				
C¹²H₃C¹²N¹⁵	$B_0 = 8921.81$ [446]						
C₂H₃N¹⁴ (CH₃NC, methyl isocyanide)	$B_0 = 10052.90$ [480]	$\omega_8 = 290$ [402]	$\alpha_8 = 1.2$ [503]; $D_{JK} = 0.22$ [480]; $q_8 = 28$ [503]	[N¹⁴] < 0.5 [229],[480],[543]	3.83 [803]	[480],[543],[974]	
C¹³H₃NC¹³	$B_0 = 9695.91$ [480]						
C¹²H₃DNC¹³	$B_0 = 9578.27$ [974]						
C¹²HD₂NC¹³	$C_0 = 9397.88$ [974]						
C¹²D₃NC¹²	$C_0 = 9096.80$ [974]; $C_0 = 8910.61$ [974]		$D_{JK} = 0.14$ [480]				
C¹²D₃NC¹³	$B_0 = 8582.06$ [480]; $B_0 = 8278.79$ [480]		$D_{JK} = 0.13$ [480]				

Molecule	Rotational constants (Mc)	ω (torsional) / quadrupole coupling	μ (D)	Ref.	Remarks
$C_3H_5NS^{22}$ (CH₃NCS, methyl isothiocyanate)	$(B_0 + C_0)/2 = 2527.1$ [345]; $A_0 = 78.2 \times 10^3$ [345]		3.18	[345]	
$C_3H_5NS^4$ (CH₃NCS)	$(B_0 + C_0)/2 = 2462.6$ [345]		3.16	[345]	
C_2H_5NS (CH₃SCN, methyl thiocyanate)	$(B_0 + C_0)/2 = 2837$ [345]				
C_2HF_2 (CH₂CHF₂)	$A_0 = 9491.95$ [978]; $B_0 = 8962.65$ [978]; $C_0 = 5170.43$ [978]			[978]	Height of potential barrier = 1250 cm⁻¹ [978]
$C^{12}C^{13}H_4O$ (ethylene oxide)	$A_0 = 25,484$ [316],[264],[360],[566]; $B_0 = 22,121$ [316],[264],[360],[566]; $C_0 = 14,098$ [316],[264],[360],[566]		1.88 [566]	[316],[360],[566]	
$C^{12}C^{13}H_4O$	$A_0 = 25,291.2$ [566]; $B_0 = 21,597.4$ [566]; $C_0 = 13,825.2$ [566]				
$C_4^{11}D_4O$	$A_0 = 20,399$ [360],[566]; $B_0 = 15,457$ [360],[566]; $C_0 = 11,544$ [360],[566]				
$C_2H_4S^{22}$ (ethylene sulfide)	$A_0 = 21,974$ [566]; $B_0 = 10,824.9$ [566]; $C_0 = 8026.3$ [566]		1.84 [566]	[566]	
$C_2D_4S^{22}$	$A_0 = 15,471$ [566]; $B_0 = 9197.6$ [566]; $C_0 = 6819.0$ [566]				
$C_2H_4S^{34}$	$A_0 = 21,974$ [566]; $B_0 = 10,551.0$ [566]; $C_0 = 7874.7$ [566]				
$C_2H_5Cl^{35}$ (ethyl chloride)	$B_0 = 5493.76$ [995]; $C_0 = 4962.24$ [995]	ω (torsional) = 215 [995]; $e\dfrac{\partial^2 V}{\partial a^2} Q_{Cl} = -48.44$ [995]; $e\dfrac{\partial^2 V}{\partial b^2} Q_{Cl} = 13.9$ [995]		[995]	Potential barrier height = 1050 cm⁻¹ [995]

MOLECULAR CONSTANTS INVOLVED IN MICROWAVE SPECTRA

Chemical symbol	Rotational constants A, B, and C, Mc	Vibrational frequencies ω in wave numbers (cm⁻¹); d indicates a degenerate vibration	Rotation-vibration constants, Mc (α, D, or q_i)	Quadrupole coupling constant eqQ, Mc	Dipole moment μ in 10^{-18} esu (Debye)	Reference for structure	Remarks
C_2HCl^{37}	$B_0 = 5397.29$ [995] $C_0 = 4812.22$ [995]			$e\dfrac{\partial^2 V}{\partial a^2} Q_{Cl} = -36.9$ [995] $e\dfrac{\partial^2 V}{\partial \delta^2} Q_{Cl} = 10.1$ [995]		[995]	
C_2H_5N (ethylenimine)	$A_0 = 22{,}736.1$ [877], [889],[820] $B_0 = 21{,}192.3$ [877], [889],[820] $C_0 = 13{,}383.3$ [877], [889],[820]				$\mu_a = 1.89$ [820] $\mu_b = 1.67$ [820] $\mu_c = 0.89$ [820]	[877]	Discrepancy between A, B, and C given by [877],[889], and [820]
C_2H_6O (ethyl alcohol)					1.7		9 lines measured [344]
C_2HF_3 (CF_3CCH)	$B_0 = 2877.95$ [557]		$\alpha_{10} = -6.51$ [557] $D_J = 0.0002$ [557] $D_{JK} = 0.0063$ [557] $q_{10} = 3.62$ [557] $D_J = 0.0002$ [557] $D_{JK} = 0.0062$ [557]			[557]	
CF_3CCD	$B_0 = 2696.07$ [557]						
C_3HN^{14} (HCCCN, cyanoacetylene)	$B_0 = 4549.07$ [548]			$N^{14} = -4.2$ [548]	3.6	[548]	
$HC^{13}C^{12}C^{12}N^{14}$	$B_0 = 4408.47$ [548]						
$HC^{12}C^{13}C^{12}N^{14}$	$B_0 = 4529.84$ [548]						
$HC^{12}C^{12}C^{13}N^{14}$	$B_0 = 4530.23$ [548]						
$DC^{12}C^{12}C^{12}N^{14}$	$B_0 = 4221.60$ [548]						
$DC^{13}C^{12}C^{12}N^{14}$	$B_0 = 4107.21$ [548]						
$DC^{12}C^{13}C^{12}N^{14}$	$B_0 = 4207.59$ [548]						
$DC^{12}C^{12}C^{13}N^{14}$	$B_0 = 4202.54$ [548]						
$HC^{12}C^{12}C^{12}N^{15}$	$B_0 = 4416.91$ [548]						
$DC^{12}C^{12}C^{12}N^{15}$	$B_0 = 4100.41$ [548]						
$C_3H_6O_3$ (vinylene carbonate)	$A_0 = 9346.79$ [977] $B_0 = 4188.46$ [977] $C_0 = 2891.54$ [977]				$\mu_a = 4.51$ [977] $\mu_b = \mu_c = 0$	[977]	
$C_2H_3Br^{79}$ (H_3CCCBr, methyl bromoacetylene)	$B_0 = 1561.11$ [744]		$D_{JK} = 0.0114$ [744]	$Br^{79} = 647$ [744]		[744]	

Molecule	Rotational constants	Vibrational	Distortion constants	Quadrupole coupling	Dipole moment	Notes / refs
$C_2H_5Br^{81}$	$B_0 = 1550.42$ [744]	....	$D_{JK} = 0.0111$ [744]	$Br^{81} = 539$ [744]	....	Molecular g factor parallel to axis = 0.31 [784]; perpendicular to axis = 0 [784] [744]
C_2H_5I (H_3CCCl, methyl iodoacetylene)	$B_0 = 1259.02$ [744]	....	$D_{JK} = 0.0072$ [744]	$I^{127} = -2230$ [744]	....	
C_3H_3N (vinyl cyanide)	$A_0 = 49,076.2$ [1000] $B_0 = 4971.33$ [1000] $C_0 = 4514.05$ [1000]	....	....	$e\dfrac{\partial^2 V}{\partial a^2} Q_N = -3.0$ [1000]	$\mu = 3.89$ [1000] $\mu_a = 3.68$ [1000] $\mu_b = 1.25$ [1000]	[1000]
$C^{13}H_4$ (CH_3CCH, methyl acetylene)	$B_0 = 8545.84$ [544], [595]	$\omega_{10} = 336$ [544]	$\alpha_{10} = -23.92$ [544] $D_J = 0.0031$ [544],[595] $D_{JK} = 0.16$ [544],[595] $q_{10} = 16.7$ [544]	....	0.75 [803]	[544],[988c]
$C^{13}H_3C^{12}C^{12}H$	$B_0 = 8542.28$ [544]	....	$D_{JK} = 0.1$ [544]			
$C^{13}H_3C^{12}C^{12}H$	$B_0 = 8313.23$ [544]	....	$D_{JK} = 0.1$ [544]			
$C^{12}H_3C^{13}C^{12}H$	$B_0 = 8290.24$ [544]	....	$D_{JK} = 0.1$ [544]			
$C^{12}H_3C^{12}C^{12}D$	$B_0 = 7788.14$ [544]	....	$D_{JK} = 0.1$ [544]			
$C^{13}H_2DC^{12}C^{12}H$	$B_0 = 8155.67$ [974] $C_0 = 8025.46$ [974]	....	$D_J \approx 0.003$ [988c] $D_{JK} = 0.13$ [988c]			
$C^{13}HD_2C^{12}C^{12}H$	$B_0 = 7765.73$ [974] $C_0 = 7630.99$ [974]	....	$D_J \approx 0.002$ [988c] $D_{JK} = 0.13$ [988c]			
$C^{13}H_2DC^{12}C^{12}D$	$B_0 = 7440.77$ [974] $C_0 = 7331.96$ [974]	....	$D_J \approx 0.001$ [988c] $D_{JK} = 0.12$ [988c]			
$C^{13}HD_2C^{12}C^{12}D$	$B_0 = 7095.09$ [974] $C_0 = 6982.56$ [974]	....	$D_J \approx 0.004$ [988c] $D_{JK} = 0.11$ [988c]			
$C^{12}D_3C^{12}C^{12}H$	$B_0 = 7355.75$ [974]	....	$D_J \approx 0.002$ [974]			
$C^{12}D_3C^{12}C^{12}D$	$B_0 = 6734.31$ [544]	....	$D_J = 0.102$ [974] $D_{JK} = 0.09$ [544]			
$C_3H_5Cl^{35}$ (cyclopropyl chloride)	$A_0 = 17,695.1$ [787] $B_0 = 3905.4$ [787] $C_0 = 3622.4$ [787]	....	....	$e\dfrac{\partial^2 V}{\partial a^2} Q_{Cl}^{35} = -55.8$ [(787) and private comm.] $e\dfrac{\partial^2 V}{\partial b^2} Q_{Cl}^{35} = 24.4$ [(787) and private comm.]		
$C_3H_5Cl^{37}$	$A_0 = 17,930$ [787] $B_0 = 3810.0$ [787] $C_0 = 3405.5$ [787]					

MOLECULAR CONSTANTS INVOLVED IN MICROWAVE SPECTRA

Chemical symbol	Rotational constants A, B, and C, Mc	Vibrational frequencies ω in wave numbers (cm^{-1}); d indicates a degenerate vibration	Rotation-vibration constants, Mc (α, D, or q_i)	Quadrupole coupling constant eqQ, Mc	Dipole moment μ in 10^{-18} esu (Debye)	Reference for structure	Remarks
C_3H_6O [$(CH_3)_2CO$, acetone]	....	....	....	....	2.8	....	Approx. 20 lines measured [344], [762]
$C_3^{12}H_6O_3$ (trioxane)	$B_0 = 5273.6$ [553]	....	....	....	2.08 [553]	[553]	
$C_3^{12}C^{14}H_6O_3$	$B_0 = 5225.0$ [553]	....	....	....	....		
$C_3H_6Cl^{35}Si$ [$(CH_3)_3SiCl$] [848]	$B_0 = 2197.44$ [847], [848]	....	....	....	....	[847],[848]	
$C_3H_6Cl^{37}Si$ [848]	$B_0 = 2147.88$ [847], [848]	....	....	....	....		
C_3H_9FSi [$(CH_3)_3SiF$]	$B_0 = 3411.0$ [931a]	....	....	....	[931a]	[712b]	
C_3N_3P [$P(CN)_3$]	$B_0 = 2326$ [712b]	....	....	....	....		
C_4H_4 (vinyacetylene)	$A_0 = 4262 \times 10$ [731] $B_0 = 4744.85$ [731] $C_0 = 4329.73$ [731]	....	....	....	....		
C_4H_4O (furan)	$A_0 = 9447.04$ [648] $B_0 = 9246.76$ [648] $C_0 = 4670.84$ [648]	....	....	....	0.661 [648]	[648]	
C_4H_5N (pyrrol)	....	....	....	....	....	[763]	16 lines measured and identified [763]
$C_4H_9Br^{79}$ [$(CH_3)_3CBr$, tertiary butyl bromide]	$B_0 = 2044$ [550]	....	....	....	2.21	[550]	
$C_4H_9Br^{81}$	$B_0 = 2028$ [550]	....	....	....	....		
$C_4H_9Cl^{35}$ [$(CH_3)_3CCl$, tertiary butyl chloride]	$B_0 = 3016$ [550]	....	....	....	2.15	[550]	
$C_4H_9Cl^{37}$	$B_0 = 2954$ [550]	....	....	....	....		
C_4H_9I [$(CH_3)_3CI$, tertiary butyl...]	$B_0 = 1562$ [550]	....	....	....	2.13	[550]	
$C_4H_{10}O$ [$(C_2H_5)_2O$, diethyl ether]	....	....	....	....	....		28 lines measured [714]
C_4H_4 [$(CHC{\equiv}C{-}C{\equiv}CH)$]	$B_0 = 2035.73$ [934]	....	$D_J \leq 0.0002$ [934] $D_{JK} = 0.020$ [934]	....	....	[934]	

Molecule	Rotational / molecular constants	D / α constants	eQq constants	μ·10¹⁸	Ref.	Remarks
C_5H_5N (pyridine)	$A_0 = 6039.13$ [772b],[893],[914] $B_0 = 5804.72$ [772b],[893],[914] $C_0 = 2959.25$ [772b],[893],[914]	$D_J = -0.0036$ [961] $D_K = 0.0059$ [961] $D_{JK} = -0.0019$ [961]		2.15 [914]	[772b],[893],[914]	
C_7H_6O ($\overline{COCH_2CH_2CH_2CH_2}$)	$A_0 = 6618.9$ [917] $B_0 = 3350.8$ [917] $C_0 = 2409.9$ [917]					
C_6H_5Br (bromobenz.)	$A_0 = 5666.7$ [918a]			1.6	[918a]	3 lines measured [344]
C_6H_5Cl (chlorobenz.)	$A_0 = 5686.1$ [796]				[796]	
C_6H_5F (fluorobenzene)	$B_0 = 2571.0$ [796] $C_0 = 1767.6$ [796]					$B_0 = 1576.9$, $C_0 = 1233.3$ [918a]
C_7H_5N (benzonitrile, phenyl cyanide)	$A_0 = 5655.35$ [919],[955] $B_0 = 1546.88$ [919],[955] $C_0 = 1214.43$ [919],[955]			4.14 [955]	[955]	
$C_8H_{13}Br^{79}$ [1-bromo-bicyclo (2,2,2) octane]	$B_0 = 725.9$ [851]				[851]	
$C_8H_{13}B^{81}$	$B_0 = 718.6$ [851]					
$C_8H_{13}Cl^{35}$ [1-chloro-bicyclo (2,2,2) octane]	$B_0 = 1090.90$ [851]				[851]	
$C_8H_{13}Cl^{37}$						
$Cl^{35}Cs$ (CsCl)	$B_e = 2161.195$ [606a],[751],[938]	240 [772c] $\alpha = 10.085$ [606a],[751],[938]	$|Cl^{35}| < 3$ [606a]	10.5 [606a]	[606a],[751],[938]	Rotation-vibration constant $\gamma_e = 0.0071$ Mc [938]
$Cl^{37}Cs$	$B_e = 2068.761$ [751],[938]	$\alpha = 9.46$ [751],[938]	$|Cs^{133}| < 4$ [606a]			
$Cl^{35}F$ (FCl)	$B_e = 15{,}483.69$ [366]	$\alpha = 130.67$ [366] $D_e = 0.026$ [366]	$Cl^{35} = -145.99$ [366]	0.88 [366]	[366]	Magnetic h.f.s. = 0.03 I·J [366]
$Cl^{37}F$	$B_e = 15{,}189.22$ [366]	$\alpha = 126.96$ [366]	$Cl^{37} = -114.92$ [366]			Magnetic h.f.s. = 0.02 I·J [366]
$Cl^{35}F_3$	$A_0 = 13{,}747.7$ [867] $B_0 = 4611.7$ [867] $C_0 = 3448.7$ [867]		$e \dfrac{\partial^2 V}{\partial a^2} Q_{Cl^{35}} = -81$ [867] $e \dfrac{\partial^2 V}{\partial b^2} Q_{Cl^{35}} = -64$ [867]		[867]	

MOLECULAR CONSTANTS INVOLVED IN MICROWAVE SPECTRA

Chemical symbol	Rotational constants A, B, and C, Mc	Vibrational frequencies ω in wave numbers (cm^{-1}); d indicates a degenerate vibration	Rotation-vibration constants, Mc (α, D, or q_i)	Quadrupole coupling constant eqQ, Mc	Dipole moment μ in 10^{-18} esu (Debye)	Reference for structure	Remarks
$Cl^{37}F_3$	$A_0 = 13{,}653.2$ [867] $B_0 = 4611.9$ [867] $C_0 = 3442.8$ [867]			$e\frac{\partial^2 V}{\partial a^2} Q_{Cl^{37}} = -65$ [867] $e\frac{\partial^2 V}{\partial b^2} Q_{Cl^{37}} = -51$ [867]			
$Cl^{35}F_3Ge^{70}$ (GeF_3Cl)	$B_0 = 2168.52$ [555]		$D_J = 0.0006$ [555] $|D_{JK}| < 0.001$ [555]			[555]	
$Cl^{37}F_3Ge^{70}$	$B_0 = 2108.13$ [555]						
$Cl^{35}F_3Ge^{72}$	$B_0 = 2167.53$ [555]						
$Cl^{37}F_3Ge^{72}$	$B_0 = 2107.04$ [555]						
$Cl^{35}F_3Ge^{74}$	$B_0 = 2166.60$ [555]						
$Cl^{37}F_3Ge^{74}$	$B_0 = 2105.98$ [555]						
$Cl^{35}F_3Si$ (SiF_3Cl)	$B_0 = 2477.7$ [525], [641]		$D_{JK} = 0.0018$ [525],[641]	$Cl^{35} = -43$ [525],[641]		[641]	
$Cl^{37}F_3Si$	$B_0 = 2413.0$ [525], [641]			$Cl^{37} = -34$ [525],[641]			
$Cl^{35}Ge^{70}H_3$ (GeH_3Cl)	$B_0 = 4401.71$ [423]			$Cl^{35} = -46$ [423] $Ge^{73} = -95$ [423],[614]	2.148 [727]	[423],[727]	
$Cl^{35}Ge^{72}H_3$	$B_0 = 4333.91$ [423]						
$Cl^{35}Ge^{74}H_3$	$B_0 = 4177.90$ [423]						
$Cl^{37}Ge^{74}H_3$	$B_0 = 4146.5$ [423]			$Cl^{37} = -36$ [423]			
$Cl^{37}Ge^{76}H_3$							
$Cl^{35}H_3Si^{28}$ (SiH_3Cl)	$B_0 = 6673.81$ [315], [423],[727],[772]			$Cl^{35} = -40.0$ [315],[772]	1.303 [423]	[315],[423], [727],[772]	
$Cl^{37}H_3Si^{28}$	$B_0 = 6512.40$ [315], [423],[727],[772]			$Cl^{37} = -30.8$ [315]			
$Cl^{35}D_3Si^{28}$	$B_0 = 5917.7$ [772]			$Cl^{35} = -39.4$ [772]			
$Cl^{37}D_3Si^{28}$	$B_0 = 5772.8$ [772]						
$Cl^{35}H_3Si^{29}$	$B_0 = 5850.6$ [772]						
$Cl^{35}H_3Si^{30}$	$B_0 = 6485.8$ [423], [727],[772]						
$Cl^{35}D_3Si^{30}$	$B_0 = 5787.0$ [772]						
$Cl^{35}I$ (ICl)	$B_0 = 3422.300$ [337], [330]	384.18	$\alpha = 16.06$ [330]	$Cl^{35} = -82.5$ [330] $I^{127} = -2930.0$ [330]	0.65 [330]	[330]	$\Delta\nu = 5.5$ Mc/mm Hg for $J = 3 \rightarrow 4$ [330] $= 3.15$ Mc/mm Hg for $J = 0 \rightarrow 1$ [337]

Molecule	B_e		α_e	eQq	μ	Ref.	Notes
$Cl^{37}I$	$B_e = 3277.365$ [330]	....	$\alpha = 15.05$ [330]	$[Cl^{36}] < 0.04$ [835]	10.48 [835],[988]	[835],[938]	Quadrupole coupling constants and dipole moments in several vibrational states [835]; rotation-vibration constants $\gamma_e = 0.050$ [835]
$Cl^{35}K^{39}$ (KCl)	$Y_{01}[\approx B_e] = 3856.370$ [750],[835]	305 [772c]	$\alpha = 23.680$ [750],[835]	$K^{39} = -5.656$ [835]	....	....	
$Cl^{35}K^{39}$	$Y_{01}[\approx B_e] = 3746.583$ [835]	....	$\alpha = 22.676$ [835]	....	....	....	Rotation-vibration constant $\gamma_e = 0.047$ [835]
$Cl^{35}K^{41}$	$Y_{01}[\approx B_e] = 3767.394$ [835]	....	$\alpha = 22.865$ [835]	$K^{41} = -6.899$ [835]	....	....	Quadrupole coupling constant in excited vibrational state [835]; rotation-vibration constant $\gamma_e = 0.048$ [835]
$Cl^{35}Na$ (NaCl)	$Y_{01}[\approx B_e] = 6536.86$ [751],[938]	380	$\alpha = 48.1$ [751]	$Na^{23} = -5.40$ [722c]	8.5 [988]	[751],[938]	
$Cl^{35}NO$ (NOCl)	$A_0 = 85{,}290$ [632]; $B_0 = 5738.3$ [632]; $C_0 = 5376.4$ [632]			$e\,\dfrac{\partial^2 V}{\partial a^2}\,Q_{Cl^{35}} = 30$ [632]; $e\,\dfrac{\partial^2 V}{\partial b^2}\,Q_{Cl^{35}} = 19.6$ [632]	1.83 [632] parallel to a axis $= 1.28$ [632]	[436]	
$Cl^{37}NO$	$A_0 = 85{,}560$ [632]; $B_0 = 5600.7$ [632]; $C_0 = 5259.2$ [632]			$e\,\dfrac{\partial^2 V}{\partial a^2}\,Q_{Cl^{37}} = 23$; $e\,\dfrac{\partial^2 V}{\partial b^2}\,Q_{Cl^{37}} = 14$			
$Cl^{35}Rb^{85}$ (RbCl)	$B_e = 2627.414$ [993], [938]	270 [772c]	$\alpha = 13.601$ [993],[938]	$Cl^{35} = 0.774$ [993]; $Rb^{85} = -52.675$ [993]		[938],[993]	$\gamma_e = 0.021$ Mc [993] Magnetic h.f.s. for $Rb^{85} = (C.3 \pm 0.3)I\cdot J$ Mc [993] $\gamma_e = 0.021$ Mc [993] $\Delta\nu = 60$ Mc/mm Hg [665]
$Cl^{35}Rb^{87}$	$B_e = 2609.779$ [993]	....	$\alpha = 13.464$ [993]	$Rb^{87} = -25.485$ [993]			
$Cl^{35}Re^{185}O_3$ (ReO$_3$Cl)	$B_0 = 2094.23$ [665]	....		$Cl^{35} = -34$ [818],[943a]; $Re^{185} = 270$ [818],[943a]		[665]	
$Cl^{37}Re^{185}O_3$	$B_0 = 2025.02$ [665]						
$Cl^{35}Re^{187}O_3$	$B_0 = 2093.59$ [665]			$Re^{187} = 253$ [818],[943a]			
$Cl^{37}Re^{187}O_3$	$B_0 = 2024.36$ [665]						
$Cl^{35}Tl^{203}$ (TlCl)	$B_e = 2743.94$ [561a], [958a]	287.47	$\alpha_e = 11.96$ [958a]	$Cl^{35} = -15.795$ [561a]		[561a]	Magnetic h.f.s. for Tl, $0.073I\cdot J$; for Cl^{35}, $0.0012I\cdot J$; [561a]
$Cl^{37}Tl^{203}$	$B_e = 2617.5$ [958a]	....	....	$Cl^{37} = -12.446$ [561a]			
$Cl_3^{35}Ge^{70}H$ (GeCl$_3$H)	$B_0 = 2172.75$ [881]		$|D_J| < 0.002$ [881]; $|D_{JK}| < 0.004$ [881]			[881]	Spectra for excited vibrational states [881]
$Cl_3^{35}Ge^{72}H$	$B_0 = 2169.26$ [881]		....				
$Cl_3^{35}Ge^{74}H$	$B_0 = 2165.84$ [881]		....				

MOLECULAR CONSTANTS INVOLVED IN MICROWAVE SPECTRA

Chemical symbol	Rotational constants A, B, and C, Mc	Vibrational frequencies ω in wave numbers (cm^{-1}); d indicates a degenerate vibration	Rotation-vibration constants, Mc (ω, D, or q_i)	Quadrupole coupling constant eqQ, Mc	Dipole moment μ in 10^{-18} esu (Debye)	Reference for structure	Remarks
$Cl_3{}^{73}Ge^{70}H$	$B_0 = 2063.74$ [881]						
$Cl_3{}^{73}Ge^{72}H$	$B_0 = 2060.43$ [881]						
$Cl_3{}^{73}Ge^{74}H$	$B_0 = 2057.20$ [881]						
$Cl_3{}^{28}SiH$ (SiCl$_3$H)	$B_0 = 2472.49$ [847]					[730a],[847]	
$Cl_3{}^{30}SiH$	$B_0 = 2346.07$ [847]						
$Cl_3{}^{30}OP$ (POCl$_3$)	$B_0 = 2015.20$ [765]					[765]	
$Cl_3{}^{30}OP$	$B_0 = 1932.38$ [765]					[765]	
$Cl_3{}^{35}P$ (PCl$_3$)	$B_0 = 2617.1$ [481]	$\omega_1 = 510$ $\omega_2 = 257$ $\omega_3 = 480d$ $\omega_4 = 190d$	$\alpha_2 = 1.9$ [481] $\alpha_4 = -1.9$ [481]		0.80	[481]	Lines of PCl$_2{}^{35}$Cl^{37} and PCl^{35}Cl$_2{}^{37}$ measured [481]
$Cl_3{}^{31}P$	$B_0 = 2487.5$ [481]						
$Cl_3{}^{32}PS^{32}$ (PSCl$_3$)	$B_0 = 1402.65$ [765]	$\omega_1 = 360$ $\omega_2 = 165$ $\omega_3 = 320d$ $\omega_4 = 134d$				[765]	
$Cl_3{}^{32}PS^{32}$	$B_0 = 1355.72$ [765]						
$Cl_3{}^{32}PS^{34}$	$B_0 = 1370.13$ [765]						
$Cl_3{}^{35}Sb^{121}$ (SbCl$_3$)	$B_0 = 1753.9$ [597]				3.93	[597],[947]	Ratio quadrupole coupling constants [597]
$Cl_3{}^{35}Sb^{123}$	$B_0 = 1750.7$ [597]						
CsF	$B_e = 5527.27$ [938]	385 [772c]	$\alpha = 35.07$ [938]		7.874 [160], [423a], [938]	[938]	$\gamma_e = 0.009$ [938]
CsI	$B_e = 708.36$ [938]	120 [772c]	$\alpha = 2.044$ [938]		12.1 [938]	[938]	$\gamma_e = 0.0015$ Mc [938]
FH_3Si^{28} (SiH$_3$F)	$B_e = 14327.9$ [522]				1.268 [522]	[938] [522],[888]	
FD_3Si^{28}	$B_0 = 12,253.114$ [771], [888]					[888]	
FH_3Si^{29}	$B_0 = 14,196.7$ [522]						
FD_3Si^{29}	$B_0 = 12,175.580$ [771], [888]						
FH_3Si^{30}	$B_0 = 14,072.6$ [522]						
FD_3Si^{30}	$B_0 = 12,101.949$ [771], [888]						

Molecule	Rotational constants	ω	α	eQq	eQq	μ	Ref	Remarks
FLi^7 (LiF)				$Li^7 = 0.408$ [722a]				Magnetic h.f.s. for $F = 0.037I\cdot J$ Me [753]; μ^2/B_0 measured [753]
$FMnO_3$ (MnO_3F)	$B_0 = 4129.11$ [702]	$\omega_3 \sim 400$ [943a] $\omega_4 \sim 600d$ [943a] $\omega_5 \sim 470d$ [943a] $\omega_6 \sim 350d$ [943a]	$\alpha_3 = 7.77$ [702],[943a] $\alpha_4 = 14.38;\ q_4 = 5.90$ $\alpha_5 = -12.80;\ q_5 = 16.20$ $\alpha_6 = 5.87;\ q_6 = 9.81$ [702],[943a]	$Mn^{55} = 16.8$ [702]	$Mn^{55} = 16.1$ [943a]	1.5 [943a]	[943a]	Rotational constants for $FMn(O^{16})_3O^{18}$ [943a]
FNO (NOF)	$A_0 = 95{,}191.7$ [609] $B_0 = 11{,}843.9$ [609] $C_0 = 10{,}508.5$ [609]	$\omega_1 = 1844.03$ [609] $\omega_3 = 765.85$ [609]				1.81 [609] μ parallel to $a =$ 1.70 μ parallel to $b =$ 0.62 [609]	[609]	
$FReO_3$	$B_0 = 3566.75$ [957]		$\alpha_3 = 12.5$ [957] $\alpha_5 = -10.5;\ q_5 = 16.35$ [957] $\alpha_6 = 2.64$ [957]; $q_6 = 4.82$ [957]	$Re = -53$ [957]				Two isotopes of Re were not resolved
F_2OS (thionyl fluoride)	$A_0 = 8614.75$ [800] $B_0 = 8356.98$ [800] $C_0 = 4923.55$ [800]					1.618 [800]	[922]	
$F_2O^{38}S$	$A_0 = 8582.33$ [922] $B_0 = 7843.37$ [922] $C_0 = 4777.90$ [922]							
$F_2S^{32}O_2$ (sulfuryl fluoride)	$A_0 = 5139.77$ [684] $B_0 = 5077.81$ [684] $C_0 = 5057.22$ [684]					0.228 [684]	[684]	
$F_2S^{34}O_2$	$A_0 = 5139.77$ [684] $B_0 = 5073.00$ [684] $C_0 = 5052.51$ [684]							
F_3HSi^{28} (SiF_3H)	$B_0 = 7207.98$ [525],[641]		$D_J = 0.005$ [933]			1.26 [933]	[641],[933]	
F_3HSi^{29}	$B_0 = 7195.66$ [641]							
F_3HSi^{30}	$B_0 = 7183.70$ [641]							
F_3DSi^{28}	$B_0 = 6890.08$ [933]		$D_J = 0.004$ [933]					
F_3DSi^{29}	$B_0 = 6880.15$ [933]							
F_3DSi^{30}	$B_0 = 6870.53$ [933]							
F_3N^{14} (NF_3)	$B_0 = 10{,}680.96$ [527]		$D_{JK} = -0.025$ [527]	$N^{14} = -7.07$ [527]		0.234 [712a],[803], [947]	[527]	
FN^{15}	$B_0 = 10{,}629.35$ [527]		$D_J \approx 0.001$ [765] $|D_{JK}| < 0.002$ [765]					
$F_3O^{38}P$ (POF_3)	$B_0 = 4594.25$ [518], [765],[696]					1.77 [518],[696], [803]	[765],[696]	

MOLECULAR CONSTANTS INVOLVED IN MICROWAVE SPECTRA

Chemical symbol	Rotational constants A, B, and C, Mc	Vibrational frequencies ω in wave numbers (cm⁻¹); d indicates a degenerate vibration	Rotation-vibration constants, Mc (α, D, or q_l)	Quadrupole coupling constant eqQ, Mc	Dipole moment μ in 10^{-18} esu (Debye)	Reference for structure	Remarks
F₃O³¹P	$B_0 = 4395.27$ [765], [696]						
F₃P (PF₃)	$B_0 = 7819.90$ [367], [712a]	$\omega_1 = 890$ $\omega_2 = 531$ $\omega_3 = 840d$ $\omega_4 = 486d$	$\alpha_1 = 38$ [712a] $\alpha_2 = 10.8$ [712a] $\alpha_3 = -13.8$ [712a] $\alpha_4 = -3.5$ [712a] $q_4 = 33.8$ [712a]	$\cdots\cdots\cdots\cdots$	1.025 [528], [803]	[367], [712a]	$\Delta\nu = 16$ Mc/mm Hg [367]
F₃PS³² (PSF₃)	$B_0 = 2657.63$ [765], [696]	$\cdots\cdots\cdots\cdots$	$D_J \approx 0.0003$ [765] $D_{JK} = 0.0018$ [765]	$\cdots\cdots\cdots\cdots$	0.633 [696]	[765], [696]	
F₃PS³³	$B_0 = 2614.73$ [765], [696]						
F₃PS³⁴	$B_0 = 2579.77$ [765], [696]						
H⁷⁹Br⁷⁹ (DBr)	$B_0 = 127{,}358.2$ [927]	$\cdots\cdots\cdots\cdots$	$\alpha_e = 1257$ [821a] $D_e = 2.8$ [821a]	Br⁷⁹ = 533 [927]	0.79	[927]	
DBr⁸¹	$B_0 = 127{,}280.0$ [927]		$\alpha_e = 1258$ [821a] $D_e = 2.8$ [821a]	Br⁸¹ = 455 [927]			HBr constants known from I.R. measurements [471]
HI¹²⁷ (DI)	$B_0 = 97{,}537.2$ [827a], [782b]	$\cdots\cdots\cdots\cdots$		I¹²⁷ = -1823 [827a], [782b]	0.38		HI molecular constants known from I.R. measurements [471]; magnetic h.f.s. for $I = 0.141 \cdot \text{J}$ Mc [782b]
HN₃¹⁴	$A_0 = 609{,}850$ [427], [633] $B_0 + C_0 = 23{,}815.7$ [427], [633]			N¹⁴ (end nitrogen) = -4.67 [740]	0.847 [427]	[427]	
DN₃¹⁴	$B_0 + C_0 = 22{,}316.1$ [427]						
HN¹⁴N¹⁴N¹⁵	$B_0 + C_0 = 23{,}048.2$ [427]						
HN¹⁵N¹⁴N¹⁴	$B_0 + C_0 = 23{,}096.7$ [427]						
HN¹⁴N¹⁵N¹⁴	$B_0 + C_0 = 23{,}814$ [427]						

HO (OH)	$B_e = 5658 \times 10^2$ [471]	3735.21	$\alpha = 214 \times 10^2$ [471]				Fine structure [471]; Λ-doubling constants and h.f.s. [861],[971d]
DO	$B_e = 3004 \times 10^2$ [471]	2720.9	$\alpha = 885 \times 10$ [471]				Λ-doubling constants and h.f.s. [916],[971d]
H₂O	$A_0 = 8332 \times 10^2$ [130] $B_0 = 4347 \times 10^2$ [130] $C_0 = 2985 \times 10^2$ [130]	$\omega_1 = 3693.8$ $\omega_2 = 1614.5$ $\omega_3 = 3801.7$			1.94 [276]	[130]	$\Delta\nu = 14$ Mc/mm Hg [173]; $g_{aa} = 0.585$, $g_{bb} = 0.742$, and $g_{cc} = 0.666$ [592],[742]
HDO	$A_0 = 7039.6 \times 10^2$ [969] $B_0 = 2736.0 \times 10^2$ [969] $C_0 = 1918.6 \times 10^2$ [969]	$\omega_1 = 2719$ $\omega_2 = 1402$ $\omega_3 = 3363$	$D_J = 12$ [856] $D_{JK} = 36.8$ [856] $D_K = 287$ [856]		1.84 [415]	[415]	$\Delta\nu = 15$ Mc/mm Hg [173],[415]; centrifugal distortion [856]
D₂O	$A_0 = 46{,}149 \times 10$ [774] $B_0 = 21{,}774 \times 10$ [774] $C_0 = 14{,}546 \times 10$ [774]	$\omega_1 = 2666$ $\omega_2 = 1178.3$ $\omega_3 = 2787.2$			1.87 [323],[670]		
H₂O₂	$B_0 = 26{,}180$ [960]				2.26 [960]	[960]	Height of potential barrier = 113 cm⁻¹ [960]
HDO₂		$\omega_1 = 2610.8$ $\omega_2 = 1290$ $\omega_3 = 2684$					Series of $\Delta J = 0$ transitions observed [959]
H₂S³²	$A_0 = 316{,}304$ [783] $B_0 = 276{,}512$ [783] $C_0 = 147{,}536$ [783]				1.02 [587]	[783]	$\varrho_i = 0.24$ [783]
H₂S³³	$A_0 = 315{,}735$ [783] $B_0 = 276{,}512$ [783] $C_0 = 147{,}412$ [783]			$e\,\dfrac{\partial^2 V}{\partial a^2}\,Q_s = -32$ [783] $e\,\dfrac{\partial^2 V}{\partial b^2}\,Q_s = -8$ [783]			
H₂S³⁴	$A_0 = 315{,}201$ [783] $B_0 = 276{,}512$ [783] $C_0 = 147{,}296$ [783]	$\omega_1 = 1924$ $\omega_2 = 1090$ $\omega_3 = 2684$					
HDS³²	$A_0 = 290{,}257$ [587] $B_0 = 145{,}218$ [587] $C_0 = 94{,}134$ [587]		$D_{JK} = 27.38$ [587] $D_K = -4.917$ [587]	$e\,\dfrac{\partial^2 V}{\partial a^3}\,Q_s = -31.0$ [903] $e\,\dfrac{\partial^2 V}{\partial b^3}\,Q_s = -10.0$ [903]	1.02 [587]	[587],[903]	Centrifugal distortion [587]
HDS³³							
HDS³⁴							4 lines measured [587]

MOLECULAR CONSTANTS INVOLVED IN MICROWAVE SPECTRA

Chemical symbol	Rotational constants A, B, and C, Mc	Vibrational frequencies ω in wave numbers (cm^{-1}); d indicates a degenerate vibration	Rotation-vibration constants, Mc (α, D, or g_l)	Quadrupole coupling constant eqQ, Mc	Dipole moment μ in 10^{-18} esu (Debye)	Reference for structure	Remarks
H$_3$N^{14} (NH$_3$)	$C_0 = 189 \times 10^3$ [130] $B_0 = 298 \times 10^3$ [130]	$\omega_1 = 3335$ $\omega_2 = 950$ $\omega_3 = 3414d$ $\omega_4 = 1627.5d$		N$^{14} = -4.084$ [317], [662],[930]	1.468 [172],[562]	[130]	Formulas for inversion frequencies p. 310; magnetic h.f.s. for N^{14} = $\left[6.1 + \dfrac{0.4K^2}{J(J+1)}\right]$ I·J kc [930], for H see p. 220; variation of eqQ with rotation [930]; $\Delta\nu = 28$ Mc/mm Hg [172],[179]; $\Delta\nu$ at high pressure p. 370; variation of $\Delta\nu$ with temperature [906], [952]; g-factor p. [294].
H$_3$N^{15}							Inversion frequencies p. 309
H$_2$DN^{14}	$\frac{1}{2}(A_0 - C_0) = 74,350$ [662] $\dfrac{2B_0 - A_0 - C_0}{A_0 - C_0} = -0.315$ [662]						Rotation-inversion spectrum [662]
HD$_2$N^{14}	$\frac{1}{2}(A_0 - C_0) = 55,200$ [662] $\dfrac{2B_0 - A_0 - C_0}{A_0 - C_0} = -0.1385$ [662]					[662]	Rotation-inversion spectrum [662]
D$_3$N^{14}	$A_0 = 946 \times 10^3$ [130] $B_0 = 154 \times 10^3$ [130]		$D_J = 3.15$ [909]				Inversion frequencies p. 309
H$_3$P (PH$_3$)	$B_0 = 133,478.3$ [909]	$\omega_1 = 1700$ [762a] $\omega_2 = 892$ [762a]			0.579 [606],[866]	[909]	Inversion frequency <0.5 Mc
H$_2$DP	$\frac{1}{2}(A_0 - A_0 - C_0) = 23,292.6$ [909] $\dfrac{2B_0 - A_0 - C_0}{A_0 - C_0} = -0.74138$ [866]	$\omega_4 = 1097$ [762a] $\omega_3 = 2320$ [762a]				[606],[736], [866]	
HD$_2$P	$\dfrac{B_0 - C_0}{2} = 8533.81$ [866] $\dfrac{2A_0 - B_0 - C_0}{C_0 - B_0} = -2.40671$ [866]	$\omega_4 = 906$ [762a]			0.565 [866]	[606],[736], [866]	Inversion frequency <0.5 Mc in ground state [606]

Molecule	Rotational constants (Mc)	ω (cm⁻¹)	α, D (Mc)	eQq (Mc)	(Mc)	Ref.	Remarks
D₃P	B_e = 69,470.41 [909]		D_J = 0.71 [909]				
H₂DSb¹²¹	· · · · · ·	· · · · · ·		Sb¹²¹ = 455 [606]	0.116	[606]	
H₂DSb¹²³	· · · · · ·	· · · · · ·		Sb¹²³ = 575 [606]			
IK³⁹ (KI)	B_e = 1825.01 [938]	200 [772c]	α = 8.034 [938] D = 0.0010 [938]	I¹²⁷ = −60 [938]	11.05	[938]	γ_e = 0.0122 Mc [938]
IK⁴¹	B_e = 1756.90 [938] $Y_{01}(\approx B_e)$ = 13,286.39 [938]	450	α = 122.6 [938]	I¹²⁷ = −198.2 [938]	6.25	[938]	γ_e = 0.455 [938]
ILi⁷ (LiI)	$Y_{01}(\approx B_e)$ = 15,381.45 [938]	· · · · · ·	α = 152.6 [938]	I¹²⁷ = −259.9 [938]			
ILi⁶							
INa (NaI)	B_e = 3531.76 [938]	286	α = 19.44 [938]	· · · · · ·		[938]	γ_e = 0.047 [938]
IRb⁸⁵ (RbI)	B_e = 984.31 [938]	147 [938]	α = 3.281 [938] D = 0.00023 [938]	· · · · · ·		[938]	γ_e = 0.0030 [938]
IRb⁸⁷	B_e = 970.76 [938] B_0 = 51,084.5 [782b],[924]	· · · · · ·	α = 3.214 [938]				
NO		1904	α_e = 534 [782b],[924]	N¹⁴ = −1.9 [435], [782b], [899], [962]	0.16		Λ-doubling [671],[872b],[924], magnetic h.f.s. for N¹⁴ [435], [899],[872b],[924]
NO		· · · · · ·	· · · · · ·	· · · · · ·	0.29		$J = 6_{-6} \longleftrightarrow 5_{-4}$ transition [497], [615]; magnetic h.f.s. not completely understood [340],[615]
NO₂F (nitryl fluoride)	A_0 = 13203 [748] B_0 = 11447 [748] C_0 = 6120 [748]	· · · · · ·	· · · · · ·	$e\dfrac{\partial^2 V}{\partial a^2} Q_{N^{14}}$ = 0.7 [748] $e\dfrac{\partial^2 V}{\partial b^2} Q_{N^{14}}$ = 1.5 [748]	0.47	[748]	
N¹⁴₂O¹⁶	B_0 = 12,561.66 [184],[357],[595]	ω_1 = 1285.0 ω_2 = 588.8d ω_3 = 2223.5	α_1 = 52 [915] α_2 = −13 [915] α_3 = 104 [915] D_J = 0.0057 [595],[755] q_1 = 26 [915]	N¹⁴ (end atom) = −0.8 [357],[755] N¹⁴ (central atom) = −0.3 [184]	0.166	[184],[357],[915]	$\Delta\nu$ = 4.2 Mc/mm Hg [357] $\|\mu_J\|$ = 0.086[592]
N¹⁵N¹⁴O¹⁶	B_0 = 12,137.30 [184],[357]	· · · · · ·	α_1 = 46[915] α_2 = −11[915] α_3 = 101[915]				
N¹⁴N¹⁵O¹⁶	B_0 = 12,560.78 [357]						
N¹⁵N¹⁵O¹⁶	B_0 = 12,137.39 [357]						
N¹⁴N¹⁴O¹⁸	B_0 = 11,859.11 [714]						
N¹⁵N¹⁴O¹⁸	B_0 = 11,855.82 [714]						
N¹⁴N¹⁵O¹⁸	B_0 = 11,449.66 [714]						
N¹⁵N¹⁵O¹⁸	B_0 = 11,448.04 [714]						

MOLECULAR CONSTANTS INVOLVED IN MICROWAVE SPECTRA

Chemical symbol	Rotational constants A, B, and C, Mc	Vibrational frequencies ω in wave numbers (cm^{-1}); d indicates a degenerate vibration	Rotation-vibration constants, Mc (α, D, or q_i)	Quadrupole coupling constant eqQ, Mc	Dipole moment μ in 10^{-18} esu (Debye)	Reference for structure	Remarks
O_2^{16}	$B_0 = 43{,}102$ [667],[963]	1580.36				[842]	ρ-type triplet frequencies see p. 184; $\Delta\nu = 1.9$ Mc/mm Hg [139],[441],[669],[667],[892b], [937]; temperature dependence of $\Delta\nu$[937]; Zeeman effect [937], [990]
$O^{16}O^{17}$							Magnetic h.f.s. for $O^{17} =$ $-101I\cdot S + 140I_2S_2$ [841]
$O^{16}O^{18}$							Lines compared with theory [841]
$O^{18}O^{18}$							Lines and alternating intensities [616]
O_2S^{32} (SO$_2$)	$A_0 = 60{,}778.79$ [187], [565],[647],[948] $B_0 = 10{,}318.10$ [187], [565],[647],[948] $C_0 = 8799.96$ [187], [565],[647],[948]	$\omega_1 = 1151.2$ $\omega_2 = 519$ $\omega_3 = 1361$			1.59 [565]	[187],[565], [647],[948]	Centrifugal distortion [647],[948] $[\mu I] = 0.084$ [592]
O_2S^{33}				$e\dfrac{\partial^2 V}{\partial a^2} = -1.7$ [903] $e\dfrac{\partial^2 V}{\partial b^2} = 25.71$ [903]			
O_3	$A_0 = 106{,}530.0$ [875] $B_0 = 13{,}349.1$ [875] $C_0 = 11{,}834.3$ [875]	$\omega_1 = 1043.4$ $\omega_2 = 710$ $\omega_3 = 1740$			0.53 [699],[875]	[699],[875]	Effective molecular g factors for 1_{11} and 2_{22} rotational states [875]

Properties of the Stable Nuclei
(Abundance, Mass, and Moments)

Relative abundances of isotopes are taken from the table of Hollander, Perlman, and Seaborg [813]. Masses are principally those listed by E. Segre [*Experimental Nuclear Physics*, Vol. I, John Wiley & Sons, Inc., (1953)] supplemented by more recent work. For isotopes with masses which have not been accurately measured, calculated values are given based on the semiempirical formula of Green and Engler [A. E. S. Green and N. Engler, *Phys. Rev.*, **91**, 40 (1953)]. These are given only to three decimal places. The basic compilations of nuclear moment data used are those of Poss [H. L. Poss, Brookhaven National Laboratories report (Oct. 1, 1949)], Mack [J. E. Mack, *Rev. Mod. Phys.* **22**, 64 (1952)] and Walchli [H. E. Walchli, Oak Ridge National Laboratory report *ORNL*—1469, (Apr. 1, 1953)] with the addition of some recent results not included in the above tables.

The nuclear magnetic moment values have not been corrected for the effects of atomic or molecular diamagnetism. These diamagnetic effects increase with increasing Z from about 0.01 per cent to somewhat more than 1 per cent (see Walchli, *ibid.*). Quadrupole moments have been corrected for screening by inner electron shells [539], [749], [979].

Atomic number	Element	Mass number	Atomic mass in atomic mass units	Abundance, per cent	Spin	Magnetic moment, nuclear magnetons	Quadrupole moment $\times 10^{24}$ cm²	Ratio of quadrupole moments
1	H	1	1.008142	99.9851	$\frac{1}{2}$	+2.792670		
		2	2.014735	0.0149	1	+0.857392	+0.002738	
2	He	3	3.016977	1.3×10^{-4}	$\frac{1}{2}$	−2.12741		
		4	4.003873	99.9999	0			
3	Li	6	6.017021	7.52	1	+0.82193		$Q_6/Q_7 = 1.9 \times 10^{-2}$
		7	7.018223	92.48	$\frac{3}{2}$	+3.25598		
4	Be	9	9.015043	100	$\frac{3}{2}$	−1.1772	±0.02	
5	B	10	10.016114	18.98–18.45	3	+1.8004	+0.086	$Q_{10}/Q_{11} = 2.084$
		11	11.012789	81.02–81.55	$\frac{3}{2}$	+2.68798	+0.042	
6	C	12	12.003804	98.892	0			
		13	13.007473	1.108	$\frac{1}{2}$	+0.7021		
7	N	14	14.007515	99.635	1	+0.4036	+0.02	
		15	15.004863	0.365	$\frac{1}{2}$	−0.2830		
8	O	16	16.000000	99.758	0			
		17	17.004533	0.0373	$\frac{5}{2}$	−1.89295	−0.005	
		18	18.004874	0.2039	0			
9	F	19	19.004456	100	$\frac{1}{2}$	+2.62728		
10	Ne	20	19.998860	90.92	0			
		21	21.000589	0.257	$\frac{3}{2}$			
		22	21.998270	8.82	0			
11	Na	23	22.997139	100	$\frac{3}{2}$	+2.2161	+0.10	
12	Mg	24	23.992696	78.60				
		25	24.993815	10.11	$\frac{5}{2}$	−0.8547		
		26	25.990871	11.29				
13	Al	27	26.990140	100	$\frac{5}{2}$	+3.63853	+0.150	
14	Si	28	27.985837	92.27	0			
		29	28.985719	4.68	$\frac{1}{2}$	(−)0.5547		
		30	29.983313	3.05	0			
15	P	31	30.983622	100	$\frac{1}{2}$	+1.1305		
16	S	32	31.982236	95.02	0			
		33	32.98197	0.75	$\frac{3}{2}$	+0.6427	−0.067	
		34	33.97860	4.22				
17	Cl	35	34.97993	75.4	$\frac{3}{2}$	+0.82086	−0.085	$Q_{35}/Q_{37} = 1.2688$
		37	36.97754	24.6	$\frac{3}{2}$	+0.68330	−0.067	
18	A	36	35.97892	0.337				
		38	37.97479	0.063				
		40	39.97502	99.600				
19	K	39	38.97593	93.08	$\frac{3}{2}$	+0.39094		$Q_{39}/Q_{41} = 1.220$
		40	39.97658	0.0119	4	−1.2964		
		41	40.97476	6.91	$\frac{3}{2}$	+0.21506		
20	Ca	40	39.97534	96.97	0			
		42	41.97202	0.64				
		43	42.97237	0.145	$\frac{7}{2}$	−1.3160		
		44	43.96920	2.06				
		46	45.968	0.0033				
		48	47.96763	0.185				
21	Sc	45	44.97000	100	$\frac{7}{2}$	+4.7491		
22	Ti	46	45.96697	7.95				
		47	46.96668	7.75	$\frac{5}{2}$	−0.78710		
		48	47.96317	73.45				
		49	48.96358	5.51	$\frac{7}{2}$	−1.1022		
		50	49.96077	5.34				
23	V	50	49.96210	0.24	6	+3.3413		
		51	50.96052	99.76	$\frac{7}{2}$	+5.138	+0.3	
24	Cr	50	49.96210	4.31				
		52	51.95707	83.76				
		53	52.95772	9.55	$\frac{3}{2}$	−0.47354		
		54	53.9563	2.38				

Atomic number	Element	Mass number	Atomic mass in atomic mass units	Abundance, per cent	Spin	Magnetic moment, nuclear magnetons	Quadrupole moment $\times 10^{24}$ cm^2	Ratio of quadrupole moments
25	Mn	55	54.95581	100	$\frac{5}{2}$	+3.4611	+0.4	
26	Fe	54	53.95704	5.84				
		56	55.95272	91.68				
		57	56.95359	2.17				
		58	57.9520	0.31				
27	Co	59	58.95182	100	$\frac{7}{2}$	+4.6389	+0.5	
28	Ni	58	57.95345	67.76				
		60	59.94901	26.16				
		61	60.94907	1.25				
		62	61.94681	3.66				
		64	63.94755	1.16				
29	Cu	63	62.94926	69.1	$\frac{3}{2}$	+2.2213	−0.16	$Q_{63}/Q_{65} = 1.0806$
		65	64.94835	30.9	$\frac{3}{2}$	+2.3790	−0.14	
30	Zn	64	63.94955	48.89				
		66	65.94722	27.81				
		67	66.94815	4.11	$\frac{5}{2}$	+0.8735		
		68	67.94686	18.56				
		70	69.94779	0.62				
31	Ga	69	68.94778	60.2	$\frac{3}{2}$	+2.0108	+0.24	$Q_{69}/Q_{71} = 1.5867$
		71	70.94752	39.8	$\frac{3}{2}$	+2.5549	+0.15	
32	Ge	70	69.94637	20.55	0			
		72	71.94462	27.37	0			
		73	72.94669	7.67	$\frac{9}{2}$	−0.87680	−0.22	
		74	73.94466	36.74	0			
		76	75.94559	7.67	0			
33	As	75	74.94570	100	$\frac{3}{2}$	+1.43491	+0.3	
34	Se	74	73.94620	0.87	0			
		76	75.94357	9.02	0			
		77	76.94459	7.58	$\frac{1}{2}$	+0.53248		
		78	77.94232	23.52	0			
		80	79.94205	49.82	0			
		82	81.94285	9.19				
35	Br	79	78.94365	50.52	$\frac{3}{2}$	+2.0990	+0.33	$Q_{79}/Q_{81} = 1.19707$
		81	80.94232	49.48	$\frac{3}{2}$	+2.2626	+0.28	
36	Kr	78	77.94513	0.354				
		80	79.94194	2.27				
		82	81.93967	11.56	0			
		83	82.94059	11.55	$\frac{9}{2}$	−0.966	+0.16	
		84	83.93836	56.90	0			
		86	85.93828	17.37	0			
37	Rb	85	84.93920	72.15	$\frac{5}{2}$	+1.3482	+0.28	$Q_{87}/Q_{85} = 2.07$
		87	86.93709	27.85	$\frac{3}{2}$	+2.7414	+0.13	
38	Sr	84	83.94011	0.56				
		86	85.93684	9.86	0			
		87	86.93677	7.02	$\frac{9}{2}$	−1.0892		
		88	87.93408	82.56	0			
39	Y	89	88.93421	100	$\frac{1}{2}$	−0.136825		
40	Zr	90	89.93311	51.46				
		91	90.934	11.23	$\frac{5}{2}$	+1.3		
		92	91.933	17.11				
		94	93.934	17.40				
		96	95.936	2.80				
41	Nb	93	92.93540	100	$\frac{9}{2}$	+6.1435		
42	Mo	92	91.935	15.86	0			
		94	93.93522	9.12	0			
		95	94.936	15.70	$\frac{5}{2}$	−0.9290		
		96	95.93558	16.50				
		97	96.93693	9.45	$\frac{5}{2}$	−0.9485		
		98	97.937	23.75	0			
		100	99.93829	9.62	0			

Atomic number	Element	Mass number	Atomic mass in atomic mass units	Abundance, per cent	Spin	Magnetic moment, nuclear magnetons	Quadrupole moment $\times 10^{24}$ cm^2	Ratio of quadrupole moments
44	Ru	96	95.941	5.68				
		98	97.9363	2.22				
		99	98.938	12.81	$\frac{5}{2}$			
		100	99.9378	12.70				
		101	100.937	16.98	$\frac{5}{2}$			
		102	101.936	31.34				
		104	103.937	18.27				
45	Rh	103	102.937	100	$\frac{1}{2}$	$(-)0.11$		
46	Pd	102	101.939	0.8				
		104	103.93690	9.3				
		105	104.938	22.6	$\frac{5}{2}$	-0.6		
		106	105.936	27.2				
		108	107.93690	26.8				
		110	109.94098	13.5				
47	Ag	107	106.937	51.35	$\frac{1}{2}$	-0.11303		
		109	108.937	48.65	$\frac{1}{2}$	-0.12994		
48	Cd	106	105.943	1.215				
		108	107.940	0.875				
		110	109.93911	12.39				
		111	110.941	12.75	$\frac{1}{2}$	-0.59216		
		112	111.93999	24.07				
		113	112.94206	12.26	$\frac{1}{2}$	-0.61947		
		114	113.94013	28.86				
		116	115.94212	7.58				
49	In	113	112.942	4.23	$\frac{9}{2}$	$+5.4962$	$+1.18$	$Q_{115}/Q_{113} = 1.0146$
		115	114.94207	95.77	$\frac{9}{2}$	$+5.5074$	$+1.20$	
50	Sn	112	111.944	0.95				
		114	113.94109	0.65				
		115	114.94154	0.34	$\frac{1}{2}$	-0.91320		
		116	115.93806	14.24				
		117	116.94171	7.57	$\frac{1}{2}$	-0.9949		
		118	117.938	24.01				
		119	118.940	8.58	$\frac{1}{2}$	-1.0409		
		120	119.93904	32.97				
		122	121.94260	4.71				
		124	123.945	5.98				
51	Sb	121	120.942	57.25	$\frac{5}{2}$	$+3.3416$	-1.3	$Q_{123}/Q_{121} = 1.27475$
		123	122.944	42.75	$\frac{7}{2}$	$+2.5334$	-1.8	
52	Te	120	119.940	0.089				
		122	121.9391	2.46				
		123	122.9422	0.87	$\frac{1}{2}$	-0.7319		
		124	123.9393	4.61				
		125	124.9427	6.99	$\frac{1}{2}$	-0.8825		
		126	125.9417	18.71				
		128	127.9438	31.79				
		130	129.9475	34.49				
53	I	127	126.946	100	$\frac{5}{2}$	$+2.7938$	-0.61	
54	Xe	124	123.944	0.096				
		126	125.943	0.090				
		128	127.944	1.919				
		129	128.94533	26.44	$\frac{1}{2}$	-0.77244		
		130	129.945	4.08				
		131	130.947	21.18	$\frac{3}{2}$	$+0.704$	-0.12	
		132	131.94618	26.89				
		134	133.94804	10.44				
		136	135.95046	8.87				
55	Cs	133	132.948	100	$\frac{7}{2}$	$+2.5642$	-0.03	

Atomic number	Element	Mass number	Atomic mass in atomic mass units	Abundance, per cent	Spin	Magnetic moment, nuclear magnetons	Quadrupole moment $\times 10^{24}$ cm^2	Ratio of quadrupole moments
56	Ba	130	129.943	0.101				
		132	131.942	0.097				
		134	133.944	2.42				
		135	134.945	6.59	$\frac{3}{2}$	+0.830		
		136	135.946	7.81				
		137	136.948	11.32	$\frac{3}{2}$	+0.927		
		138	137.9498	71.66				
57	La	138	137.947	0.089	5	+3.68	+3	$Q_{138}/Q_{139} = 3.0$
		139	138.949	99.911	$\frac{7}{2}$	+2.7615	+0.9	
58	Ce	136	135.946	0.193				
		138	137.947	0.250				
		140	139.9488	88.48				
		142	141.9528	11.07				
59	Pr	141	140.9509	100	$\frac{5}{2}$	+3.8	−0.054	
60	Nd	142	141.959	27.13				
		143	142.956	12.20	$\frac{7}{2}$	−1.0		
		144	143.9562	23.87				
		145	144.959	8.30	$\frac{7}{2}$	−0.62		
		146	145.959	17.18				
		148	147.9642	5.72				
		150	149.9676	5.60				
62	Sm	144	143.9567	3.16				
		147	146.961	15.07	$\frac{5}{2}$	−0.68		
		148	147.9616	11.27				
		149	148.963	13.84	$\frac{5}{2}$	−0.55		
		150	149.9632	7.47				
		152	151.9677	26.63				
		154	153.9712	22.53				
63	Eu	151	150.963	47.77	$\frac{5}{2}$	+3.4	+1.2	
		153	152.965	52.23	$\frac{5}{2}$	+1.5	+2.6	
64	Gd	152	151.970	0.20				
		154	153.9694	2.15				
		155	154.970	14.73				
		156	155.9715	20.47				
		157	156.973	15.68				
		158	157.9736	24.87				
		160	159.9785	21.90				
65	Tb	159	158.972	100	$\frac{3}{2}$			
66	Dy	156	155.972	0.0524				
		158	159.975	0.0902				
		160	159.9752	2.294				
		161	160.977	18.88	$\frac{7}{2}$			
		162	161.9779	25.53				
		163	162.980	24.97	$\frac{7}{2}$			
		164	163.9814	28.18				
67	Ho	165	164.9822	100	$\frac{7}{2}$			
68	Er	162	161.980	0.136				
		164	163.9827	1.56				
		166	165.982	33.41				
		167	166.983	22.94	$\frac{7}{2}$		+10	
		168	167.9849	27.07				
		170	169.9907	14.88				
69	Tm	169	168.985	100	$\frac{1}{2}$			
70	Yb	168	167.983	0.140				
		170	169.985	3.03				
		171	170.987	14.31	$\frac{1}{2}$	+0.45		
		172	171.988	21.82				
		173	172.989	16.13	$\frac{5}{2}$	−0.66	+4.0	
		174	173.991	31.84				
		176	175.995	12.73				

Atomic number	Element	Mass number	Atomic mass in atomic mass units	Abundance, per cent	Spin	Magnetic moment, nuclear magnetons	Quadrupole moment $\times 10^{24}$ cm^2	Ratio of quadrupole moments
71	Lu	175	174.993	97.40	$\frac{7}{2}$	+2.9	+6.5	
		176	175.995	2.60	7	+4.2	+8	
72	Hf	174	173.992	0.18				
		176	175.9957	5.15				
		177	176.998	18.39	$(\frac{1}{2},\frac{3}{2})$			
		178	177.9988	27.08				
		179	179.002	13.78	$(\frac{1}{2},\frac{3}{2})$			
		180	180.0031	35.44				
73	Ta	181	180.999	100	$\frac{7}{2}$	+2.1	+7	
74	W	180	180.001	0.135				
		182	182.0041	26.4				
		183	183.0066	14.4	$\frac{1}{2}$	+0.087		
		184	184.0074	30.6				
		186	186.010	28.4				
75	Re	185	185.009	37.07	$\frac{5}{2}$	+3.1438	+2.9	$Q_{185}/Q_{187} = 1.06$
		187	187.012	62.93	$\frac{5}{2}$	+3.1760	+2.7	
76	Os	184	184.010	0.018				
		186	186.012	1.59				
		187	187.014	1.64				
		188	188.0157	13.3				
		189	189.017	16.1	$\frac{3}{2}$	+0.65066	+2.0	
		190	190.0174	26.4				
		192	192.0225	41.0				
77	Ir	191	191.020	38.5	$\frac{3}{2}$	0.17	+1.5	
		193	193.024	61.5	$\frac{3}{2}$	+0.17	+1.5	
78	Pt	190	190.018	0.012				
		192	192.021	0.78				
		194	194.0241	32.8				
		195	195.0265	33.7	$\frac{1}{2}$	+0.6036		
		196	196.0267	25.4				
		198	198.0327	7.23				
79	Au	197	197.030	100	$\frac{3}{2}$	+0.14		
80	Hg	196	196.029	0.146				
		198	198.032	10.02				
		199	199.034	16.84	$\frac{1}{2}$	+0.49930		
		200	200.036	23.13				
		201	201.038	13.22	$\frac{3}{2}$	−0.607	+0.5	
		202	202.040	29.80				
		204	204.045	6.85				
81	Tl	203	203.03499	29.50	$\frac{1}{2}$	+1.5960		
		205	205.03792	70.50	$\frac{1}{2}$	+1.6116		
82	Pb	204	204.0363	1.48				
		206	206.0388	23.6				
		207	207.0405	22.6	$\frac{1}{2}$	+0.58367		
		208	208.0416	52.3				
83	Bi	209	209.0446	100	$\frac{9}{2}$	+4.388	−0.4	
90	Th	232	232.11034	100				
92	U	234	234.11379	0.0058				
		235	235.11704	0.715	$(\frac{5}{2})$			
		238	238.12493	99.28				

BIBLIOGRAPHY

The bibliography has been arranged by year and within each year alphabetically according to the name of the first author. A brief statement of the contents of each reference will be found in the right hand column of the bibliography except when a book is listed, in which case the title is expected to serve as a statement of contents.

In addition to references concerning microwave spectroscopy directly, there are others listed because they contain closely related material, or because they are referred to in the text. References which are not directly connected with microwave spectroscopy are identified by numbers in italics.

Prior to 1929

1. Lebedew, P., *Wied. Ann.*, **56**, 1 (1895). — Mm-wave spark generator
2. Lorentz, H. A., *Proc. Amst. Akad. Sci.*, **8**, 591 (1906). — Th. pressure broadening
3. Nicholls, E. F., and J. D. Tear, *Phys. Rev.*, **21**, 587 (1923). — Mm waves
4. Nicholls, E. F., and J. D. Tear, *Proc. Natl. Acad. Sci. U.S.*, **9**, 221 (1923). — Mm waves
5. Glagolewa-Arkadiewa, A., *Nature*, **113**, 640 (1924). — Mm-wave spark generator
6. Lewitzky, M., *Phys. Zeits.*, **25**, 107 (1924); **27**, 177 (1926). — Mm-wave spark generator
7. Thomas, L. H., *Nature*, **117**, 514 (1926). — Thomas precession
8. Born, M., and J. R. Oppenheimer, *Ann. Physik*, **4–84**, 457 (1927). — Separation of motions
8a. Kramers, H. A., *Atti del congr. intern. dei fisici, Como*, **2**, 545 (1927). — Kramers-Kronig relations
9. Grotrian, W., *Graphische Darstellung der Spektren von Atomen*, Springer-Verlag-OHG, Berlin (1928).
10. Hill, E. L., and J. H. Van Vleck, *Phys. Rev.*, **32**, 250 (1928). — Λ-doubling
11. Mulholland, H. P., *Proc. Cambridge Phil. Soc.*, **24**, 280 (1928). — Partition function

1929

12. Debye, P., *Polar Molecules*, Chemical Catalog Company, Inc., New York.
13. Glagolewa-Arkadiewa, A., *Z. Physik*, **55**, 234. — Mm-wave spark generator
14. Hill, E. L., *Phys. Rev.*, **34**, 1507. — Th. mol. Zeeman effect
15. Kramers, H. A., *Z. Physik*, **53**, 422. — ρ tripling
16. Morse, P. M., *Phys. Rev.*, **34**, 57. — Mol. potential
17. Van Vleck, J. H., *Phys. Rev.*, **33**, 467. — Λ doubling
18. Wang, S. C., *Phys. Rev.*, **34**, 243. — Asymmetric rotator

1930

19. Breit, G., *Phys. Rev.*, **35**, 1447. — Atomic h.f.s.
19a. Breit, G., and I. I. Rabi, *Phys. Rev.*, **38**, 2082L. — Zeeman effect in atoms

20. Brouwer, F., Dissertation, Amsterdam.   Stark effect diatomic molecules

21. Kronig, R. de L., *Band Spectra and Molecular Structure*, Cambridge University Press, New York.

22. Mulliken, R. S., *Revs. Mod. Phys.*, **2**, 60.   Review mol. spectra

23. Pauling, L., and S. A. Goudsmit, *The Structure of Line Spectra*, McGraw-Hill Book Company, Inc., New York.

1931

24. Casimir, H. B. G., *Rotation of a Rigid Body in Quantum Mechanics*, J. B. Wolter's, The Hague.

25. Dennison, D. M., *Revs. Mod. Phys.*, **3**, 280.   Review mol. spectra

26. Mulliken, R. S., *Revs. Mod. Phys.*, **3**, 89.   Review mol. spectra

27. Mulliken, R. S., and A. Christy, *Phys. Rev.*, **38**, 87. Λ doubling

28. Racah, G., *Z. Physik*, **71**, 431.   Atomic h.f.s.

29. Weizel, W., *Bandenspektren*, Akad. Verlagsges., Leipzig.

1932

30. Bacher, R., and S. A. Goudsmit, *Atomic Energy States*, McGraw-Hill Book Company, Inc., New York.

31. Betz, O., *Ann. Physik*, **15**, 321.   Exp. H fine structure

32. Condon, E. U., *Phys. Rev.*, **41**, 759.   Induced dipole transitions

33. Dennison, D. M., and G. E. Uhlenbeck, *Phys. Rev.*, **41**, 313.   Th. NH_3

34. Dunham, J. L., *Phys. Rev.*, **41**, 721.   Rotation-vibration interaction

36. Nielsen, H. H., *Phys. Rev.*, **40**, 445.   Th. hindered rotation

37. Ray, B. S., *Z. Physik*, **78**, 74.   Asymmetric rotator

38. Van Vleck, J. H., *Theory of Electric and Magnetic Susceptibilities*, Clarendon Press, Oxford.

1933

39. Blockinzew, D., *Phys. Zeit. U.S.S.R.*, **4**, 501.   High-frequency modulation

40. Fermi, E., and E. Segre, *Z. Physik*, **82**, 729; *Reale Accad. D'Italia—Scienze Fiziche, Mat. E. Naturali Memorie*, **4**, 131.   Atomic h.f.s.

41. Goudsmit, S. A., *Phys. Rev.*, **43**, 636.   Atomic h.f.s.

42. Kronig, R. de L., *Physica*, **1**, 617.   Masses from mol. spectra

43. Placzek G., and E. Teller, *Z. Physik*, **81**, 209.   Mol. symmetries and statistical weights

44. Weisskopf, V. F., *Phys. Zeits.*, **34**, 1.   Th. pressure broadening

44a. Wright, N., and H. M. Randall, *Phys. Rev.*, **44**, 391. NH_3 I.R. spectrum

1934

45. Cleeton, C. E., and N. H. Williams, *Phys. Rev.*, **45**, 234. NH_3 microwave spectrum

46. Cosens, C. R., *Proc. Phys. Soc. London*, **46**, 818.   Lock-in amplifier

47. Crawford, F. H., *Revs. Mod. Phys.*, **6**, 90.   Survey mol. Zeeman effect

47a. Dieke, G. H., and G. B. Kistiakowsky, *Phys. Rev.*, **45**, 4.   Asymmetric rotor

48. Gordon, A. R., *J. Chem. Phys.*, **2**, 65.   Partition function

49. Glagolewa-Arkadiewa, A., *Compt. rend. U.S.S.R.*, **3**, 415.   Mm-wave spark generator

50. Kuhn, H., and F. London, *Phil. Mag.*, **18**, 983.   Th. pressure broadening
51. Kuhn, H., *Phil. Mag.*, **18**, 987.   Th. pressure broadening
52. Pekeris, C. L., *Phys. Rev.*, **45**, 98.   Rotation-vibration
                                                                 diatomic molecule

53. White, H. E., *Introduction to Atomic Spectra*,
    McGraw-Hill Book Company, Inc., New York.

1935

54. Bartunek, P. F., and E. F. Barker, *Phys. Rev.*, **48**,   OCS I.R. spectrum
    516.
55. Budó, A., *Z. Physik*, **96**, 219.   Fine-structure $^3\Pi$
                                                                 molecules

56. Condon, E. U., and G. H. Shortley, *The Theory of
    Atomic Spectra*, The Macmillan Company, New
    York.
57. Crawford, F. H., and T. Jorgensen, Jr., *Phys. Rev.*,   Rotation-vibration in
    47, 358.   LiH
58. Crawford, F. H., and T. Jorgensen, Jr., *Phys. Rev.*,   Rotation-vibration in
    47, 932.   LiH
59. Cross, P. C., *Phys. Rev.*, **47**, 7.   I.R. spectrum H_2S
60. Haase, T., *Ann. Physik*, **23**, 675.   Exp. H fine structure
61. Manning, M. F., *J. Chem. Phys.*, **3**, 136.   Theory NH_3
62. Pauling, L., and E. B. Wilson, Jr., *Introduction to
    Quantum Mechanics*, McGraw-Hill Book Company,
    Inc., New York.
63. Page, L., *Introduction to Theoretical Physics*, D. Van
    Nostrand Company, Inc., New York.
64. Wilson, E. B., Jr., *J. Chem. Phys.*, **3**, 276.   Statistical weights
65. Whittaker, E. T., and G. N. Watson, *Modern
    Analysis*, Cambridge University Press, New York.
66. Wilson, E. B., Jr., *J. Chem. Phys.*, **3**, 818.   Symmetry and
                                                                 statistical weights

1936

67. Bethe, H. A., and R. F. Bacher, *Revs. Mod. Phys.*,   Review nuclei also
    8, 82.   atomic h.f.s.
68. Brandt, W. H., *Phys. Rev.*, **50**, 778.   Electronic quartet
                                                                 states
69. Budó, A., *Z. Physik*, **98**, 437. $^3\Pi$ states
70. Casimir, H. B. G., *On the Interaction between Atomic
    Nuclei and Electrons*, Teyler's Tweede Genootschap,
    E. F. Bohn, Haarlem.
71. Crawford, F. H., and T. Jorgensen, Jr., *Phys. Rev.*,   Rotation-vibration
    49, 745.   LiH
72. Gilbert, C., *Phys. Rev.*, **49**, 619.   Fine-structure triplet
                                                                 molecules
73. Hebb, M. H., *Phys. Rev.*, **49**, 610. ρ-type tripling
74. Schmid, R., A. Budó, and J. Zemplén, *Z. Physik*, **103**,   Zeeman effect O_2
    250.
75. Van Vleck, J. H., *J. Chem. Phys.*, **4**, 327.   Isotope shift molecular
                                                                 spectra
76. Wilson, E. B., Jr., and J. B. Howard, *J. Chem. Phys.*,   General formulation
    4, 260.   rotation-vibration

1937

77. Almy, G. M., and R. B. Horsfall, *Phys. Rev.*, **51**, 491. $^2\Pi$ states
78. Budó, A., *Z. Physik*, **105**, 73.   Mol. quartet states
79. Candler, A. C., *Atomic Spectra and the Vector Model*,
    Cambridge University Press, New York.

80. Houston, W. V., *Phys. Rev.*, **51**, 446.   H fine structure
81. Kemble, E. C., *The Fundamental Principles of Quantum Mechanics*, McGraw-Hill Book Company, Inc., New York.
82. Margenau, H., and D. T. Warren, *Phys. Rev.*, **51**, 748.   Theory mol. interaction
83. Nevin, T. E., *Nature*, **140**, 1101.   Mol. quartet states
84. Randall, H. M., D. M. Dennison, N. Ginsburg, and H_2O I.R. spectrum
    L. R. Weber, *Phys. Rev.*, **52**, 160.
85. Schlapp, R., *Phys. Rev.*, **51**, 342. ρ-type tripling

1938

86. Kennard, E. H., *Kinetic Theory of Gases*, McGraw-Hill Book Company, Inc., New York.
86a. Meacham, L. A., *Proc. IRE*, **26**, 1278.   Stabilized crystal
                                                                  oscillator
87. Nevin, T. E., *Phil. Trans. Roy. Soc. London*, **237**, 471. Mol. quartet states
88. Pasternack, S., *Phys. Rev.*, **54**, 1113.   H fine structure
89. Williams, R. C., *Phys. Rev.*, **54**, 558.   H fine structure
90. Wilson, E. B., Jr., *J. Chem. Phys.*, **6**, 740.   Symmetry and
                                                                  statistical weights

1939

91. Kellogg, J. M. B., I. I. Rabi, N. F. Ramsey, and H_2 R.F. spectrum
    J. R. Zacharias, *Phys. Rev.*, **56**, 728.
91a. Kellogg, J. M. B., N. F. Ramsey, Jr., I. I. Rabi, and   HD and D_2 R.F.
    J. R. Zacharias, *Phys. Rev.*, **57**, 677.   spectrum
92. Kalugina, A., *J. Exp. Theor. Phys. U.S.S.R.*, **9**, 362. Mm-wave spark
                                                                  generator
93. Margenau, H., *Revs. Mod. Phys.*, **11**, 1.   Review mol.
                                                                  interactions
94. Pauling, L., *Nature of the Chemical Bond*, Cornell University Press, Ithaca, N.Y.
94a. Slawsky, Z. I., and D. M. Dennison, *J. Chem. Phys.*,   Theory centrifugal
    7, 509.   distortion
95. Wu, T. Y., *Vibrational Spectra and Structure of Polyatomic Molecules*, National University of Peking, Kun-ming, China.

1940

96. Dennison, D. M., *Revs. Mod. Phys.*, **12**, 175.   Review mol. spectra
97. Drinkwater, J. W., O. Richardson, and W. E.   H fine structure
    Williams, *Proc. Roy. Soc. London A*, **174**, 164.
98. Kellogg, J. M. B., I. I. Rabi, N. F. Ramsey, and H_2 R.F. spectrum
    J. R. Zacharias, *Phys. Rev.*, **57**, 677.
99. Koehler, J. S., and D. M. Dennison, *Phys. Rev.*, **57**,   Th. CH_3OH
    1006.
100. Mulliken, R. S., *Phys. Rev.*, **57**, 500.   Hund's case (*c*)
101. Nordsieck, A., *Phys. Rev.*, **60**, 310.   *q* for H_2
102. Ramsey, N. F., *Phys. Rev.*, **58**, 226. H_2 R.F. spectrum
103. Sandeman, I., *Proc. Roy. Soc. Edinburgh*, **60**, 210.   Rotation-vibration
                                                                  diatomic molecules
104. Spitzer, L., Jr., *Phys. Rev.*, **58**, 348.   Th. pressure broadening
105. Wills, A. P., *Vector Analysis*, Prentice-Hall, Inc., New York.

1941

106. Hulburt, H. M., and J. O. Hirschfelder, *J. Chem.   Molecular potential
     Phys.*, **8**, 61.
107. Michaels, W. C., and N. L. Curtis, *Rev. Sci. Instr.*,   Lock-in amplifier
     12, 444.

108. Nielsen, H. H., *Phys. Rev.*, **60**, 794.
Rotation-vibration polyatomic molecules

109. Schomaker, V., and D. P. Stevenson, *J. ACS*, **63**, 37.
Bond lengths

110. Sheng, H.-Y., E. F. Barker, and D. M. Dennison, *Phys. Rev.*, **60**, 786.
Th. and exp. NH_3

110a. Shrader, J. H., and E. C. Pollard, *Phys. Rev.*, **59**, 277.
Cl masses

111. Torrey, H. C., *Phys. Rev.*, **59**, 293.
Theory intensities R.F. spectra

1942

112. Pitzer, K. S., and W. D. Gwinn, *J. Chem. Phys.*, **10**, 428.
Internal rotation

113. Racah, G., *Phys. Rev.*, **62**, 438.
Theory complex spectra

114. Shaffer, W. H., *J. Chem. Phys.*, **10**, 1.
Rotation-vibration XY_3Z molecules

115. Silver, S., *J. Chem. Phys.*, **10**, 565.
Rotation-vibration XY_2Z molecules

116. Silver, S., and E. S. Ebers, *J. Chem. Phys.*, **10**, 559.
Rotation-vibration XYZ molecules

117. Slater, J. C., *Microwave Transmission*, McGraw-Hill Book Company, Inc., New York.

1943

118. King, G. W., R. M. Hainer, and P. C. Cross, *J. Chem. Phys.*, **11**, 27.
Energy levels asymmetric rotor

119. Nielsen, A. H., *J. Chem. Phys.*, **11**, 160.
Rotation-vibration linear XYZ molecules

120. Nielsen, H. H., and W. H. Shaffer, *J. Chem. Phys.*, **11**, 140. *Errata: Phys. Rev.*, **75**, 1961 (1949).
l-type doubling

121. Terman, F. E., *Radio Engineers' Handbook*, McGraw-Hill Book Company, Inc., New York.

1944

122. Cross, P. C., R. M. Hainer, and G. W. King, *J. Chem. Phys.*, **12**, 210.
Intensities asymmetric rotor

123. Eyring, H., J. Walter, and G. E. Kimball, *Quantum Chemistry*, John Wiley & Sons, Inc., New York

124. Herzberg, G., *Atomic Spectra and Atomic Structure*, 2d ed., Dover Publications, New York.

125. Nielsen, H. H., *J. Opt. Soc. Amer.*, **34**, 521.
Rotation-vibration interactions

126. Nielsen, H. H., *Phys. Rev.*, **66**, 282.
Th. linear molecules

127. Shaffer, W. H., and R. P. Schuman, *J. Chem. Phys.*, **12**, 504.
Rotation-vibration XYZ molecules

1945

128. Cooley, J. P., and J. H. Rohrbaugh, *Phys. Rev.*, **67**, 296.
Spark radiator

129. Feld, B. T., and W. E. Lamb, Jr., *Phys. Rev.*, **76**, 15.
Quadrupole h.f.s. in molecules

130. Herzberg, G., *Infrared and Raman Spectra*, D. Van Nostrand Company, Inc., New York.

130a. Jabłónski, A., *Phys. Rev.*, **68**, 78. *Errata: Phys. Rev.*, **69**, 31 (1946).
Th. pressure broadening

131. Lindholm, E., *Ark. Mat. Astron. Fysik*, **32**, 17.
Th. pressure broadening

132. Nielsen, H. H., *Phys. Rev.*, **68**, 181.
Rotation-vibration polyatomic molecules

133. Pauling, L., *Nature of the Chemical Bond*, Cornell University Press, Ithaca, N.Y.

134. Shaffer, W. H., and R. C. Herman, *J. Chem. Phys.,* **13,** 83. — Rotation-vibration X_2XZ_2 molecules

134a. Southworth, G. C., *J. Franklin Inst.,* **239,** 285. — Microwaves from the sun

135. Van Vleck, J. H., *Report 664 from MIT Radiation Laboratory* (Div. 14, N.D.R.C). — Atmospheric absorption microwaves

135a. Van Vleck, J. H., *Report 735 from MIT Radiation Laboratory.* — Kramers-Kronig relations

136. Van Vleck, J. H., and V. F. Weisskopf, *Revs. Mod. Phys.,* **17,** 227. — Th. pressure broadening

1946

137. Becker, G. E., and S. H. Autler, *Phys. Rev.,* **70,** 300. — H_2O pressure broadening

139. Beringer, R., *Phys. Rev.,* **70,** 53. — O_2 high pressure

140. Bleaney, B., *Physica,* **12,** 595. — Review

141. Bleaney, B., and R. P. Penrose, *Nature,* **157,** 339. — NH_3

142. Bleaney, B., and R. P. Penrose, *Phys. Rev.,* **70,** 775L. — NH_3

143. Coles, D. K., and W. E. Good, *Phys. Rev.,* **70,** 979L. — NH_3 quadrupole structure

144. Coon, J., *J. Chem. Phys.,* **14,** 665. — ClO_2 electronic spectrum

145. Dailey, B. P., R. L. Kyhl, M. W. P. Strandberg, J. H. Van Vleck, and E. B. Wilson, Jr., *Phys. Rev.,* **70,** 984L. — NH_3 quadrupole structure

146. Dakin, T. W., W. E. Good, and D. K. Coles, *Phys. Rev.,* **70,** 560L. — OCS, Stark effect

148. Dicke, R. H., *Rev. Sci. Instr.,* **17,** 268. — Microwave radiometer

149. Dicke, R. H., and R. Beringer, *Astrophys. J.,* **103,** 375. — Microwaves from sun

150. Dicke, R. H., R. Beringer, R. L. Kyhl, and A. B. Vane, *Phys. Rev.,* **70,** 340. — Atmospheric absorption

151. Fisk, J. B., H. D. Hagstrum, and P. L. Hartman, *Bell System Tech. J.,* **25,** 167. — Review magnetrons

152. Fiske, M. D., *Rev. Sci. Instr.,* **17,** 478. — Waveguide windows

153. Foley, H. M., *Phys. Rev.,* **69,** 616. — Th. pressure broadening

154. Fröhlich, H., *Nature,* **157,** 478. — Th. pressure broadening

155. Good, W. E., *Phys. Rev.,* **70,** 109A; 213; *Phys. Rev.,* **69,** 539L. — NH_3

156. Haar, D. ter, *Phys. Rev.,* **70,** 222. — Vibration anharmonic oscillator

157. Hadley, L. N., and D. M. Dennison, *Phys. Rev.,* **70,** 780. — Th. NH_3 inversion

158. Hershberger, W. D., *J. Appl. Phys.,* **17,** 495. — Absorption at medium pressures

159. Hershberger, W. D., E. T. Bush, and G. W. Leck, *RCA Rev.,* **7,** 422. — Thermal, acoustic effects microwave absorption

160. Hughes, H. K., *Phys. Rev.,* **70,** 570. *Errata: Phys. Rev.,* **70,** 909. — Electric field technique molecular beam

161. Jablónski, A., *Phys. Rev.,* **69,** 31. — Th. pressure broadening

162. Jablónski, A., *Physica's Grav.,* **7,** 541. — Th. pressure broadening

163. Kellogg, J. M. B., and S. Millman, *Revs. Mod. Phys.,* **18,** 323. — Review mol. beams

164. Kyhl, R. L., R. H. Dicke, and R. Beringer, *Phys. Rev.,* **69,** 694A. — H_2O atmospheric absorption

165. Lamb, W. E., Jr., *Phys. Rev.,* **70,** 308. — Th. large cavity

166. Lamont, H. R. L., and A. G. D. Watson, *Nature,* **158,** 943. — 6 mm atm. absorption

137. Malter, L., R. L. Jepsen, and L. R. Bloom, *RCA Rev.*, **7**, 622. — Mica windows

168. Nielsen, H. H., *Phys. Rev.*, **70**, 184. — Rotation-vibration symmetric top

169. Pound, R. V., *Rev. Sci. Instr.*, **17**, 490. — Stabilized microwave oscillator

170. Roberts, A., Y. Beers, and A. G. Hill, *Phys. Rev.*, **70**, 112A. — Atomic Cs h.f.s.

171. Sproull, R. L., and E. G. Linder, *Proc. IRE*, **34**, 305. — Resonant cavities

172. Townes, C. H., *Phys. Rev.*, **70**, 109A, 665. — NH_3

173. Townes, C. H., and F. R. Merritt, *Phys. Rev.*, **70**, 558L. — H_2O, HDO

174. Walter, J. E., and W. D. Hershberger, *J. Appl. Phys.*, **17**, 814. — Absorption at medium pressures

175. Wilson, I. G., C. W. Schramm, and J. P. Kinzer, *Bell System Tech. J.*, **25**, 408. — Resonant cavities

1947

176. Beard, C. I., and B. P. Dailey, *J. Chem. Phys.*, **15**, 762L. — HNCS, DNCS

177. Bethe, H. A., *Phys. Rev.*, **72**, 339. — H fine-structure theory

178. Bleaney, B., J. H. N. Loubser, and R. P. Penrose, *Proc. Phys. Soc.*, **59**, 185. — Cavity resonators

179. Bleaney, B., and R. P. Penrose, *Proc. Roy. Soc.*, **A189**, 358. — NH_3

180. Bleaney, B., and R. P. Penrose, *Proc. Phys. Soc.*, **59**, 418. — NH_3 pressure broadening

181. Bracewell, R. N., *Proc. IRE*, **35**, 830. — Cavity resonators

182. Carter, R. L., and W. V. Smith, *Phys. Rev.*, **72**, 1265L. — Frequency markers

183. Coles, D. K., *Proc. Nat. Electronics Conf.*, **1947**, 180. — Review

184. Coles, D. K., E. S. Elyash, and J. G. Gorman, *Phys. Rev.*, **72**, 973L. — N_2O

185. Coles, D. K., and W. E. Good, *Phys. Rev.*, **72**, 157A. — Stark, Zeeman effects

186. Dailey, B. P., *Phys. Rev.*, **72**, 84L. — CH_3OH

187. Dailey, B. P., S. Golden, and E. Bright Wilson, Jr., *Phys. Rev.*, **72**, 871L. — SO_2

188. Dailey, B. P., and E. Bright Wilson, Jr., *Phys. Rev.*, **72**, 522A. — SO_2, CH_3NO_2, CH_3OH

189. Dakin, T. W., W. E. Good, and D. K. Coles, *Phys. Rev.*, **71**, 640L. — OCS

190. Feld, B. T., *Phys. Rev.*, **72**, 1116L. — Th. quad. interactions

191. Foley, H. M., *Phys. Rev.*, **71**, 747. — Th. quad. interactions

192. Foley, H. M., *Phys. Rev.*, **72**, 504. — Magnetic effects $^1\Sigma$ mols

194. Ginsburg, V. L., *Bull. Acad. Sci. U.S.S.R., Ser. Phys.*, **11**, 165. — Survey mm waves

195. Golay, M. J. E., *Rev. Sci. Instr.*, **18**, 357. — Pneumatic detector

196. Golay, M. J. E., *Rev. Sci. Instr.*, **18**, 347. — Th. pneumatic detector

197. Good, W. E., and D. K. Coles, *Phys. Rev.*, **71**, 383L. — $N^{14}H_3$ $N^{15}H_3$

198. Good, W. E., and D. K. Coles, *Phys. Rev.*, **72**, 157A. — $N^{14}H_3$ $N^{15}H_3$

199. Gordy, W., *J. Chem. Phys.*, **15**, 305. — Bond lengths

200. Gordy, W., *J. Chem. Phys.*, **15**, 81. — Bond lengths

201. Gordy, W., and M. Kessler, *Phys. Rev.*, **71**, 640L. — NH_3 h.f.s.

202. Gordy, W., and M. Kessler, *Phys. Rev.*, **72**, 644L. — Double modulation spectrograph

203. Gordy, W., J. W. Simmons, and A. G. Smith, *Phys. Rev.*, **72**, 344L. — CH_3Cl, CH_3Br

204. Gordy, W., A. G. Smith, and J. W. Simmons, *Phys. Rev.*, **71**, 917L.   CH₃I

204. Gordy, W., A. G. Smith, and J. W. Simmons, *Phys. Rev.*, **71**, 917L. CH$_3$I
205. Gordy, W., A. G. Smith, and J. W. Simmons, *Phys. Rev.*, **72**, 249L. CH$_3$I
206. Gordy, W., W. V. Smith, A. G. Smith, and H. Ring, *Phys. Rev.*, **72**, 259. BrCN, ICN
207. Hershberger, W. D., and J. Turkevich, *Phys. Rev.*, **71**, 554L. CH$_3$OH, CH$_3$NH$_2$
208. Hillger, R. E., M. W. P. Strandberg, T. Wentink, and R. Kyhl, *Phys. Rev.*, **72**, 157A. OCS
209. Hughes, H. K., *Phys. Rev.*, **72**, 614. CsF molecular beam
210. Hughes, R. H., and E. B. Wilson, Jr., *Phys. Rev.*, **71**, 562L. Stark spectrograph
211. Hunt, L. E., *Proc. IRE*, **35**, 979. Frequency measurement
212. Jauch, J. M., *Phys. Rev.*, **72**, 715; 535A. NH$_3$ theory h.f.s. and Stark effect
213. Jen, C. K., *Phys. Rev.*, **72**, 986L. Zeeman; NH$_3$
214. Kahan, T., *J. Phys. Radium*, **8**, 192. Microwave interferometer
215. King, G. W., *J. Chem. Phys.*, **15**, 820. Asymmetric rotor correspondence principle
216. King, G. W., and R. M. Hainer, *Phys. Rev.*, **71**, 135. HDO
217. King, G. W., R. M. Hainer, and P. C. Cross, *Phys. Rev.*, **71**, 433; **70**, 108A (1946). Predicted absorption H$_2$O, etc.
218. Kinzer, J. P., and I. G. Wilson, *Bell System Tech. J.*, **410.** Cavity resonators
219. Lamb, W. E., Jr., and R. C. Retherford, *Phys. Rev.*, **72**, 241. H fine structure
220. Miller, P. H., *Proc. IRE*, **35**, 252. Crystal detector noise
221. Montgomery, C. G., *Technique of Microwave Measurements*, MIT Radiation Laboratory Series, Vol. 11, McGraw-Hill Book Company, Inc., New York.
222. Mumford, W. W., *Proc. IRE*, **35**, 160. Directional couplers
223. Nielsen, H. H., and D. M. Dennison, *Phys. Rev.*, **72**, 86L; 1101. NH$_3$ anomalies
224. Nierenberg, W. A., and N. F. Ramsey, *Phys. Rev.*, **72**, 1075. NaCl, NaBr, NaI mol. beams
225. Pierce, J. R., and W. G. Shepherd, *Bell System Tech. J.*, **26**, 460. Reflex klystrons
226. Pond, T. A., and W. F. Cannon, *Phys. Rev.*, **72**, 1121L. NH$_3$ saturation
227. Pound, R. V., *Proc. IRE*, **35**, 1405. Stabilization of oscillators
228. Rideout, V. C., *Proc. IRE*, **35**, 767. Stabilization of oscillators
229. Ring, H., H. Edwards, M. Kessler, and W. Gordy, *Phys. Rev.*, **72**, 1262. CH$_3$CN, CH$_3$NC
230. Roberts, A., *Nucleonics*, **1**, 10. Review nuclear effects
231. Saxton, J. A., *Rept. Phys. Soc., London; Rept. Met. Soc.*, **1947**, 215. Water-vapor absorption
232. Sherbin, L. E., *Electronics*, **20**, 122. Waveguide data
233. Smith, W. V., *Phys. Rev.*, **71**, 126L. Spin from intensities
234. Smith, W. V., and R. L. Carter, *Phys. Rev.*, **72**, 638L. NH$_3$, saturation
235. Smith, W. V., J. L. Garcia de Quevedo, R. L. Carter, and W. S. Bennett, *J. Appl. Phys.*, **18**, 1112. Frequency stabilization
236. Strandberg, M. W. P., R. Kyhl, T. Wentink, and R. E. Hillger, *Phys. Rev.*, **71**, 326L. *Errata: Phys. Rev.*, **71**, 639L. NH$_3$

237. Talpey, R. G., and H. Goldberg, *Proc. IRE*, **35**, 965.   Frequency standard
238. Tolansky, S., *High Resolution Spectroscopy*, Methuen & Co., Ltd., London.
239. Townes, C. H., *Phys. Rev.*, **71**, 909L.   Quadrupole moments, q
240. Townes, C. H., A. N. Holden, J. Bardeen, and F. R.   Br, Cl, N moments
   Merritt, *Phys. Rev.*, **71**, 644L.   *Errata:* 829L.
241. Townes, C. H., A. N. Holden, and F. R. Merritt,   Linear molecules
   Phys. Rev., **71**, 64L, 479A.
242. Townes, C. H., A. N. Holden, and F. R. Merritt,   Linear molecules, mass
   Phys. Rev., **72**, 513, 740A.   ratios, l doubling
243. Townes, C. H., and F. R. Merritt, *Phys. Rev.*, **72**,   High-frequency Stark
   1266L; *Phys. Rev.*, **73**, 1249A.   effect
244. Van Vleck, J. H., *Phys. Rev.*, **71**, 413. O_2
245. Van Vleck, J. H., *Phys. Rev.*, **71**, 425. H_2O
246. Van Vleck, J. H., *Phys. Rev.*, **71**, 468A.   Symmetric-top quad.
     coupling
247. Watts, R. J., and D. Williams, *Phys. Rev.*, **71**, 639L; NH_3 quad.
   Phys. Rev., **72**, 157A; 263.
248. Watts, R. J., and D. Williams, *Phys. Rev.*, **72**, 1122L.   Double-modulation
     spectrograph
249. Watts, R. J., and D. Williams, *Phys. Rev.*, **72**, 980L.   Stark spectrograph
250. Weidner, R. T., *Phys. Rev.*, **72**, 1268L.   ICl
251. Williams, D., *Phys. Rev.*, **72**, 974L. NH_3

1948

252. Bardeen, J., and C. H. Townes, *Phys. Rev.*, **73**, 97.   Th. h.f.s. in molecules
253. Bardeen, J., and C. H. Townes, *Phys. Rev.*, **73**, 627.   Second-order
   Errata: Phys. Rev., **73**, 1204.   quadrupole h.f.s.
254. Bleaney, B., *Repts. Progr. Phys.*, **11**, 178 (1946–1947).   Review
255. Bleaney, B., and J. H. N. Loubser, *Nature*, **161**, 522L. NH_3 pressure
     broadening
256. Bleaney, B., and R. P. Penrose, *Proc. Phys. Soc.*   Power saturation
   London, **60**, 83.
257. Bleaney, B., and R. P. Penrose, *Proc. Phys. Soc.* NH_3 pressure
   London, **60**, 540.   broadening
258. Bragg, J. K., *Phys. Rev.*, **74**, 533.   Th. quad. h.f.s.
     asymmetric molecules
259. Carter, R. L., and W. V. Smith, *Phys. Rev.*, **73**, 1053;   Power saturation
   74, 123A.
260. Coates, R. J., *Rev. Sci. Instr.*, **19**, 586.   Grating for mm waves
261. Coles, Donald K., *Phys. Rev.*, **74**, 1194L. CH_3OH
262. Collins, G. B., *Microwave Magnetrons*, MIT Radiation Laboratory Series, Vol. 6, McGraw-Hill Book
   Company, Inc., New York.
263. Crain, C. M., *Phys. Rev.*, **74**, 691.   Dielectric constants of
     gases
264. Cunningham, G. L., W. I. LeVan, and W. D. Gwinn,   Ethylene oxide
   Phys. Rev., **74**, 1537L.
265. Culshaw, W., *Proc. Phys. Soc.*, **61**, 562.   Microwave Michelson
     interferometer
266. Dailey, B. P., K. Rusinow, R. G. Shulman, and C. H. AsF_3
   Townes, *Phys. Rev.*, **74**, 1245A.
267. Dumond, J. W. M., and E. R. Cohen, *Revs. Mod.*   Survey physical
   Phys., **20**, 82.   constants
268. Edgell, W. F., and A. Roberts, *J. Chem. Phys.*, **16**, CF_3CH_3
   1002L.
269. Fano, U., *J. Research Natl. Bur. Standards*, **40**, 215.   Th. Stark effect with
     h.f.s.
270. Foley, H. M., *Phys. Rev.*, **73**, 259L.   Th. pressure broadening

271. LeBot, Jean, *J. Phys. Radium* **9**, 1D; Freymann, M.,   Review
     and R. Freymann, 29D; M. Freymann, R. Freymann,
     and Jean LeBot, 45D.
272. Gilliam, O. R., H. D. Edwards, and W. Gordy, *Phys.*   H.f.s. in CH_3I, ICN
     Rev., **73**, 635L.
273. Ginsburg, N., *Phys. Rev.*, **74**, 1052.   I.R. spectrum H_2O, D_2O
274. Golden, S., *J. Chem. Phys.*, **16**, 78.   Th. rotational energy
                                                   asymmetric rotor
275. Golden, S., *J. Chem. Phys.*, **16**, 250. *Errata:* **17**,   Th. rotational energy
     586L.   asymmetric rotor
276. Golden, S., T. Wentink, R. Hillger, and M. W. P.   Stark effect H_2O
     Strandberg, *Phys. Rev.*, **73**, 92.
277. Golden, S., and E. Bright Wilson, Jr., *J. Chem. Phys.*,   Th. Stark effect
     16, 669.   asymmetric rotor
278. Gordy, W., *Revs. Mod. Phys.*, **20**, 668.   Review
279. Gordy, W., H. Ring, and A. B. Burg, *Phys. Rev.*, **74**,   Nuclear moments
     1191L. *Errata:* **75**, 208L. B^{10}, B^{11}
280. Gordy, W., James W. Simmons, and A. G. Smith, CH_3Cl, CH_3Br, CH_3I
     Phys. Rev., **74**, 243; 1246A.
281. Hamilton, D. R., J. K. Knipp, and J. B. H. Kuper,
     Klystrons and Microwave Triodes, MIT Radiation
     Laboratory Series, Vol. 7, McGraw-Hill Book Com-
     pany, Inc., New York.
282. Henderson, R. S., *Phys. Rev.*, **74**, 107L. *Errata:* **74**, NH_3 h.f.s.
     626L.
283. Henderson, R. S., and J. H. Van Vleck, *Phys. Rev.*,   Fine-structure
     74, 106L.   polyatomic molecules
284. Herman, H., and R. J. Coates, *N.R.L. Rept.* R-3223.   Mm-wave components
285. Herman, R. C., and W. H. Shaffer, *J. Chem. Phys.*,   Rotation-vibration
     16, 453.   polyatomic molecules
286. Hershberger, W. D., *J. Appl. Phys.*, **19**, 411; *Phys.*   Sensitivity Stark
     Rev., **73**, 1249A.   spectrometer
287. Hershberger, W. D., and L. E. Norton, *RCA Rev.*,   Frequency stabilization
     9, 38.   on microwave lines
288. Husten, B. F., and H. Lyons, *Trans. AIEE*, Part 1,   Microwave frequency
     67, 321.   measurement
289. Jablónski, A., *Phys. Rev.*, **73**, 258L.   Th. pressure broadening
290. Jauch, J. M., *Phys. Rev.*, **74**, 1262A.   Th. NH_3 h.f.s.
291. Jen, C. K., *Phys. Rev.*, **73**, 1248A; **74**, 1246A; **74**,   Zeeman effect NH_3,
     1396. CH_3Cl, SO_2
292. Jen, C. K., *J. Appl. Phys.*, **19**, 649.   Dielectric constant
                                                    gases
293. Karplus, R., *Phys. Rev.*, **73**, 1027.   High-frequency
                                                  modulation
294. Karplus, R., *Phys. Rev.*, **73**, 1120L.   Power saturation NH_3
295. Karplus, R., *Phys. Rev.*, **74**, 223.   Power saturation
296. Karplus, R., *J. Chem. Phys.*, **16**, 1170.   Th. molecular rotational
                                                     energy
297. Karplus, R., and J. Schwinger, *Phys. Rev.*, **73**, 1020.   Th. power saturation
298. Kessler, M., and Gordy, W., *Phys. Rev.*, **74**, 354A.   Microwave
                                                                 spectrometer
299. Klinger, H. H., *Funk und Ton*, **2**, 135.   Review mm waves
300. Klinger, H. H., *Funk und Ton*, **2**, 183.   Review microwaves
301. Lamont, H. R. L., *Phys. Rev.*, **74**, 353L.   Atmospheric absorption
                                                      6 mm
302. Lamont, H. R. L., *Proc. Phys. Soc. London*, **61**, 562.   Atmospheric absorption
                                                                  6 mm
303. Mizushima, M., *Phys. Rev.*, **74**, 705L.   Th. NH_3 pressure
                                                    broadening

304. Montgomery, C. G., R. H. Dicke, and E. M. Purcell, *Principles of Microwave Circuits*, MIT Radiation Laboratory Series, Vol. 8, McGraw-Hill Book Company, Inc., New York.

305. Mott, N. F., and I. N. Sneddon, *Wave Mechanics and Its Applications*, Oxford University Press, London.

306. Newton, R. R., and L. H. Thomas, *J. Chem. Phys.*, **16,** 310. — Th. NH_3 inversion

307. Nierenberg, W. A., I. I. Rabi, and M. Slotnick, *Phys. Rev.*, **73,** 1430; **74,** 1246A. — Stark effect with h.f.s.

308. Pollard, E. C., and J. M. Sturtevant, *Microwaves and Radar Electronics*, John Wiley & Sons, Inc., New York.

309. Pound, R. V., *Microwave Mixers*, MIT Radiation Laboratory Series, Vol. 16, McGraw-Hill Book Company, Inc., New York.

310. de Quevedo, J. L., and W. V. Smith, *J. Appl. Phys.*, **19,** 831. — Frequency stabilization on microwave lines

311. Raev, A., *Annuaire Univ. Sofia, Fac. Sci.*, *Livre* 1, **45,** 303. — Review

312. Ragan, G. L., *Microwave Transmission Circuits*, MIT Radiation Laboratory Series, Vol. 9, McGraw-Hill Book Company, Inc., New York.

313. Richards, P. I., and H. S. Snyder, *Phys. Rev.*, **73,** 269L. — Th. power saturation

314. Roberts, A., *Phys. Rev.*, **73,** 1405L. — C^{14} in OCS

315. Sharbaugh, A. H., *Phys. Rev.*, **74,** 1870L. — SiH_3Cl

316. Shulman, R. G., B. P. Dailey, and C. H. Townes, *Phys. Rev.*, **74,** 846L. — Ethylene oxide

317. Simmons, J. W., and W. Gordy, *Phys. Rev.*, **73,** 713; **74,** 123A. — NH_3 h.f.s.

318. Skinner, H. A., *J. Chem. Phys.*, **16,** 553L. — Structure methyl halides

319. Smith, A. G., H. Ring, W. V. Smith, and W. Gordy, *Phys. Rev.*, **73,** 633. — H.f.s. ICN, N_2O

320. Smith, A. G., H. Ring, W. V. Smith, and W. Gordy, *Phys. Rev.*, **74,** 370, 123A. — ClCN, BrCN, ICN

321. Smith, D. F., *Phys. Rev.*, **74,** 506L. — NH_3 pressure broadening

322. Snyder, H. S., and P. I. Richards, *Phys. Rev.*, **73,** 1178. — Th. power saturation

323. Strandberg, M. W. P., *Phys. Rev.*, **74,** 1245A. — D_2O

324. Strandberg, M. W. P., T. Wentink, R. E. Hillger, G. H. Wannier, and M. L. Deutsch, *Phys. Rev.*, **73,** 188L. — Stark effect HDO

325. Torrey, H. C., and C. A. Whitmer, *Crystal Rectifiers*, MIT Radiation Laboratory Series, Vol. 15, McGraw-Hill Book Company, Inc., New York.

326. Townes, C. H., H. M. Foley, and W. Low, *Phys. Rev.*, **76,** 1415L. — Th. nuclear quad. moments

327. Townes, C. H., and S. Geschwind, *Phys. Rev.*, **74,** 626L. — S^{33} in OCS

328. Townes, C. H., and S. Geschwind, *J. Appl. Phys.*, **19,** 795L. — Th. sensitivity of spectrometer

329. Townes, C. H., A. N. Holden, and F. R. Merritt, *Phys. Rev.*, **74,** 1113. — OCS, ClCN, BrCN, ICN

330. Townes, C. H., F. R. Merritt, and B. D. Wright, *Phys. Rev.*, **73,** 1334; 1249A. — ICl

331. Trischka, J. W., *Phys. Rev.*, **74**, 718.　　　H.f.s. CsF

332. Trischka, J. W., *Phys. Rev.*, **76**, 1365.　　CsF mol. beam

333. Tuller, W. G., W. C. Galloway, and F. P. Zaffarano,　Frequency stabilization
Proc. IRE, **36**, 794.

334. Turner, T. E., Thesis, McGill University.　　CH_2Cl_2, CH_2Br_2

335. Unterberger, R. R., and Smith, W. V., *Rev. Sci.*　Microwave frequency
Instr., **19**, 580.　　　　　　　　　　　　　standard

336. Watts, R. J., W. J. Pietenpol, J. D. Rogers, and D.　Power saturation
Williams, *Phys. Rev.*, **74**, 1246A.

336a. Wesson, L. G., *Tables of Electric Dipole Moments,*
Technology Press, Cambridge, Mass.

337. Weidner, R. T., *Phys. Rev.*, **73**, 254L.　　　ICl

338. Weingarten, I. R., Thesis, Columbia University.　NH_3 high pressure

339. Wick, G. C., *Phys. Rev.*, **73**, 51.　　　　Mol. magnetic effects

340. Witmer, E. E., *Phys. Rev.*, **74**, 1247A; **74**, 1250A.　Th. asymmetric rotor

1949

341. Anderson, P. W., *Phys. Rev.*, **75**, 1450L.　　Th. NH_3 pressure
broadening

342. Anderson, P. W., *Phys. Rev.*, **76**, 647, 471A.　Th. pressure broadening

343. Anderson, P. W., Thesis, Harvard.　　　　Th. pressure broadening

344. Bak, B., E. S. Knudsen, and E. Madsen, *Phys. Rev.*,　C_6H_5Br, C_2H_5OH,
75, 1622L.　　　　　　　　　　　　　　CH_3OH, $(CH_3)_2CO$,
CH_3NO_2

345. Beard, C. I., and B. P. Dailey, *J. ACS*, **71**, 929.　CH_3NCS, CH_3SCN

346. Benedict, W. S., *Phys. Rev.*, **75**, 1317A.　　Centrifugal distortion

347. Beringer, R., and J. G. Castle, Jr., *Phys. Rev.*, **75**,　Zeeman effect O_2
1963L.

348. Beringer, R., and J. G. Castle, Jr., *Phys. Rev.*, **76**,　Zeeman effect NO
868L.

349. Bianco, D., G. Matlack, and A. Roberts, *Phys. Rev.*,　OCS, CH_3Cl
76, 473A.

350. Birks, J. B., *J. Brit. Inst. Radio Engrs.*, **9**, 10.　　Review microwave
physics

351. Bitter, F., *Phys. Rev.*, **76**, 833.　　　　　Resonant modulation

352. Bragg, J. K., and S. Golden, *Phys. Rev.*, **75**, 735.　Th. h.f.s. asymmetric
rotor

353. Bragg, J. K., and A. H. Sharbaugh, *Phys. Rev.*, **75**,　CH_2O
1774L.

354. Burgess, J. S., *Phys. Rev.*, **76**, 1267L.　　　I.R. spectrum
deuteroammonia

355. Carrara, N., P. Lombardini, R. Cine, and L. Sacconi,　NH_3 inversion
Nuovo Cimento, **6**, 552.

356. Cohen, V. W., W. S. Koski, and T. Wentink, Jr.,　S^{35} in OCS
Phys. Rev., **76**, 703L.

357. Coles, D. K., and R. H. Hughes, *Phys. Rev.*, **76**, 178A.　N_2O

358. Coles, D. K., and R. H. Hughes, *Phys. Rev.*, **76**, 858L.　CF_3Cl

359. Crawford, M. F., and A. L. Schawlow, *Phys. Rev.*,　Atomic h.f.s.
76, 1310.

360. Cunningham, G. L., A. W. Boyd, W. D. Gwinn, and　Ethylene oxide
W. I. LeVan, *J. Chem. Phys.*, **17**, 211L.

361. Dailey, B. P., *Anal. Chem.*, **21**, 540.　　　Chemical analysis by
microwave
spectroscopy

362. Dailey, B. P., J. M. Mays, and C. H. Townes, *Phys.*　CH_3Cl, SiH_3Cl, GeH_3Cl
Rev., **76**, 136L, 472A.

363. Dailey, B. P., H. Minden, and R. G. Shulman, *Phys.*　CH_3CF_3
Rev., **75**, 1319(A).

364. Davis, L., B. T. Feld, C. W. Zabel, and J. R. Zach-　H.f.s. atomic Cl
arias, *Phys. Rev.*, **76**, 1076.

365. Edwards, H. D., O. R. Gilliam, and W. Gordy, *Phys. Rev.*, **76**, 196A. — CH_3OH, CH_3NH_2

366. Gilbert, D. A., A. Roberts, and P. A. Griswold, *Phys. Rev.*, **76**, 1723L; *Phys. Rev.*, **77**, 742A (1950). — FCl

367. Gilliam, O. R., H. D. Edwards, and W. Gordy, *Phys. Rev.*, **75**, 1014; *Phys. Rev.*, **76**, 195A. — CH_3F, CHF_3, PF_3

368. Golay, M. J. E., *Rev. Sci. Instr.*, **20**, 816. — I.R. detector

369. Golden, S., and J. K. Bragg, *J. Chem. Phys.*, **17**, 439. — Th. asymmetric rotor

370. Goldstein, J. H., and J. K. Bragg, *Phys. Rev.*, **75**, 1453L. — H.f.s. vinyl chloride

371. Gordy, W., O. R. Gilliam, and R. Livingston, *Phys. Rev.*, **75**, 443. — Magnetic moment I^{127}, I^{129}

372. Hainer, R. M., P. C. Cross, and G. W. King, *J. Chem. Phys.*, **17**, 826. — Th. asymmetric rotor

373. Hedrick, L. C., *Rev. Sci. Instr.*, **20**, 781. — Square-wave generator

374. Hicks, B. L., E. Ossofsky, and R. N. Jones, *Ballistics Res. Lab., Tech. Note* 130. — Free radicals

375. Hughes, H. K., *Phys. Rev.*, **76**, 1675. — Stark effect strong fields

376. Jen, C. K., *Phys. Rev.*, **76**, 1494; **75**, 1319A. — Paschen-Back effect; NH_3, N_2O

377. Jen, C. K., *Phys. Rev.*, **76**, 471A. — Magnetic effects H_2O, HDO

378. Karplus, R., and A. H. Sharbaugh, *Phys. Rev.*, **75**, 889L. *Errata:* 1449L. — Stark effect CH_3Cl

379. Knight, G., and B. T. Feld, *MIT Research Lab., Rept.* 123; *Phys. Rev.*, **74**, 354A (1948). — Th. h.f.s. asymmetric rotor

380. Kusch, P., *Phys. Rev.*, **75**, 887. — Mol. h.f.s. Li^6

381. Lamb, W. E., Jr., and M. Skinner, *Phys. Rev.*, **75**, 1325. — Fine-structure He^+

382. Lassettre, E. N., and L. B. Dean, Jr., *J. Chem. Phys.*, **17**, 317. — Mol. quad. moments

383. Lawrance, R. B., *Research Lab. Electronics, MIT, Prog. Rept., Jan.* 15 (1949). — H.f.s. in H

384. Lenard, A., *Tables for Calculation of Stark and Zeeman effects*, Department of Physics, State University of Iowa. — Tables Stark and Zeeman effects

385. Lengyel, B. A., *Proc. IRE*, **37**, 1242. — Microwave interferometer

386. Lengyel, B. A., and A. J. Simmons, *N.R.L. Rept.*, 3562. — Microwave interferometer

387. Lew, H., *Phys. Rev.*, **76**, 1086. — H.f.s. atomic Al

388. Lines, A. W., *T.R.E. J.*, July, 1949, p. 1. — Survey mm waves

389. Livingston, R., O. R. Gilliam, and W. Gordy, *Phys. Rev.*, **76**, 149L. — I^{129} in CH_3I

390. Loubser, J. H. N., and C. H. Townes, *Phys. Rev.*, **76**, 178A. — 1.5 to 2-mm magnetron harmonics

391. Low, W., and C. H. Townes, *Phys. Rev.*, **76**, 1295; **75**, 1319A. — Th. Stark effect symmetric rotor

392. Low, W., and C. H. Townes, *Phys. Rev.*, **75**, 529L; 1318A. — O^{17} and S^{36} in OCS

393. Margenau, H., *Phys. Rev.*, **76**, 121; 585A. — Th. NH_3 pressure broadening

394. Margenau, H., *Phys. Rev.*, **76**, 1423. — Th. NH_3 pressure broadening

395. Matossi, F., *Phys. Rev.*, **76**, 1845. — I.R. pressure broadening

396. McAfee, K. B., Jr., R. H. Hughes, and E. B. Wilson, Jr., *Rev. Sci. Instr.*, **20**, 821. — Stark spectrometer

397. Millman, G. H., and R. C. Raymond, *J. Appl. Phys.*, **20**, 413L. — Absorption at high pressure

398. Mizushima, M., *J. Phys. Soc. Japan*, **4**, 191 (1949).   Th. NH₃

400. Moore, C. E., Atomic Energy Levels, *Natl. Bur. Standards Circ.* 467.

401. Mulliken, R. S., C. A. Rieke, D. Orloff, and H. Orloff, *J. Chem. Phys.*, **17**, 510L.   Overlap integrals

402. Nielsen, H. H., *Phys. Rev.*, **75**, 1961L.   Th. *l*-type doubling

403. Pietenpol, W. J., and J. D. Rogers, *Phys. Rev.*, **76**, 690L.   CH₂Br₂

404. Pippard, A. B., *J. Sci. Instr.*, **26**, 296.   Microwave interferometer

405. Roberts, A., Y. Beers, and A. G. Hill, *MIT Research Lab. Electronics, Tech. Rept.* 120.   H.f.s. atomic Cs

406. Roberts, A., and W. F. Edgell, *J. Chem. Phys.*, **17**, 742L; *Phys. Rev.*, **76**, 178A.   CF₂CH₂

407. Robinson, D. Z., *J. Chem. Phys.*, **17**, 1022.   Th. HCl structure

408. Schiff, L. I., *Quantum Mechanics*, McGraw-Hill Book Company, Inc., New York.

409. Sharbaugh, A. H., J. K. Bragg, T. C. Madison, and V. G. Thomas, *Phys. Rev.*, **76**, 1419L.   SiH₃Br

410. Sharbaugh, A. H., T. C. Madison, and J. K. Bragg, *Phys. Rev.*, **76**, 1529L.   NH₃ inversion

411. Sharbaugh, A. H., and J. Mattern, *Phys. Rev.*, **75**, 1102L.   CH₃Br

412. Simmons, J. W., *Phys. Rev.*, **76**, 686L.   CD₃Cl, CD₃I

413. Smith, A. G., W. Gordy, J. W. Simmons, and W. V. Smith, *Phys. Rev.*, **75**, 260.   Techniques 3 to 5 mm

414. Smith, W. V., and R. R. Unterberger, *J. Chem. Phys.*, **17**, 1348L.   CHCl₃

415. Strandberg, M. W. P., *J. Chem. Phys.*, **17**, 901.   HDO

416. Strandberg, M. W. P., C. Y. Meng, and J. G. Ingersoll, *Phys. Rev.*, **75**, 1524.   O₂

417. Strandberg, M. W. P., C. S. Pearsall, and M. T. Weiss, *J. Chem. Phys.*, **17**, 429L.   H₃B¹⁰CO

418. Strandberg, M. W. P., T. Wentink, Jr., and A. G. Hill, *Phys. Rev.*, **75**, 827; *Phys. Rev.*, **73**, 1249A.   OCSe

419. Strandberg, M. W. P., T. Wentink, Jr., and R. L. Kyhl, *Phys. Rev.*, **75**, 270.   OCS

420. Stutt, C. A., *MIT Research Lab. Electronics, Tech. Rept.* 105.   Lock-in amplifier

421. Townes, C. H., and L. C. Aamodt, *Phys. Rev.*, **76**, 691L.   Cl³⁶ in ClCN

422. Townes, C. H., and B. P. Dailey, *J. Chem. Phys.*, **17**, 782; *Phys. Rev.*, **74**, 1245A.   Th. quad. coupling

423. Townes, C. H., J. M. Mays, and B. P. Dailey, *Phys. Rev.*, **76**, 700L, 137A.   Nuclear moments Ge, Si

423a. Trischka, J. W., *Phys. Rev.*, **76**, 1365.   CsF, mol. beam

424. Van Vleck, J. H., and H. Margenau, *Phys. Rev.*, **76**, 1211, 585A.   Th. pressure broadening

425. Westenberg, A. A., J. H. Goldstein, and E. B. Wilson, Jr., *J. Chem. Phys.*, **17**, 1319; *Phys. Rev.*, **76**, 472A.   HCCCl

1950

426. Allen, P. W., and L. E. Sutton, *Acta Cryst.*, **3**, Part 1, 46.   Table mol. structure from electron diffraction

427. Amble, E., and B. P. Dailey, *J. Chem. Phys.*, **18**, 1422L.   HN₃

428. Anderson, P. W., *Phys. Rev.*, **80**, 511.   NH₃ pressure broadening quad.-induced dipole

429. Autler, S. H., and C. H. Townes, *Phys. Rev.*, **78**, 340A.   Resonant modulation
430. Baird, D. H., R. M. Fristrom, and M. H. Sirvetz, *Rev. Sci. Instr.*, **21**, 881L.   Stark cells
431. Bak, B., E. S. Knudsen, E. Madsen, and J. Rastrup-Andersen, *Phys. Rev.*, **79**, 190L. CH_2CO
432. Bak, B., R. Sloan, and D. Williams, *Phys. Rev.*, **80**, 101L.   SCSe
433. Barriol, J., *J. phys. radium*, **11**, 52.   Stark effect th.
434. Beard, C. I., and B. P. Dailey, *J. Chem. Phys.*, **18**, 1437; *Phys. Rev.*, **75**, 1318A. *Errata:* **19**, 975L (1951).   HNCS
435. Beringer, R., and J. G. Castle, Jr., *Phys. Rev.*, **78**, 581, 340A.   NO
436. Bernstein, H. J., *J. Chem. Phys.*, **18**, 1514L.   NOCl
437. Bersohn, R., *J. Chem. Phys.*, **18**, 1124L.   Th. quad. coupling, three nuclei
438. Birnbaum, G., *Phys. Rev.*, **77**, 144L. NH_3 dispersion
439. Bleaney, B., and J. H. N. Loubser, *Proc. Phys. Soc.*, **63A**, 483. NH_3, CH_3Cl, CH_3Br, high pressures
440. Bragg, J. K., T. C. Madison, and A. H. Sharbaugh, *Phys. Rev.*, **77**, 148L. *Errata:* 571L. CH_2CFCl
441. Burkhalter, J. H., R. S. Anderson, W. V. Smith, and W. Gordy, *Phys. Rev.*, **79**, 651; *Phys. Rev.*, **77**, 152L (1950); **79**, 224A. O_2 fine structure
442. Casimir, H. B. G., *Ned. Tijdschr. Natuurk.*, **16**, 198.   Th. h.f.s.
443. Castle, J. G., Jr., and R. Beringer, *Phys. Rev.*, **80**, 114L. NO_2
444. Coester, F., *Phys. Rev.*, **77**, 454.   Stark-Zeeman th.
445. Coles, D. K., *Advances in Electronics*, **2**, 299.   Review
446. Coles, D. K., W. E. Good, and R. H. Hughes, *Phys. Rev.*, **79**, 224A. CH_3CN
447. Cornwell, C. D., *J. Chem. Phys.*, **18**, 1118L. C_2H_3Br, B_2H_5Br
448. Crain, C. M., *Rev. Sci. Instr.*, **21**, 456.   Atmosphere refractive index
449. Crawford, B. L., Jr., and D. E. Mann, *Ann. Rev. Phys. Chem.*, **1**, 151.   Review
450. Culshaw, W., *Proc. Phys. Soc.*, **B63**, 939.   Michelson interferometer
451. Epprecht, G. W., *Z. angew. Math. Phys.*, **1**, 138.   Dielectric constants of gases
452. Eshbach, J. R., R. E. Hillger, and C. K. Jen, *Phys. Rev.*, **80**, 1106; *Phys. Rev.*, **78**, 339A. S^{33} nuclear magnetic moment
453. Essen, L., and K. D. Froome, *Proc. Phys. Soc.*, **B64**, 862.   Dielectric constants of gases
454. Fletcher, E. W., and S. P. Cooke, *Cruft Laboratory O.N.R. Rept.*, 64.   Frequency standard
455. Freymann, M. R., *L'Onde electrique*, **30**, 416.   Review
456. Geschwind, S., H. Minden, and C. H. Townes, *Phys. Rev.*, **78**, 174L; *Phys. Rev.*, **79**, 226A.   OCSe nuclear properties
457. Gilliam, O. R., C. M. Johnson, and W. Gordy, *Phys. Rev.*, **78**, 140.   2- to 3-mm spectroscopy
458. Girdwood, B. M., *Can. J. Research*, **28**, 180. CH_3OH
459. Goldstein, J. H., and J. K. Bragg, *Phys. Rev.*, **78**, 347A.   Asymmetric molecules with quad. coupling
460. Good, W. E., *Proc. Natl. Elec. Conf.*, **6**, 29.   Techniques
461. Gordy, W., H. Ring, and A. B. Burg, *Phys. Rev.*, **78**, 512; *Phys. Rev.*, **75**, 1325A. BH_3CO
462. Gordy, W., and J. Sheridan, *Phys. Rev.*, **79**, 224A.   Methyl mercuric halides
463. Grabner, L., and V. Hughes, *Phys. Rev.*, **79**, 819.   KF mol. beams

464. Griffing, V., *J. Chem. Phys.*, **18**, 744. — Power saturation
465. Hartz, T. R., and A. van der Ziel, *Phys. Rev.*, **78**, 473L. — Square-wave double modulation
467. Henry, A. F., *Phys. Rev.*, **80**, 396; **79**, 213A. — O_2 Zeeman th.
468. Henry, A. F., *Phys. Rev.*, **80**, 549. — NO Zeeman h.f.s.
469. Herman, R. C., and W. H. Shaffer, *J. Chem. Phys.*, **18**, 1207. — Vibration-rotation $X_2Y_2Z_2$ molecules
470. Hershberger, W. D., and L. E. Norton, *J. Franklin Inst.*, **249**, 359. — Stabilization servo theory
471. Herzberg, G., *Spectra of Diatomic Molecules*, D. Van Nostrand Company, Inc., New York.
472. Holstein, T., *Phys. Rev.*, **79**, 744L. — Th. pressure broadening
473. Howard, R. R., and W. V. Smith, *Phys. Rev.*, **77**, 840L. — Pressure broadening, temperature dependence
474. Howard, R. R., and W. V. Smith, *Phys. Rev.*, **79**, 128; 225A. — Collision diameters
476. Hughes, V., and L. Grabner, *Phys. Rev.*, **79**, 314. — RbF mol. beams
477. Hughes, V., and L. Grabner, *Phys. Rev.*, **79**, 829. — Diatomic molecule theory, mol. beams
478. Jones, L. C., *Phys. Rev.*, **77**, 741A. — Pressure broadening
479. Jones, L. H., J. N. Shoolery, R. G. Shulman, and D. M. Yost, *J. Chem. Phys.*, **18**, 990L. — HNCO
480. Kessler, W., H. Ring, R. Trambarulo, and W. Gordy, *Phys. Rev.*, **79**, 54. — CH_3CN, CH_3NC
481. Kisliuk, P., and C. H. Townes, *Phys. Rev.*, **78**, 347A; *J. Chem. Phys.*, **18**, 1109. — PCl_3, $AsCl_3$
482. Kisliuk, P., and C. H. Townes, *J. Research Natl. Bur. Standards*, **44**, 611. — Table microwave lines
483. Klages, G., *Experientia*, **6**, 321. — Review
483a. Kusch, P., and A. G. Prodell, *Phys. Rev.*, **79**, 1009. — H.f.s. in H and D
484. Lamb, W. E., Jr., and R. C. Retherford, *Phys. Rev.*, **79**, 549. — Atomic H fine structure
485. Lamb, W. E., Jr., and M. Skinner, *Phys. Rev.*, **78**, 539. — Fine structure He^+
486. Lamont, H. R. L., *Wave Guides*, 3d ed., Methuen & Co., Ltd., London.
488. Lide, D. R., and D. K. Coles, *Phys. Rev.*, **80**, 911L. — CH_3SiH_3 internal rotation
489. Loubser, J. H. N., and J. A. Klein, *Phys. Rev.*, **78**, 348A. — ND_3 mm waves
490. Low, W., and C. H. Townes, *Phys. Rev.*, **80**, 608; **79**, 198A. — Nuclear masses
491. Low, W., and C. H. Townes, *Phys. Rev.*, **79**, 224A. — OCS, OCSe Fermi resonance
492. Lyons, H., *J. Appl. Phys.*, **21**, 59L. — Frequency dividers
493. Maier, W., *Z. Elektrochem.*, **54**, 521. — Review
494. Margenau, H., and S. Bloom, *Phys. Rev.*, **79**, 213A. — Th. pressure broadening
495. Margenau, H., and A. Henry, *Phys. Rev.*, **78**, 587. — Th. NO
496. Matlack, G., G. Glockler, D. R. Bianco, and A. Roberts, *J. Chem. Phys.*, **18**, 332. — CH_3Cl
497. McAfee, K. B., Jr., *Phys. Rev.*, **78**, 340A. — NO_2
498. Minden, H. T., J. M. Mays, and B. P. Dailey, *Phys. Rev.*, **78**, 347A. — CH_3SiF_3
499. Mizushima, M., *Research Chem. Phys.*, **29**, 25. — Th. pressure broadening
500. Morgan, H. W., G. W. Keilholtz, W. V. Smith, and A. L. Southern, *O.R.N.L. Rept.* Y-621. — ClCN isotopic analysis
501. Mulliken, R. S., *J. ACS*, **72**, 4493. — Th. chemical binding
502. Murphy, J., and R. C. Raymond, *J. Appl. Phys.*, **21**, 1064. — Dielectric constants of gases

503. Nielsen, H. H., *Phys. Rev.*, **77**, 130. CH₃CN, CH₃NC
l doubling

503a. Nethercot, A. H., Ph.D. Thesis, University of Michigan. Mm-wave spark generator

504. Nethercot, A. H., and C. W. Peters, *Phys. Rev.*, **79**, 225A. NH₃ infrared line widths

505. Nielsen, H. H., *Phys. Rev.*, **78**, 296L. *l*-doubling OCS, HCN

506. Nielsen, H. H., *Phys. Rev.*, **78**, 415. Centrifugal stretching

507. Nierenberg, W. A., *Phys. Rev.*, **80**, 1102L. H.f.s. mol. beams

509. Pietenpol, W. J., J. D. Rogers, and D. Williams, *Phys. Rev.*, **78**, 480L. HCOOH, NOCl

510. Pierce, J. R., *Physics Today*, **3**, 24. Review mm waves

511. Pierce, J. R., *Travelling Wave Tubes*, D. Van Nostrand Company, Inc., New York.

512. Pryce, M. H. L., *Phys. Rev.*, **77**, 136. Th. resonant modulation

513. Rainwater, J., *Phys. Rev.*, **79**, 432. Th. nuclear quad.

514. Ramsey, N. F., *Phys. Rev.*, **78**, 221. Mol. quad. moment

515. Ramsey, N. F., *Phys. Rev.*, **78**, 699. "Chemical" effects, magnetic h.f.s.

516. Rogers, J. D., H. L. Cox, and P. G. Braunschweiger, *Rev. Sci. Instr.*, **21**, 1014. Frequency measurement

517. Rouse, A. G., A. V. Bushkovitch, L. C. Jones, C. A. Potter, and W. F. Sullivan, *Phys. Rev.*, **78**, 347A. Pressure broadening and shift

518. Senatore, S. J., *Phys. Rev.*, **78**, 293L. POF₃

519. Sharbaugh, A. H., *Rev. Sci. Instr.*, **21**, 120. Stark spectrograph

520. Sharbaugh, A. H., B. S. Pritchard, and T. C. Madison, *Phys. Rev.*, **77**, 302. CF₃Br

521. Sharbaugh, A. H., B. S. Pritchard, V. G. Thomas, J. M. Mays, and B. P. Dailey, *Phys. Rev.*, **79**, 189L. GeH₃Br, SiH₃Br

522. Sharbaugh, A. H., V. G. Thomas, and B. S. Pritchard, *Phys. Rev.*, **78**, 64L. SiH₃F

523. Shaull, J. M., *Proc. IRE*, **38**, 6. Frequency standards

524. Sheridan, J., and W. Gordy, *Phys. Rev.*, **77**, 292L. CF₃Br, CF₃I, CF₃CN

525. Sheridan, J., and W. Gordy, *Phys. Rev.*, **77**, 719L. SiF₃H, SiF₃CH₃, SiF₃Cl, SiF₃Br

526. Sheridan, J., and W. Gordy, *Phys. Rev.*, **79**, 224A. CH₃CCBr

527. Sheridan, J., and W. Gordy, *Phys. Rev.*, **79**, 513. NF₃

528. Shulman, R. G., B. P. Dailey, and C. H. Townes, *Phys. Rev.*, **78**, 145; **75**, 472A. Dipole moments

529. Shulman, R. G., and C. H. Townes, *Phys. Rev.*, **77**, 421L; **78**, 347A. OCS, HCN *l*-type doubling transitions

530. Shulman, R. G., and C. H. Townes, *Phys. Rev.*, **77**, 500; **75**, 1318A. OCS Stark effect

531. Simmons, J. W., and W. E. Anderson, *Phys. Rev.*, **80**, 338. CH₃Cl, CH₃Br, CH₃I, ICN centrifugal distortion

532. Simmons, J. W., W. E. Anderson, and W. Gordy, *Phys. Rev.*, **77**, 77. *Errata*: **86**, 1055 (1952). HCN

533. Simmons, J. W., and W. O. Swan, *Phys. Rev.*, **80**, 289L. CH₃Br

534. Smith, D. F., M. Tidwell, and D. V. P. Williams, *Phys. Rev.*, **77**, 420L. BrF

535. Smith, D. F., M. Tidwell, and D. V. P. Williams, *Phys. Rev.*, **79**, 1007L. BrCl

536. Smith, W. V., and R. R. Howard, *Phys. Rev.*, **79**, 132; **76**, 473A. Mol. quad. moments

537. Smythe, W. R., *Static and Dynamic Electricity*, 2d ed., McGraw-Hill Book Company, Inc., New York.

538. Southern, A. L., H. W. Morgan, G. W. Keilholtz, and W. V. Smith, *Phys. Rev.*, **78**, 629A. — Isotopic analysis

539. Sternheimer, R., *Phys. Rev.*, **80**, 102. — Theory quad. coupling

539a. Takahashi, I., A. Okaya, T. Ogawa, and T. Hashi, *Mem. College Science, Univ. Kyoto, A*, **26**, 113. — Microwave spectrometer

540. Tomassini, M., *Nuovo Cimento*, **7**, 1. — NH_3

541. Torkington, P., *J. Chem. Phys.*, **18**, 407. — Internal rotation

542. Townes, C. H., and B. P. Dailey, *Phys. Rev.*, **78**, 346A. — Quad. coupling and ionic character

543. Trambarulo, R., and W. Gordy, *Phys. Rev.*, **79**, 224A. — CD_3NC, CD_3CN

544. Trambarulo, R., and W. Gordy, *J. Chem. Phys.*, **18**, 1613. — CH_3CCH

545. Unterberger, R. R., R. Trambarulo, and W. V. Smith, *J. Chem. Phys.*, **18**, 565L. — $CHCl_3$

546. Weiss, M. T., M. W. P. Strandberg, R. B. Lawrance, and C. C. Loomis, *Phys. Rev.*, **78**, 202. — B^{10} spin

547. Wells, A. F., *Structural Inorganic Chemistry*, Clarendon Press, Oxford.

548. Westenberg, A. A., and E. B. Wilson, Jr., *J. ACS*, **72**, 199. — CHCCN

549. Whiffen, D. H., *Quart. Rev.*, **4**, 131. — Review rotation spectra

550. Williams, J. Q., and W. Gordy, *J. Chem. Phys.*, **18**, 994. — Tertiary butyl halides

551. Williams, J. Q., and W. Gordy, *Phys. Rev.*, **79**, 225A. — $CHBr_3$, PBr_3

552. Wilson, E. B., Jr., *Faraday Soc. Discussion*, **9**, 108. — Review

1951

553. Amble, E., *Phys. Rev.*, **83**, 210A. — Trioxane

554. Anderson, R. S., C. M. Johnson, and W. Gordy, *Phys. Rev.*, **83**, 1061. — O_2 2.5 mm

555. Anderson, W. E., J. Sheridan, and W. Gordy, *Phys. Rev.*, **81**, 819. — GeF_3Cl

556. Anderson, R. S., W. V. Smith, and W. Gordy, *Phys. Rev.*, **82**, 264L. — O_2 line widths

557. Anderson, W. E., R. Trambarulo, J. Sheridan, and W. Gordy, *Phys. Rev.*, **82**, 58. — CF_3CCH

557a. Aslakson, C. I., *Trans. Am. Geophys. Union*, **32**, 813. — Velocity of microwaves

557b. Barrow, R. F., *et al.*, *Données spectroscopiques concernant les molécules diatomiques*, Hermann & Cie, Paris.

558. Beringer, R., and J. G. Castle, Jr., *Phys. Rev.*, **81**, 82. — O_2

559. Birnbaum, G., *Phys. Rev.*, **82**, 110L. — Atmospheric refractive index

560. Birnbaum, G., S. J. Kryder, and H. Lyons, *J. Appl. Phys.*, **22**, 95. — Dielectric constant gases

561. Burkhard, D. G., and D. M. Dennison, *Phys. Rev.*, **84**, 408. — CH_3OH

561a. Carlson, R. O., C. A. Lee, and B. P. Fabricand, *Phys. Rev.*, **85**, 784. — TlCl

562. Coles, D. K., W. E. Good, J. K. Bragg, and A. H. Sharbaugh, *Phys. Rev.*, **82**, 877. — NH_3 Stark

562a. Collins, T. L., A. O. Nier, and W. H. Johnson, Jr., *Phys. Rev.* **84**, 717. — Masses near A = 40

563. Cornwell, C. D., *O.N.R. Rept.*, Iowa State, Jan. 1, 1951. — C_2H_3Br

564. Costain, C. C., *Phys. Rev.*, **82**, 108L. — NH_3

565. Crable, G. F., and W. V. Smith, *J. Chem. Phys.*, **19**, 502L. — SO_2

566. Cunningham, G. L., Jr., A. W. Boyd, R. J. Myers, W. D. Gwinn, and W. I. LeVan, *J. Chem. Phys.*, **19**, 676. — Ethylene oxide, ethylene sulfide

567. Dayhoff, E. S., *Rev. Sci. Instr.*, **12**, 1025L. — Klystron frequency controller

568. DeHeer, J., *Phys. Rev.*, **83**, 741. — Theory *l* doubling

568*a.* Deutsch, M., *Phys. Rev.*, **82**, 455L. — Positronium

569. Dickinson, W. C., *Phys. Rev.*, **81**, 717. — "Chemical effects," magnetic resonance

570. Essen, L., and K. D. Froome, *Proc. Phys. Soc. B*, **64**, 862. — Dielectric constant of air and constituents

571. Ewen, H. I., and E. M. Purcell, *Phys. Rev.*, **83**, 881A; *Nature*, **168**, 356. — Interstellar hydrogen radiation

572. Freymann, R., *Physica*, **17**, 328. — CH_3CH_2Cl

572a. Friedburg, H., and W. Paul, *Naturwiss.*, **38**, 159. — Molecular beam focussing

573. Geschwind, S., Thesis, Columbia University. — High-resolution microwave spectroscopy

574. Geschwind, S., and R. Gunther-Mohr, *Phys. Rev.*, **81**, 882L; **82**, 346A. — Ge, Si, S masses

575. Geschwind, S., R. Gunther-Mohr, and C. H. Townes, *Phys. Rev.*, **81**, 288L; **82**, 343A. — Cl^{35}/Cl^{37} quad. moment ratio

575a. Gobau, G., *Proc. IRE*, **39**, 319. — Surface wave propagation

576. Gokhale, B. V., and M. W. P. Strandberg, *Phys. Rev.*, **84**, 844L; **82**, 327A. — O_2 line widths

577. Gokhale, B. V., H. R. Johnson, and M. W. P. Strandberg, *Phys. Rev.*, **83**, 881A. — *L* uncoupling

577*a.* Good, W. E., D. K. Coles, G. R. Gunther-Mohr, A. L. Schawlow, and C. H. Townes, *Phys. Rev.*, **83**, 880A. — NH_3 h.f.s.

578. Gordy, W., *J. Chem. Phys.*, **19**, 792. — Quad. coupling interpretation

579. Gorter, C. J., *Physica*, **17**, 169. — Review r.f. spectroscopy

580. Gozzini, A., *Nuovo Cimento*, **8**, 361. — Dielectric constant of gases

581. Grabner, L., and V. Hughes, *Phys. Rev.*, **82**, 561. — Two-quantum transition-molecular beams

582. Greenhow, C., and W. V. Smith, *J. Chem. Phys.*, **19**, 1298. — Mol. quad. N_2, O_2

583. Gunther-Mohr, G. R., S. Geschwind, and C. H. Townes, *Phys. Rev.*, **81**, 289L. — Polarization of nucleus

585. Hedrick, L. C., *Rev. Sci. Instr.*, **22**, 537L. — Square-wave generator

586. Hill, R. M., and W. V. Smith, *Phys. Rev.*, **82**, 451L. — Mol. quad. moments

587. Hillger, R. E., and M. W. P. Strandberg, *Phys. Rev.*, **83**, 575; **82**, 327A. — HDS centrifugal distortion

588. Honerjäger, R., *Naturwiss.*, **38**, 34. — Review

589. Hughes, R. H., *Instruments*, **24**, 1352. — Analytical applications

590. Hughes, R. H., W. E. Good, and D. K. Coles, *Phys. Rev.*, **84**, 418; **77**, 741A (1950). — CH_3OH

591. Hurd, F. K., and W. D. Hershberger, *Phys. Rev.*, **82**, 95L. — CH_3SH

592. Jen, C. K., *Phys. Rev.*, **81**, 197. — Mol. magnetic moments

593. Jen, C. K., *Physica*, **17**, 378. — Magnetic moments

594. Johnson, C. M., W. Gordy, and R. Livingston, *Phys. Rev.*, **83**, 1249L. — Cl^{36} moments

595. Johnson, C. M., R. Trambarulo, and W. Gordy, *Phys. Rev.*, **84**, 1178. — 2–3 mm spectroscopy

596. Johnson, K. C., *Proc. Inst. Elec. Engrs. London*, Part III, **98**, 77.   Frequency stabilization

597. Kisliuk, P., and C. H. Townes, *Phys. Rev.*, **83**, 210A. $AsCl_3$, $SbCl_3$

598. Koch, B., *Ergeb. exakt. Naturw.*, **24**, 222.   Review techniques

599. Lamb, W. E., and R. C. Retherford, *Phys. Rev.*, **81**, 222.   Atomic H fine structure

600. Lamont, H. R. L., *Physica*, **17**, 446.   Frequency stabilization

601. Lawrance, R. B., and M. W. P. Strandberg, *Phys. Rev.*, **83**, 363; **78**, 347A (1950).   Centrifugal distortion H_2CO

602. Leslie, D. C. M., *Phil. Mag.*, **42**, 37.   Th. pressure broadening

603. Lide, D. R., *J. Chem. Phys.*, **19**, 1605. CH_3SnH_3

604. Lide, D. R., Jr., and D. K. Coles, *Phys. Rev.*, **80**, 911L. CH_3SiH_3 internal rotation

605. Logan, R. A., R. E. Cote, and P. Kusch, *Phys. Rev.*, **85**, 280.   Quad. interaction-mol. beams

606. Loomis, C. C., and M. W. P. Strandberg, *Phys. Rev.*, **81**, 798. PH_3, AsH_3, SbH_3

606a. Luce, R. G., and J. W. Trischka, *Phys. Rev.*, **83**, 851L; **82**, 323A.   CsCl

607. Lyons, H., L. J. Rueger, R. G. Nuckolls, and M. Kessler, *Phys. Rev.*, **81**, 630; **81**, 297A.   Deuterated ammonias

608. Magnuson, D. W., *J. Chem. Phys.*, **19**, 1614L.   Dielectric constant UF_6

609. Magnuson, D. W., *J. Chem. Phys.*, **19**, 1071L; *Phys. Rev.*, **83**, 485A.   NOF, dipole moments

610. Maier, W., *Ergeb. exakt. Naturw.*, **24**, 275.   Review

611. Maier, W., *Landolt-Börnstein Tabellen*, Aufl. 6, B. 1, T. 2.   Tables of mol. constants

612. Margenau, H., *Phys. Rev.*, **82**, 156.   Th. pressure broadening

613. Marshall, W. F., *Electronics*, **24**, 92.   Frequency standard

614. Mays, J. M., and C. H. Townes, *Phys. Rev.*, **81**, 940.   Ge isotopes

615. McAfee, K. B., Jr., *Phys. Rev.*, **82**, 971L. NO_2

616. Miller, S. L., A. Javan, and C. H. Townes, *Phys. Rev.*, **82**, 454; **83**, 209A. O^{18} spin

617. Millman, S., *Proc. IRE*, **39**, 1035.   Mm-wave tube

618. Minden, H. T., and B. P. Dailey, *Phys. Rev.*, **82**, 338A. CH_3CF_3, CH_3SiF_3 hindered rotation

619. Mizushima, M., *Phys. Rev.*, **83**, 94; *Physica*, **17**, 453A. *Errata: Phys. Rev.*, **84**, 363.   Th. pressure broadening

620. Mizushima, M., and T. Ito, *J. Chem. Phys.*, **19**, 739.   Th. quad. coupling three nuclei

621. Muller, C. A., and J. N. Oort, *Nature*, **108**, 357.   Interstellar hydrogen

622. Newell, G., Jr., and R. H. Dicke, *Phys. Rev.*, **83**, 1064L; **81**, 297A.   Doppler width reduction

623. Nielsen, H. H., *Physica*, **17**, 432.   *l* doubling

624. Nielsen, H. H., *Revs. Mod. Phys.*, **23**, 90.   Vibration-rotation energies

625. Nierenberg, W. A., *Phys. Rev.*, **82**, 932.   Spin-orbit coupling mol. beams

626. Nuckolls, R. G., L. J. Rueger, and H. Lyons, *Phys. Rev.*, **83**, 880. ND_3

626a. Post, E. J., and H. F. Pit, *Proc. IRE*, **39**, 169.   Stabilized crystal oscillator

627. Potter, C. A., A. V. Bushkovitch, and A. G. Rouse, *Phys. Rev.*, **83**, 987; **82**, 323A. NH_3 pressure broadening

628. Pierce, J. R., *Electronics*, **24**, 66.   Mm wave review

629. Poynter, R. L., *O.N.R. Rept.*, Iowa State, Jan. 1, 1951. H_2CCHI

630. Ramsey, N. F., *Phys. Rev.*, **87**, 1075.   Magnetic effects of vibration and rotation

631. Reesor, G. E., *Can. J. Phys.*, **29**, 87.   Absorption in excited H

632. Rogers, J. D., W. J. Pietenpol, and D. Williams, *Phys. Rev.*, **83**, 431; **77**, 741A; **82**, 323A.　NOCl
633. Rogers, J. D., and D. Williams, *Phys. Rev.*, **82**, 131A.　HN₃ — HN$_3$
634. Rogers, J. D., and D. Williams, *Phys. Rev.*, **83**, 210A.　HCOOH
635. Rogers, T. F., *Phys. Rev.*, **83**, 881A.　Line shapes
636. Roubine, E., *Rev. Tech. C.F.T.H.*, **16**, 21.　Stark spectrometer
637. Rueger, L. J., H. Lyons, and R. G. Nuckolls, *Rev. Sci. Instr.*, **22**, 428L.　High-temperature Stark cell
638. Sawyer, K. A., and J. D., Kierstead, *MIT Research Lab. Electronics Tech.*, *Rept.* 188.　ND₂H — ND$_2$H
639. Schuster, N. A., *Rev. Sci. Instr.*, **22**, 254.　Phase sensitive detector
640. Shaw, T. M., and J. J. Windle, *J. Chem. Phys.*, **19**, 1063.　CH₃SH — CH$_3$SH
641. Sheridan, J., and W. Gordy, *J. Chem. Phys.*, **19**, 965.　Trifluorosilane derivatives
642. Shimoda, K., and T. Nishikawa, *J. Phys. Soc. Japan*, **6**, 512.　Atomic Na h.f.s.
643. Shimoda, K., and T. Nishikawa, *J. Phys. Soc. Japan*, **6**, 516.　Zeeman spectrograph
644. Shoolery, J. N., and A. H. Sharbaugh, *Phys. Rev.*, **82**, 95L.　Dipole moments OCS, HNCO, H₂CO, CHF₃ — HNCO, H$_2$CO, CHF$_3$
645. Shoolery, J. N., R. G. Shulman, W. F. Sheehan, Jr., V. Schomaker, and D. M. Yost, *J. Chem. Phys.*, **19**, 1364; *Phys. Rev.*, **82**, 323A.　CF₃CCH — CF$_3$CCH
646. Shoolery, J. N., R. G. Shulman, and D. M. Yost, *J. Chem. Phys.*, **19**, 250.　HNCO, HNCS
647. Sirvetz, M. H., *J. Chem. Phys.*, **19**, 938.　SO₂ — SO$_2$
648. Sirvetz, M. H., *J. Chem. Phys.*, **19**, 1609.　Furan
649. Smith, D. F., M. Tidwell, D. V. P. Williams, and S. J. Senatore, *Phys. Rev.*, **83**, 485A.　CF₂O — CF$_2$O
650. Southern, A. L., H. W. Morgan, G. W. Keilholtz, and W. V. Smith, *Anal. Chem.*, **23**, 1000.　N and C isotopic determination
651. Swartz, J. C., and J. W. Trischka, *Phys. Rev.*, **88**, 1085.　LiF mol. beam
652. Talley, R. M., and A. H. Nielsen, *J. Chem. Phys.*, **19**, 805.　C₂D₂ vibration-rotation — C$_2$D$_2$ vibration-rotation
653. Townes, C. H., *J. Appl. Phys.*, **22**, 1365.　Frequency stabilization
654. Townes, C. H., *Physica*, **17**, 354.　Nuclear properties
655. Trischka, J., *J. Chem. Phys.*, **20**, 1811L.　LiF mol. beam
656. Van Vleck, J. H., *Phys. Rev.*, **83**, 880A.　Theory NH₃ h.f.s. — Theory NH$_3$ h.f.s.
657. Van Vleck, J. H., *Revs. Mod. Phys.*, **23**, 213; *Phys. Rev.*, **82**, 320A.　Coupling of angular momenta
658. Weber, J., *Phys. Rev.*, **83**, 1058L; 881A.　Pressure broadening
659. Weber, J., and K. J. Laidler, *J. Chem. Phys.*, **19**, 381L.　Kinetics NH₃-D₂ exchange — Kinetics NH$_3$-D$_2$ exchange
660. Weber, J., and K. J. Laidler, *J. Chem. Phys.*, **19**, 1089.　Kinetics NH₃-D₂ exchange — Kinetics NH$_3$-D$_2$ exchange
661. Weiss, M. T., and M. W. P. Strandberg, *Phys. Rev.*, **81**, 286L; **82**, 326A.　Deuteroammonias
662. Weiss, M. T., and M. W. P. Strandberg, *Phys. Rev.*, **83**, 567.　Deuteroammonias
663. Wentink, T., Jr., W. S. Koski, and V. W. Cohen, *Phys. Rev.*, **81**, 948; **81**, 296A; **77**, 742A (1950).　S³⁵ mass — S^{35} mass
664. Wilson, E. B., Jr., *Ann. Rev. Phys. Chem.*, **2**, 151.　Review

1952

665. Amble, E., S. L. Miller, A. L. Schawlow, and C. H. Townes, *J. Chem. Phys.*, **20**, 192L; *Phys. Rev.*, **82**, 328A (1951).　ReO₃Cl — ReO$_3$Cl
666. Anderson, P. W., *Phys. Rev.*, **86**, 809L.　Th. pressure broadening

666a. Anderson, J. R., *Trans. Instruments and Meas. Conf.*,   Square-wave generator
Stockholm, **5**.

667. Anderson, R. S., W. V. Smith, and W. Gordy, *Phys.* O_2 pressure broadening
Rev., **87**, 561.

668. *Annual Review of Nuclear Science*, Vol. I, Annual   Review nuclear
Reviews, Inc., Stanford, Calif.   moments

669. Artman, J. O., and J. P. Gordon, *Phys. Rev.*, **87**, O_2 pressure broadening
227A.

669a. Bak, B., *Trans. Instruments and Meas. Conf.*, *Stock-*   Review
holm, **8**.

670. Beard, C. I., and D. R. Bianco, *J. Chem. Phys.*, **20**, D_2O
1488L.

671. Beringer, R., and E. B. Rawson, *Phys. Rev.*, **86**, 607A.   NO Λ doubling

672. Beringer, R., and E. B. Rawson, *Phys. Rev.*, **87**, 228A.   Zeeman effect H

673. Beringer, R., *Ann. N.Y. Acad. Sci.*, **55**, 814.   Zeeman effect
   paramagnetic gases

674. Biedenharn, L. C., J. M. Blatt, and M. E. Rose, *Rev.*   Tables Racah
Mod. Phys., **24**, 249.   coefficients

675. Birnbaum, G., and S. K. Chatterjee, *J. Appl. Phys.*, H_2O dielectric constant
23, 220.

675a. Birnbaum, G., H. E. Bussey, and R. R. Larson,   Atmospheric refractive
Trans. IRE Prof. Group on Antennas and Propagation,   index
No. 3, 74.

676. Bloom, S., and H. Margenau, *Phys. Rev.*, **85**, 717A.   Th. pressure broadening

676a. Bolef, D. I., and H. J. Zeiger, *Phys. Rev.*, **85**, 799. $Rb^{87}F$, $Rb^{87}Cl$ mol.
   beam

676b. Carlson, R. O., C. A. Lee, and B. P. Fabricand, *Phys.*   TlCl, mol. beam
Rev., **85**, 784.

676c. Boyd, D. R. J., and H. W. Thompson, *Spectrochim.*   HBr (I.R.)
Acta, **5**, 308.

677. Cohen, V. W., *Ann. N.Y. Acad. Sci.*, **55**, 904.   Determination moments
   radioactive nuclei

678. Costain, C. C., and G. B. B. M. Sutherland, *Phys.*   Th. inversion
Chem., **56**, 321.

678a. Deutsch, M., and S. C. Brown, *Phys. Rev.*, **85**, 1047.   Positronium

679. Duchesne, J., *J. Chem. Phys.*, **20**, 1804.   Th. quad. coupling

679a. Duchesne, J., *Nuovo Cimento*, **9**, Suppl. 3, 270.   Comparison I.R. and
   microwaves

680. Dailey, B. P., *Ann. N.Y. Acad. Sci.*, **55**, 915.   Survey hindered
   rotation

681. Eshbach, J. R., R. E. Hillger, and M. W. P. Strand-   Magnetic moment S^{33}
berg, *Phys. Rev.*, **85**, 532.

682. Eshbach, J. R., and M. W. P. Strandberg, *Phys. Rev.*,   Mol. g factors
85, 24; **82**, 327A.

683. Eshbach, J. R., and M. W. P. Strandberg, *Rev. Sci.*   Zeeman effect apparatus
Instr., **23**, 623.

683a. Essen, L., and K. D. Froome, *Nuovo Cimento*, **9**,   Refractive index of air
Suppl. 3, 277.

684. Fristrom, R. M., *J. Chem. Phys.*, **20**, 1; *Phys. Rev.*, SO_2F_2
85, 717A.

685. Froome, K. D., *Proc. Roy. Soc.*, **213**, 123.   Microwave
   measurement of c

686. Frosch, R. A., and H. M. Foley, *Phys. Rev.*, **88**, 1337.   Mol. magnetic h.f.s.

687. Gabriel, W. F., *Proc. IRE*, **40**, 940.   Frequency stabilization

688. Geschwind, S., G. R. Gunther-Mohr, and G. Silvey, O^{17} in OCS
Phys. Rev., **85**, 474; **83**, 209A.

689. Geschwind, S., *Ann. N.Y. Acad. Sci.*, **55**, 751.   Survey high resolution

690. Ghosh, S. N., R. Trambarulo, and W. Gordy, *J.* CHF_3, $CHCl_3$, CH_3CF_3,
Chem. Phys., **20**, 605; *Phys. Rev.*, **87**, 172A. CH_3CCl_3

691. Gilbert, D. A., *Phys. Rev.*, **85**, 716A. — Cl^{36} in CH_3Cl

692. Golay, M. J. E., *Proc. IRE*, **40**, 1161. — Mm waves

693. Gordy, W., *Physics Today*, **7**, 5. — Elementary review

694. Gordy, W., *Ann. N.Y. Acad. Sci.*, **55**, 774. — Mm waves

695. Hardy, W. A., G. Silvey, and C. H. Townes, *Phys. Rev.*, **85**, 494L; **86**, 608A. — Se^{79} in OCSe

695a. Harrick, N. J., and N. F. Ramsey, *Phys. Rev.*, **88**, 228. — R.F. spectrum of H_2

696. Hawkins, N. J., V. W. Cohen, and W. S. Koski, *J. Chem. Phys.*, **20**, 528L. — POF_3, PSF_3

696a. Hogan, C. L., *Bell System Tech. J.*, **31**, 1. — Applications of ferrites

697. Hrostowski, H. J., R. J. Myers, and G. C. Pimentel, *J. Chem. Phys.*, **20**, 518L. — Pentaborane

698. Hughes, J. V., and H. L. Armstrong, *J. Appl. Phys.*, **23**, 501. — Dielectric constant air

699. Hughes, R. H., *Phys. Rev.*, **85**, 717A. — O_3

700. Hughes, R. H., *Ann. N.Y. Acad. Sci.*, **55**, 872. — Chemical analysis

701. Ince, C. R. S., *J. Appl. Phys.*, **23**, 1408L. — Atomic clock

702. Javan, A., and A. V. Grosse, *Phys. Rev.*, **87**, 227A. — MnO_3F

703. Javan, A., and C. H. Townes, *Phys. Rev.*, **86**, 608A. — ICN h.f.s. anomalies

704. Jen, C. K., J. W. B. Borghausen, and R. W. Stanley, *Phys. Rev.*, **85**, 717A. — Mol. *g* factor

705. Jen, C. K., *Ann. N.Y. Acad. Sci.*, **55**, 822. — Survey Zeeman effects

706. Johnson, C. M., and D. M. Slager, *Phys. Rev.*, **87**, 677L. — OCS pressure broadening

707. Johnson, H. R., *Phys. Rev.*, **85**, 764A. — Th. spectrograph

708. Johnson, H. R., and M. W. P. Strandberg, *J. Chem. Phys.*, **20**, 687; *Phys. Rev.*, **82**, 327A. — CH_2CO

709. Johnson, H. R., and M. W. P. Strandberg, *Phys. Rev.*, **85**, 503L. — Beam spectrometer

710. Johnson, H. R., and M. W. P. Strandberg, *Phys. Rev.*, **86**, 811L. — Wall-collision broadening

711. Jones, L. C., A. V. Bushkovitch, C. A. Potter, and A. G. Rouse, *Phys. Rev.*, **87**, 227A. — Pressure broadening

712. Kagarise, R. E., H. D. Rix, and D. H. Rank, *J. Chem. Phys.*, **20**, 1437. — I.R. spectrum HCN

712a. Karplus, R., and A. Klein, *Phys. Rev.*, **86**, 257. — Theory positronium

712b. Kisliuk, P., Thesis, Columbia University. — Halides of N, P, As, Sb; CH_3HgCN, $P(CN)_3$

713. Kisliuk, P., and G. A. Silvey, *J. Chem. Phys.*, **20**, 517. — CF_3SF_5

714. Kisliuk, P., and C. H. Townes, *Natl. Bur. Standards Circ.* 518. — Tables microwave spectra

715. Kivelson, D., and E. B. Wilson, Jr., *J. Chem. Phys.*, **20**, 1575; *Phys. Rev.*, **87**, 214A. — Th. centrifugal distortion asymmetric rotor

716. Klein, J. A., J. H. N. Loubser, A. H. Nethercot, and C. H. Townes, *Rev. Sci. Instr.*, **23**, 78. — Techniques 1 to 3 mm

716a. Kolsky, H. G., T. E. Phipps, N. F. Ramsey, and H. B. Silsbee, *Phys. Rev.*, **87**, 395. — H_2 and D_2 R.F. spectrum

717. Kojima, S., K. Tsukada, S. Hagiwara, M. Mizushima, and T. Ito, *J. Chem. Phys.*, **20**, 804. — $CHBr_3$

718. Lamb, W. E., and R. C. Retherford, *Phys. Rev.*, **86**, 1014. — Fine-structure H

719. Lamont, H. R. L., and E. M. Hickin, *Brit. J. Appl. Phys.*, **3**, 182. — Frequency stabilization microwave lines

720. Lide, D. R., *J. Chem. Phys.*, **20**, 1761. — Intensity slightly asymmetric rotor

721. Lide, D. R., *J. Chem. Phys.*, **20**, 1812L.; *Erratum*, **21**, 571 (1953). — CH_3NH_2

722. Lide, D. R., Jr., *J.ACS*, **74**, 3548; *Phys. Rev.*, **87**, 227A. CH_2F_2

722a. Logan, R. A., R. E. Cote, and P. Kusch, *Phys. Rev.*, **86**, 280. eqQ in alkali halides

723. Lord, R. C., and R. E. Merrifield, *J. Chem. Phys.*, **20**, 1348.   Rotation-vibration symmetric rotors

724. Lyons, H., *Ann. N.Y. Acad. Sci.*, **55**, 831.   Frequency standards

725. Magnuson, D. W., *J. Chem. Phys.*, **20**, 229.   Dielectric constant ClF_3

726. Massey, J. T., and D. R. Bianco, *Phys. Rev.*, **85**, 717A. H_2O_2

727. Mays, J. M., and B. P. Dailey, *J. Chem. Phys.*, **20**, 1695. XYH_3 molecules

728. Mays, J. M., *Ann. N.Y. Acad. Sci.*, **55**, 789.   Survey high temp., free radicals

729. Miller, S. L., L. C. Aamodt, G. Dousmanis, C. H. Townes, and J. Kraitchman, *J. Chem. Phys.*, **20**, 1112; *Phys. Rev.*, **82**, 327A. CH_3Cl, CH_3Br, CH_3I

730. Minden, H. T., *J. Chem. Phys.*, **20**, 1964.   Statistical weights CH_3CF_3

730a. Mockler, R., J. H. Bailey, and W. Gordy, *Phys. Rev.*, **87**, 172A. $HSiCl_3$, CH_3SiCl_3

731. Morgan, H. W., and J. H. Goldstein, *J. Chem. Phys.*, **20**, 1981L. CH_2CHCN

732. Myers, R. J., and W. D. Gwinn, *J. Chem. Phys.*, **20**, 1420. CH_2Cl_2

733. Nethercot, A. H., J. A. Klein, and C. H. Townes, *Phys. Rev.*, **86**, 798L. HCN

734. Nethercot, A. H., J. A. Klein, J. H. N. Loubser, and C. H. Townes, *Nuovo Cimento*, **9**, Suppl. 3, 358.   1 to 2 mm

735. Nielsen, H. H., *J. Chem. Phys.*, **20**, 1955.   I.R. structure AsH_3

736. Nielsen, H. H., *J. Chem. Phys.*, **20**, 759L.   I.R. structure PH_3, AsH_3, SbH_3

737. Nuckolls, R. G., and L. J. Rueger, *Phys. Rev.*, **85**, 731A.   Lock-in detector

737a. Prodell, A. G., and P. Kusch, *Phys. Rev.*, **88**, 184.   Atomic H h.f.s. mol beam

738. Rank, D. H., *J. Chem. Phys.*, **20**, 1975L.   I.R. mol. constants

739. Rank, D. H., R. P. Ruth, and J. L. Vander Sluis, *Phys. Rev.*, **86**, 799L.   I.R. microwave measurement c

739a. Rawson, E. B., and R. Beringer, *Phys. Rev.*, **88**, 677L.   Atomic oxygen, Zeeman effect

740. Rogers, J. D., and D. Williams, *Phys. Rev.*, **86**, 654A.   H.f.s. HN_3

741. Rueger, L. J., and R. G. Nuckolls, *Rev. Sci. Instr.*, **23**, 635.   Stark cell

742. Schwarz, R. F., *Phys. Rev.*, **86**, 606A; Thesis, Harvard University.   Molecular g factors

743. Sheridan, J., and W. Gordy, *J. Chem. Phys.*, **20**, 591. CF_3Br, CF_3I, CF_3CN

744. Sheridan, J., and W. Gordy, *J. Chem. Phys.*, **20**, 735. CH_3CCBr, CH_3CCI

745. Silvey, G., W. A. Hardy, and C. H. Townes, *Phys. Rev.*, **87**, 236A. $TeCS$

746. Simmons, J. W., and J. H. Goldstein, *J. Chem. Phys.*, **20**, 122; *Phys. Rev.*, **83**, 485A (1951). CD_3Cl, CD_3Br, CD_3I

747. Sinton, W. M., *Phys. Rev.*, **86**, 424L.   Mm solar radiation

748. Smith, D. F., and D. W. Magnuson, *Phys. Rev.*, **87**, 226A. NO_2F

748a. Smith, E. K., Jr., and S. Weintraub, *N.B.S. Rept.*, 1938.   Refractive index of air

748b. Smith, W. V., *Ann. N.Y. Acad. Sci.*, **55**, 891.   Survey pressure broadening

749. Sternheimer, R., *Phys. Rev.*, **86**, 316; 595A. — Th. magnetic h.f.s.
750. Stitch, M. L., A. Honig, and C. H. Townes, *Phys. Rev.*, **86**, 607A. — KCl, TlCl
751. Stitch, M. L., A. Honig, and C. H. Townes, *Phys. Rev.*, **86**, 813L. — NaCl, CsCl
752. Strandberg, M. W. P., *Ann. N.Y. Acad. Sci.*, **55**, 808. — Centrifugal distortion
753. Swartz, J. C., and J. W. Trischka, *Phys. Rev.*, **88**, 1085; **86**, 606A. — LiF molecular beams
754. Tetenbaum, S. J., *Phys. Rev.*, **86**, 440; **82**, 323A. — BrCN 6 mm
755. Tetenbaum, S. J., *Phys. Rev.*, **88**, 772. — OCS, N_2O 6 mm
755a. Thompson, H. W., R. L. Williams, and H. J. Calloman, *Spectrochim. Acta*, **5**, 311. — HI (I.R.)
756. Townes, C. H., *Ann. N.Y. Acad. Sci.*, **55**, 745. — Brief survey
756a. Townes, C. H., *Am. Scientist*, **40**, 270. — Elementary review
757. Townes, C. H., and B. P. Dailey, *J. Chem. Phys.*, **20**, 35. — Th. quad. coupling solids
758. Twiss, R. Q., *S.E.R.L. Tech. J.*, **2**, 10. — Mm-wave generation
758a. Van den Bosch, J. C., and F. Bruin, *Nuovo Cimento*, **9**, Suppl. 3, 238. — Spectrometer
758b. Van den Bosch, J. C., and F. Bruin, *Nuovo Cimento*, **9**, Suppl. 3, 245. — Interferometers
759. Van Kranendonk, J., Thesis, Amsterdam. — Th. pressure broadening
760. Wang, T. C., C. H. Townes, A. L. Schawlow, and A. N. Holden, *Phys. Rev.*, **86**, 809. — Ratio Cl^{35}/Cl^{37} quadrupole
760a. Warner, A. W., *Proc. IRE*, **40**, 1030. — Quartz-crystal oscillators
761. Weatherly, T. L., and D. Williams, *Phys. Rev.*, **87**, 517; with Y. Ting, **83**, 210A (1951); with E. R. Manring, **85**, 717A (1951). — l doubling HCN, DCN
762. Weatherly, T. L., and D. Williams, *J. Chem. Phys.*, **20**, 755L. — Acetone
762a. Weston, R. E., Jr., and M. H. Sirvetz, *J. Chem. Phys.*, **20**, 1820. — Vibration frequency PH_2D, PHD_2
763. Wilcox, W. S., and J. H. Goldstein, *J. Chem. Phys.*, **20**, 1656. — Pyrrole
763a. Wilcox, W. S., J. H. Goldstein, and J. W. Simmons, *Phys. Rev.*, **87**, 172. — Vinyl cyanide
764. Williams, Q., J. T. Cox, and W. Gordy, *J. Chem. Phys.*, **20**, 1524. — $CHBr_3$
765. Williams, Q., J. Sheridan, and W. Gordy, *J. Chem. Phys.*, **20**, 164. — POF_3, PSF_3, $POCl_3$, $PSCl_3$
766. Wilson, E. B., Jr., *Ann. N.Y. Acad. Sci.*, **55**, 943. — Survey mol. structure from microwave spectroscopy
767. Zeiger, H. J., and D. I. Bolef, *Phys. Rev.*, **85**, 788. — TlCl mol. beam
768. Zieman, C. M., *J. Appl. Phys.*, **23**, 154L. — Dielectric constants gases

1953

768a. Abragam, A., and J. H. Van Vleck, *Phys. Rev.*, **92**, 1448. — Th. O Zeeman effect
769. Anderson, F., J. R. Andersen, B. Bak, O. Bastiensen, E. Risberg, and L. Smedvik, *J. Chem. Phys.*, **21**, 373L. — $(CH_3)_3CF$
769a. Arendale, W. F., and W. H. Fletcher, *J. Chem. Phys.*, **21**, 1898. — CH_2CO
770. Artman, J. O., and J. P. Gordon, *Phys. Rev.*, **90**, 338A. — O_2 pressure broadening
770a. Artman, J. O., *Rev. Sci. Inst.* **24**, 873L. — Interferometer
771. Bak, B., J. Bruhn, and J. Rastrup-Andersen, *J. Chem. Phys.*, **21**, 752L. — SiD_3F

772. Bak, B., J. Bruhn, and J. Rastrup-Andersen, *J.* SiD_3Cl
    Chem. Phys., **21**, 753L.
772a. Bak, B., L. Hansen, and J. Rastrup-Andersen, *J.* CH_3CCCF_3
    Chem. Phys., **21**, 1612.
772b. Bak, B., and J. Rastrup-Andersen, *J. Chem. Phys.*,   Pyridine
    21, 1305.
772c. Barrow, R. F., and Caunt, A. D., *Proc. Roy. Soc. A,*   Alkali halides
    219, 120.   (U.V. spectrum)
773. Beers, Y., and S. Weisbaum, *Phys. Rev.*, **91**, 1014L.   HDO
773a. Bedard, F. D., J. J. Gallagher, and C. M. Johnson, D_0 for CO
    Phys. Rev., **92**, 1440.
774. Benedict, W. S., N. Gailar, and E. K. Plyler, *J.* D_2O (I.R.)
    Chem. Phys., **21**, 1301.
775. Benedict, W. S., N. Gailar, and E. K. Plyler, *J. Chem.*   HDO (I.R.)
    Phys., **21**, 1302.
776. Benesch, W., and T. Elder, *Phys. Rev.*, **91**, 308.   Pressure broadening
777. Bird, R., and Richard C. Mockler, *Phys. Rev.*, **91**,   CS
    222A.
778. Birnbaum, G., *J. Chem. Phys.*, **21**, 57. H_2O dispersion mm.
                                                  wavelengths
779. Birnbaum, G., and A. A. Maryott, *Phys. Rev.*, **92**, ND_3 high pressure
    270; **89**, 895A.
780. Birnbaum, G., and A. A. Maryott, *J. Chem. Phys.*, NH_3 pressure
    21, 1774.   broadening
781. Bloom, S., and H. Margenau, *Phys. Rev.*, **90**, 791.   Th. pressure broadening
781a. Brossel, J., B. Cagnac, and A. Kastler, *Compt. rend.*,   Atomic Zeeman effect
    237, 984.
781b. Braunstein, R., and J. W. Trischka, *Phys. Rev.*, **90**,   LiF $I \cdot J$ interaction
    348A.
782. Burke, B. F., and M. W. P. Strandberg, *Phys. Rev.*,   Zeeman effect
    90, 303, 338A.   asymmetric rotor
782a. Burkhard, D. G., *J. Chem. Phys.*, **21**, 1541.   Th. hindered rotation
782b. Burrus, C. A., and W. Gordy, *Phys. Rev.*, **92**, 1437.   NO and DI
783. Burrus, C. A., and W. Gordy, *Phys. Rev.*, **92**, 274. H_2S
783a. Chang, T. S., and D. M. Dennison, *J. Chem. Phys.*,   Centrifugal distortion
    21, 1293. CH_3Cl
783b. Cox, H. L., Jr., *Rev. Sci. Inst.*, **24**, 307.   Lock-in amplifier
784. Cox, J. T., P. B. Peyton, Jr., and W. Gordy, *Phys.* CH_3F, CH_3CCH
    Rev., **91**, 222A.
785. Crawford, H. D., *J. Chem. Phys.*, **21**, 2099L. D_2O
786. Culshaw, W., *Proc. Phys. Soc. London*, **B66**, 597.   Fabry-Perot
                                                  interferometer
787. Dailey, B. P., *Phys. Rev.*, **90**, 337A.   Cyclopropyl chloride
788. Danos, M., and S. Geschwind, *Phys. Rev.*, **91**, 1159.   Wall-collision
                                                  broadening
789. Danos, M., S. Geschwind, H. Lashinsky, and A. Van   Microwave Cerenkov
    Trier, *Phys. Rev.*, **92**, 828L.   effect
790. Dayhoff, E. S., S. Triebwasser, and W. E. Lamb,   H fine structure
    Phys. Rev., **89**, 106.
791. Dehmelt, H. G., *Phys. Rev.*, **91**, 313.   Quad. coupling sulfur
792. Dicke, R. H., *Phys. Rev.*, **89**, 472.   Th. line width
793. Ditchfield, C. R., *Proc. Inst. Elec. Engrs. London,*   Mm-wave crystal mixer
    Part III, **68**, 365.
794. Dousmanis, G. C., T. M. Sanders, C. H. Townes, and   HNCS
    H. J. Zeiger, *J. Chem. Phys.*, **21**, 1416.
795. DuMond, J. W. M., and E. R. Cohen, *Revs. Mod.*   Atomic constants
    Phys., **25**, 691.
796. Erlandsson, G., *Arkiv. Fysik.*, **6**, Paper 45, 477; **7**,   Fluorobenzene
    Paper 17, 189.

796a. Erlandsson, G., *Arkiv. Fysik.*, **6**, 69. CH_3OH, CH_3NO_2
796b. Erlandsson, G., *Arkiv. Fysik.*, **6**, 491.   HCOOH
797. Essen, L., *Proc. Phys. Soc.*, **66B**, 189.   Refractive index H_2O,
                                                                      air, O_2, N_2, H_2, D_2, He

798. Essen, L., *Proc. Inst. Elec. Engrs. London*, Part III,   Refractive index
     100, 19.   atmosphere
799. Fabricand, B. P., R. O. Carlson, C. A. Lee, and I. I.   KBr
     Rabi, *Phys. Rev.*, **91**, 1403.
800. Ferguson, R. C., and E. B. Wilson, Jr., *Phys. Rev.*, SOF_2
     90, 338A.
801. Ferigle, S. M., and A. Weber, *Am. J. Phys.*, **21**, 102.   Rotation-vibration
                                                                      polyatomic molecules
803. Ghosh, S. N., R. Trambarulo, and W. Gordy, *J.*   Dipole moments NF_3,
     Chem. Phys., **21**, 308; *Phys. Rev.*, **87**, 172A. PF_3, POF_3, HCN,
                                                                      CH_3CN, CH_3NC,
                                                                      CH_3F, CH_3CCH,
                                                                      SiF_3H

804. Goldman, I. I., *Doklady Akad. Nauk SSSR*, **88**, 241.   Th. quad. coupling
804a. Gordy, W., *J. chim. phys.*, **50**, C114. H_2O
805. Gordy, W., W. V. Smith, and R. F. Trambarulo,   Microwave
     John Wiley & Sons, Inc., New York.   spectroscopy
805a. Gozzini, A., and E. Polacco, *Compt. rend.* **237**, 1497.   Dielectric constant
                                                                      gases
806. Gorter, C. J., *Experimenta*, **9**, 161.   Review
806a. Hardy, W. A., G. Silvey, C. H. Townes, B. F. Burke, $OCSe^{79}$
     M. W. P. Strandberg, G. W. Parker, and V. W.
     Cohen, *Phys. Rev.*, **92**, 1532.
807. Harrick, N. J., R. G. Barnes, P. J. Bray, and N. F. H_2, D_2
     Ramsey, *Phys. Rev.*, **90**, 260.
808. Hawkins, W. B., and R. H. Dicke, *Phys. Rev.*, **91**,   Na atoms
     1008L.
809. Hedrick, L. C., *Rev. Sci. Instr.*, **24**, 565.   Frequency standards
811. Hicks, B. L., T. E. Turner, and W. W. Widule, *J.*   Th. asymmetric rotors
     Chem. Phys., **21**, 564L.
812. Hill, R. M., and W. Gordy, *Phys. Rev.*, **91**, 222A. O_2 temperature
                                                                      dependence of line
                                                                      widths
813. Hollander, J. M., I. Perlman, and G. T. Seaborg,   Isotopes table
     Revs. Mod. Phys., **25**, 469.
814. Honig, A., M. L. Stitch, and M. Mandel, *Phys. Rev.*,   CsF, CsCl, CsBr
     92, 901.
814a. Huggins, M. L., *J.A.C.S.*, **75**, 4123.   Electronegativities
815. Hughes, R. H., *J. Chem. Phys.*, **21**, 959. O_3
816. Hughes, V., G. Tucker, E. Rhoderick, and G. Wein-   He atom
     reich, *Phys. Rev.*, **91**, 828.
817. Ivash, E. V., and D. M. Dennison, *J. Chem. Phys.*, CH_3OH
     21, 1804; *Phys. Rev.*, **89**, 895A.
818. Javan, A., G. Silvey, C. H. Townes, and A. V. Grosse, Mn^{55}, Re^{185}, Re^{187}
     Phys. Rev., **91**, 222A.   quad. moments
819. Jen, C. K., D. R. Bianco, and J. T. Massey, *J. Chem.* D_2O
     Phys., **21**, 520.
820. Johnson, R. D., R. J. Myers, and W. D. Gwinn, *J.*   Ethylenimine
     Chem. Phys., **21**, 1425L.
821. Jones, L. C., A. V. Bushkovitch, C. A. Potter, and   Pressure broadening
     A. G. Rouse, *Phys. Rev.*, **89**, 895A.
821a. Keller, F. L., and A. H. Nielsen, *Phys. Rev.*, **91**, 235. DBr IR spectrum
822. Kendrick, W. M., and T. E. Turner, *Ballistic Re-*   Spectrograph
     search Lab. Rept. 660.
823. King, W. C., and W. Gordy, *Phys. Rev.*, **90**, 319.   OCS mm wave

824. Kisliuk, P., and S. Geschwind, *J. Chem. Phys.*, **21**, 828. AsF$_3$

825. Kivelson, D., *J. Chem. Phys.*, **21**, 536. Th. asymmetric rotor

826. Kivelson, D., and E. Bright Wilson, Jr., *J. Chem. Phys.*, **21**, 1229; *Phys. Rev.*, **90**, 338A. Th. centrifugal distortion

827. Kivelson, D., and E. Bright Wilson, Jr., *J. Chem. Phys.*, **21**, 1236. Th. mol. parameters from rotational constants

827a. Klein, J. A., and A. H. Nethercot, *Phys. Rev.*, **91**, 1018L. DI

828. Klinger, H. H., *J. Franklin Inst.*, **256**, 353. Review radio astronomy

829. Klinger, H. H., *J. Franklin Inst.*, **256**, 129. Review microwaves

830. Kompfner, R., *Proc. I.R.E.*, **41**, 1602. Mm-wave tube

831. Kraitchman, J., *Am. J. Phys.*, **21**, 17–24. Th. mol. structure determination

832. Krishnaji, and P. Swarup, *J. Sci. Ind. Research*, **12B**, 1–3. NH$_3$ pressure broadening

833. Krishnaji, and P. Swarup, *Z. Physik.*, **136**, 374. NH$_3$ dispersion

833a. Krishnaji, and P. Swarup, *J. Appl. Phys.*, **24**, 1525. Absorption in gases

834. Kusch, P., *Phys. Rev.*, **29**, 268. LiCl Li quad. interaction

835. Lee, C. A., B. P. Fabricand, R. O. Carlson, and I. I. Rabi, *Phys. Rev.*, **86**, 607A; **91**, 1395. KCl

836. Livingston, R., B. M. Benjamin, J. T. Cox, and W. Gordy, *Phys. Rev.*, **92**, 1271. I^{131} spin, quad. mom.

837. Loubser, J. H. N., *J. Chem. Phys.*, **21**, 2231L. CH$_3$COOH

838. Luce, R. G., and J. W. Trischka, *J. Chem. Phys.*, **21**, 105. CsCl mol. beams

839. McCulloh, K. E., and G. F. Pollnow, *J. Chem. Phys.*, **21**, 2082L. Pyridine

840. Meier, R., *Ann. Physik*, **12**, 26. Sub-mm waves

841. Miller, S. L., and C. H. Townes, *Phys. Rev.*, **90**, 537. O^{17} and O^{18} in O$_2$

842. Miller, S. L., C. H. Townes, and M. Kotani, *Phys. Rev.*, **90**, 542; **86**, 607A (1952). Magnetic h.f.s. O$_2$

843. Mizushima, M., *Phys. Rev.*, **91**, 222A. Th. O$_2$

844. Mizushima, M., *J. Chem. Phys.*, **21**, 1222; *Phys. Rev.*, **91**, 464A. Th. allene-type molecules

845. Mizushima, M., *J. Chem. Phys.*, **21**, 539. Stark effect asymmetric rotor h.f.s.

846. Mizushima, M., and P. Venkateswarlu, *J. Chem. Phys.*, **21**, 705; *Phys. Rev.*, **89**, 893A. Rotation-vibrational dipole moment

847. Mockler, R. C., J. H. Bailey, and W. Gordy, *J. Chem. Phys.*, **21**, 1710. H SiCl$_3$, CH$_3$SiCl$_3$, (CH$_3$)$_3$SiCl

848. Mockler, R. C., and W. Gordy, *Phys. Rev.*, **91**, 222A. (CH$_3$)$_3$SiCl

849. Motz, H., W. Thon, and R. N. Whitehurst, *J. Appl. Phys.*, **24**, 826. Microwave generation

850. Muller, N., *J. ACS*, **75**, 860. CH$_2$FCl

851. Nethercot, A. H., and A. Javan, *J. Chem. Phys.*, **21**, 363; and (with C. H. Townes) *Phys. Rev.*, **87**, 226A (1952). C$_8$H$_{13}$Cl, C$_8$H$_{13}$Br

852. Nielsen, H. H., *J. Chem. Phys.*, **21**, 142. Th. XY$_3$ molecules

853. Nishikawa, T., and K. Shimoda, *J. Phys. Soc. Japan*, **8**, 426. NH$_3$ inversion spectrum

854. Nuckolls, R. G., L. J. Rueger, and H. Lyons, *Phys. Rev.*, **89**, 1101. ND$_3$ inversion

854a. Obi, S. Y., T. Ishidzu, S. Yanagawa, Y. Tanabe, and M. Sato, *Ann. Tokyo Astron. Observatory*, **3**, 89. Tables Racah coefficients

855. Ochs, S. A., R. E. Cote, and P. Kusch, *J. Chem. Phys.*, **21**, 459. NaCl mol. beam

855a. Ogata, K., and H. Matsuda, *Phys. Rev.*, **89**, 27.　　Masses of light atoms
856. Posener, D. W., and M. W. P. Strandberg, *J. Chem.*　HDO
　　 Phys., **21**, 1401L.
856a. Potok, H. N., *J. Brit. Inst. Radio Engrs.*, **13**, 490.　Mm-wave spark
　　　　　　　　　　　　　　　　　　　　　　　　　　　　generators
856b. Ramsey, N. F., *Phys. Rev.*, **89**, 527L.　　　　　　Pseudo quad. effect

857. Ramsey, N. F., in *Experimental Nuclear Physics*,　Review nuclear
　　 John Wiley & Sons, Inc., New York.　　　　　　　　moments
858. Ramsey, N. F., *Phys. Rev.*, **91**, 303.　　　　　　　Interactions between
　　　　　　　　　　　　　　　　　　　　　　　　　　　　nuclei

859. Reich, H. J., P. F. Ordung, H. L. Krauss, and J. G.
　　 Skalnick, *Microwave Theory and Techniques*, D.
　　 Van Nostrand Company, Inc., New York.
859a. Robinson, G. W., and C. D. Cornwell, *J. Chem.*　Th. quadrupole h.f.s.
　　 Phys., **21**, 1436.　　　　　　　　　　　　　　　　two nuclei
859b. Robinson, G. W., *J. Chem. Phys.*, **21**, 1741.　　　CCl_2O
860. Rowen, J. H., *Bell System Tech. J.*, **32**, 1333.　　Ferrite microwave
　　　　　　　　　　　　　　　　　　　　　　　　　　　　applications
861. Sanders, T. M., A. L. Schawlow, G. C. Dousmanis,　OH radical
　　 and C. H. Townes, *Phys. Rev.*, **89**, 1158L.
862. Satomura, S., *Mem. Inst. Sci. Ind. Research, Osaka*
　　 Univ., **10**, 34.
863. Scheibe, A., *Z. angew. Phys.*, **5**, 307.　　　　　　Review time standards
864. Sharbaugh, A. H., G. A. Heath, L. F. Thomas, and　SiH_3I
　　 J. Sheridan, *Nature*, **171**, 87.
865. Shimoda, K., and T. Nishikawa, *J. Phys. Soc. Japan*,　Methylamine
　　 8, 133; **8**, 425.
866. Sirvetz, M. E., and R. E. Weston, *J. Chem. Phys.*, **21**,　PHD_2, PH_2D
　　 898.
867. Smith, D. F., *J. Chem. Phys.*, **21**, 609; *Phys. Rev.*, **86**,　ClF_3
　　 608A.
868. Smith, E. K., and S. Weintraub, *J. Research Natl.*　Dielectric constant air
　　 Bur. Standards, **50**, 39.
870. Solimene, N., and B. P. Dailey, *Phys. Rev.*, **91**, 464A.　CH_3SH
871. Strandberg, M. W. P., *Microwave Spectroscopy*, John
　　 Wiley & Sons, Inc., New York.
872. Sverdlov, L. M., *Doklady Akad. Nauk SSSR*, **88**, 249.　Th. isotope shifts
874. Tate, P., and M. W. P. Strandberg, *Phys. Rev.*, **91**,　High-temp.
　　 464A.　　　　　　　　　　　　　　　　　　　　　　spectrometer
875. Trambarulo, R., S. N. Ghosh, C. A. Burrus, and W.　O_3
　　 Gordy, *J. Chem. Phys.*, **21**, 851; *Phys. Rev.*, **91**, 222A.
876. Triebwasser, S., E. S. Dayhoff, and W. E. Lamb,　Fine-structure H
　　 Phys. Rev., **89**, 98.
877. Turner, T. E., V. C. Fiora, W. M. Kendrick, and　Ethylenimine
　　 B. L. Hicks, *J. Chem. Phys.*, **21**, 564; *Phys. Rev.*, **90**,
　　 338A.
877a. Turner, T. E., B. L. Hicks, and G. Reitwiesner, *Re-*　Asym. rotor tables
　　 port 878, Ballistics Research Laboratory, Aberdeen, Md.
878. van den Bosch, J. C., and F. Bruin, *Physica*, **19**, 705.　Cavity wavemeter
879. Van Kranendonk, J., Thesis, Amsterdam.　　　　　　Th. pressure broadening
880. Van Vleck, J. H., in Quantum-mechanical Methods　Th. pressure broadening
　　 in Valence Theory, *Nat. Acad. Sci.*, 117.
881. Venkateswarlu, P., R. C. Mockler, and W. Gordy,　$GeHCl_3$
　　 Phys. Rev., **91**, 222A; *J. Chem. Phys.*, **21**, 1713.
882. Warner, A. W., *Bell Labs. Record*, **31**, 205.　　　　Quartz-crystal
　　　　　　　　　　　　　　　　　　　　　　　　　　　　oscillators
883. Weber, J., *Trans. Inst. Radio Engrs. Prof. Group on*　Amplification of
　　 Electron Devices, **3**, 1.　　　　　　　　　　　　　microwaves

884. Weber, D., S. and S. Penner, *J. Chem. Phys.*, **21**,   NO, HCl, HBr line
     1503.   widths
885. Weinreich, G., and V. Hughes, *Phys. Rev.*, **90**, 377A.   He³ mol. beam
886. Weisbaum, S., Y. Beers, and G. Herrmann, *Phys.*   HDO
     Rev., **90**, 338A.
886a. Wessel, G., and H. Lew, *Phys. Rev.*, **92**, 641.   Molecular beam
                                                                detector
887. White, R. L., *Phys. Rev.*, **91**, 1014L.   DCCCl, DCN deuteron
                                                                quad. coupling
888. White, R. L., and C. H. Townes, *Phys. Rev.*, **92**, 1256. SiD₃F h.f.s.
889. Wilcox, W. S., K. C. Brannock, W. DeMore, and   Ethylenimine
     J. H. Goldstein, *J. Chem. Phys.*, **21**, 563L.
891. Yergin, P. F., W. E. Lamb, Jr., E. Lipworth, and R.   Fine-structure He⁺
     Novick, *Phys. Rev.*, **90**, 377A.

                              1954

892. Aamodt, L. C., P. C. Fletcher, G. Silvey, and C. H.   OCSe⁷⁵
     Townes, *Phys. Rev.*, **94**, 789A.
892a. Althoff, K., and H. Kruger, *Naturwiss.*, **41**, 368.   Cs h.f.s. P₃ state
892b. Artman, J. O., and J. P. Gordon, *Phys. Rev.*, **96**, 1237. O₂
     J. O. Artman, Thesis, Columbia Univ. (1953).
892c. Autler, S. H., and C. H. Townes (to be published).   High frequency Stark
                                                                effects
893. Bak, B., L. Hansen, and J. Rastrup-Andersen, *J.*   Pyridine
     Chem. Phys., **22**, 565L; **22**, 2013.
894. Baird, D. H., and G. R. Bird, *Rev. Sci. Instr.*, **25**, 319. Meas. of intensities
895. Barnes, R. G., and W. V. Smith, *Phys. Rev.*, **93**, 95.   Quad. coupling
                                                                constants atoms
896. Barnes, R. G., P. J. Bray, and N. F. Ramsey, *Phys.*   Magnetic moment of H₂
     Rev., **94**, 893.
896a. Bassov, N. G., and A. M. Prokhorov, *J. Exp. Theor.*   Mol. beams techniques
     Phys., U.S.S.R., **27**, 431.
898. Beringer, R., and M. A. Heald, *Phys. Rev.*, **95**, 1474. Atomic H
899. Beringer, R., E. B. Rawson, and A. F. Henry, *Phys.*   NO
     Rev., **94**, 343.
900. Bernstein, R. B., F. F. Cleveland, and F. L. Voelz,   IR spectrum CH₃I
     J. Chem. Phys., **22**, 193.
901. Bird, G. R., *Rev. Sci. Instr.*, **25**, 324.   Meas. of intensities
902. Bird, G. R., *Phys. Rev.*, **95**, 1686L.   CH₃Cl saturation
903. Bird, G. R., and C. H. Townes, *Phys. Rev.*, **94**, 1203. Quad. coupling of S
904. Birnbaum, G., and A. A. Maryott, *Phys. Rev.*, **95**,   Absorption in
     622A.   compressed gases
905. Birnbaum, G., A. A. Maryott, and P. F. Wacker,   CO₂, high pressure
     J. Chem. Phys., **22**, 1782L.
906. Birnbaum, G., and A. A. Maryott, *J. Chem. Phys.*,   NH₃, line width
     22, 1457L.   versus T
907. Bruin, F., *Proc. K. Ned. Akad. Wetensch. B*, **56**, 515. Interferometers
908. Burke, B. F., M. W. P. Strandberg, V. W. Cohen, and   S³⁵ magnetic moment
     W. S. Koski, *Phys. Rev.*, **93**, 193.
908a. Burkhard, D. G., and J. C. Irvin, *Tech. Rep. 2, Uni-*   Th. hindered motions
     versity of Colorado.
909. Burrus, C. A., A. Jache, and W. Gordy, *Phys. Rev.*,   PH₃
     95, 299A and *Phys. Rev.*, **95**, 706.
910. Burrus, C. A., and W. Gordy, *Phys. Rev.*, **93**, 897.   Mm waves OCS
911. Collier, R. J., *Phys. Rev.*, **95**, 1201.   HCN l-doubling
911a. Collins, T. L., W. H. Johnson, Jr., and A. O. Nier,   Mass spectra
     Phys. Rev., **94**, 398.
912. Cornwell, C. D., and R. L. Poynter, *J. Chem. Phys.*,   Vinyl iodide
     22, 1257L.

913. Cox, J., W. J. O. Thomas, and W. Gordy, *Phys. Rev.*, **95**, 299A. — CH_3Cl

913a. Daly, R. T., Jr., and J. H. Holloway, *Phys. Rev.*, **96**, 539L. — Ga^{69}, Ga^{71} octupole moments

913b. Dehmelt, H. G., *Phys. Rev.* (to be published). — Atomic P

913c. Dailey, B. P., and C. H. Townes, *J. Chem. Phys.* (to be published). — Th. quadrupole coupling

914. DeMore, B. B., W. S. Wilcox, and J. H. Goldstein, *J. Chem. Phys.*, **22**, 876. — Pyridine

915. Douglas, A. E., and C. K. Møller, *J. Chem. Phys.*, **22**, 275. — IR spectrum N_2O

916. Dousmanis, G. C., *Phys. Rev.*, **94**, 789A. — OD

916a. Dousmanis, G. C., *Phys. Rev.* (to be published). — NO h.f.s.

917. Erlandsson, G. *J. Chem. Phys.*, **22**, 563L. — Cyclopentenone

918. Erlandsson, G., (private communication, to be published). — Fluorobenzene

918a. Erlandsson, G., *Ark. Fys.*, **8**, 341. — Chlorobenzene

919. Erlandsson, G., *J. Chem. Phys.*, **22**, 1152L. — Benzonitrile

920. Farrands, J. L., and J. Brown, *Wireless Engr.*, **31**, 81. — Interferometer for mm waves

921. Feeny, H., H. Lackner, P. Moser, and W. V. Smith, *J. Chem. Phys.*, **22**, 79. — Press. broad. linear mols.

922. Ferguson, R. C., *J. ACS*, **76**, 850. — F_2OS

923. Foley, H. M., R. M. Sternheimer, and D. Tycko, *Phys. Rev.*, **93**, 734. — Th. quad. coupling

924. Gallagher, J. J., F. D. Bedard, and C. M. Johnson, *Phys. Rev.*, **93**, 729. — NO

924a. Geschwind, S., G. R. Gunther-Mohr, and C. H. Townes, *Rev. Mod. Phys.*, **26**, 444. — Mass determinations

925. Gordon, J. P., H. J. Zeiger, and C. H. Townes, *Phys. Rev.*, **95**, 282L; thesis, Columbia Univ. (1955). — NH_3 h.f.s.; molecular oscillator

926. Gordy, W., *J. Chem. Phys.*, **22**, 1276L. — Th. quad. coupling and chemical bonds

926a. Gordy, W., *J. Phys. Radium*, **15**, 521. — Mm. wave spectroscopy

926b. Gordy, W., *J. Chem. Phys.*, **22**, 1470L. — Theory quad. coupling

927. Gordy, W., and C. A. Burrus, *Phys. Rev.*, **93**, 419. — DBr

928. Gordy, W., and J. Sheridan, *J. Chem. Phys.*, **22**, 92. — CH_3HgCl, CH_3HgBr

929. Gross, E. P., *Phys. Rev.*, **94**, 1424A. — Th. press. broadening

930. Gunther-Mohr, G. R., R. L. White, A. L. Schawlow, W. E. Good, and D. K. Coles, *Phys. Rev.*, **94**, 1184; *Phys. Rev.*, **83**, 880A (1951). — NH_3 h.f.s.

931. Gunther-Mohr, G. R., C. H. Townes, and J. H. Van Vleck, *Phys. Rev.*, **94**, 1191. — NH_3 h.f.s.

931a. Gunton, R. C., J. F. Ollom, and H. N. Rexroad, *J. Chem. Phys.*, **22**, 1942L. — $(CH_3)_3SiF$

931b. Hardy, W. A., P. Fletcher, and V. Suarez, *Rev. Sci. Inst.*, **25**, 1135L. — Stark cell

932. Hardy, W. A., and G. Silvey, *Phys. Rev.*, **95**, 385. — TeCS

932a. Heald, M. A., and R. Beringer, *Phys. Rev.*, **96**, 645. — N

933. Heath, G. A., L. F. Thomas, and J. Sheridan, *Trans. Faraday Soc.*, **50**, 779. — F_3HSi

934. Heath, G. A., L. F. Thomas, and J. Sheridan, (private communication, to be published). — C_5H_4 (penta-1:3-Diyne)

934a. Hecht, K. T., and D. M. Dennison (private communication). — Hindered torsional motions

935. Heineken, F. W., and F. Bruin, *Physica*, **20**, 350. — Microwave refractive index of gases

936. Heitler, W., *The Quantum Theory of Radiation*, 3d ed., Oxford University Press, London.
937. Hill, R. M., and W. Gordy, *Phys. Rev.*, **93**, 1019. O_2 Zeeman effect and Δv
938. Honig, A., M. Mandel, M. L. Stitch, and C. H. Townes, *Phys. Rev.*, **93**, 953A; **96**, 629. Alkali halides
939. Hrostowski, H. J., and R. J. Myers, *J. Chem. Phys.*, **22**, 262. Pentaborane
940. Ishiguro, E., and S. Koide, *Phys. Rev.*, **94**, 350. H_2, magnetic properties
941. Ito, T., Y. Tanabe, and M. Mizushima, *Phys. Rev.*, **93**, 1242. Th. line width
942. Jaccarino, V., J. G. King, R. A. Satten, and H. H. Stroke, *Phys. Rev.*, **94**, 1798L. I^{127} nuclear octupole moment
943. Jache, A., G. Blevins, and W. Gordy, *Phys. Rev.*, **95**, 299A. AsH_3
943a. Javan, A., and A. Engelbrecht, *Phys. Rev.*, **96**, 649. MnO_3F, ReO_3Cl
943b. Jen, C. X., *Am. J. Phys.*, **22**, 553. Rotational magnetic moments
944. Johnson, C. M., D. M. Slazer, and D. D. King, *Rev. Sci. Instr.*, **25**, 213. Mm-wave harmonic generators
945. Kambe, K., and J. H. Van Vleck, *Bull. Am. Phys. Soc.*, **29**, 10. O Zeeman effect
946. King, W. C., and W. Gordy, *Phys. Rev.*, **93**, 407. Mm waves, OCS, CH_3F, H_2O
947. Kisliuk, P., *J. Chem. Phys.*, **22**, 86. Group V trihalides
948. Kivelson, D., *J. Chem. Phys.*, **22**, 904. SO_2
948a. Kivelson, D., *J. Chem. Phys.*, **22**, 1733. Hindered motions
949. Klinger, H. H., *Introduction to Microwaves and Their Scientific Application*, S. Hirzel, Stuttgart.
949a. Kojima, T., and T. Nishikawa, *J. Phys. Soc. Japan* (to be published). CH_3SH
949b. Kojima, S., and K. Tsukada, *J. Chem. Phys.*, **22**, 2093L. $CHBr_3$
950. Kraitchman, J. A., and B. P. Dailey, *Phys. Rev.*, **94**, 788A. C_2H_5F
950a. Kraitchman, J. A., and B. P. Dailey, *J. Chem. Phys.*, **22**, 1477. eqQ for methyl halides
951. Krishnaji and P. Swarup, *J. Chem. Phys.*, **22**, 568. CH_3Br, high pressure
952. Krishnaji and P. Swarup, *J. Chem. Phys.*, **22**, 1456L. Temp. dependence NH_3 absorption
952a. Kusch, P., *J. Chem. Phys.*, **22**, 1203. $KClFeCl_2$, $KBrFeBr_2$ g. factor
953. Kusch, P., and T. G. Eck, *Phys. Rev.*, **94**, 1799. In^{115} octupole moment
954. Lide, D. R., Jr., *Phys. Rev.*, **94**, 788A; *J. Chem. Phys.* **22**, 1613L. CH_3NH_2
955. Lide, D. R., Jr., *J. Chem. Phys.*, **22**, 1577. Benzonitrile
955a. Liuima, F. A., A. V. Bushkovitch, and A. G. Rouse, *Phys. Rev.*, **96**, 434. Press. broadening
956. Lovell, R. J., and E. A. Jones, *Phys. Rev.*, **95**, 300A. COF_2
957. Lotspeich, J. F., and A. Javan, *Phys. Rev.*, **94**, 789A. ReO_3F
958. Low, W., *Phys. Rev.* (to be published). Fermi resonance
958a. Mandel, M., and A. H. Barrett (to be published). $TlCl$
959. Massey, J. T., C. I. Beard, and C. K. Jen, *Phys. Rev.*, **95**, 622A. HDO_2
960. Massey, J. T., and D. R. Bianco, *J. Chem. Phys.*, **22**, 442. H_2O_2
960a. Matricon, M., and Bonnet, *J. Phys. Radium*, **15**, 647. Ethylamine
961. McCulloh, K. E., and G. F. Pollnow, *J. Chem. Phys.*, **22**, 681. Pyridine

962. Mizushima, M., *Phys. Rev.*, **94**, 789A; *Phys. Rev.*, **94**, 569.   Th. NO h.f.s.

963. Mizushima, M., and R. M. Hill, *Phys. Rev.*, **93**, 745. O_2

964. Morgan, H. W., and J. H. Goldstein, *J. Chem. Phys.*, **22**, 1427. C_2H_3I

965. Muller, N., *J. ACS*, **75**, 860. CH_2ClF

965a. Nethercot, A. H., Jr. *Trans. IRE MTT2*, 17 (1954)   Mm-wave techniques

966. Nielsen, H. H., *J. Chem. Phys.*, **22**, 1383.   IR spectrum PH_3

967. Ogg, R. A., Jr., and J. D. Ray, *J. Chem. Phys.*, **22**, 147. Si^{29} spin

967a. Osipov, B. D., *J. Exp. Theor. Phys.*, *U.S.S.R.*, **27**, 115. NH_3 dispersion

968. Peter, M., and M. W. P. Strandberg, *Phys. Rev.*, **95**, 622A.   OCS

968a. Prokhorov, A. M., and A. I. Barchukov, *J. Exp. Theor. Phys.*, *U.S.S.R.*, **26**, 761.   Absorption measurements

969. Poscncr, D. W., and M. W. P. Strandbcrg, *Phys. Rev.*, **95**, 374.   Centr. distor H_2O, D_2O, HDO

969a. Ramsey, N. F., *Nuclear Moments*, John Wiley & Sons, Inc., New York.

970. Rogers, T. F., *Phys. Rev.*, **95**, 622A. H_2O in atm.

971. Sanders, T. M., Jr., A. L. Schawlow, G. C. Dousmanis, and C. H. Townes, *J. Chem. Phys.*, **22**, 245.   OH

971a. Sanders, T. M., Jr., G. C. Dousmanis, and C. H. Townes (to be published).   OH and OD

972. Satomura, S., *Inst. Sci. and Ind. Res.*, *Memoirs*, **10**, 34. Review microwaves

973. Schatz, P. N., *J. Chem. Phys.*, **22**, 755L.   Th. quad. coupling

973a. Sheridan, J., and Thomas, L. F., *Nature*, **174**, 798L.   Methyl-cyanoacetylene

974. Sherrard, E. I., L. F. Thomas, and J. Sheridan (private communication, to be published). CH_3CN, CH_3NC, CH_3F

975. Shimoda, K., *J. Phys. Soc.*, Japan, **9**, 378; **9**, 558; **9**, 567. NH_3 frequency standard

976. Shimoda, K., T. Nishikawa, and T. Itoh, *J. Chem. Phys.*, **22**, 1456L; *J. Phys. Soc. Japan*, **9**, 974. CH_3NH_2

977. Slayton, G. R., J. W. Simmons, and J. H. Goldstein, *Phys. Rev.*, **95**, 299A; *J. Chem. Phys.*, **22**, 1678.   Vinylene carbonate

978. Solimene, N., and B. P. Dailey, *Phys. Rev.*, **94**, 789A; *J. Chem. Phys.*, **22**, 2042.   1,1 Difluoroethane

978a. Solimene, N., and B. P. Dailey, *J. Chem. Phys.* (to be published). CH_3SH

979. Sternheimer, R. M., *Phys. Rev.*, **95**, 736.   Quad. screening **theory**

979a. Sterzer, F., *J. Chem. Phys.*, **22**, 2094L. CF_3I

980. Sterzer, F., and Y. Beers, *Phys. Rev.*, **94**, 1410A. CH_3I quad. transitions

980a. Stitch, M. L., A. Honig, and C. H. Townes, *Rev. Sci. Instr.*, **25**, 759.   High temp. spectrograph

981. Strandberg, M. W. P., *Microwave Spectroscopy*, Methuen & Co., Ltd., London; John Wiley & Sons, Inc., New York.

982. Strandberg, M. W. P., and H. Dreicer, *Phys. Rev.*, **94**, 1393.   Beam absorption spectrometer

982a. Strandberg, M. W. P., H. R. Johnson and J. R. Eshbach, *Rev. Sci. Inst.*, **25**, 776.   Microwave spectrometers

983. Strandberg, M. W. P., and M. Tinkham, *Phys. Rev.*, **95**, 623A.   Th. O_2

983a. Swarup, P., *J. Sci. Ind. Res.*, **13B**, 311.   Ethyl chloride absorption

984. Swarup, P., *J. Sci. Ind. Res.*, **13B**, 389.   Temp. dependence **of** absorption

985. Takahashi, I., A. Okaya, and T. Ogawa, *J. Inst. Elec. Commun. Engrs. Japan*, **35**, 462.   Frequency measurement

986. Tannenbaum, E., R. D. Johnson, R. J. Myers, and W. D. Gwinn, *J. Chem. Phys.*, **22**, 949L. CH_3NO_2

987. Tate, P. A., and M. W. P. Strandberg, *J. Chem. Phys.*,   KCl, NaCl
     22, 1380.
988. Tate, P. A., and M. W. P. Strandberg, *Rev. Sci.*   High temp.
     Instr., **25**, 956.   spectrograph
988a. Thomas, L. F., E. I. Sherrard, and J. Sheridan (pri- CH_3CN, CH_3CCH
     vate communication, to be published).
989. Thomas, W. J. O., J. T. Cox, and W. Gordy, *J. Chem.* D_J, D_{JK}, methyl halides
     Phys., **22**, 1718.
990. Tinkham, M., and M. W. P. Strandberg, *Phys. Rev.*, O_2
     95, 622A; thesis, M.I.T.
991. Trambarulo, R., and P. M. Moser, *J. Chem. Phys.*,   HCOOH
     22, 1622L.
992. Trambarulo, R., H. Lackner, P. Moser, and H. Feeny,   Press. broadening BrCN
     Phys. Rev., **95**, 622A.
993. Trischka, J. W., and R. Braunstein, *Phys. Rev.*, **96**, RbCl (mol. beam)
     968.
994. Van Winter, C., *Physica*, **20**, 274.   Th. asymmetric rotor
995. Wagner, R. S., and B. P. Dailey, *J. Chem. Phys.*, **22**, C_2H_5Cl
     1459L.
996. Weissman, H. B., R. B. Bernstein, S. E. Rosser, A. G.   IR spectrum CH_3Br
     Meister, and F. F. Cleveland, (private communica-
     tion, to be published).
997. Westerkamp, J. F., *Phys. Rev.*, **93**, 716.   HCN *l*-doubling
997a. Weinstein, R., M. Deutsch, and S. C. Brown, *Phys.*   Positronium
     Rev. (to be published).
998. White, R. L., *Phys. Rev.*, **94**, 789A. eqQ in DCN, HCN,
                                                                 DCCCl
998a. White, R. L., *J. Chem. Phys.* (to be published). N^{14} coupling in HCN
999. White, R. L., *Bul. Am. Phys. Soc.*, **20**, 11; thesis, $I·J$ interactions
     Columbia Univ.
1000. Wilcox, W. S., J. H. Goldstein, and J. W. Simmons,   Vinyl cyanide
     J. Chem. Phys., **22**, 516.
1000a. Wilson, E. B., Jr., C. C. Lin, and D. R. Lide (to be   Th. hindered rotation
     published).
1001. Wolfe, P. N., and Dudley Williams, *Phys. Rev.*, **93**, O_3
     360A.

AUTHOR INDEX

SUBJECT INDEX

In addition to subject headings, this index contains the more important symbols used and all molecules mentioned in the text and tables. Molecules are listed alphabetically according to their empirical formulas by the following procedures:

1. All empirical formulas beginning with a given letter are listed at the end of the alphabetical section for that letter.

2. Symbols for the elements in the empirical formula for a molecule are in alphabetical order except that within a molecular formula:

   a. C for carbon precedes all other symbols. This groups organic compounds together.

   b. In organic compounds, H precedes all other symbols except C.

   c. Isotopes are not distinguished, so that D (for deuterium) is regarded as H for purposes of listing.

3. All molecules with formulas of the form $X_n Y$ precede those of the form $X_{n+1}YZ$, etc.

To aid in identification, in some cases the usual chemical formula or the name of a compound is given in parentheses. The names are not usually given separate entries. Greek letters are listed as if spelled out.

A (argon), 364–366

A, rotational constant, 48, 50, 83, 613–642

Absorption cells, high temperature, 443–445

  outgassing, 498

  for reactive materials, 447, 448, 498

  resonant cavity, 435–439, 498

  Stark modulation, 264–266, 418, 419, 498

  untuned cavity, 439–441

  waveguide, 264, 266, 411–415, 424, 443–446, 462, 463, 497, 498

    for analysis, 497, 498

    reflections in, 415, 445, 446, 496, 497

    windows, 444, 447

  Zeeman modulation, 424

Absorption intensity (see Intensity of absorption)

Abundance of isotopes, 495, 643–648

Accidentally symmetric rotor, 49, 50, 52, 92, 155

Alkali atom energy levels, 119, 120

α, attenuation, 412–414, 436, 437, 440, 441, 445, 446

  minimum detectable, 412–415, 437

α, fine-structure constant, 123

α_e, diatomic molecule, 9–16, 613–642

α_i, linear molecule, 25–30, 33, 39–42, 45–47, 613–642

  symmetric-top, 79, 613–642

Ammonia (NH_3) (see H_3N)

  absorption-frequencies table, 311, 312

Amplifiers, lock-in, 421–424

  modulation-frequency, 420–424

Analysis, chemical, 486–498

  equipment, 497, 498

  qualitative, 488–492

  quantitative, 492–497

Anharmonicity of potential, 15, 16, 27–29, 107

Astronomical sources of microwaves, 146–148, 448–450

Asymmetric rotor, centrifugal distortion, 105–109

  energy levels, 83–92, 522–556

  hindered torsional motions, 324–335

  intensities and selection rules, 92–102, 557–612

  inversion, 314–315

  matrix elements, 92–102

  quadrupole coupling, 241–245

Fundamental Constants and Conversion Factors

This listing is based on the report of J. W. M. DuMond and E. R. Cohen [795].

Velocity of light, $c = (2.997929 \pm 0.000008) \times 10^{10}$ cm/sec

Avogadro's constant (physical scale), $N = (6.02472 \pm 0.00036) \times 10^{23}$ mol/g-mol

Loschmidt's constant (physical scale), $L_0 = N/V_0 = (2.68713 \pm 0.00016)$
$\times 10^{19}$ molecules/cm³

Electronic change, $e = (4.80288 \pm 0.00021) \times 10^{-10}$ esu
$e' = e/c = (1.60207 \pm 0.00007) \times 10^{-20}$ emu

Electron rest mass, $m = (9.1085 \pm 0.0006) \times 10^{-28}$ g

Proton rest mass, $m_p = (1.67243 \pm 0.0001) \times 10^{-24}$ g

Planck's constant, $h = (6.6252 \pm 0.0005) \times 10^{-27}$ erg-sec
$\hbar = h/2\pi = (1.05444 \pm 0.00009) \times 10^{-27}$ erg-sec

Fine structure constant, $\alpha = e^2/\hbar c = (7.29726 \pm 0.00008) \times 10^{-3}$
$1/\alpha = 137.0377 \pm 0.002$

Bohr magneton, $\mu_0 = he/(4\pi mc) = (0.92732 \pm 0.00006) \times 10^{-20}$ erg/gauss
$= 1.39967 \pm 0.00004$ Mc/gauss

Nuclear magneton, $\mu_n = he/(4\pi m_p c) = (0.505038 \pm 0.000036) \times 10^{-23}$ erg/gauss
$= (7.6230 \pm 0.0006) \times 10^{-4}$ Mc/gauss

Unit atomic mass, $M(O_{16})/16 = (1.65983 \pm 0.0001) \times 10^{-24}$ g

Rotational constant, $B \times I = h/8\pi^2 = (8.39091 \pm 0.0005) \times 10^{-35}$ Mc/g-cm²
$= 5.05531 \pm 0.0003 \times 10^5$ Mc/atomic mass units A²

Boltzmann's constant, $k = (1.38042 \pm 0.0001) \times 10^{-16}$ erg/deg
$= (2.0836 \pm 0.0004) \times 10^4$ Mc/deg
$= (0.69501 \pm 0.0001)$ cm⁻¹/deg

Stark effect constant, $\mu E = 0.50348$ Mc/(debye)(volt/cm)

1 electron volt/particle $= (2.306 \pm 0.001) \times 10^4$ cal/mole $= (8065.98 \pm 0.30)$ cm⁻¹